OLDSMOBILE CUTLASS

1970-1982 REAR DRIVE, GAS & DIESEL

SHOP MANUAL

ERIC JORGENSEN
Editor

JEFF ROBINSON
Publisher

CLYMER PUBLICATIONS

World's largest publisher of books devoted exclusively to automobiles and motorcycles

12860 MUSCATINE STREET · P.O. BOX 20 · ARLETA, CALIFORNIA 91331

FIRST EDITION
First Printing September, 1980

SECOND EDITION
Revised by Kalton C. Lahue to include 1981 models
First Printing December, 1981

THIRD EDITION
Revised by Kalton C. Lahue to include 1982 models
First Printing September, 1982

Printed in U.S.A.

ISBN: 0-89287-324-8

Marina Lerique; production coordinator

COVER:
Photographed by Mike Brown/Visual Imagery, Los Angeles, Calif.

CONTENTS

OLDSMOBILE CUTLASS

1970-1982 REAR DRIVE, GAS & DIESEL

SHOP MANUAL

QUICK REFERENCE DATA

DISTRIBUTOR

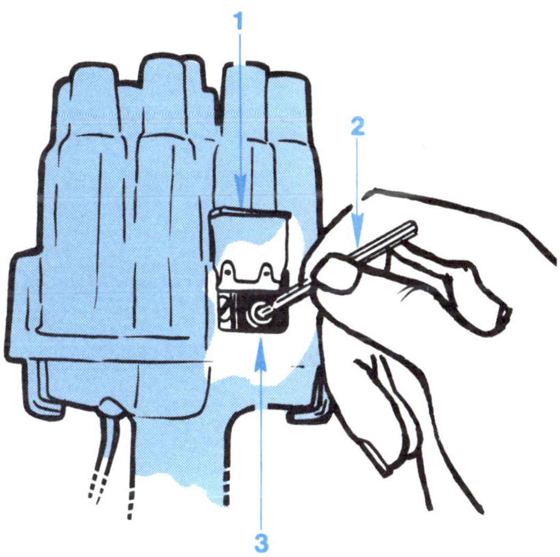

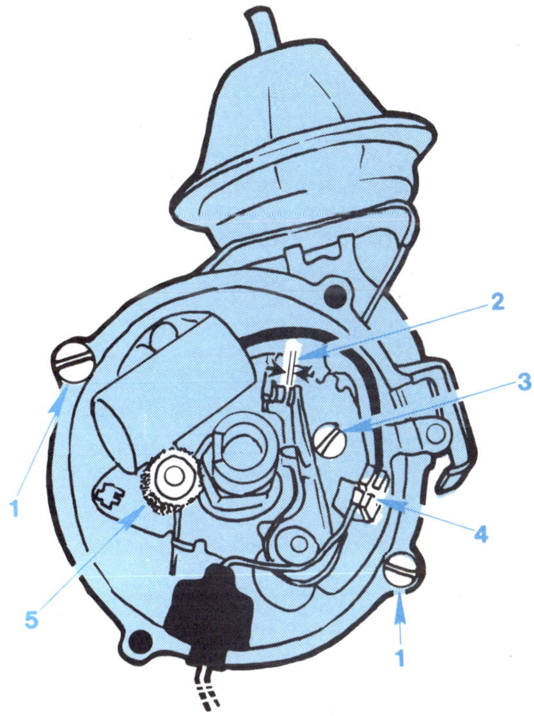

1967-1974 V8 ENGINES

1. Window
2. Hexagon wrench
3. Adjusting screw

1967-1974 L6 ENGINES

1. Breaker plate attaching screws
2. Point gap
3. Contact set attaching screws
4. Quick disconnect terminal
5. Cam lubricator

TIMING MARKS

L6 ENGINES

Marks are 2° increments

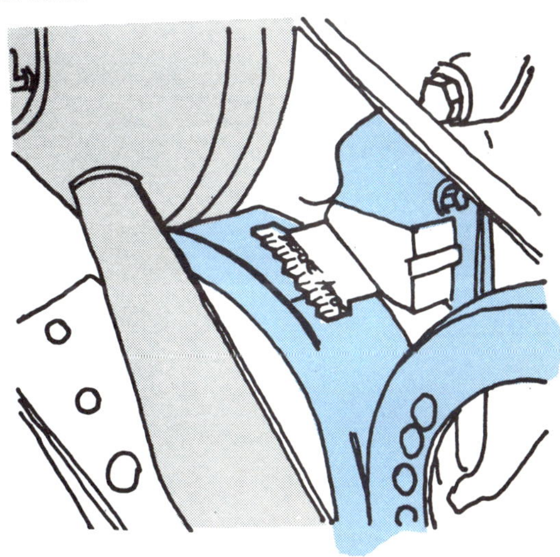

V8 ENGINES

TUNE-UP SPECIFICATIONS

Cylinder Head Bolt Torque	
L6	95 ft.-lb.
Chevy V8	65 ft.-lb.
V6	80 ft.-lb.
Olds V8 (1970-1971)	80 ft.-lb.
Olds V8 (1972-1976)	85 ft.-lb.
Olds 4.3 liter V8 (1977-on)	85 ft.-lb.
Olds other V8s (1977-on)	130 ft.-lb.
Valve Clearance	Hydraulic lifters-no adjustment required.
Spark Plugs (type and gap)	See Table 6, Chapter Three
Point gap (1970-1974, all)	0.016 in. (used), 0.019 in. (new).
Dwell angle (1970-1974)	
L6	31-34°
V8	30°
Ignition timing	See Table 7, Chapter Three
Idle speed	See VECI sticker or Table 7, Chapter Three
Compression	
L6	130 psi.
V6 and V8	Lowest cylinder within 70 percent of highest, no cylinder under 100 psi
Firing order	
L6	1-5-3-6-2-4 (Cylinder No.1 — front)
V6	1-6-5-4-3-2 (Cylinder No.1 — left front)
V8	1-8-4-3-6-5-7-2 (Cylinder No.1 — left front)

RECOMMENDED LUBRICANTS AND FLUIDS

Use	Recommendation
Engine oil	Product bearing letter designation SE (gasoline engine); SE/CD (1978 diesel); or SE/CC (1979-on diesel)
Power steering	GM power steering fluid or, if not available, DEXRON® or DEXRON® II automatic transmission fluid
Differential—standard	SAE 80 or SAE 90 GL-5 gear lubricant (SAE 80 GL-5 in Canada)
Differential—Positraction	GM part No. 1051022 (lubricant) or equivalent
Steering gear—manual	GM part No. 1051052 (lubricant) or equivalent
Transmission—manual	SAE 80 or SAE 90 GL-5 gear lubricant (SAE 80 GL-5 in Canada)
Brake master cylinder	DOT 3 or Delco Supreme 11 brake fluid
Clutch linkage (manual transmission) 1. Pivot points 2. Cross-shaft grease fitting and pushrod to clutch fork joint	 Engine oil EP chassis lubricant meeting requirements of GM specification GM 6031-M
Shift linkage (all)	Engine oil
Hood latch and hinge assemblies 1. Pivots and spring anchor 2. Release pawl 3. Hinges	 Engine oil Chassis lubricant Engine oil
Chassis lubrication (front suspension, steering linkage, etc.)	Chassis lubricant meeting requirements of GM specification GM 6031-M
Constant velocity universal joint	Lubricant, GM part No. 1050679, or meeting requirements of GM specification GM 6040-M
Automatic transmission	DEXRON® or DEXRON® II automatic transmission fluid
Parking brake cables	Chassis lubricant
Front wheel bearings	Chassis lubricant meeting the requirements of GM specification GM 6031-M
Door, tailgate, seat and trunk hinges	Engine oil
Convertible door-to-lock wedge plates	Stick-type lubricant
Windshield washer	GM Optikleen washer solvent, part No. 1050001 or equivalent
Battery	Clean, odorless drinking water
Engine cooling system	Mixture of water and high quality Ethylene Glycol base antifreeze, meeting requirements of GM specification 1899-M

AUTOMATIC TRANSMISSION FLUID CAPACITY

Transmission	Capacity
Turbo Hydramatic 400	7½ U.S. pints
Turbo Hydramatic 375	7½ U.S. pints
Turbo Hydramatic 350	5 U.S. pints
Turbo Hydramatic 250	5 U.S. pints
Turbo Hydramatic 200	6 U.S. pints

OIL FILTER USAGE

Engine	Filter Brand			
	AC	Lee	Fram	Purolator
250 cid L6	PF-25	LF-25	PH-30	PER-49
231 cid V6	PF-47	LF-111	PH-3387A	PER-111
1970-1977 V8	PF-24	LF-24	PH-25	PER-33
1978-1980 V8				
except 260 cid	PF-25	LF-25	PH-30	PER-49
260 cid	PF-45	LF-45	PH-3429	PER-123
1981-1982 gasoline V8	PF-45	LF-45	PH-3429	PER-123
All diesel				
except 260 cid	PF-45	LF-45	PH-3429	PER-123
260 cid diesel	PF-51	— —	PH-3535	— —

—NOTES—

INTRODUCTION

This detailed, comprehensive manual covers 1970-1982 rear-wheel drive Oldsmobile Cutlass automobiles. The expert text gives complete information on maintenance, repair and overhaul. Hundreds of photos and drawings guide you through every step. The book includes all you need to know to keep your car running right.

Chapters One through Twelve contain general information on all models and specific information on 1970-1980 models. The Supplement at the end of the book contains specific information on 1981 and later models which differs from that in the main book.

Where repairs are practical for the owner/mechanic complete procedures are given. Equally important, difficult jobs are pointed out. Such operations are usually more economically performed by a dealer or independent garage.

A shop manual is a reference. You want to be able to find information fast. As in all Clymer book, this one is designed with this in mind. All chapters are thumb tabbed. Important items are indexed at the rear of the book. Finally, all the most frequently used specifications and capacities are summarized on the *Quick Reference* pages at the front of the book.

Keep the book handy. Carry it in your glove box. It will help you to better understand your car, lower repair and maintenance costs and generally improve your satisfaction with your vehicle.

CHAPTER ONE

GENERAL INFORMATION

The troubleshooting, tune-up, maintenance, and step-by-step repair procedures in this book are written for the owner and home mechanic. The text is accompanied by useful photos and diagrams to make the job as clear and correct as possible.

Troubleshooting, tune-up, maintenance, and repair are not difficult if you know what tools and equipment to use and what to do. Anyone not afraid to get their hands dirty, of average intelligence, and with some mechanical ability can perform most of the procedures in this book.

In some cases, a repair job may require tools or skills not reasonably expected of the home mechanic. These procedures are noted in each chapter and it is recommended that you take the job to your dealer, a competent mechanic, or machine shop.

MANUAL ORGANIZATION

This chapter provides general information and safety and service hints. Also included are lists of recommended shop and emergency tools as well as a brief description of troubleshooting and tune-up equipment.

Chapter Two provides methods and suggestions for quick and accurate diagnosis and repair of problems. Troubleshooting procedures discuss typical symptoms and logical methods to pinpoint the trouble.

Chapter Three explains all periodic lubrication and routine maintenance necessary to keep your vehicle running well. Chapter Three also includes recommended tune-up procedures, eliminating the need to constantly consult chapters on the various subassemblies.

Subsequent chapters cover specific systems such as the engine, transmission, and electrical systems. Each of these chapters provides disassembly, repair, and assembly procedures in a simple step-by-step format. If a repair requires special skills or tools, or is otherwise impractical for the home mechanic, it is so indicated. In these cases it is usually faster and less expensive to have the repairs made by a dealer or competent repair shop. Necessary specifications concerning a particular system are included at the end of the appropriate chapter.

When special tools are required to perform a procedure included in this manual, the tool is illustrated either in actual use or alone. It may be possible to rent or borrow these tools. The inventive mechanic may also be able to find a suitable substitute in his tool box, or to fabricate one.

The terms NOTE, CAUTION, and WARNING have specific meanings in this manual. A NOTE provides additional or explanatory information. A CAUTION is used to emphasize areas where equipment damage could result if proper precautions are not taken. A WARNING is used to stress those areas where personal injury or death could result from negligence, in addition to possible mechanical damage.

SERVICE HINTS

Observing the following practices will save time, effort, and frustration, as well as prevent possible injury.

Throughout this manual keep in mind two conventions. "Front" refers to the front of the vehicle. The front of any component, such as the transmission, is that end which faces toward the front of the vehicle. The "left" and "right" sides of the vehicle refer to the orientation of a person sitting in the vehicle facing forward. For example, the steering wheel is on the left side. These rules are simple, but even experienced mechanics occasionally become disoriented.

Most of the service procedures covered are straightforward and can be performed by anyone reasonably handy with tools. It is suggested, however, that you consider your own capabilities carefully before attempting any operation involving major disassembly of the engine.

Some operations, for example, require the use of a press. It would be wiser to have these performed by a shop equipped for such work, rather than to try to do the job yourself with makeshift equipment. Other procedures require precision measurements. Unless you have the skills and equipment required, it would be better to have a qualified repair shop make the measurements for you.

Repairs go much faster and easier if the parts that will be worked on are clean before you begin. There are special cleaners for washing the engine and related parts. Brush or spray on the cleaning solution, let it stand, then rinse it away with a garden hose. Clean all oily or greasy parts with cleaning solvent as you remove them.

WARNING

Never use gasoline as a cleaning agent. It presents an extreme fire hazard. Be sure to work in a well-ventilated area when using cleaning solvent. Keep a fire extinguisher, rated for gasoline fires, handy in any case.

Much of the labor charge for repairs made by dealers is for the removal and disassembly of other parts to reach the defective unit. It is frequently possible to perform the preliminary operations yourself and then take the defective unit in to the dealer for repair, at considerable savings.

Once you have decided to tackle the job yourself, make sure you locate the appropriate section in this manual, and read it entirely. Study the illustrations and text until you have a good idea of what is involved in completing the job satisfactorily. If special tools are required, make arrangements to get them before you start. Also, purchase any known defective parts prior to starting on the procedure. It is frustrating and time-consuming to get partially into a job and then be unable to complete it.

Simple wiring checks can be easily made at home, but knowledge of electronics is almost a necessity for performing tests with complicated electronic testing gear.

During disassembly of parts keep a few general cautions in mind. Force is rarely needed to get things apart. If parts are a tight fit, like a bearing in a case, there is usually a tool designed to separate them. Never use a screwdriver to pry apart parts with machined surfaces such as cylinder head and valve cover. You will mar the surfaces and end up with leaks.

Make diagrams wherever similar-appearing parts are found. You may think you can remember where everything came from — but mistakes are costly. There is also the possibility you may get sidetracked and not return to work for days or even weeks — in which interval, carefully laid out parts may have become disturbed.

Tag all similar internal parts for location, and mark all mating parts for position. Record number and thickness of any shims as they are removed. Small parts such as bolts can be identified by placing them in plastic sandwich bags that are sealed and labeled with masking tape.

Wiring should be tagged with masking tape and marked as each wire is removed. Again, do not rely on memory alone.

When working under the vehicle, do not trust a hydraulic or mechanical jack to hold the vehicle up by itself. Always use jackstands. See **Figure 1**.

Disconnect battery ground cable before working near electrical connections and before disconnecting wires. Never run the engine with the battery disconnected; the alternator could be seriously damaged.

Protect finished surfaces from physical damage or corrosion. Keep gasoline and brake fluid off painted surfaces.

Frozen or very tight bolts and screws can often be loosened by soaking with penetrating oil like Liquid Wrench or WD-40, then sharply striking the bolt head a few times with a hammer and punch (or screwdriver for screws). Avoid heat unless absolutely necessary, since it may melt, warp, or remove the temper from many parts.

Avoid flames or sparks when working near a charging battery or flammable liquids, such as brake fluid or gasoline.

No parts, except those assembled with a press fit, require unusual force during assembly. If a part is hard to remove or install, find out why before proceeding.

Cover all openings after removing parts to keep dirt, small tools, etc., from falling in.

When assembling two parts, start all fasteners, then tighten evenly.

The clutch plate, wiring connections, brake shoes, drums, pads, and discs should be kept clean and free of grease and oil.

When assembling parts, be sure all shims and washers are replaced exactly as they came out.

Whenever a rotating part butts against a stationary part, look for a shim or washer. Use new gaskets if there is any doubt about the condition of old ones. Generally, you should apply gasket cement to one mating surface only, so the parts may be easily disassembled in the future. A thin coat of oil on gaskets helps them seal effectively.

Heavy grease can be used to hold small parts in place if they tend to fall out during assembly. However, keep grease and oil away from electrical, clutch, and brake components.

High spots may be sanded off a piston with sandpaper, but emery cloth and oil do a much more professional job.

Carburetors are best cleaned by disassembling them and soaking the parts in a commercial carburetor cleaner. Never soak gaskets and rubber parts in these cleaners. Never use wire to clean out jets and air passages; they are easily damaged. Use compressed air to blow out the carburetor, but only if the float has been removed first.

Take your time and do the job right. Do not forget that a newly rebuilt engine must be broken in the same as a new one. Refer to your owner's manual for the proper break-in procedures.

1

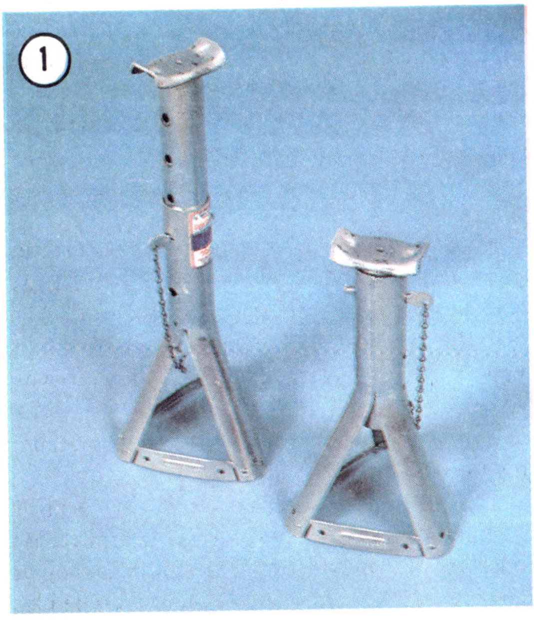

SAFETY FIRST

Professional mechanics can work for years and never sustain a serious injury. If you observe a few rules of common sense and safety, you can enjoy many safe hours servicing your vehicle. You could hurt yourself or damage the vehicle if you ignore these rules.

1. Never use gasoline as a cleaning solvent.

2. Never smoke or use a torch in the vicinity of flammable liquids such as cleaning solvent in open containers.

3. Never smoke or use a torch in an area where batteries are being charged. Highly explosive hydrogen gas is formed during the charging process.

4. Use the proper sized wrenches to avoid damage to nuts and injury to yourself.

5. When loosening a tight or stuck nut, be guided by what would happen if the wrench should slip. Protect yourself accordingly.

6. Keep your work area clean and uncluttered.

7. Wear safety goggles during all operations involving drilling, grinding, or use of a cold chisel.

8. Never use worn tools.

9. Keep a fire extinguisher handy and be sure it is rated for gasoline (Class B) and electrical (Class C) fires.

EXPENDABLE SUPPLIES

Certain expendable supplies are necessary. These include grease, oil, gasket cement, wiping rags, cleaning solvent, and distilled water. Also, special locking compounds, silicone lubricants, and engine cleaners may be useful. Cleaning solvent is available at most service stations and distilled water for the battery is available at most supermarkets.

SHOP TOOLS

For proper servicing, you will need an assortment of ordinary hand tools (**Figure 2**).

As a minimum, these include:

a. Combination wrenches
b. Sockets
c. Plastic mallet
d. Small hammer
e. Snap ring pliers
f. Gas pliers
g. Phillips screwdrivers
h. Slot (common) screwdrivers
i. Feeler gauges
j. Spark plug gauge
k. Spark plug wrench

Special tools necessary are shown in the chapters covering the particular repair in which they are used.

2

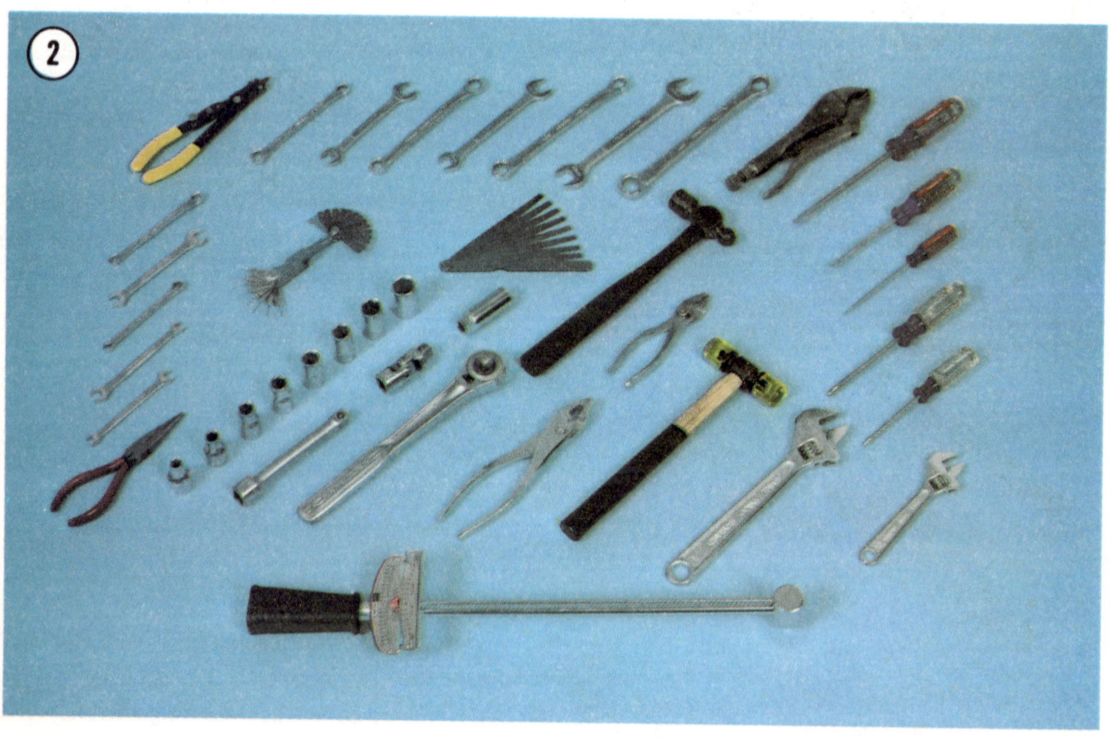

Engine tune-up and troubleshooting procedures require other special tools and equipment. These are described in detail in the following sections.

EMERGENCY TOOL KIT

A small emergency tool kit kept in the trunk is handy for road emergencies which otherwise could leave you stranded. The tools listed below and shown in **Figure 3** will let you handle most roadside repairs.

a. Combination wrenches
b. Crescent (adjustable) wrench
c. Screwdrivers — common and Phillips
d. Pliers — conventional (gas) and needle nose
e. Vise Grips
f. Hammer — plastic and metal
g. Small container of waterless hand cleaner
h. Rags for clean up
i. Silver waterproof sealing tape (duct tape)
j. Flashlight
k. Emergency road flares — at least four
l. Spare drive belts (water pump, alternator, etc.)

TROUBLESHOOTING AND TUNE-UP EQUIPMENT

Voltmeter, Ohmmeter, and Ammeter

For testing the ignition or electrical system, a good voltmeter is required. For automotive use, an instrument covering 0-20 volts is satisfac-

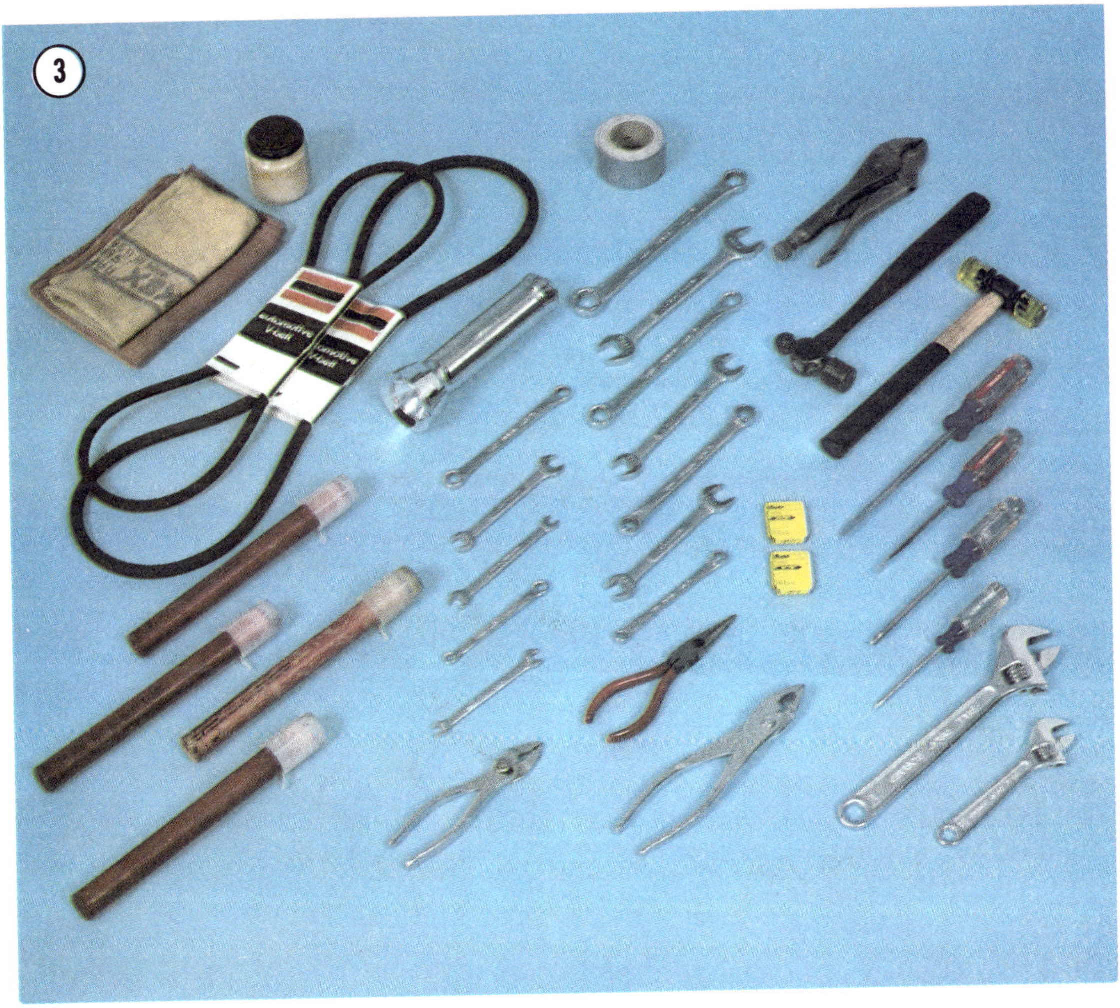

3

tory. One which also has a 0-2 volt scale is necessary for testing relays, points, or individual contacts where voltage drops are much smaller. Accuracy should be ± ½ volt.

An ohmmeter measures electrical resistance. This instrument is useful for checking continuity (open and short circuits), and testing fuses and lights.

The ammeter measures electrical current. Ammeters for automotive use should cover 0-50 amperes and 0-250 amperes. These are useful for checking battery charging and starting current.

Several inexpensive VOM's (volt-ohm-milliammeter) combine all three instruments into one which fits easily in any tool box. See **Figure 4**. However, the ammeter ranges are usually too small for automotive work.

4

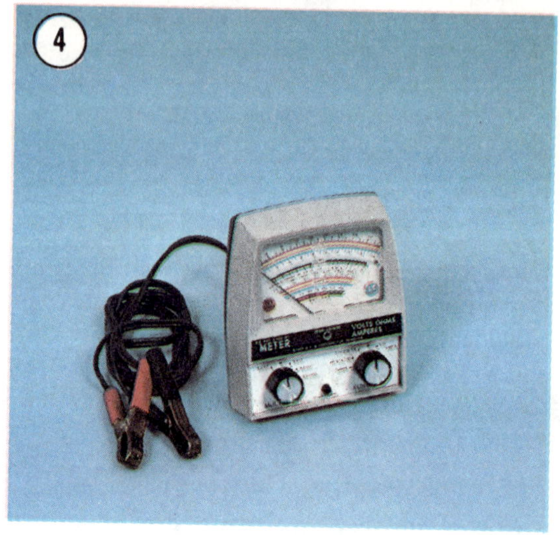

Hydrometer

The hydrometer gives a useful indication of battery condition and charge by measuring the specific gravity of the electrolyte in each cell. See **Figure 5**. Complete details on use and interpretation of readings are provided in the electrical chapter.

5

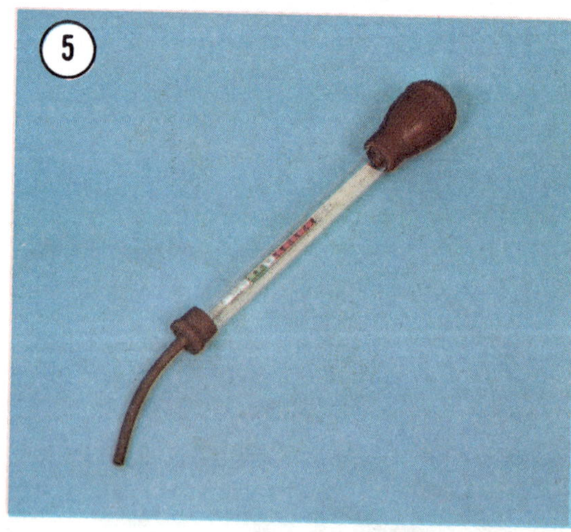

Compression Tester

The compression tester measures the compression pressure built up in each cylinder. The results, when properly interpreted, can indicate general cylinder and valve condition. See **Figure 6**.

6

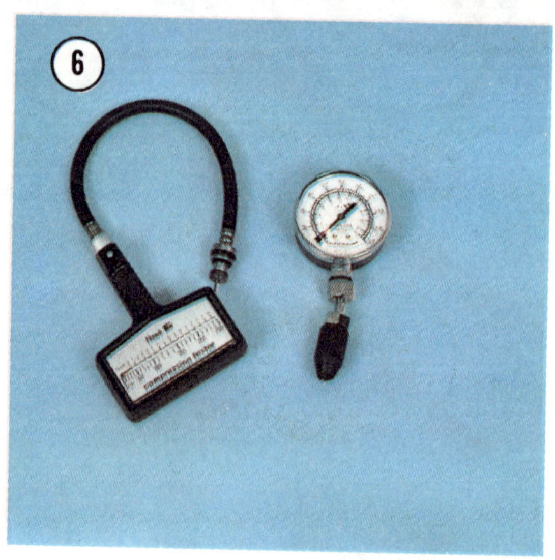

Vacuum Gauge

The vacuum gauge **(Figure 7)** is one of the easiest instruments to use, but one of the most difficult for the inexperienced mechanic to interpret. The results, when interpreted with other findings, can provide valuable clues to possible trouble.

To use the vacuum gauge, connect it to a vacuum hose that goes to the intake manifold. Attach it either directly to the hose or to a T-fitting installed into the hose.

NOTE: *Subtract one inch from the reading for every 1,000 ft. elevation.*

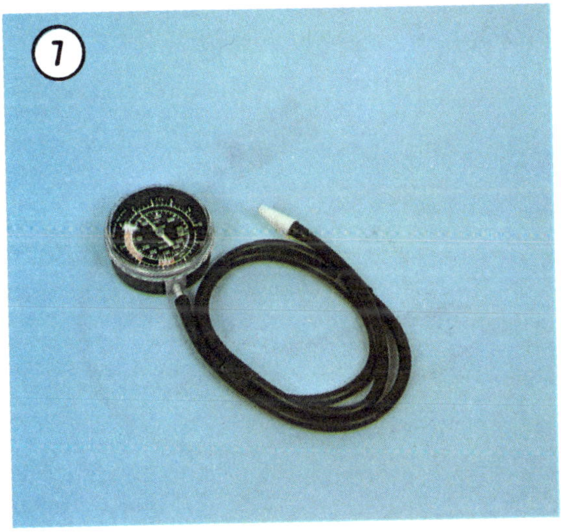

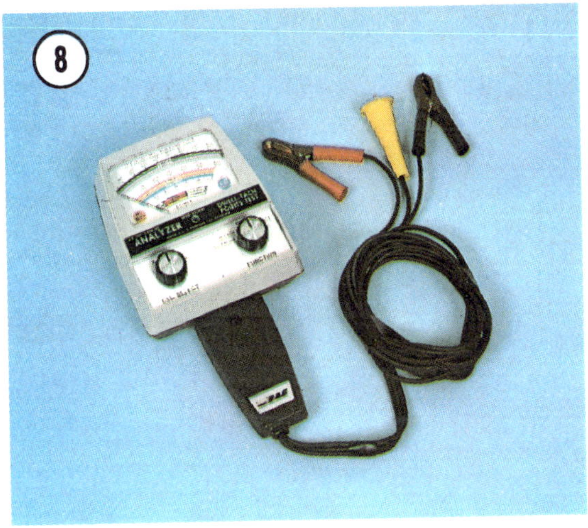

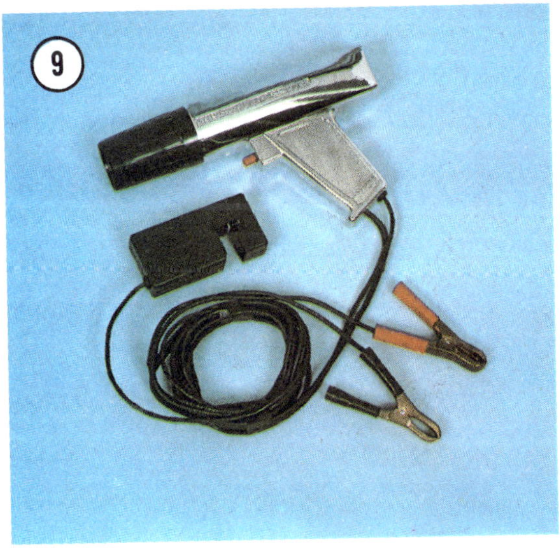

Fuel Pressure Gauge

This instrument is invaluable for evaluating fuel pump performance. Fuel system troubleshooting procedures in this manual use a fuel pressure gauge. Usually a vacuum gauge and fuel pressure gauge are combined.

Dwell Meter (Contact Breaker Point Ignition Only)

A dwell meter measures the distance in degrees of cam rotation that the breaker points remain closed while the engine is running. Since this angle is determined by breaker point gap, dwell angle is an accurate indication of breaker point gap.

Many tachometers intended for tuning and testing incorporate a dwell meter as well. See **Figure 8**. Follow the manufacturer's instructions to measure dwell.

Tachometer

A tachometer is necessary for tuning. See **Figure 8**. Ignition timing and carburetor adjustments must be performed at the specified idle speed. The best instrument for this purpose is one with a low range of 0-1,000 or 0-2,000 rpm for setting idle, and a high range of 0-4,000 or more for setting ignition timing at 3,000 rpm. Extended range (0-6,000 or 0-8,000 rpm) instruments lack accuracy at lower speeds. The instrument should be capable of detecting changes of 25 rpm on the low range.

Strobe Timing Light

This instrument is necessary for tuning, as it permits very accurate ignition timing. The light flashes at precisely the same instant that No. 1 cylinder fires, at which time the timing marks on the engine should align. Refer to Chapter Three for exact location of the timing marks for your engine.

Suitable lights range from inexpensive neon bulb types ($2-3) to powerful xenon strobe lights ($20-40). See **Figure 9**. Neon timing lights are difficult to see and must be used in dimly lit areas. Xenon strobe timing lights can be used outside in bright sunlight. Both types work on this vehicle; use according to the manufacturer's instructions.

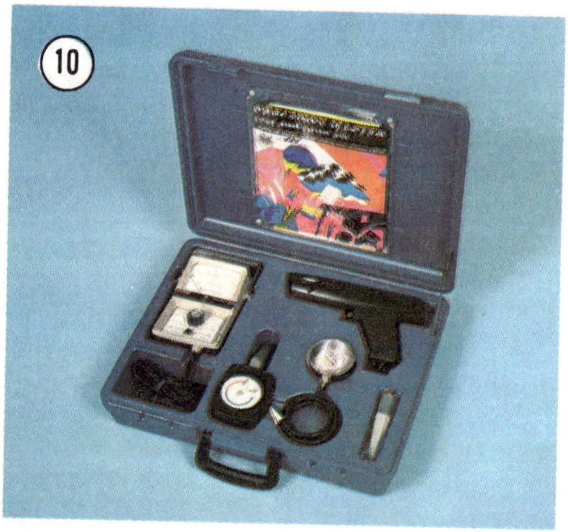

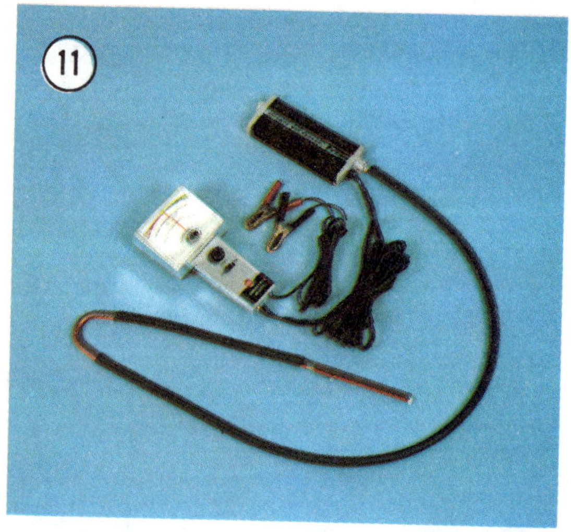

Tune-up Kits

Many manufacturers offer kits that combine several useful instruments. Some come in a convenient carry case and are usually less expensive than purchasing one instrument at a time. **Figure 10** shows one of the kits that is available. The prices vary with the number of instruments included in the kit.

Exhaust Gas Analyzer

Of all instruments described here, this is the least likely to be owned by a home mechanic. This instrument samples the exhaust gases from the tailpipe and measures the thermal conductivity of the exhaust gas. Since different gases conduct heat at varying rates, thermal conductivity of the exhaust is a good indication of gases present.

An exhaust gas analyzer is vital for accurately checking the effectiveness of exhaust emission control adjustments. They are relatively expensive to buy ($70 and up), but must be considered essential for the owner/mechanic to comply with today's emission laws. See **Figure 11**.

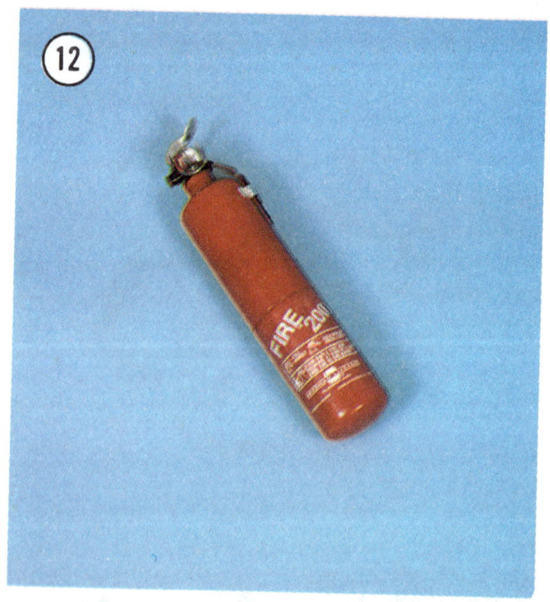

Fire Extinguisher

A fire extinguisher is a necessity when working on a vehicle. It should be rated for both *Class B* (flammable liquids — gasoline, oil, paint, etc.) and *Class C* (electrical — wiring, etc.) type fires. It should always be kept within reach. See **Figure 12**.

CHAPTER TWO

2

TROUBLESHOOTING

Troubleshooting can be a relatively simple matter if it is done logically. The first step in any troubleshooting procedure must be defining the symptoms as closely as possible. Subsequent steps involve testing and analyzing areas which could cause the symptoms. A haphazard approach may eventually find the trouble, but in terms of wasted time and unnecessary parts replacement, it can be very costly.

The troubleshooting procedures in this chapter analyze typical symptoms and show logical methods of isolation. These are not the only methods. There may be several approaches to a problem, but all methods must have one thing in common — a logical, systematic approach.

STARTING SYSTEM

The starting system consists of the starter motor and the starter solenoid. The ignition key controls the starter solenoid, which mechanically engages the starter with the engine flywheel, and supplies electrical current to turn the starter motor.

Starting system troubles are relatively easy to find. In most cases, the trouble is a loose or dirty electrical connection. **Figures 1 and 2** provide routines for finding the trouble.

CHARGING SYSTEM

The charging system consists of the alternator (or generator on older vehicles), voltage regulator, and battery. A drive belt driven by the engine crankshaft turns the alternator which produces electrical energy to charge the battery. As engine speed varies, the voltage from the alternator varies. A voltage regulator controls the charging current to the battery and maintains the voltage to the vehicle's electrical system at safe levels. A warning light or gauge on the instrument panel signals the driver when charging is not taking place. Refer to **Figure 3** for a typical charging system.

Complete troubleshooting of the charging system requires test equipment and skills which the average home mechanic does not possess. However, there are a few tests which can be done to pinpoint most troubles.

Charging system trouble may stem from a defective alternator (or generator), voltage regulator, battery, or drive belt. It may also be caused by something as simple as incorrect drive belt tension. The following are symptoms of typical problems you may encounter.

1. ***Battery dies frequently, even though the warning lamp indicates no discharge*** — This can be caused by a drive belt that is slightly too

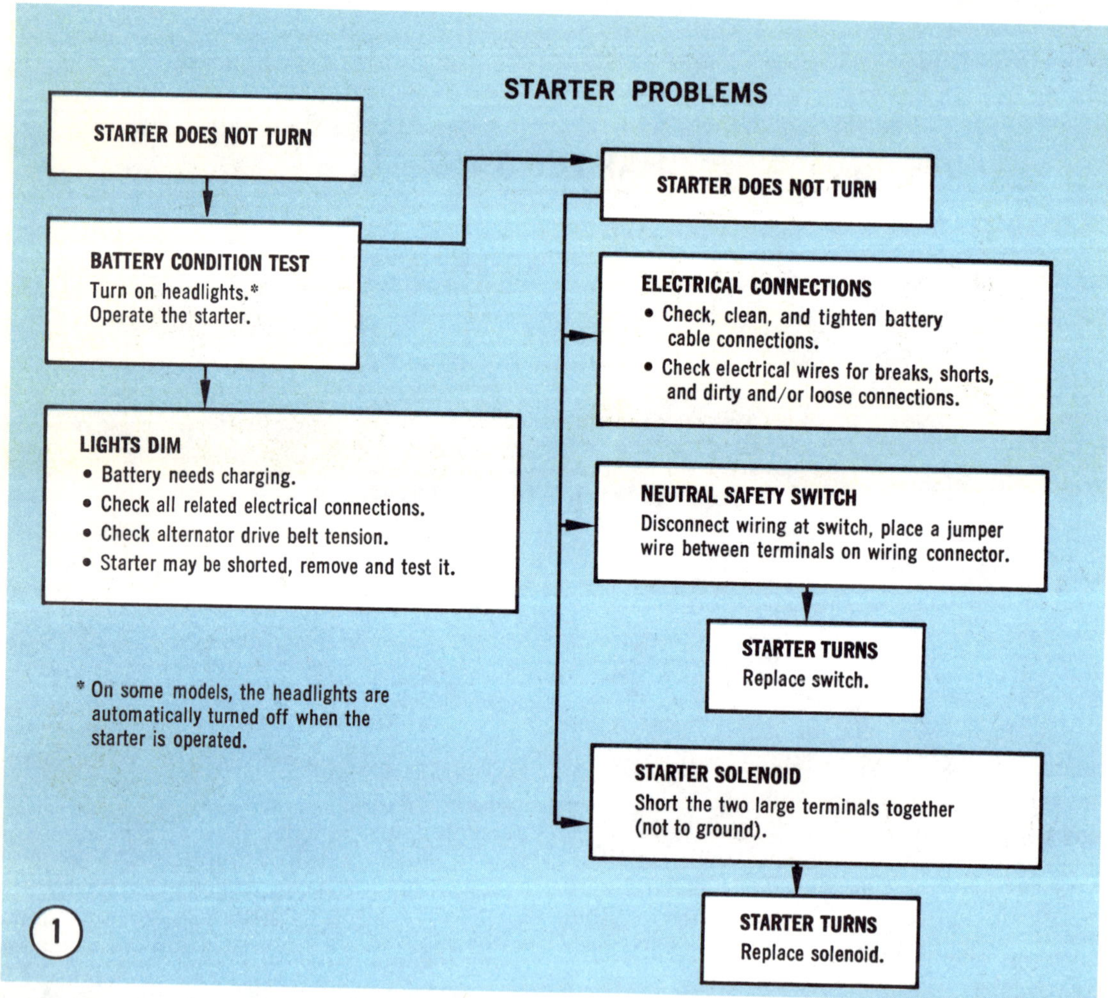
STARTER PROBLEMS
STARTER DOES NOT TURN
BATTERY CONDITION TEST
Turn on headlights.*
Operate the starter.
LIGHTS DIM
• Battery needs charging.
• Check all related electrical connections.
• Check alternator drive belt tension.
• Starter may be shorted, remove and test it.
* On some models, the headlights are automatically turned off when the starter is operated.
STARTER DOES NOT TURN
ELECTRICAL CONNECTIONS
• Check, clean, and tighten battery cable connections.
• Check electrical wires for breaks, shorts, and dirty and/or loose connections.
NEUTRAL SAFETY SWITCH
Disconnect wiring at switch, place a jumper wire between terminals on wiring connector.
STARTER TURNS
Replace switch.
STARTER SOLENOID
Short the two large terminals together (not to ground).
STARTER TURNS
Replace solenoid.

1

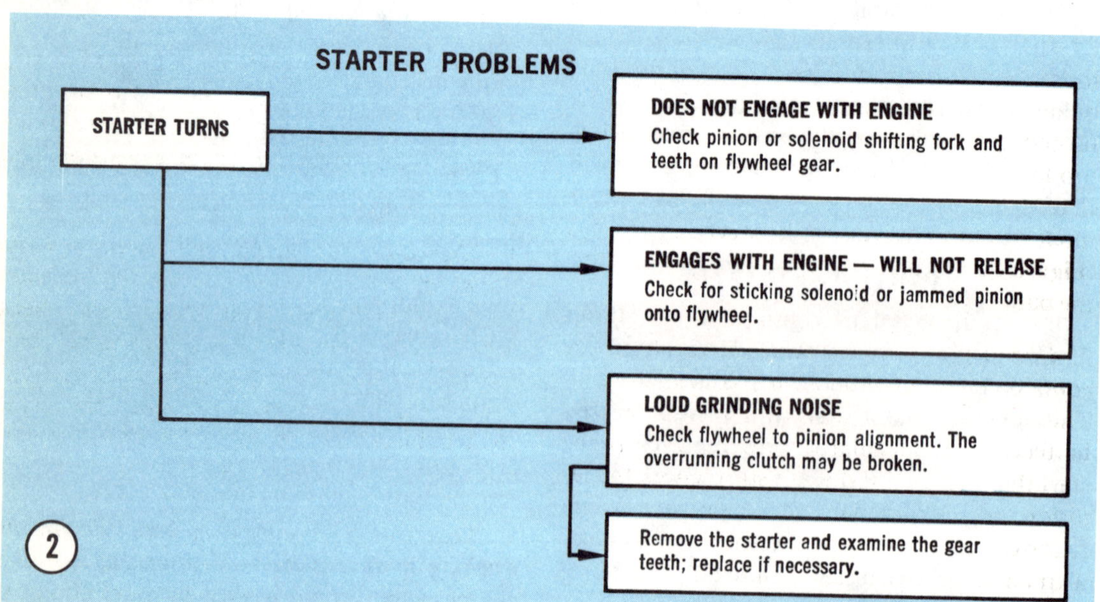
STARTER PROBLEMS
STARTER TURNS
DOES NOT ENGAGE WITH ENGINE
Check pinion or solenoid shifting fork and teeth on flywheel gear.
ENGAGES WITH ENGINE — WILL NOT RELEASE
Check for sticking solenoid or jammed pinion onto flywheel.
LOUD GRINDING NOISE
Check flywheel to pinion alignment. The overrunning clutch may be broken.
Remove the starter and examine the gear teeth; replace if necessary.

2

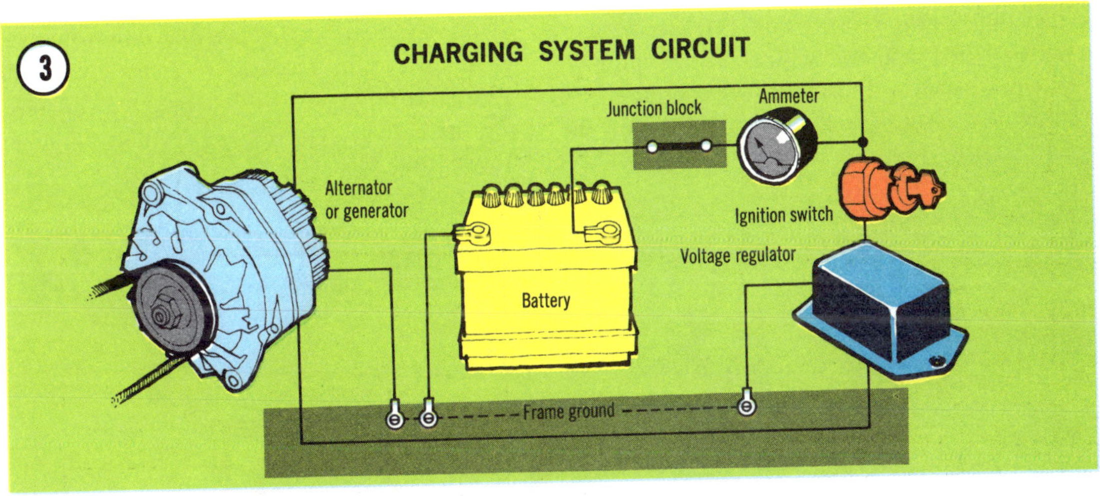

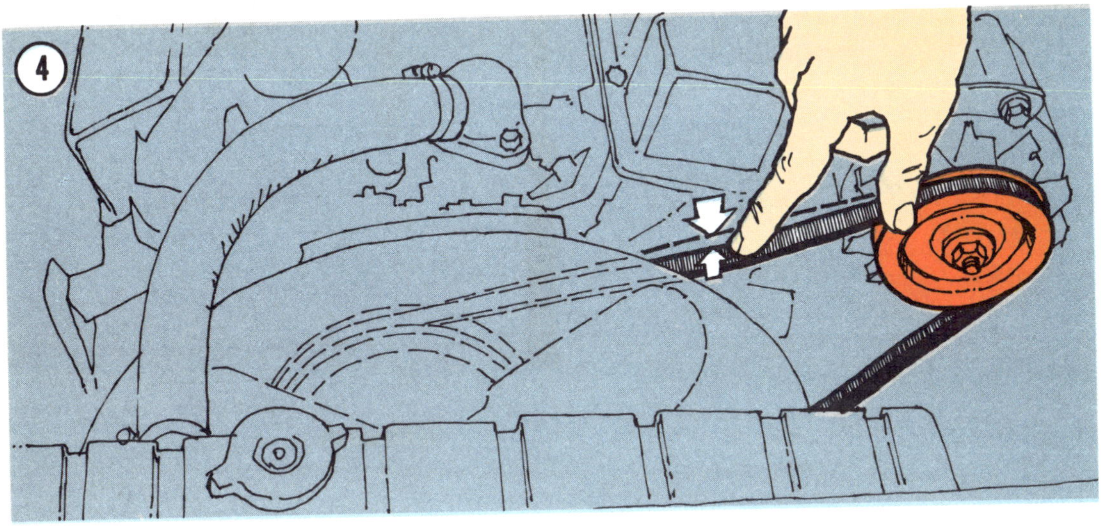

loose. Grasp the alternator (or generator) pulley and try to turn it. If the pulley can be turned without moving the belt, the drive belt is too loose. As a rule, keep the belt tight enough that it can be deflected about ½ in. under moderate thumb pressure between the pulleys (**Figure 4**). The battery may also be at fault; test the battery condition.

2. ***Charging system warning lamp does not come on when ignition switch is turned on*** — This may indicate a defective ignition switch, battery, voltage regulator, or lamp. First try to start the vehicle. If it doesn't start, check the ignition switch and battery. If the car starts, remove the warning lamp; test it for continuity with an ohmmeter or substitute a new lamp. If the lamp is good, locate the voltage regulator and make sure it is properly grounded (try tightening the mounting screws). Also the alternator (or generator) brushes may not be making contact. Test the alternator (or generator) and voltage regulator.

3. ***Alternator (or generator) warning lamp comes on and stays on*** — This usually indicates that no charging is taking place. First check drive belt tension (**Figure 4**). Then check battery condition, and check all wiring connections in the charging system. If this does not locate the trouble, check the alternator (or generator) and voltage regulator.

4. ***Charging system warning lamp flashes on and off intermittently*** — This usually indicates the charging system is working intermittently.

Check the drive belt tension (**Figure 4**), and check all electrical connections in the charging system. Check the alternator (or generator). *On generators only*, check the condition of the commutator.

5. ***Battery requires frequent additions of water, or lamps require frequent replacement*** — The alternator (or generator) is probably overcharging the battery. The voltage regulator is probably at fault.

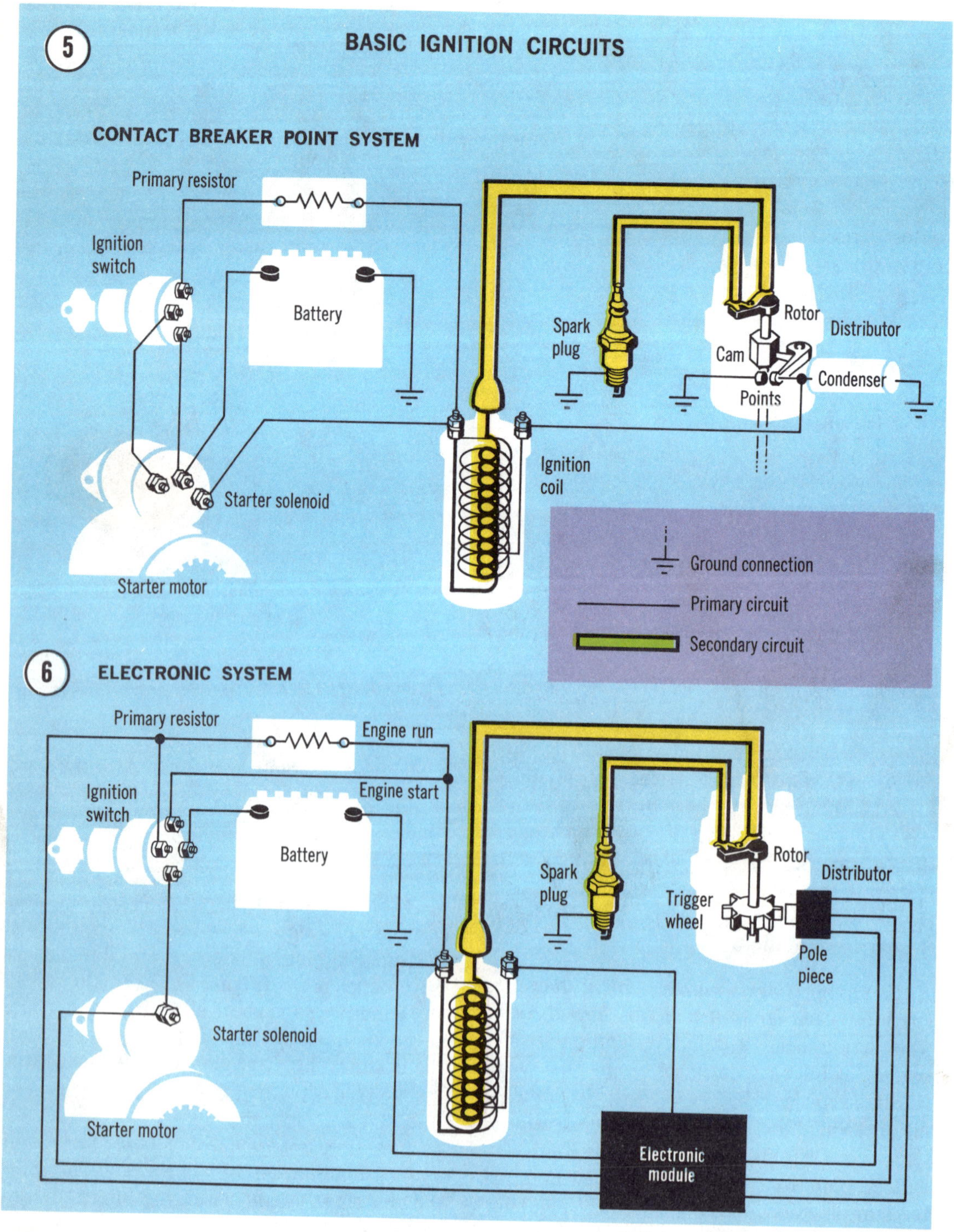

6. ***Excessive noise from the alternator (or generator)*** — Check for loose mounting brackets and bolts. The problem may also be worn bearings or the need of lubrication in some cases. If an alternator whines, a shorted diode may be indicated.

IGNITION SYSTEM

The ignition system may be either a conventional contact breaker type or an electronic ignition. See electrical chapter to determine which type you have. **Figures 5 and 6** show simplified diagrams of each type.

Most problems involving failure to start, poor performance, or rough running stem from trouble in the ignition system, particularly in contact breaker systems. Many novice troubleshooters get into trouble when they assume that these symptoms point to the fuel system instead of the ignition system.

Ignition system troubles may be roughly divided between those affecting only one cylinder and those affecting all cylinders. If the trouble affects only one cylinder, it can only be in the spark plug, spark plug wire, or portion of the distributor associated with that cylinder. If the trouble affects all cylinders (weak spark or no spark), then the trouble is in the ignition coil, rotor, distributor, or associated wiring.

The troubleshooting procedures outlined in **Figure 7** (breaker point ignition) or **Figure 8**

7

BREAKER POINT IGNITION PROBLEMS

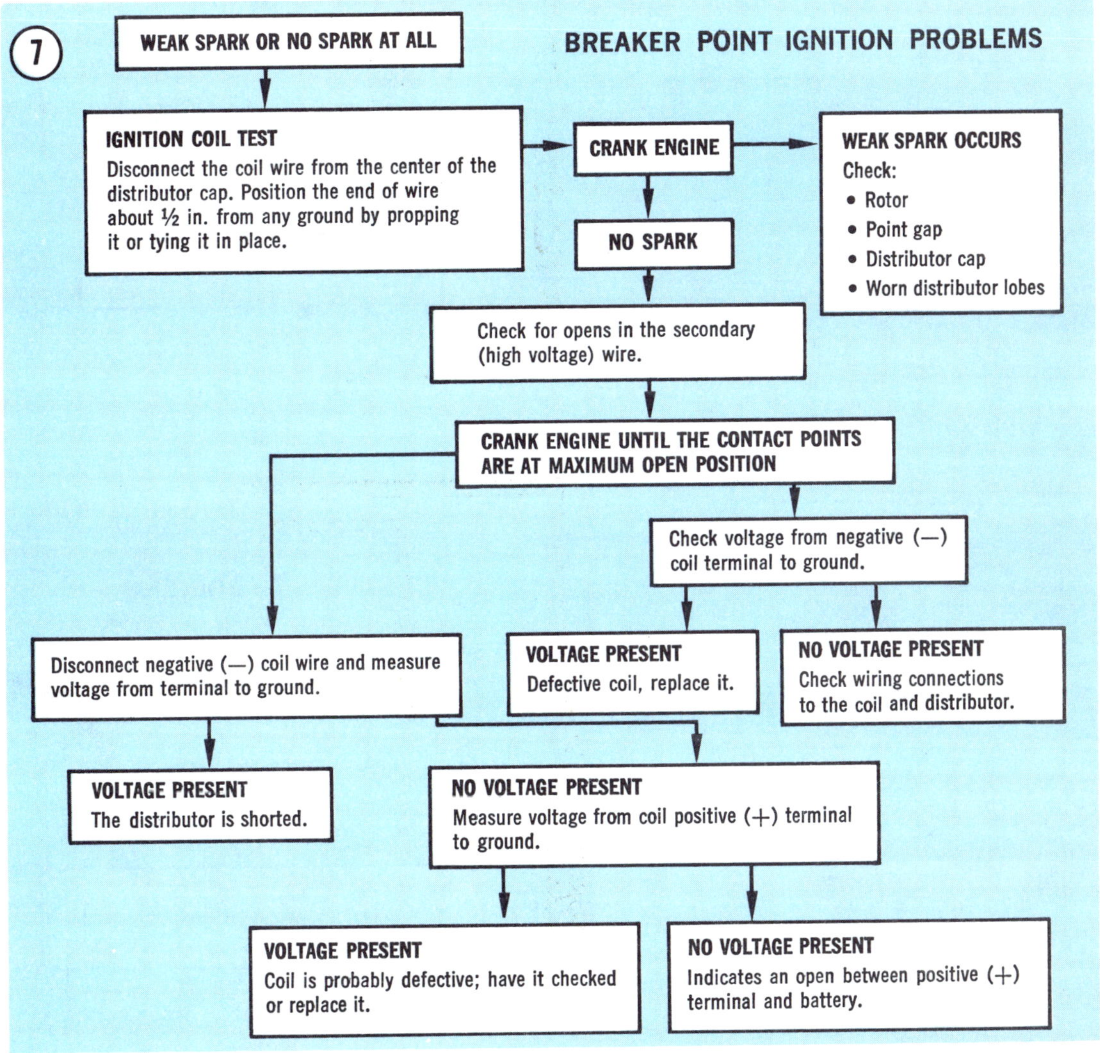

(electronic ignition) will help you isolate ignition problems fast. Of course, they assume that the battery is in good enough condition to crank the engine over at its normal rate.

ENGINE PERFORMANCE

A number of factors can make the engine difficult or impossible to start, or cause rough running, poor performance and so on. The majority of novice troubleshooters immediately suspect the carburetor or fuel injection system. In the majority of cases, though, the trouble exists in the ignition system.

The troubleshooting procedures outlined in **Figures 9 through 14** will help you solve the majority of engine starting troubles in a systematic manner.

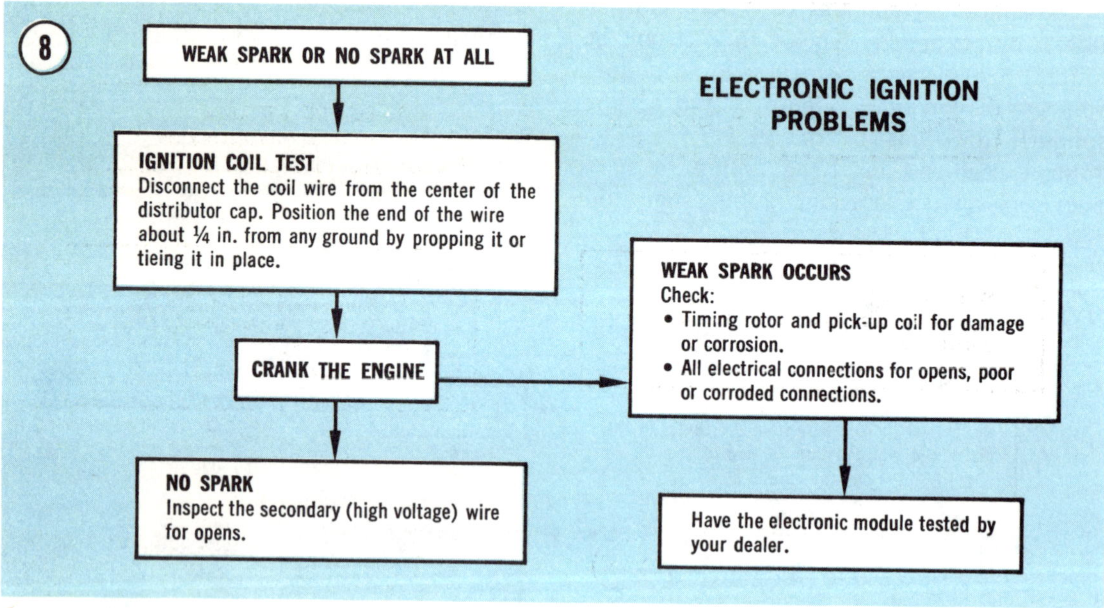

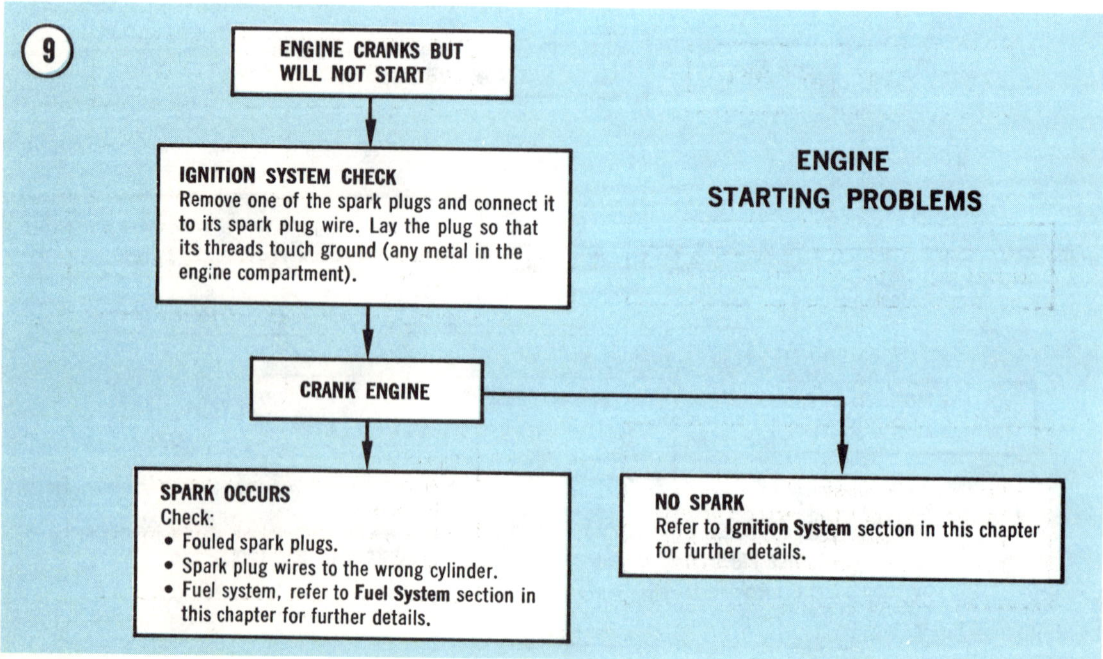

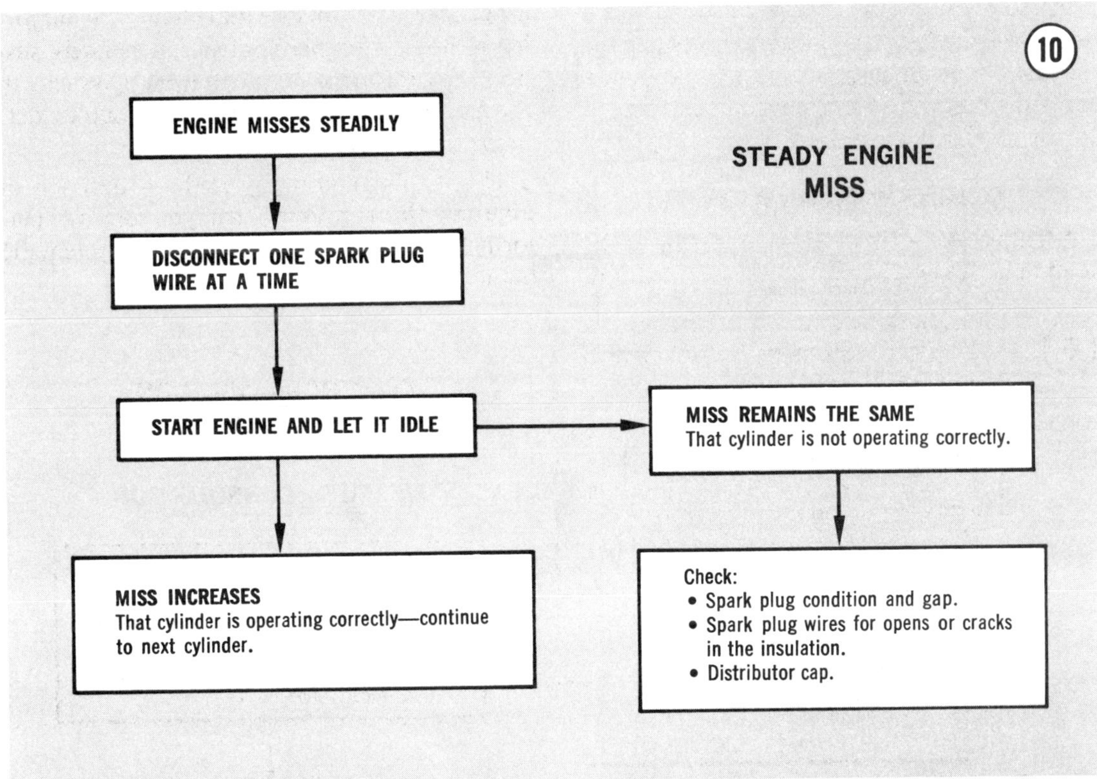
10
STEADY ENGINE MISS
ENGINE MISSES STEADILY
DISCONNECT ONE SPARK PLUG WIRE AT A TIME
START ENGINE AND LET IT IDLE
MISS REMAINS THE SAME
That cylinder is not operating correctly.
Check:
• Spark plug condition and gap.
• Spark plug wires for opens or cracks in the insulation.
• Distributor cap.
MISS INCREASES
That cylinder is operating correctly—continue to next cylinder.

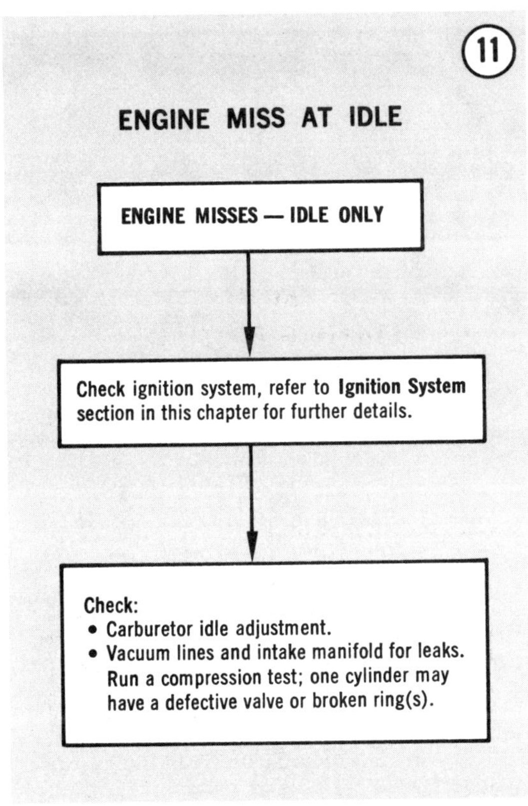
11
ENGINE MISS AT IDLE
ENGINE MISSES — IDLE ONLY
Check ignition system, refer to Ignition System section in this chapter for further details.
Check:
• Carburetor idle adjustment.
• Vacuum lines and intake manifold for leaks.
Run a compression test; one cylinder may have a defective valve or broken ring(s).

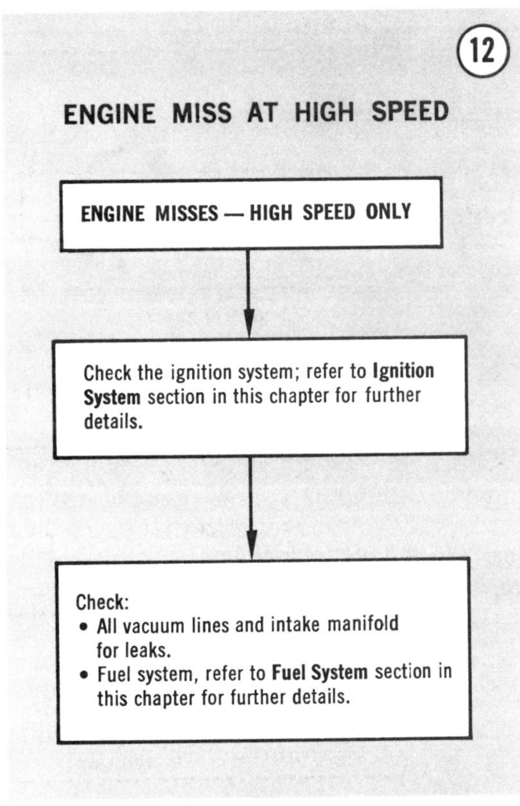
12
ENGINE MISS AT HIGH SPEED
ENGINE MISSES — HIGH SPEED ONLY
Check the ignition system; refer to Ignition System section in this chapter for further details.
Check:
• All vacuum lines and intake manifold for leaks.
• Fuel system, refer to Fuel System section in this chapter for further details.

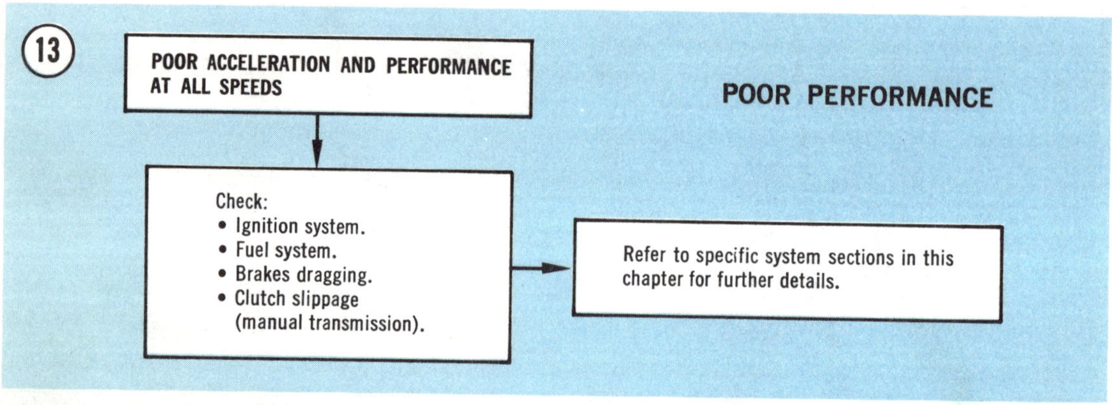

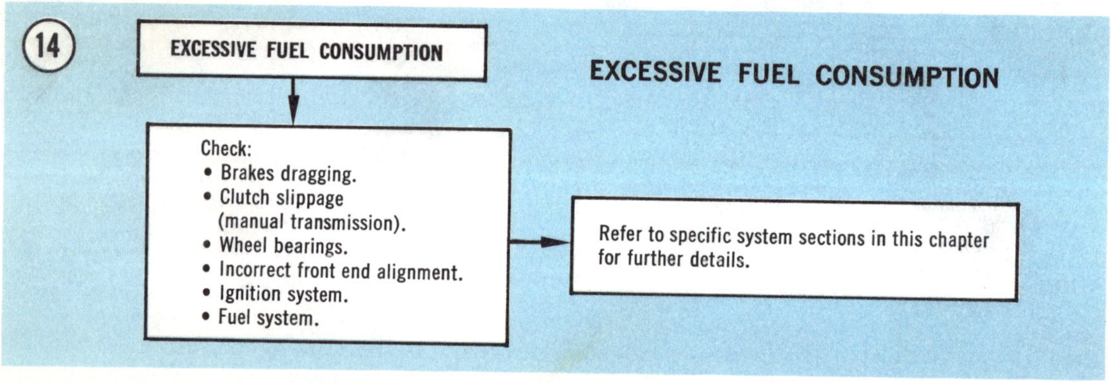

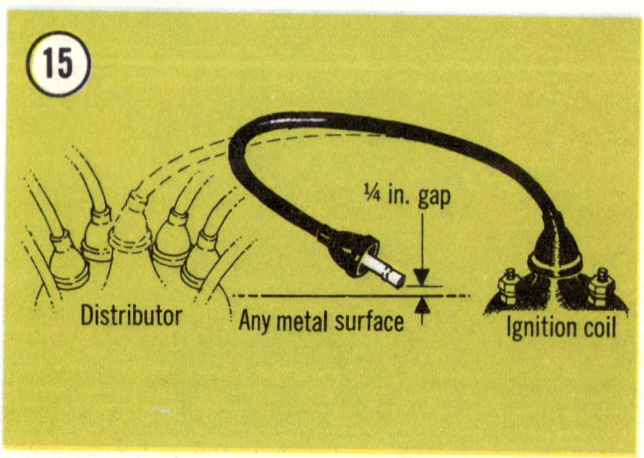

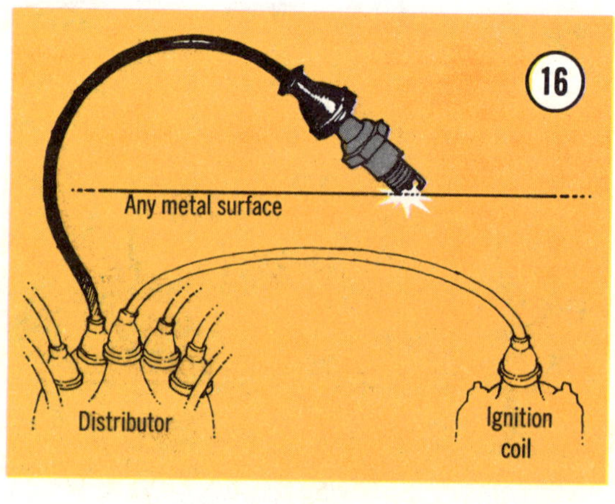

Some tests of the ignition system require running the engine with a spark plug or ignition coil wire disconnected. The safest way to do this is to disconnect the wire with the engine stopped, then prop the end of the wire next to a metal surface as shown in **Figures 15 and 16**.

WARNING

Never disconnect a spark plug or ignition coil wire while the engine is running. The high voltage in an ignition system, particularly the newer high-energy electronic ignition systems could cause serious injury or even death.

Spark plug condition is an important indication of engine performance. Spark plugs in a properly operating engine will have slightly pitted electrodes, and a light tan insulator tip. **Figure 17** shows a normal plug, and a number of others which indicate trouble in their respective cylinders.

(17)

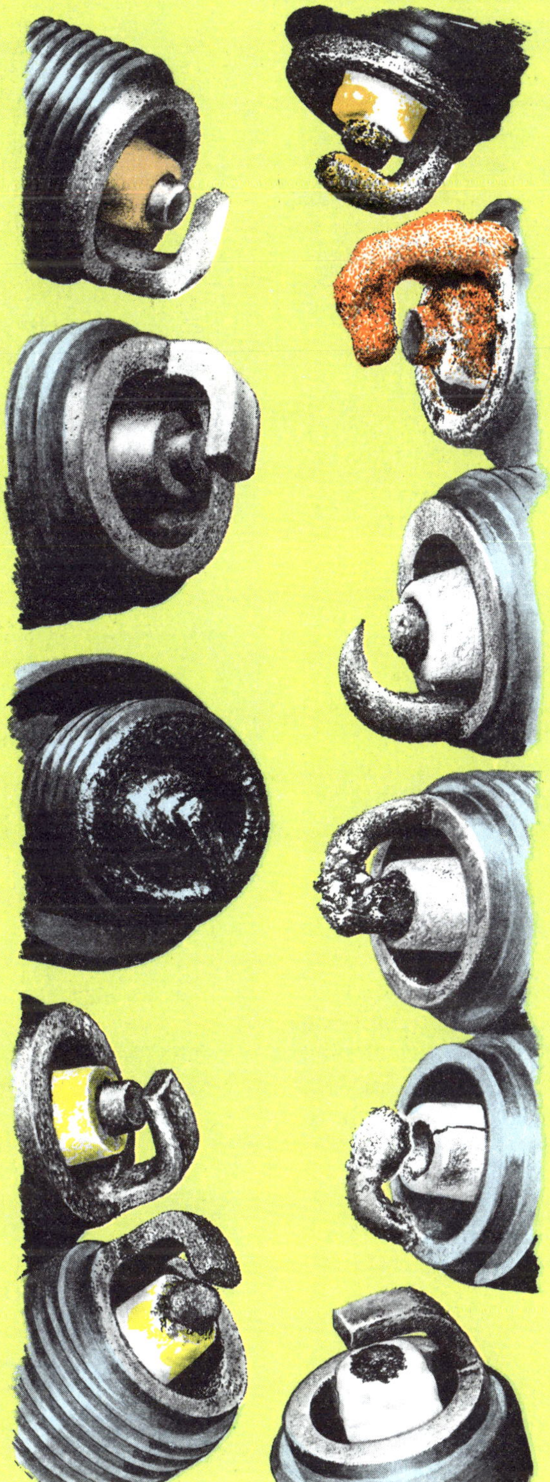

NORMAL

- Appearance—Firing tip has deposits of light gray to light tan.
- Can be cleaned, regapped and reused.

CARBON FOULED

- Appearance—Dull, dry black with fluffy carbon deposits on the insulator tip, electrode and exposed shell.
- Caused by—Fuel/air mixture too rich, plug heat range too cold, weak ignition system, dirty air cleaner, faulty automatic choke or excessive idling.
- Can be cleaned, regapped and reused.

OIL FOULED

- Appearance—Wet black deposits on insulator and exposed shell.
- Caused by—Excessive oil entering the combustion chamber through worn rings, pistons, valve guides or bearings.
- Replace with new plugs (use a hotter plug if engine is not repaired).

LEAD FOULED

- Appearance — Yellow insulator deposits (may sometimes be dark gray, black or tan in color) on the insulator tip.
- Caused by—Highly leaded gasoline.
- Replace with new plugs.

LEAD FOULED

- Appearance—Yellow glazed deposits indicating melted lead deposits due to hard acceleration.
- Caused by—Highly leaded gasoline.
- Replace with new plugs.

OIL AND LEAD FOULED

- Appearance—Glazed yellow deposits with a slight brownish tint on the insulator tip and ground electrode.
- Replace with new plugs.

FUEL ADDITIVE RESIDUE

- Appearance — Brown colored hardened ash deposits on the insulator tip and ground electrode.
- Caused by—Fuel and/or oil additives.
- Replace with new plugs.

WORN

- Appearance — Severely worn or eroded electrodes.
- Caused by—Normal wear or unusual oil and/or fuel additives.
- Replace with new plugs.

PREIGNITION

- Appearance — Melted ground electrode.
- Caused by—Overadvanced ignition timing, inoperative ignition advance mechanism, too low of a fuel octane rating, lean fuel/air mixture or carbon deposits in combustion chamber.

PREIGNITION

- Appearance—Melted center electrode.
- Caused by—Abnormal combustion due to overadvanced ignition timing or incorrect advance, too low of a fuel octane rating, lean fuel/air mixture, or carbon deposits in combustion chamber.
- Correct engine problem and replace with new plugs.

INCORRECT HEAT RANGE

- Appearance—Melted center electrode and white blistered insulator tip.
- Caused by—Incorrect plug heat range selection.
- Replace with new plugs.

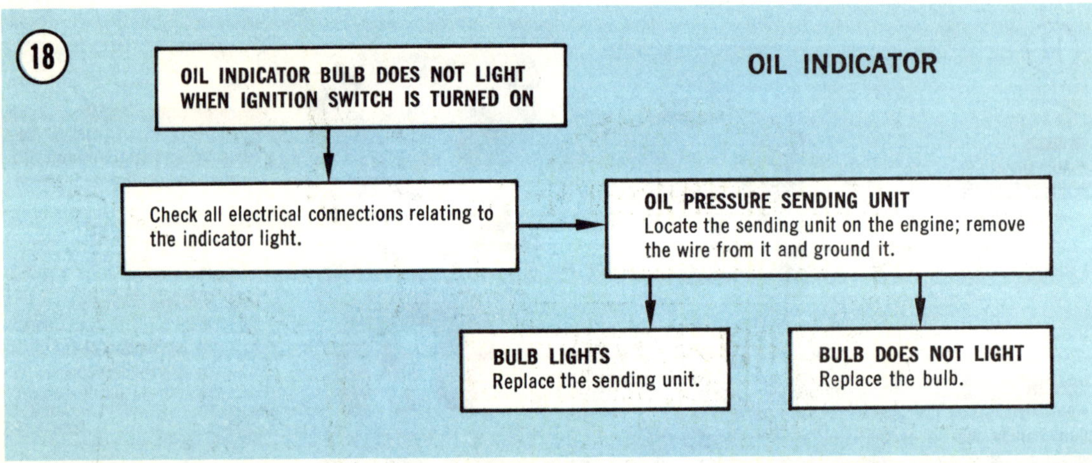

ENGINE OIL PRESSURE LIGHT

Proper oil pressure to the engine is vital. If oil pressure is insufficient, the engine can destroy itself in a comparatively short time.

The oil pressure warning circuit monitors oil pressure constantly. If pressure drops below a predetermined level, the light comes on.

Obviously, it is vital for the warning circuit to be working to signal low oil pressure. Each time you turn on the ignition, but before you start the car, the warning light should come on. If it doesn't, there is trouble in the warning circuit, not the oil pressure system. See **Figure 18** to troubleshoot the warning circuit.

Once the engine is running, the warning light should stay off. If the warning light comes on or acts erratically while the engine is running there is trouble with the engine oil pressure system. *Stop the engine immediately*. Refer to **Figure 19** for possible causes of the problem.

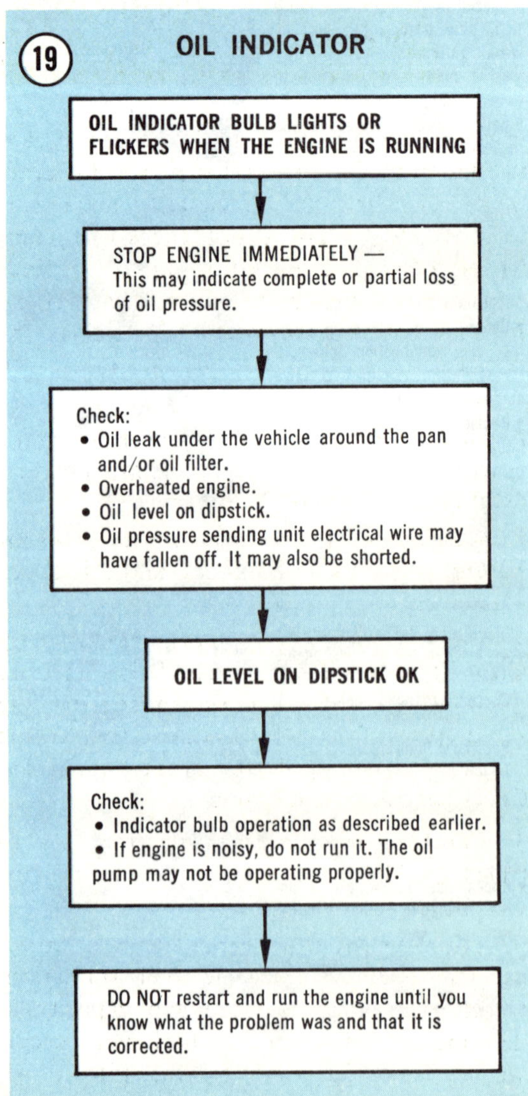

FUEL SYSTEM (CARBURETTED)

Fuel system problems must be isolated to the fuel pump (mechanical or electric), fuel lines, fuel filter, or carburetor. These procedures assume the ignition system is working properly and is correctly adjusted.

1. ***Engine will not start*** — First make sure that fuel is being delivered to the carburetor. Remove the air cleaner, look into the carburetor throat, and operate the accelerator

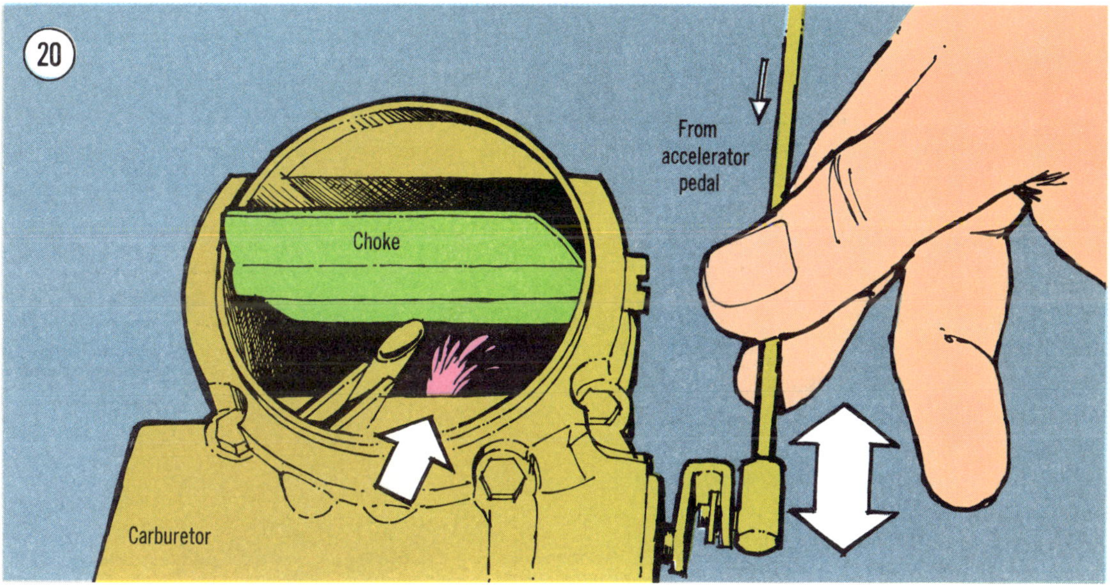

2

linkage several times. There should be a stream of fuel from the accelerator pump discharge tube each time the accelerator linkage is depressed (**Figure 20**). If not, check fuel pump delivery (described later), float valve, and float adjustment. If the engine will not start, check the automatic choke parts for sticking or damage. If necessary, rebuild or replace the carburetor.

2. ***Engine runs at fast idle*** — Check the choke setting. Check the idle speed, idle mixture, and decel valve (if equipped) adjustment.

3. ***Rough idle or engine miss with frequent stalling*** — Check idle mixture and idle speed adjustments.

4. ***Engine "diesels" (continues to run) when ignition is switched off*** — Check idle mixture (probably too rich), ignition timing, and idle speed (probably too fast). Check the throttle solenoid (if equipped) for proper operation. Check for overheated engine.

5. ***Stumbling when accelerating from idle*** — Check the idle speed and mixture adjustments. Check the accelerator pump.

6. ***Engine misses at high speed or lacks power*** — This indicates possible fuel starvation. Check fuel pump pressure and capacity as described in this chapter. Check float needle valves. Check for a clogged fuel filter or air cleaner.

7. ***Black exhaust smoke*** — This indicates a badly overrich mixture. Check idle mixture and idle speed adjustment. Check choke setting. Check for excessive fuel pump pressure, leaky floats, or worn needle valves.

8. ***Excessive fuel consumption*** — Check for overrich mixture. Make sure choke mechanism works properly. Check idle mixture and idle speed. Check for excessive fuel pump pressure, leaky floats, or worn float needle valves.

FUEL SYSTEM (FUEL INJECTED)

Troubleshooting a fuel injection system requires more thought, experience, and know-how than any other part of the vehicle. A logical approach and proper test equipment are essential in order to successfully find and fix these troubles.

It is best to leave fuel injection troubles to your dealer. In order to isolate a problem to the injection system make sure that the fuel pump is operating properly. Check its performance as described later in this section. Also make sure that fuel filter and air cleaner are not clogged.

FUEL PUMP TEST (MECHANICAL AND ELECTRIC)

1. Disconnect the fuel inlet line where it enters the carburetor or fuel injection system.

2. Fit a rubber hose over the fuel line so fuel can be directed into a graduated container with about one quart capacity. See **Figure 21**.

3. To avoid accidental starting of the engine, disconnect the secondary coil wire from the coil.

4. Crank the engine for about 30 seconds.

5. If the fuel pump supplies the specified amount (refer to the fuel chapter later in this book), the trouble may be in the carburetor or fuel injection system. The fuel injection system should be tested by your dealer.

6. If there is no fuel present or the pump cannot supply the specified amount, either the fuel pump is defective or there is an obstruction in the fuel line. Replace the fuel pump and/or inspect the fuel lines for air leaks or obstructions.

7. Also pressure test the fuel pump by installing a T-fitting in the fuel line between the fuel pump and the carburetor. Connect a fuel pressure gauge to the fitting with a short tube (**Figure 22**).

8. Reconnect the primary coil wire, start the engine, and record the pressure. Refer to the fuel chapter later in this book for the correct pressure. If the pressure varies from that specified, the pump should be replaced.

9. Stop the engine. The pressure should drop off very slowly. If it drops off rapidly, the outlet valve in the pump is leaking and the pump should be replaced.

EMISSION CONTROL SYSTEMS

Major emission control systems used on nearly all U.S. models include the following:

a. Positive crankcase ventilation (PCV)
b. Thermostatic air cleaner
c. Air injection reaction (AIR)
d. Fuel evaporation control
e. Exhaust gas recirculation (EGR)

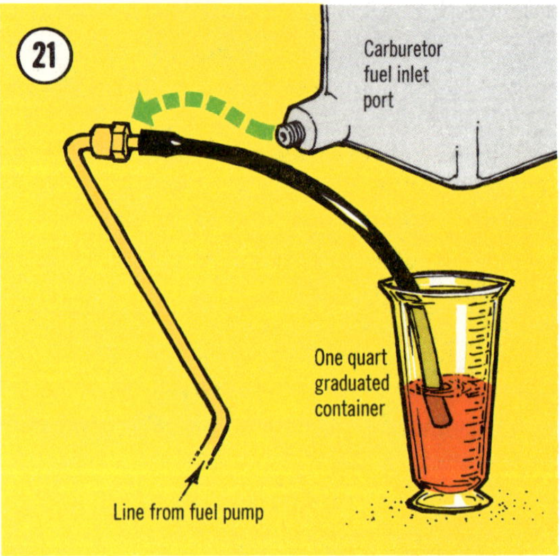

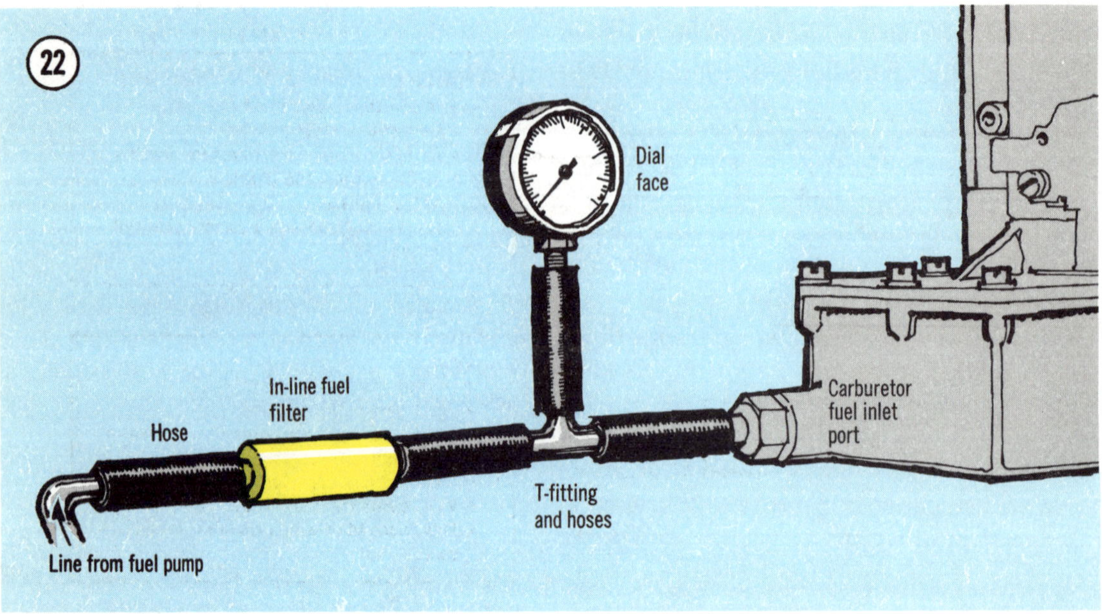

Emission control systems vary considerably from model to model. Individual models contain variations of the four systems described here. In addition, they may include other special systems. Use the index to find specific emission control components in other chapters.

Many of the systems and components are factory set and sealed. Without special expensive test equipment, it is impossible to adjust the systems to meet state and federal requirements.

Troubleshooting can also be difficult without special equipment. The procedures described below will help you find emission control parts which have failed, but repairs may have to be entrusted to a dealer or other properly equipped repair shop.

With the proper equipment, you can test the carbon monoxide and hydrocarbon levels. **Figure 23** provides some sources of trouble if the readings are not correct.

Positive Crankcase Ventilation

Fresh air drawn from the air cleaner housing scavenges emissions (e.g., piston blow-by) from thc crankcasc, thcn thc intakc manifold vacuum draws emissions into the intake manifold. They can then be reburned in the normal combustion process. **Figure 24** shows a typical system. **Figure 25** provides a testing procedure.

Thermostatic Air Cleaner

The thermostatically controlled air cleaner maintains incoming air to the engine at a predetermined level, usually about 100°F or higher. It mixes cold air with heated air from the exhaust manifold region. The air cleaner in-

(23) EXHAUST EMISSION PROBLEMS

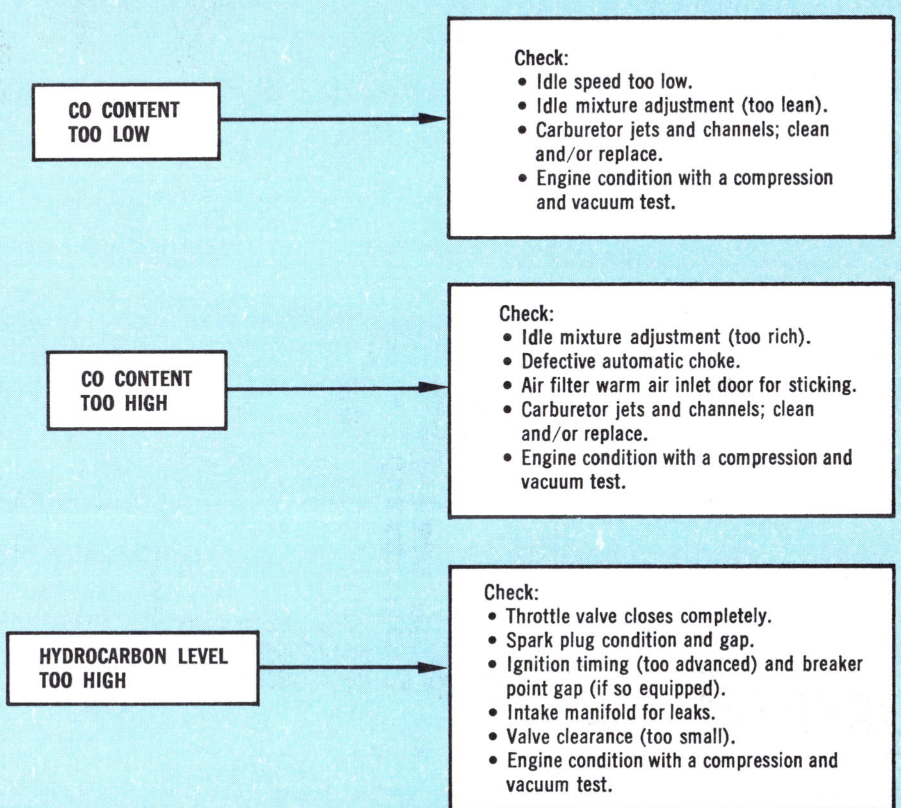

cludes a temperature sensor, vacuum motor, and a hinged door. See **Figure 26**.

The system is comparatively easy to test. See **Figure 27** for the procedure.

Air Injection Reaction System

The air injection reaction system reduces air pollution by oxidizing hydrocarbons and carbon monoxide as they leave the combustion chamber. See **Figure 28**.

The air injection pump, driven by the engine, compresses filtered air and injects it at the exhaust port of each cylinder. The fresh air mixes with the unburned gases in the exhaust and promotes further burning. A check valve prevents exhaust gases from entering and damaging the air pump if the pump becomes inoperative, e.g., from a fan belt failure.

Figure 29 explains the testing procedure for this system.

Fuel Evaporation Control

Fuel vapor from the fuel tank passes through the liquid/vapor separator to the carbon canister. See **Figure 30**. The carbon absorbs and

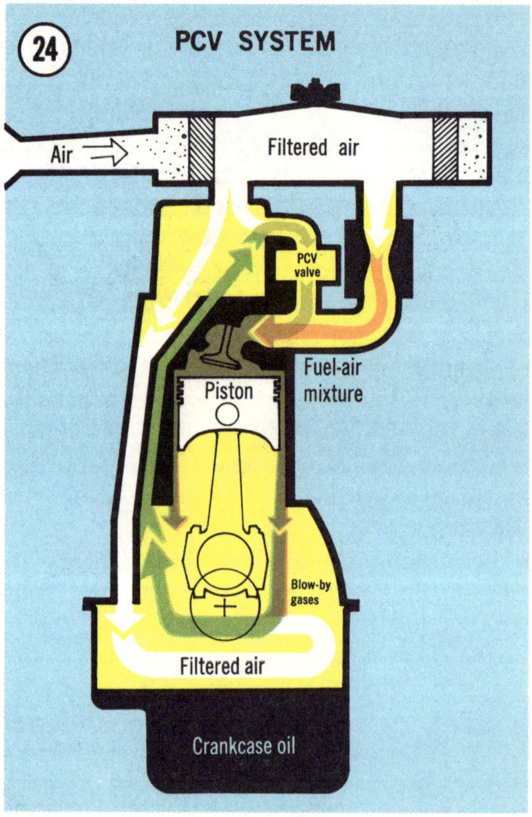

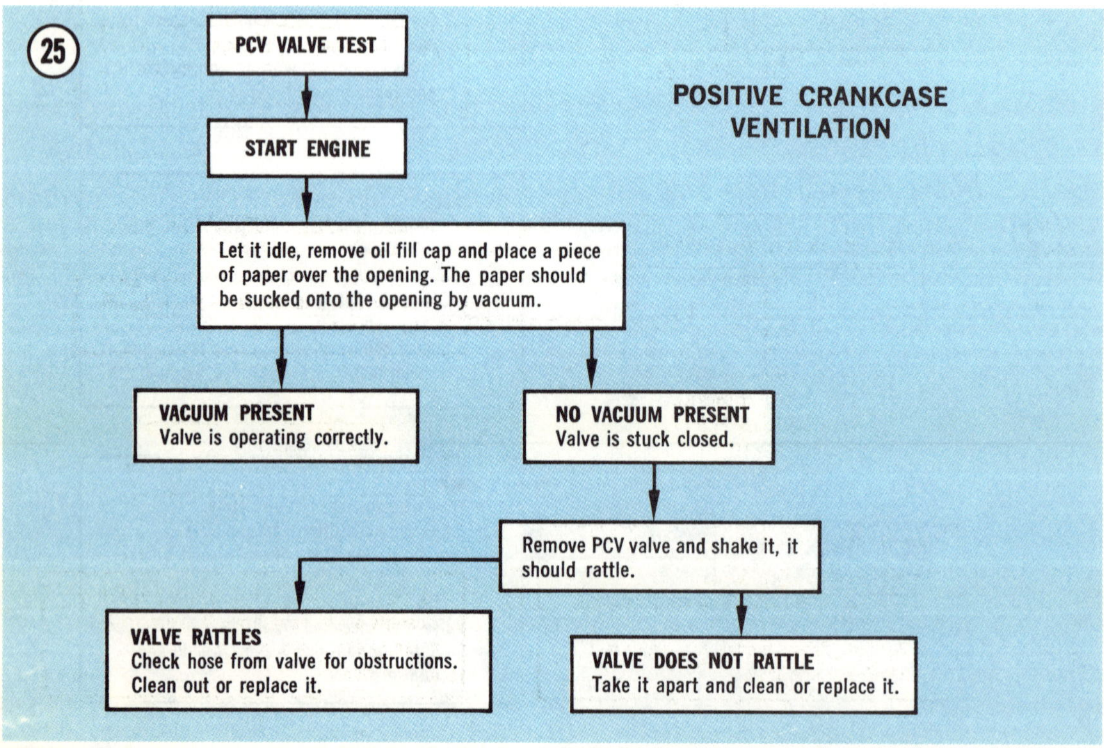

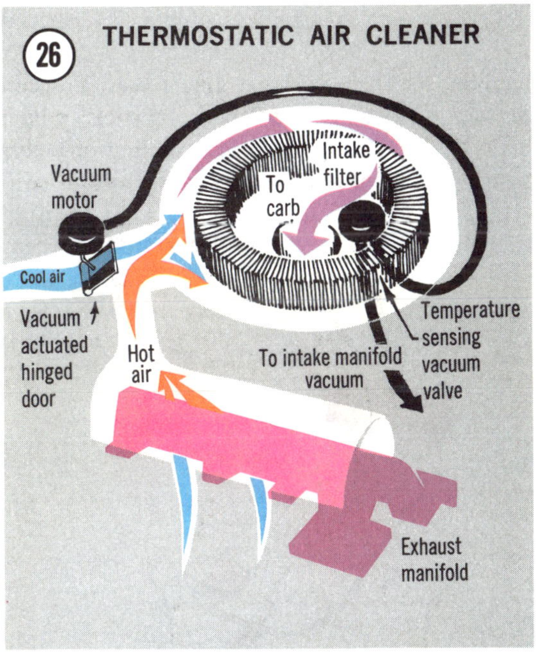

(26) THERMOSTATIC AIR CLEANER

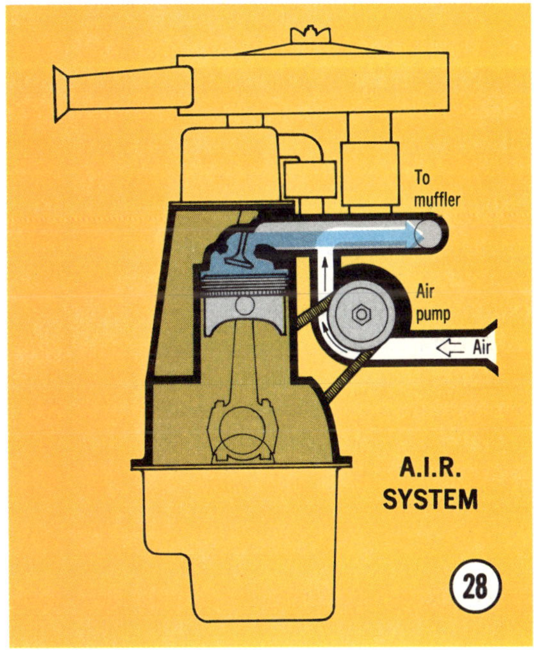

(28) A.I.R. SYSTEM

(27)

THERMOSTATIC AIR CLEANER

THERMOSTATIC AIR CLEANER

↓

Normal operation — Closed for cold engine.
— Open for warm engine.

↓

OPENS AND CLOSES
Is operating correctly.

DOES NOT OPEN OR CLOSE
Check for binding linkage or a leak in the vacuum line.

(29)

AIR INJECTION REACTOR

AIR INJECTION REACTOR

↓

PUMP NOT RUNNING AT PROPER SPEED

↓

Check:
- Drive belt tension.
- Oil the bearings (if there are provisions to do so).
- Air filter (some models have their own small air filter).

PUMP NOT PRODUCING AIR PRESSURE
(Approximately 1 psi)

↓

REMOVE THE AIR FILTER
(if so equipped)

↓

Let the engine idle, place a burning match or cigarette at the inlet. The pump should draw in the smoke. If not, have it serviced by your dealer.

stores the vapor when the engine is stopped. When the engine runs, manifold vacuum draws the vapor from the canister. Instead of being released into the atmosphere, the fuel vapor takes part in the normal combustion process.

Exhaust Gas Recirculation

The exhaust gas recirculation (EGR) system is used to reduce the emission of nitrogen oxides (NOx). Relatively inert exhaust gases are introduced into the combustion process to slightly reduce peak temperatures. This reduction in temperature reduces the formation of NOx.

Figure 31 provides a simple test of this system.

ENGINE NOISES

Often the first evidence of an internal engine trouble is a strange noise. That knocking, clicking, or tapping which you never heard before may be warning you of impending trouble.

While engine noises can indicate problems, they are sometimes difficult to interpret correctly; inexperienced mechanics can be seriously misled by them.

Professional mechanics often use a special stethoscope which looks similar to a doctor's stethoscope for isolating engine noises. You can do nearly as well with a "sounding stick" which can be an ordinary piece of doweling or a section of small hose. By placing one end in contact with the area to which you want to listen and the other end near your ear, you can hear

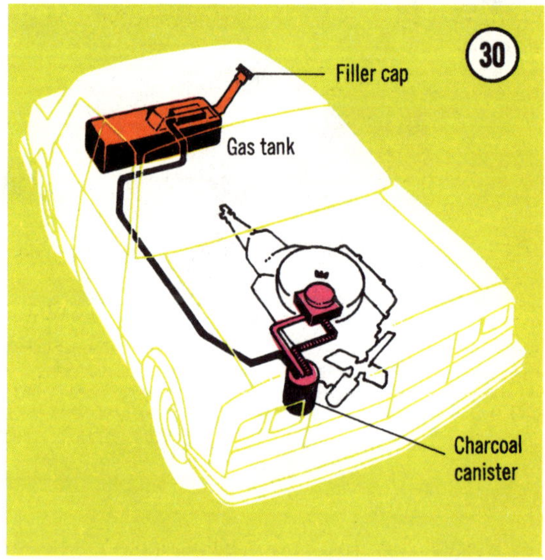

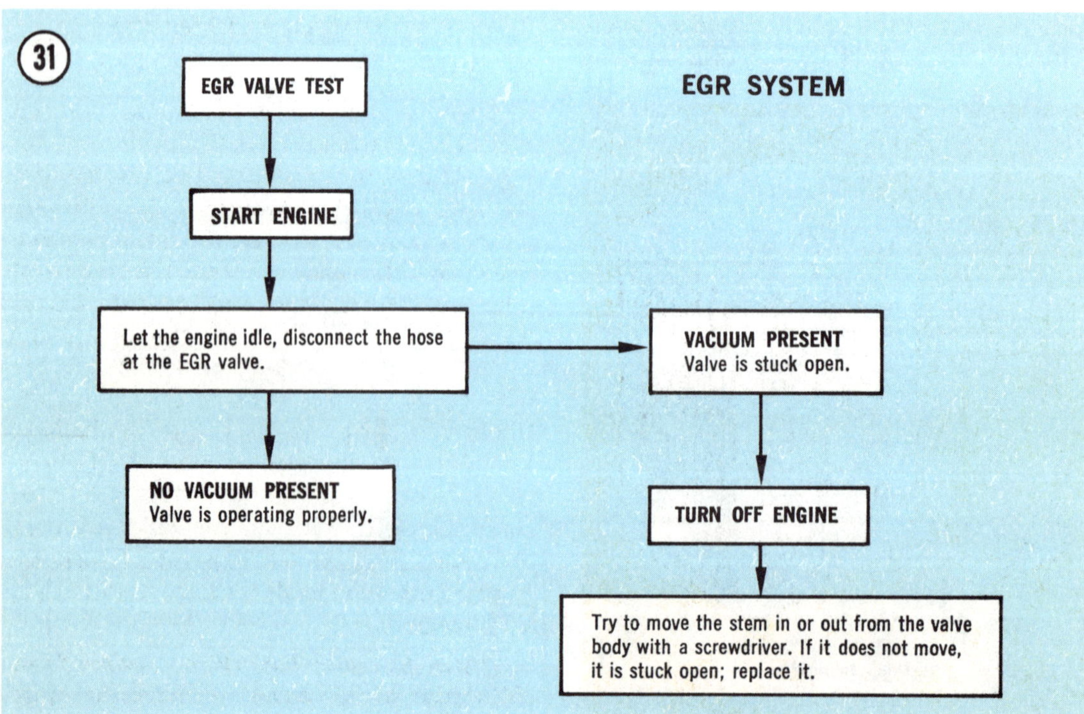

sounds emanating from that area. The first time you do this, you may be horrified at the strange noises coming from even a normal engine. If you can, have an experienced friend or mechanic help you sort the noises out.

Clicking or Tapping Noises

Clicking or tapping noises usually come from the valve train, and indicate excessive valve clearance.

If your vehicle has adjustable valves, the procedure for adjusting the valve clearance is explained in Chapter Three. If your vehicle has hydraulic lifters, the clearance may not be adjustable. The noise may be coming from a collapsed lifter. These may be cleaned or replaced as described in the engine chapter.

A sticking valve may also sound like a valve with excessive clearance. In addition, excessive wear in valve train components can cause similar engine noises.

Knocking Noises

A heavy, dull knocking is usually caused by a worn main bearing. The noise is loudest when the engine is working hard, i.e., accelerating hard at low speed. You may be able to isolate the trouble to a single bearing by disconnecting the spark plugs one at a time. When you reach the spark plug nearest the bearing, the knock will be reduced or disappear.

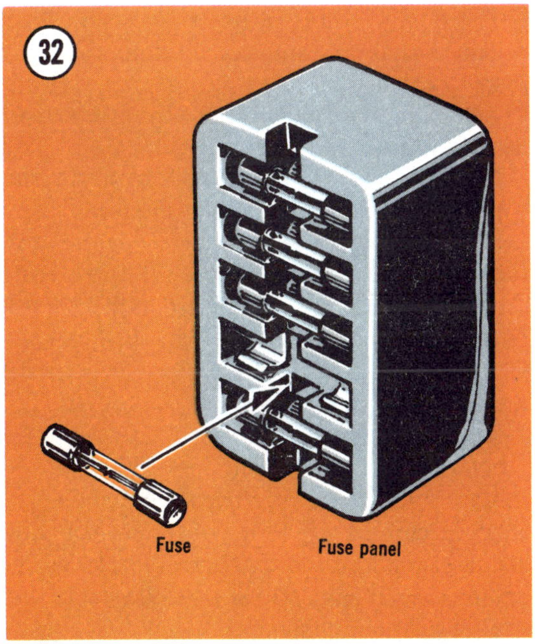

Worn connecting rod bearings may also produce a knock, but the sound is usually more "metallic." As with a main bearing, the noise is worse when accelerating. It may even increase further just as you go from accelerating to coasting. Disconnecting spark plugs will help isolate this knock as well.

A double knock or clicking usually indicates a worn piston pin. Disconnecting spark plugs will isolate this to a particular piston, however, the noise will *increase* when you reach the affected piston.

A loose flywheel and excessive crankshaft end play also produce knocking noises. While similar to main bearing noises, these are usually intermittent, not constant, and they do not change when spark plugs are disconnected.

Some mechanics confuse piston pin noise with piston slap. The double knock will distinguish the piston pin noise. Piston slap is identified by the fact that it is always louder when the engine is cold.

ELECTRICAL ACCESSORIES

Lights and Switches (Interior and Exterior)

1. ***Bulb does not light*** — Remove the bulb and check for a broken element. Also check the inside of the socket; make sure the contacts are clean and free of corrosion. If the bulb and socket are OK, check to see if a fuse has blown or a circuit breaker has tripped. The fuse panel (**Figure 32**) is usually located under the instrument panel. Replace the blown fuse or reset the circuit breaker. If the fuse blows or the breaker trips again, there is a short in that circuit. Check that circuit all the way to the battery. Look for worn wire insulation or burned wires.

If all the above are all right, check the switch controlling the bulb for continuity with an ohmmeter at the switch terminals. Check the switch contact terminals for loose or dirty electrical connections.

2. ***Headlights work but will not switch from either high or low beam*** — Check the beam selector switch for continuity with an ohmmeter

at the switch terminals. Check the switch contact terminals for loose or dirty electrical connections.

3. ***Brake light switch inoperative*** — On mechanically operated switches, usually mounted near the brake pedal arm, adjust the switch to achieve correct mechanical operation. Check the switch for continuity with an ohmmeter at the switch terminals. Check the switch contact terminals for loose or dirty electrical connections.

4. ***Back-up lights do not operate*** — Check light bulb as described earlier. Locate the switch, normally located near the shift lever. Adjust switch to achieve correct mechanical operation. Check the switch for continuity with an ohmmeter at the switch terminals. Bypass the switch with a jumper wire; if the lights work, replace the switch.

Directional Signals

1. ***Directional signals do not operate*** — If the indicator light on the instrument panel burns steadily instead of flashing, this usually indicates that one of the exterior lights is burned out. Check all lamps that normally flash. If all are all right, the flasher unit may be defective. Replace it with a good one.

2. ***Directional signal indicator light on instrument panel does not light up*** — Check the light bulbs as described earlier. Check all electrical connections and check the flasher unit.

3. ***Directional signals will not self-cancel*** — Check the self-cancelling mechanism located inside the steering column.

4. ***Directional signals flash slowly*** — Check the condition of the battery and the alternator (or generator) drive belt tension (**Figure 4**). Check the flasher unit and all related electrical connections.

Windshield Wipers

1. ***Wipers do not operate*** — Check for a blown fuse or circuit breaker that has tripped; replace or reset. Check all related terminals for loose or dirty electrical connections. Check continuity of the control switch with an ohmmeter at the switch terminals. Check the linkage and arms for loose, broken, or binding parts. Straighten out or replace where necessary.

2. ***Wiper motor hums but will not operate*** — The motor may be shorted out internally; check and/or replace the motor. Also check for broken or binding linkage and arms.

3. ***Wiper arms will not return to the stowed position when turned off*** — The motor has a special internal switch for this purpose. Have it inspected by your dealer. Do not attempt this yourself.

Interior Heater

1. ***Heater fan does not operate*** — Check for a blown fuse or circuit breaker that has tripped. Check the switch for continuity with an ohmmeter at the switch terminals. Check the switch contact terminals for loose or dirty electrical connections.

2. ***Heat output is insufficient*** — Check the heater hose/engine coolant control valve usually located in the engine compartment; make sure it is in the open position. Ensure that the heater door(s) and cable(s) are operating correctly and are in the open position. Inspect the heat ducts; make sure that they are not crimped or blocked.

COOLING SYSTEM

The temperature gauge or warning light usually signals cooling system troubles before there is any damage. As long as you stop the vehicle at the first indication of trouble, serious damage is unlikely.

In most cases, the trouble will be obvious as soon as you open the hood. If there is coolant or steam leaking, look for a defective radiator, radiator hose, or heater hose. If there is no evidence of leakage, make sure that the fan belt is in good condition. If the trouble is not obvious, refer to **Figures 33 and 34** to help isolate the trouble.

Automotive cooling systems operate under pressure to permit higher operating temperatures without boil-over. The system should be checked periodically to make sure it can withstand normal pressure. **Figure 35** shows the equipment which nearly any service station has for testing the system pressure.

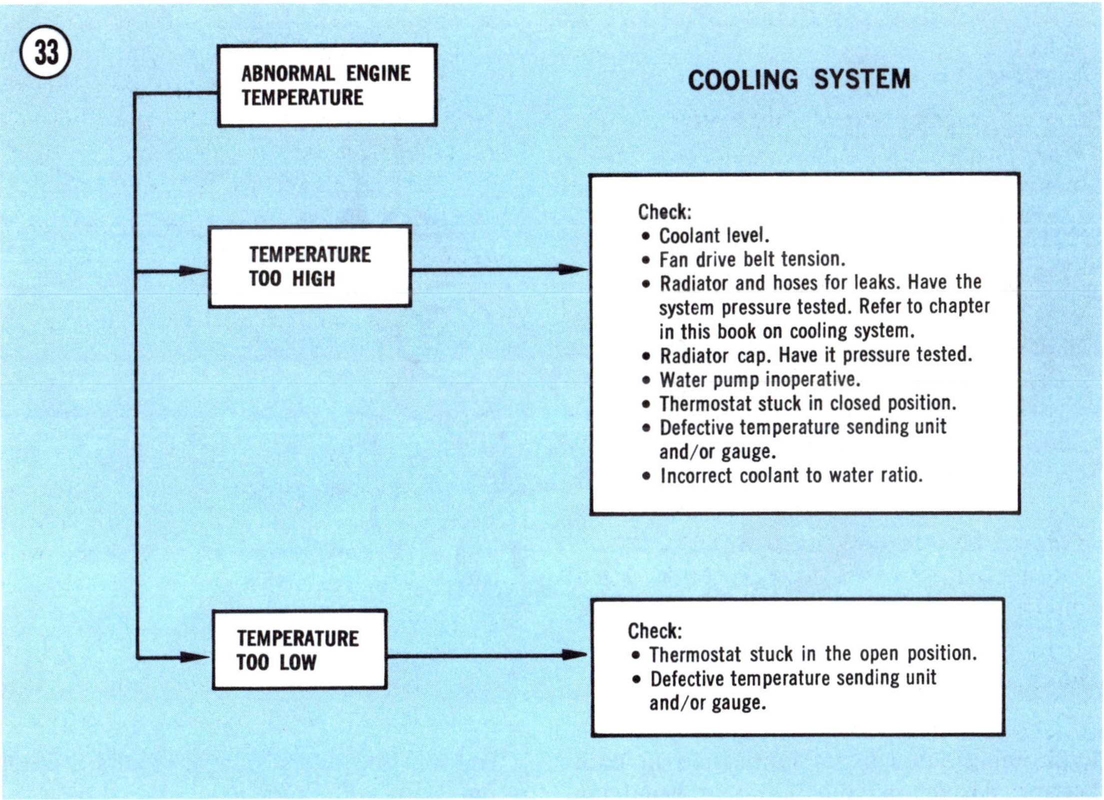

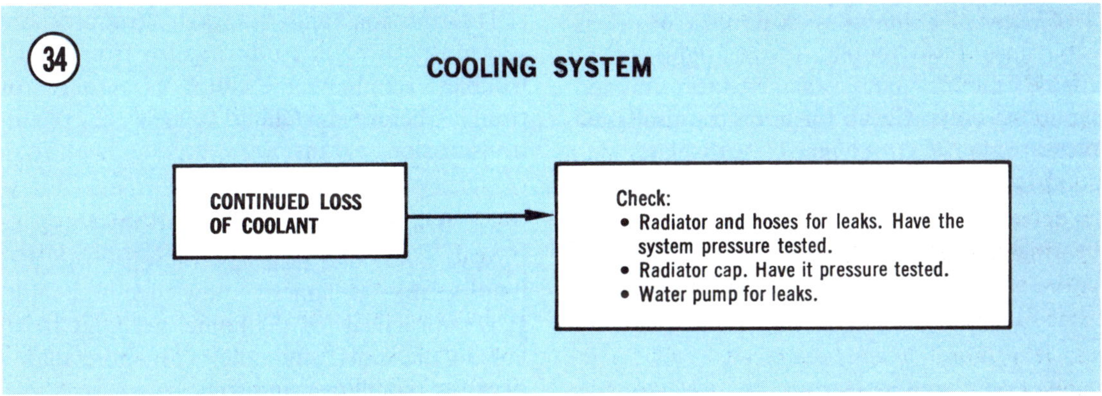

CLUTCH

All clutch troubles except adjustments require transmission removal to identify and cure the problem.

1. *Slippage* — This is most noticeable when accelerating in a high gear at relatively low speed. To check slippage, park the vehicle on a level surface with the handbrake set. Shift to 2nd gear and release the clutch as if driving off. If the clutch is good, the engine will slow and stall. If the clutch slips, continued engine speed will give it away.

Slippage results from insufficient clutch pedal free play, oil or grease on the clutch disc, worn pressure plate, or weak springs.

2. ***Drag or failure to release*** — This trouble usually causes difficult shifting and gear clash, especially when downshifting. The cause may be excessive clutch pedal free play, warped or bent pressure plate or clutch disc, broken or

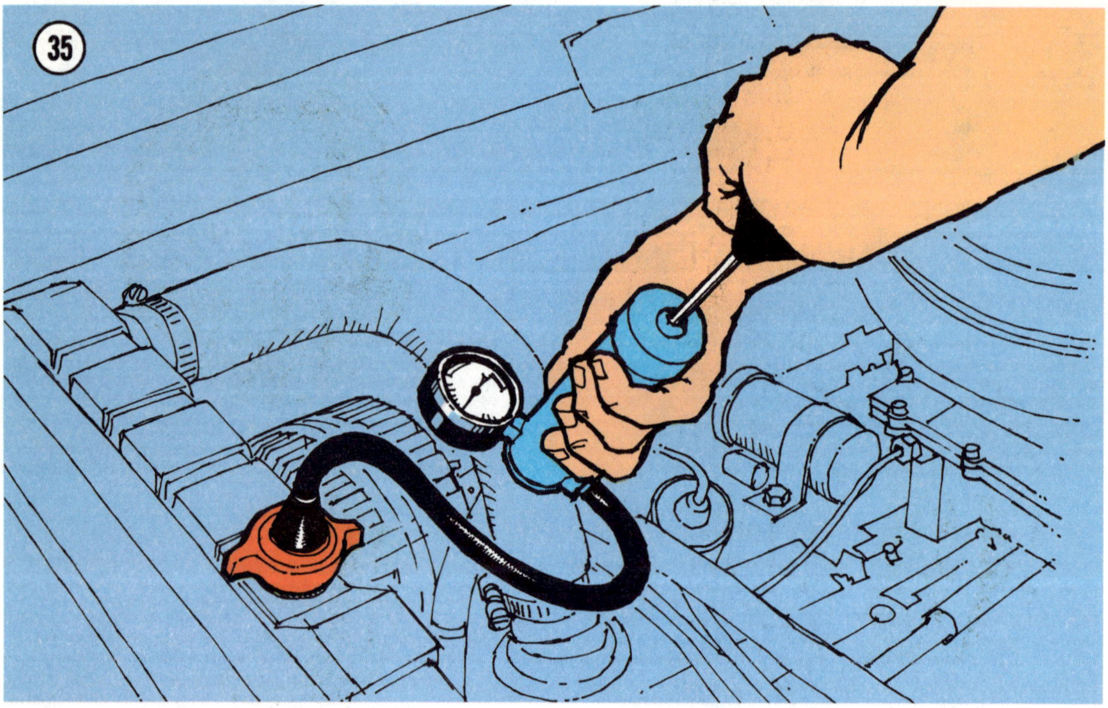

loose linings, or lack of lubrication in pilot bearing. Also check condition of transmission main shaft splines.

3. ***Chatter or grabbing*** — A number of things can cause this trouble. Check tightness of engine mounts and engine-to-transmission mounting bolts. Check for worn or misaligned pressure plate and misaligned release plate.

4. ***Other noises*** — Noise usually indicates a dry or defective release or pilot bearing. Check the bearings and replace if necessary. Also check all parts for misalignment and uneven wear.

MANUAL TRANSMISSION/TRANSAXLE

Transmission and transaxle troubles are evident when one or more of the following symptoms appear:

a. Difficulty changing gears
b. Gears clash when downshifting
c. Slipping out of gear
d. Excessive noise in NEUTRAL
e. Excessive noise in gear
f. Oil leaks

Transmission and transaxle repairs are not recommended unless the many special tools required are available.

Transmission and transaxle troubles are sometimes difficult to distinguish from clutch troubles. Eliminate the clutch as a source of trouble before installing a new or rebuilt transmission or transaxle.

AUTOMATIC TRANSMISSION

Most automatic transmission repairs require considerable specialized knowledge and tools. It is impractical for the home mechanic to invest in the tools, since they cost more than a properly rebuilt transmission.

Check fluid level and condition frequently to help prevent future problems. If the fluid is orange or black in color or smells like varnish, it is an indication of some type of damage or failure within the transmission. Have the transmission serviced by your dealer or competent automatic transmission service facility.

BRAKES

Good brakes are vital to the safe operation of the vehicle. Performing the maintenance speci-

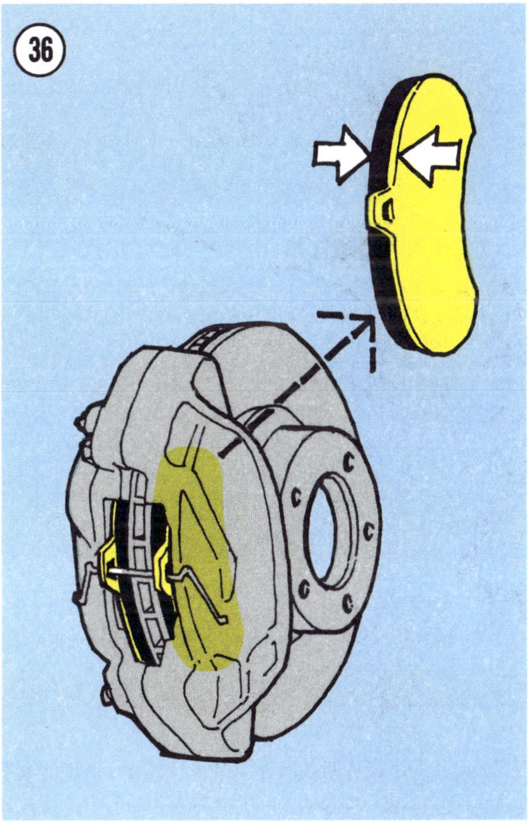

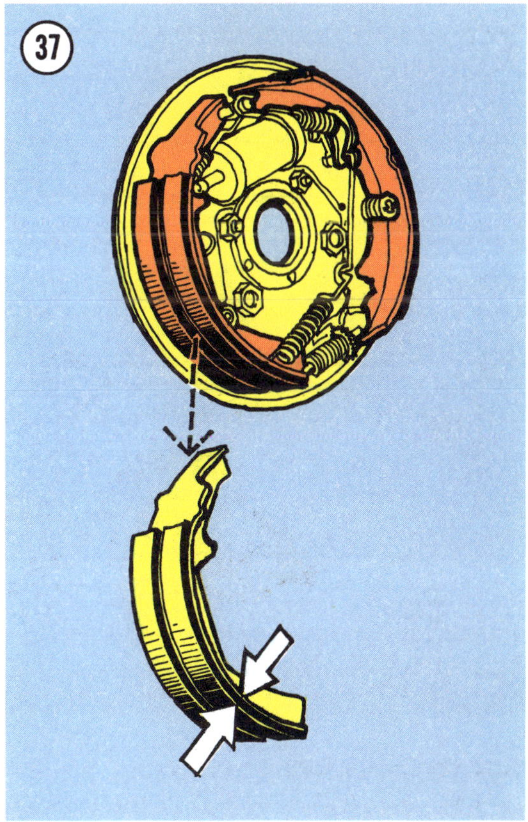

fied in Chapter Three will minimize problems with the brakes. Most importantly, check and maintain the level of fluid in the master cylinder, and check the thickness of the linings on the disc brake pads (**Figure 36**) or drum brake shoes (**Figure 37**).

If trouble develops, **Figures 38 through 40** will help you locate the problem. Refer to the brake chapter for actual repair procedures.

STEERING AND SUSPENSION

Trouble in the suspension or steering is evident when the following occur:

a. Steering is hard
b. Car pulls to one side
c. Car wanders or front wheels wobble
d. Steering has excessive play
e. Tire wear is abnormal

Unusual steering, pulling, or wandering is usually caused by bent or otherwise misaligned suspension parts. This is difficult to check without proper alignment equipment. Refer to the suspension chapter in this book for repairs that you can perform and those that must be left to a dealer or suspension specialist.

If your trouble seems to be excessive play, check wheel bearing adjustment first. This is the most frequent cause. Then check ball-joints as described below. Finally, check tie rod end ball-joints by shaking each tie rod. Also check steering gear, or rack-and-pinion assembly to see that it is securely bolted down.

TIRE WEAR ANALYSIS

Abnormal tire wear should be analyzed to determine its causes. The most common causes are the following:

a. Incorrect tire pressure
b. Improper driving
c. Overloading
d. Bad road surfaces
e. Incorrect wheel alignment

38

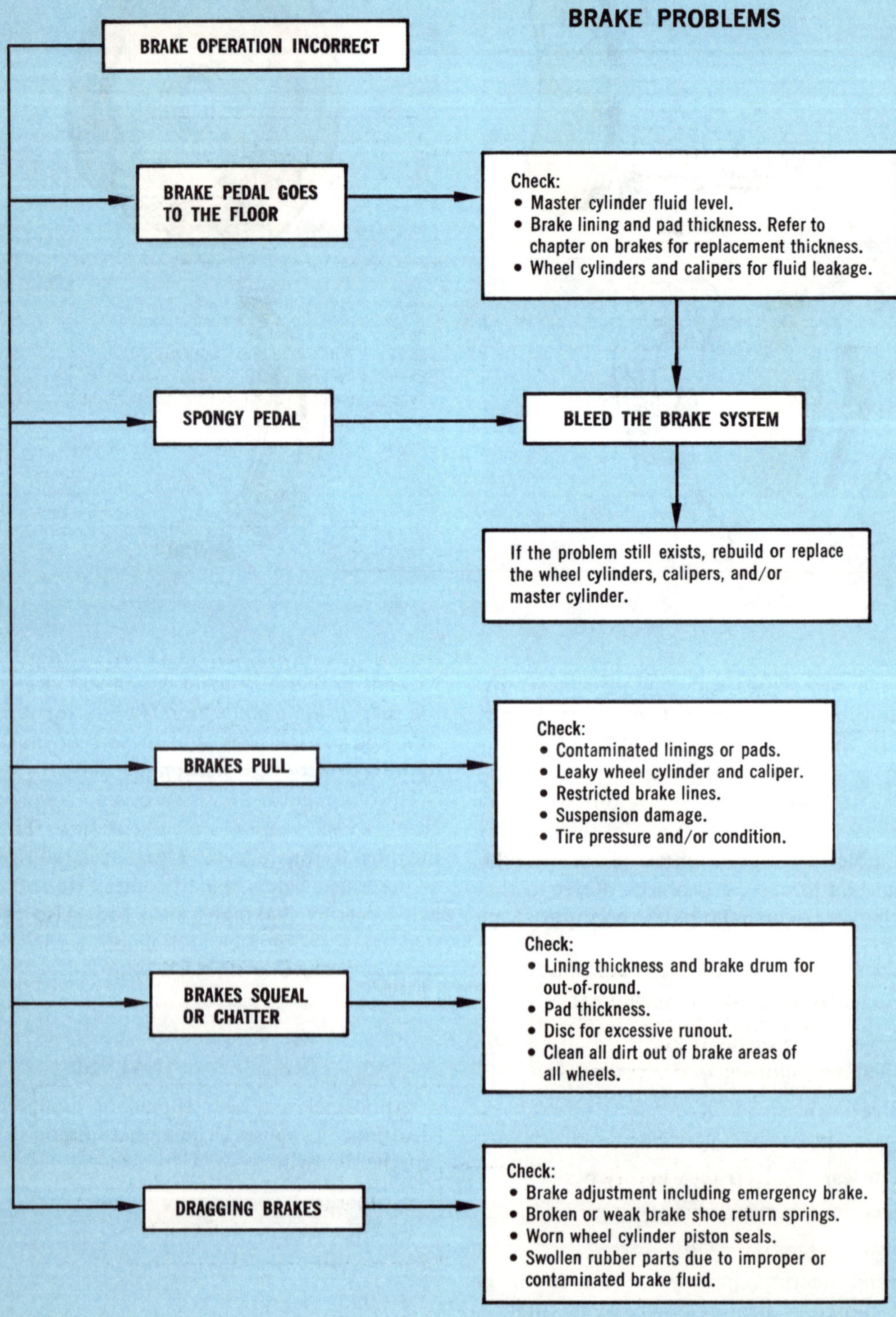
BRAKE PROBLEMS
BRAKE OPERATION INCORRECT
BRAKE PEDAL GOES TO THE FLOOR
Check:
• Master cylinder fluid level.
• Brake lining and pad thickness. Refer to chapter on brakes for replacement thickness.
• Wheel cylinders and calipers for fluid leakage.
SPONGY PEDAL
BLEED THE BRAKE SYSTEM
If the problem still exists, rebuild or replace the wheel cylinders, calipers, and/or master cylinder.
BRAKES PULL
Check:
• Contaminated linings or pads.
• Leaky wheel cylinder and caliper.
• Restricted brake lines.
• Suspension damage.
• Tire pressure and/or condition.
BRAKES SQUEAL OR CHATTER
Check:
• Lining thickness and brake drum for out-of-round.
• Pad thickness.
• Disc for excessive runout.
• Clean all dirt out of brake areas of all wheels.
DRAGGING BRAKES
Check:
• Brake adjustment including emergency brake.
• Broken or weak brake shoe return springs.
• Worn wheel cylinder piston seals.
• Swollen rubber parts due to improper or contaminated brake fluid.

39

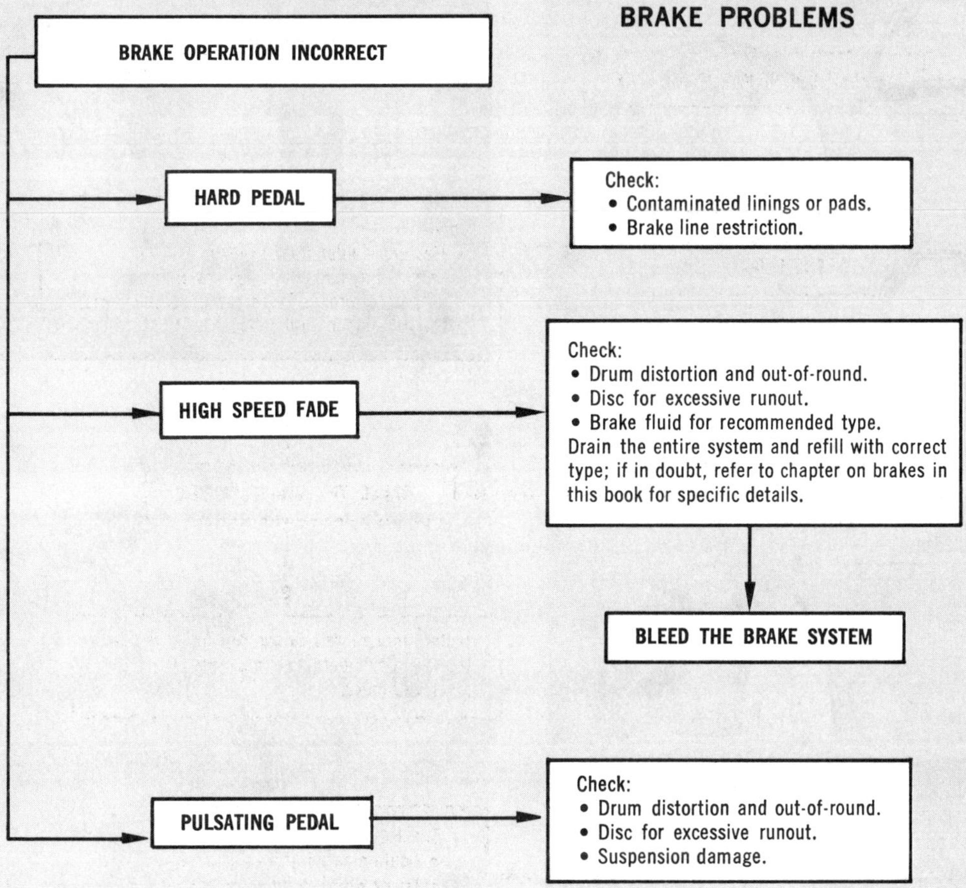
BRAKE PROBLEMS
BRAKE OPERATION INCORRECT
HARD PEDAL
Check:
• Contaminated linings or pads.
• Brake line restriction.
HIGH SPEED FADE
Check:
• Drum distortion and out-of-round.
• Disc for excessive runout.
• Brake fluid for recommended type.
Drain the entire system and refill with correct type; if in doubt, refer to chapter on brakes in this book for specific details.
BLEED THE BRAKE SYSTEM
PULSATING PEDAL
Check:
• Drum distortion and out-of-round.
• Disc for excessive runout.
• Suspension damage.

40

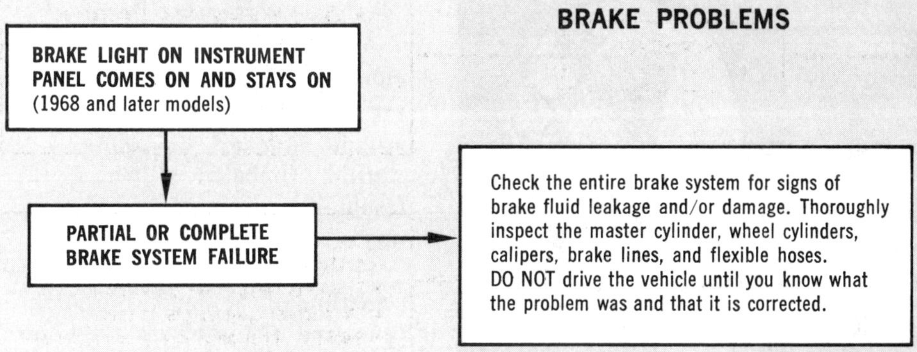
BRAKE PROBLEMS
BRAKE LIGHT ON INSTRUMENT PANEL COMES ON AND STAYS ON (1968 and later models)
PARTIAL OR COMPLETE BRAKE SYSTEM FAILURE
Check the entire brake system for signs of brake fluid leakage and/or damage. Thoroughly inspect the master cylinder, wheel cylinders, calipers, brake lines, and flexible hoses.
DO NOT drive the vehicle until you know what the problem was and that it is corrected.

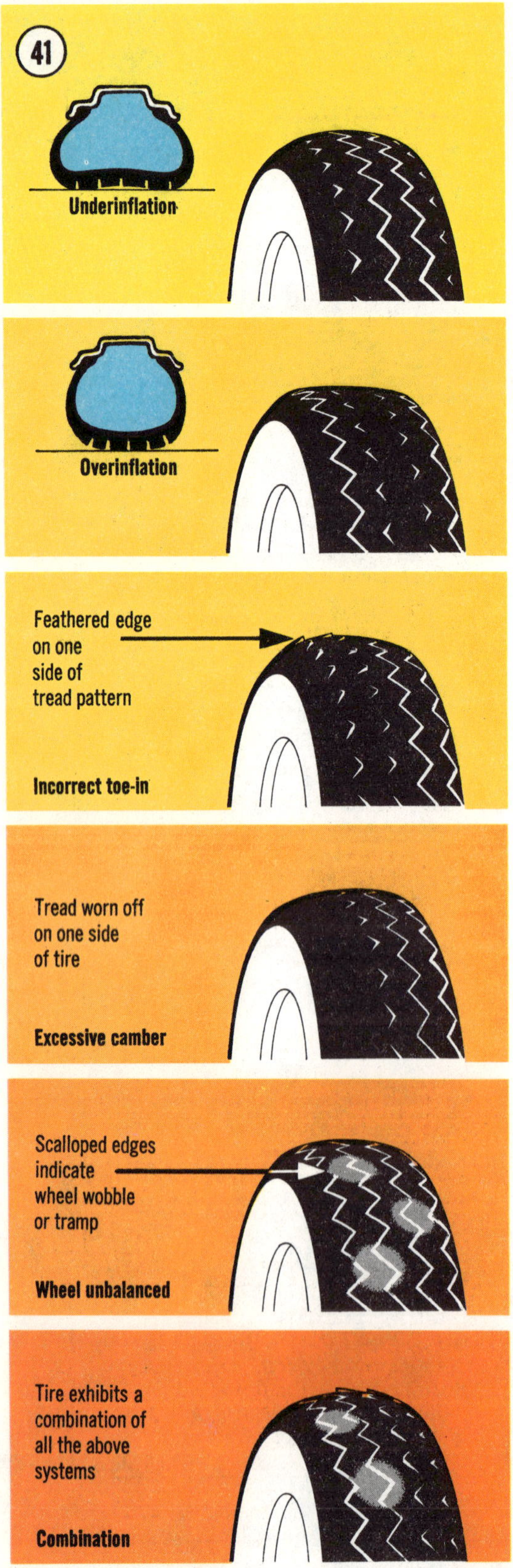

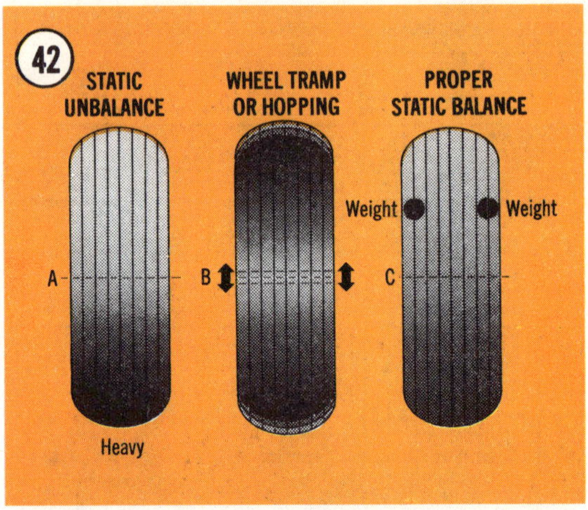

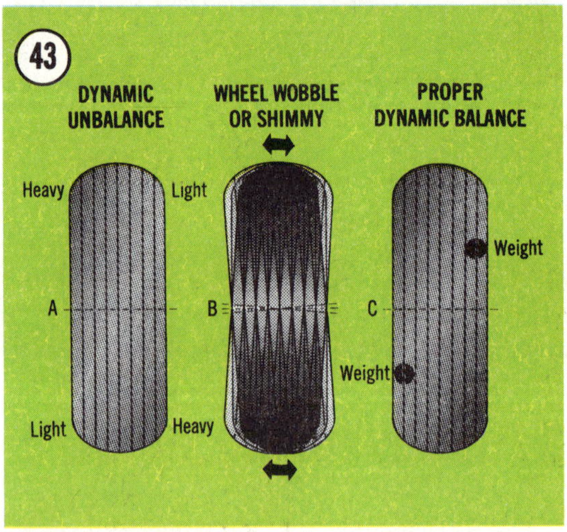

Figure 41 identifies wear patterns and indicates the most probable causes.

WHEEL BALANCING

All four wheels and tires must be in balance along two axes. To be in static balance (**Figure 42**), weight must be evenly distributed around the axis of rotation. (A) shows a statically unbalanced wheel; (B) shows the result — wheel tramp or hopping; (C) shows proper static balance.

To be in dynamic balance (**Figure 43**), the centerline of the weight must coincide with the centerline of the wheel. (A) shows a dynamically unbalanced wheel; (B) shows the result — wheel wobble or shimmy; (C) shows proper dynamic balance.

NOTE: If you own a 1981 and later model, first check the Supplement at the back of the book for any new service information.

CHAPTER THREE

3

LUBRICATION, MAINTENANCE, AND TUNE-UP

A carefully followed program of maintenance will pay dividends in longer engine and vehicle life and fewer repair bills. Such a program is especially important if the vehicle is used on unpaved roads or in out-of-the-way places. Breakdowns and failures are much less likely to happen if the car has been well-maintained.

Certain maintenance jobs should be done at each gas stop. Others should be done at specified time or mileage intervals. Still others should be done whenever certain symptoms develop. This chapter describes all maintenance necessary to keep your car in good condition.

Tables 1-7 are found at the end of the chapter.

GAS STOP CHECKS

In recent years more and more gas stations have converted to self service. It seems likely that this trend will continue to grow. The result is that many of the checks that formerly were made routinely by service station attendants now must be made by car owners, if they are to be done at all. Even if you still patronize full-service stations, you may want to develop the habit of looking over the attendant's shoulder or of doing the tasks yourself. Even though the checks are simple, they are important to the care and trouble-free driving of any automobile. Sometimes they give an indication of the need for other maintenance.

Items to be checked at every gas stop include:

a. Engine oil level
b. Coolant level
c. Battery electrolyte level (unless your car is equipped with the maintenance-free Freedom battery)
d. Windshield washer fluid level

Engine Oil Level

NOTE

You may want to check the oil as the last step at a fuel stop. This will allow the oil in the upper part of the engine to drain back into the crankcase and provide a truer reading of the actual level.

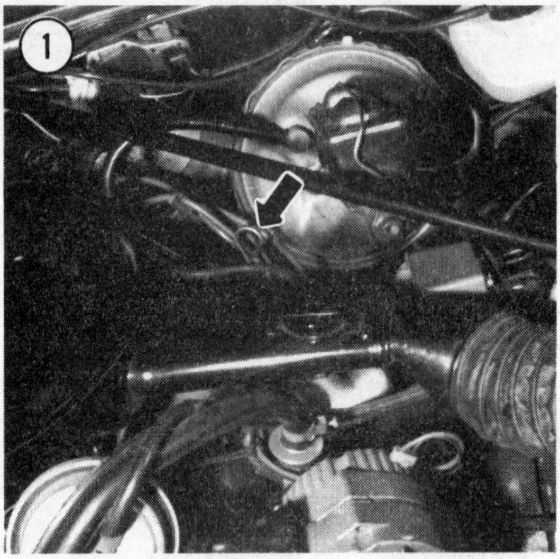

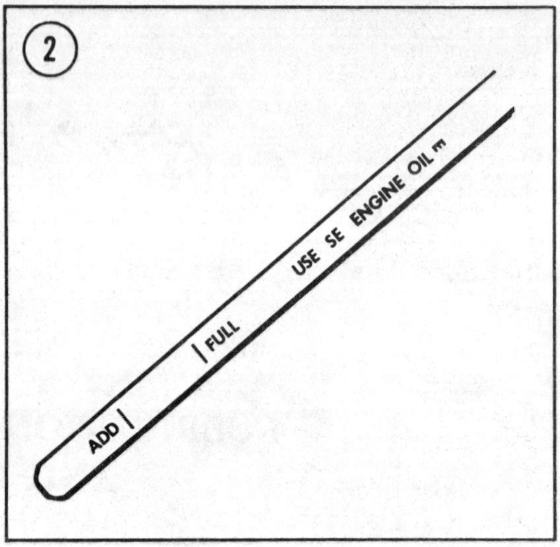

To check engine oil level, remove the dipstick (**Figure 1**, typical), wipe it clean with a cloth or paper towel, replace the dipstick until it seats firmly, then remove it again and read the level of the oil on the lower end. Reinsert the dipstick after taking the reading. The oil level should be maintained between the ADD or ADD OIL mark and the FULL mark on the dipstick. See **Figure 2**. Oil should be added whenever the level drops below the ADD mark. It takes one quart of oil to bring the level from ADD to FULL. Do not overfill the engine. Too much oil sometimes can be as harmful to the engine as too little. As long as the level is somewhere between ADD and FULL there is adequate oil in the engine. See **Table 1** for type of oil recommended.

Coolant Level

Almost all late model Oldsmobiles are equipped with coolant recovery systems. This system consists of a coolant recovery tank, usually made of translucent plastic, which is connected to the radiator filler neck by a hose. See **Figure 3**. When the coolant heats up it expands and flows into the recovery tank. Then when the engine is stopped and/or the coolant temperature drops, the coolant is drawn back into the cooling system. This helps prevent the loss of expensive coolant.

On vehicles equipped with coolant recovery systems, check the coolant level by observing the liquid level in the coolant rcovery tank. This see-through tank has self-explanatory markings, usually FULL HOT and FULL COLD. The level should be somewhere between the 2 marks when the coolant is at normal operating temperature. If the level is not up to the FULL COLD mark after the engine cools to ambient temperature, bring the level up to the mark by adding a 50/50 mixture of ethylene glycol type antifreeze and water to the tank (not the radiator).

WARNING

Do not remove the radiator cap to check fluid level while the engine is hot, especially if the air conditioner has been in use or if the engine has been overheating. The coolant is under pressure and could cause severe burns. If your car does not have a coolant recovery system, wait as long as possible before removing the cap from a hot radiator. If possible, run cool water over the radiator, then insulate the cap with rags and carefully turn the cap to the first detent. Allow all possible steam and coolant to escape with the cap in this position before completely removing the cap.

The fluid level on vehicles without coolant recovery systems should be kept 3 inches below the bottom of the filler neck when the cooling system is cold.

3

4

5

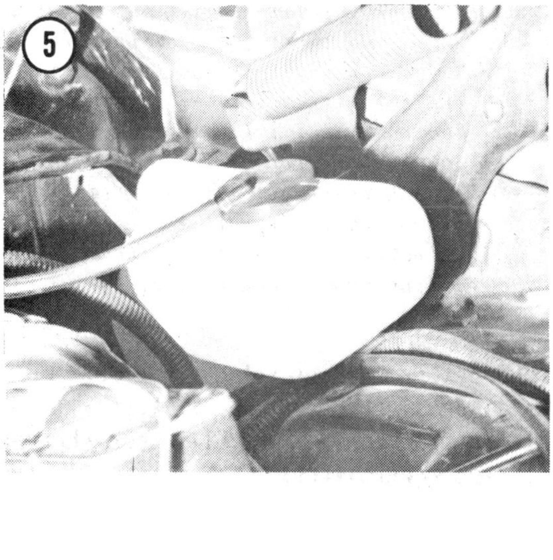

NOTE
Any significant loss of coolant could mean a leak in the cooling system. If the system consistently loses coolant, see Chapter Seven for check procedure.

Battery Electrolyte Level

NOTE
Late models are equipped with a sealed, maintenance-free battery. This battery does not require periodic electrolyte level checks.

Check battery electrolyte level by removing the vent plug from each cell and observing the level in each vent well. The level in each cell should reach the bottom of the split vent well. See **Figure 4**. If the level is low, add clear, odorless drinking water until the level contacts the bottom of the vent well. Do not overfill, as this will result in loss of electrolyte and shorter battery life.

Windshield Washer Fluid

The windshield washer is an important safety feature and an adequate supply of fluid should be present in the reservoir (**Figure 5**) at all times. A quick glance at the reservoir while performing other under-the-hood service should become a habit. If the reservoir is less than 1/2 full, it should be refilled. If an additive is used, follow the manufacturer's instructions.

PERIODIC MAINTENANCE (GASOLINE AND DIESEL ENGINES)

A complete maintenance schedule for vehicles with gasoline engines is provided in **Table 2**, and a similar schedule for diesel-powered cars is shown in **Table 3**. The schedules are intended only as guides. If your vehicle is subjected to conditions such as heavy dust, continuous short trips, or pulling trailers (or use of leaded fuel in 1974 and later models), more frequent servicing will be required.

The following is a brief explanation of each of the services listed in **Tables 2 and 3**. Use only those services that apply to your vehicle.

1. *Chassis*—Lubricate all grease fittings in front suspension, steering linkage, and universal joint. Shift linkage and cable guides should also be lubricated.

2. *Fluid levels*—Check levels in brake master cylinder, power steering pump, battery, engine crankcase, rear axle (differential), transmissions, and windshield washer. Also check engine coolant for proper level and freeze protection to -20 degrees F or the lowest temperature expected (whichever is lower). Any significant loss of fluid in any of these systems should be investigated as it could mean that trouble is developing and that immediate corrective action is required. Low level in the front reservoir of the brake master cylinder could be an indicator that disc brake pads (on cars so equipped) need replacing.

3. *Engine oil*—Change oil at the intervals indicated if vehicle was operated under normal conditions. Change oil every 2 or 3 months or every 3,000 miles when the vehicle is operated under any of the following conditions:

a. Heavy dust
b. Trailer towing
c. Extensive idling
d. Short trips, especially under freezing temperatures with the engine not thoroughly warmed up

4. *Engine oil filter*—Replace at first oil change, and then every second oil change thereafter. If driving under severe conditions, as described in Step 3 above, change oil filter with every oil change.

5. *Tires*—Radial tires should be rotated (see **Figure 6**) at first 7,500 miles (6,000, diesel) and every 15,000 miles (12,000, diesel) thereafter. Bias-belted tires should be rotated every 7,500 miles (6,000, diesel). Inflate tires to pressures shown on tire placard on rear face of driver's door (late models) or to values given in owner's manual.

6. *Rear axle*—Check at intervals shown in **Table 2 or 3**. Change lubricant every 12,000 miles (15,000 miles on 1975 and later models) on Anti-Spin axles. Change lubricant every 12,000 miles (7,500 miles on 1975 and later models) on all types of rear axles or final drives on vehicles used extensively to pull trailers.

7. *Air conditioning*—Check condition of air conditioner hoses and have them replaced if need is indicated. Check refrigerant charge at sight glass. Flowing refrigerant should be clear, although a few bubbles may be present in mild weather. Foam in the flow indicates that refrigerant charge is low. Absence of flow and no temperature difference between compressor inlet and outlet lines indicates no charge.

8. *Cooling system*—At intervals shown in **Table 2 or 3**, test system for leaks (see Chapter Seven) and check or replace pressure cap. Tighten hose clamps and inspect condition of all cooling and heater hoses. Replace all hoses every 24 months, or earlier if when checked, swollen or otherwise deteriorated. Clean exterior of radiator and air conditioner every 12 months (preferably using compressed air). Drain and flush radiator and replace coolant every 24 months or 24,000 miles (30,000 miles on 1975 and later models).

9. *Wheel bearings*—Clean and repack wheel bearings at intervals shown on chart. See **Table 1** for recommended lubricant.

10. *Manual steering gear*—Check around pitman arm and housing for seal leakage and oozing. Leakage (solid grease, not just oily film) should be corrected immediately. Lubricate clutch cross shaft lever.

11. *Clutch cross shaft*—On the manual transmission models, lubricate the clutch cross shaft lever.

12. *Automatic transmission fluid*—Under normal driving conditions, change fluid and service sump filter every 24,000 miles (100,000 miles on 1975 and later models). Cut intervals in half if vehicle has been operated under severe conditions, such as trailer pulling, and so forth.

13. *Owner safety checks*—Items listed below should be checked at the intervals given in **Table 2 or 3** and deficiencies should be corrected as soon as possible.

a. *Steering column lock (if so equipped)*—Key should turn to LOCK postition only when transmission selector is in PARK (automatic transmission) or REVERSE (manual transmission).

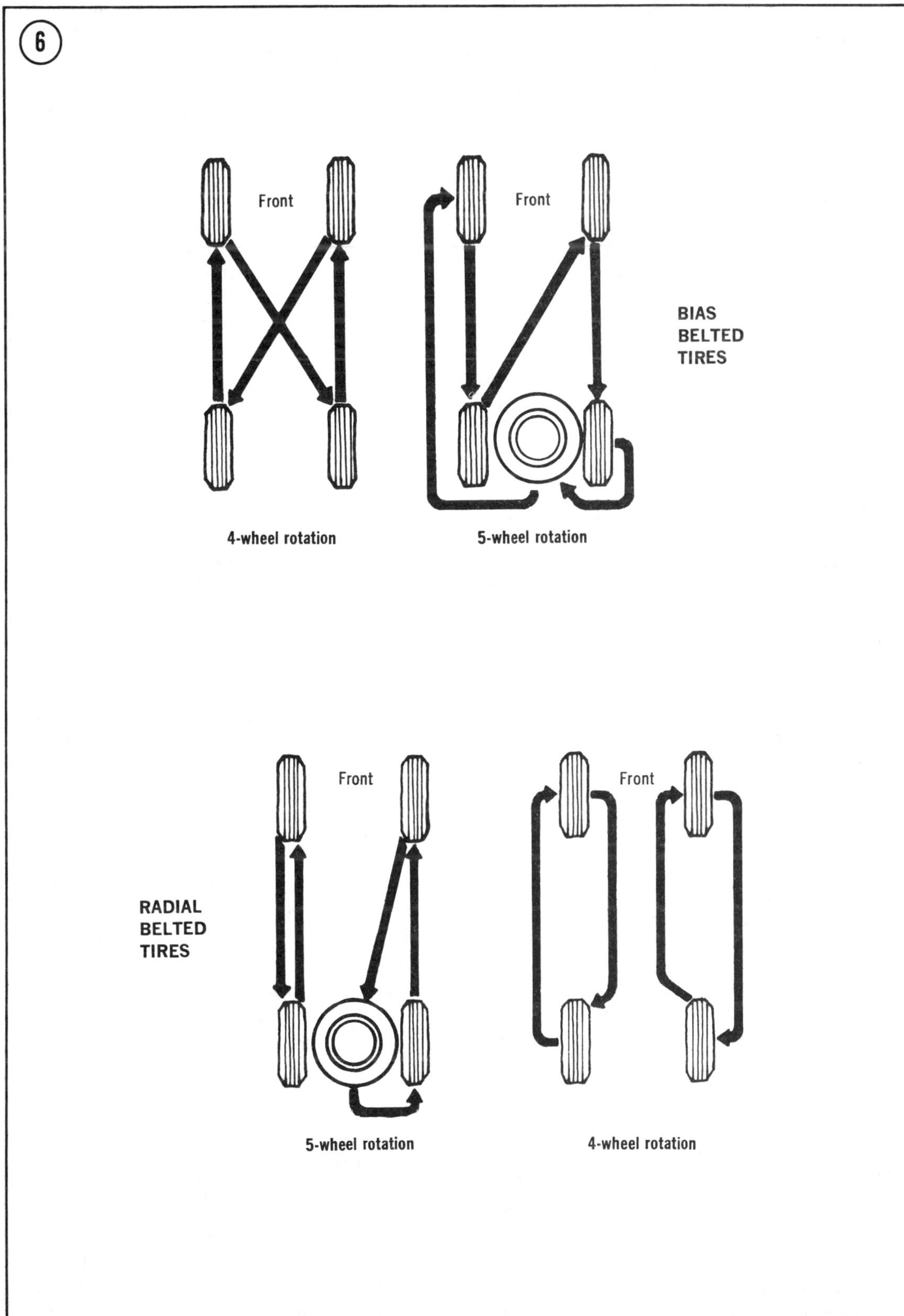
6
Front
Front
BIAS
BELTED
TIRES
4-wheel rotation
5-wheel rotation
Front
Front
RADIAL
BELTED
TIRES
5-wheel rotation
4-wheel rotation

b. *Parking brake and transmission PARK mechanism*—Check holding ability by restraining vehicle with parking brake only on a fairly steep hill. Then check the PARK mechanism (automatic transmission) by placing transmission selector in PARK (P) and releasing all brakes.

c. *Starter safety switch*—Starter should operate only in PARK (P) or NEUTRAL (N) positions (automatic transmissions) or in NEUTRAL with the clutch fully depressed (manual transmissions, if equipped with starter safety switch).

d. *Transmission starter switch*—Verify that automatic shift indicator accurately indicates the shift position selected.

e. *Steering*—Verify that the steering mechanism operates freely and does not have excessive play or make harsh sounds when turning or parking.

f. *Wheel alignment and balance*—Check for abnormal tire wear, then road test the vehicle. The need for alignment may be indicated by a pull to either the right or left on a straight, level road. The need for wheel balancing is usually indicated by excessive vibration of the steering wheel or the front of the vehicle while driving at normal highway speeds.

g. *Brakes*—Observe brake warning light (if so equipped) during braking action. Also check for changes in braking action, such as pulling to one side, unusual sounds, or increased brake pedal travel. See Chapter Ten for additional checks.

h. *Exhaust system*—Be alert to any smell of fumes in the vehicle, or to any change in the sound of the exhaust system that might indicate leakage.

i. *Windshield wipers and washers*—Check operation of wipers and condition and alignment of wiper blades. Check operation of washers (amount and direction of sprayed fluid) and fill fluid reservoir.

j. *Defrosters*—Turn on heater, then move control to DEFROST (DEF) and check amount of air directed to windshield.

k. *Rear view mirrors and sun visors*—Verify that friction joints are adjusted so that mirrors and visors stay in selected position.

l. *Horn*—Verify that horn still works.

m. *Lap and shoulder belts*—Check all components for proper operation. Verify that all anchor bolts are tight. Check belts for fraying.

n. *Head restraints*—Verify that head restraints, if so equipped, adjust up and down properly and that no components are missing, loose or damaged.

o. *Seat back latches*—Verify that latches are holding by pulling forward on seat backs with doors closed (if equipped with automatic seat back latches).

p. *Lights and buzzers*—Verify that all lights and buzzers are working. These include seat belt reminder light and buzzer, ignition key buzzer, interior lights, instrument panel illuminating and warning lights, headlights, license plate lights, side marker lights, parking lights, turn direction signals, back-up lights, and hazard warning flashers.

q. *Glass*—Check for any condition that could obscure vision or be a safety hazard. Correct as required.

r. *Door latches*—Verify positive closing, latching and locking action.

s. *Hood latches*—Verify that hood closes firmly by lifting up on hood after closing it. Check for missing or damaged parts.

t. *Fluid leaks*—Check under vehicle after it has been parked for a while for evidence of fuel, water, or oil leaks. (Water dripping from air conditioner after use is normal.) Immediately determine and correct cause of any gasoline fumes or liquid to avoid possible fire or explosion.

14. *Tires and wheels*—Check tires for nails, cuts, excessive wear or other damage. Verify that tread wear indicators are not visible. Verify that tires are inflated to pressures given on tire placard (on rear face of driver's door, late models) or in owner's manual.

15. *Exhaust system*—Check entire exhaust system, from exhaust manifold to tailpipe,

including catalytic converter, if so equipped, and nearby areas of body and trunk lid for broken, damaged, missing, or mispositioned parts, open seams, holes, loose connections, or any other defect that could permit exhaust fumes to enter the trunk or passenger compartment. The presence of dust or water in the trunk could be an indication of a problem in one of these areas. Whenever a muffler is replaced, all system components to the rear of the muffler (exhaust pipes, resonators, etc.) should also be replaced to maintain the integrity of the system.

16. *Drive belts*—Check all engine drive belts (fans, Delcotron, AIR pump, power steering pump, and air conditioner compressor belts) for cracks, fraying and tension. Adjust or replace, as necessary.

17. *Suspension and steering*—Check for loose, damaged, or missing parts, signs of excessive wear, and lack of lubrication. Repair or correct, as required.

18. *Brakes and power steering*—Check all hoses and lines for proper connections and leaks or deterioration. Immediately replace any questionable parts. If abrasion or other wear is evident on hoses, find and correct the cause.

19. *Disc brakes*—Check brake pads and rotor condition while wheels are removed for tire rotation. Condition should be checked more often if driving conditions and habits result in frequent brake applications.

20. *Drum brakes and parking brakes*—Remove drums, springs, etc. Verify that cylinders are not leaking. Check parking brake adjustment for drag, and lubricate parking brake cable passages and lever mechanism at chassis lube period. More frequent inspections and service should be made if driving conditions and habits result in frequent brake application.

21. *Throttle linkage*—Inspect for damaged or missing parts, interference or binding. Correct deficiencies immediately.

22. *Headlights*—Have headlights checked for proper aim. More frequent checks should be made if illumination area ahead of car seems inadequate.

23. *Underbody*—If car is operated in areas where salt is used on roads for snow removal, or other corrosive materials are used, underside of car should be flushed and inspected yearly, preferably in the spring. Particular attention should be given to areas where mud and other materials have collected.

24. *Bumpers*—Check front and rear bumper system for loose bolts and alignment.

25. *Thermostatically controlled air cleaner*—Verify that all hoses and ducts are properly connected and that valve is working properly.

26. *Carburetor choke*—Verify that choke mechanism is operating freely and remove any petroleum gum formation on choke shaft or correct any other damage or condition that may be causing binding.

27. *Timing, dwell, carburetor idle, distributor and coil*—Tune-up engine at the intervals indicated, using procedures given in this chapter.

28. *Carburetor mounting*—Tighten carburetor mounting bolts at the interval shown, then every 24,000 miles, to compensate for gasket compression and/or vibration.

29. *Early Fuel Evaporation (EFE) Valve* — Check for proper operation. Any binding condition must be corrected. Check switch for operation. Check for and replace damaged hoses.

30. *Spark plugs*—Replace at the intervals shown. If misfiring occurs earlier, remove and clean spark plug if condition permits, otherwise replace.

31. *Exhaust Gas Recirculation System (EGR)*—At the interval shown (and if vehicle is so equipped), remove, inspect and clean EGR valve. Inspect and clean, if required EGR passages in inlet manifold. If EGR valve is damaged, it must be repaired or replaced.

32. *Carburetor fuel inlet filter*—Replace at the intervals indicated, or more frequently if filter becomes clogged.

33. *Thermal vacuum switch and hoses*—Check for proper operation and replace switch if not functioning properly. Check hoses for proper connection and damage. Replace if damaged.

34. *Vacuum advance solenoid and hoses*—Check both vacuum and electrical functions and replace if not operating properly. Check all electrical wires and vacuum hoses and replace those found defective or reconnect all improper connections.

35. *Transmission control switch*—Verify switch operation and check all wires and connections. Replace defective parts.
36. *Positive Crankcase Ventilation System (PCV)*—Check operation at the intervals given in **Table 2** and replace every 24,000 miles. Clean and inspect hoses, replacing any damaged ones. Replace PCV filter (if so equipped) at 12 month/12,000 mile intervals.
37. *Spark plug and ignition coil wires*— Inspect and test at intervals shown in **Table 2**. See procedure in this chapter. Replace faulty or deteriorated wires.
38. *Engine compression*—Check compression at the intervals listed in **Table 2**, using the procedure given in this chapter.
39. *Evaporation Control System (ECS)*—Check all hoses and lines for proper routing, connections and condition. Check vapor canister for cracks or other damage. Replace damaged parts as required. Replace filter in canister lower section.
40. *Fuel cap, lines and tank*—Inspect for damage that could cause leakage. Inspect cap for sealing ability. Replace damaged parts.
41. *Air Injection Reactor (AIR) system*—Check hoses for deterioration and improper connections. Correct or replace, as required.
42. *Air cleaner element*— Replace at the intervals shown in **Table 2**, or more frequently if vehicle is being operated in dusty areas.
43. *Idle stop or speed-up solenoid*—Verify proper operation and replace if defective.

44. *Carburetor vacuum break*—Inspect linkage for proper operation. Correct any binding of linkage. Check for deterioration and correct connection, and replace any damaged parts.

45. *Oxygen change sensor*—Replace sensor at the intervals shown (or when sensor warning signal is visible in speedometer face). Reset signal. See Chapter Five for procedures.

46. *Vacuum Differential Valve (VDV) and Differential Vacuum Delay and Separator Valve (DVDSV)*—Check valve and connecting hoses for correct routing and proper connections, and damage and deterioration. Replace faulty parts as required.

47. *Diesel crankcase ventilation system service*—Clean the filter assemblies from both valve covers, as well as the ventilation pipes and tubes. Also inspect the rubber fittings and replace those in bad condition. Every 30,000 miles the ventilation regulator valve should be cleaned and the breather cap assembly should be replaced.
48. *Engine idle speed adjustment (diesel)*—Adjust diesel engine idle speed to specifications (see decal in engine compartment) at the specified interval (see **Table 3**) by turning the slow idle screw on the injection pump.

49. *Diesel fuel filter replacement*—The fuel filter should be replaced at the interval stated in **Table 3**, or sooner if operating in dusty areas (or if the filter becomes clogged). The filter is located in the fuel line at the top rear of the engine.

LUBRICATION AND LUBRICANTS

The selection of lubricants and fluids is as important to automobile care as the actual maintenance program itself. Only those types of lubricants and fluids recommended by the manfacturers should be used. This is especially true if the vehicle is still under warranty, as the use of non-recommended types could cause cancellation of the warranty.

Table 1 lists the various lubricants and fluids recommended by General Motors Corporation.

ENGINE

Crankcase Capacity

All gasoline engines use 4 qts., plus 1 qt. at filter change. Diesel engines require 7 qts.

Engine Oil and Filters

Engine oil should be selected to meet the demands of the temperatures and driving conditions anticipated. General Motors recommends the use of oil with the letter designation SE for all gasoline engines regardless of the model year or previous oil recommendations. Designation SE/CD is

7

8

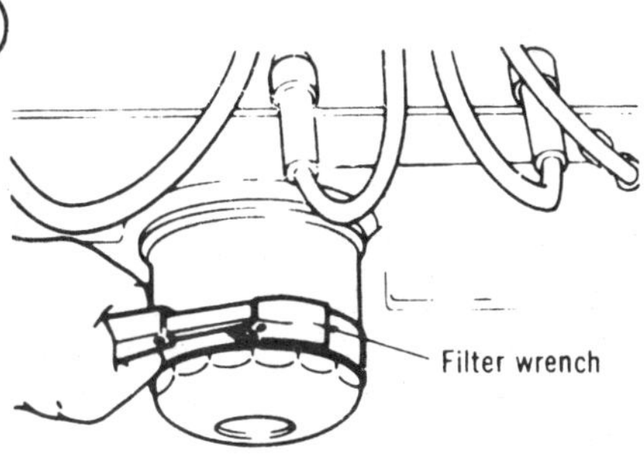

required for 1978 diesel engines and SE/CC for 1979 and later diesels. This designation is plainly marked on the oil can. See **Figure 7**.

If the temperatures expected during the period the oil will be used are between -30 and +20 degrees F, the use of 5W-20 or 5W-30 viscosity oil is recommended. However, 5W-20 oil is not recommended for sustained high-speed driving.

If the expected temperatures will be between 0 and +60 degrees F, 10W, 5W-30, 10W-30 or 10W-40 viscosity oils are recommended. SAE 30 oil may be used if the temperature is not expected to drop below 40 degrees F.

If the temperature is not expected to drop below +20 degrees F, 20W-20, 10W-30, 10W-40, 20W-40, or 20W-50 viscosity oils are recommended.

If the vehicle is to be operated exclusively in Canada, 5W-30 viscosity oil should be used in all seasons.

The recommended frequency of oil and filter changes are given in **Tables 2 and 3**. The recommended periods should be cut in half under any of the following conditions:

1. Driving in dusty areas. Operation in a dust storm may require an immediate oil change.
2. Trailer pulling.
3. Extensive engine idling.
4. Short trip operation in sub-freezing weather.

CAUTION

*General Motors specifically warns against the use of non-detergent, low quality oil, or any oil not having the SE designation. The regular use of oil additives is specifically not recommended. Occasional use of additives to correct specific problems is authorized. For instance, the use of Super Engine Oil Supplement (available at Oldsmobile dealers) is recommended to reduce varnish and sludge deposits in older engines. As a rule, an Oldsmobile dealer should be consulted before using additives. Oil filters should be changed at the intervals recommended in **Tables 2 and 3**. Use a high quality filter.*

To drain the oil and change the filter, you will need:

a. Drain pan
b. Funnel
c. Can opener or pour spout
d. Filter wrench
e. Wrench for removing drain plug
f. 4 or 5 quarts of oil (7 quarts for diesel engine)
g. Oil filter (see **Table 4**)

There are a number of ways of discarding the old oil safely. The easiest way is to pour it from a drain pan into a gallon bleach bottle. Tighten the cap and throw it in your household trash (if not prohibited).

1. Warm engine to operating temperature, then shut it off.
2. Put drain pan under oil drain plug, and remove plug with a wrench.
3. Let oil drain for at least 10 minutes.
4. Unscrew oil filter counterclockwise by hand or use a filter wrench. See **Figure 8** (typical).

5. Wipe the gasket surface of the engine block with a clean, lint-free cloth.
6. Coat the neoprene gasket on the new filter with clean oil. See **Figure 9**.
7. Screw the filter onto the engine *by hand* until the filter gasket just touches the base, i.e., until you feel the slightest resistance when turning the filter. Then tighten the filter *by hand* 2/3 turn or more.

CAUTION
Do not overtighten and do not use a filter wrench or the filter will leak.

8. Install oil drain plug, and tighten securely.
9. Remove oil filler cap.
10. Pour in the required amount of oil.
11. Start engine and let it idle. The oil pressure light on instrument panel will remain on for a short time (15-30 seconds), then it will go out.

CAUTION
Do not rev engine to make oil light go out. It takes time for the oil to reach all areas of the engine and excessive engine speed could damage dry parts.

12. While the engine is running, make sure that the drain plug and oil filter are not leaking.
13. Turn the engine off and check the oil level with the dipstick. See **Figures 1 and 2**. Add oil if necessary to bring oil up to FULL mark, but *do not overfill.*

Air Cleaners

Two types of air cleaners were used during the period covered by this book. One type was the semi-permanent polyurethane foam type and the other was the replaceable paper element type.

Service to the polyurethane type used in some early models consists of removal, washing in solvent, soaking in engine oil, squeezing to wring out excess oil, and replacing the cleaner on the vehicle.

Service to the paper element type used in later models consists only of replacement. Elements must not be cleaned with air hose, tapped, washed or oiled.

To remove the filter element, remove the wing nut holding the air cleaner cover in place (**Figure 10**, typical) and lift out the filter element. Installation is the reverse. Securely tighten the wing nut.

Battery Terminal Washers

Some older models may have batteries equipped with felt washers on the terminals. The purpose of these oil-soaked washers is to control corrosive action. The washers should be saturated with engine oil every 6,000 miles.

Fuel Filter

A clogged fuel filter can cause "stumbling" or "cutting out" at high speeds and will eventually lead to complete fuel starvation. Filters should be changed at the intervals recommended in **Table 2** and **Table 3** (sooner if they become clogged). The fuel filter on all gasoline models is located in the carburetor fuel inlet fitting. See **Figure 11** (typical).

To replace, disconnect the fuel line from the fuel inlet and remove the fuel filter nut. Remove the old filter and filter spring. Install spring and new filter element in carburetor.

9

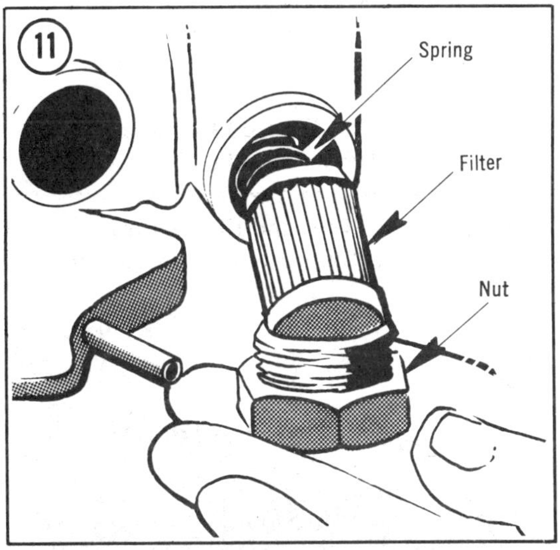

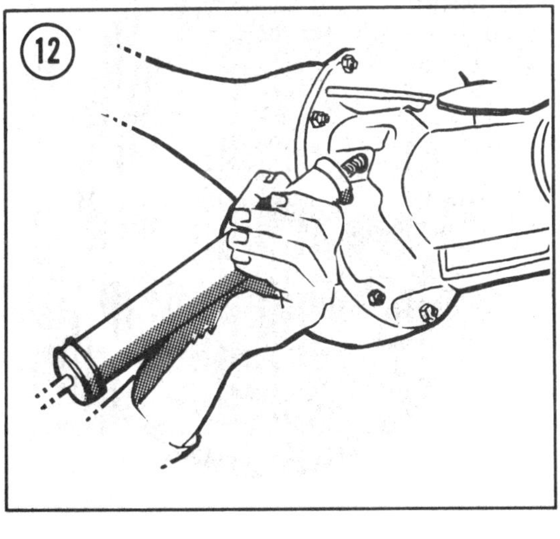

Install a new gasket on the filter nut and install. Securely tighten nut. Install fuel line and tighten connector.

The fuel filter on all diesel models is located in a line between the mechanical fuel pump and the injection pump. To replace, disconnect the filter from the line and install a new one. Diesel fuel filters should be installed when engine temperature is below 125° F to reduce initial cranking time.

Rear Axle and Manual Transmission Lubrication

CAUTION

Operating conditions at high speeds are very severe and damage to differential and transmission gears may occur if improper lubricants are used. Straight mineral oil gear lubricant must not be used, especially in hypoid rear axles. Instead, use SAE 80 or SAE 90 GL-5 Gear Lubricant. Use SAE 80 GL-5 in vehicles used in Canada or other areas where the temperatures are expected to be consistently below 32 degrees F.

Rear axle and manual transmission lubricants normally do not require changing during the life of the vehicle. However, if the vehicle is used consistently for trailer pulling, rear axle lubricant should be changed at 12,000-mile intervals (15,000 miles for 1975 and later models).

Anti-spin rear axles require a special lubricant (General Motors Part No.1052271 or equivalent). Lubricant levels should checked at operating temperatures and at the intervals recommended in **Tables 2 or 3**. Add only lubricant of the same type already in the housing. When at operating temperature, lubricant level should be even with bottom of filler plug hole. Rear axle filler hole is located on the right side of the differential carrier (**Figure 12**). Transmission filler hole is located on the right side of the transmission housing.

Transmission Shift Linkage (Manual and Automatic)

Shift linkage and manual transmission floor control lever contacting faces should be lubricated with engine oil every 4 months or 6,000 miles, whichever comes first.

13

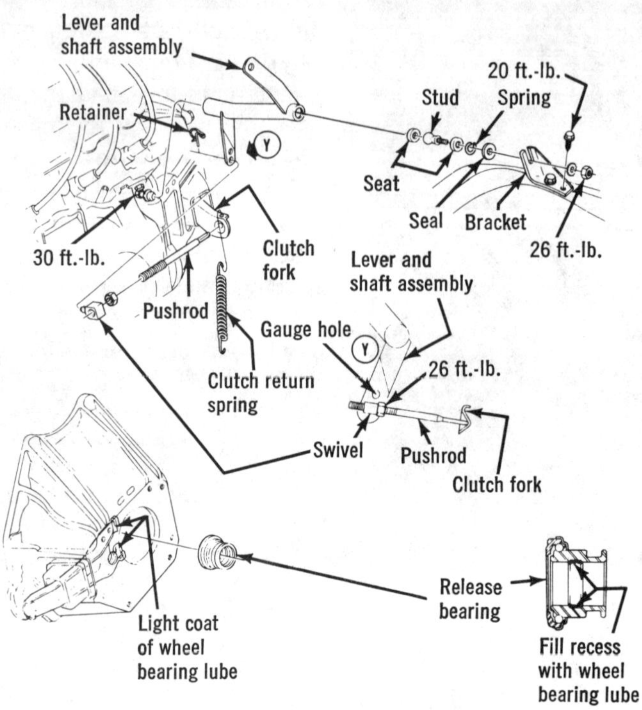

NOTE

If chassis lubricant has been used previously on linkage, continue using this lubricant.

Clutch Cross Shaft

Periodic lubrication of the clutch cross shaft is not required. However, at 36,000 miles (or sooner, if required), remove plug, install grease fitting and apply chassis lubricant with a grease gun.

An exploded view of the cross shaft is shown in **Figure 13**.

AUTOMATIC TRANSMISSIONS

Automatic transmission fluids identified with the names DEXRON II or DEXRON are recommended for use in all Oldsmobile automatic transmissions. These fluids are available at Oldsmobile dealers and also at service stations.

Although several models of automatic transmissions were supplied during the period covered herein, the service procedure is basically the same for all models.

1. Check fluid level at the interval recommended in **Table 2 or 3**.
2. Operate the automobile for several miles to bring transmission up to normal operating temperature.
3. Park in a level space and leave engine running, with the parking brake set and transmission selector in PARK.
4. Remove transmission dipstick, wipe clean, and reinsert until cap fully seats. Remove dipstick and note reading.
5. If fluid is at or below the ADD mark, (**Figure 14**), add sufficient fluid to bring level to FULL mark.

NOTE

One pint of fluid is required to bring the level from the ADD mark to the FULL mark in all models.

CAUTION

Do not overfill the transmission. To do so could cause damage.

Under normal driving conditions the automatic transmission should be drained and refilled at the intervals indicated in **Tables 2 and 3**. If the car is driven extensively in heavy

14

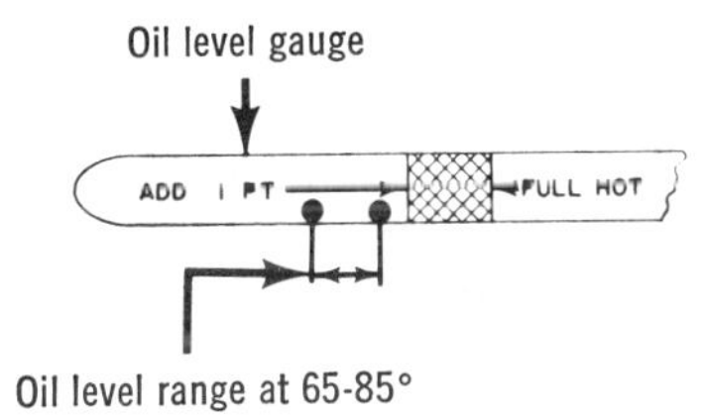

city traffic, in hot weather, or is used to pull a trailer, cut the recommended intervals in half. See **Table 5** for the fluid capacities of automatic transmissions.

After replacing automatic transmission fluid, always recheck the fluid level, using the procedure given above.

To change the transmission fluid, you will need the following:

a. The required amount of DEXRON II ATF (see **Table 5**)
b. Funnel
c. Pour spout
d. Container large enough to hold the old fluid.

CAUTION

Do not attempt to change the transmission fluid if you must work outdoors. Cleanliness is very important when working with an automatic transmission. Blowing dust could settle on internal parts and cause serious damage. If you do not have access to a clean, enclosed garage, let your dealer or local repair shop change the fluid.

1. Drive the vehicle for at least 5 miles to warm up the transmission fluid.
2. Raise the vehicle and support it on jackstands. Also place a suitable jack under the transmission. Remove the transmission crossmember support, if necessary.
3. Place a suitable drain pan under the transmission and remove the attaching bolts from the front and side of the oil pan.
4. Unscrew the rear oil pan attaching bolts about 4 turns and then carefully pry the oil pan loose and allow fluid to drain.
5. Remove the rear oil pan attaching screws and remove the oil pan and gasket. Discard the gasket.
6. Drain the remaining fluid from the oil pan and clean the pan with solvent. Make sure the pan is thoroughly clean and then dry with compressed air.
7. Remove the attaching screws and then remove the strainer and gasket from the valve body. Discard the gasket.
8. If the strainer is to be reused, thoroughly clean it in solvent and blow it dry with compressed air.
9. Install the strainer (or new filter), using a new gasket and the 2 attaching screws.
10. Install the oil pan, using a new gasket. Tighten the attaching screws to 12 ft.-lb.

CAUTION

Do not overtighten the transmission oil pan attaching screws, as this could damage the gasket and cause transmission fluid leaks. This, in turn, could cause serious transmission damage.

11. Lower the vehicle and fill the transmission with the proper amount of DEXRON II automatic transmission fluid. Use a clean funnel inserted in the filler tube (inside the engine compartment), and a clean pouring spout installed in the fluid container to help prevent spillage.
12. Place the transmission selector lever in PARK position; set the handbrake, start the engine, and let it idle.

CAUTION

Do not race the engine at this time, as the transmission has not yet had an opportunity to circulate properly and damage could result from lack of lubrication.

13. Move the transmission selector lever through each position and then return it to PARK. Check the transmission fluid level with the dipstick (**Figure 14**) and add enough fluid, if necessary, to bring the level to 1/4 inch *below* the ADD mark.

CAUTION
Do not overfill the transmission, as this could cause foaming of the fluid and inadequate lubrication, resulting in damage. Remember that the transmission fluid you have just added is cold, and that the marks on the transmission dipstick are intended for measuring hot fluid (the fluid expands as it heats).

14. Test drive the car for several miles to allow the transmission to reach operating temperature and then recheck the level with the transmission in PARK. The level now should be between the ADD and FULL marks on the dipstick. Add fluid as required to bring the level up to, but not above, the FULL mark. The distance between the ADD and FULL marks represents about one pint.

CHASSIS

Chassis Lubricant

General Motors recommends a water resistant EP chassis lubricant meeting the requirements of GM Specification GM 6031-M. This lubricant is designed to be applied by commercial pressure gun equipment. Chassis should be lubricated at the intervals shown in **Tables 2 and 3**.

Front Wheel Bearings

Front wheel bearings must be lubricated every 30,000 miles (1975 and later models) or 24,000 miles (1970-1974 models) according to model year. The lubricant recommended is a high-melting point, water resistant, front wheel bearing lubricant meeting the requirements of GM Specification GM 6031-M or equivalent. If the vehicle is equipped with disc brakes, use wheel bearing lubricant, GM Part No. 1051344, or equivalent. This, according to General Motors, is a premium, high-melting point lubricant.

CAUTION
General Motors warns against the use of "long fiber" or "viscous" type lubricants on front wheel bearings, and also against the mixing of lubricant types. To avoid bearing damage, use the recommended lubricants and thoroughly clean bearings and hubs of old lubricants before repacking.

To lubricate bearings, remove front wheels and hub assemblies (see Chapter Ten for disc brake removal and replacement procedure). Remove, clean, and inspect bearing assemblies and replace with new ones if necessary. Pack bearings with lubricant, clean hub assembly, and reinstall bearings. Reinstall wheel and hub assemblies and adjust bearings, using the procedure given in Chapter Ten.

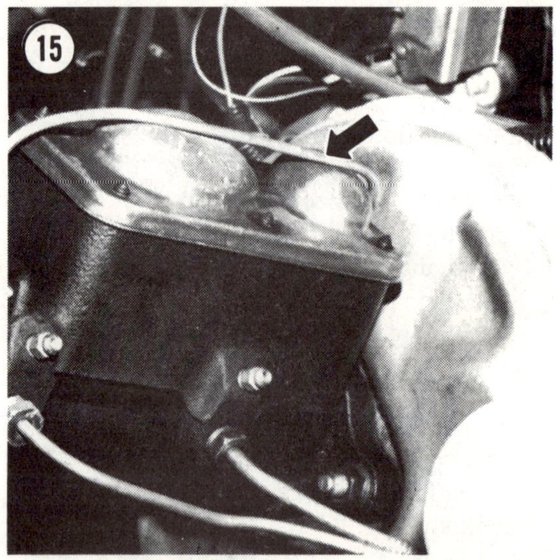

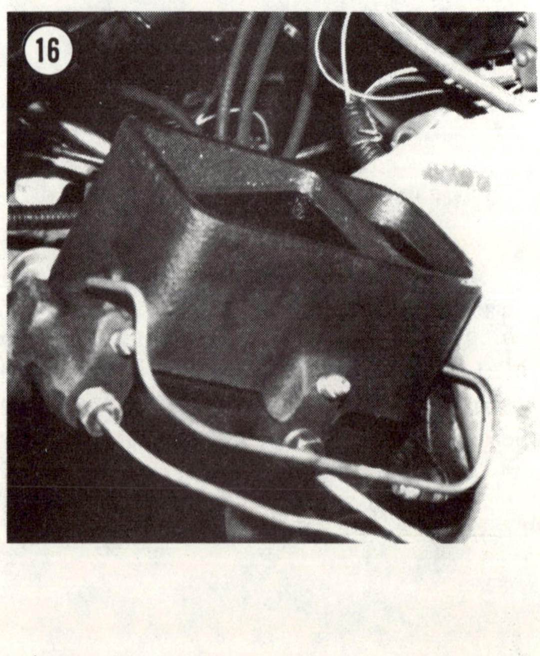

CAUTION
Do not pack hub between inner and outer bearings or the dust caps as excess lubricant may destroy seals and work into the brake drums and linings, causing brake deterioration or failure.

CAUTION
The proper adjustment of wheel bearings is important to both safety and proper operation. The procedure given in Chapter Ten should be read, understood, and closely followed before attempting adjustment.

Brake Master Cylinder

At the intervals shown in **Tables 2 and 3**, check the fluid level in the master cylinder reservoir. Clean dirt and grime from edge of cover so that it will not fall into the reservoir. Pry the wire retainers back with a screwdriver (**Figure 15**) and lift the cover off. The level should be 1/4 inch below the lowest edge of the reservoir (see **Figure 16**). Top up if necessary with hydraulic brake fluid clearly marked DOT-3 or GM Hydraulic Brake Fluid Supreme No. 11.

Parking Brake

Apply chassis lubricant to parking brake cable, cable guides, and all operating linkages and levers at the intervals recommended in **Tables 2 and 3**.

17

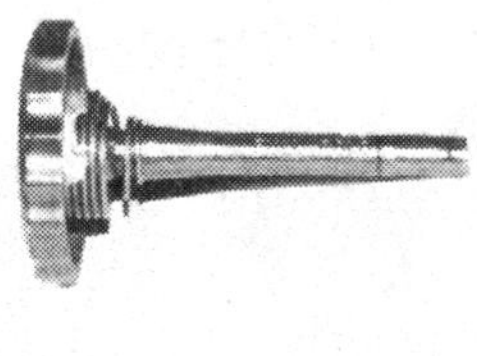

Manual Steering Gear

NOTE
The steering gear is filled at the factory. Seasonal change is not required and no lubrication is required for the life of the steering gear.

Every 36,000 miles (30,000 on 1975 and later models), the steering gear should be inspected for seal linkage. (Worry about actual solid grease; an oily film can be ignored.) If the steering gear is overhauled or a seal is replaced, refill housing with GM Part No. 1051052 Steering Gear Lubricant.

CAUTION
Do not use chassis lubricant to fill the steering gear housing. Also, do not overfill the housing.

Power Steering System

Check power steering fluid at each oil change period. Add GM Power Steering Fluid (or DEXRON II or DEXRON Automatic Transmission Fluid) as required to bring level into proper range on filler cap dipstick, depending on fluid temperature (see **Figure 17**). If at operating temperature (1972-on models), fluid should be between HOT and COLD marks. If at room temperature, fluid should be between ADD and COLD marks. On earlier models, fluid should be brought up to the FULL mark. The fluid does not require periodic changing.

Hood Latches and Hinges

At each oil change period, lubricate hood latch and hood hinge assemblies as follows:

1. Wipe latch and hinge areas clean of accumulated dirt or contamination.
2. Apply Lubriplate or equivalent to latch pilot bolts and locking plate.
3. Apply engine oil to all pivot points in hood release mechanism and primary and secondary latch mechanism.
4. Lubricate hinges with engine oil.
5. Make a functional check to verify that all parts of the hinge and latch assemblies are functioning properly.

3

GASOLINE ENGINE TUNE-UP

The tune-up consists of a series of inspections, adjustments and parts replacements to compensate for wear and deterioration of engine components. Regular tune-ups are especially important to modern, high-powered, high-performance engines. Emission control systems, improved electrical systems, and other advances make these engines especially sensitive to improperly operating or incorrectly adjusted parts.

Since proper engine operation depends upon a number of inter-related system functions, a tune-up consisting of only one or 2 corrections will seldom get lasting results. Instead, a thorough, systematic procedure of analysis and correction will pay dividends in improved power, performance and operating economy.

The procedure presented in this chapter requires a series of visual and mechanical checks and adjustments, followed by an instrument checkout. The instruments required are described in detail in Chapter One. Tune-up specifications are given on a Vehicle Emission Control Information sticker located in the engine compartment. If this sticker is missing or cannot be read, use the information given in **Tables 6 and 7**.

The tune-up for most 1974 and older models consists of the following, in the order listed:

1. Engine compression check
2. Ignition adjustment and timing, consisting of:
   a. Spark plug replacement
   b. Breaker point replacement and/or adjustment and condenser replacement
   c. Distributor cap and rotor inspection (and replacement, if required)
   d. Spark plug wire inspection (and replacement, if required)
   e. Ignition timing adjustment
3. Carburetor adjustment

18

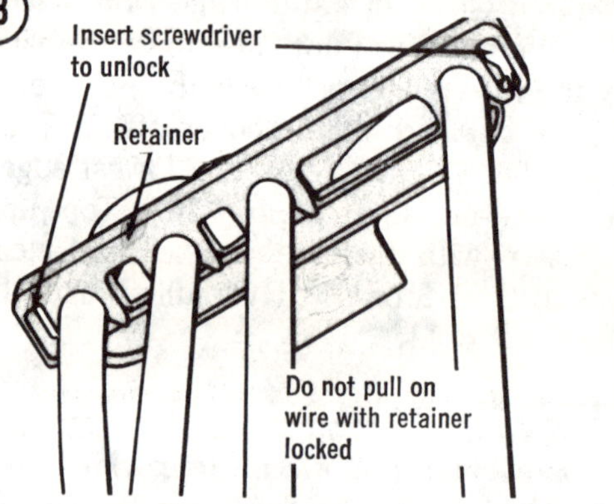

19

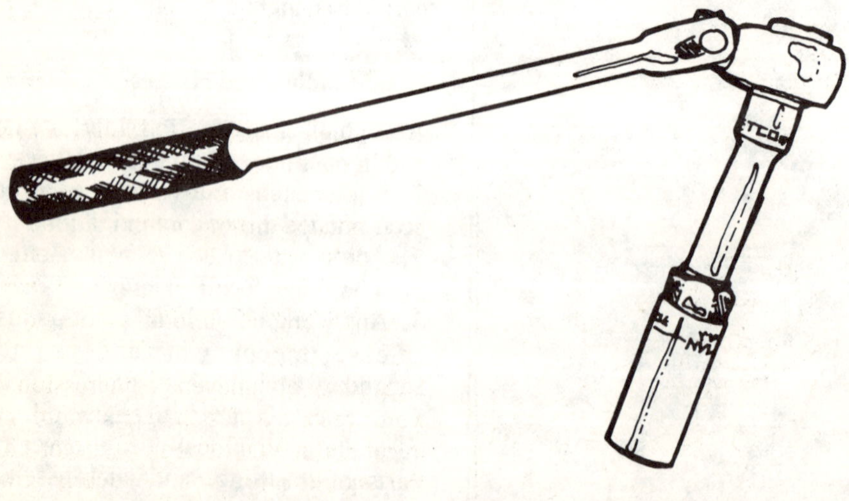

NOTE
The 1974 455 cid engine and all 1975 and later engines have breakerless ignition systems.

The above steps also apply to 1975 and later cars, except for breaker point replacement and adjustment.

Firing Order

The cylinder firing order for in-line 6-cylinder engines is 1-5-3-6-2-4 (cylinder No. 1 front). The order for V6 engines is 1-6-5-4-3-2, and the order for V8 engines is 1-8-4-3-6-5-7-2 (cylinder No. 1 left front).

SPARK PLUG REMOVAL

1. Remove all foreign matter from around spark plugs, using compressed air, if available. A tire pump, vacuum cleaner, or a small paintbrush can also be used.
2. Disconnect spark plug wires by grasping the *boot* portion of the wire and applying only enough twisting force to remove the wire from the plug.

CAUTION
Yanking on wires can lead to damage, especially on late models with HEI ignition. Also, do not force wires out of cable retainers. First open retainer. See ***Figure 18****.*

NOTE
Tag each spark plug wire—a strip of masking tape will do—and mark the proper cylinder number. The cylinder closest to the front of the car is No. 1.

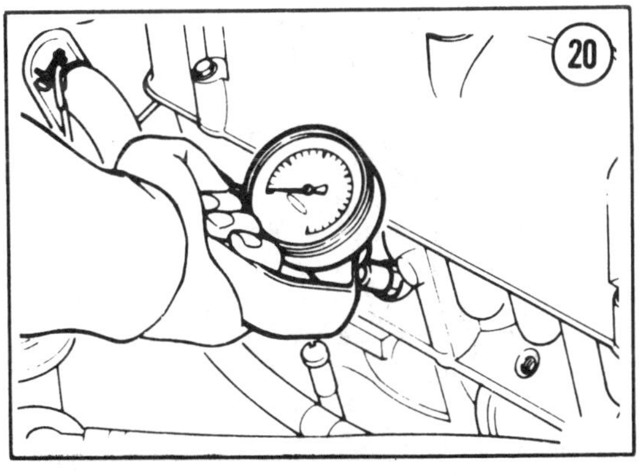

Note that on V6 and V8 engines one bank of cylinders is slightly offset from the other. The bank which is farthest forward contains cylinders 1, 3, 5, (and 7) and the opposite bank cylinders 2, 4, 6, (and 8) from front to back.

3. Using a spark plug wrench (**Figure 19**), remove the spark plugs in order.
4. Inspect each spark plug as it is removed and compare its appearance with Figure 17, Chapter Two. Electrode appearance is a good indicator of performance in each cylinder and permits early recognition of trouble.

COMPRESSION TEST

An engine with low or uneven compression cannot be properly tuned. In view of this, the engine should be given a compression test before proceeding with the tune-up.

1. Remove air cleaner from carburetor and block throttle and choke in wide open position.
2. Disconnect the distributor primary lead from the coil negative post on breaker point ignitions. With HEI ignitions, unplug the outermost harness connector (BAT terminal) from the distributor cap. Remove the spark plugs.
3. Hook up remote starter per manufacturer's instructions.
4. Firmly insert a compression gauge in each spark plug hole, in order, and crank the engine through at least 4 compression strokes to obtain the highest possible reading. See **Figure 20**. Record the reading for each cylinder.

NOTE
If there is more than 20 pounds difference between the high and low reading cylinders, the engine cannot be properly tuned until repairs have been made. The remainder of this procedure may be performed to help determine the kind of repairs that are needed.

5. If low or uneven compression readings were recorded for one or more cylinders, inject about a tablespoon of motor oil through the spark plug hole of each low-reading cylinder. Crank the engine through several

compression strokes and recheck compression. If compression improves, the problem is probably worn rings. If no improvement is noted, in all probability the valves are burned, sticking, or not seated properly. If 2 adjacent cylinders read low and oil injection does not increase compression, the problem may be a defective head gasket.

SPARK PLUG REPLACEMENT

Spark plugs should be replaced at the intervals shown in **Tables 2 and 3**. If misfiring occurs earlier than these intervals, spark plugs in good condition can often be cleaned, regapped and reinstalled with acceptable results. If all new plugs are being installed, skip to Step 3 in the following procedure.

New plugs should be carefully gapped to ensure a reliable, consistent spark. You must use a special spark plug gapping tool with a wire gauge. See **Figure 21**.

21

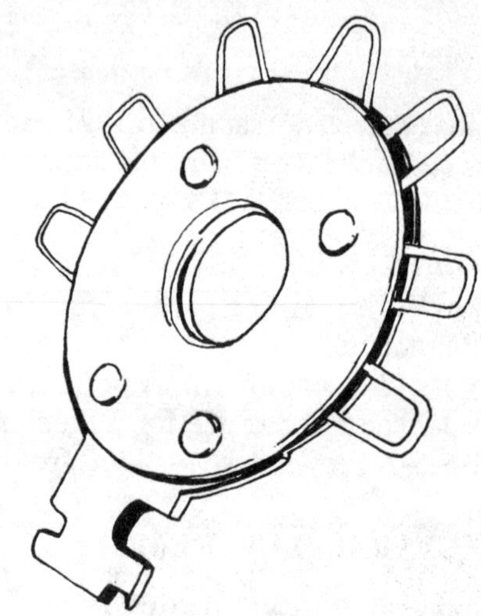

1. Inspect plugs and discard and replace those with badly worn electrodes and/or glazed, blistered, or broken porcelain insulators.

2. Clean the serviceable plugs with an abrasive cleaner, such as a sandblast. File the center electrode flat.

3. Verify that all plugs to be installed are of the same make and of the proper heat range number (**Table 6**).

4. Remove new plugs from box and screw on the small end pieces that are loose in each box. See **Figure 22**.

5. Insert the correct gauge wire (**Table 6**) between the center and side electrode of each spark plug. See **Figure 23**. If the gap is correct, you will feel a slight drag as you pull the wire through. If there is no drag, or the gauge won't pass through, bend the side electrode *with the gapping tool* (see **Figure 24**) to set proper gap.

6. Inspect spark plug hole threads and clean before installing the plugs. If required, corrosion can be removed with a 14 mm x 1.25 SAE spark plug thread chaser (use grease on the chaser to catch chips).

CAUTION
Use extreme care when using thread chaser to avoid cross-threading. Also crank engine several times to blow out any dislodged material from the engine.

7. Apply a thin film of oil—a drop from engine dipstick will do—to spark plug threads. Install plug in hole and torque to 15 ft.-lb. (20 ft.-lb. on 231 cid V6 and all 1978 and later models).

NOTE
If a torque wrench is not available, tighten spark plugs as tight as possible by hand, then, using wrench, tighten another 1/2 turn. Do not overtighten. Excessive torque may change gap setting or squash the gasket so badly it cannot seal.

8. Reconnect the spark plug wires.

NOTE
If spark plug wires have been in service for a year or longer, or appear cracked, oil soaked, or brittle, check them with an ohmmeter. Any wire with a resistance over 20,000 ohms or "infinity" should be replaced. Use only brand name (Delco, Belden, etc.) replacement resistance cable.

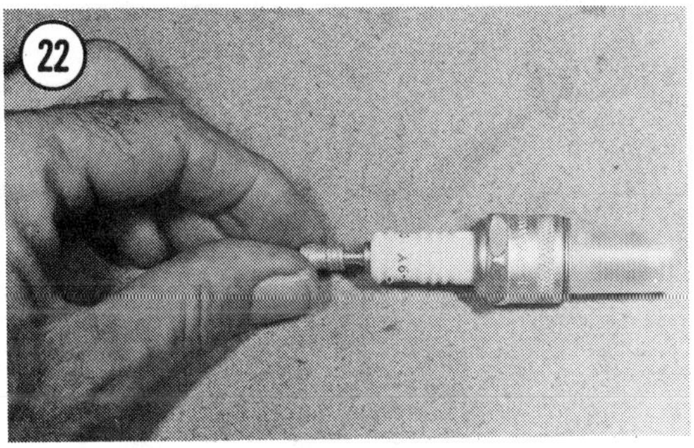

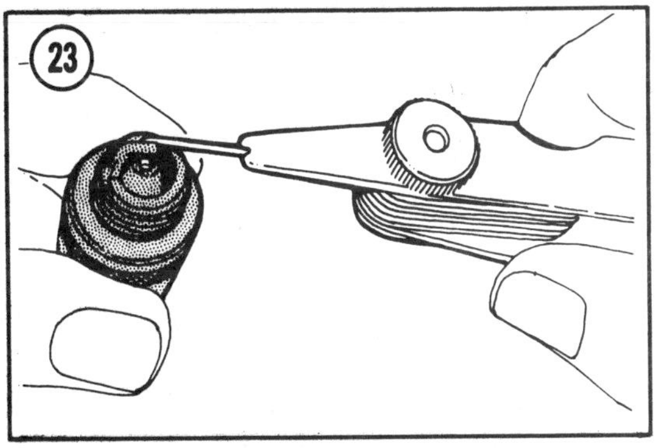

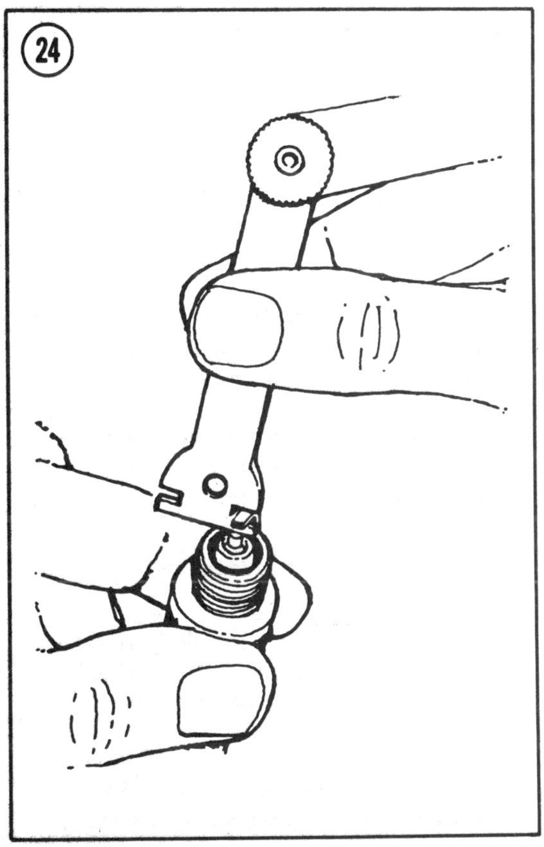

BREAKER POINT IGNITION SYSTEMS

Ignition System Inspection

1. Remove and clean distributor cap. Inspect for cracks, carbon tracks, and burned, worn or corroded terminals, etc. (**Figure 25**). Replace cap if necessary.
2. Clean rotor and inspect for damage or deterioration (**Figure 25**). Replace if necessary.
3. Inspect spark plug wiring (see NOTE above). Verify that all wires are installed to proper plugs. Properly position wires in support to avoid contact with engine and to avoid cross-firing.
4. Tighten all ignition system connections and replace any loose, frayed or damaged wires.

Distributor Service

Refer to **Figure 26** (6-cylinder) or **Figure 27** (V8) for this procedure.

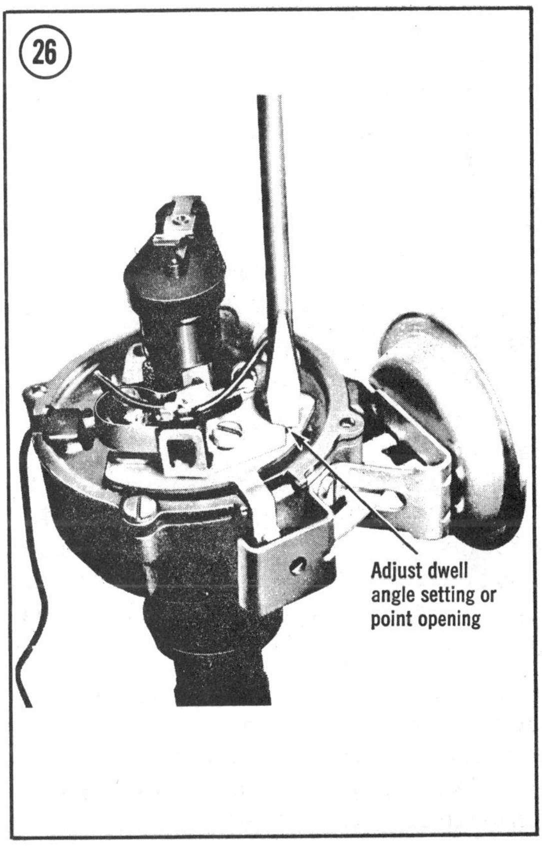

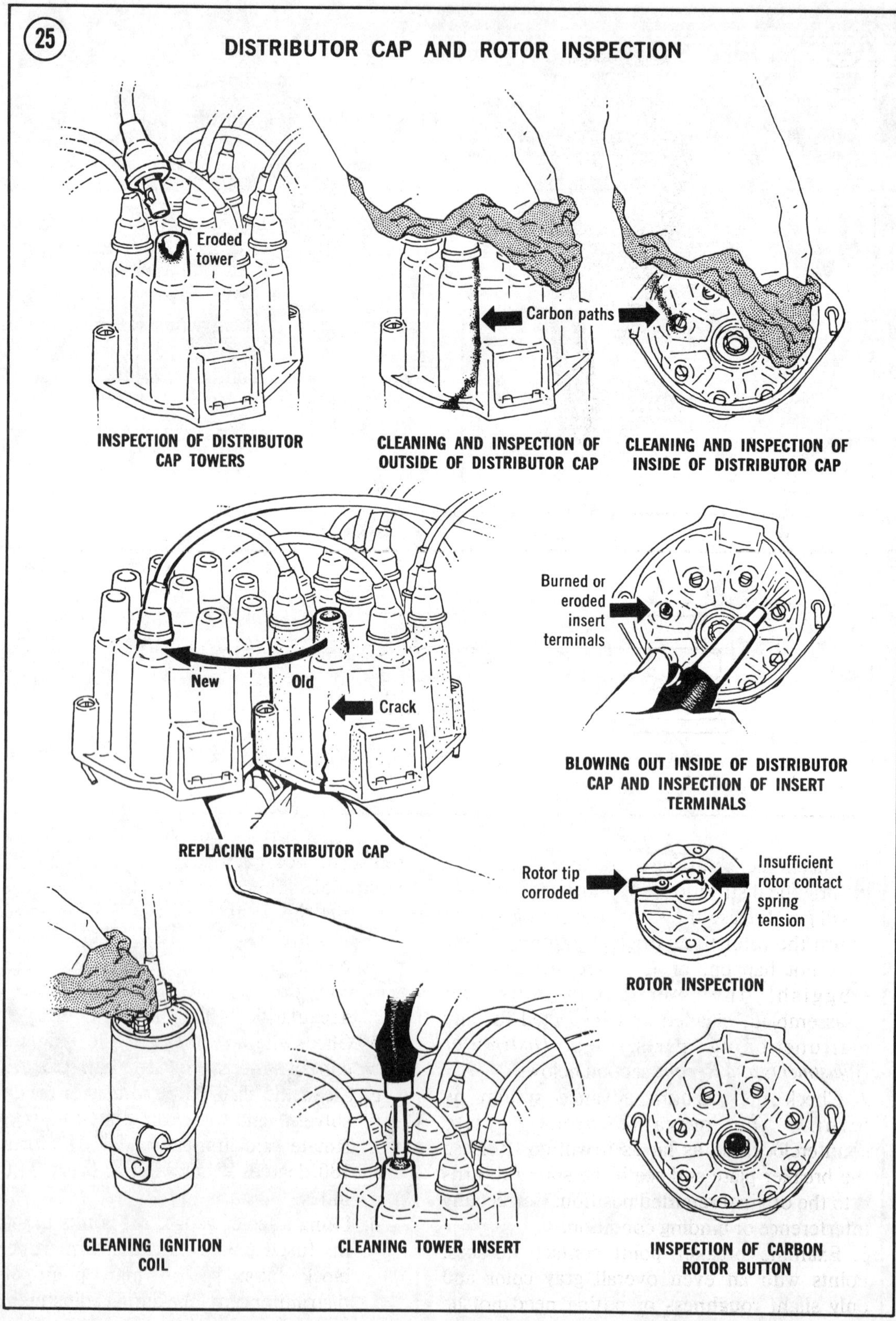

25
DISTRIBUTOR CAP AND ROTOR INSPECTION
Eroded tower
Carbon paths
INSPECTION OF DISTRIBUTOR CAP TOWERS
CLEANING AND INSPECTION OF OUTSIDE OF DISTRIBUTOR CAP
CLEANING AND INSPECTION OF INSIDE OF DISTRIBUTOR CAP
Burned or eroded insert terminals
New
Old
Crack
BLOWING OUT INSIDE OF DISTRIBUTOR CAP AND INSPECTION OF INSERT TERMINALS
REPLACING DISTRIBUTOR CAP
Rotor tip corroded
Insufficient rotor contact spring tension
ROTOR INSPECTION
CLEANING IGNITION COIL
CLEANING TOWER INSERT
INSPECTION OF CARBON ROTOR BUTTON

27

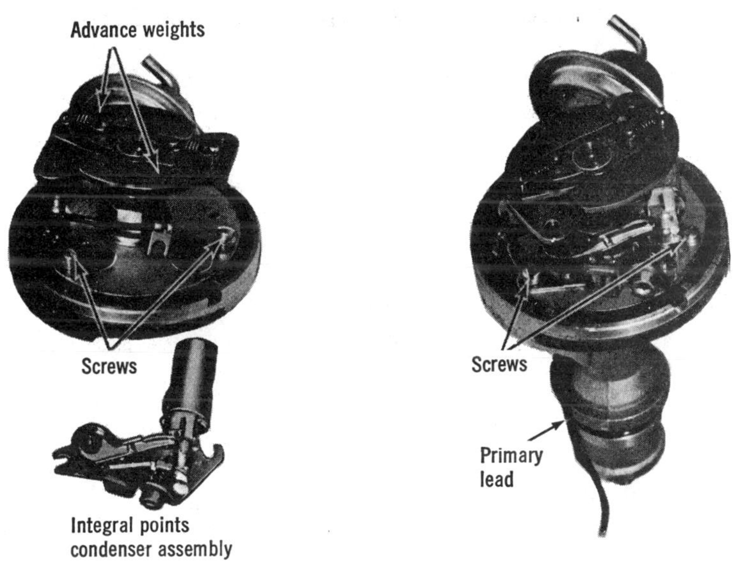

28

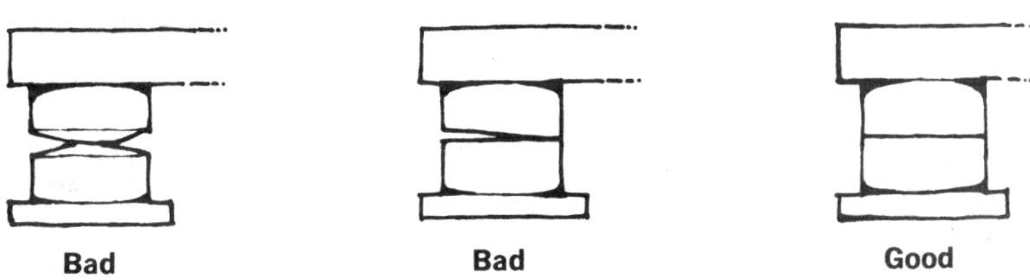

1. Check the centrifugal advance system by turning the distributor rotor clockwise as far as it will go. Release the rotor. The spring should return the rotor to its original position. If this does not happen, or if the return action is sluggish, the mechanism must be disassembled, cleaned and inspected for any malfunctioning parts. See *Distributor Adjustments and Repairs* section below.
2. Check the vacuum advance system by turning the movable breaker plate counterclockwise as far as it will go. Release the breaker plate and see if the spring returns it to the original retarded position. Correct any interference or binding condition.
3. Examine breaker point contact surfaces. Points with an even overall gray color and only slight roughness or pitting need not be replaced, but may be dressed with a clean point file as follows:
   a. File the points, using only a few strokes and a clean fine-cut, contact point file. Never use sandpaper or emery cloth, and do not attempt to remove all irregularities—just the scale or dirt.
   b. Check the alignment of the points and correct as necessary. See **Figure 28**.
   c. Clean the distributor cam with cleaning solvent and a clean, lint-free cloth. Rotate cam lubricator wick 1/2 turn or 180 degrees. Replace wick every 24,000 miles.
   d. Using a feeler gauge, set points to 0.016 in. (used points). Breaker arm rubbing block must be on high point of a distributor cam lobe during adjustment.

4. If points are badly pitted or burned, they must be replaced. However, these conditions are usually caused by improper conditions in other parts of the ignition system or by dirt or other contamination in the distributor. The cause must be corrected (see *Distributor Adjustments and Repairs* section below) when installing new points, or the same condition will rapidly develop again. To replace points (for a 6-cylinder engine) proceed as follows:

a. Remove distributor cap by releasing hold-down screws. Move cup out of the way.
b. Remove rotor.
c. Remove condenser and primary lead wires from quick disconnect terminal.
d. Remove contact set attaching screw and remove point set.
e. If condenser is to be replaced, remove attaching screw and remove condenser from breaker plate.
f. Using a clean, lint-free cloth, wipe breaker plate free of oil and smudge.
g. Install new point set on breaker plate and attach with screw.

NOTE
Pilot on point set must engage matching hole in breaker plate.

h. If condenser was removed, install new condenser on breaker plate with attaching screw.
i. Connect primary and condenser lead wires to quick disconnect terminal on point set.
j. Using a spring gauge, measure the tension of the breaker arm spring. Pressure must be between 19 and 23 ounces.

NOTE
Hook spring gauge over the breaker lever (movable arm of the point set) and read tension just as points separate. Spring tension can be adjusted by carefully bending the breaker lever spring. Decrease pressure by carefully pinching the spring. To increase pressure, remove point set from distributor and bend spring away from lever. Excessive distortion of the spring must be avoided.

k. Using a feeler gauge, set the new points to 0.019 inch. Rubbing block of points must be resting on the highest point of a cam lobe when this adjustment is made (See *Dwell Angle* procedure below).
l. Rotate cam lubricator 180 degrees at 12,000-mile intervals and replace lubricator every 24,000 miles.
m. Replace rotor and distributor cap. Lock cap to distributor housing with hold-down screws.
n. Start engine and check dwell angle (see below).

To replace points on a 8-cylinder engine, proceed as follows:

NOTE
The contact point set (and condenser on some models) is replaced as a single unit. Breaker lever tension and point gap are preset at the factory, and only dwell angle requires adjustment after replacement.

a. Remove distributor cap by placing a screwdriver blade in the slotted head of each latch screw, pressing down, and rotating a quarter turn in either direction. Move cap out of the way.
b. Remove rotor attaching screws, then remove rotor (see **Figure 29**).
c. Remove the RFI shield, if so equipped.
d. Loosen the 2 point set assembly attaching screws and remove the set from the breaker plate. Disconnect the primary.
e. Remove the primary and condenser lead wires from the nylon insulated connector in the point set.
f. If the condenser is to be replaced, remove attaching screw and then remove condenser.
g. Install new condenser, using attaching screw (if old condenser was removed).
h. Install new point set, using attaching screws.
i. Connect primary and condenser lead wires to nylon insulated connector on point set. Make certain lead wires are routed so they will not interfere with point set operation.
j. Replace rotor and install and tighten the attaching screws.

k. Rotate cam lubricator 180 degrees at 12,000-mile intervals. Replace every 24,000 miles.
l. Replace distributor cap and lock into place with latches.
m. Start engine and check dwell angle (see below).

DWELL ANGLE

The preferred method for setting dwell angle is to first set the breaker point gap with a feeler gauge (6-cylinder engines only—points for 8-cylinder engines are factory pre-set) and then check the setting with a dwell meter. It is very important that breaker points be set to the proper gap. Points set too closely tend to burn and pit rapidly. Points with an excessive gap result in a weak spark at high speeds. New points must be set to a wider gap than used points to compensate for wear of the rubbing block while seating to the distributor cam.

Adjusting Dwell Angle (6-Cylinder Engines)

1. Remove distributor cap and rotor.
2. Verify alignment of points as per **Figure 28**. If necessary, align by bending fixed contact arm. Do not bend the breaker lever (movable arm). Do not attempt to align used points. Instead, replace them if serious misalignment is present.

29

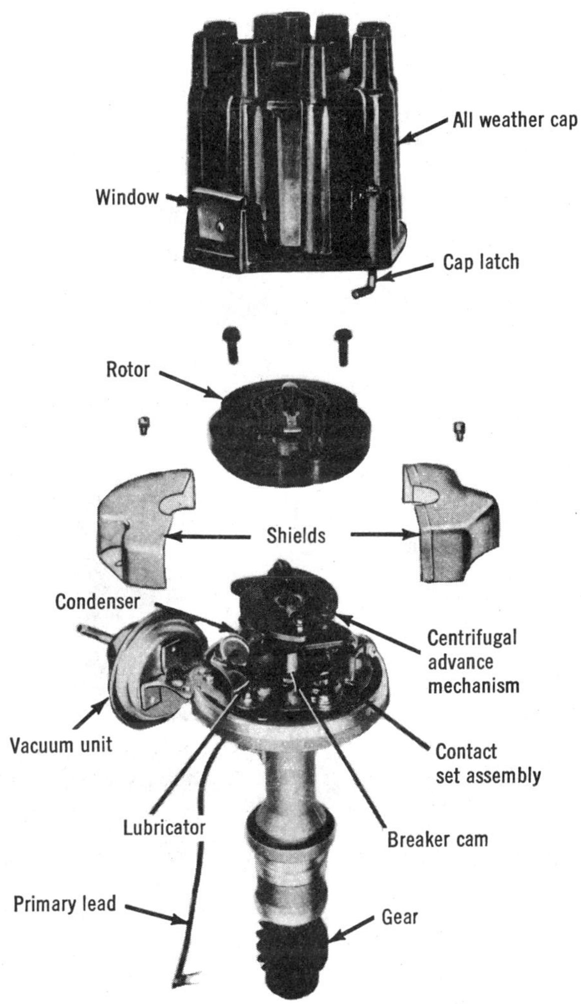

3

3. Crank or turn engine until rubbing block on breaker lever is resting on high point of a distributor cam lobe (this provides maximum point gap).
4. Loosen the point set attaching screw (see **Figure 26**) and insert a screwdriver in the opposing slots in the point set and the breaker plate. Move the point set to obtain a setting of 0.019 in. (new points) or 0.016 in. (used points), measured with blade-type feeler gauge.
5. Tighten the attaching screw and remeasure the gap. Readjust if setting was moved during the tightening of the screw.
6. Reassemble the rotor and distributor cap.
7. Connect dwell meter and tachometer to engine, using manufacturer's instructions.
8. Disconnect vacuum line from distributor and plug line with a sharp end of a pencil, a golf tee, or other similar pointed object.
9. Operate engine at normal idle speed and check dwell angle. For 6-cylinder engines, angle should be between 31 and 34 degrees.

NOTE

If dwell angle is not within these limits, points should be reset. Check for misalignment of points or worn distributor cam lobes.

10. Accelerate engine to 1,750 rpm. Variation in dwell angle should not exceed 3 degrees. Higher variation indicates excessive distributor wear or loose breaker plate.
11. Reconnect vacuum line to distributor.

Adjusting Dwell Angle (8-Cylinder Engine)

1. After installing point set and reassembling distributor, start engine and allow to warm up to normal operating temperature. Disconnect and plug vacuum line to distributor.
2. Attach a dwell meter and tachometer to the engine, using the manufacturer's instructions.
3. Raise window in distributor cap and insert a hex wrench in the adjusting screw head (see **Figure 30**).
4. Turn the adjusting screw until a dwell angle of 29-31 degrees (30 degrees preferred) is reached.
5. If a dwell meter is not available, turn the adjusting screw clockwise until the engine begins to misfire. Then turn screw 1/2 turn in opposite direction. This method will give an approximate dwell angle setting, but should be used only when a dwell meter is not available.
6. Accelerate the engine to 1,750 rpm and check for variation in dwell angle. The angle should not vary more than 3 degrees. Variation in excess of this indicates excessive distributor wear or loose breaker plate.
7. Remove hex wrench and close access window in distributor cap. Unplug and replace vacuum line to distributor.

BREAKER POINT IGNITION TIMING ADJUSTMENT

Ignition timing should be checked and adjusted (if required) after point replacement and dwell angle adjustment have been completed.

1. Connect timing light in accordance with the manufacturer's instructions. **Figure 31** shows a typical hook-up.
2. Determine the timing specifications from the VECI sticker (engine compartment) or **Table 7**.

30

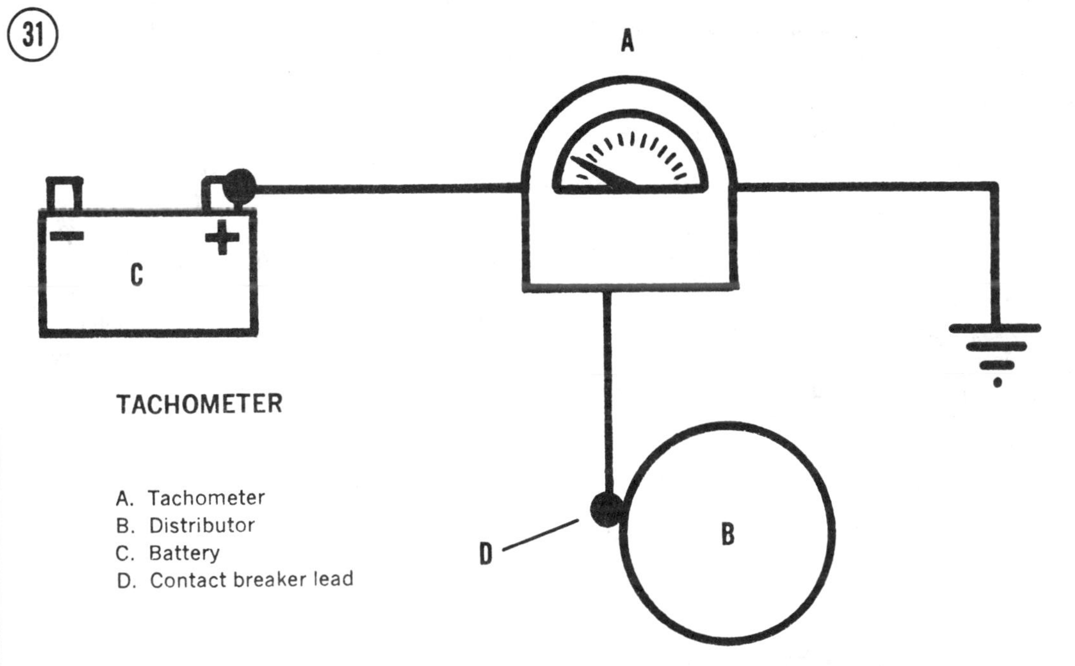

TACHOMETER

A. Tachometer
B. Distributor
C. Battery
D. Contact breaker lead

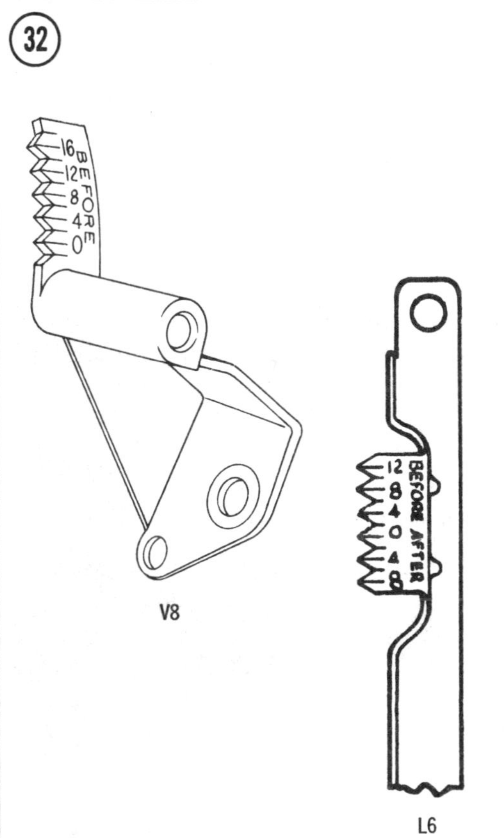

V8

L6

NOTE

Use chalk or white paint, etc., to emphasize the proper markings on the timing tab and the harmonic balancer.

3. Disconnect the vacuum line to the distributor and plug line with sharp end of a pencil or golf tee.
4. Operate the engine at normal idle speed.
5. Aim the timing light at the timing tab (see **Figure 32**, typical).
6. Adjust timing by loosening distributor clamp bolt and rotating distributor body as required while watching timing marks on engine tab and harmonic balancer and operation of timing light. When the marks appear to stand still directly opposite each other, retighten the distributor clamp bolt. Recheck timing and readjust if required.
7. Stop engine, disconnect timing light, and reconnect vacuum hose.

HIGH ENERGY IGNITION SYSTEM

Routine maintenance is not required for the HEI system itself. If parts or components fail, they are not repairable and must be replaced. However, engine timing should be checked and the distributor components visually checked for cracks, wear, dust, moisture, burns, etc., every 18 months or 22,500 miles, whichever comes first. At the same time, the secondary wiring (spark plug wires) should be inspected, checked out with an ohmmeter, and replaced if necessary.

NOTE

The HEI system has larger (8 mm) diameter, silicone-insulated, spark plug wires. While these gray-colored wires are more heat resistant and less vulnerable to deterioration than conventional wires, they should not be mistreated. When removing wires from spark plugs, grasp only on the boots. Twist boot 1/2 turn in either direction to break seal, then pull to remove.

Spark plugs also should be replaced at the intervals given in **Table 2**.

The procedure for ignition timing is identical to that given above for conventional ignition systems, except that timing light connections should be made in parallel using an adapter at the No. 1 spark plug wire terminal on the distributor. The distributor cap also has a special terminal marked TACH. Connect one lead of the tachometer to this terminal and the other to ground.

NOTE

Some tachometers must connect from the TACH terminal to the battery positive terminal. Check the manufacturer's instructions before making connections.

CARBURETOR ADJUSTMENTS

Idle Mixture Adjustment

The idle mixture screws on all 1971-1978 carburetors are equipped with limiter caps. These caps, which should not be altered or removed during a routine tune-up, limit or prevent the adjustment of the air/fuel mixture. This mixture is set during a flow test at the factory, and in most instances a satisfactory idle can be obtained within the limits imposed by the caps.

On all 1979-1980 models, the idle mixture screws are recessed into the carburetor body and covered with plugs. Idle mixture adjustment is not a part of the routine tune-up on any 1979-1980 model. Removal of the plugs covering the adjustment screws requires chiselling away a part of the carburetor body. This job should be left to a mechanic who is familiar with the correct procedure.

On all 1975 and later models (and some earlier models), the carburetor is designed to work and interact with other emission control systems. Tampering with carburetor adjustments by an unskilled mechanic who does not understand how all of the systems work together can result in poor performance, shortened engine and component life, and possible violation of the emission control laws.

In all cases, the ignition system should be tuned and timed as described earlier in this chapter before attempting any carburetor adjustments. If idle mixture adjustment is required following carburetor overhaul, see procedures in Chapter Five. Correct engine idle speed is obtained by a combination of idle mixture and throttle linkage adjustments. As stated above, in a routine tune-up, mixture adjustment should be made within the limits imposed by the limiter caps. Throttle linkage adjustment is made, depending upon model, by turning the idle stop screw, the idle stop solenoid plunger, or both.

When making any idle adjustment, always refer to the Vehicle Emission Control Information (VECI) sticker in the engine compartment of your car. The sticker lists the correct idle speeds and other pertinent information for your specific vehicle. If this sticker is missing or cannot be read, use the information provided in **Table 7**.

Idle Speed Adjustment (1975-On 1 BBL-Carburetor)

1. Start the engine and allow it to reach normal operating temperature. Remove the air cleaner and verify that the choke valve is fully open. If equipped with air conditioner, make sure it is turned off.
2. Connect a tachometer to the distributor, following the manufacturer's instructions.
3. Set the parking brake and block the rear wheels.

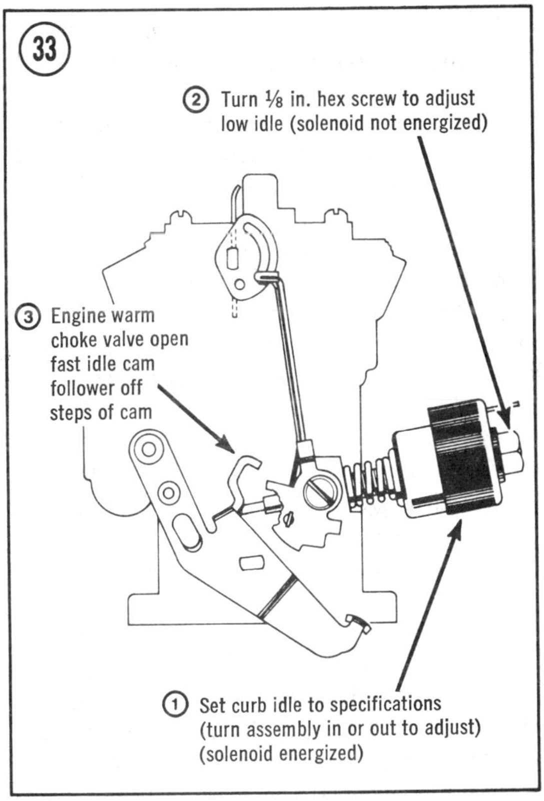

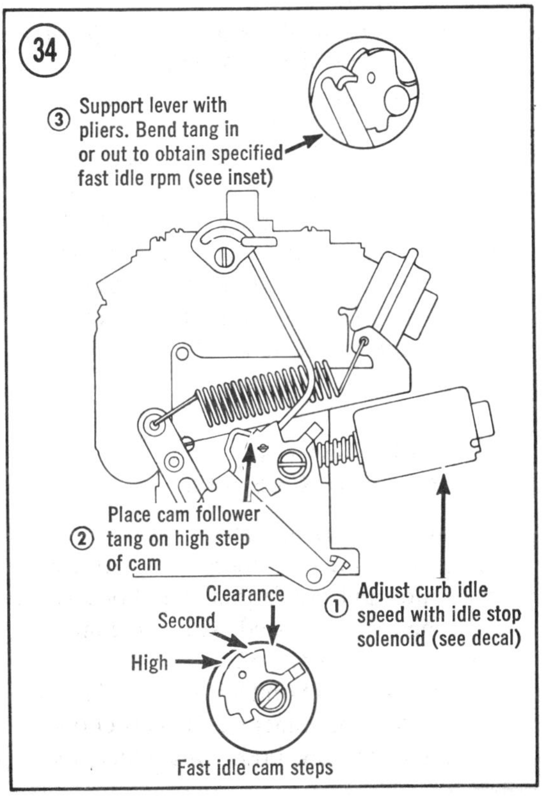

4. Disconnect the fuel tank hose from the vapor canister. Disconnect and plug the vacuum hose at the distributor. Check the VECI sticker in the engine compartment and disconnect and plug other hoses as directed.
5. Check the engine timing with a timing light (see *Ignition Timing Adjustment* this chapter) and correct timing to specifications, if required. See VECI sticker or **Table 7**. Reconnect vacuum hose to distributor.
6. If the vehicle is equipped with automatic transmission, start engine and place the selector in DRIVE. If equipped with manual transmission, start engine and place selector in NEUTRAL.
7. With the idle stop solenoid energized, adjust the curb idle speed to specifications by turning the solenoid assembly. See **Figure 33**.
8. Disconnect the electrical leads to the idle stop solenoid and turn the 1/8 in. hex screw in the end of the solenoid to obtain the specified low idle speed.
9. Reconnect the solenoid leads. Advance the throttle slightly and release it, then recheck to verify that the specified curb idle speed is obtained when the throttle is released.
10. Reconnect the hoses that were disconnected in Step 4. Remove the tachometer.

NOTE

It is not necessary to adjust idle mixture as a part of a routine tune-up. If the carburetor has been given a major overhaul, see ***Idle Mixture Adjustment*** *for your carburetor in Chapter Five.*

Fast Idle Adjustment
(1975-On 1 BBL Carburetor)

Refer to **Figure 34**.
1. Verify that low and curb idle speeds are correctly set.
2. Start engine and allow it to reach normal operating temperature. Air cleaner should be installed, choke should be fully open, EGR valve signal line should be removed and plugged, and air conditioner should be off. Connect tachometer to engine, using manufacturer's instructions.

3. Disconnect vacuum hose from distributor and plug hose.
4. With transmission in NEUTRAL, set fast idle cam follower so that tang is on high step of cam.
5. Bend cam follower tang to arrive at fast idle speed listed on emissions control decal.

Idle Speed Adjustment (1977-1978 2GC-2GE Carburetor)

1. Start engine and allow it to warm up to normal operating temperature. Air cleaner should be installed, choke should be fully open, and air conditioner should be off. Connect tachometer to engine, using manufacturer's instructions.
2. Set parking brake and block rear wheels.
3. Disconnect fuel tank hose at vapor canister.
4. Disconnect vacuum hose from distributor and plug hose. Disonnect and plug vacuum hose at EGR valve on 231 cid engine.
5. Verify that timing is properly set or adjust if required, using the procedure given above. Reconnect vacuum hose.
6. Place the transmission selector in DRIVE (automatic transmision) or NEUTRAL (manual transmission).
7. If the carburetor does not have an idle stop solenoid, refer to **Figure 35** and place the idle speed screw on the low step of the fast idle cam. Turn the idle stop screw as required to obtain the specified curb idle speed.
8. If the carburetor has an idle stop solenoid, but no air conditioning, refer to **Figure 36**. Open the throttle slightly to allow the solenoid plunger to fully extend, then allow the speed to return to idle. Turn the solenoid screw as required to adjust the speed to the specified curb idle speed. Then disconnect the electrical connecter from the solenoid and turn the idle speed screw to the specified low idle speed. Reconnect the solenoid and recheck the curb idle speed with the solenoid energized and extended.
9. If the vehicle is equipped with air conditioning, make sure the air conditioner is turned off and then set the engine at the specified curb idle speed, using the idle speed screw. See **Figure 37**. Disconnect the electrical lead at the air conditioner compressor and turn the air conditioner on. Open the throttle slightly to allow the solenoid plunger to fully extend. Release throttle and turn the solenoid screw to adjust engine speed to the specified rpm. Reconnect the air conditioner electrical lead.
10. Stop engine and remove tachometer. Connect fuel tank hose to vapor canister and vacuum hose to EGR valve. Remove blocks from wheels.

35

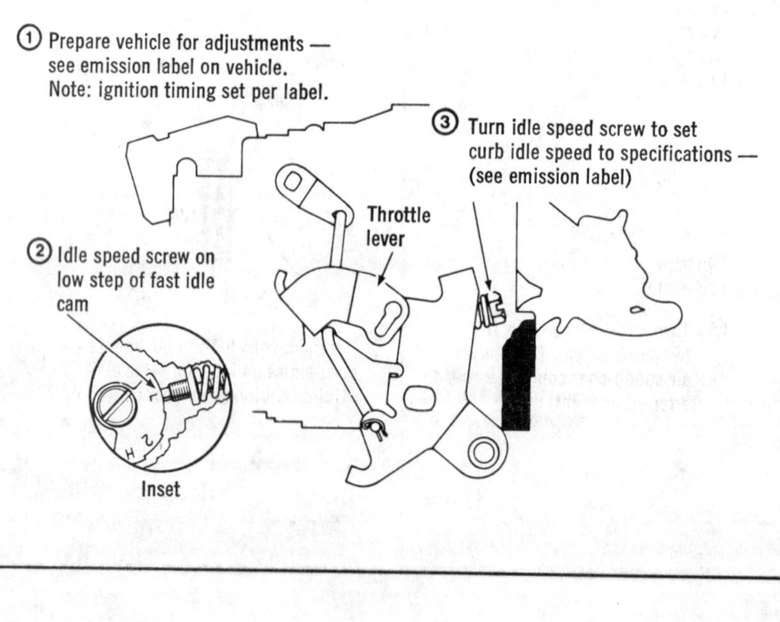

36

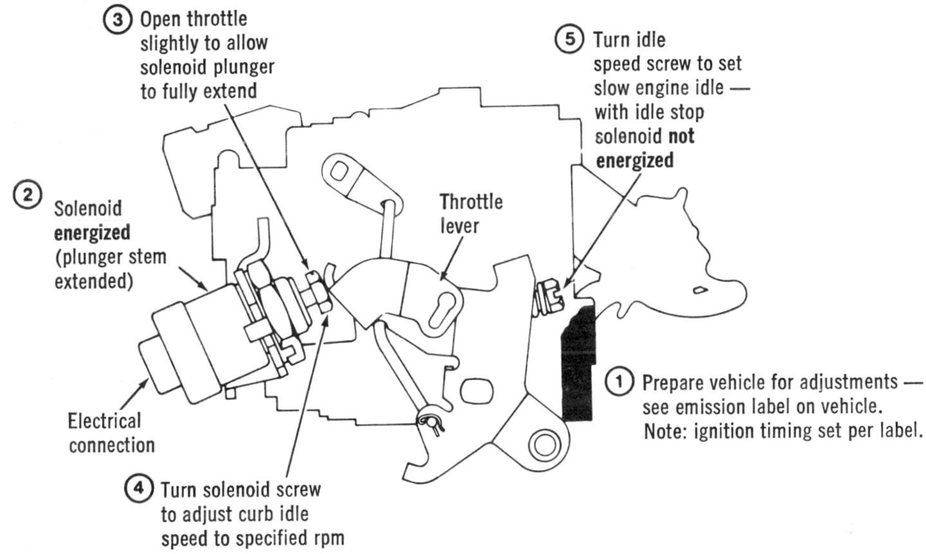
③ Open throttle slightly to allow solenoid plunger to fully extend
⑤ Turn idle speed screw to set slow engine idle — with idle stop solenoid **not energized**
② Solenoid **energized** (plunger stem extended)
Throttle lever
① Prepare vehicle for adjustments — see emission label on vehicle. Note: ignition timing set per label.
Electrical connection
④ Turn solenoid screw to adjust curb idle speed to specified rpm

37

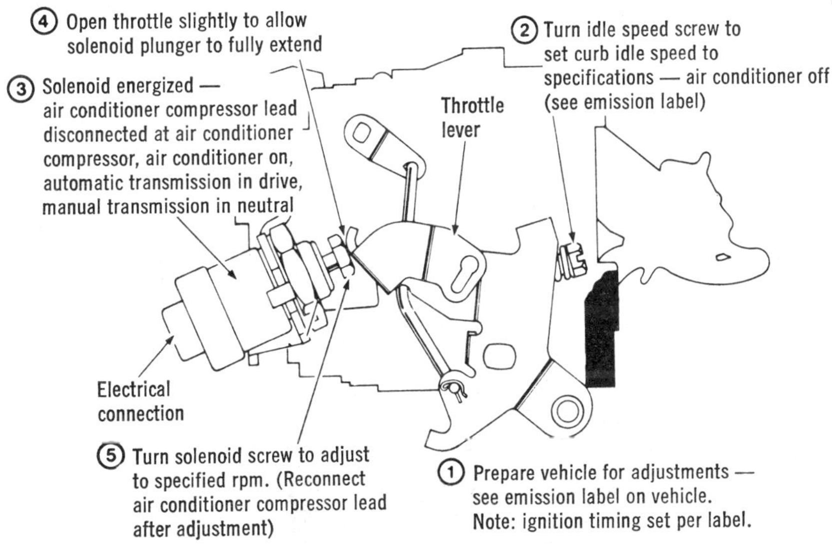
④ Open throttle slightly to allow solenoid plunger to fully extend
② Turn idle speed screw to set curb idle speed to specifications — air conditioner off (see emission label)
③ Solenoid energized — air conditioner compressor lead disconnected at air conditioner compressor, air conditioner on, automatic transmission in drive, manual transmission in neutral
Throttle lever
Electrical connection
⑤ Turn solenoid screw to adjust to specified rpm. (Reconnect air conditioner compressor lead after adjustment)
① Prepare vehicle for adjustments — see emission label on vehicle. Note: ignition timing set per label.

Idle Speed Adjustment (1978-On M2MC/E2ME Carburetors)

Refer to **Figure 38** (M2MC) or **Figure 39** (E2ME).

1. Start engine and allow to warm up to normal operating temperature. Air cleaner should be installed, choke should be fully open, and air conditioner should be off. Connect tachometer to engine, using manufacturer's instructions.
2. Set parking brake and block rear wheels.
3. Disconnect fuel tank hose at vapor canister.
4. Disconnect vacuum hose or electrical connector from distributor and plug hose.
5. Verify that timing is properly set or adjust if required, using the procedure given above. Reconnect vacuum hose or electrical connector.
6. Place the transmission selector in DRIVE (automatic transmmssion) or NEUTRAL (manual transmission).
7. On models without air conditioning, adjust the idle rpm to the specified curb idle speed, using the idle speed screw.
8. On air conditioned models, make sure the air conditioner is off and adjust idle rpm to specified curb idle speed, using the idle speed screw. Then disconnect the electrical lead to the air conditioner compressor, and turn the air conditioner on. Open the throttle slightly to let the solenoid plunger fully extend, then release the throttle. Turn the solenoid screw as required to adjust the engine idle speed to the specified solenoid screw rpm. Reconnect the electrical lead to the air conditioner compressor.
9. Refer to **Figure 40** (M2MC) or **Figure 41** (E2ME). Place automatic transmission in PARK and manual transmission in NEUTRAL. Hold cam follower on the specified step of the fast idle cam (see VECI sticker) and turn the fast idle screw to obtain the specified fast idle speed. Stop the engine and remove the tachometer.
10. Connect the fuel tank hose to the vapor canister and remove the blocks from the wheels.

Low and Curb Idle Adjustment (1975-On 4 BBL-Carburetor)

NOTE

*Obtain low and curb idle speeds from Vehicle Emission Control Information decal in engine compartment. If decal is missing or cannot be read, speeds given in **Table** 7 can be used.*

38

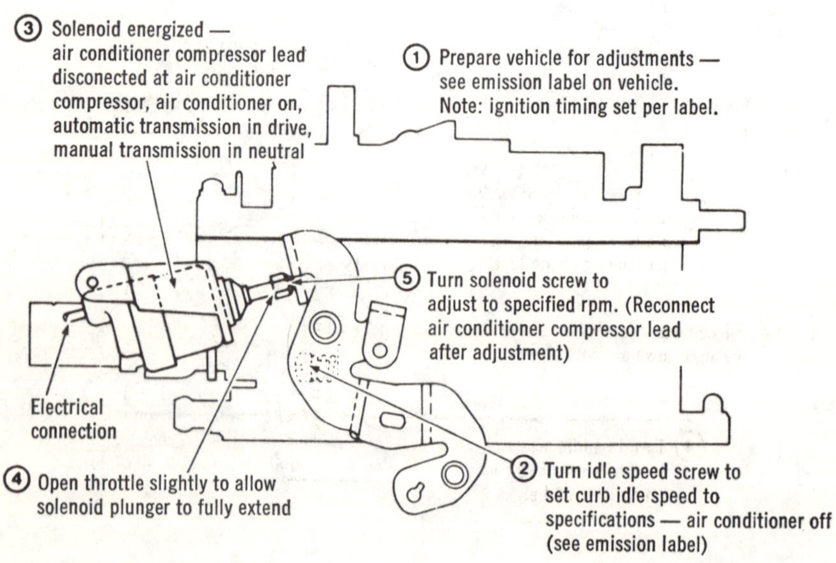

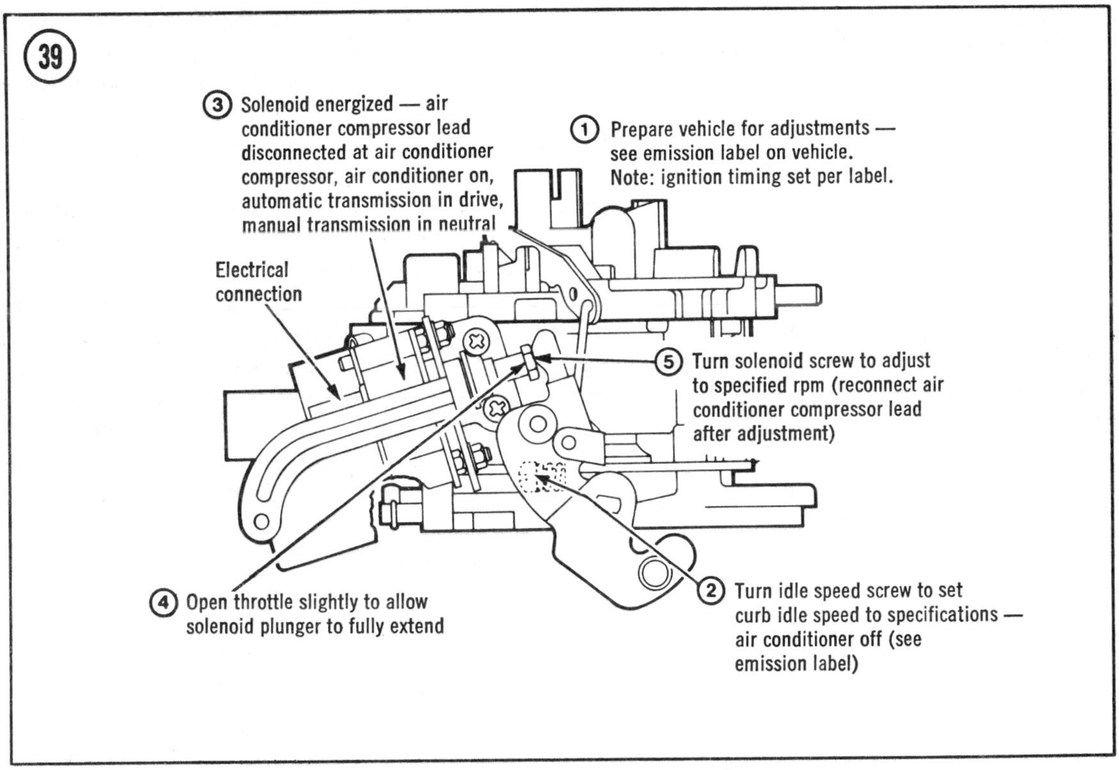
39
3 Solenoid energized — air conditioner compressor lead disconnected at air conditioner compressor, air conditioner on, automatic transmission in drive, manual transmission in neutral
1 Prepare vehicle for adjustments — see emission label on vehicle. Note: ignition timing set per label.
Electrical connection
5 Turn solenoid screw to adjust to specified rpm (reconnect air conditioner compressor lead after adjustment)
4 Open throttle slightly to allow solenoid plunger to fully extend
2 Turn idle speed screw to set curb idle speed to specifications — air conditioner off (see emission label)

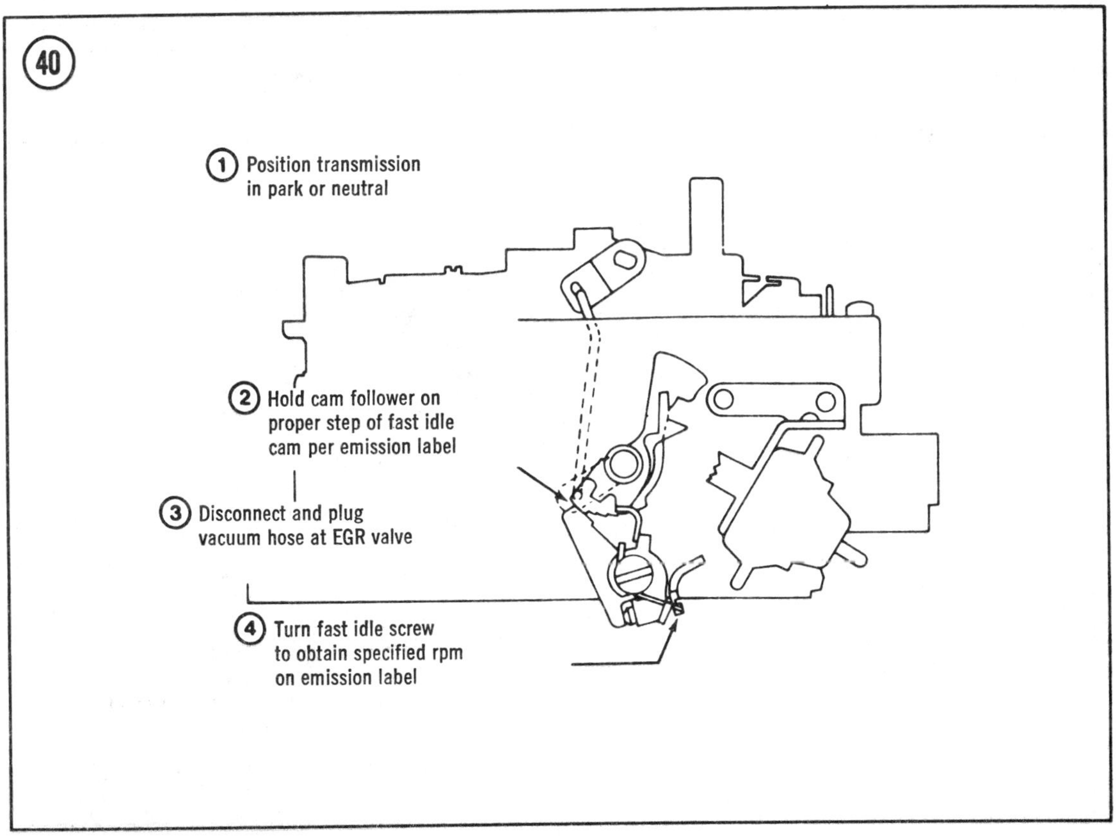
40
1 Position transmission in park or neutral
2 Hold cam follower on proper step of fast idle cam per emission label
3 Disconnect and plug vacuum hose at EGR valve
4 Turn fast idle screw to obtain specified rpm on emission label

41

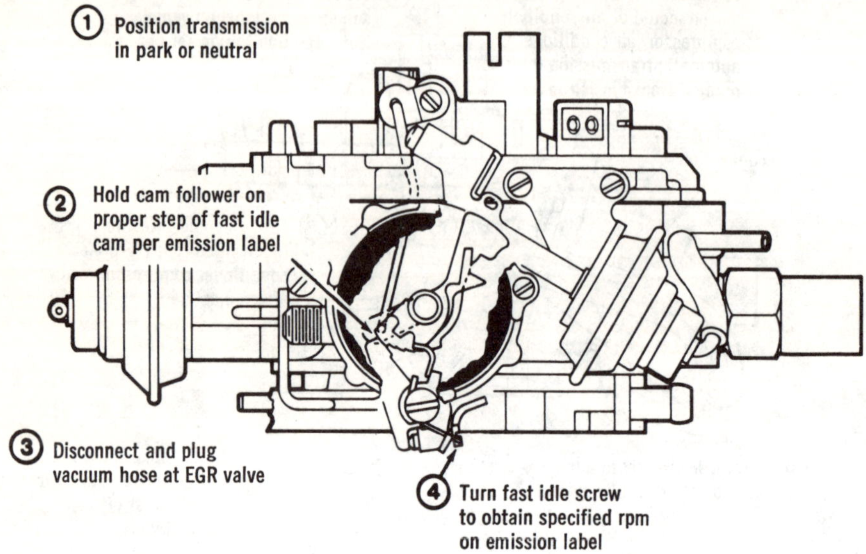
① Position transmission in park or neutral
② Hold cam follower on proper step of fast idle cam per emission label
③ Disconnect and plug vacuum hose at EGR valve
④ Turn fast idle screw to obtain specified rpm on emission label

42

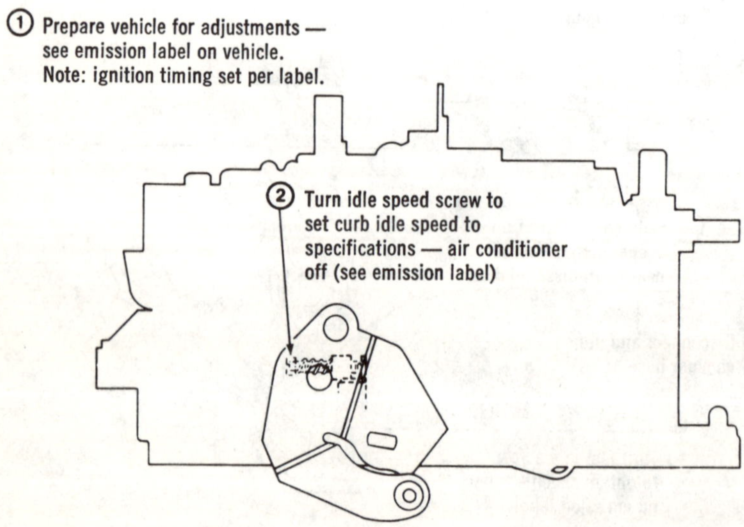
① Prepare vehicle for adjustments — see emission label on vehicle. Note: ignition timing set per label.
② Turn idle speed screw to set curb idle speed to specifications — air conditioner off (see emission label)

43

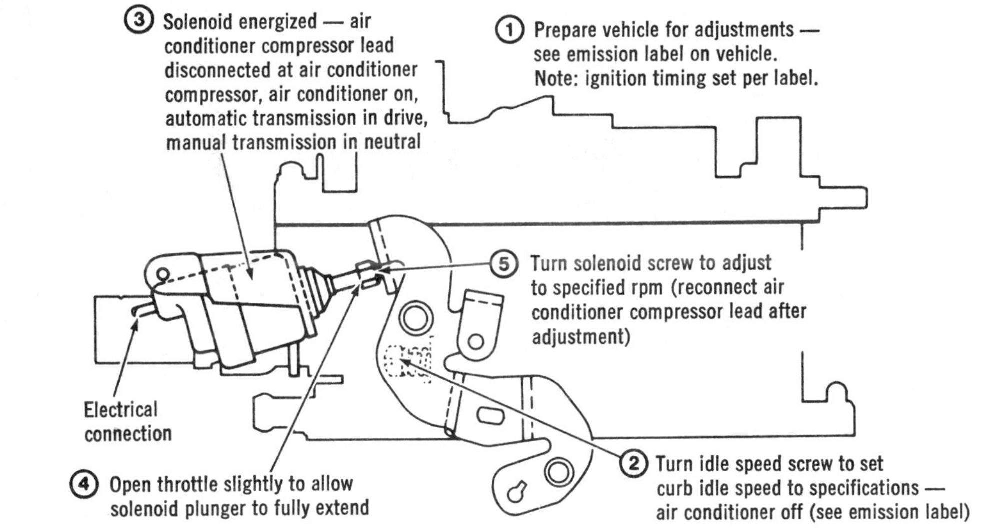

44

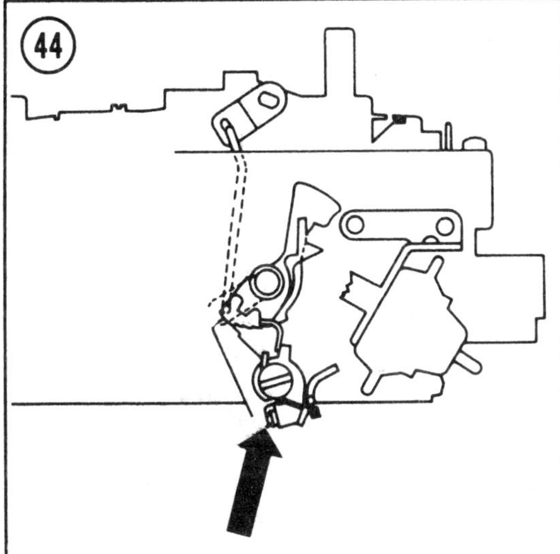

Refer to **Figure 42** or **Figure 43**.

1. Start the engine and let it warm up to operating temperature. The air cleaner should be installed, the choke should be fully open, and the air conditioner should be turned off. Connect the tachometer to the engine, using the manufacturer's instructions.
2. Set parking brake and block rear wheels.
3. Disconnect fuel tank hose from vapor canister. Disconnect distributor vacuum hose and plug hose. Disconnect other hoses and wires as specified on VECI sticker.
4. Check timing, using procedure given above, and adjust as required. Reconnect distributor vacuum hose.
5. Disconnect the electrical lead at idle stop solenoid.
6. Place transmission selector in DRIVE (automatic transmission) or NEUTRAL (manual transmission).
7. Turn the idle speed screw as required to obtain the specified curb or carburetor screw speed. See **Figure 42** (no solenoid) or **Figure 43** (with air conditioner idle speed solenoid), and VECI sticker.
8. Place automatic transmission in PARK or manual transmission in NEUTRAL. Hold cam follower on specified step of fast idle cam (see VECI sticker) and adjust fast idle screw to obtain the specified fast idle rpm. See **Figure 44**.
9. If the carburetor is equipped with an idle speed solenoid, disconnect the electrical lead to the air conditioner compressor, place automatic transmission in PARK or manual transmission in NEUTRAL, and turn the air conditioner ON. Open the throttle slightly to allow the solenoid plunger to fully extend and then release throttle. Turn the solenoid screw

45

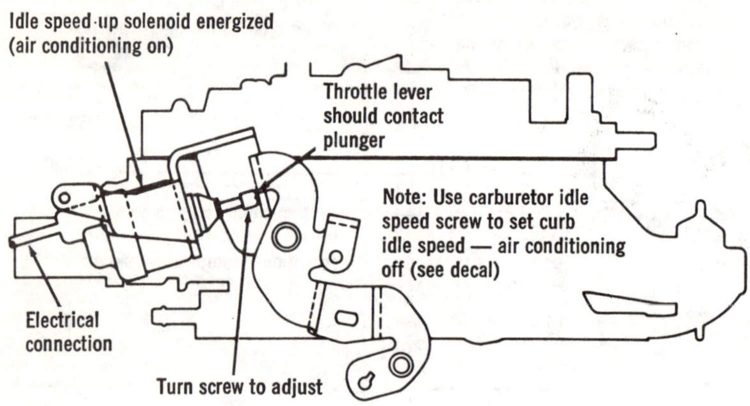

(see **Figure 43**) to obtain the specific solenoid screw speed. See VECI sticker. Reconnect the lead to the electrical sticker.

10. Stop the engine and remove the tachometer. Reconnect hoses and wires disconnected in Step 3.

Idle Speed Adjustment
(1975-1977 2MC Carburetor)

Refer to **Figure 45**.

1. Start the engine and warm it up to normal operating temperature. Remove the air cleaner. Disconnect the air cleaner vacuum hose at the intake manifold and plug the fitting so that additional air is not drawn into the manifold.

2. Verify that the choke valve is fully open and that the air conditioner, if so equipped, is off. Set the parking brake and block the rear wheels.

3. If the car has an automatic transmission, disconnect and plug the vapor canister and the spark plug advance vacuum modulator hoses at the carburetor. If equipped with manual transmission, disconnect and plug the vapor canister, EGR valve, and distributor hoses at the carburetor.

4. Check timing, using the procedure elsewhere in this chapter, and adjust to specifications if required. Reinstall the distributor vacuum line.

5. Place automatic transmission selector in DRIVE or manual transmission selector in NEUTRAL and adjust carburetor slow idle screw to obtain specified slow idle rpm (750 rpm for manual transmission and 550 rpm for automatic).

6. If the car has an automatic transmission and air conditioning, disconnect the electrical lead from the air conditioner compressor and turn the air conditioner on. Adjust the idle speed-up solenoid screw (**Figure 45**) to obtain an engine speed of 650 rpm. Turn the air conditioner off and reconnect the electrical leads to the compressor.

7. With automatic transmission in PARK or manual in NEUTRAL, place the fast idle cam follower on the low step of the cam and against the shoulder of the next higher step. Turn fast idle screw (see **Figure 44** for approximate location) to obtain a fast idle speed of 900 rpm.

8. Reconnect all hoses that were previously disconnected, remove the tachometer and install the air cleaner and its vacuum hose.

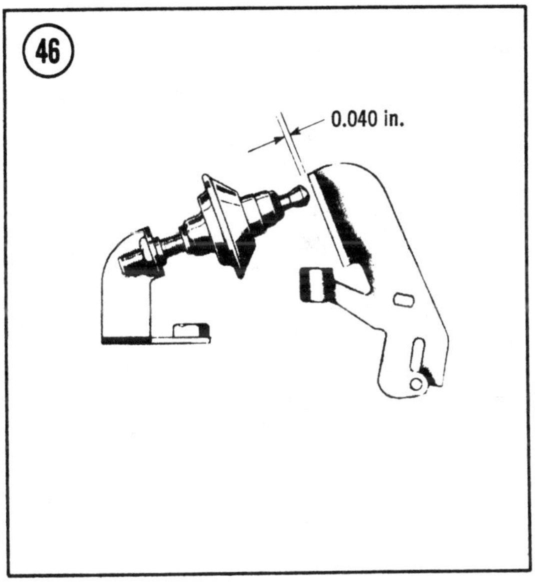

9. To check and adjust dashpot clearance, adjust idle speeds as described in the preceeding steps and make sure the choke is fully open and the fast idle cam follower is off the cam steps. Depress the dashpot plunger and measure the clearance between the end of the plunger and the throttle lever. See **Figure 46**. If the clearance is not 0.040 in., turn the dashpot to make the necessary adjustment.

Idle Speed Adjustment (1970-1971 1-BBL Carburetor)

NOTE

The 1971 model carburetors are equipped with limiter caps on idle mixture screws. These caps, which should not be removed, do not permit adjustment of the idle mixture.

1. Start the engine and allow it to warm up to normal operating temperature. Remove the air cleaner and make sure the choke valve is fully open.
2. On 1970 models, remove the distributor vacuum valve from the carburetor and plug the fitting. Also plug the carburetor hot idle compensator so it is closed. On 1971 models, disconnect the distributor vacuum hoses from the combination emissions control (CEC) valve and plug the hose leading to the carburetor. Disconnect and plug vapor canister hose to fuel tank.

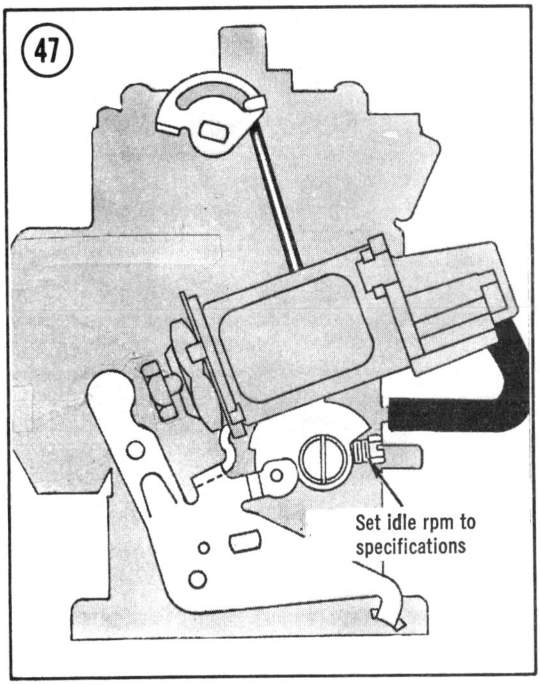

3. If equipped with air conditioning, make sure the air conditioner is turned off. Set the parking brake and block the rear wheels. Place automatic transmission in DRIVE or manual tansmission in NEUTRAL.
4. Connect a tachometer to the engine, following the manufacturer's instructions. Check dwell and timing, and adjust to specifications if necessary. See procedures elsewhere in this chapter.
5. On 1970 models, make sure the throttle solenoid wires are connected and that the idle stop screw is not touching the throttle lever. Adjust idle mixture screw to obtain the highest possible steady rpm. Then adjust the throttle solenoid plunger to obtain a speed of 25 rpm higher than the highest of the 2 idle speeds specified on the VECI sticker. Then turn the idle mixture screw in to the specified higher speed. Disconnect the solenoid wires and adjust the throttle stop screw to obtain the lower of the 2 specified idle speeds. Reconnect solenoid wires.
6. On 1971 models, do not attempt to adjust idle mixture as a part of a routine tune-up, as this mixture is preset at the factory. Set engine speed to the specified rpm, using the idle speed screw. See **Figure 47**.

3

7. Place cam follower on lowest step of fast idle cam and verify that fast idle speed (in PARK or NEUTRAL) is 900 rpm for automatic transmission or 750 rpm for manual transmission. If not, adjust by bending cam follower as shown in **Figure 48**.
8. Remove tachometer and reconnect all hoses that were disconnected in Step 2.

Idle Speed Adjustment (1970-1974 2-BBL and 4-BBL Carburetors)

1. Start the engine and allow it to warm up to normal operating temperature. Remove the air cleaner and disconnect the air cleaner vacuum hose at the intake manifold and plug the fitting on the manifold. Make sure the choke is fully open and the air conditioning is off.
2. Disconnect and plug the carburetor hose from the vapor canister. Also disconnect and plug the vacuum advance hose at the distributor and the EGR valve hose, if so equipped.
3. Connect a tachometer to the engine and check dwell and timing. Adjust to specifications if necessary. See procedures elsewhere in this chapter.
4. If the carburetor does not have an idle solenoid, turn the carburetor idle screw to obtain the specified slow idle speed. See VECI sticker or **Table** 7.
5. If the carburetor is equipped with an idle solenoid, adjust the energized solenoid screw to obtain the specified higher slow idle speed. Then de-energize the solenoid by disconnecting the electrical leads and use the carburetor idle screw to adjust engine speed to the lower specified slow idle speed.
6. Place cam follower on the specified step of the fast idle cam (see VECI sticker) and turn the fast idle adjustment screw to obtain the specified fast idle speed.
7. Reconnect the hoses that were disconnected in Step 2 and install the air cleaner and air cleaner vacuum hose. Remove the tachometer.

DRIVE BELT REPLACEMENT

Belts driven by the crankshaft pulley are used to drive various accessories, including the alternator, power steering pump, air induction reactor pump, water pump, and air conditioner compressor. Belt replacement usually is a matter of loosening the attaching and adjusting bolts of an accessory, removing the belt, installing a new belt, and adjusting the belt tension to a specified value. It can be seen from **Figure 49** (typical) that it may be necessary to remove one or more other belts before certain belts can be removed.

48

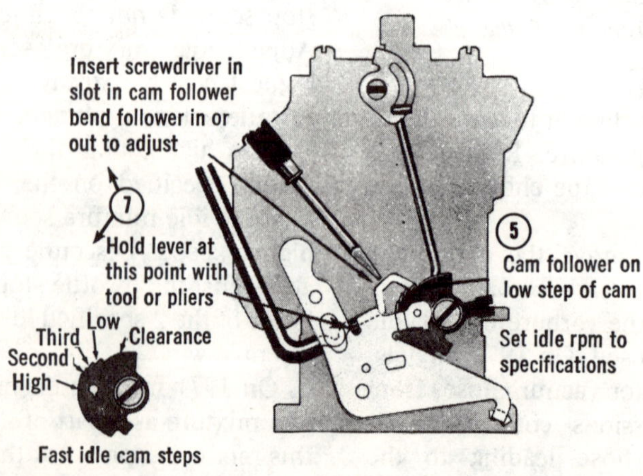

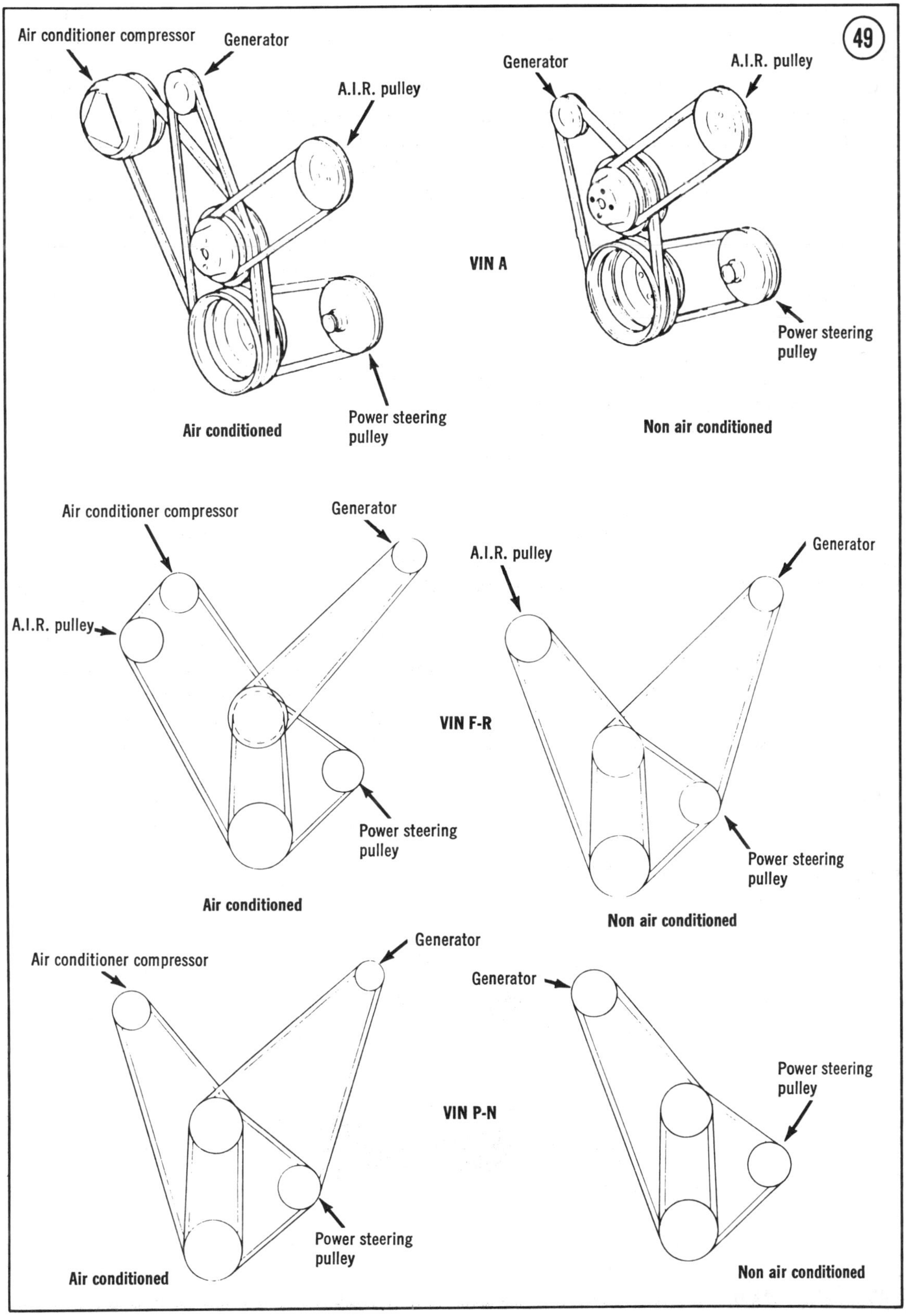
49
Air conditioner compressor
Generator
A.I.R. pulley
Power steering pulley
Air conditioned
Generator
A.I.R. pulley
Power steering pulley
Non air conditioned
VIN A
Air conditioner compressor
Generator
A.I.R. pulley
Power steering pulley
Air conditioned
A.I.R. pulley
Generator
Power steering pulley
Non air conditioned
VIN F-R
Air conditioner compressor
Generator
Power steering pulley
Air conditioned
Generator
Power steering pulley
Non air conditioned
VIN P-N

Belt tension is very important. A belt that is too loose will wear quickly and, if loose enough to slip, will result in improper operation of the accessory or accessories that it drives. On the other hand, a belt that is too tight will also wear quickly and will result in undue strain and wear on the bearings of the driven accessory or accessories.

The only recommended method of belt tension adjustment is the use of a tension gauge. See **Figure 50**. The gauge is attached to the belt as shown in the figure and pressure is applied to the accessory that was loosened for belt replacement until the specified tension is reached. Then the accessory adjustment and attaching bolts are securely tightened while maintaining the belt tension. If you do not have a tension gauge, you may be able to rent one from a tool rental business or a service station. A time-honored method for checking belt tension when a gauge is not available is to depress the belt halfway between 2 pulleys with your thumb. The belt should deflect 1/2 inch under about 12 lb. of pressure. This method is not recommended, however, and should be used only in emergencies.

A belt is considered used after it has been in service (engine running) for 15 minutes or longer. If a new belt is installed, the tension should be rechecked and adjusted, if necessary, after 30 minutes of service. General Motors specifies the following tensions for drive belts: 5/16 inch wide belts, 80 lb. maximum (new) or 50 lb. minimum (used); 3/8 inch wide belts, 140 lb. maximum (new) or 70 lb. minimum (used); and 15/32 inch wide belts, 165 lb. maximum (new) or 90 lb. minimum (used).

50

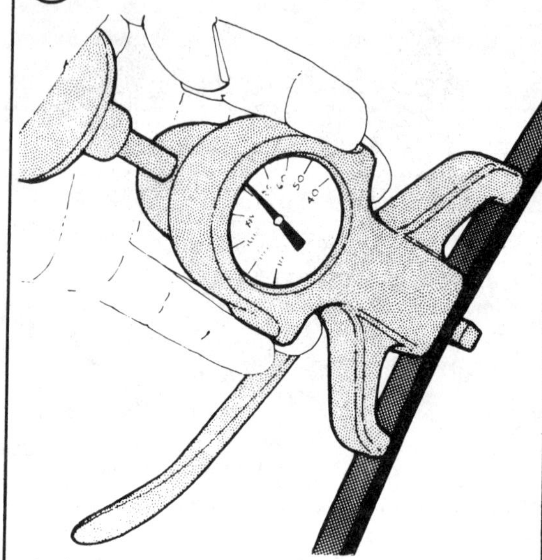

DIESEL ENGINE TUNE-UP

Diesel engines do not require a tune-up in the same sense as a gasoline engine, primarily because the diesel engine uses a compression ignition instead of an electrical ignition. Fewer maintenance tasks are thus required on a diesel engine, but the required tasks are just as important, if not more so, as the more extensive gasoline engine maintenance. The required tasks and the intervals at which they should be performed are given in **Table 3**.

Owner maintenance on diesel engines should be limited to the tasks listed in **Table 3**. Tampering, especially with the injection system, by an unskilled mechanic can lead to serious (and expensive) damage.

Engine Oil Change

Use only oil intended for diesel engine service. This will be plainly marked on the oil can label and/or top by a letter code (**Figure 51**). Use only oils marked SE/CD (1978 diesel) or SE/CC (1979-80 diesel). Oils not marked with these letter codes will not provide the protection necessary for diesel engines and their use could lead to engine damage. See *Engine Oil and Filters* in this chapter for oil change instructions. Also note that the oil should be changed every 3,000 miles in diesel engines. If the temperature in your area is consistently under 85° F, use 10W-30 viscosity grade oil. If it is consistently above 85° F, use 30W.

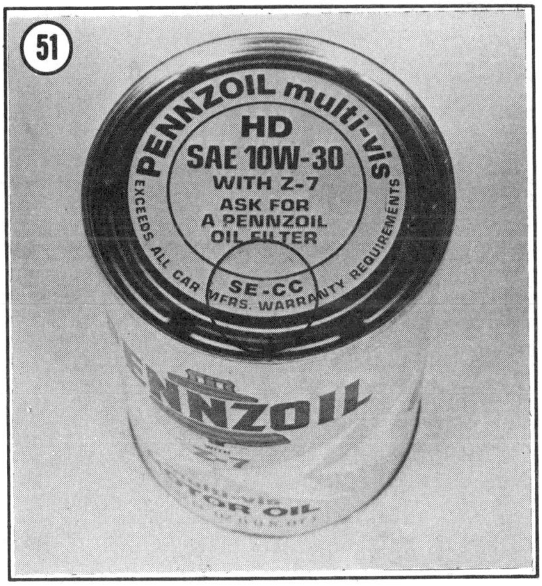

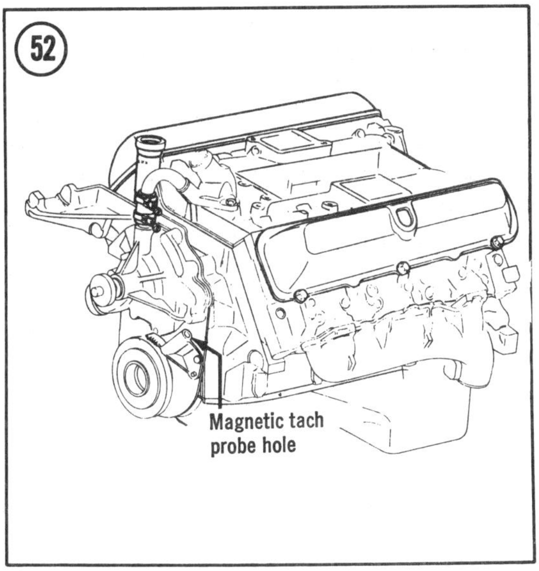

Engine Oil Filter Replacement

The engine oil filter should be replaced every time the engine oil is changed. See **Table 4** for the proper replacement and *Engine Oil and Filters* in this chapter for instructions.

Fuel Filter

The fuel filter is located in the line between the fuel pump and injection pump. It should be changed at the interval stated in **Table 3**. See *Fuel Filters* in this chapter for replacement instructions.

Idle Speed Adjustment

NOTE
A special tachometer with a magnetic probe (J-26925 or equivalent) must be used on diesel engines. Most tachometers for gasoline engines operate from the electrical ignition system impulses and will not work on diesel engines.

1. Connect the tachometer probe to the engine by inserting it in the timing indicator hole. See **Figure 52**.

2. Start the engine and allow it to reach normal operating temperature.

3. Block the rear wheels and set the parking brake. Place the transmission selector in DRIVE and make sure the air conditioner, if so equipped, is off.

4. Turn the slow idle speed screw on the injection pump (see **Figure 53**) to obtain the slow idle speed (rpm) listed on the VECI sticker.

5. If the car has an air conditioner, turn it on and disconnect the wires to the air conditioning compressor. If there is no air conditioner, disconnect the solenoid connector. Connect a jumper wire between ground and one of the solenoid terminals. Connect a second jumper wire between the second solenoid terminal and the battery positive terminal to energize the solenoid. Adjust the solenoid plunger (**Figure 54**) to obtain an idle speed of 650 rpm.

6. Stop the engine, reconnect all wires and connectors to their original locations, and remove all test equipment.

Other Maintenance Tasks

Other maintenance tasks listed in **Table 3** are described under *Periodic Maintenance, Gasoline and Diesel* in this chapter.

53

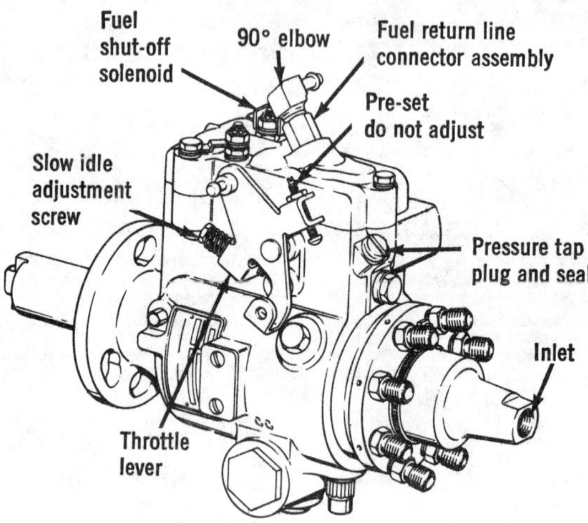
Fuel
shut-off
solenoid
90° elbow
Fuel return line
connector assembly
Pre-set
do not adjust
Slow idle
adjustment
screw
Pressure tap
plug and seal
Inlet
Throttle
lever

54

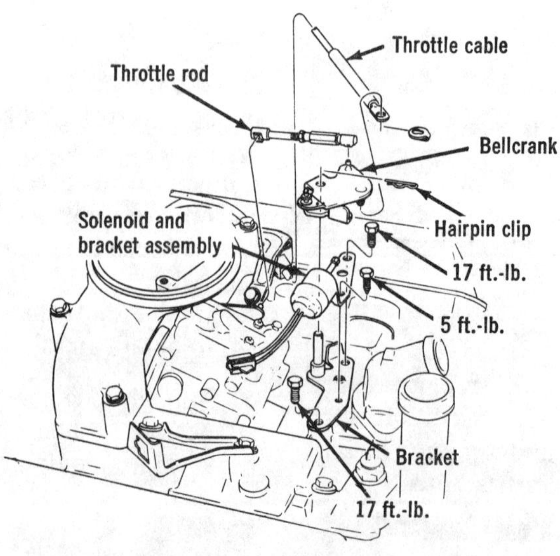
Throttle cable
Throttle rod
Bellcrank
Hairpin clip
Solenoid and
bracket assembly
17 ft.-lb.
5 ft.-lb.
Bracket
17 ft.-lb.

Table 1 RECOMMENDED LUBRICANTS AND FLUIDS

Use	Recommendation
Engine oil	Product bearing letter designation SE (gasoline engine); SE/CD (1978 diesel); or SE/CC (1979-on diesel)
Power steering	GM power steering fluid or, if not available, DEXRON® or DEXRON® II automatic transmission fluid
Differential—standard	SAE 80 or SAE 90 GL-5 gear lubricant (SAE 80 GL-5 in Canada)
Differential—Positraction	GM part No. 1051022 (lubricant) or equivalent
Steering gear—manual	GM part No. 1051052 (lubricant) or equivalent
Transmission—manual	SAE 80 or SAE 90 GL-5 gear lubricant (SAE 80 GL-5 in Canada)
Brake master cylinder	DOT 3 or Delco Supreme 11 brake fluid
Clutch linkage (manual transmission) 1. Pivot points 2. Cross-shaft grease fitting and pushrod to clutch fork joint	 Engine oil EP chassis lubricant meeting requirements of GM specification GM 6031-M
Shift linkage (all)	Engine oil
Hood latch and hinge assemblies 1. Pivots and spring anchor 2. Release pawl 3. Hinges	 Engine oil Chassis lubricant Engine oil
Chassis lubrication (front suspension, steering linkage, etc.)	Chassis lubricant meeting requirements of GM specification GM 6031-M
Constant velocity universal joint	Lubricant, GM part No. 1050679, or meeting requirements of GM specification GM 6040-M
Automatic transmission	DEXRON® or DEXRON® II automatic transmission fluid
Parking brake cables	Chassis lubricant
Front wheel bearings	Chassis lubricant meeting the requirements of GM specification GM 6031-M
Door, tailgate, seat and trunk hinges	Engine oil
Convertible door-to-lock wedge plates	Stick-type lubricant
Windshield washer	GM Optikleen washer solvent, part No. 1050001 or equivalent
Battery	Clean, odorless drinking water
Engine cooling system	Mixture of water and high quality Ethylene Glycol base antifreeze, meeting requirements of GM specification 1899-M

Table 2 MAINTENANCE SCHEDULE (GASOLINE ENGINE)

Frequency	Item No.	Service
		Section A — Lubrication and General Maintenance
1975-on — every 12 months or 7,500 miles Others — every 4 months or 6,000 miles	1	Chassis lubrication
	2	Fluid levels check
	3	Engine oil change
At first oil change, then every second oil change	4	Oil filter change
See explanation in text	5	Tire rotation
	6	Rear axle lubrication check
Every 12 months	7	Air conditioner charge and hose condition check
1975-on — every 12 months or 15,000 miles Others — every 12 months or 12,000 miles	8	Cooling system service
1975-on — every 30,000 miles Others — every 24,000 miles	9	Front wheel bearing lubrication
	10	Manual steering gear seals check
	11	Clutch crossshaft lubrication
1975-on — every 60,000 miles Others — every 24,000 miles	12	Automatic transmission fluid change
		Section B — Safety Maintenance
1975-on — every 12 months or 7,500 miles Others — every 4 months or 6,000 miles	13	Owner safety checks
	14	Tires and wheels condition check
	15	Exhaust system condition check
	16	Drive belts condition and adjustment check
	17	Suspension and steering systems check
	18	Brakes and power steering hoses check
	19	Disc brakes condition check
1975-on — every 12 months or 15,000 miles Others — every 12 months or 12,000 miles	20	Drum and parking brakes condition check
	21	Throttle linkage operation and condition check
	22	Headlights check
	23	Underbody condition check
	24	Bumpers condition check

(continued)

Table 2 MAINTENANCE SCHEDULE (GASOLINE ENGINE) (continued)

Frequency	Item No.	Service
		Section C — Emission Control Maintenance
1975-1976 and 1977-on Schedule I — at 6 months or 7,500 miles, then at 18 month/22,500 mile intervals 1977-on Schedule II — at 6 months or 7,500 miles, then at 24 month/30,000 mile intervals Others — at 4 months or 6,000 miles, then at 12 month/12,000 mile intervals	25	Thermostatically controlled air cleaner operations check
	26	Carburctor choke operations check
	27	Timing, dwell, idle speed, distributor and coil checks and adjustments
	28	Carburetor mounting bolts torque check
	29	EFE system operations check
1975-1976 and 1977-on Schedule I — every 22,500 miles 1977-on Schedule II — every 30,000 miles Others — every 12,000 miles	30	Spark plug replacement
1975-on — every 12 months or 15,000 miles Others — every 12 months or 12,000 miles NOTE: 1977-on Schedule I, perform item 34 at 7,500 miles, at 22,500 miles, and every 22,500 miles thereafter. Perform item 36 every 22,500 miles.	31	EGR system service (1973-1974)
	32	Carburetor fuel inlet filter service
	33	Thermal vacuum switch and hose check (if so equipped)
	34	Vacuum advance solenoid and hoses check
	35	Transmission control switch (1974 and earlier)
	36	PCV system check
	37	Spark plug and ignition coil wires check
1975-on — every 30,000 miles Others — every 24,000 miles	38	Engine compression check
	39	ECS system check
	40	Fuel cap, tank, and hoses check
	41	AIR system check
	42	Air cleaner filter element replacement
1975-1976 and 1977-on Schedule I — every 22,500 miles 1977-on Schedule II — every 30,000 miles Others — first 24,000 miles, then every 12,000 miles	43	Idle stop or speed up solenoid check
	44	Carburetor vacuum break adjustment
	45	Oxygen change sensor check (if so equipped)
	46	DVDS valve check (if so equipped)

Table 3 MAINTENANCE SCHEDULE (DIESEL ENGINE)

Frequency	Item No.	Service
Section A — Lubrication and General Maintenance		
Every 3,000 miles	3 4	Engine oil change Oil filter change
Every 12 months or 9,000 miles	1 2	Chassis lubrication Fluid levels check
See explanation in text	5 6	Tire rotation Rear axle lubrication check
Every 12 months or 15,000 miles	7 8 47	Air conditioner charge and hose condition check Cooling system service Crankcase ventilation system service
Every 30,000 miles Every 100,000 miles	9 12	Front wheel bearing lubrication Automatic transmission fluid change
Section B — Safety Maintenance		
Every 12 months or 9,000 miles	13 14 15 17 18 19	Owner safety checks Tires and wheels condition check Exhaust system condition check Suspension and steering systems check Brake and power steering hoses check Disc brake condition check
Every 12 months or 15,000 miles	16 20 21 24	Drive belts condition and adjustment check Drum and parking brakes condition check Throttle linkage operation and condition check Bumper check
Section C — Emission Control Maintenance		
First 3,000 miles, then every 30,000 miles	48	Engine idle speed adjustment
Every 30,000 miles	42 49	Air cleaner filter element replacement Fuel filter replacement

Table 4 OIL FILTER USAGE

Engine	Filter Brand: AC	Lee	Fram	Purolator
231 cid V6	PF-40/PF-47	LF-111	PH-3387/PH-3387A	PER-111
250 cid L6	PF-25	LF-25	PH-30	PER-49
1970-77 V8	PF-24	LF-24	PH-25	PER-33
1978-80 V8 except 260 cid	PF-25	LF-25	PH-30	PER-49
260 cid	PF-45	LF-45	PH-3429	PER-123
All diesel	PF-45	LF-45	PH-3429	PER-123

Table 5 AUTOMATIC TRANSMISSION FLUID CAPACITY

Transmission	Capacity
Turbo Hydramatic 400	7½ U.S. pints
Turbo Hydramatic 375	7½ U.S. pints
Turbo Hydramatic 350	5 U.S. pints
Turbo Hydramatic 250	5 U.S. pints
Turbo Hydramatic 200	6 U.S. pints

Table 6 SPARK PLUG SPECIFICATIONS

Engine	Spark Plug Type (AC)	Gap Setting (in.)
1970		
250 cid 6	R46T	0.035
350 cid 8, 250 hp	R46S	0.030
350 cid 8, 310 hp	R45S	0.030
350 cid 8, 325 hp	R43S	0.030
455 cid 8, 310 hp	R46S	0.030
455 cid 8, 320 hp	R45S	0.030
455 cid 8, 365 hp	R44S	0.030
455 cid 8, 370 hp	R44S	0.030
455 cid 8, 375 hp	R44S	0.030
455 cid 8, 390 hp	R45S	0.030
455 cid 8, 390 hp air induction	R44S	0.030
455 cid 8, 400 hp	R44S	0.030
1971		
250 cid 6	R46TS	0.035
350 cid 8, 155 hp	R46S	0.040
350 cid 8, 180 hp/M	R45S	0.040
350 cid 8, 180 hp/A	R46S	0.040
350 cid 8, 260 hp	R45S	0.040
455 cid 8, 185/225 hp	R46S	0.040
455 cid 8, 260 hp	R45S	0.040
455 cid 8, 265 hp	R46S	0.040
1972		
350 cid 8, 2 bbl carb.	R46S	0.040
350 cid 8, 4 bbl/M	R45S	0.040
350 cid 8, 4 bbl/A	R46S	0.040
455 cid 8, 250/270 hp	R46S	0.040
455 cid 8, 300 hp	R45S	0.040
455 cid 8, others/M	R45S	0.040
455 cid 8, others/A	R46S	0.040
1973		
250 cid 6	R46TS	0.035
350 cid 8, 2 bbl carb.	R46S	0.040
350 cid 8, 4 bbl/M	R45S	0.040
350 cid 8, 4 bbl/A	R46S	0.040
455 cid 8	R45S	0.040

(continued)

Table 6 **SPARK PLUG SPECIFICATIONS** (continued)

Engine	Spark Plug Type (AC)	Gap Setting (in.)
1974		
250 cid 6	R46TS	0.035
350 cid 8	R46S	0.040
455 cid 8, breaker ign.	R46S	0.040
455 cid 8, HEI ign.	R46SX	0.080
1975		
250 cid 6	R46TX	0.060
260 cid 8	R46SX	0.080
350 cid 8	R46SX	0.080
400 cid 8, 2 bbl carb.	R46TSX	0.060
400 cid 8, 4 bbl	R45TSX	0.060
455 cid 8	R46SX	0.080
1976		
250 cid 6	R46TS	0.035
All others	Early production, use R46SX gapped to 0.080 in. Later production, R46SZ gapped to 0.060. See VECI decal in engine compartment for your car.	
1977		
231 cid V6	R46TS	0.060
All others	R46SZ	0.060
1978		
231 cid V6	R46TSX	0.060
All others	R45TS	0.060
1979		
231 cid V6	R46TSX	0.060
260 cid 8	R46SZ	0.060
305 cid 8	R45TS	0.045
350 cid 8	R46SZ	0.060
1980		
231 cid V6	R45TS	0.040
260 cid 8	R46SX	0.080
305 cid 8	R45TS	0.045
350 cid 8	R46SX	0.080

Table 7 TUNE-UP SPECIFICATIONS

Engine Description and Usage[2]	Timing (°BTDC @ rpm)		Slow Idle (rpm)[1]		Fast Idle (rpm)	
	Manual Transmission	Automatic Transmission	Manual Transmission (in Neutral)	Automatic Transmission (in Drive)	Manual Transmission (in Neutral)	Automatic Transmission (in Park)[3]
1970						
250 cid 6 (all)	TDC @ 750	4 @ 600	830/750	630-600	900	750
350 cid 8, 250 hp (all), 2 bbl	10 @ 1,100	10 @ 1,100	750	575	1,000	1,000
350 cid 8, 310 hp (all), 4 bbl	10 @ 1,100	10 @ 1,100	650	575	1,050	1,050
350 cid 8, 325 hp (all), 4 bbl	14 @ 1,100	14 @ 1,100	750	625	1,050	1,050
455 cid 8, 320 hp (all), 2 bbl	—	8 @ 1,100	—	575	—	1,000
455 cid 8, 365 hp (Cut), 4 bbl	—	12.5 @ 1,100	—	600	—	1,050
455 cid 8, 365 hp (VC), 4 bbl	—	8 @ 1,100	—	575	—	1,050
455 cid 8, 365 hp (442), 4 bbl	12 @ 1,100	12 @ 1,100	700	650	1,050	1,050
455 cid 8, 370 hp (442), 4 bbl	8 @ 1,100	8 @ 1,100	700	650	1,050	1,050
1971						
250 cid 6 (all)	4 @ 550	4 @ 500	550	500	900	750
350 cid 8, 240 hp (all), 2 bbl	10 @ 1,100	10 @ 1,100	750	600	1,000	1,000
350 cid 8, 260 hp (all), 4 bbl	10 @ 1,100	12 @ 1,100	750	600	1,050	1,050
455 cid 8, 280 hp (all), 2 bbl	—	8 @ 1,100	—	600	—	1,000
455 cid 8, 320 hp (all), 4 bbl	—	8 @ 1,100	—	600	—	1,050
455 cid 8, 340 hp (442), 4 bbl	10 @ 1,100	10 @ 1,100	750	600	1,050	1,050
455 cid 8, 350 hp (442), 4 bbl	12 @ 1,100	12 @ 1,100	750	600	1,050	1,050
1972						
350 cid 8, 160 hp (all), 2 bbl	8 @ 1,100	—	750	—	1,000	—
350 cid 8, 160 hp (Fed), 2 bbl	—	8 @ 1,100	—	650/550	—	1,000
350 cid 8, 160 hp (Cal), 2 bbl	—	6 @ 1,100	—	650/550	—	1,000
350 cid 8, 180 hp (all), 4 bbl	8 @ 1,100	12 @ 1,100	750	600	1,050	1,050
455 cid 8, 250/270 hp (all), 4 bbl	10 @ 1,100	8 @ 1,100	750	600	1,050	1,050
455 cid 8, 300 hp (all), 4 bbl	12 @ 1,100	10 @ 1,100	750	650	1,050	1,050
1973						
350 cid 8, 160 hp (all), 2 bbl	—	14 @ 1,100	—	650/550	1,000	1,000
350 cid 8, 180 hp (all), 4 bbl	12 @ 1,100	12 @ 1,100	1,000/600	650/550	1,050	1,050
455 cid 8, 225 hp (all), 4 bbl	10 @ 1,100	8 @ 1,100	1,000/750	650/550	1,050	1,050
1974						
350 cid 8 (all), 4 bbl	—	12 @ 1,100	—	650/550	—	1,000
455 cid 8 (all), 4 bbl	—	8 @ 1,100	—	650/550	—	1,000

(continued)

Table 7 TUNE-UP SPECIFICATIONS (continued)

Engine Description and Usage[2]	Timing (°BTDC @ rpm)		Slow Idle (rpm)[1]		Fast Idle (rpm)	
	Manual Transmission	Automatic Transmission	Manual Transmission (in Neutral)	Automatic Transmission (in Drive)	Manual Transmission (in Neutral)	Automatic Transmission (in Park)[3]
1975						
250 cid 6 (Fed)	10 @ 850	10 @ 550	850/425	550/425	Preset	Preset
250 cid 6 (Cal)	—	10 @ 600	—	600/425	Preset	Preset
260 cid 8 (Fed)	16 @ 1,100	18 @ 1,100	750	650/550	900	900
260 cid 8 (Cal)	—	16 @ 1,100	—	650/600	—	900
350 cid 8 (Fed)	—	20 @ 1,100	—	650/550	—	900
350 cid 8 (Cal)	—	20 @ 1,100	—	650/600	—	1,000
455 cid 8 (Fed)	—	16 @ 1,100	—	650/550	—	900
455 cid 8 (Cal)	—	16 @ 1,100	—	650/600	—	800
1976						
250 cid 6 (Fed)	6 @ 850	10 @ 550	850/425	550/425	Preset	Preset
250 cid 6 (Cal)	—	10 @ 600	—	600/425	Preset	Preset
260 cid 8 (Fed)	16 @ 1,100	18 @ 1,100	750	650/550	900	900
260 cid 8 (Cal)	14 @ 1,100	14 @ 1,100	750	650/550	900	900
350 cid 8 (Fed)	—	20 @ 1,100	—	650/550	—	900
350 cid 8 (Cal)	—	20 @ 1,100	—	650/600	—	1,000
455 cid 8 (Fed)	—	16 @ 1,100	—	650/550	—	900
455 cid 8 (Cal)	—	16 @ 1,100	—	650/600	—	800
1977						
231 cid 6 (Fed)	12 @ 600	12 @ 600	800	670/600	Preset	Preset
231 cid 6 (Cal)	12 @ 600	12 @ 600	800/600	670/600	Preset	Preset
231 cid 6 (High Altitude)	—	12 @ 600	—	670/600	—	Preset
260 cid 8 (Fed)	16 @ 1,100	18 @ 1,100	750	650/550	900	900
350 cid 8 (Fed)	—	20 @ 1,100	—	650/550	—	900
350 cid 8 (Cal)	—	20 @ 1,100	—	650/550	—	1,000
350 cid 8 (High Altitude)	—	20 @ 1,100	—	700/600	—	1,000
403 cid 8 (Fed except VC)	—	20 @ 1,100	—	650/550	—	900
403 cid 8 (Fed VC)	—	22 @ 1,100	—	650/550	—	900
403 cid 8 (Cal)	—	20 @ 1,100	—	650/550	—	1,000
1978						
231 cid 6 (Fed)	15 @ 800	15 @ 600	800	600	1,850	1,850
231 cid 6 (Cal and High Alt.)	—	15 @ 600	—	600	—	1,850
260 cid 8 (Fed)	18 @ 1,100	20 @ 1,100	800/650	650/500	750	800
260 cid 8 (Cal)	—	18 @ 1,100	—	650/500	—	800
260 cid 8 (High Altitude)	—	20 @ 1,100	—	650/550	—	900
305 cid 8 (Fed)	4 @ 600	4 @ 500	700/600	600/500	1,600	1,600
305 cid 8 (Cal)	—	8 @ 500	—	650/500	—	1,600
305 cid 8 (High Altitude)	—	8 @ 600	—	700/600	—	1,600
350 cid 8 (High Altitude)	—	8 @ 600	—	650/600	—	1,600

(continued)

Table 7 TUNE-UP SPECIFICATIONS (continued)

Engine Description and Usage[2]	Timing (°BTDC @ rpm) Manual Transmission	Timing (°BTDC @ rpm) Automatic Transmission	Slow Idle (rpm)[1] Manual Transmission (in Neutral)	Slow Idle (rpm)[1] Automatic Transmission (in Drive)	Fast Idle (rpm) Manual Transmission (in Neutral)	Fast Idle (rpm) Automatic Transmission (in Park)[3]
1979						
231 cid 6 (Fed)	15 @ 800	15 @ 600	800/600	670/550	2,200	2,200
231 cid 6 (Cal and High Alt.)	—	15 @ 600	—	600	—	2,200
260 cid 8 (Fed)	18 @ 1,100	20 @ 1,100	800/650	625/500	750	800
260 cid 8 (Cal)	—	18 @ 1,100	—	625/500	—	800
260 cid 8 (High Altitude)	—	20 @ 1,100	—	650/550	—	900
260 cid 8 (Diesel)	—	—	—	650/590	—	—
305 cid 8 (Fed)	4 @ 700	4 @ 500	700	600/500	1,300	1,600
305 cid 8 (Cal)	—	4 @ 500	—	600/500	—	1,600
305 cid 8 (High Altitude)	—	8 @ 600	—	650/600	—	1,750
350 cid 8 (Fed)	—	18 @ 1,100	—	650/550	—	900
350 cid 8 (Cal)	—	20 @ 1,100	—	600/500	—	1,000
350 cid 8 (High Altitude)	—	8 @ 600	—	650/600	—	1,750
350 cid 8 (Diesel)	—	—	—	650/570	—	—
1980						
231 cid 6 (Fed)	15 @ 800	15 @ 550	800/600	600/500	2,200	2,000
231 cid 6 (Cal)	—	15 @ 550	—	620/550	—	2,200
260 cid 8 (Fed)	—	See VECI sticker	—	625/500	—	700
305 cid 8 (Fed)	—	4 @ 500	—	600/500	—	1,850
305 cid 8 (Cal)	—	4 @ 550	—	650/550	—	2,200
350 cid 8 (Fed)	—	18 @ 1,100	—	600/500	—	700DR
350 cid 8 (Diesel)	—	—	—	750/600	—	—

1. If two speeds are listed, higher speed is with the solenoid energized and lower speed is with the solenoid de-energized.
2. All - all models; Cut - Cutlass; VC - Vista Cruiser; 442 - 4-4-2 series; Fed - cars first sold in any state except California; Cal - Cars first sold in California; 2 bbl - 2 barrel carburetor; 4 bbl - 4 barrel carburetor.
3. In PARK unless stated otherwise (DR - in DRIVE).

NOTE: If you own a 1981 and later model, first check the Supplement at the back of the book for any new service information.

CHAPTER FOUR

ENGINE

The Cutlass has been offered with a wide range of power options. Fortunately, all of these engines can be grouped into a few general classes. The inline 6-cylinder engine, produced by Chevrolet and offered as standard equipment on some Cutlass models through 1975, will be referred to in this chapter as the L6 engine. The Buick-built 231 cid V6 engine, used on many 1977 and later Cutlasses, will be referred to as the V6 engine. The Oldsmobile V8 engines used on all years, including the 260, 350, 403, and 455 cid engines, will be called Olds 260, etc. The Chevrolet 305 and 350 cid engines, used on many 1978 and later models, will be called Chevy 305 and Chevy 350. Some 1979 and later Cutlasses were supplied with 260 and 350 cid diesel engines, and these will be called 260 diesel and 350 diesel.

All of the gasoline engines are remarkably similar, and the diesel engines were developed from the Olds 260 and 350 engines. There is little difference in the service procedures for the different classes of gasoline engines. Service by amateur mechanics to diesel engines, except for the maintenance procedures given in Chapter Three, is not recommended at this time.

Procedures are provided for removal and installation of the engine, removal and replacement of sub-assemblies, and inspection, adjustments, and repairs of some sub-assemblies. The procedures are general in nature and apply to all engines, except where special instructions are necessary for a specific model. Even though the illustrations usually show workbench operations, most single procedures, when not a part of a major overhaul, can be performed successfully with the engine in the car. Engine specifications are given in **Table 1** and tightening torques in **Table 2**, both at the end of the chapter.

ENGINE REMOVAL

If at all possible, have the engine steamed cleaned before removing it. This will save time later on in cleaning individual parts, and it will make your work much more enjoyable. If steam cleaning equipment is not available, clean the exterior with a commercial engine cleaning solvent, following the manufacturer's instructions.

Consider renting an engine work stand if you are planning a major overhaul. These are available at many tool and equipment rental stores, and you will find them a great convenience. The engine can be left in an upright position for removal and installation of the intake and exhaust manifold, cylinder head(s), piston and rod assemblies, etc. It can then be turned over for removal of the oil pan, bearing caps, crankshaft, etc.

Do your work under the cleanest possible conditions. Dirt and dust introduced to bearing surfaces can lead to trouble later on.

Keep track of the parts you remove, especially nuts, bolts, and washers. Keep them in special containers (clean coffee cans, etc) that are plainly marked with the part of the engine from which they were removed. Also, if one or more attaching bolts for a certain part are longer than the others, make a note of where they were removed from. This may sound like a lot of trouble, but remember that your work could be interrupted and it might be some time before you reassemble the engine. Searching for lost parts, and/or installing the wrong parts, can be both time consuming and frustrating.

People tend to make mistakes when they are tired. In view of this, it is a good idea to be fresh when you start working on a engine, and to quit working when you become tired. The job may take a little longer, but you will do better work. Before starting any procedure, read it thoroughly and make sure you understand what is required. If you do not fully understand, ask an expert. If you still do not understand, do not start. Mechanics tend to charge heavily for working on basket cases.

Before disconnecting any hoses or electrical connector, it is a good idea to identify both sides of the connection so that hoses and connectors can be returned to their original positions during installation. A good way to do this is to wrap a piece of masking tape around each side of the connection to be broken, and then to mark each of the tapes with matching numbers. Small tags with strings attached also can be used. These tags are available at most stationery stores.

1. Make scribe marks around the hood hinges and remove the hood.

2. Disconnect both battery cables, and then drain the cooling system. If the coolant is to be reused, drain it into a suitable clean container.

3. If the car has air conditioning, disconnect the electrical lead to the compressor, then remove the compressor-to-bracket bolts and place the compressor to one side, out of the way, with the hoses attached.

4. Remove the fan belts, blade, and pulley. Remove the air cleaner.

5. Remove the heater and radiator hoses from the engine. Although radiator removal is not necessary on some models, additional clearance is gained and the possibility of damage to the radiator is lessened if the fan shroud, upper radiator support, and the radiator are removed. If the radiator is left in the car, either remove the hoses or tie them back out of the way and remove the shroud.

6. Remove the power steering pump from its bracket, leaving the hoses attached, and tie it out of the way.

7. Remove the fuel pump hoses and plug them to prevent loss of fuel.

8. Disconnect the engine ground strap and battery ground cable from the engine.

9. Identify and disconnect all vacuum hoses and electrical wires leading from the engine to parts not attached to the engine.

10. Disconnect the throttle cable from the carburetor. If the vehicle has automatic transmission, disconnect the downshift or throttle valve cable.

11. Disconnect oil and coolant sending unit connections at the engine.

12. Raise the vehicle and disconnect the starter cables and cable shield, if so equipped, from the engine.

13. Disconnect the exhaust pipe(s) from the exhaust manifold(s). Be sure to properly support the exhaust system.

14. Remove the lower flywheel or torque converter cover on car with automatic transmission, and remove the bolts attaching the torque converter to the flywheel or drive plate, first making matching chalk marks on the converter and flywheel for alignment during installation.

15. Remove the transmission-to-engine attaching bolts.

16. On V8 engines, remove the starter.

17. Remove the motor mount bolts. See **Figure 1** (typical).

18. Lower the vehicle and support the transmission.

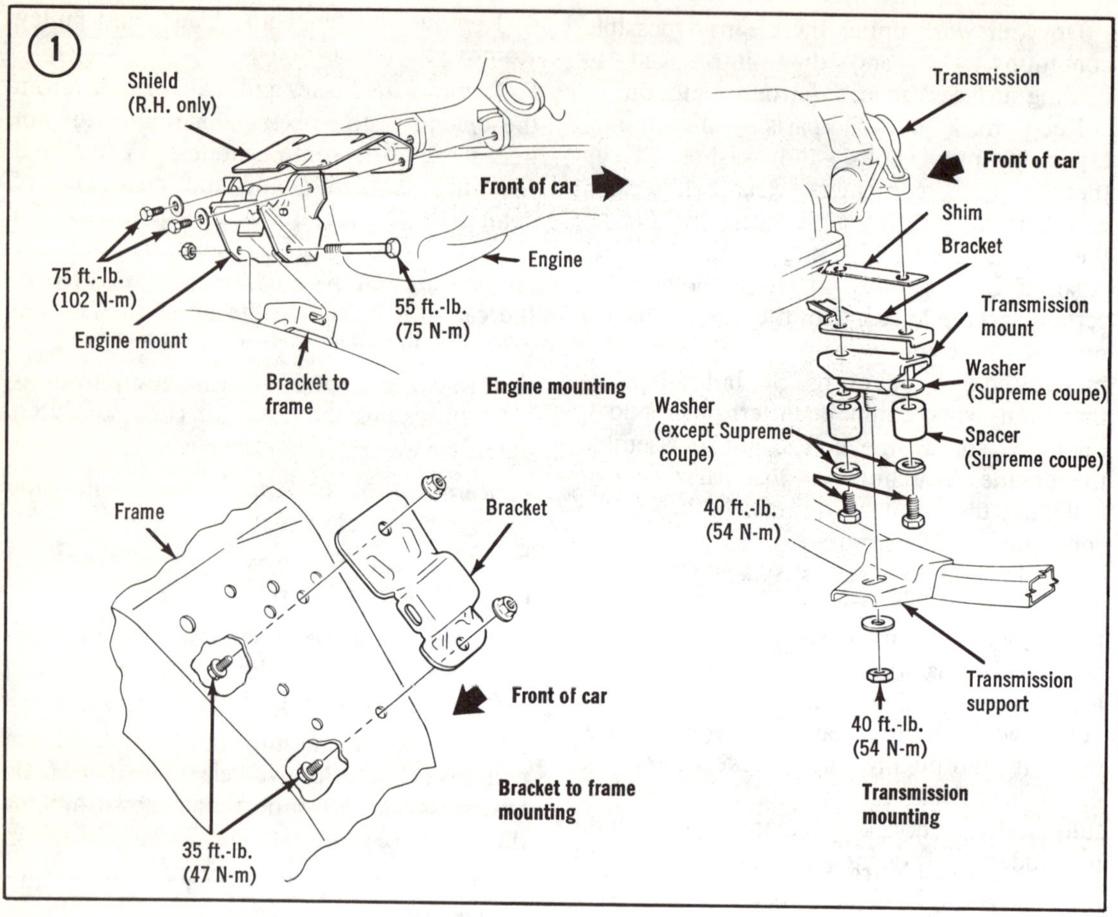
1
Shield
(R.H. only)
75 ft.-lb.
(102 N-m)
Engine mount
Front of car
Engine
55 ft.-lb.
(75 N-m)
Bracket to
frame
Engine mounting
Transmission
Front of car
Shim
Bracket
Transmission
mount
Washer
(Supreme coupe)
Washer
(except Supreme
coupe)
Spacer
(Supreme coupe)
40 ft.-lb.
(54 N-m)
Transmission
support
40 ft.-lb.
(54 N-m)
Transmission
mounting
Frame
Bracket
Front of car
Bracket to frame
mounting
35 ft.-lb.
(47 N-m)

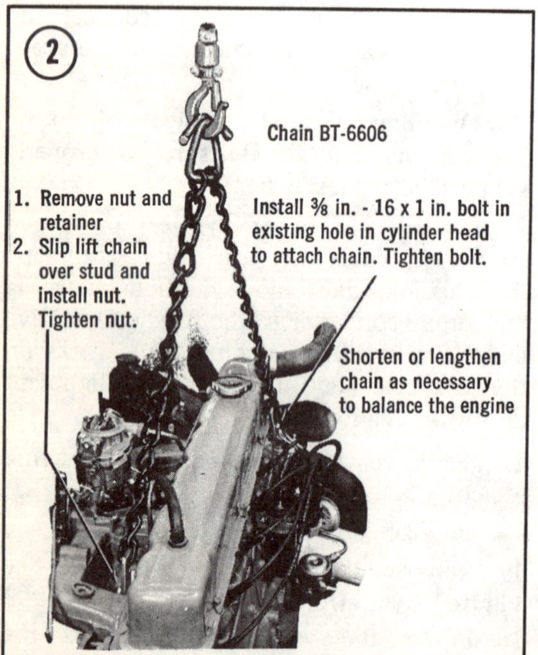
2
Chain BT-6606
1. Remove nut and retainer
2. Slip lift chain over stud and install nut. Tighten nut.
Install 3/8 in. - 16 x 1 in. bolt in existing hole in cylinder head to attach chain. Tighten bolt.
Shorten or lengthen chain as necessary to balance the engine

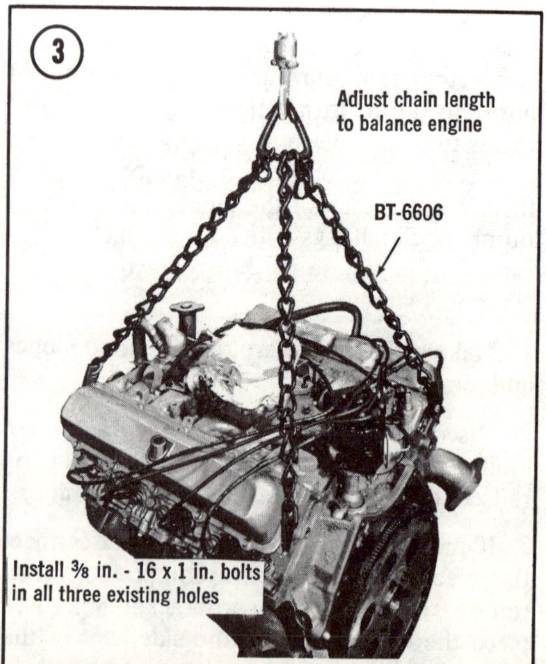
3
Adjust chain length to balance engine
BT-6606
Install 3/8 in. - 16 x 1 in. bolts in all three existing holes

4

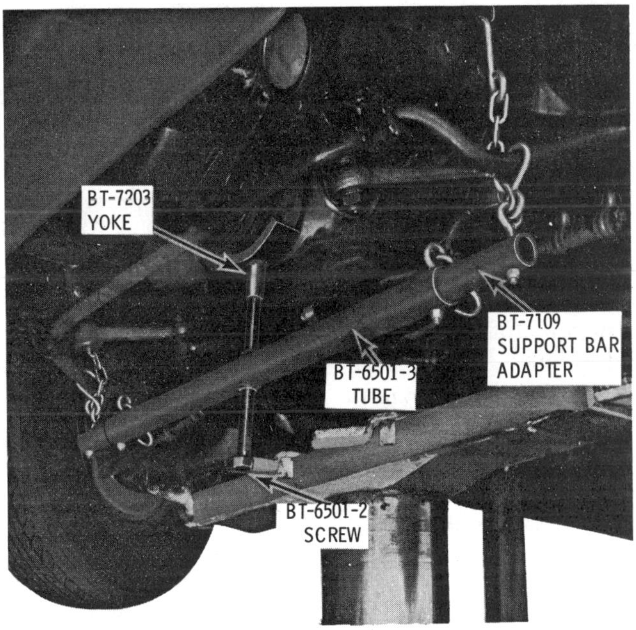

19. Install a lifting fixture on the engine. The specified tool for the L6 and V8 engines is BT-6606. **Figure 2** shows the fixture installed on the L6 and **Figure 3** on the V8. A cradle type lifting fixture (BT-6501) is recommended for the V6 engine. See **Figure 4**.

20. Make a final check to make sure all electrical wires and vacuum and other hoses (and all other parts) are free and clear of the engine.

21. Attach the lifting fixture to a lifting device. Raise the engine and transmisssion slightly to clear the engine mounts, then pull the engine forward to clear the transmission. Lift the engine out of the vehicle and place it on a work stand.

ENGINE INSTALLATION

1. Position the engine over the engine compartment and carefully lower it into the vehicle. If the car has an automatic transmission, make sure the scribe marks on the flywheel and the torque converter are aligned and install the attaching bolts and tighten them to 35 ft.-lb. If the car has a manual transmission, make sure the transmission input shaft splines mesh properly with those of the clutch. Install the transmission housing-to-engine attaching bolts and tighten them to the specifications given in **Table 2**.

2. Raise the car, install the engine mount bolts, and tighten them to specifications. See **Table 2**.

3. Install the lower cover and tighten the attaching bolts to about 5 ft.-lb.

4. Install the exhaust pipes to the exhaust manifold(s) and tighten the attaching bolts to the specified torque. See **Table 2**.

5. Lower the car and reconnect all electrical connectors, vacuum hoses, and fuel hoses to their original positions. If these parts were marked as described, in *Engine Removal*, this chapter, this is simply a matter of finding connectors, etc., marked with identical numbers and reconnecting them. Make sure the proper wires are reconnected to the oil pressure and coolant temperature senders, and that the engine-to-body ground strap is properly installed.

6. Install the power steering pump on its bracket and securely tighten the attaching bolts.

7. If the radiator was removed, install the radiator and upper support and securely tighten the attaching bolts. Position the fan shroud, if so equipped, and install the fan blade and pulley. Install the shroud and the fan belts.
8. Install the air conditioner compressor, if so equipped.
9. Make sure all accessory drive belts are installed and properly tension the belts. See *Drive Belt Replacement*, Chapter Three.
10. Install all coolant and radiator hoses to the engine and radiator and fill the cooling system with the proper solution of ethylene glycol-type coolant and water, following the coolant manufacturer's recommendations.
11. Install the air cleaner and verify that the engine is filled with the proper amount of the recommended engine oil. See *Engine Oil and Filters*, Chapter Three. Make sure a new filter is installed before adding the oil.
12. Install the hood, making sure the scribe marks made during removal are properly aligned.
13. Make a last check to make sure that all wires and hoses are properly connected, then reconnect the battery cables.
14. Start the engine and check for coolant and oil leaks. Tune the engine as described in *Gasoline Engine Tune-up*, Chapter Three.

GENERAL OVERHAUL SEQUENCE

NOTE
For partial disassembly of the engine, see the applicable procedure later in this chapter. For complete disassembly, follow the sequence outlined below. This sequence presumes the engine has been removed from the vehicle and mounted on an engine stand or suitable workbench.

1. Remove the following accessories or components from the engine (if present):
   a. Air injection reactor system (and brackets)
   b. Alternator (and brackets)
   c. Accessory drive pulleys
   d. Fuel pump (and pushrod, V8 engines)
   e. Later pump and hoses
   f. Spark plug wires and distributor cap
   g. Carburetor and fuel lines
   h. Oil filter
   i. Starter
   j. Clutch pressure plate and disc
   k. Ground strap
   l. Oil dipstick and tube

2. Remove intake and exhaust manifolds.
3. Remove pushrod covers (inline engines).
4. Remove the rocker arm assemblies (and shafts on V6 or pivots on Olds V8 engines); remove pushrods and valve lifters. Store removed items in racks so they can be reinstalled in the same locations from which they were removed.
5. Remove cylinder head(s).
6. Remove torsional damper.
7. Remove oil pan.
8. Remove front cover from crankcase.
9. Remove oil baffle, if so equipped.
10. Remove oil pump and screen. Remove extension shaft (V8 engines).
11. If connecting rods and caps are not marked with cylinder number identification, mark them. Also check cylinder bores for ridges. If necessary, remove ridges.
12. Remove connecting rod caps, then remove connecting rod/piston assemblies from engine.

NOTE
Turn crankshaft as required to remove rods and pistons.

CAUTION
Use care when removing camshaft to avoid damaging bearings.

13A. Remove camshaft from V6 and all V8 models as follows:
   a. Remove attaching bolts from camshaft sprocket, tap lower edge of sprocket lightly with plastic hammer to dislodge, and remove sprocket and timing chain. On V6 engines, remove camshaft, sprockets, and timing chain after aligning timing marks on sprockets.
   b. Install 2 bolts in camshaft sprocket bolt holes (except on V6) and carefully pull out the camshaft.

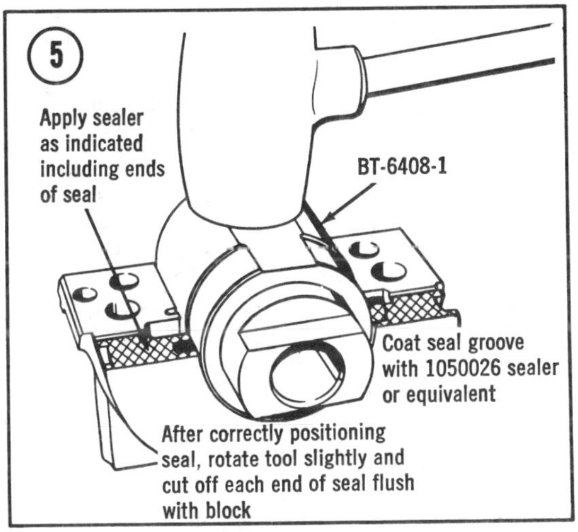

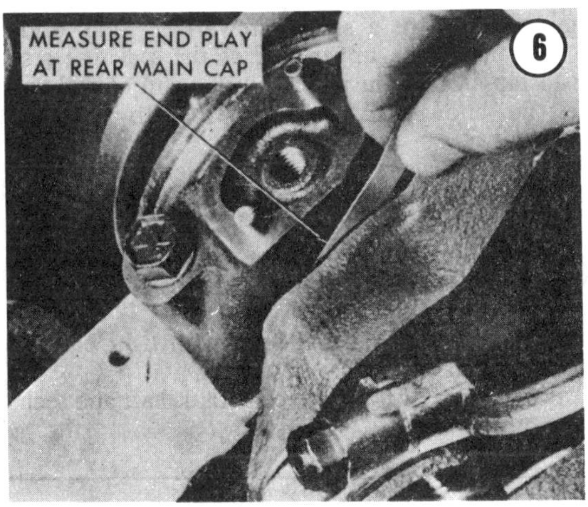

13B. Remove camshaft from L6 engines as follows:

a. Remove camshaft thrust plate screws, using holes in camshaft gear for access.
b. Remove camshaft and gear as an assembly.

14. Remove flywheel.
15. Remove caps from main bearings and remove crankshaft from cylinder block.
16. Remove rear main bearing oil seal from cylinder block and bearing cap.
17. Discard all gaskets and seals removed during the above steps.

NOTE

See procedures later in this chapter for further disassembly, cleaning, and inspection of the parts or subassemblies removed during this procedure.

GENERAL REASSEMBLY SEQUENCE

NOTE

Use only new gaskets and seals on engine during assembly.

1. Install crankshaft as follows:
   a. Place rear main bearing oil seal in grooves in cylinder block and bearing cap. Make certain seal lip faces toward front of engine. If seal has 2 lips, one with helix should face front of engine.

   b. Lubricate seal lips with engine oil; keep oil off parting line surface.

   c. Install main bearings in block and bearing caps and lubricate bearing surface with engine oil.

   d. Carefully install crankshaft in cylinder block.

   e. Brush a thin coat of oil sealing compound on block mating surface (see **Figure 5**) and on corresponding surface on bearing cap.

   CAUTION

   Do not allow sealant on crankshaft or seal.

   f. Install main bearing caps, with arrows pointing to front of engine; torque all bolts (except rear cap) to specifications (see **Table 2** at the end of the chapter). Torque rear bearing cap bolts to 10-12 ft.-lb., then tap end of crankshaft, first to the rear and then to the front, with a lead hammer to line up rear main bearing and crankshaft thrust surfaces. Retorque all bolts to specifications.
   g. Measure crankshaft end play with feeler gauge (see **Figure 6**). Force crankshaft forward as far as it will go and measure between front of rear main bearing and crankshaft thrust surface. See **Table 1**, Engine Specifications (at the end of the chapter), for allowable end play.

2. Install a wood block between crankshaft and cylinder block to prevent rotation and install flywheel. Torque to specifications (see **Table 2**).

NOTE
Align dowel holes in flywheel and crankshaft. If vehicle has automatic transmission, make certain converter attaching pads on flywheel are facing toward transmission.

NOTE
If camshaft or lifters are new, lubricate cam lobes and lifter feet with Molykote or equivalent.

3A. Install camshaft on V6 and V8 models as follows:

a. Thread bolts into crankshaft bolt holes (except V6) lubricate camshaft journals with engine oil, and install camshaft. See **Figure** 7, typical. Take care not to damage bearings. Remove bolts.
b. Install timing chain on camshaft sprocket. Align marks on camshaft and crankshaft sprockets (see **Figure 8**) and install chain on crankshaft sprocket. Align dowel on camshaft with hole in sprocket and install sprocket on camshaft.

CAUTION
Do not hammer on camshaft sprocket during installation, as camshaft rear expansion plug could be loosened.

c. Use mounting bolts to draw sprocket onto camshaft. Torque to specifications (see **Table 2**), and lubricate the chain with engine oil.

3B. Install camshaft on inline engines as follows:

a. Lubricate journals on camshaft with engine oil.
b. Position crankshaft and camshaft so that timing marks are aligned (see **Figure 9**) and install camshaft and gear assembly. Take care not to damage the camshaft bearings.
c. Using a dial indicator, check gear runout (see **Figure 10**). Camshaft gear runout should not exceed 0.004 in. and crankshaft gear runout should not exceed 0.003 in.

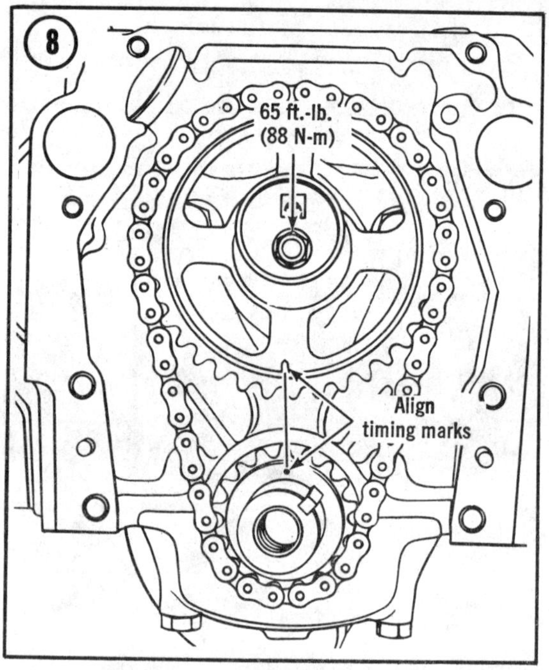

d. Use a dial indicator to check backlash between timing gear teeth (see **Figure 11**). Backlash should be not less than 0.004 in. and not more than 0.006 in.
e. Lubricate timing gears with engine oil.

4. Install connecting rod and piston assemblies as follows:
a. Install bearings in connecting rods and caps. Lubricate bearings, pistons, piston rings, bolts and cylinder walls with engine oil.

10

11

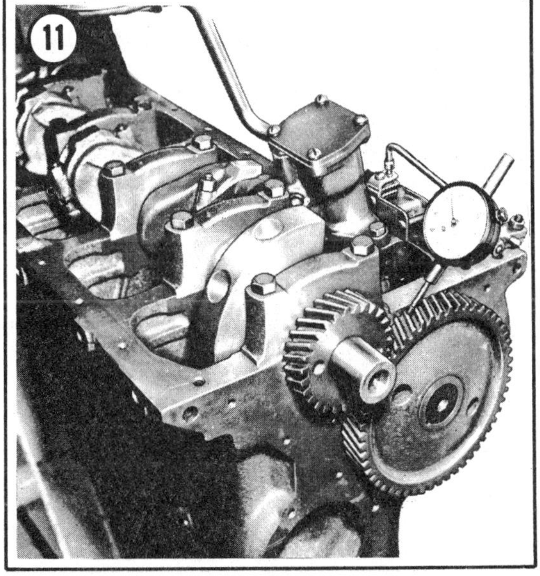

b. Using a ring compressor to compress the piston rings, insert No. 1 piston assembly in No. 1 cylinder, using light taps with a wooden hammer handle. Hold ring compressor firmly against cylinder block until all rings have entered the bore.

CAUTION
*Make certain ring gaps are properly positioned on piston and piston is properly positioned in cylinder. See **Figure 12** (top of **Figure 12** faces the front of the engine).*

NOTE
A connecting rod guide tool set will greatly help in installing piston and rod assemblies. If not available, use care to avoid damage to the bearing surfaces, threads, and crankshaft journals. Masking tape over threads and sharp projections will help. However, make certain all adhesive from the tape is removed before assembling rods to the crankshaft.

c. If installed, remove rod guide set. Install connecting rod cap and torque nuts to specifications (see **Table 2**).
d. Repeat Steps b and c for remaining rod and piston assemblies.

12

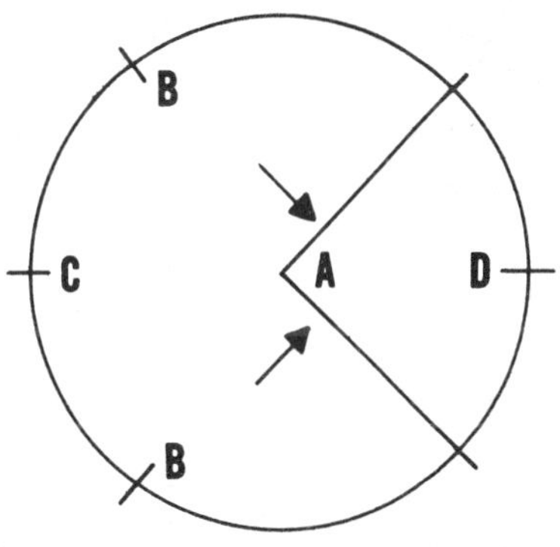

A. Oil ring spacer gap (tang in hole or slot within arc)
B. Oil ring rail gaps
C. 2nd compression ring gap
D. Top compression ring gap

4

e. Measure side clearance of each rod (see **Table 1**). Make measurement between connecting rod cap and side of crankpin (L6 engines) or between connecting rod caps (V6 and V8 engines). Refer to **Figure 13 or 14**.

5. Install oil pump and screen assembly, extension shaft (V6 and V8 engines), and oil baffle, if so equipped.

6A. Install crankcase front cover on L6 engines as follows:

a. Lubricate lip seal with engine oil and install crankcase front cover aligning tool (GM tool No. J-23042 or equivalent) in seal (see **Figure 15**).

13

14

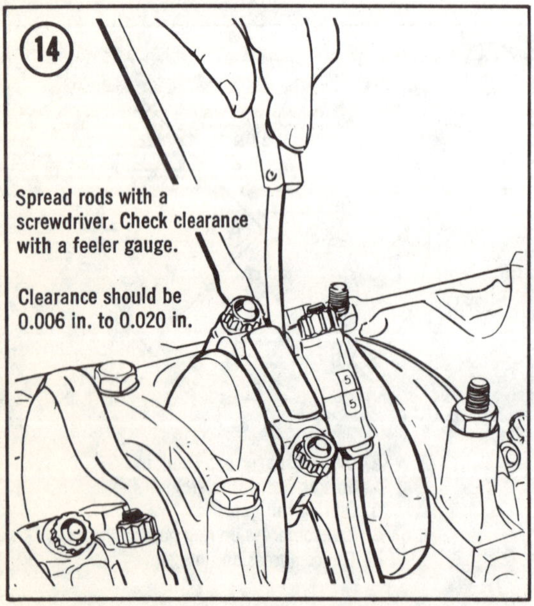

15

CENTERING TOOL MUST BE INSTALLED AS SHOWN BEFORE TIGHTENING TIMING COVER TO CENTER SEAL SURFACE AROUND C/S.

16

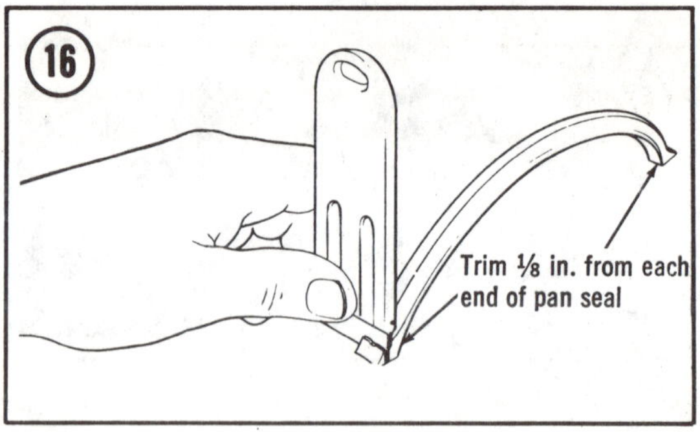

17

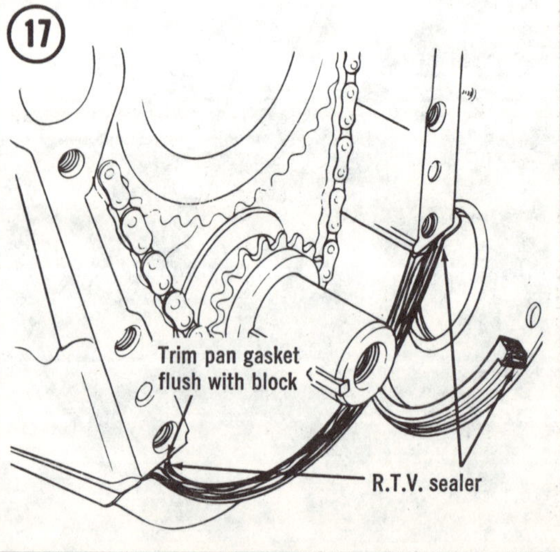

b. Position gasket on cover and install cover on cylinder block. Torque to specifications (see **Table 2**) and remove tool.

6B. Install crankcase front cover on Chevy V8 engines as follows:

a. Place gasket in position on dowels on cylinder block.

b. Lubricate lip seal with engine oil. Place cover in position on dowel pins. Install bolts and torque to specifications (see **Table 2**).

6C. Install front cover on V6 engines as follows:

a. Remove the oil pump cover and completely pack the oil pump cavity with petroleum jelly, leaving no air spaces. Then reinstall the cover, using a new gasket.

b. Clean all gasket surfaces thoroughly and install a new gasket on the engine block, making sure it is properly positioned.

18

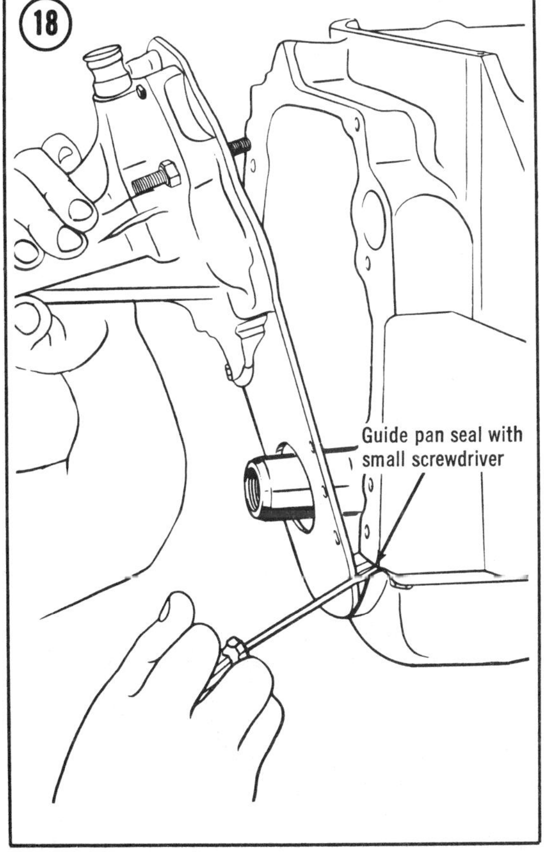

c. Install the front cover over the gasket, making sure that the locating dowels are properly engaged in their holes.

d. Dip the attaching bolt threads in sealer (Loctite, etc.), install them, and tighten them to 30 ft.-lb.

6D. Install front cover on Olds V8 engines as follows:

a. Put excess material from the front of the oil pan gasket on each side of the engine block.

b. Clean all mating surfaces with a suitable solvent.

c. Trim about 1/8 in. of material from each side of the front oil seal as shown in **Figure 16**. Use a razor blade or other sharp tool.

d. Install a new gasket on the engine block, applying a sealer (1050026 or equivalent) around the coolant holes. Install a new seal on the front cover.

e. Apply a room temperature vulcanizing (RTV) sealer to the junction of the engine block, front cover, and oil pan. See **Figure 17**.

f. Install the cover on the block, pressing downward to compress the seal. Turn the cover as required and use a small screwdriver to guide the pan seal into the cavity. See **Figure 18**.

g. Dip the threads of the attaching bolts in engine oil and loosely install (finger-tight) 2 bolts to hold the cover in place.

h. Chamfer the ends of the dowels as shown in **Figure 19** and install the dowels chamfered ends first in their holes.

19

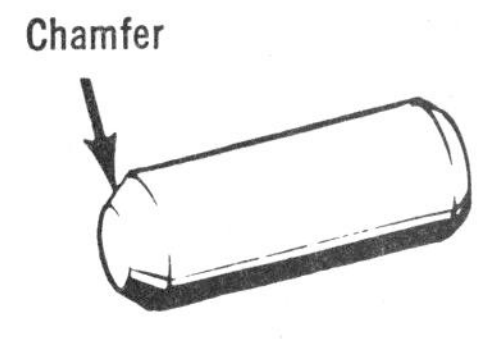

4

20

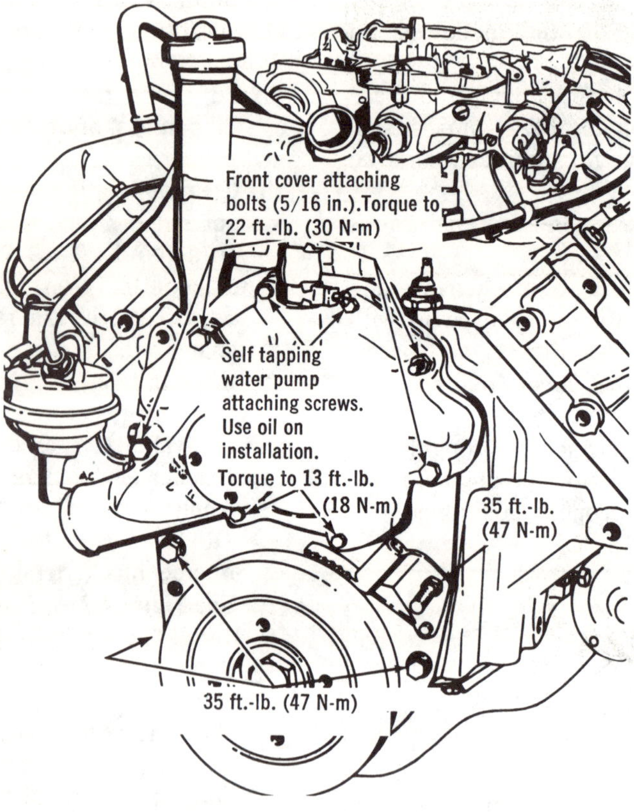

i. Install the timing indicator and the water pump assembly, and then install all remaining attaching bolts. Tighten the bolts evenly to the torque values shown in **Figure 20**.

7. Install oil pan as follows:
   a. Place side gaskets on cylinder block sealing surfaces and apply sealer at the intersection of the end seals and side gaskets.
   b. Place oil pan rear seal in groove in rear main bearing cap. Ends of seal should butt against the side gaskets.
   c. Place oil pan front seal in crankcase front cover. Ends of seal should butt against side gaskets.
   d. Install oil pan and torque to specifications (see **Table 2**).

CAUTION

The installation procedures in Step 8 must be followed, using the proper tool, or movement of the inertial weight section on the hub (assembled to hub with rubber-like material) may destroy the tuning of the torsional damper.

8A. Install drive-on torsional damper as follows:

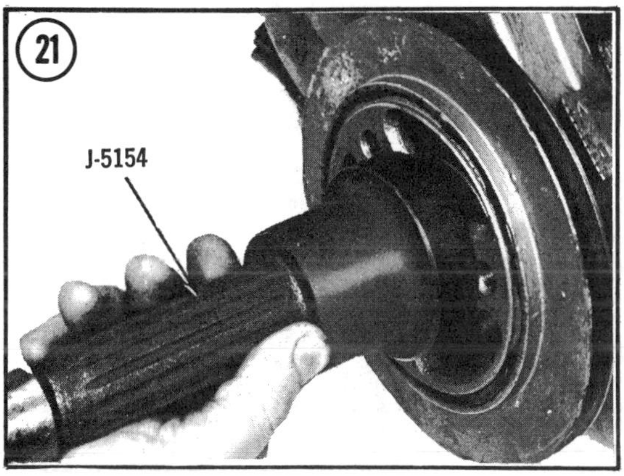

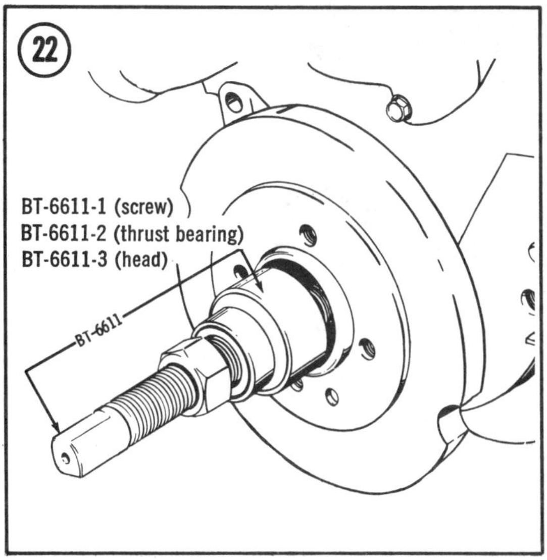

a. Coat seal contact area on damper with engine oil.

b. Position damper on crankshaft and drive into position, using hollow drift bar (GM part No. J-5154), until the damper bottoms against the crankshaft gear or sprocket. See **Figure 21**. Remove tool.

8B. Install pull-on type torsional damper as follows:

a. Coat seal contact area on damper with engine oil.

b. Position damper over key on crankshaft.

c. On V8 engines, use tool BT-6611 on Olds V8 (**Figure 22**) or J-23523 on Chevy V8 (**Figure 23**) to pull damper into position. Tool is not required on V6; install damper and use attaching bolt to pull it into place.

CAUTION

Bolt should be installed in crankshaft with a minimum of 1/2 in. thread engagement on Chevy V8.

d. Remove tool and install retaining bolt. Torque to specifications (see **Table 2**).

9. Install cylinder head(s) as follows:

CAUTION

Make certain gasket surfaces on head(s) and cylinder block are clean and free of nicks and deep scratches. Threads in the block and on cylinder head bolts must be clean, as dirty threads will affect bolt torques.

a. Apply a thin coat of sealer on both sides of steel gaskets.

CAUTION

Too much sealer will affect the sealing quality of the gasket.

CAUTION

Do not use sealer on either side of a composite steel-asbestos gasket.

b. Place gasket over dowel pins on cylinder block with bead side up.

c. Carefully lower cylinder head into place.

d. Apply sealing compound to bolt threads and install them finger-tight.

4

24

Intake manifold to exhaust manifold bolts
30 ft.-lb. (except 10, 11, 12 & 13)
(10, 11, 12 & 13) 20 ft.-lb.

25

26

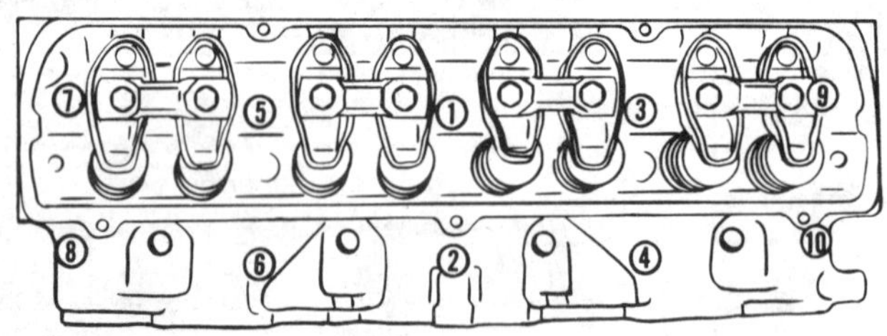

85 ft.-lb. (115 N-m) - 4.3 liter engine
130 ft.-lb. (176 N-m) - 5.0 and 5.7 liter engines

e. Following the sequence shown in **Figure 24** (L6), **Figure 25** (V6), **Figure 26** (Olds V8) or **Figure 27** (Chevy V8), tighten each cylinder head bolt a small amount at a time until all bolts have been tightened to specification (see **Table 2**).

10. Install valve lifters and pushrods, making certain they are replaced in the same locations from which they were removed.

NOTE

If new lifters and/or rocker arms and balls are installed, coat feet of valve lifters and surfaces of rocker arm and balls with Molykote or equivalent.

27

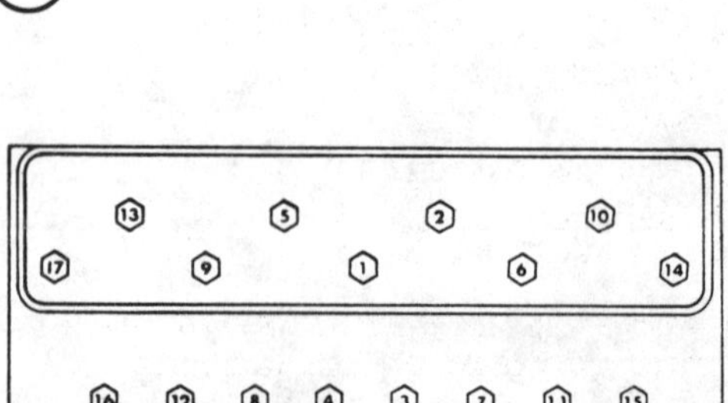

28

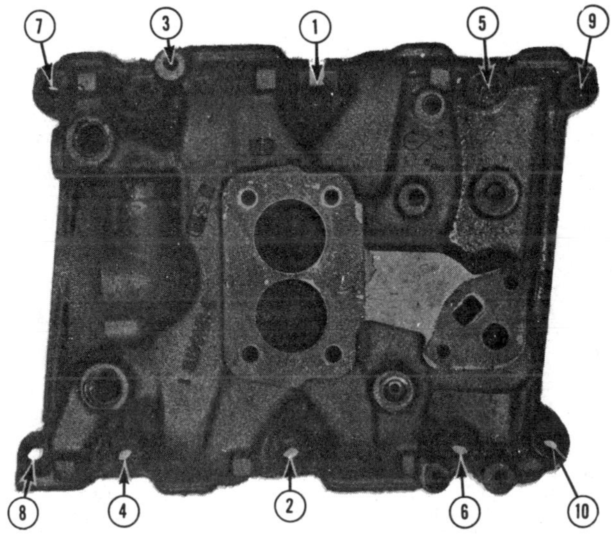

11. Install rocker arms, balls and nuts. Tighten nuts to take up all pushrod end play.
12. Install pushrod covers (inline engines).
13. Install exhaust and intake manifolds. Tighten L6 exhaust-to-intake bolts to the pattern shown in **Figure 24**. Tighten V6 intake manifold bolts in pattern shown in **Figure 28**. On Olds V8 engines use pattern shown in **Figure 29**. Use pattern in **Figure 30** for Chevy V8 intake manifold. Tighten all bolts to specifications given in **Table 2**.
14. Install the following applicable sub-assemblies or components, using the appropriate procedures given later in this chapter.
   a. Engine oil dipstick tube and dipstick
   b. Clutch pressure plate and disc
   c. Ground strap
   d. Starter
   e. New oil filter
   f. Fuel lines and carburetor
   g. Distributor
   h. Distributor cap and spark plug wire
   i. Accessory drive pulleys and belts
   j. Fuel pump (and pushrod on V8 engines)
   k. Water pump and bypass hose
   l. Accessory drive pulleys and belts
   m. Alternator and bracket
   n. Air injector reactor system with brackets

29

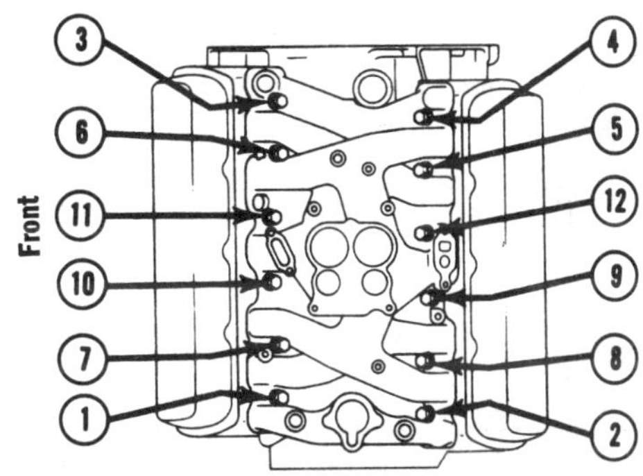

30

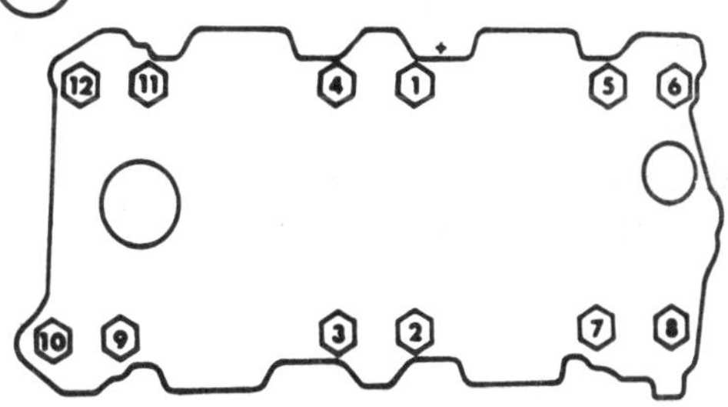

31

32

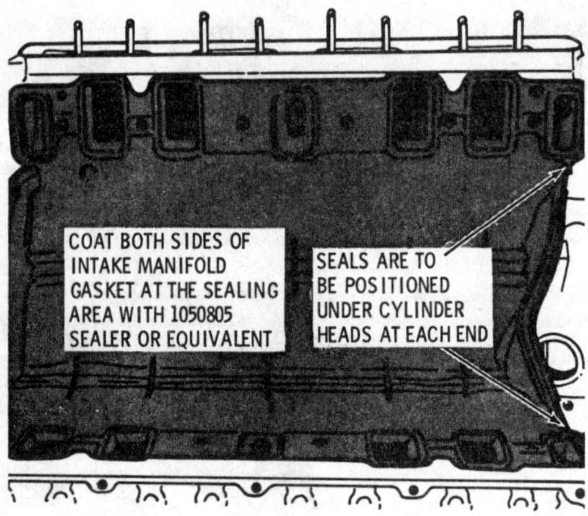

15. Adjust all belts to proper tension.
16. Adjust valves, using the procedure given elsewhere in this chapter.
17. Connect lifting device to engine lifting brackets and install in car, using the procedure given elsewhere in this chapter.

MANIFOLD ASSEMBLY (L6 ENGINES)

Removal

1. Remove air cleaner and disconnect both throttle rods at bellcrank and remove throttle spring.
2. Disconnect vacuum and fuel lines from carburetor.
3. Disconnect ventilation hose from rocker arm cover and hose at vapor canister.
4. Disconnect exhaust pipe from manifold flange and discard packing.
5. Remove attaching bolts and clamps and then remove manifold assembly. Discard all gaskets.
6. Check for cracks in manifold assembly.
7. Separate manifolds by removing bolts and nuts at center.

Installation

1. Clean all gasket surfaces on manifold and cylinder head.
2. Check manifold exhaust port for gaps by placing a straightedge along the full length of the port faces. If at any point a gap of 0.030 in. exists, the manifold will not, in all probability, seal and the manifold should be replaced.
3. Rejoin the exhaust and intake manifolds by installing the center bolt and nuts (finger-tight).
4. Place new gasket on cylinder head manifold end studs. Place manifold assembly on cylinder head and install bolts and clamps. Torque to 25-30 ft.-lb. (inner bolts) and 20 ft.-lb. (outer bolts).
5. Torque exhaust-to-intake manifold bolt and nuts to 30 ft.-lb.
6. Using new packing, connect exhaust pipe to manifold flange.
7. Connect ventilation hose to rocker arm cover and connect hoses to vapor canister.
8. Connect vacuum and fuel lines to carburetor.
9. Reconnect throttle linkage and install throttle spring.
10. Install air cleaner, start engine, and check for manifold leaks and adjust carburetor idle speed.

33

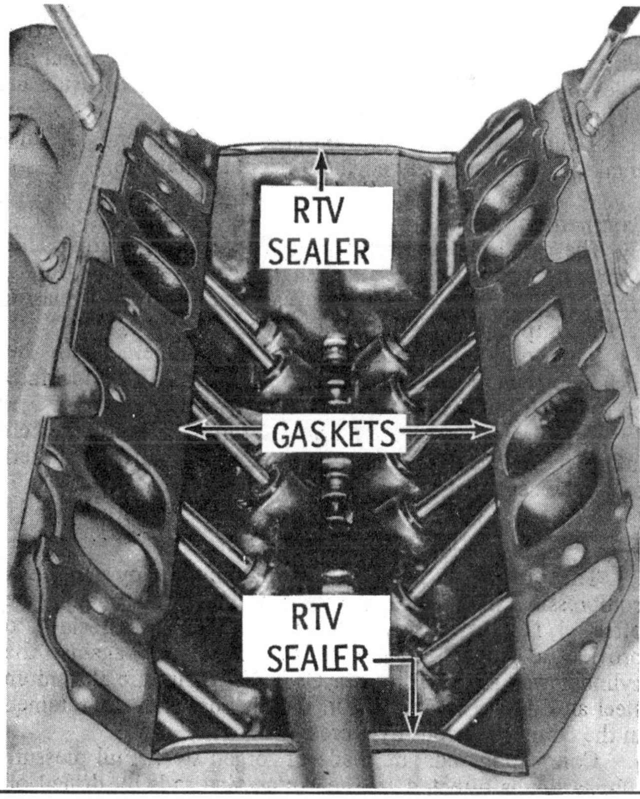

4

INTAKE MANIFOLD (V6 AND V8 ENGINES)

Removal

NOTE

If engine has been removed from vehicle, skip to Step 5.

1. Remove air cleaner and drain radiator.
2. Disconnect the following:
   a. Battery cables (at battery)
   b. Radiator hoses (at manifold)
   c. Accelerator linkage (at pedal lever)
   d. Fuel line (at carburetor)
   e. All coil and temperature sender wires
   f. Power brake vacuum hose
   g. Crankcase ventilation hoses (if so equipped)
   h. Vacuum hose (at distributor and carburetor)
   i. Idle stop solenoid wire (if so equipped)
3. Remove cap from distributor and mark position of rotor. First remove distributor clamp, then remove distributor from vehicle. Position distributor cap rearward so it is clear of manifold.
4. Remove alternator upper bracket.
5. Remove coil bracket and coil.
6. Remove attaching bolts and then remove intake manifold and carburetor from vehicle. Discard all gaskets and seals.
7. If the manifold is to be replaced, transfer the following to the new manifold, using new gaskets:
   a. Carburetor (and attaching bolts)
   b. Water outlet and thermostat
   c. Heater hose adapter
   d. Choke coil
   e. EGR valve (if so equipped)

Installation

1. Clean all sealing and gasket surfaces on cylinder block, cylinder heads, and intake manifold.
2. Install new end seals and side gaskets as shown in **Figure 31** (V6), **Figure 32** (Olds V8), and **Figure 33** (Chevy V8). Use sealants as directed in the figures.

3. Install attaching bolts and torque to specifications (see **Table 2**), using the torquing patterns shown in **Figure 28** (V6), **Figure 29** (Olds V8), or **Figure 30** (Chevy V8).
4. Install coil bracket and coil.
5. Install distributor and distributor clamp. Make certain rotor points toward mark made during removal. Install distributor cap.

NOTE
If crankshaft was rotated while distributor was removed, see procedure later in this chapter for installation instructions.

6. Install alternator upper bracket and adjust belt tension and tighten bolts.
7. Connect the following components:
   a. Battery cables
   b. Radiator hoses
   c. Accelerator linkage
   d. Fuel line
   e. Wires to coil
   f. Power brake vacuum hose
   g. Distributor vacuum hose
   h. Crankcase ventilation hoses
8. Fill cooling system with coolant, start engine, and check manifold and cooling system for leaks. Adjust timing and idle speed, using the procedures given in Chapter Three.

EXHAUST MANIFOLD (V8 ENGINES)

Removal

NOTE
If engine is equipped with Air Injection Reactor System, remove AIR manifold and tubes, using procedure given in Chapter Five.

1. Disconnect battery ground cable and remove pre-heater air stove from air cleaner. Remove alternator, if required.
2. Remove flange nuts and lower and support crossover (if so equipped) and exhaust pipe assembly. Disconnect EFE pipe, if so equipped.
3. Remove bolts (end bolts first) and remove exhaust manifold from engine. Discard gaskets.

Installation

1. Clean mating surfaces on manifolds and cylinder heads. Install manifolds and secure with center bolts.
2. Install end bolts and snug up all bolts. Torque center bolts to specifications (see **Table 2**), then torque end bolts to specification.
3. Using new gaskets, install crossover and exhaust pipes to manifold flanges. Connect EFE pipe, if so equipped.
4. Install alternator (if removed), and connect battery ground cable.
5. Install air cleaner pre-heater, and AIR manifold tubes, if so equipped.
6. Start engine and check for leaks.

ROCKER ARM COVER (ALL MODELS)

Removal/Installation

1. Remove air cleaner and disconnect crankcase ventilation hose(s) at rocker arm cover(s).
2. Remove all wires from clips on the rocker arm cover. Remove air injection hose from check valve of AIR pipe, if so equipped.
3. Remove attaching bolts and remove rocker arm cover. If so equipped, rotate cover out from under AIR pipe.
4. To install, clean all gasket surfaces on cylinder head and rocker cover, install new gasket, and reverse Steps 1 through 3.

VALVE MECHANISM

Removal

1. Remove rocker arm cover(s), using procedure above.
2. On L6 and all V8 engines, remove rocker arm nuts, balls, arms, and pushrods. On V6 engines, remove shaft and arm assemblies and pushrods. Remove nylon rocker arm retainers with a pair of channel lock pliers to separate rocker arms from shaft.

NOTE
Store nuts, balls, arms, and pushrods in a rack, properly identified, so they can be reinstalled in the same locations from which they were removed. A rack can be built by drilling a series of 12 or 16 holes in a piece of wood and marking them with the cylinder numbers.

34A

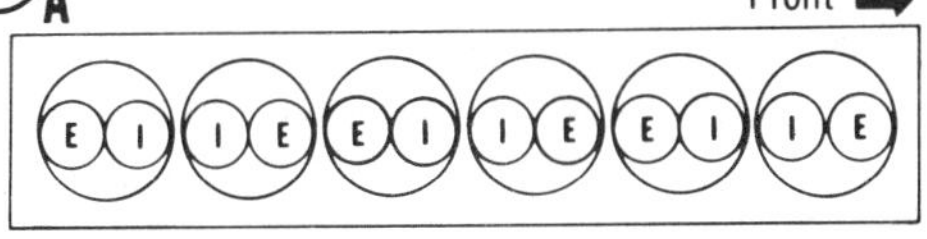

34B

Installation and Adjustment (Inline Engines)

CAUTION

Whenever new rocker arms and/or rocker arm balls are installed, coat bearing surfaces with Molykote or equivalent.

1. Install pushrods, making certain rods seat in lifter sockets.
2. Install rocker arms, balls, and nuts. Tighten nuts until all valve lash is eliminated.
3. Adjust valves when lifter is on base circle of camshaft lobe as follows:
   a. Mark distributor housing with chalk at No. 1 and No. 6 cylinder positions, then disconnect wires from spark plugs and coil. Remove distributor cap and wires.
   b. Crank engine until distributor rotor points to No. 1 cylinder position and breaker points are open, and adjust the following valves:

      No. 1 cylinder exhaust and intake
      No. 2 cylinder intake
      No. 3 cylinder exhaust
      No. 4 cylinder intake
      No. 5 cylinder exhaust

      NOTE

      *See **Figure 34A** for valve locations.*

   c. Back out adjusting nut until lash is felt at pushrod, then turn nut to eliminate all lash. Verify by checking pushrod end play while turning nut (see **Figure 34B**). When all play has been removed, tighten nut one additional full turn (to center lifter plunger).
   d. Crank engine until rotor points to No. 6 cylinder position and breaker points are open. Adjust the following valves:

      No. 2 cylinder exhaust
      No. 3 cylinder intake
      No. 4 cylinder exhaust
      No. 5 cylinder intake
      No. 6 cylinder intake and exhaust

4. Reinstall distributor cap and spark plug wires.
5. Install rocker arm cover, using the procedure given above.
6. Adjust idle speed.

Installation and Adjustment (V6 and V8 Engines)

NOTE

If new rocker arms and/or rocker arm balls are being installed, coat bearing surfaces with Molykote or equivalent.

1. Install pushrods in same locations from which they were removed. Verify that rods are seated in lifter sockets.
2. On Olds V8 engines, install rocker arms in cylinder pairs and tighten bolts to 25 ft.-lb. On V6 engines, install rocker arm and shaft assemblies and tighten nuts to 30 ft.-lb. On Chevy V8 engines, install rocker arms, balls, and nuts. Tighten nuts until all valve lash is removed and adjust as described in Step 3.
3. Adjust valves (Chevy V8 engines) as follows:

NOTE

Valve lash adjustment is neither possible nor required on Olds V8 and V6 engines.

a. Crank engine until mark on torsional damper is aligned with zero mark on timing tab and engine is in No. 1 firing position.

NOTE

When marks are aligned as described above, engine may be in either No. 1 or No. 6 firing position. To determine firing position, hold fingers on No. 1 cylinder valves while cranking engine. If valves are not moving as mark on damper comes near the zero mark, engine is in No. 1 firing position. If valves move, engine is in No. 6 firing position.

b. With engine in No. 1 position, adjust following valves (see Step d):
Exhaust: 1, 3, 4, 8
Intake: 1, 2, 5, 7

c. Crank engine one revolution to No. 6 firing position and adjust the following valves (see Step d):
Exhaust: 2, 5, 6, 7
Intake: 3, 4, 6, 8

NOTE

Intake valves are those closer to the intake manifold. Exhaust valves are those closer to the exhaust manifold.

d. Adjust valves by backing off adjusting nut (rocker arm stud nut) until play can be felt in pushrod. Tighten nut to the point that all pushrod-to-rocker arm clearance is removed. This can be determined by rotating pushrod while tightening nut (see **Figure 35**). When pushrod does not turn easily, clearance has been eliminated. Now tighten nut one additional turn to center hydraulic lifter plunger. No other adjustment is required.

4. Install rocker arm cover(s) using new gaskets and procedure given elsewhere in this chapter.

HYDRAULIC VALVE LIFTERS

All engines covered in this book are equipped with hydraulic valve lifters, which seldom require attention and do not require readjustment. Exercise care and cleanliness when handling parts, however.

Valve Lifter Removal (Inline Engine)

1. Remove valve mechanism, using procedure given above.
2. Mark No. 1 and No. 6 cylinder positions on the distributor housing, then disconnect coil and spark plug wire and remove wires and distributor cap.
3. Turn engine over until distributor rotor points to No. 1 cylinder position. Disconnect primary lead at coil and remove distributor from vehicle.
4. Remove pushrod covers and discard gaskets.
5. Remove valve lifters.

NOTE

Store valve lifters in order in a rack so they may be reinstalled in the same locations from which they were removed.

Valve Lifter Installation (Inline Engines)

1. Install valve lifters in the engine.

NOTE

If new valve lifters are being installed, coat the foot of each lifter with Molykote or equivalent.

36

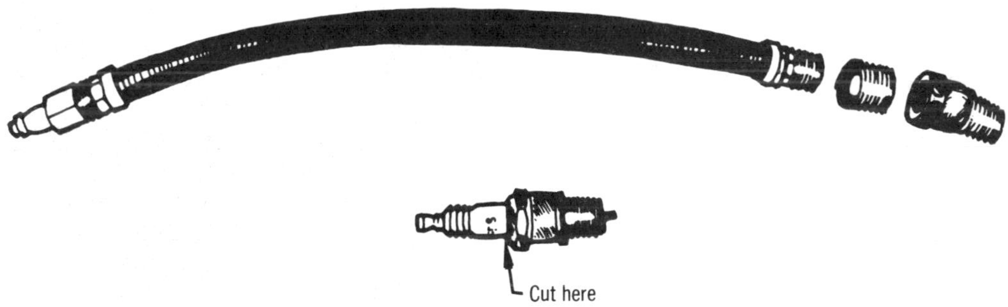

2. Install pushrod covers, using new gaskets. Torque to 80 in.-lb. (1970 models) or 50 in.-lb. (1971 and later models).
3. Install distributor and secure with clamp and bolt. Make certain rotor is pointing to No. 1 cylinder position. Connect primary lead to coil.
4. Reinstall and adjust valve mechanism and distributor cap and wires, using the procedure given above.
5. Set ignition timing and idle speed, using the procedure given in Chapter Three.

Valve Lifter Removal (V6 and V8 Engines)

1. Remove intake manifold, using procedure given above.
2. Remove valve mechanism, using procedure given above.
3. Remove valve lifters. Store lifters in a rack so they can be reinstalled in the locations from which they were removed.

Valve Lifter Installation (V6 and V8 Engines)

1. Install valve lifters.

NOTE
If new lifters are being installed, coat feet of lifters with Molykote or equivalent.

2. Install intake manifold, using procedure given above.
3. Install and adjust (if required) valve mechanism, using the procedure given above.

VALVE STEM OIL SEAL AND/OR VALVE SPRING (ALL ENGINES)

Removal/Installation

1. Remove rocker arm cover(s), using procedure given above.
2. Remove spark plug, rocker arm and pushrod on valve to be serviced.
3. Install air line adapter tool in spark plug hole and apply compressed air to hold valves.

NOTE
*Tool is GM part No. J-23590 or BT-72-1B. A suitable substitute can be made by removing the porcelain and side electrode from an old spark plug (see **Figure 36**). Drive out porcelain, after cutting as shown, by tapping on center electrode. Using a 3/8 in. pipe tap, cut thread in remaining body of plug and assemble air connection as shown. A suitable tool also may be available at an auto parts dealer.*

4. Using a valve spring compressor tool (GM J-5892 on Chevy V8, BT-64-13 on all other engines), applied as shown in **Figure 37** (typical), remove valve locks, cap, spring, and damper, if so equipped.
5. Remove valve stem oil seal (if replacement is required).
6. On L6 and Chevy V8 engines, replace as follows:
   a. Set valve spring and damper, valve shield and valve cup in place.
   b. Compress the spring and install oil seal in lower groove of stem, taking care that seal is flat and not twisted.

NOTE

A light coat of oil on seal will help avoid twisting.

   c. Install valve locks and release spring compression. Make sure locks seat properly in upper groove of valve stem.

NOTE

A small amount of grease can be used to hold locks in place.

7. On Olds V8 engines, install new seals over the valve stems, using special tool BT-6804 or equivalent. See **Figure 38**. Standard to 0.005 in. oversize intake valve seals are gray and exhaust seals are ivory colored. Larger oversize (0.010-0013 in.) intake valve seals are orange and the same oversize exhaust seals are blue colored. After installation, inspect seals for cracks. Position seals over valves and install valve rotators. Compress springs and install keys. Make sure springs and keys are properly seated.
8. On V6 engines, install seals on intake valves only. They are not required on exhaust valves. Install the seal over the valve stem and push it down until it just touches the top of the valve guide. Then use a special tool (J-22509 or equivalent) to push the seal down over the guide. Install the valve springs, caps, and retainers. Compress springs carefully to avoid damage to intake valve seals.
9. Remove air line adapter tool and replace spark plug.
10. Install and adjust the valve mechanism, using the procedure given above.

CYLINDER HEAD

Removal

1. Drain cooling system, including block.
2. Remove intake and exhaust manifolds, using the rpocedures given above.
3. Remove rocker arm cover(s) and valve mechanisms, using the procedure given above.
4. On inline engines, remove fuel and vacuum lines from retaining clip at water outlet and disconnect wires from temperature sending units.

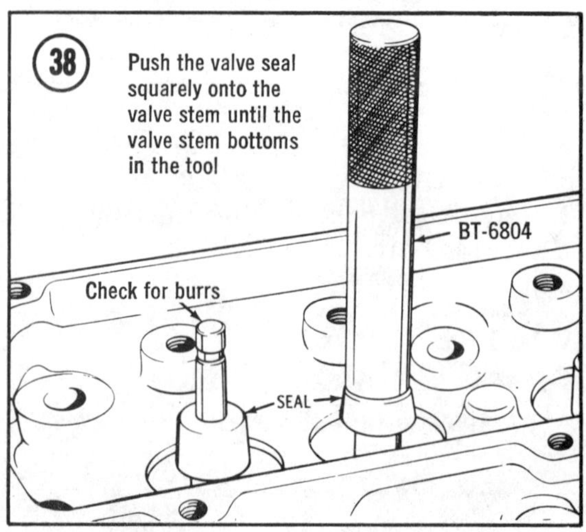

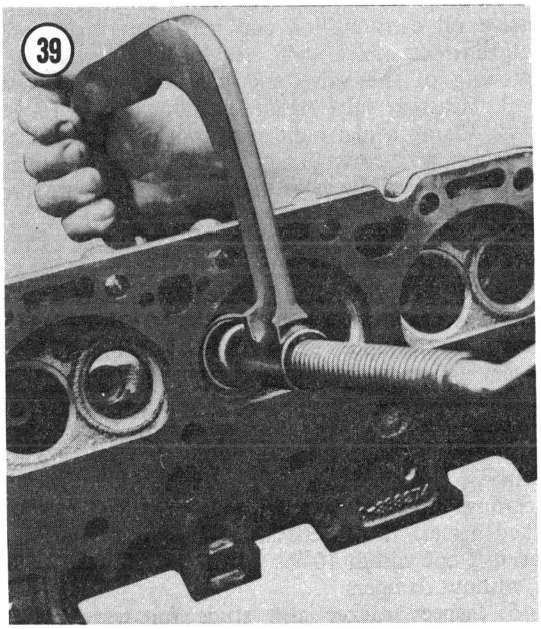

5. On vehicles so equipped, disconnect Air Injection Reactor System hoses at check valve, vapor canister hose, and EGR valve.
6. On inline engines, disconnect radiator upper hose at water outlet and battery ground strap at cylinder head.
7. On inline engines, remove coil.
8. Remove cylinder head bolts, cylinder head(s) and gasket(s). Support heads on wood blocks to prevent damage.

Installation

CAUTION
Clean gasket sealing surfaces on both head and block. These surfaces must be free of all foreign matter and nicks or heavy scratches. All threads on bolts and in cylinder blocks must be clean, as dirt will affect bolt torque. If an all-steel head gasket is used, give both sides a thin coat of sealing compound. DO NOT use sealer on combination metal and asbestos gaskets.

1. Place gasket in position over dowel pins with bead side up.
2. Using care, lower and guide cylinder head into position.
3. Lightly coat cylinder head bolt threads with sealing compound and install finger-tight.
4. Tighten bolts, applying a small amount of torque at a time and following the sequence shown in **Figure 24** (L6), **Figure 25** (V6), **Figure 26** (Olds V8), or **Figure 27** (Chevy V8), until specified torque is reached (see **Table 2**).
5. On inline engines, perform the following:
   a. Install coil.
   b. Connect engine ground strap and upper radiator hose.
   c. Connect fuel and vacuum lines in clip at water outlet.
   d. Connect temperature sender wires.
6. Connect vapor canister hose, if so equipped.
7. Fill cooling system and check for leaks.
8. Install intake and exhaust manifolds, using the procedures given above.
9. Install and adjust valve mechanisms, using the procedures given elsewhere in this chapter.
10. Install and torque rocker arm cover.
11. Connect AIR pipe, if so equipped.
12. Connect EGR valve, if so equipped.
13. Start engine and allow to warm up to operating temperature. Stop engine and retorque cylinder head bolts to specification.
14. Adjust timing and idle speed, using the procedures given in Chapter Three.

Disassembly

NOTE
This procedure is performed after rocker arm nuts, balls, and rocker arms have been removed and the cylinder head has been removed from the engine.

1. Using a spring compressor (GM Part No. J-8062 or equivalent), compress valve springs and remove valve keys (see **Figure 39**). Release compressor and remove rotators or spring caps, shields (if so equipped) and spring damper. Then remove oil seals and valve spring shims.
2. Remove valves from cylinder head and store in sequence in a rack so they can be replaced in their original positions.

Cleaning

1. Clean all carbon from combustion chambers and valve ports, using a wire brush (see **Figure 40**).
2. Clean valve guides, using a suitable tool (GM Part No. BT-6415 or equivalent). Refer to **Figure 41**.
3. Clean all carbon and sludge from pushrods, rocker arms, and rocker arm guides.
4. Clean valve stems and heads, using a buffing wheel.
5. Clean all foreign material from head gasket mating surface.

40

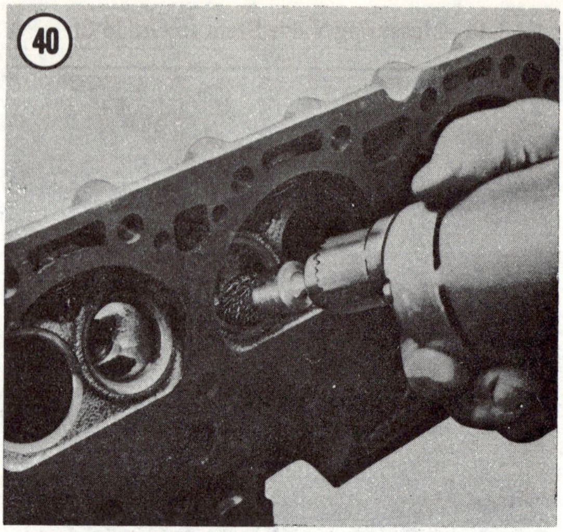

41

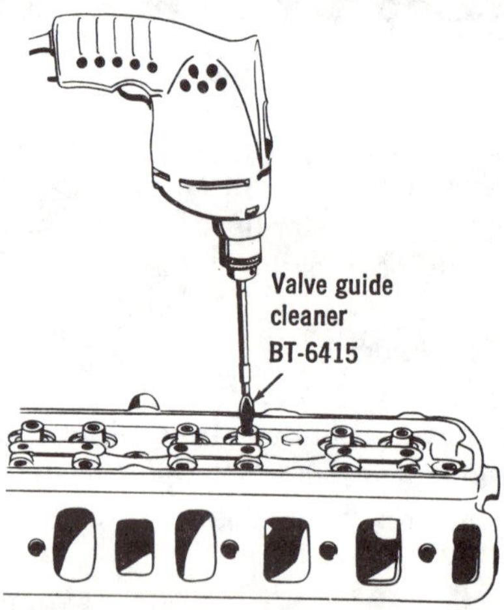

Inspection

1. Inspect for cracks in exhaust ports, combustion chambers and water chambers.
2. Check valves for damaged stems, cracked faces or burned heads.

NOTE
Excessive valve stem clearance will result in excessive oil usage and could cause stems to break. Insufficient clearance can result in noisy operation and sticking of valves.

3. Measure the valve stem clearance as follows:

NOTE
Valve stem clearance measurement and valve guide reaming, if required, are tasks which should be performed by competent automotive machinists.

a. Install the dial indicator as shown in **Figure 42**, so that the side-to-side (crosswise-to-head) movement of valve stem will cause direct movement of the indicator stem. Indicator stem must contact valve stem just above valve guide.

b. Drop valve head about 1/16 inch off valve seat. Move valve stem from side to side, using light pressure, and take indicator reading. If reading exceeds specified tolerance (see **Table 1**), valve guide must be reamed and oversized valves installed.

42

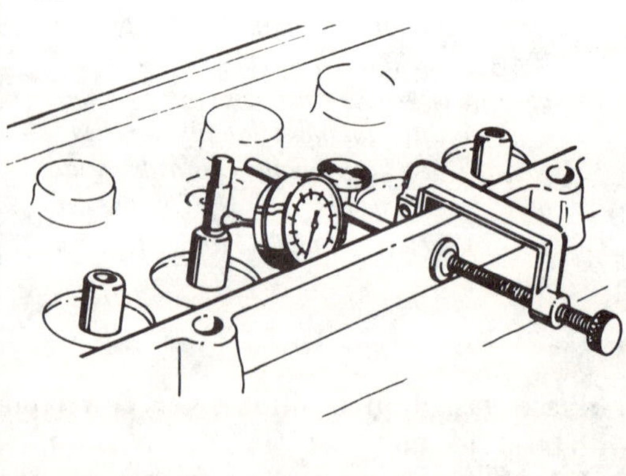

43

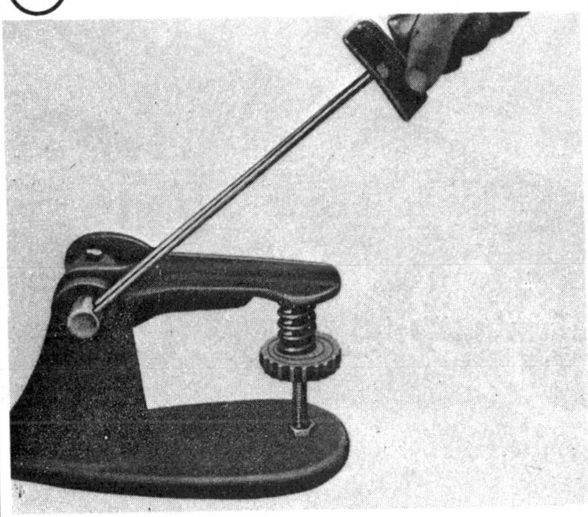

44

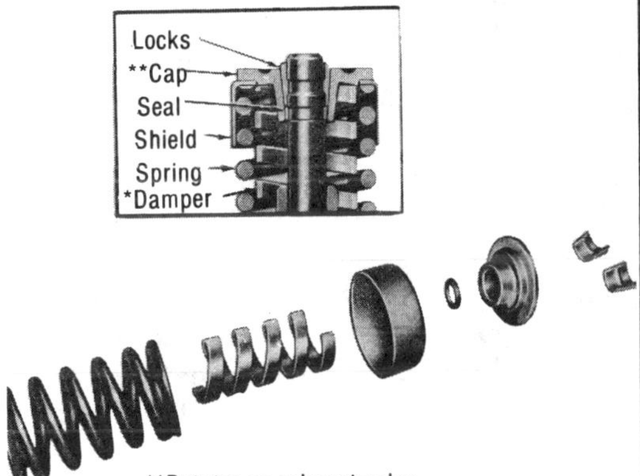

4. Check valve spring tension, using a tension tester (GM Part No. J-8056 or equivalent). Refer to **Figure 43**.

NOTE

*If a tool is not available, have tension checked by an automotive machine shop. If tool is available, compress spring to specified height (see **Table 1**) and measure tension. Spring should be replaced if not within 10 lb. of specified load (without dampers).*

5. Inspect rocker arm studs for wear or damage. Inspect pushrod guides (if used) for wear or damage.

Repairs

On L6 and Chevy V8 engines, damaged or loose rocker arm studs should be replaced with oversized studs (available in 0.003 in. and 0.03 in. oversize). This task should be performed by an experienced automotive machinist, as holes must be reamed before the oversized studs can be installed and special tools and skills are required.

The reconditioning of valve guide bores, valve seats and valves requires the use of precision special equipment and should not be attempted by the home mechanic. These tasks can be performed competently and economically by a well-equipped automotive machine shop. Specifications are given in **Table 1**.

Assembly

1. Insert a valve in the port from which it was removed.

2. For L6 and Chevy V8 engines, position valve spring shim, spring, damper (if used), shield and cap or rotator on valve stem (see **Figure 44**). Using a spring compressor tool (GM Part No. J-8062 or equivalent), compress the spring. Install oil seal in stem lower groove. Make sure seal is not twisted. Install valve locks and remove compressor. Make sure that lock seats in upper valve stem groove.

3. On Olds V8 engines, install new valve seals as described in *Valve Stem Oil Seal And/Or Valve Spring*, this chapter. Then install valve spring and rotators, compress the spring (taking care not to damage seals), and install the retaining keys. Make sure spring and keys are properly seated.

4. On V6 engines, install valve spring, cap, and cap retainer, using the same equipment used during disassembly. See **Figure 39**.

5. Install remaining valves.

6. On L6 and Chevy V8 engines, check height of installed springs against specification (see **Table 1**). A cutaway scale (see **Figure 45**) will help. Measure from shim or spring seat top to valve spring or valve spring shield top (see **Figure 46**). If height exceeds specification, install a valve seat shim (approximately 1/16 in. thick). Do not shim spring to give an installed height under the specified minimum.
7. On Olds V8 engines, a special gauge is required to measure valve stem and installed spring height. See **Figure 47**. There should be at least 0.015 in. clearance between the gauge and the tip of the valve stem. If clearance is less, remove the valve and grind off the tip of the stem, using V-blocks and a precision valve grinder. Both measurement and grinding should be left entrusted to your dealer or an automotive machine shop.
8. On V6 engines, the valve stem tip should be precision ground or the valve should be replaced if the tip is 0.050 in. or more above the normal height. Both measurement and grinding are jobs for your dealer or an automotive machine shop.

45

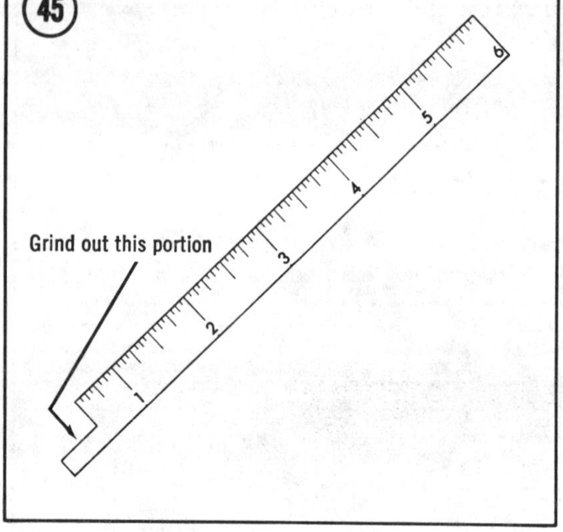

46

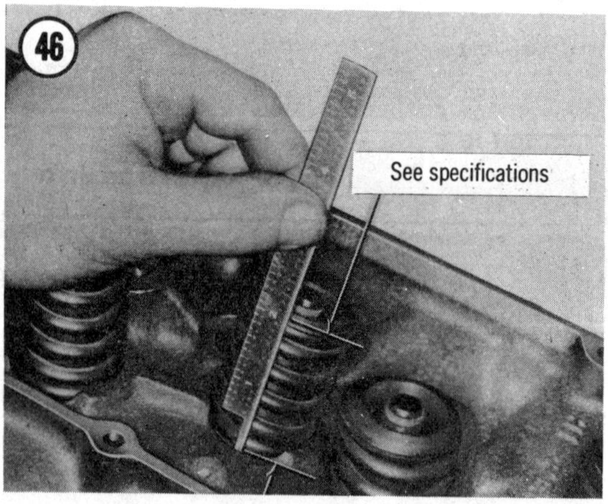

47

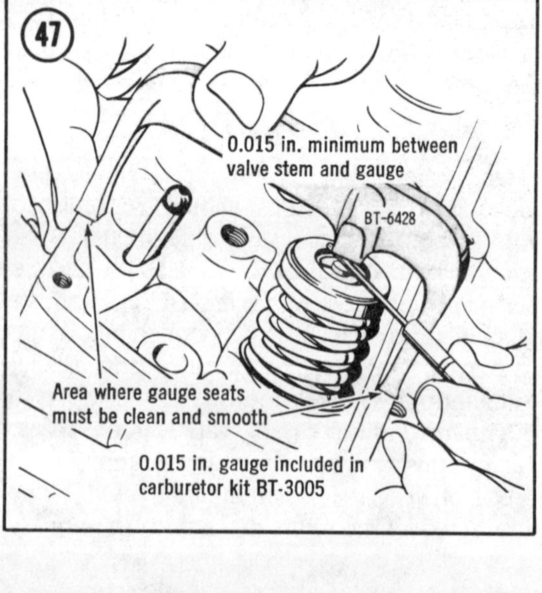

OIL PAN REPLACEMENT

L6 Engine

1. Disconnect the positive cable from the battery and then disconnect the fuel line at the fuel pump. Plug the line to prevent loss of fuel.
2. Disconnect the electrical leads at the starter motor.
3. Remove the upper radiator support bracket-to-side panel bolts. Drain the cooling system and remove the radiator hoses.
4. If the car is equipped with air conditioning, remove the fan and clutch assembly.
5. Remove the front engine mount bolts (bracket-to-mount bolts).
6. Raise the car and place jackstands under the front lifting points, then drain the engine oil.
7. If equipped with automatic transmission, remove the transmission manual linkage.
8. Remove the flywheel cover.
9. Remove the starter, then disconnect the exhaust pipe from the exhaust manifold.

10. Turn the crankshaft so that the timing mark on the harmonic balancer is in the 6 o'clock position (facing down).
11. Using a suitable jack, raise the engine at the harmonic balancer. Then remove the right engine mount and bracket.
12. Remove the attaching screws and loosen the oil pan. Tap the oil pan with a rubber hammer, if necessary, to break it loose.
13. Raise the car just as much as is required to remove the oil pan, taking care not to damage parts mounted on the cowl. Remove the oil pan.

14. If further work is to be done on the engine before the oil pan is reinstalled, block the engine as shown in **Figures 48 and 49**.

15. Installation is the reverse of the above steps. Make sure the engine is filled with the proper amounts and kinds of coolant and engine oil before starting. Use a new gasket and tighten attaching bolts to 7 ft.-lb.

48

49

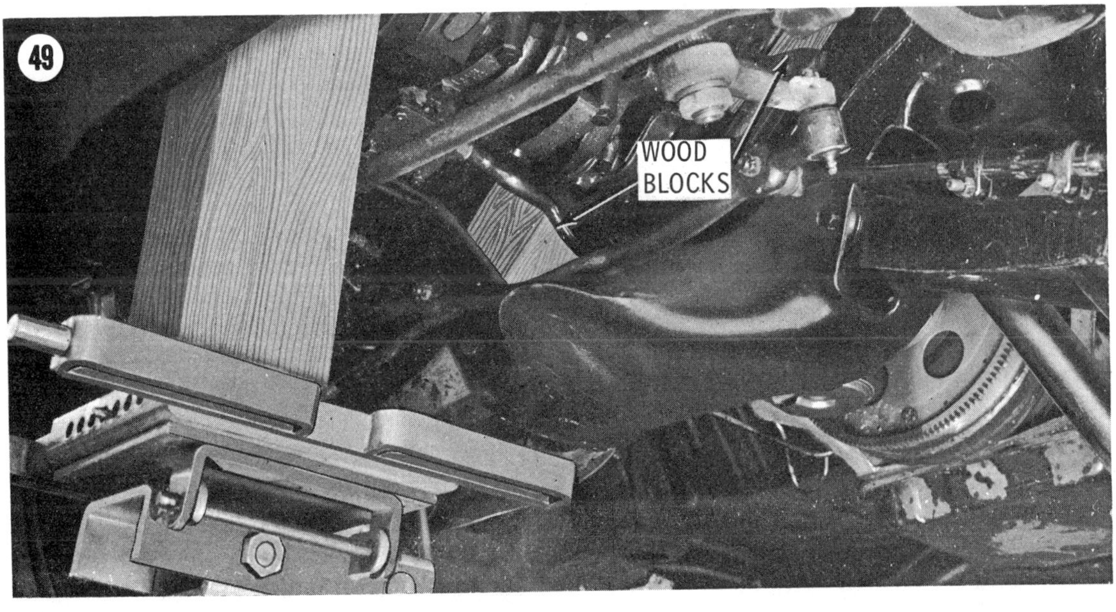

50

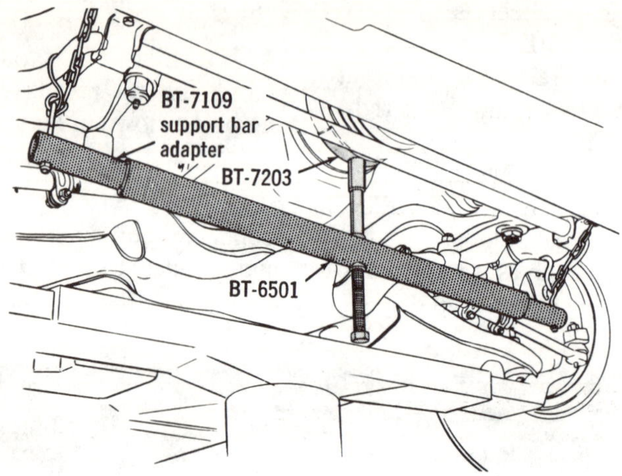

V6 Engine

1. Raise the car and support it on jackstands.
2. Drain the engine oil.
3. Remove the flywheel cover and the exhaust crossover pipe.
4. Remove the oil pan.
5. Installation is the reverse of these steps. Install a new gasket on the block, using a few spots of room temperature vulcanizing (RTV) sealer to hold it in place. Make sure all sealing surfaces on the block and pan are clean and free of all old gasket material. Tighten oil pan attaching bolts to 14 ft.-lb.

V8 Engines

1. Turn the crankshaft until the timing marks are aligned (No. 1 cylinder in top dead center position). Disconnect the positive cable from the battery.
2. Remove the oil dipstick. On Chevy V8 engines, remove the dipstick tube.
3. Remove the upper half of the fan shroud.
4. Raise the front of the car and support it on jackstands. Drain the engine oil.
5. Remove the exhaust crossover pipe.
6. On Olds V8 engines, remove the starter motor. Attach engine support tool BT-6501. See **Figure 50**. A suitable jack could be substituted for this tool, if care is taken. Remove the engine mount pivot bolts and raise the engine, then remove the engine mounts. Remove the attaching bolts and remove the oil pan.
7. On Chevy V8 engines, remove the torque converter cover and the engine mount through bolts. Then remove the oil pan attaching bolts and lower, but do not remove, the pan. Make sure crankshaft throw is in the horizontal position (turn crankshaft if necessary), then raise the engine and loosely install the through bolts. Remove the oil pan.
8. Installation is the reverse of the above steps. Use a new gasket, and make sure all sealing surfaces are clean and free of old gasket material. Coat both sides of gasket with sealer (Part No. 1050026 or equivalent). On Olds V8 engines, install new rubber seals at front and rear, and tighten bolts to 10 ft.-lb. On Chevy V8 engines, tighten 1.4 in. bolts to 7 ft.-lb., and 5/16 in. bolts to 22 ft.-lb. Make sure engine is filled with proper amount of engine oil before starting. Check sealing surfaces for leaks after starting engine.

OIL PUMP (ALL ENGINES EXCEPT V6)

Removal/Installation

1. Remove oil pan, using the procedure given above.
2. On L6 engines, remove 2 flange mounting bolts, then remove pump and screen as an assembly. Remove the extension shaft.

(51) OIL PUMP (L6 ENGINE)

1. Pressure regulator valve
2. Spring
3. Retaining pin
4. Screws
5. Pump cover
6. Cover gasket
7. Idler gear
8. Drive gear and shaft
9. Pump body
10. Pick-up screen and pipe

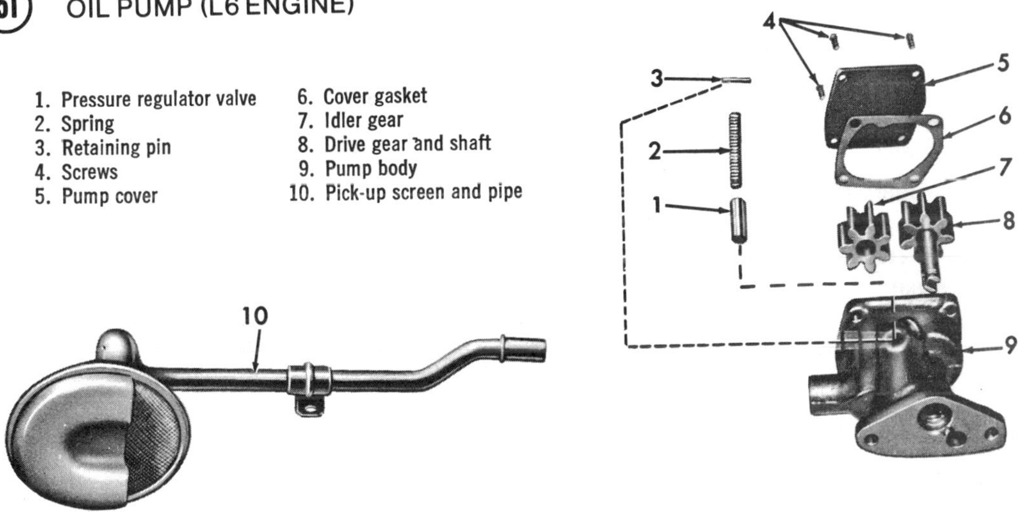

(52) OIL PUMP (OLDS V8)

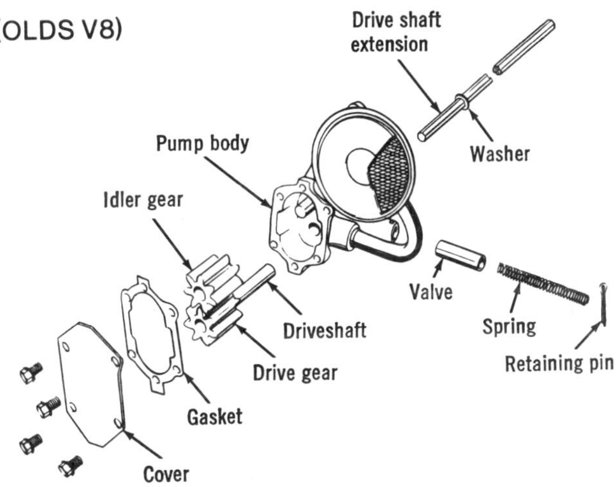

3. To install oil pump, proceed as follows:
   a. On L6 engines, align oil pump drive shaft to match with distributor tang and install oil pump to block. Position flange over lower distributor bushing. No gasket is required.

      NOTE
      If oil pump does not slide easily into place, remove and realign slot with distributor tang.

   b. On V8 engines, assemble pump and extension shaft to rear main bearing cap, aligning slot in shaft top end with tang on lower end of distributor shaft. Install bolt and torque to specification (see **Table 2**).

      NOTE
      Bottom edge of oil pump screen should be parallel to oil pan rails.

Disassembly

Refer to **Figure 51** (L6), **Figure 52** (Olds V8), or **Figure 53** (Chevy V8).

1. Remove pump cover attaching screws and pump cover. On L6 engines, remove gasket.

NOTE

Mark gear teeth so same indexing of teeth can be duplicated when reassembling pump.

2. Remove idler and drive gears and drive shaft from pump body.
3. Remove retaining pin, then remove pressure regulator valve and related parts.
4. If replacement is indicated, mount pump in soft-jawed vise and extract pickup screen and pipe from pump body.

Cleaning and Inspection

1. Wash all parts in solvent. Dry with compressed air (if available).
2. Inspect pump body for cracks and signs of wear.
3. Inspect gears for excessive wear or damage.
4. Check shaft of drive gear for looseness in pump body.
5. Inspect inside of cover for wear which would allow oil to leak past ends of gears.
6. Inspect pickup screen and pipe for damage.
7. Check pressure regulator valve for fit.

NOTE

Pump body and gears cannot be replaced separately. If damage or excessive wear is present, entire pump must be replaced.

Assembly

Refer to **Figure 51, 52, or 53** for this procedure.

CAUTION

If pickup screen and pipe assembly were removed, install new assembly. Do not attempt to install old assembly, as loss of press fit could result in loss of oil pressure.

1. To install new pickup screen and pipe assembly, mount pump in soft-jawed vise. Apply sealer to end of pipe and, using a suitable tool (GM Part No. J-8369), tap pipe into place with plastic hammer.

CAUTION

Avoid twisting, shearing or collapsing pipe while installing in pump. Screen on inline engines must be parallel to bottom of oil pan when pump is installed.

53

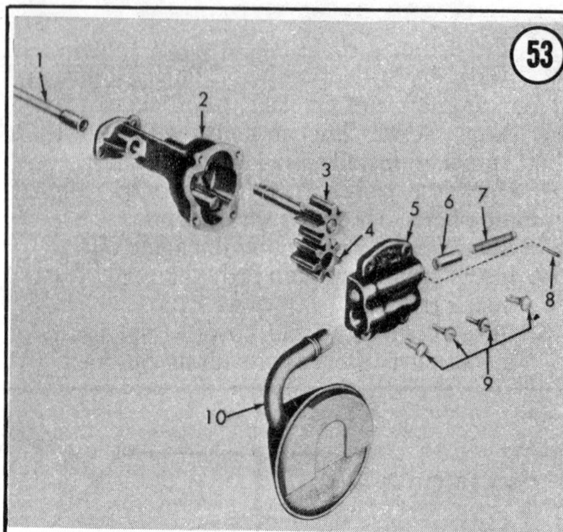

OIL PUMP (CHEVY V8)

1. Shaft extension
2. Pump body
3. Drive gear and shaft
4. Idler gear
5. Pump cover
6. Pressure regulator valve
7. Pressure regulator spring
8. Retaining pin
9. Screws
10. Pickup screen and pipe

2. Install pressure regulator valve and related parts with retaining pin.
3. Install drive gear and shaft in body, then install idler gear so that teeth mesh in original position and smooth side of gear is facing pump cover opening.
4. Install pump cover (using new gasket on inline engine) and torque screws to 70 in.-lb. (L6 engine), 80 in.-lb. (Chevy V8), or 96 in.-lb. (Olds V8).
5. Check for smooth operation by turning drive shaft by hand.

OIL PUMP (V6 ENGINE)

Pipe and Screen Assembly Removal/Installation

1. Remove the oil pan as described elsewhere in this chapter.
2. Remove the bolts attaching the pipe and screen assembly to the cylinder block and

remove the assembly from the engine. See **Figure 54**.

3. Clean the assembly in solvent, paying particular attention to the screen and the pipe passage, and blow dry with compressed air.

4. Clean the gasket surface on the cylinder block flange and install the pipe and screen assembly on the flange, using a new gasket. Securely tighten the attaching bolts.

5. Install the oil pan, using a new gasket, as described elsewhere in this chapter.

Pump Removal/Installation

Figure 55 is a diagram of the lubrication system and shows the location of the oil pump

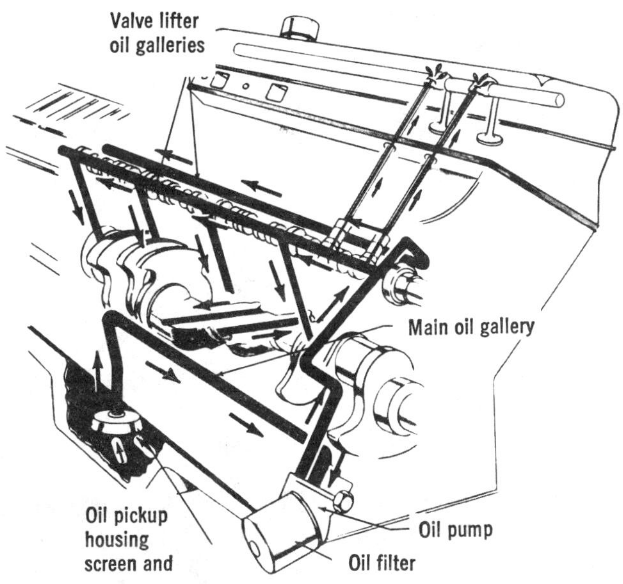

and filter assembly. **Figure 56** is a cut-away view of the oil pump and filter assembly. **Figure 57** is an exploded view of the oil pump cover and relief valve assembly.

1. Remove the oil filter and then remove the oil pump cover assembly by removing the attaching screws. Slide out the oil pump gears.
2. Clean the gears in solvent and check them for wear, scoring, or other damage. Gears in poor or doubtful condition should be replaced.
3. Remove the valve cap (**Figure 57**) and remove the spring and valve from the oil pump cover. Clean the parts in solvent and check the valve for wear, scoring, or other damage. Also check the spring for side wear or other damage. Replace parts that are questionable.

NOTE
The oil filter bypass valve and spring are permanently staked in place in the oil pump cover and should not be removed.

4. Check the fit of the relief valve in its bore in the oil pump cover. It should be an easy slip fit, without side movement. If the valve fits too loosely, or does not slide easily, install a new valve and recheck the fit. If the valve still does not fit properly, replace the pump cover.
5. Inspect the bypass valve for warping or other damage. The valve should be flat and should be free of nicks and scratches. Replace the cover assembly if the bypass valve appears to be damaged.
6. Lubricate the relief valve and spring with engine oil and install them in their bore in the pump cover. Install the valve cap, using a new gasket, and tighten the cap to 35 ft.-lb., using a reliable torque wrench. Do not over-tighten.
7. Install the gears and shaft in the oil pump cavity in the timing chain cover. Check the end clearance of the gears, using a straightedge and feeler gauge as shown in **Figure 58**. Measure between the straightedge and the gasket surface of the pump cavity as shown in the figure. If the clearance is not between 0.002-0.006 in., measure the gears as shown in **Figure 59** and the cavity as shown in **Figure 60** to determine which is out of tolerance. Specifications are given in the figures. Replace the defective part(s).

56

57

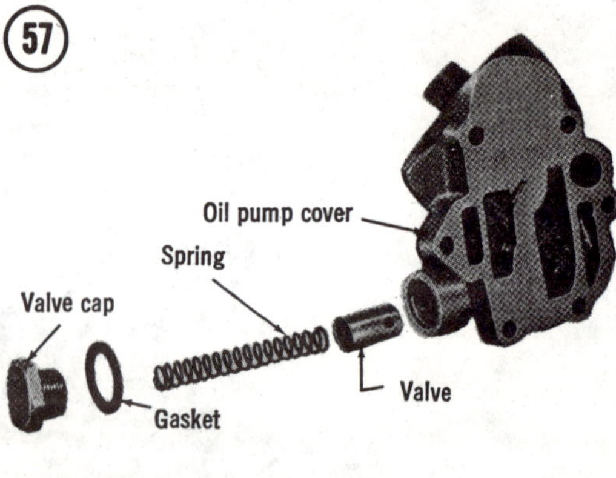

58

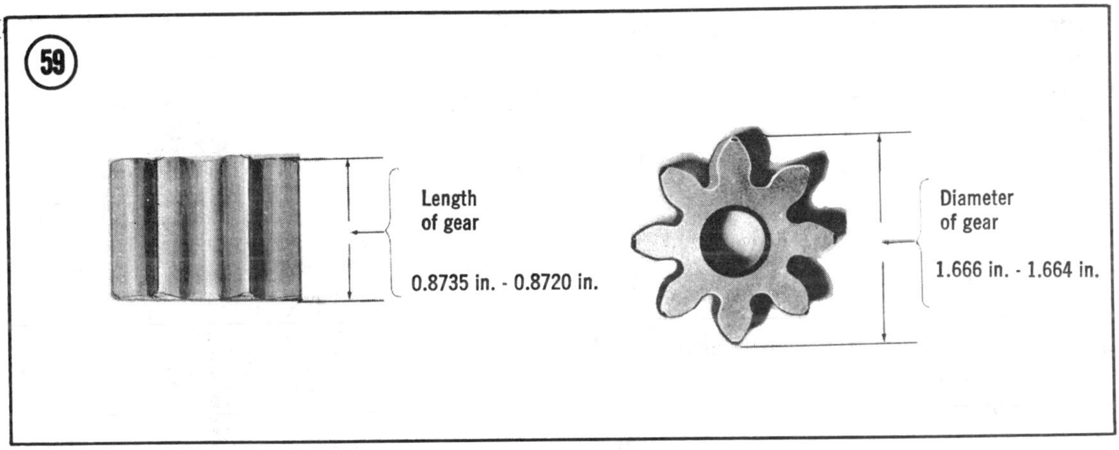

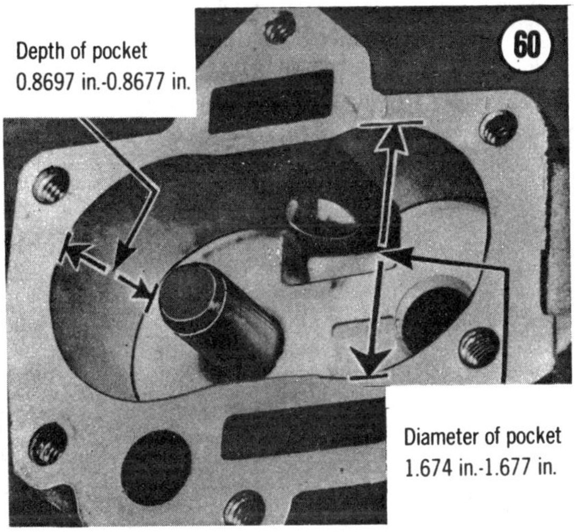

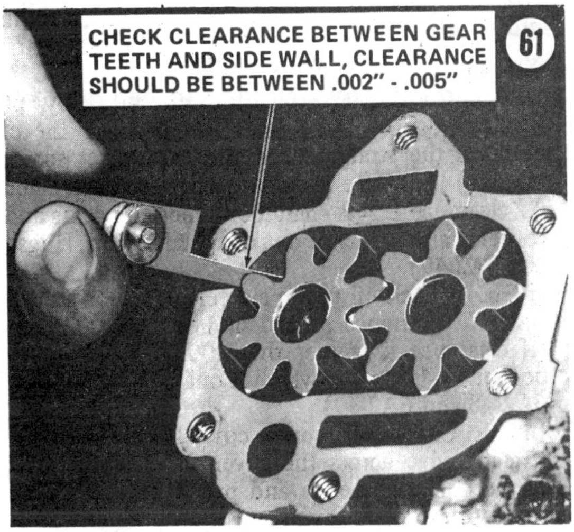

8. Measure the gear side clearance as shown in **Figure 61**. If not between 0.0025-0.0050 in., measure the diameter of the gears and the diameter of the pockets as shown in **Figures 59 and 60** to determine which is out of tolerance. Replace the defective part(s).

9. Check the pump cover flatness by trying to insert a 0.001 in. feeler gauge between the cover face and a straightedge. If the clearance is 0.001 in. or larger at any point, replace the cover.

10. After all clearances are within specifications, remove the gears and pack the pump cavity with petroleum jelly (Vaseline, etc.). Do not use chassis lube or any other type of lubricant.

11. Reinstall the gears in the pump cavity, making sure that the petroleum jelly fills all space and that no air pockets are left.

CAUTION

This step is very important. If the pump cavity is not properly packed with petroleum jelly, it could fail to prime itself and serious damage to engine parts could result from the lack of lubrication.

12. Using a new gasket, install the pump cover. Tighten the attaching screws evenly and alternately to 10 ft.-lb.

13. Install a new oil filter.

REAR MAIN OIL SEAL REPLACEMENT

L6 and Chevy V8 Engines

NOTE

Upper and lower seal must be replaced as a unit. Install seal with lip facing front of engine.

The rear main bearing seal can be replaced without crankshaft removed. However, care must be taken to protect the sealing bead on the outside diameter of the seal. An installation tool (see **Figure 62**), can be used to protect the bead.

1. Remove oil pan, using the procedure given elsewhere in this chapter.
2. Remove oil pump, using the procedure given above.
3. Remove rear main bearing cap. Use a small screwdriver to remove seal half from bearing cap (see **Figure 63**).
4. Tap one end of upper half of seal with a small hammer and brass pin punch (see **Figure 64**) until other end protrudes enough to grasp and remove with pliers.
5. Use non-abrasive cleaner to remove all traces of sealant and other foreign matter from the cylinder block, crankshaft and bearing cap.
6. Inspect all components for defects on sealing surfaces, case assembly, and crankshaft. Coat seal lips and bead with engine oil, keeping oil off seal mating ends.
7. Position tool (**Figure 62**) tip between crankshaft and cylinder block seal seat.
8. Position seal between crankshaft and tip of tool so that sealing bead contacts tip of tool, making certain that seal lip points toward front of engine (see **Figure 65**).
9. Roll seal around crankshaft using tool as a "shoehorn" to protect seal bead.

CAUTION

Installation tool must remain in position until seal is positioned with both ends flush with block to protect seal bead from sharp edge of seal seat surface in cylinder block.

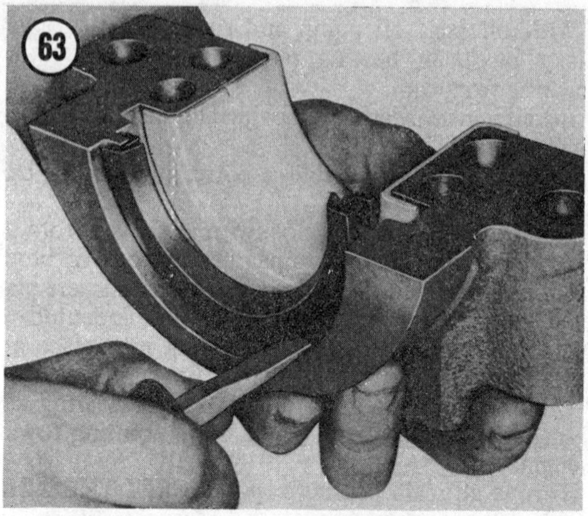
63

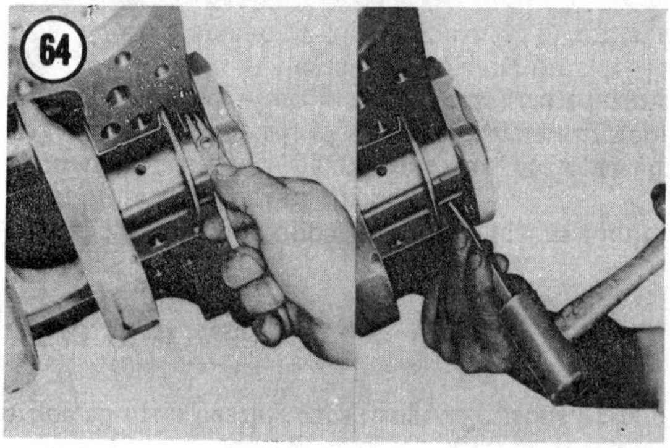
64

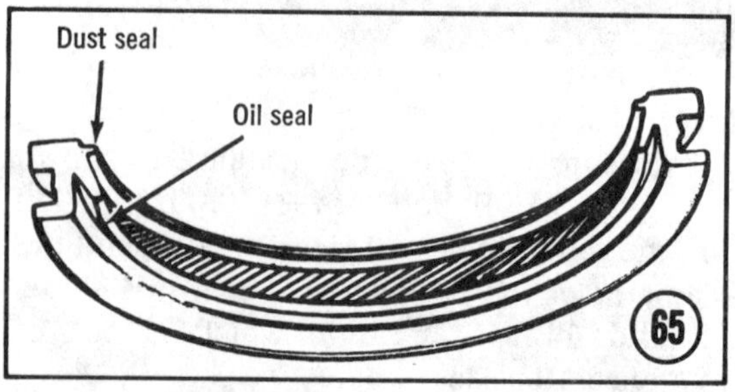

65

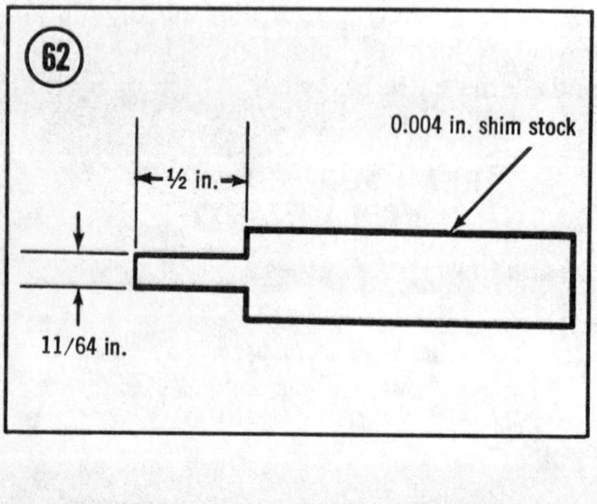

62

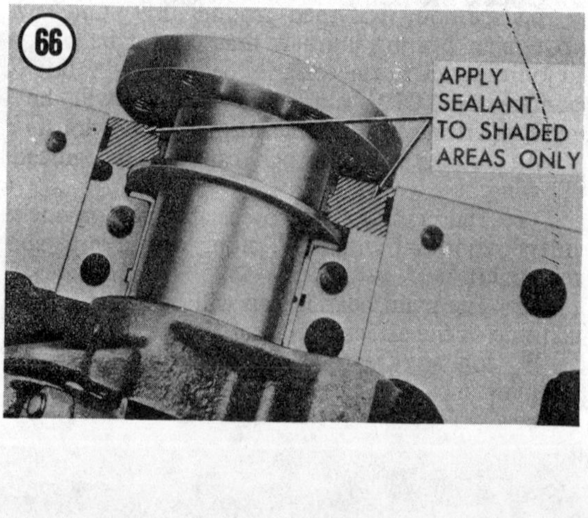

66

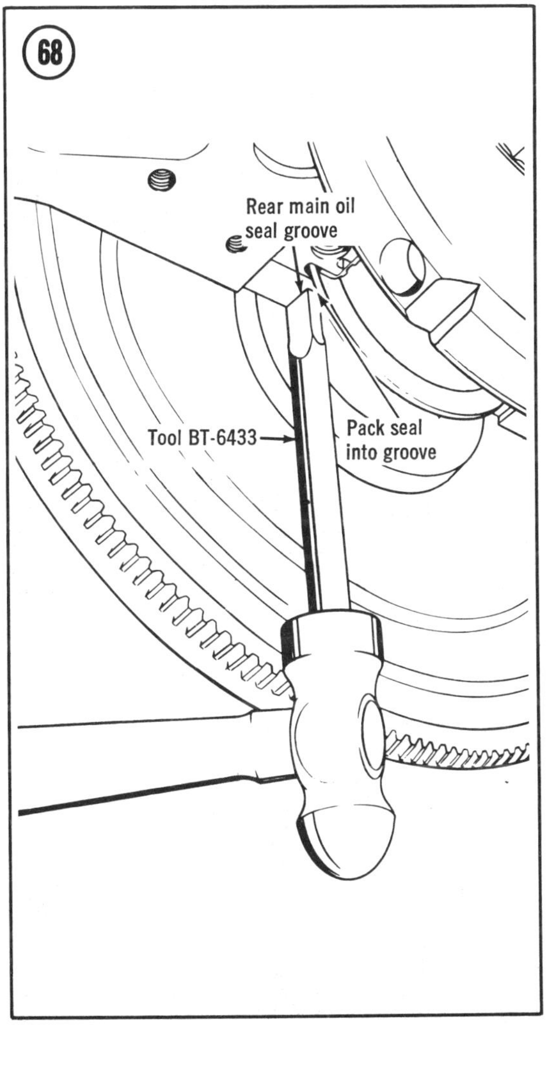

10. Remove tool. Use care not to remove or displace seal.
11. Again using tool as "shoehorn" install seal half in bearing cap. Feed seal into cap, using light pressure with thumb and finger.
12. Apply sealant as shown in **Figure 66** and install bearing cap. Take care to avoid getting sealant on the seal split line. Torque to specifications (see **Table 2**).

REAR MAIN OIL SEAL REPLACEMENT

V6 and Olds V8 Engines

Replacement of this seal requires that the crankshaft be removed from the engine. However, the old seal can be repaired in some cases without crankshaft removal. Repair consists of removing the oil pan and rear main bearing cap, and packing the seal into the block with a special tool (J-21526-2 on V6 engines, or BT-6433 on Olds V8, or equivalent), as shown in **Figure 67** (V6) or **Figure 68** (V8), about 1/4 to 3/4 in. on both ends. Then a guide tool (J-21526-1) is installed (**Figure 69**) on the V6 engine. A piece of the old seal removed from the bearing cap is cut for each side, using a razor blade. Each piece should be about 1/16 in. longer than the depth the old seal was driven into the block. Use the bearing cap as a holding fixture so the seal pieces can be cut squarely. Then use the packing tool to work the short pieces

4

of seal into the block on the sides of the rear main bearing. See **Figure 70**. Remove the guide tool, if installed. Install a new fabric seal in the bearing cap, install the bearing cap, and tighten the attaching bolts to 100 ft.-lb. Then reinstall the oil pan.

To install new seals, remove the crankshaft as described elsewhere in this chapter. Remove the old seals from the grooves in the rear main bearing cap and the cylinder block, and install new seals. On V6 engines, use a hammer handle or similar tool to force the fabric seal into the grooves (see **Figure 71**), and then cut the ends of the seal off squarely and flush. On V8 engines, use a special tool as shown in **Figure 72**. On V6 engines, install new Neoprene composition seals in the side grooves of the bearing cap (**Figure 71**). These seals, which swell when subjected to oil and heat, are a loose fit and are slightly longer than the grooves. Do not trim them or attempt to pack them. A small amount of leakage is normal when these seals are newly installed, but should disappear in a short time. This time can be shortened by soaking the seals in oil or kerosene for a couple of minutes just prior to installation. Before installing the bearing cap, coat the surfaces shown in **Figure 73** (V6) or **Figure 72** (V8) with silastic sealer. Install the crankshaft and then install the bearing caps.

MAIN BEARINGS

Precision insert type bearings are used in all engines, and shims are not required for fitting. If at any time bearing clearance is found to be excessive, a new bearing (both upper and lower halves) is required. Bearings are available in standard and 0.001, 0.002, 0.009, 0.010, 0.020, and 0.030 in. undersize. Close tolerances are obtained in production by selective fitting of rod and main bearing inserts. Thus one half of a bearing may be standard and the other half 0.001 in. undersize (this decreases clearance by 0.0005 in.). If a production crankshaft cannot be properly fitted in this manner it is ground 0.009 in. undersize on the main journals only and again is selectively fitted. If your engine has a 0.009 in. undersize crankshaft, the counterweight forward of the center main journal will be stamped with ".009", and the number 9 will be stamped on the block at the left front oil pan rail.

NOTE

In repair work on L6 and Chevy V8 engines, shimming may be required if bearing caps are replaced for any reason. Laminated shims are available for service work.

Inspection

As a rule the lower bearing half shows greater wear and distress from fatigue. If inspection shows lower half is suitable for use, it can usually be assumed that the upper half also is satisfactory without removing it. If the lower half is unsatisfactory, replace both halves (never replace only one half of a main bearing).

70

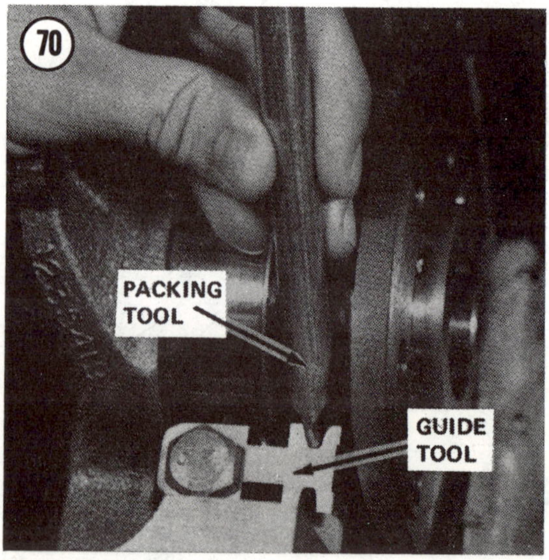

71

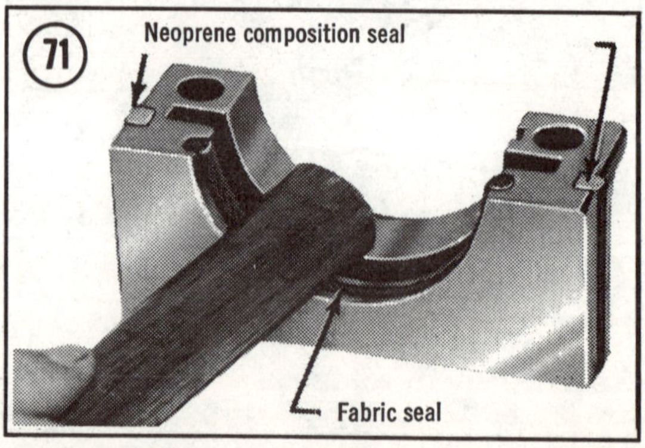

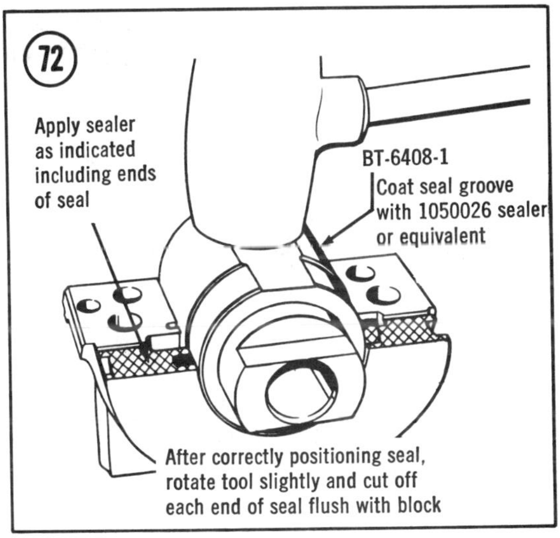

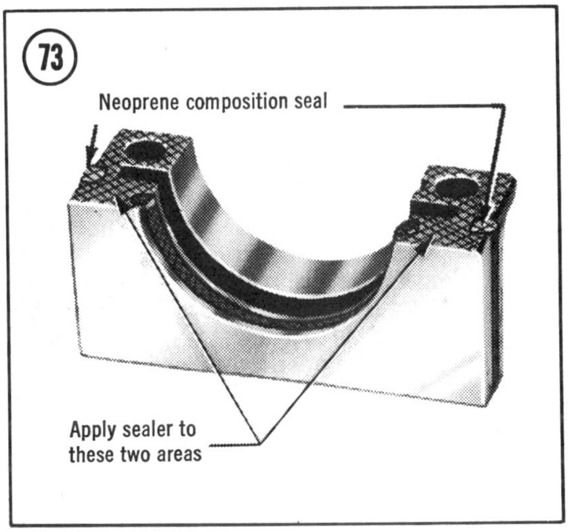

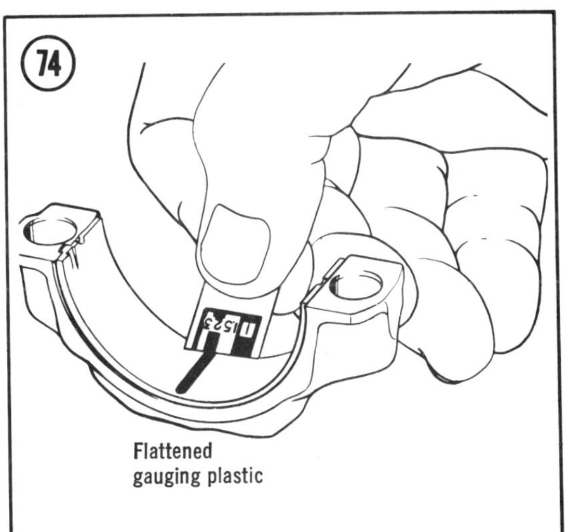

Checking Bearing Clearance

NOTE

Use Plastigage or equivalent for checking bearing clearances. All bearing cap bolts should be torqued to specification, and surfaces to be measured should be wiped free of oil. If engine is still installed in vehicle, support crankshaft at flywheel and tortional damper to remove clearance from upper bearing half.

1. With oil pan and pump removed (see procedures above), and starting with rear main bearing, remove cap and wipe oil from journal and bearing half in cap.
2. Place a piece of Plastigage the full width of the bearing on the journal, parallel to the crankshaft.

CAUTION

Do not rotate crankshaft while Plastigage is between journal and bearing, as false reading will be obtained.

3. Install bearing cap and evenly torque bolts to specification (see **Table 2**).
4. Remove bearing cap. Flattened Plastigage will be found either on journal or bearing.
5. Using the scale printed on the Plastigage package, measure width of compressed plastic at its widest point. Do not remove plastic before measurement is made (see **Figure 74**).

NOTE

As a rule, main bearings wear evenly and are not out-of-round. If a bearing is fitted to an out-of-round journal (0.001 in. max.), the bearing must be fitted to the maximum diameter of the journal. Otherwise rapid wear of the bearing can be expected. If the flattened plastic tapers toward the center or ends, and a difference of more than 0.001 in. is indicated, the crankshaft journal should be checked for roundness, and corrected if required, before proceeding.

6. If clearance is within specifications (see **Table 1**), insert is satisfactory. If not, replace both halves of insert.

4

7. If a standard or 0.001 in. or 0.002 in. undersize bearing fails to produce proper clearance, journal must be ground for the next undersize bearing.
8. Repeat the above steps for the remaining bearings, then rotate crankshaft to verify that excessive drag is not present.
9. Measure crankshaft end play (see specifications, **Table 1**), by forcing crankshaft to extreme front. Measure at front end of rear main bearing. See **Figure 75**.
10. Install new rear main bearing oil seal, using the procedure given above.

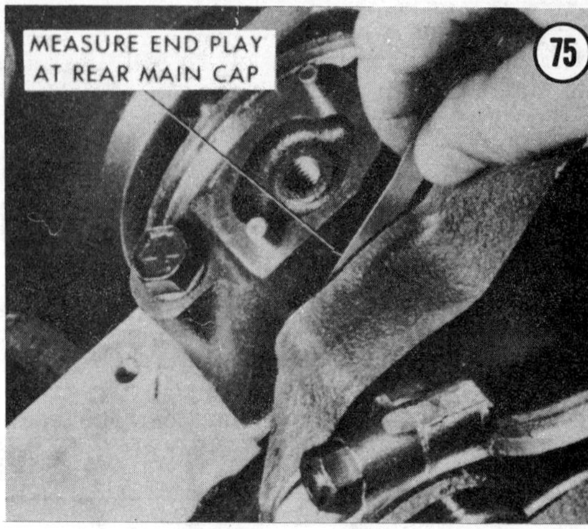

Main Bearing Replacement (With Crankshaft Removal)

1. Remove and inspect crankshaft.
2. Remove old bearings from cylinder block and caps.
3. Select new bearings of correct size (see *Checking Bearing Clearance* procedure above) and coat bearings with engine oil. Install in cylinder blocks and bearing caps.
4. Install crankshaft in cylinder block and install bearing caps with arrows pointing to front of engine.
5. Install bolts and torque to specification (see **Table 2**).

Main Bearing Replacement (Without Crankshaft Removal)

1. Remove oil pan, oil pump, and spark plugs from engine (see procedure above). Remove cap from main bearing to be replaced and remove bearing from cap.
2. On L6 engine crankshaft, rear main bearing journal has no oil hole. Rear main bearing upper half is removed as follows:
   a. Use hammer and small drift punch to start bearing half out of block.
   b. Grasp bearing thrust surface and oil slinger with a pair of pliers (with taped jaws) and rotate crankshaft to remove bearing half (see **Figure 76**).
   c. Select new bearing of correct size and coat with oil. Insert unnotched end between crankshaft and notched side of block.
   d. Using pliers as in Step b, rotate bearing half into place. During last fourth of movement, pliers may be used on slinger only, or, if care is taken, bearing may be tapped into place with a soft drift punch.

3. All remaining mainbearing journals on inline crankshafts and all V6 crankshaft journals have oil holes. Upper bearing halves may be installed as follows:

   a. Install a long cotter pin in oil hole in crankshaft journal and rotate crankshaft clockwise (viewed from front of engine) to roll out insert.

b. Select and oil proper size upper bearing (see *Checking Bearing Clearance* above) and insert unnotched end between crankshaft and notched side of block. Rotate into place and remove cotter pin from journal oil hole.

4. Oil new lower bearing insert and install in bearing cap. Install cap with arrows pointing to engine front. Torque cap bolts to specifications (see **Table 2**).

CONNECTING ROD BEARING REPLACEMENT

CAUTION

***Do not** file rods or rod caps to obtain specified clearance. If clearance is excessive, new bearings are required. Bearings are available in standard sizes and 0.001 in. and 0.002 in. undersize for use with new and used standard size crankshafts, and in 0.010 in. and 0.020 in. undersize for use with reconditioned crankshafts.*

1. Remove oil pan and oil pump, using procedures given above.
2. Remove connecting rod cap and bearing, and inspect bearing for wear or damage. Do not reuse bearing if either is in evidence.
3. Clean bearings and crankpin to remove all oil.
4. Using a micrometer, measure crankpin for out-of-round or taper. If not within specifications (see **Table 1**), replace or recondition crankshaft. If within tolerances and a new bearing is to be installed, measure maximum diameter of crankpin to determine required bearing size.

CAUTION

If a new bearing is fitted to the minimum diameter of an out-of-round crankpin, rapid bearing failure will result.

5. Install new or used bearing and check clearance with Plastigage or equivalent as follows:

   a. Place a piece of Plastigage on crankpin (parallel to crankshaft). Plastigage should cover the full width contacted by the bearing.
   b. Install bearing halves in connecting rod and bearing.
   c. Install connecting rod and cap on crankpin and torque nuts evenly to specification (see **Table 2**).

CAUTION

***Do not** turn crankshaft while Plastigage is installed. To do so will distort Plastigage which will result in incorrect measurement.*

   d. Remove cap and measure Plastigage at widest point, using scale on Plastigage package (see **Figure 74**).

6. If clearance is not within specified tolerances, select and install a different, correct size bearing and remeasure clearance.
7. Coat bearing surfaces with oil, install rod and cap on crankpin, and torque nuts to specification.
8. When all bearings have been installed, tap each connecting rod lightly to assure clearance, and measure side clearances (see **Table 1**) as shown in **Figure 13** (L6 engines) or **Figure 14** (V6 and V8 engines).
9. Replace oil pump and oil pan, using the procedures given above.

CONNECTING ROD AND PISTON ASSEMBLIES

Removal

1. Remove oil pan, oil pump, and cylinder head, using the procedures given earlier in this chapter.
2. If ridges or deposits are found in the upper end of the cylinder bore, remove them as follows:

   a. Turn crankshaft until piston is at bottom of stroke and place a cloth on top of piston to catch cuttings.
   b. Use a ridge reamer to remove ridge or deposit.
   c. Turn crankshaft until piston is at top of stroke and remove cloth and all cuttings.

3. Inspect rods and bearing caps for cylinder identification markings. If none are present, mark rods and caps so they can be replaced in proper cylinders.

4

4. Remove bearing caps and install lengths of 3/8 in. hose over the rod bolt threads to prevent damage to the threads and crankshaft journals. See **Figure 77**, typical. Push out rod and piston assembly with a piece of hardwood or a hammer handle.

NOTE
It will be necessary to rotate crankshaft slightly to remove some assemblies.

Disassembly

1. Remove piston rings with a ring expander tool.
2. Support rod tightly and try to rock piston along axis of piston pin. Any rocking motion (do not confuse with sliding) indicates wear in the piston pin, rod bushing, pin bore, or more likely, a combination of all three. Mark the assembly for further inspection.

NOTE
Disassembly and assembly of the piston, piston pin, and rod require an arbor press and special tools. These jobs are best performed by an automotive machine shop, as the use of makeshift tools could result in damage.

Inspection

1. Clean pistons thoroughly in solvent. Scrape carbon deposits from top and grooves. Take care to avoid damage to piston.

CAUTION
Do not use wire brush on piston skirts.

2. Examine grooves for burrs, dented edges, and side wear. Pay particular attention to top compression ring groove, as it usually wears more than the others.
3. Measure piston-to-cylinder clearance, using procedure given below.
4. If damage or wear indicates piston replacement, select a new piston as described in *Piston Clearance and Selection* procedure.
5. Measure all parts marked in Step 2 of the *Disassembly* procedure above with micrometer and dial bore gauge to determine which parts are worn (this can be done by an automotive machine shop if measuring equipment is not available). Replace piston and pin set as a unit if either or both are worn.

77

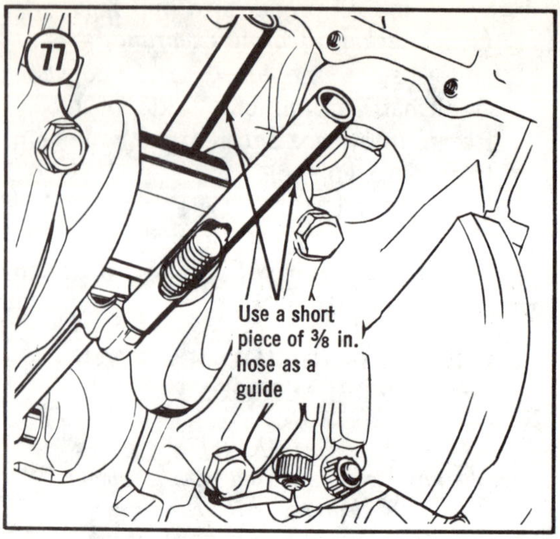

78

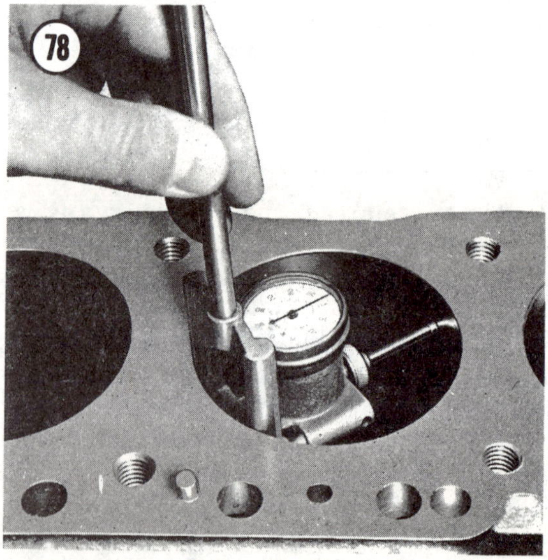

79

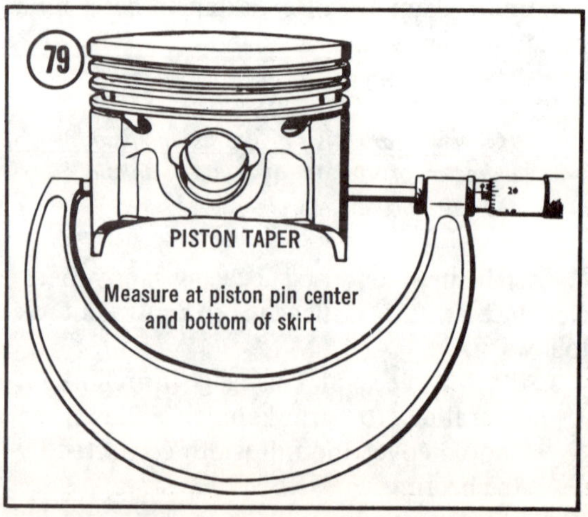

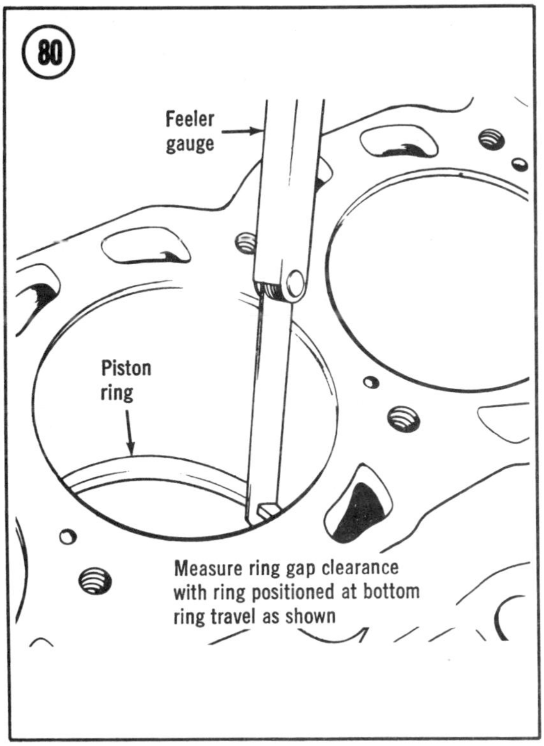

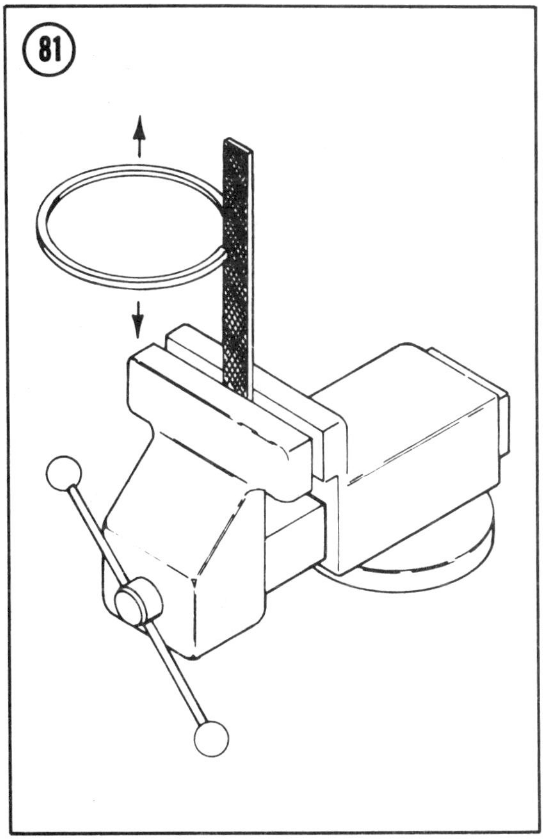

Piston Clearance Selection

1. Make sure cylinder walls are clean and dry.
2. Measure cylinder bore with a telescope bore gauge. Make measurement 2-1/2 in. from top of bore (see **Figure 78**).
3. Measure outside diameter of piston at bottom of skirt (see **Figure 79**).
4. If the difference between the 2 readings is the piston clearance. Compare difference with the specification in **Table 1**. If greater than specification, select another piston size.

Piston Ring Installation

NOTE

All compression rings have marks on their upper sides. Install these rings with marks toward top of piston. Oil control rings are made of 3 pieces, including 2 rails and a spacer.

1. Select new rings of proper size for piston being used.
2. Insert compression ring into bore of cylinder, about 1/4 in. above travel ring. Square ring with cylinder wall by tapping with a piston. Using a feeler gauge, measure the gap (see **Figure 80**) and compare to specifications (see **Table 1**). If gap is too large, fit another ring. If smaller than specified, hold a small file in a vise, grip ends of ring with fingers, and carefully enlarge gap as shown in **Figure 81**.
3. Fit each cylinder with 2 compression rings, using the instructions in Step 2.
4. Roll each ring around its piston groove as shown in **Figure 82** to check for binding.

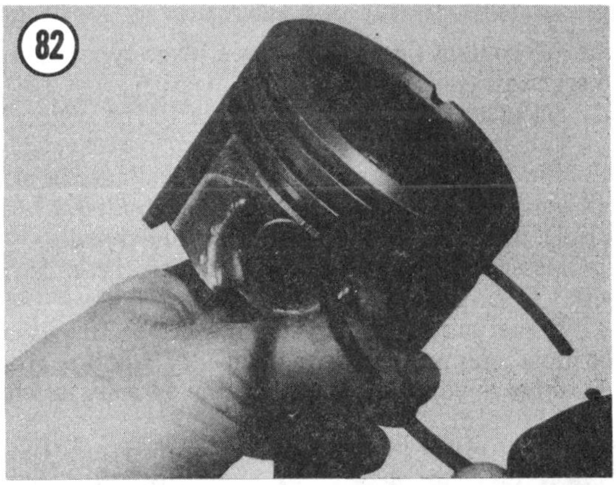

83

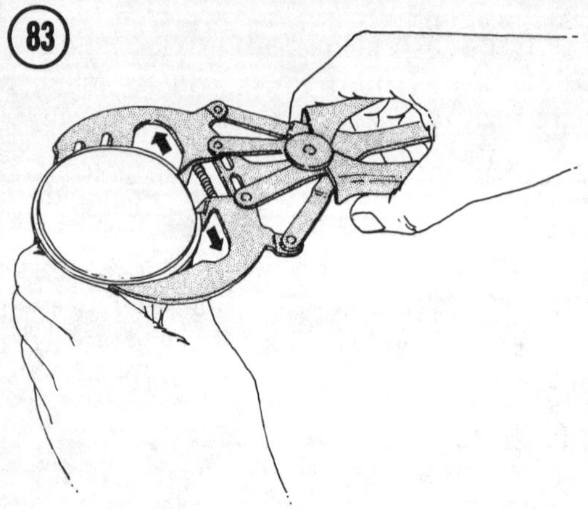

84

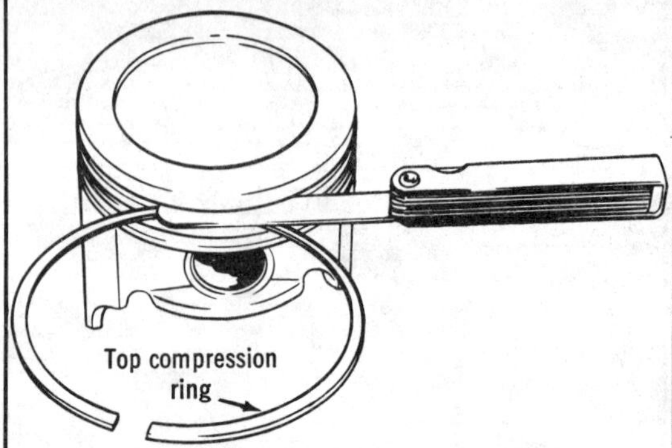

Minor binding may be cleaned up with a fine cut file, if care is used.

5. Using a ring expander tool, install oil ring spacer in groove and, except on inline engines, install anti-rotational tang in oil hole. Hold spacer ends butted and install lower steel rail with gap properly located. Then install upper steel rail with gap properly located (see **Figure 12** for gap locations).

6. Flex oil ring assembly in its groove to make sure ring is free and binding does not occur at any point. Minor binding can be cleaned up with a fine cut file.

7. Install 2 compression rings, marked side up, using ring expander tool (see **Figure 83**). Make certain the rings fitted to each cylinder are installed on the proper pistons.

8. Check side clearance of each ring as shown in **Figure 84**. Compare with specifications for your engine (see **Table 1**).

9. Verify that ring gaps are properly positioned as shown in **Figure 12**.

Piston Installation

NOTE

Used pistons must be installed in the cylinders from which they were removed, and new pistons must be installed in the cylinders to which they were fitted.

1. Coat cylinder walls, pistons and rings with light engine oil.

2. With the bearing caps removed, install short lengths of 3/8 in. hose over rod bolt

85

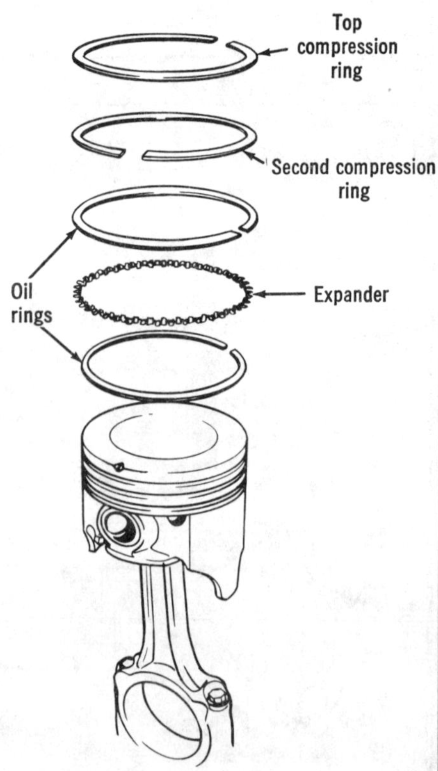

threads, and have an assistant help you to slowly guide rods into place.

3. Install each piston in its correct cylinder bore. In L6, V6, and Olds V8 engines, piston notch (**Figure 85**) must face front of engine. In Chevy V8, install the piston and rod assembly with rod bearing tang slots on opposite side from camshaft. If pistons are installed in cylinders other than the ones from which they were removed, they must have new rings and bearings installed on them.

86

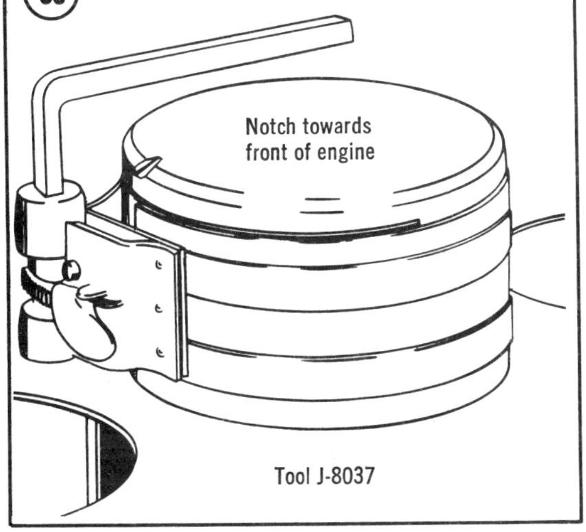

4. Compress rings with a ring compressor tool and press pistons into bores with a wooden hammer handle (**Figure 86**).

5. Guide bearing carefully onto journal. Install bearing caps and check clearances as described above. Torque nuts to specifications (**Table 2**).

CAMSHAFT

Camshaft removal and installation procedures are given in Steps 13 and 3 of the *General Overhaul* and *Reassembly* sequences, respectively.

Inspection

1. Measure camshaft bearing journals for out-of-round condition, using a micrometer. Replace camshaft if out-of-round exceeds 0.001 in.

NOTE

If measuring equipment is not available, take camshaft to an automotive machinist for the measurements described in this procedure.

2. Mount camshaft in V-blocks and, using a dial indicator (**Figure 87**), check bearing journals for alignment. Replace camshaft if out of alignment more than 0.0015 in.

87

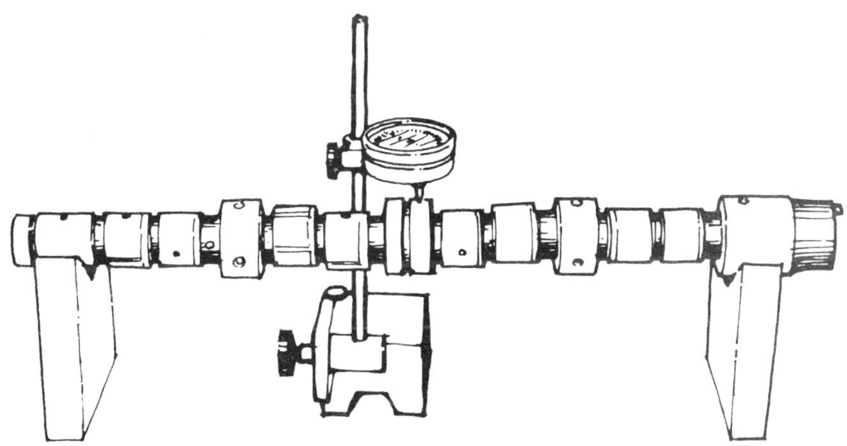

3. On L6 engines, check gear and thrust plate for wear and damage. Check end play (**Figure 88**). End play should be between 0.001 and 0.005 in.

Oil Nozzle Replacement (L6 Engines)

1. Pull out nozzle with pliers (**Figure 89**).
2. Drive new nozzle in with plastic hammer. Make sure oil hole is in vertical position.

Camshaft Gear Replacement (L6 Engines)

Replacement of camshaft gear requires an arbor press and special tools. This job should be referred to an automotive machine shop.

Camshaft Bearing Replacement

Bearing replacement requires special tools and adequate substitutes are not easily improvised. An automotive machine shop will replace the bearing for a small fee.

CRANKSHAFT

The engine must be removed from the vehicle in order to remove the crankshaft. The procedures for crankshaft removal; bearing removal, inspection, and replacement; and installation are given in Steps 15 and 1 of the *General Overhaul* and *Reassembly* sequences, respectively.

Crankshaft Inspection

1. Using a micrometer, check crankshaft journals and crankpins for out-of-round taper, or undersize. See **Table 1** for specifications.

NOTE

If measuring equipment is not available, take the crankshaft to an automotive machine shop to have the checks described in this procedure.

2. Mount front and rear main journals in V-blocks and check at intermediate main journals with a dial indicator for runout. See **Table 1** for specifications.
3. Replace crankshaft or have it reconditioned if out of specifications.

88

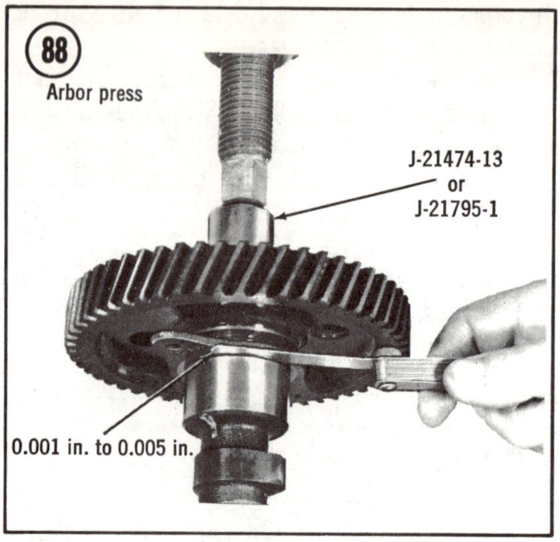

89

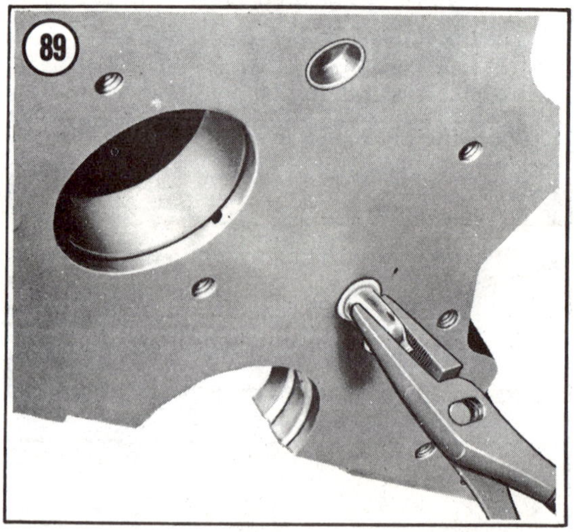

90

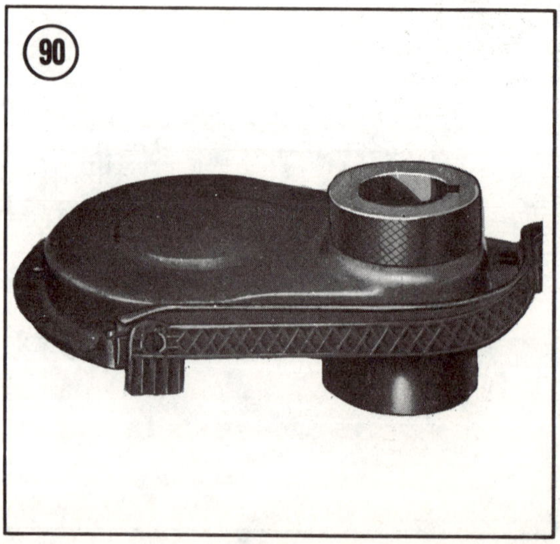

91

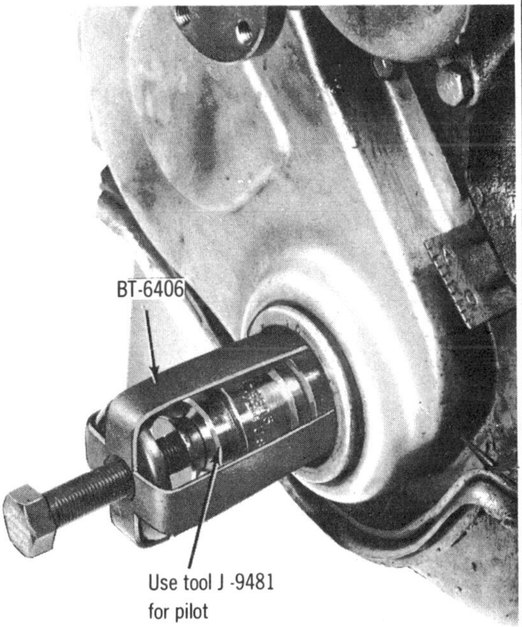

92

CYLINDER BLOCK

Cleaning and Inspection

1. Clean block with solvent and remove all foreign material from gasket surfaces.
2. Remove plugs from oil galleries and clean all oil passages.
3. Clean and inspect all water passages.
4. Check for cracks in cylinder walls, water jacket, valve lifter bores, and main bearing webs.
5. Check cylinder walls for out-of-round, taper, or excessive ridges, using a dial indicator (this job should be refered to an automotive machinist). If cylinders exceed specifications, honing or reboring will be required.

FRONT COVER OIL SEAL REPLACEMENT

L6 and Chevy V8 Engines

1. If the front cover has been removed from the engine, pry out the old seal with a screw driver or similar tool, taking care not to damage the cover. Position seal with open end facing inside of cover and drive into place with special tool (J-23042 or equivalent). See **Figure 90** (L6 shown, V8 similar).
2. If front cover has not been removed, remove torsional damper and install special tools (J-9481 and BT-6406 or equivalents) as shown in **Figure 91**. Remove the seal with the tools. Position seal with open end facing inside of cover and drive into place with a special tool (J-5154 or equivalent). See **Figure 92**.

V6 Engine

1. Remove the timing chain cover (see Step 6, *General Assembly Sequence* and use a punch or similar tool to drive out old seal and shedder.
2. Coil new packing material around opening, so that ends of the material are at the top. Use a suitable punch to drive in the shedder, then stake the shedder in at least three places.
3. Use a hammer handle or similar tool to pack in the material so the harmonic balancer hub can be inserted through the hole.

Olds V8 Engines

1. With the front cover installed on the engine, remove drive belts, crankshaft pulley, and pulley hub.
2. Use special tool (BT-6406 or equivalent) to remove old seal as shown in **Figure 93**.
3. Apply a sealer (1050026 or equivalent) to the outside diameter of the seal and position it on the front cover and use a special tool (BT-6405 or equivalent) to install the seal in the cover. See **Figure 94**. Tighten the tool until an 0.005 in. feeler gauge will just fit between the tool and the front cover.

4. Install the crankshaft hub and pulley. Install the drive belts and adjust their tension. See *Drive Belt Replacement*, Chapter Three.

93

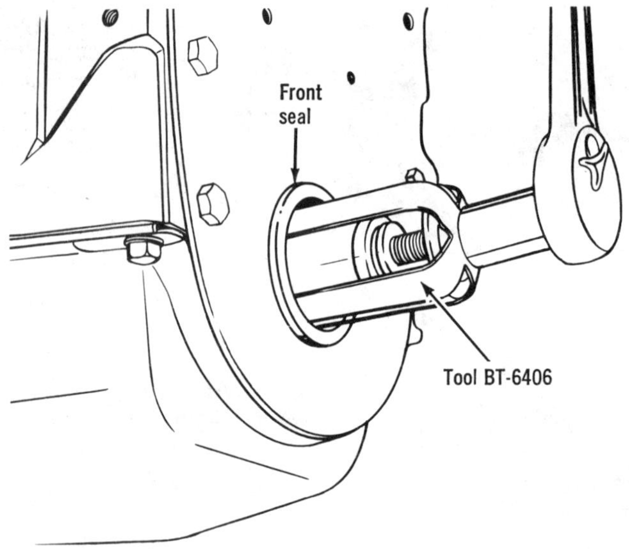

94

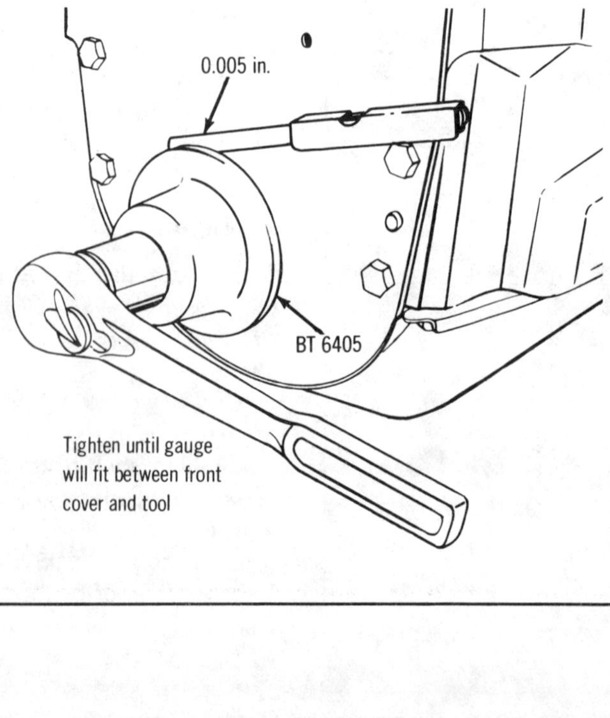

Table 1 ENGINE SPECIFICATIONS

	250 CID L6
General	
Bore	$3\frac{7}{8}$ in.
Stroke	3.53 in.
Firing order	1-5-3-6-2-4
Cylinder bore	
Diameter	3.8775 in.
Out-of-round, new (wear limit)	(0.005 in. maximum)
Piston	
Clearance in bore	0.0025 in.
Piston rings	
Number per cylinder	3
Ring end gap	
Top	0.010-0.020 in.
2nd	0.010-0.020 in.
Oil control	0.015-0.055 in.
Ring side clearance	
Top	0.0012-0.0027 in.
2nd	0.0012-0.0032 in.
Oil control	0.000-0.005 in.
Piston pin	
Diameter	0.9270-0.9273 in.
Clearance	
In piston	0.0015-0.00025 in.
In rod	0.0008-0.0016 in.
Crankshaft	
End play	0.002-0.006 in.
Main bearing journal	
Diameter (all)	2.2983-2.993 in.
Taper	0.001 in. maximum
Out-of-round	0.001 in. maximum
Main bearing clearance (all)	0.0003-0.0039 in.
Crankpin	
Diameter	1.999-2.000 in.
Taper, new (wear limit)	0.001 in. maximum
Out-of-round, new (wear limit)	0.001 in. maximum
Connecting rods	
Side clearance	0.009-0.014 in.
Bearing clearance	0.0035 in. maximum
Camshaft	
Journal diameter	1.8677-1.8697 in.
Runout	0.0015 in. maximum
Valve system	
Lifter type	Hydraulic
Rocker arm ratio	1.75:1
Valve lash (intake & exhaust)	One turn down from zero lash
Intake valve	
Face angle	45°
Seat angle	46°
Seat width	$\frac{1}{32}$-$\frac{1}{16}$ in.
Stem-to-guide clearance	0.0010-0.0027 in.
Seat runout	0.002 in. maximum
Exhaust valve	
Face angle	45°
Seat angle	46°
Seat width	$\frac{1}{15}$-$\frac{3}{32}$ in.
Stem-to-guide clearance	0.0015-0.0032 in.
Seat runout	0.002 in. maximum
Valve springs (outer)	
Free length	1.92 in. (to 1975) 1.94 in. (1976)
Load @ length (lbs. @ in.)	
Closed	55-64 @ 1.66 in. (to 1975) 78-86 @ 1.66 in. (1976)
Open	180-192 @ 1.27 in. (to 1975) 170-180 @ 1.26 in. (1976)
Installed height ($\pm \frac{1}{32}$ in.)	$1\frac{21}{32}$ in.

(continued)

Table 1 **ENGINE SPECIFICATIONS** (continued)

	231 CID V6
General	
Bore	3.800 in.
Stroke	3.400 in.
Firing order	1-6-5-4-3-2
Cylinder bore	
Diameter	3.800 in.
Out-of-round	0.0005 in. max.
Taper	0.0005 in. max.
Piston	
Clearance in bore (skirt top)	0.0008-0.0020 in.
Clearance in bore (skirt bottom)	0.0013-0.0035 in.
Piston rings	
Number per cylinder	3
Ring end gap	
Top	0.010-0.020 in.
2nd	0.010-0.020 in.
Oil control	0.015-0.035 in.
Ring side clearance	
Top	0.003-0.005 in.
2nd	0.003-0.005 in.
Oil control	0.0035 in. max.
Piston pins	
Diameter	0.9391-0.9394 in.
Clearance	
In piston	0.0004-0.0007 in.
In rod	0.0007-0.0017 in.
Crankshaft	
End play	0.004-0.008 in.
Main bearing journals	
Diameter (all)	2.4995 in.
Out-of-round	0.0015 in. maximum
Main bearing clearance (all)	0.0003-0.0018 in.
Crankpin	
Diameter	2.2495-2.2487 in.
Connecting rods	
Side clearance	0.006-0.027 in.
Bearing clearance	0.0005-0.0026 in.
Camshaft	
Journal diameter	1.785-1.786 in.
Runout	0.0015 in. maximum
Valve system	
Lifter type	Hydraulic
Rocker arm ratio	1.55:1
Valve lash (intake & exhaust)	Non-adjustable
Intake valve	
Face angle	45°
Seat angle	45°
Seat width	1/32-1/16 in.
Stem-to-guide clearance	0.0015-0.0032 in.
Exhaust valve	
Face angle	45°
Seat angle	45°
Seat width	1/16-3/32 in.
Stem-to-guide clearance	0.0015-0.0032 in.
Valve springs (outer)	
Load @ length (lb. @ in.)	
Closed	59-69 @ 1.727
Open	162-174 @ 1.3

(continued)

Table 1 **ENGINE SPECIFICATIONS** (continued)

	Chevy CID V8	
	305	**350**
General		
Bore	3.736 in.	4.000 in.
Stroke	3.48 in.	3.48 in.
Firing order	1-8-4-3-6-5-7-2	1-8-4-3-6-5-7-2
Cylinder bore		
Diameter	3.7350-3.7385 in.	3.999-4.0025 in.
Out-of-round, new (wear limit)	0.001 (0.002) in.	0.001 (0.002) in.
Piston		
Clearance in bore	0.0027 in. maximum	
Piston rings		
Number per cylinder	3	
Ring end gap		
Top	0.010-0.030 in.	
2nd	0.010-0.035 in.	
Oil control	0.015-0.065 in.	
Ring side clearance		
Top	0.0012-0.0042 in.	
2nd	0.0012-0.0042 in.	
Oil control	0.002-0.008 in.	
Piston pins		
Diameter	0.9270-0.9273 in.	
Clearance		
In piston	0.00015-0.00025 in.	
In rod	0.0008-0.0016 in. (press fit)	
Crankshaft		
End play	0.002-0.006 in.	
Main bearing journal		
Diameter	No. 1 — 2.4484-2.4493 in. No. 2-3-4 — 2.4481-2.4490 in. No. 5 — 2.4479-2.4488 in.	
Taper	0.0002 in. (0.001 in. maximum)	
Out-of-round	0.0002 in. (0.001 in. maximum)	
Main bearing clearance	No. 1 — 0.0008-0.0015 in. No. 2-3-4 — 0.0011-0.0023 in. No. 5 — 0.0025-0.0035 in.	
Crankpin		
Diameter	2.099-2.100 in.	
Taper, new (wear limit)	0.0002 in. (0.001 in. maximum)	
Out-of-round, new (wear limit)	0.0002 in. (0.001 in. maximum)	
Connecting rods		
Side clearance	0.006-0.014 in.	
Bearing clearance	0.0013-0.0030 in.	
Camshaft		
Journal diameter	1.8682-1.8692 in.	
Runout	0.0015 in. maximum	
Valve system		
Lifter type	Hydraulic	
Rocker arm ratio	1.50:1	
Valve lash — intake and exhaust	$\frac{3}{4}$ turn down from zero lash	
Intake valve		
Face angle	45°	
Seat angle	46°	
Seat width	$\frac{1}{32}$-$\frac{1}{16}$ in.	
Stem-to-guide clearance	0.0010-0.0037 in.	
Seat runout	0.002 in.	
Exhaust valve		
Face angle	45°	
Seat angle	46°	
Seat width	$\frac{1}{16}$-$\frac{3}{32}$ in.	
Stem-to-guide clearance	0.0010-0.0037 in.	
Seat runout	0.002 in.	
Valve springs (outer)		
Free length	2.03 in.	
Load @ length (lbs. @ in.)		
Closed	76-84 @ 1.61 in.	
Open	194-206 @ 1.16 in.	
Installed height	$1\frac{19}{32}$ in. exhaust, $1\frac{23}{32}$ intake	
Damper free length	1.86 in.	

(continued)

Table 1 ENGINE SPECIFICATIONS (continued)

	Olds 260 CID V8 (4.3 Liter)
General	
Bore	3.500 in.
Stroke	3.385 in.
Firing order	1-8-4-3-6-5-7-2
Cylinder bore	
Diameter	3.500 in.
Out-of-round	0.001 in.
Piston	
Clearance in bore	0.00075-0.00175 in.
Piston rings	
Number per cylinder	3
Ring end gap	
Top and 2nd	0.010-0.020 in.
Oil control	0.015-0.055 in.
Ring side clearance	
Top and 2nd	0.0020-0.0040 in.
Piston pins	
Diameter	0.9803-0.9807 in.
Clearance in piston	0.0003-0.0005 in.
Clearance in rod	0.0008-0.0018 in. (press fit)
Crankshaft	
End play	0.0035-0.0135 in.
Main bearing journals	
Diameter, No. 1	2.4998-2.4988 in.
Diameter, No. 2, 3, 4, & 5	2.4995-2.9985 in.
Out-of-round	0.0015 in.
Main bearing clearance	
No. 1, 2, 3, & 4	0.0005-0.0021 in.
No. 5	0.0015-0.0031 in.
Crankpin diameter	2.1248-2.1238 in.
Out-of-round	0.0015 in.
Connecting rods	
Side clearance	0.006-0.020 in.
Bearing clearance	0.0004-0.0033 in.
Camshaft	
Journal diameter	
No. 1	2.0365-2.0357 in.
No. 2	2.0165-2.0157 in.
No. 3	1.9965-1.9957 in.
No. 4	1.9765-1.9757 in.
No. 5	1.9565-1.9557 in.
Runout	0.011-0.077 in.
Valve system	
Lifter type	Hydraulic
Intake valve	
Face angle	44°
Seat angle	45°
Seat width	0.037-0.075 in.
Stem-to-guide clearance	0.0010-0.0027 in.
Exhaust valve	
Face angle	30°
Seat angle	31°
Seat width	0.050-0.090 in.
Stem-to-guide clearance	0.0015-0.0032 in.
Valve springs	
Free length	1.96 in.
Load @ length (lb. @ in.)	
Closed	76-84 @ 1.670
Open	180-194 @ 1.270

(continued)

Table 1 ENGINE SPECIFICATIONS (continued)

	Olds 350 CID V8 (5.7 Liter)
General	
Bore	4.05 in.
Stroke	3.385 in.
Firing order	1-8-4-3-6-5-7-2
Cylinder bore	
Diameter	4.057 in.
Out-of-round	0.001 in.
Piston	
Clearance in bore	0.00075-0.00175 in.
Piston rings	
Number per cylinder	3
Ring end gap	
Top and 2nd	0.010-0.020 in.
Oil control	0.015-0.055 in.
Ring side clearance	
Top	0.0018-0.0033 in.
2nd	0.0018-0.0038 in.
Oil control	0.0021-0.0031 in.
Piston pins	
Diameter	0.9803-0.9807 in.
Clearance in piston	0.0003-0.0005 in.
Clearance in rod	0.0008-0.0018 in.
Crankshaft	
End play	0.0035-0.0135 in.
Main bearing journals	
Diameter, No. 1	2.4998-2.4988 in.
Diameter, No. 2, 3, 4, & 5	2.4995-2.4985 in.
Out-of-round	0.0015 in.
Main bearing clearance	
No. 1, 2, 3, & 4	0.0005-0.0021 in.
No. 5	0.0015-0.0031 in.
Crankpin diameter	2.1248-2.1238 in.
Out-of-round	0.0015 in.
Connecting rods	
Side clearance	0.006-0.020 in.
Bearing clearance	0.0004-0.0033 in.
Camshaft	
Journal diameter	
No. 1	2.0365-0.0357 in.
No. 2	2.0165-2.0157 in.
No. 3	1.9965-1.9957 in.
No. 4	1.9765-1.9757 in.
No. 5	1.9565-1.9557 in.
Runout	0.011-0.077 in.
Valve system	
Lifter type	Hydraulic
Intake valve	
Face angle	45° (1970-1972); 44° (1973-on)
Seat angle	46° (1970-1972); 45° (1973-on)
Seat width	0.037-0.075 in.
Stem-to-guide clearance	0.0010-0.0027 in.
Exhaust valve	
Face angle	45° (1970-1972); 30° (1973-on)
Seat angle	46° (1970-1972); 31° (1973-on)
Seat width	0.037-0.075 in. (1970-1972); 0.050-0.090 in. (1973-on)
Stem-to-guide clearance	0.0015-0.0032 in.
Valve springs	
Free length	1.96 in.
Load @ length (lb. @ in.)	
Closed	76-84 @ 1.670
Open	180-194 @ 1.270

(continued)

4

Table 1 ENGINE SPECIFICATIONS (continued)

	Olds 403 CID V8 (6.6 Liter)
General	
Bore	4.351 in.
Stroke	3.385 in.
Firing order	1-8-4-3-6-5-7-2
Cylinder bore	
Diameter	4.351 in.
Out-of-round	0.001 in.
Piston clearance in bore	0.001-0.002 in.
Piston rings	
Number per cylinder	3
End gap	
Top and 2nd	0.010-0.023 in.
Oil control	0.015-0.055 in.
Ring side clearance	
Top and 2nd	0.0020-0.0040 in.
Oil control	0.0021-0.0031 in.
Piston pins	
Diameter	0.9803-0.9807 in.
Clearance in piston	0.0003-0.0005 in.
Clearance in rod	0.0008-0.0018 in. (press fit)
Crankshaft	
End play	0.0035-0.0135 in.
Main bearing journals	
Diameter, No. 1	2.4998-2.4988 in.
Diameter, No. 2, 3, 4, & 5	2.4995-2.4985 in.
Out-of-round	0.0015 in.
Main bearing clearance	
No. 1, 2, 3, & 4	0.0005-0.0021 in.
No. 5	0.0015-0.0031 in.
Crankpin	
Diameter	2.1248-2.1238 in.
Out-of-round	0.0015 in.
Connecting rods	
Side clearance	0.006-0.020 in.
Bearing clearance	0.0004-0.0033 in.
Camshaft	
Journal diameter	
No. 1	2.0365-2.0357 in.
No. 2	2.0165-2.0157 in.
No. 3	1.9965-1.9957 in.
No. 4	1.9765-1.9757 in.
No. 5	1.9565-1.9557 in.
Runout	0.011-0.077 in.
Valve system	
Lifter type	Hydraulic
Intake valve	
Face angle	44°
Seat angle	45°
Seat width	0.037-0.075 in.
Stem-to-guide clearance	0.0010-0.0027 in.
Exhaust valve	
Face angle	30°
Seat angle	31°
Seat width	0.050-0.090 in.
Stem-to-guide clearance	0.0015-0.0032 in.
Valve springs	
Free length	1.96 in.
Load @ length (lb. @ in.)	
Closed	76-84 @ 1.670
Open	180-194 @ 1.270

(continued)

Table 1 ENGINE SPECIFICATIONS (continued)

	Olds 455 CID V8
General	
Bore	4.126 in.
Stroke	4.250 in.
Firing order	1-8-4-3-6-5-7-2
Cylinder bore	
Diameter	4.126 in.
Out-of-round	0.001 in.
Piston clearance in bore	0.001-0.002 in.
Piston rings	
Number per cylinder	3
Ring end gap	
Top and 2nd	0.010-0.023 in.
Oil control	0.015-0.055 in.
Ring side clearance	
Top and 2nd	0.0020-0.0040 in.
Oil control	0.0021-0.0031 in.
Piston pins	
Diameter	0.9803-0.9807 in.
Clearance in piston	0.0003-0.0005 in.
Clearance in rod	0.0008-0.0018 in.
Crankshaft	
End play	0.0035-0.0135 in.
Main bearing journal diameters	3.0003-0.2993 in.
Out-of-round	0.0015 in.
Main bearing clearance	
No. 1, 2, 3, & 4	0.0005-0.0021 in.
No. 5	0.0020-0.0034 in.
Crankpin	
Diameter	2.4998-2.4988 in.
Out-of-round	0.0015 in.
Connecting rods	
Side clearance	0.006-0.020 in.
Bearing clearance	0.0004-0.0033 in.
Camshaft	
Journal diameter	
No. 1	2.0365-2.0357 in.
No. 2	2.0165-2.0157 in.
No. 3	1.9965-1.9957 in.
No. 4	1.9765-1.9757 in.
No. 5	1.9565-1.9557 in.
Runout	0.011-0.077 in.
Valve system	
Lifter type	Hydraulic
Intake valve	
Face angle	30° (1970); 45° (1971-on)
Seat angle	30° (1970); 46° (1971-on)
Seat width	0.037-0.075 in.
Stem-to-guide clearance	0.0010-0.0027 in.
Exhaust valve	
Face angle	45° (1970-1972); 30° (1973-on)
Seat angle	46° (1970-1972); 31° (1973-on)
Seat width	0.050-0.090 in.
Stem-to-guide clearance	0.0015-0.0032 in.
Valve springs	
Free length	1.96 in.
Load @ length (lb. @ in.)	
Closed	76-84 @ 1.670
Open	180-194 @ 1.270

(continued)

4

Table 1 ENGINE SPECIFICATIONS (continued)

	4.3 and 5.7 Liter Diesel V8's	
	4.3 Liter	**5.7 Liter**
General		
Bore	3.500 in.	4.057 in.
Stroke	3.385 in.	3.385 in.
Firing order	1-8-4-3-6-5-7-2	1-8-4-3-6-5-7-2
Cylinder bore		
Diameter	3.500 in.	3.500 in.
Out-of-round	0.001 in.	0.001 in.
Piston		
Clearance in bore	0.005-0.006 in.	0.005-0.006 in.
Piston rings		
Number per cylinder	3	3
Ring end gap		
Top	0.012-0.022 in.	0.015-0.025 in.
2nd	0.010-0.020 in.	0.015-0.025 in.
Oil control	0.015-0.055 in.	0.015-0.055 in.
Ring side clearance		
Top	0.004-0.006 in.	0 005-0.007 in.
2nd	0.0018-0.0038 in.	0.0018-0.0038 in.
Oil control	0.001-0.005 in.	0.001-0.005 in.
Piston pins		
Diameter	1.0949-1.0953 in.	1.0949-1.0953 in.
Clearance		
In piston	0.0003-0.0005 in.	0.0003-0.0005 in.
In rod	0.0003-0.0013 in.	0.0003-0.0013 in.
Crankshaft		
End play	0.0035-0.0135 in.	0.0035-0.0135 in.
Main bearing journals		
Diameter	3.0003-2.9993 in.	3.0003-2.9993 in.
Out-of-round	0.0015 in.	0.0015 in.
Main bearing clearance		
No. 1, 2, 3, & 4	0.0005-0.0021 in.	0.0005-0.0021 in.
No. 5	0.0015-0.0031 in.	0.0015-0.0031 in.
Crankpin		
Diameter	2.1248-2.1238 in.	2.1248-2.1238 in.
Out-of-round	0.0015 in.	0.0015 in.
Connecting rods		
Side clearance	0.006-0.020 in.	0.006-0.020 in.
Bearing clearance	0.0005-0.0026 in.	0.0005-0.0026 in.
Camshaft		
Journal diameter	1. 2.0365-2.0357	1. 2.0365-2.0357
	2. 2.0165-2.0157	2. 2.0165-2.0157
	3. 1.9965-1.9957	3. 1.9965-1.9957
	4. 1.9765-1.9757	4. 1.9765-1.9757
	5. 1.0565-1.9557	5. 1.0565-1.9557
Runout	0.011-0.077in.	0.011-0.077in.
Valve system		
Lifter type	Hydraulic	Hydraulic
Intake valve		
Face angle	44°	44°
Seat angle	45°	45°
Seat width	0.075-0.098 in.	0.075-0.098 in.
Stem-to-guide clearance	0.0010-0.0027 in.	0.0010-0.0027 in.
Exhaust valve		
Face angle	30°	30°
Seat angle	31°	31°
Seat width	0.037-0.075 in.	0.037-0.075 in.
Stem-to-guide clearance	0.0015-0.0032 in.	0.0015-0.0032 in.
Valve springs (outer)		
Free length	2.09 in.	2.09 in.
Load @ length (lb. @ in.)		
Closed	77-83 @ 1.670	77-83 @ 1.670
Open	144-158 @ 1.300	144-158 @ 1.300

Table 2 TIGHTENING TORQUES

Usage	L6	Chevy V8
Camshaft thrust plate	80 in.-lb.	—
Crankcase front cover	80 in.-lb.	80 in.-lb.
Flywheel housing cover	80 in.-lb.	80 in.-lb.
Oil filter bypass valve	—	80 in.-lb.
Oil pan (to crankcase) (¼-20)	80 in.-lb.	80 in.-lb.
Oil pan (to front cover) 1970-1973	55 in.-lb.	—
Oil pan (to front cover) 1974-on	50 in.-lb.	—
Oil pump cover	70 in.-lb.	80 in.-lb.
Rocker arm cover 1970	55 in.-lb.	—
Rocker arm cover 1971-on	80 in.-lb.	45 in.-lb.
Connecting rod cap (⅜-24)	—	45 ft.-lb.
Connecting rod cap (11/32-24)	35 ft.-lb.	—
Camshaft sprocket	20 ft.-lb.	20 ft.-lb.
Clutch pressure plate	20 ft.-lb.	—
Oil pan (to crankcase) 1970	125 in.-lb.	—
Oil pan (to crankcase) 1971-1973	75 in.-lb.	—
Oil pan (to crankcase) 1974-on (5/16-18)	75 in.-lb.	20 ft.-lb.
Oil pump	115 in.-lb.	—
Pushrod cover 1970	80 in.-lb.	—
Pushrod cover 1971-on	50 in.-lb.	—
Water pump	15 ft.-lb.	—
Clutch pressure plate	35 ft.-lb.	30 ft.-lb.
Distributor clamp	20 ft.-lb.	20 ft.-lb.
Flywheel housing	30 ft.-lb.	30 ft.-lb.
Fuel pump	22 ft.-lb.	—
Manifold (exhaust to inlet) 1970-1972	25 ft.-lb.	—
Manifold (exhaust to inlet) 1973-on	30 ft.-lb.	20 ft.-lb.
Manifold (inlet)	—	30 ft.-lb.
Manifold-to-head (outer) 1970-1971	20 ft.-lb.	—
Manifold-to-head (all others) 1970-1971	30 ft.-lb.	—
Manifold-to-head 1972-on	35 ft.-lb.	—
Thermostat housing	30 ft.-lb.	—
Water outlet 1970	20 ft.-lb.	—
Water outlet 1971-on	20 ft.-lb.	30 ft.-lb.
Water pump	—	30 ft.-lb.
Connecting rod cap	—	45 ft.-lb.
Cylinder head	—	65 ft.-lb.
Main bearing cap 1970	65 ft.-lb.	—
Main bearing cap 1971-1975	65 ft.-lb.	—
Main bearing cap 1976-on	65 ft.-lb.	70 ft.-lb.
Oil pump	65 ft.-lb.	65 ft.-lb.
Flywheel	60 ft.-lb.	60 ft.-lb.
Torsional damper	60 ft.-lb.	60 ft.-lb.
Cylinder head	95 ft.-lb.	—
Temperature sending unit	20 ft.-lb.	20 ft.-lb.
Oil filter	Hand-tight	25 ft.-lb.
Oil pan drain plug	20 ft.-lb.	20 ft.-lb.
Spark plug 1970	25 ft.-lb.	25 ft.-lb.
Spark plug 1971-1978	15 ft.-lb.	15 ft.-lb.
Spark plug 1979-on	15 ft.-lb.	22 ft.-lb.

(continued)

4

Table 2 **TIGHTENING TORQUES** (continued)

	Diesel	Olds V8	V6
Crankshaft bearing cap bolts (all, 455 cid)	—	120 ft.-lb.	—
Crankshaft bearing cap bolts (others, No. 1-2-3-4)	120 ft.-lb.	80 ft.-lb.	100 ft.-lb.
Crankshaft bearing cap bolts (others, No. 5)	120 ft.-lb.	120 ft.-lb.	—
EGR valve to intake manifold	—	25 ft.-lb.	—
Flywheel-to-crankshaft (automatic transmission)	40 ft.-lb.	60 ft.-lb.	60 ft.-lb.
Flywheel-to-crankshaft (manual transmission)	60 ft.-lb.	90 ft.-lb.	60 ft.-lb.
Oil pump-to-main bearing cap bolts	35 ft.-lb.	35 ft.-lb.	—
Oil pump cover bolts	8 ft.-lb.	8 ft.-lb.	10 ft.-lb.
Rocker arm pivot bolt to head	28 ft.-lb.	28 ft.-lb.	—
Rocker arm shaft-to-head bolt	—	—	30 ft.-lb.
Valve cover bolts	—	7 ft.-lb.	4 ft.-lb.
Oil pan bolts	10 ft.-lb.	10 ft.-lb.	14 ft.-lb.
Oil pan drain plug	30 ft.-lb.	30 ft.-lb.	30 ft.-lb.
Oil filter assembly-to-block	35 ft.-lb.	35 ft.-lb.	20 ft.-lb.
Harmonic balancer-to-crankshaft bolt	200-310 ft.-lb.	200-310 ft.-lb.	225 ft.-lb.
Fan driving pulley-to-balancer bolts	20 ft.-lb.	20 ft.-lb.	20 ft.-lb.
Fan driven pulley-to-hub bolts	20 ft.-lb.	20 ft.-lb.	20 ft.-lb.
Front cover-to-cylinder block bolts	35 ft.-lb.	35 ft.-lb.	30 ft.-lb.
Water pump-to-front cover bolts	13 ft.-lb.	13 ft.-lb.	7 ft.-lb.
Water outlet/thermostat housing attaching bolts	20 ft.-lb.	20 ft.-lb.	13 ft.-lb.
Intake manifold-to-cylinder head bolts	40 ft.-lb.	40 ft.-lb.	45 ft.-lb.
Exhaust manifold-to-cylinder head bolts	25 ft.-lb.	25 ft.-lb.	25 ft.-lb.
Carburetor to intake manifold bolts	—	10 ft.-lb.	15 ft.-lb.
Choke tube and plate-to intake manifold bolts	—	15 ft.-lb.	—
Air cleaner-to-carburetor stud	—	5 ft.-lb.	—
Fuel pump-to-block bolt and/or nut	—	25 ft.-lb.	20 ft.-lb.
Fuel pump eccentric-to-camshaft	—	65 ft.-lb.	22 ft.-lb.
Engine mount-to-cylinder block bolts	75 ft.-lb.	75 ft.-lb.	55 ft.-lb.
Engine mount-to-frame bolts	50 ft.-lb.	50 ft.-lb.	—
Starter-to-cylinder block bolts	35 ft.-lb.	35 ft.-lb.	35 ft.-lb.
Starter brace-to-cylinder block bolts	25 ft.-lb.	25 ft.-lb.	—
Starter brace-to-starter bolt	15 ft.-lb.	15 ft.-lb.	—
Starter brace-to-starter stud	—	8 ft.-lb.	—
Distributor clamp hold-down	—	17 ft.-lb.	13 ft.-lb.
Spark plugs (1970-1974)	—	35 ft.-lb.	—
Spark plugs (1975-on)	—	25 ft.-lb.	15 ft.-lb.
Cylinder head bolts (1970-1971)	—	80 ft.-lb.	—
Cylinder head bolts (1972-1976)	—	85 ft.-lb.	—
Cylinder head bolts (1977-on 4.3 liter)	—	85 ft.-lb.	—
Cylinder head bolts (1977-on exc. 4.3 liter)	130 ft.-lb.	130 ft.-lb.	80 ft.-lb.
Connecting rod nuts	42 ft.-lb.	42 ft.-lb.	40 ft.-lb.
Oil pump pressure regulator retainer	—	—	35 ft.-lb.
Oil screen housing-to-cylinder block	—	—	8 ft.-lb.
Oil gallery plugs	—	—	25 ft.-lb.

NOTE: If you own a 1981 and later model, first check the Supplement at the back of the book for any new service information.

CHAPTER FIVE

FUEL AND EXHAUST SYSTEMS

5

FUEL SYSTEM

The fuel system basically consists of a fuel tank, a mechanical fuel pump, and a carburetor, all connected by fuel lines.

A large number of carburetors are used on these cars. These are designed to meet the requirements of the particular engine/ transmission combinations. Therefore, carburetors that look alike are not always interchangeable. To obtain correct repair parts for your carburetor, copy the carburetor part number stamped on the carburetor body (or on a metal tag) and take it with you to your dealer.

This section covers removal, installation, and repair and/or replacement of carburetors, fuel pumps, and fuel tanks. Carburetor specifications (**Tables 1-4**) are at the end of the chapter.

ROCHESTER MV CARBURETOR

The Rochester MV carburetor was used on all inline 6 (L6) engines covered by this book. Specifications are given in **Table 1**.

Removal

NOTE

Acceleration stumbling and other performance problems are sometimes caused by foreign matter in the carburetor. When removing carburetor, take care to avoid draining fuel from bowl so that contents of bowl can be examined for contamination when the carburetor is disassembled. Also, check fuel filter.

Refer to **Figure 1**.

1. Remove air cleaner and gasket.
2. Disconnect fuel and vacuum lines from carburetor.

1

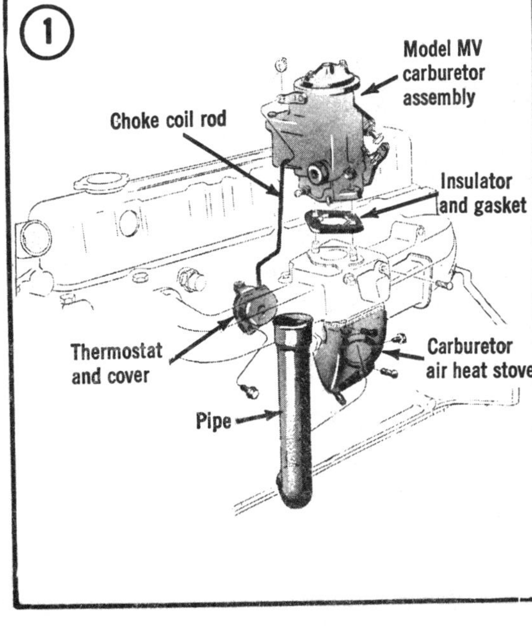

3. Disconnect choke coil rod, throttle linkage, and idle stop solenoid electrical lead (at connector).
4. Remove carburetor attaching nuts and then remove carburetor and solenoid assembly.
5. Remove insulator, air cleaner bracket, and gasket from intake manifold flange.

Installation

NOTE
For ease of starting, fill carburetor bowl before installing carburetor. Operate throttle lever several times and verify fuel discharge from pump jets prior to installation.

1. Verify that sealing flanges on carburetor and intake manifold are clean.
2. Install gasket, air cleaner bracket, and insulator over manifold studs.
3. Install carburetor over manifold studs and install and securely tighten attaching nuts.
4. Install and tighten vacuum and fuel lines at carburetor.
5. Connect accelerator linkage to carburetor.
6. Connect choke coil rod and idle stop solenoid electrical connector.

Disassembly

Air Horn

1. Note position of choke rod carefully and remove choke rod. On early models this is done by removing retaining screw from upper choke lever and removing lever from shaft and rod from lever. On later models, remove fast idle cam and remove choke rod from cam.
2. On 1975-1976 models, remove auxiliary vacuum break diaphragm assembly from air

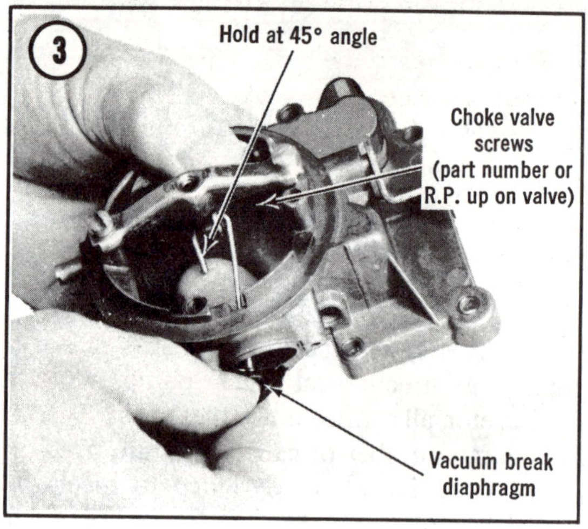

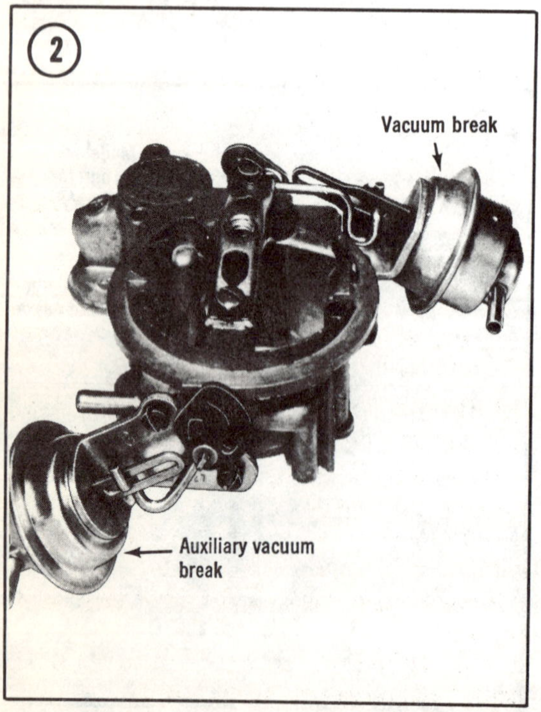

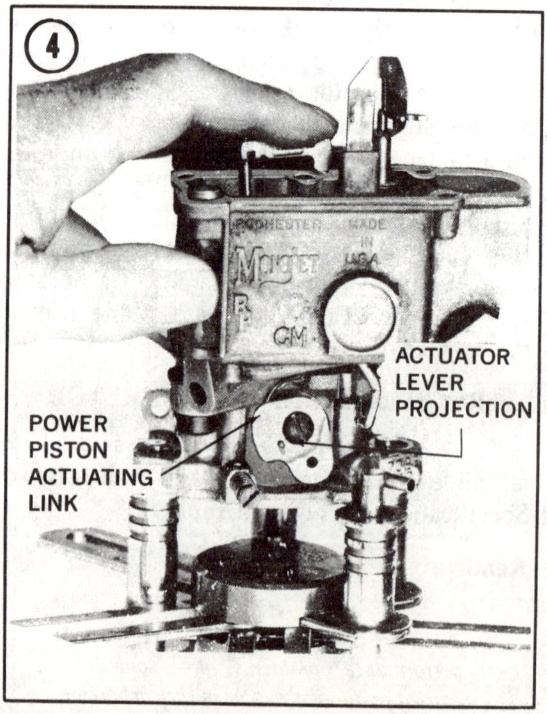

horn by removing 2 attaching screws (**Figure 2**). Remove diaphragm plunger from link which is permanently attached to choke lever. Lever and link can be removed as a unit if desired. See *Disassembly* procedure below.

3. Remove air horn-to-blow attaching screws (3 short and 3 long screws) and lockwashers.

4. On 1975-1976 models, remove primary vacuum break diaphragm unit from air horn. Remove vacuum break hose assembly and link from slot in diaphragm plunger stem.

5. Remove air horn by lifting straight up. Invert air horn and place on a clean workbench.

6. On 1971 and earlier models, remove vacuum break diaphragm cover (2 screws), hold choke wide open, and push upward on eyelet on choke valve until looped end of rod slides off wire lever attached to choke valve. Then remove diaphragm and plunger rod (**Figure 3**).

5

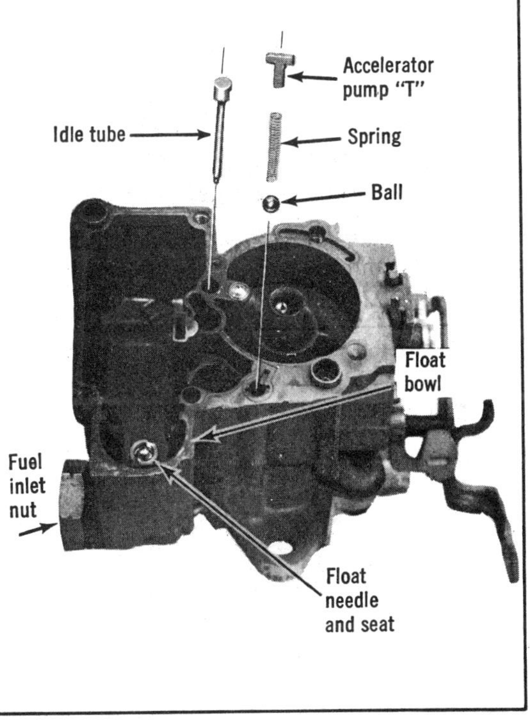

7. If desired, choke valve, vacuum break lever (if so equipped), and choke shaft can be removed by first removing thermostatic coil lever from end of choke shaft (one screw). Then remove choke valve (2 screws) and choke shaft from air horn.

NOTE

No further disassembly of air horn is required.

Float Bowl

1. Remove air horn-to-float bowl gasket.
2. Lift up on float hinge pin and remove float from bowl. Remove hinge pin from float arm.
3. Remove float needle. On 1974 and earlier models, remove float needle seat and gasket, taking care to avoid damage to needle seat.
4. Remove fuel filter element, spring, and gasket by removing fuel inlet fitting.
5. Using long-nosed pliers, remove "T" guide and pump discharge spring. Invert bowl and remove spring and ball.
6. Remove accelerator pump plunger and metering rod power piston drive assemblies as follows:

   a. Remove actuating lever from throttle shaft (one screw). See **Figure 4** for a typical example.
   b. Remove upper end of drive link from lower end of power piston rod.
   c. Rotate drive link from keyhole in actuating lever.
   d. Remove actuator lever from lower end of pump link in same manner.
   e. Push down on accelerator pump and remove actuator link by rotating until tang is aligned with slot in lever, then remove link.

7. Lift out power piston metering rod assembly and drive rod. Then remove spring from power piston cavity.
8. Remove pump plunger assembly from float bowl. Remove pump return spring from pump well.
9. Remove main metering jet from bottom of fuel bowl.
10. On 1975-1976 models, remove float needle seat and gasket. See **Figure 5**.

5

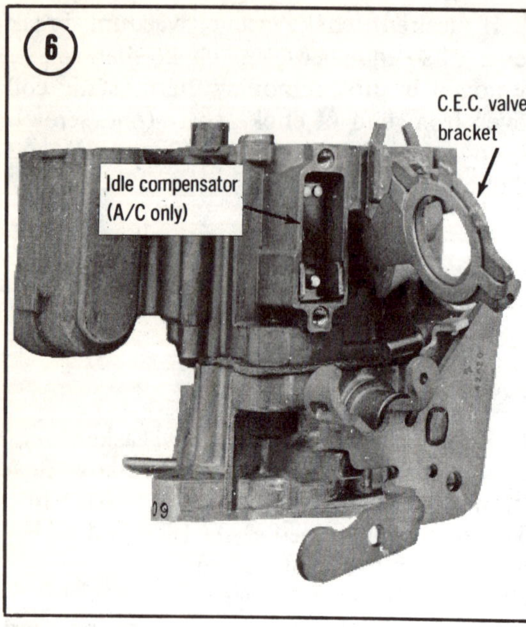

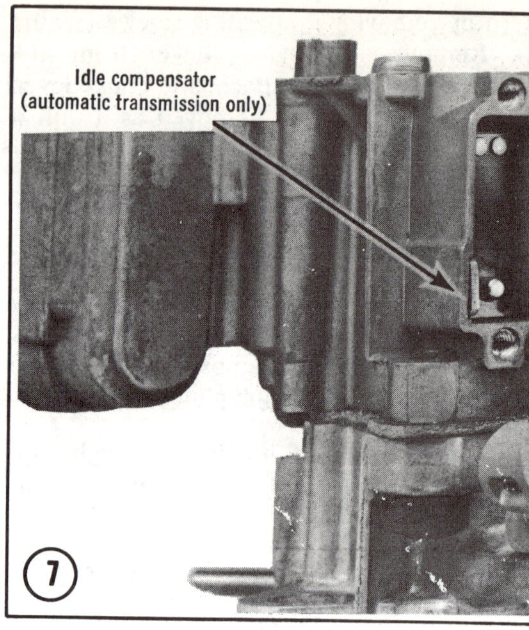

11. On 1974 and earlier models (with air conditioning) remove the screws from the throttle return spring bracket and remove bracket, hot idle compensator, and gasket from recess in bowl. See **Figure 6**.
12. On 1975-1976 models, remove 2 screws from idle compensator cover (automatic transmission only) and remove cover, compensator, and seal. See **Figure 7**.

NOTE
If desired, idle adjustment screw and fast idle cam (idle stop solenoid on 1975-1976 models) can be removed at this time. No further disassembly of float bowl is required.

Throttle Body

1. With bowl inverted, remove attaching screws and remove throttle body and gasket from bowl.

2. No further disassembly is required unless idle mixture needle is damaged or idle channels need cleaning. If necessary, cut tang from plastic limiter cap to remove needle. Do not replace limiter cap when reinstalling needle.

NOTE
Because of the close tolerances involved, do not remove throttle valve or shaft.

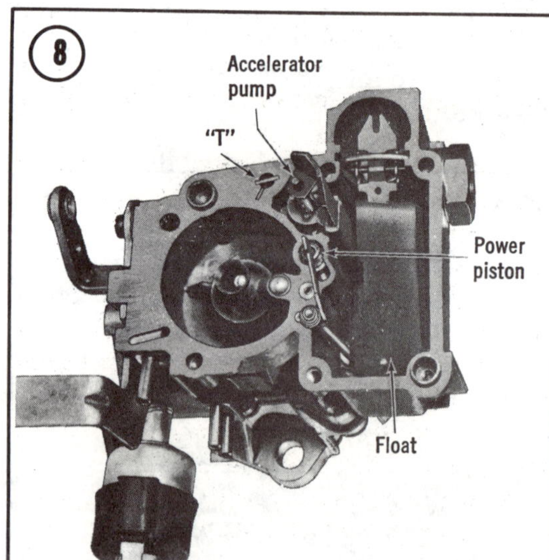

Cleaning and Inspection

1. Clean all metal parts and castings in carburetor solvent.

CAUTION
Do not immerse any rubber or plastic parts, or electrical parts (solenoids, etc.) in carburetor cleaners.

2. When clean, blow out all passages with compressed air. Do not clean jets or passages with drills or wires.

3. Inspect idle mixture needle for damage or wear. Replace if present.

4. Examine float needle and seat for wear. Replace with a new factory matched set if worn.

5. Inspect all sealing surfaces for damage.

6. Examine rods and holes in levers for excessive wear. Replace if worn.

7. Check fast idle cam for wear or damage.

8. Check levers and valve for binds and other damage.

9. Replace filter element.

Assembly

Float Bowl (1974 and Earlier)

1. If removed, install fast idle cam to float bowl boss (one screw) on 1974 and earlier models.
2. If removed, install seal in recess in idle compensator cavity and install idle compensator (2 screws). Securely tighten screws.
3. Install throttle return spring bracket (3 screws) and tighten securely.
4. Install pump return spring in pump well, making sure it is properly seated.
5. Install pump plunger assembly in pump well. Slide must protrude through bottom of bowl. Push down on slide and install drive link in hole in lower end of shaft. Drive link ends should point toward carburetor boss. Tang on link upper end holds link to slide.
6. Attach power piston and pump actuating lever to lower end of link. Lever projection must point down.
7. Install power piston in its cavity.
8. Install end of actuating rod into groove on side of power piston. Install metering rod assembly and actuating rod into float bowl. Metering rod end must enter jet orifice.
9. Install drive link in actuating lever keyhole.
10. While holding assembly down in bowl, install power piston drive link upper end in hole in power piston actuating rod lower end.

NOTE
Align "D" hole in actuating lever with flats on throttle shaft and install lever on shaft. Retain by securely tightening screw in end of shaft. Check operation of components installed for free operation before proceeding.

11. Install idle tube in float bowl cavity.
12. Install discharge ball, spring, and retainer. Retainer must be flushed with top of bowl castings.
13. Install fuel filter element and spring, using new gasket. Tighten inlet nut securely.

NOTE
Filter open end should face inlet nut hole.

14. Install float needle seat and gasket. Carefully avoid damage to seat while tightening securely. Install float needle into seat.
15. Insert straight part of float hinge pin into float arm and install float into bowl.

Float Bowl (1975-1976)

1. If removed, install idle stop solenoid.
2. If so equipped, install seal into recess in idle compensator cavity and install compensator assembly. Install cover (2 screws) and tighten securely.
3. Install main metering jet in fuel bowl and tighten securely.
4. Install float needle seat and gasket.
5. Install idle tube. Make sure tube is flush with bowl casting.
6. Install pump ball, spring, and "T" guide in pump discharge hole.
7. Press on "T" guide until it is flush with casting (**Figure 8**).
8. Install fuel filter element, spring, and gasket and tighten inlet nut securely.
9. Install accelerator pump return spring and install power piston return spring into bowl piston cavity.
10. Install right angle end of actuating rod into slot in power piston, and install piston, metering rod and actuating rod into float bowl. Make certain end of actuating rod enters hole in bowl and meter rod is located in jet orifice.
11. Install pump plunger in pump well so actuating lever protrudes through bottom of bowl. Press down on pump lever and install drive link in slot in lower end of shaft. Ends of

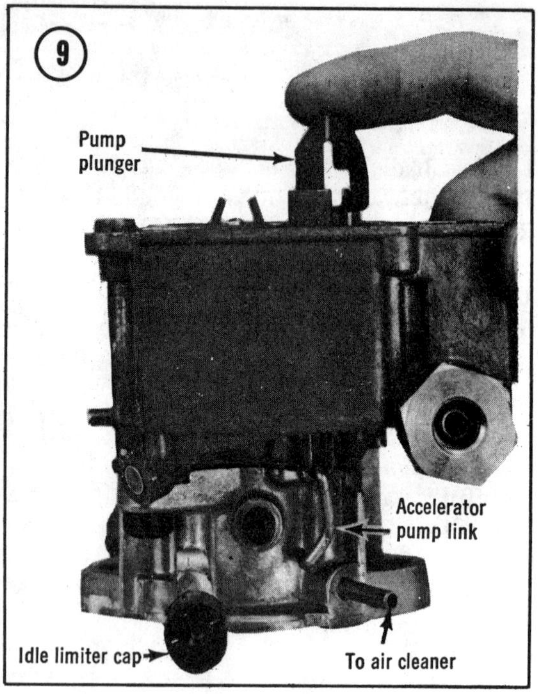

drive link must point in toward carburetor bore. Tang on link upper end holds link to pump shaft (**Figure 9**).

12. Install pump link lower end in throttle shaft actuator lever.

13. Install power piston actuator link (curved end) into throttle actuator lever. End should face outward. End has tang to retain link to lever.

14. Hold power piston assembly down and slide upper end of curved power piston actuator link into lower end of power piston actuator rod.

15. Install actuating lever on end of throttle shaft (align flats on lever with flats on shaft) with retaining screw and tighten securely.

16. Install float needle valve into seat.

17. Install hinge pin into float arm and then install the assembly into float bowl.

Air Horn

1. Install choke shaft assembly and choke valve into air horn, if removed. Align valve and tighten retaining screws. Stake screws securely or use Loctite or equivalent.

2. Carefully install air horn to float bowl with 3 long and 3 short screws and lockwashers. Tighten securely.

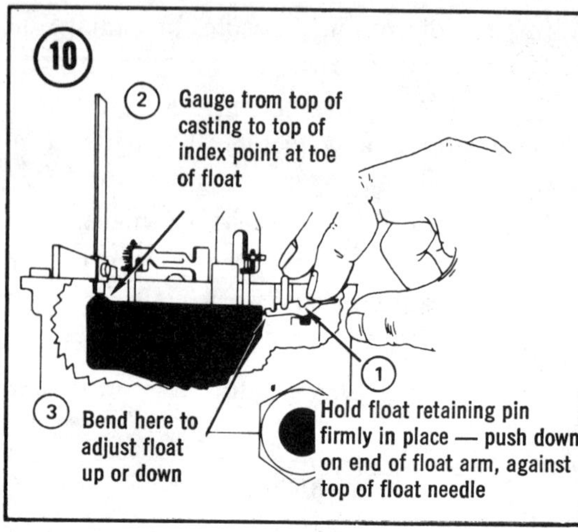

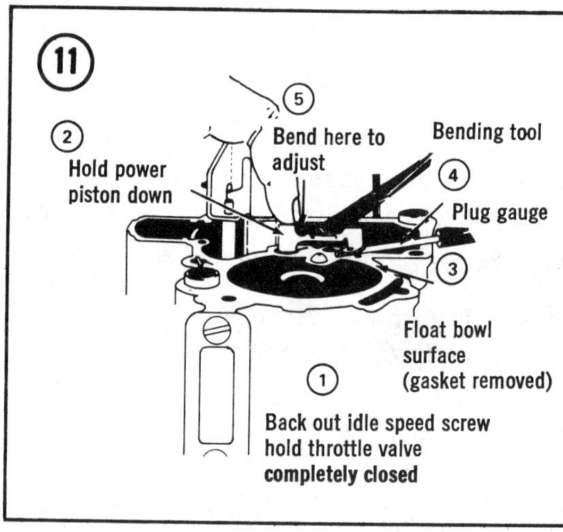

NOTE

Install primary choke vacuum break diaphragm assembly under 2 short screws next to thermostatic coil lever. Connect link to plunger stem and install lever to end of choke shaft with retaining screws. Tighten securely.

3. Install hose to vacuum break diaphragm and tube on float bowl.

4. Install choke rod in slot in upper choke lever (end of rod points away from air horn).

5. Install lower end of choke rod into fast idle cam. Cam steps should face fast idle tang on throttle lever. Install cam to boss on float bowl and securely tighten screw.

6. On 1975-1976 models, install auxiliary vacuum break diaphragm link attached to choke lever slot in diaphragm plunger stem.

12

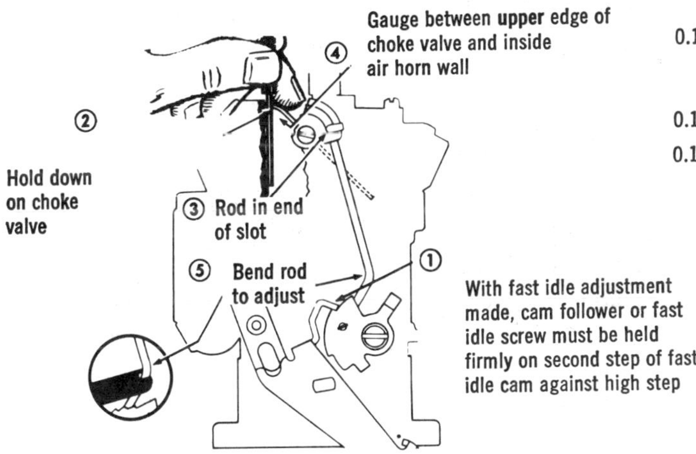

Then install auxiliary diaphragm unit to air horn and tighten screws securely.
7. If removed, install idle stop solenoid.

Adjustment

Float Level

Refer to **Figure 10** for this procedure.
1. Hold float retaining pin in place and push down on float arm at outer end.
2. Using adjustable "Y" scale, measure from top of float at index point on toe to float bowl gasket surface (gasket removed). See **Table 1** for specifications.
3. Bend float up or down, as required, at float arm junction to adjust.

Metering Rod

1. On 1975-1976 models, open throttle valve, slide metering rod out of holder and remove from main metering jet.
2. Back out idle stop solenoid (if so equipped) or idle speed screw so throttle will close. Rotate fast idle cam so follower does not contact cam steps.
3. With throttle valve completely closed, apply pressure to power piston top and hold piston down against its stop.
4. While holding pressure as described in Step 3, swing metering rod holder over flat surface of bowl next to carburetor bore until metering rod rests lightly against inside edge of bowl (**Figure 11**).
5. Insert specified gauge (see **Table 1** for specifications) between bowl casting and metering rod holder lower surface. Gauge should have a slide fit.
6. Carefully bend metering rod up or down, as required, to adjust.
7. On 1975-1976 models, install metering rod and spring assembly. Install rod in jet, then in hanger.
8. Install air horn gasket on float bowl, carefully sliding slit portion over metering rod holder. Align gasket with dowels and press firmly into place.

Fast Idle Cam

Refer to **Figure 12**.
1. Check fast idle speed, using procedure in Chapter Three, and adjust as necessary.
2. Set fast idle cam follower tang firmly on cam second stop (**Figure 12**).
3. Rotate and hold choke valve toward closed position by using light force on choke coil rod.
4. Insert specified gauge (**Table 1**) between lower edge of choke and inside of wall of air horns.
5. Bend cam-to-choke rod as required to obtain specified clearance.
6. Verify that cam-to-choke rod does not bind or meet interference.

5

Choke Unloader

Refer to **Figure 13**.

1. Hold down on choke valve by using light force on choke coil lever.
2. Rotate throttle valve to wide open position.
3. Insert gauge (**Table 1**) between upper edge of choke valve and inside wall of air horn.
4. If adjustment is required, bend tang on throttle lever.

Choke Coil Rod

Refer to **Figure 14**.

1. Pull choke coil rod up to end of travel to completely close choke valve.
2. If bottom of rod is not even with top of lever, adjust as required by bending rod.

Primary Vacuum Break

Refer to **Figure 15**.

1. Apply an outside vacuum source to primary vacuum break diaphragm until plunger is fully seated. Push up on choke coil lever rod in end of slot.
2. Insert specified gauge (**Table 1**) between lower edge of choke valve and inside wall of air horn.
3. Make adjustment, if required, by bending vacuum break rod.
4. Verify that no binding or interference is present.

Auxiliary Vacuum Break

Refer to **Figure 16**.

1. Apply an outside vacuum source to auxiliary vacuum break diaphragm until plunger is fully seated.
2. Position cam follower on highest step of fast idle cam.
3. With vacuum break diaphragm fully seated, insert specified gauge (**Table 1**) between upper edge of choke valve and inner wall of air horn.
4. Bend link between vacuum break and choke valve to adjust, if required.

Idle Mixture Adjustment

1. Start engine and allow it to reach normal operating temperature. Air cleaner should be on, choke should be fully open, and air conditioner off.
2. Connect a tachometer to the distributor, following manufacturer's instructions.
3. Set parking brake and block rear wheels.
4. Disconnect fuel tank hose from vapor canister and disconnect vacuum hose from distributor and plug hose. Check VECI decal and disconnect and plug other hoses as directed.
5. Check engine timing and correct if required. Reconnect vacuum hose to distributor.
6. If vehicle is equipped with automatic transmission, place selector in DRIVE. If equipped with manual transmission, place selector in NEUTRAL and disconnect electrical lead wire to idle stop solenoid.
7. Adjust idle stop solenoid to set idle speed to higher specified rpm.
8. Turn idle mixture screw counterclockwise until maximum idle speed is obtained. Repeat Step 7, if required, to reset idle speed to higher specified rpm on VECI decal.
9. Observe tachometer and adjust idle speed to lower specified rpm on VECI decal.
10. Stop engine and remove tachometer.
11. Connect fuel tank hose to vapor canister. Reconnect electrical lead wire to idle stop solenoid on vehicles equipped with manual transmisssion.

ROCHESTER 2GC-2GE

NOTE

Carburetors on late models (1975 and later) are a part of the emissions control system. Service by the home mechanic should be limited to adjusting idle speed and idle mixture as described in Chapter Three. Other service should be referred to an expert.

Removal/Installation

1. Remove air cleaner.
2. Disconnect vacuum and fuel lines.
3. Remove fresh air choke hoses from choke system.
4. Disconnect accelerator linkage.
5. If equipped with automatic transmission, disconnect downshift linkage.
6. Remove all vacuum and electrical connectors, including idle stop solenoid wiring (if so equipped).

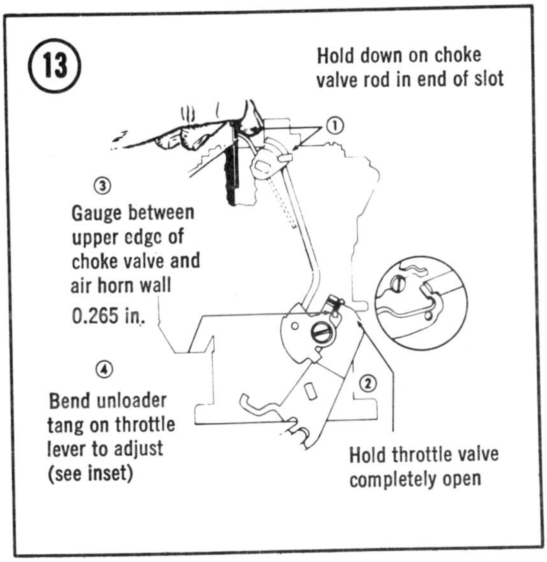
13
Hold down on choke valve rod in end of slot
①
③
Gauge between upper edge of choke valve and air horn wall
0.265 in.
④
Bend unloader tang on throttle lever to adjust (see inset)
②
Hold throttle valve completely open

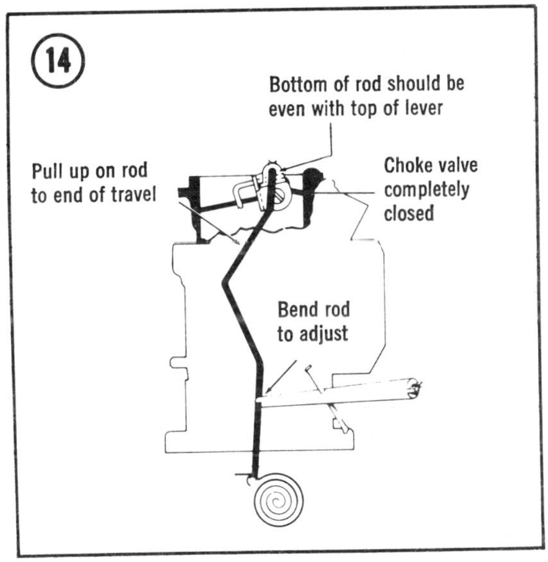
14
Bottom of rod should be even with top of lever
Pull up on rod to end of travel
Choke valve completely closed
Bend rod to adjust

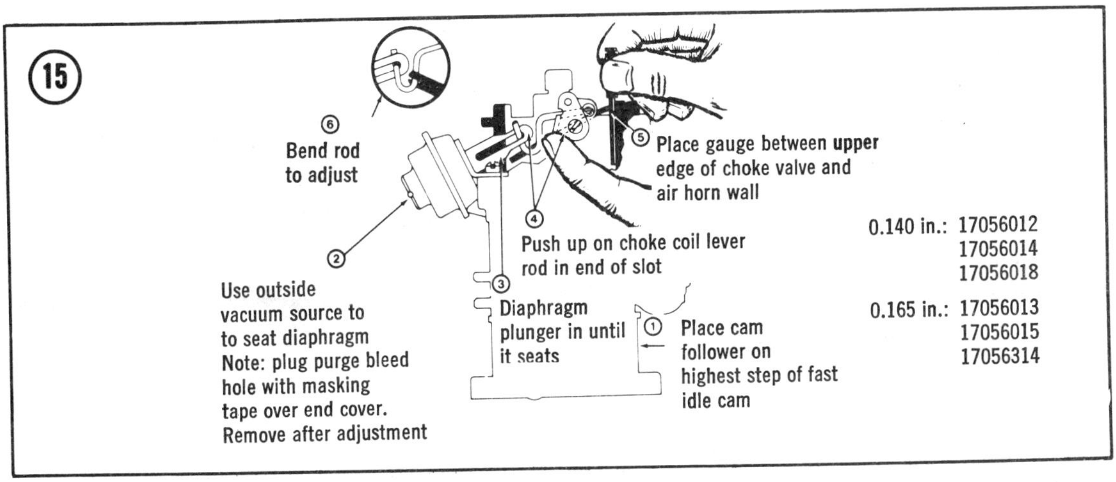
15
⑥
Bend rod to adjust
⑤ Place gauge between upper edge of choke valve and air horn wall
④
Push up on choke coil lever rod in end of slot
0.140 in.: 17056012
17056014
17056018
0.165 in.: 17056013
17056015
17056314
②
Use outside vacuum source to to seat diaphragm
Note: plug purge bleed hole with masking tape over end cover.
Remove after adjustment
③
Diaphragm plunger in until it seats
① Place cam follower on highest step of fast idle cam

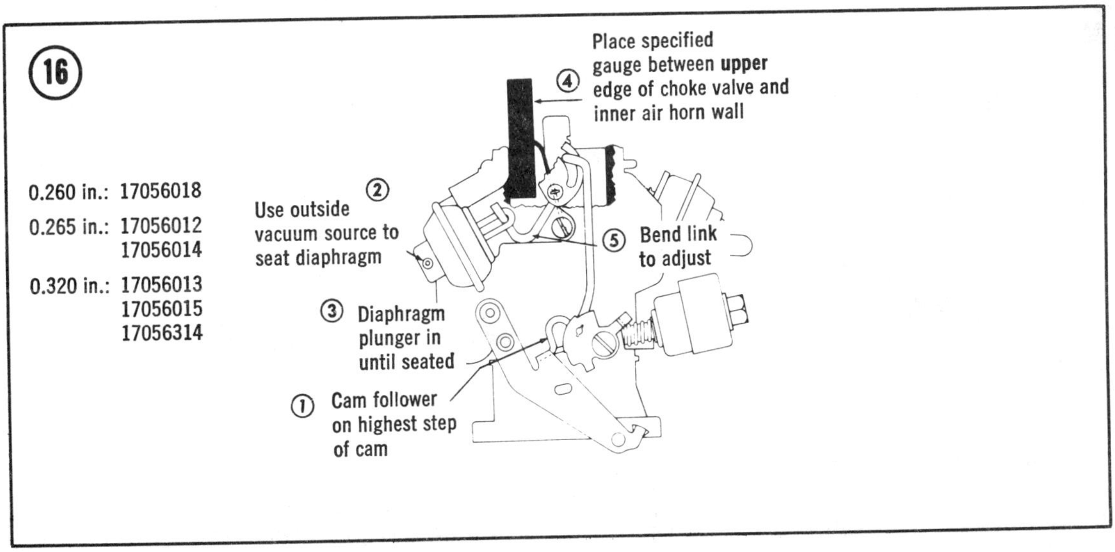
16
Place specified gauge between upper edge of choke valve and inner air horn wall
④
0.260 in.: 17056018
0.265 in.: 17056012
17056014
0.320 in.: 17056013
17056015
17056314
②
Use outside vacuum source to seat diaphragm
⑤ Bend link to adjust
③ Diaphragm plunger in until seated
① Cam follower on highest step of cam

5

7. Remove attaching nuts or bolts and remove carburetor.
8. Remove gasket or insulator.
9. To install carburetor, clean all gasket sealing surfaces and, using new gasket or insulator, reverse Steps 1 through 8.

Disassembly

1. Remove fuel filter nut, gasket, filter, and spring.
2. Remove spring clip and disconnect lower end of pump rod from throttle lever (**Figure 17**). Remove upper end of pump rod from pump lever by rotating rod out of hole in lever.
3. Remove hose connecting tube on throttle body to tube on vacuum break diaphragm unit. Remove attaching screws and remove vacuum break diaphragm assembly, detaching link assembly from choke shaft lever.
4. Remove retaining screws and then remove vacuum break lever from end of choke shaft. See **Figure 18**.
5. Remove intermediate choke rod from vacuum break lever and from lever on thermostatic coil housing. See **Figure 19**.
6. Remove fast idle cam. Remove cam from rod by rotating cam. See **Figure 20**.
7. Remove attaching screws and lift air horn from bowl. Remove upper end of choke rod from choke lever.
8. Place air horn on flat surface and remove float hinge pin, float, gasket, splash shield (if used), and float needle.
9. Remove float needle seat and gasket (use a wide blade screwdriver).
10. Depress stem of power piston and allow piston to snap free (repeat several times, if required). Remove piston, using care to avoid bending stem. See **Figure 21**.
11. Remove setscrew and remove pump plunger assembly and inner pump lever from pump shaft.

CAUTION
Do not remove pump plunger stem from inner pump lever unless the plunger requires replacement. On some models, the stem must be broken before it can be removed and on others it is rotated out. Bending of the inner lever could cause binding which could hold the throttle open.

18

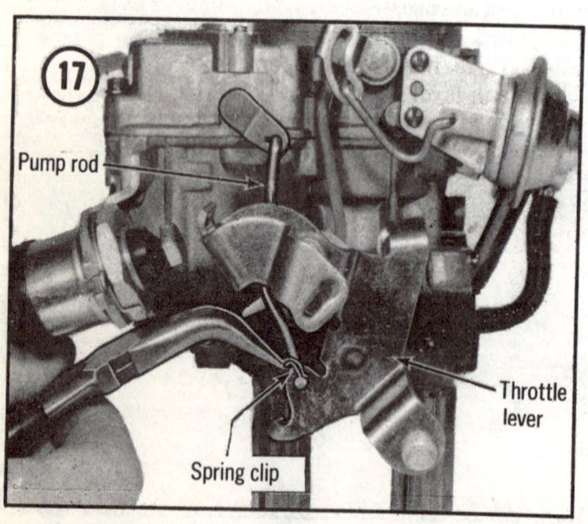

17

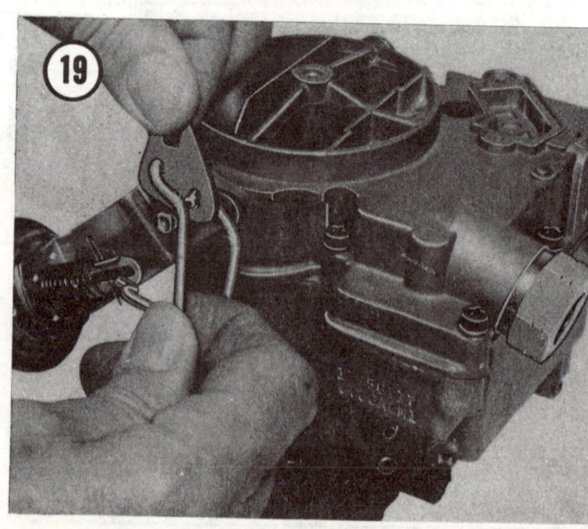
19

NOTE

If plunger is to be reused, place it in a container of kerosene or gasoline until it is required for reassembly. Do not clean plunger or other rubber or plastic parts in carburetor solvent.

12. Remove gasket from air horn and remove fuel inlet baffle, if so equipped.
13. Remove choke valve from shaft (2 screws) and remove shaft from air horn.

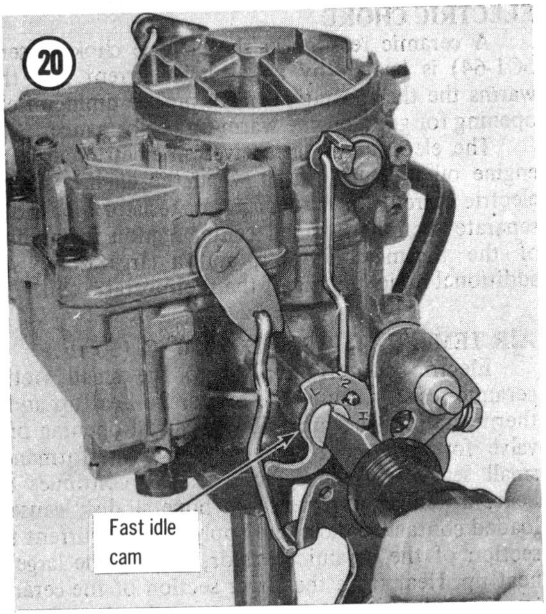

NOTE

If screws are staked, file off staked ends.

14. Remove fast idle cam rod and lever from choke shaft.

Float Bowl

1. Remove pump plunger return spring and invert bowl to remove aluminum check ball. See **Figure 22**.
2. Remove main metering jets, power valve, and gasket.
3. Remove venturi cluster and gasket (3 screws). Remove plastic main well inserts. See **Figure 23**.

4. Remove pump discharge spring retainer with long-nosed pliers (**Figure 24**). Remove spring and check ball from passage.
5. Invert the carburetor and remove bowl-to-throttle body retaining screws and separate the 2 units. See **Figure 25**.

Throttle Body

1. If replacement or cleaning is required, remove idle mixture screws. If equipped with limiter caps, cut off cap tang.
2. Remove choke cover (3 screws) and remove thermostatic coil.

CAUTION
Do not remove cup baffle beneath cover as coil distortion could result.

3. Remove baffle plate from choke housing.
4. Remove choke housing and gasket (2 screws) from throttle body.
5. Remove intermediate choke lever (one screw) from intermediate choke shaft. Remove inner lever and shaft, and rubber dust seal from choke housing.

CAUTION
Do not attempt to remove throttle valves, as relation to idle discharge orifices may be destroyed.

Cleaning and Inspection

1. Wash all metal parts in carburetor solvent and blow dry, taking care to blow out all passages.

CAUTION
Do not use drills or wire to clean jets and passages.

2. Inspect all parts for wear or damage and replace as necessary.

Assembly

NOTE
Use new gaskets in reassembly.

Throttle Body

1. Install idle speed screw and spring, if removed.

24

2. Install idle mixture needles and springs, id removed. Seat needles lightly and back out 4 turns.
3. Install new rubber dust seal in choke housing cavity. Seal lips should face toward carburetor body.
4. Install inner lever and shaft in choke housing.
5. Install intermediate lever on shaft and securely tighten retaining screw.
6. Install new gasket on choke housing and position housing on throttle body. Install and securely tighten retaining screws.
7. Using the procedure given below, adjust intermediate choke rod so that with choke valve closed, lever in choke housing is aligned with gauge.
8. Using new gaskets, install choke coil and cover with end of coil below plastic tang on inner choke housing lever. Index cover, using procedure below, and install and securely tighten cover attaching screws.
9. Using new gasket, install float bowl to throttle body assembly (3 screws) and securely tighten. Make certain holes in gasket are aligned with holes in casting.

25

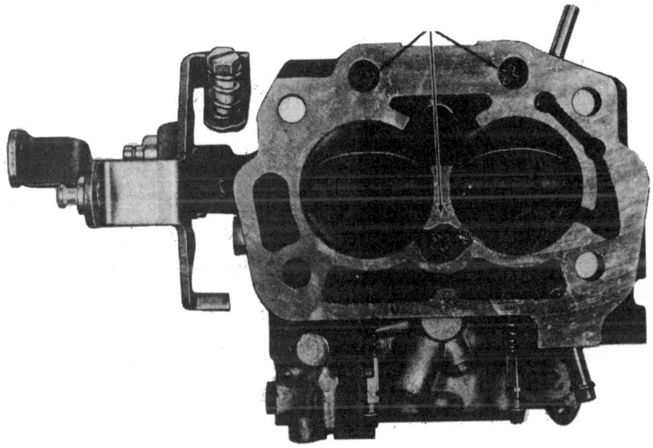

Float Bowl

1. Install main metering jets.
2. Install power valve and gasket in float bowl and securely tighten with slotted screwdriver.
3. Place aluminum inlet check ball in pump well hole and install pump return spring. Press with finger to center spring. On 2GE, make sure larger end of spring is at bottom of well.
4. Drop steel ball into pump discharge hole and install spring and retainer.

CAUTION

Do not interchange aluminum and steel balls during assembly. Steel ball is larger and heavier of the two.

5. Install plastic inserts in main fuel wells and make sure they are properly seated in the recesses.
6. Install venturi cluster and gasket. Fit center screw with gasket to avoid pump discharge leakage, and securely tighten all screws. See **Figure 23**.

Air Horn

1. Install upper choke rod lever and collar assembly on choke shaft.
2. Install choke shaft in air horn. Install shaft from throttle lever side of carburetor.
3. Center choke valve on shaft with lettered or numbered side up and install with 2 screws. Securely tighten the screws and lightly stake in place (or use Loctite, or equivalent, on threads). Check valve for free movement.
4. If removed, install outer pump shaft and lever assembly, making sure plastic washer is in place.
5. Install accelerator pump plunger to inner lever (if removed) and install retainer (provided in repair kit). End of plunger shaft should point inward.
6. Place gasket on float needle seat and install seat in air horn. Tighten securely.
7. Insert power piston into air horn and lightly stake retainer to casting. Check piston for free movement.
8. Install fuel inlet baffle or splash shield, as appropriate.
9. Install new air horn gasket, indexing gasket to casting dowel.
10. Install float needle, then install float assembly on air horn and insert hinge pin. Check for free movement of float and needle.
11. Install fuel filter, spring, gasket, and inlet fitting and securely tighten.
12. Install vacuum break diaphragm onto air horn. Part number on diaphragm bracket should face outward. Position rod in diaphragm plunger and attach to choke coil lever. Then install choke coil lever on end of choke shaft, with tang facing outward. Install and securely tighten the retaining screw.
13. Install choke rod in upper choke lever, rotating rod to align square end of rod with slot in lever.

26

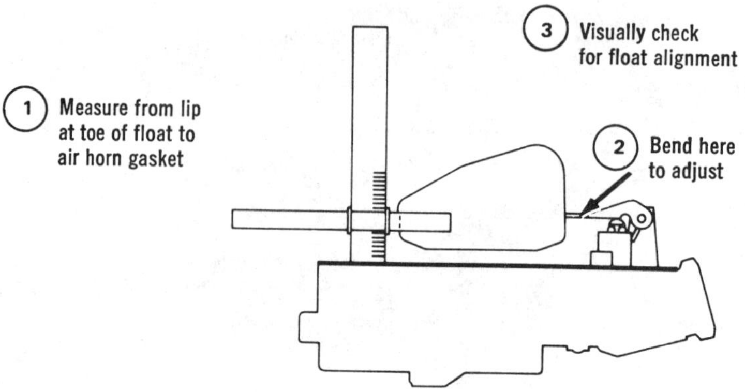

27

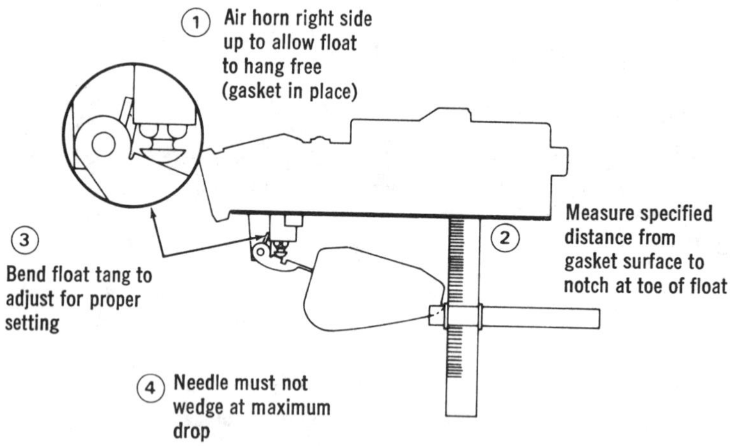

Float Level

1. Invert air horn and verify that gasket is in place and needle valve is seated. Measure distance from lip of float toe to air horn gasket. See **Figure 26**.
2. Adjust float to specifications (**Tables 2 and 3**) by bending float arm at point shown.

Float Drop

1. Hold air horn upright so float hangs free and measure distance from lip on float toe to air horn gasket. See **Figure 27**.
2. Adjust to specifications (**Tables 2 and 3**) by bending tang at rear of float hanger.

Air Horn-to-Float Bowl

1. Position air horn on bowl, making sure accelerator pump plunger is inserted in pump well and moves freely.
2. Install lockwashers and screws and securely and evenly tighten, using the pattern shown in **Figure 28**.
3. Install fast idle cam to lower end of choke rod. Fasten cam to bowl (part number or identification facing outward) with attaching screw and tighten securely. Check for free movement.
4. Install pump rod in upper pump lever and secure rod to throttle lever with spring clip. See **Figure 17**.

(28)

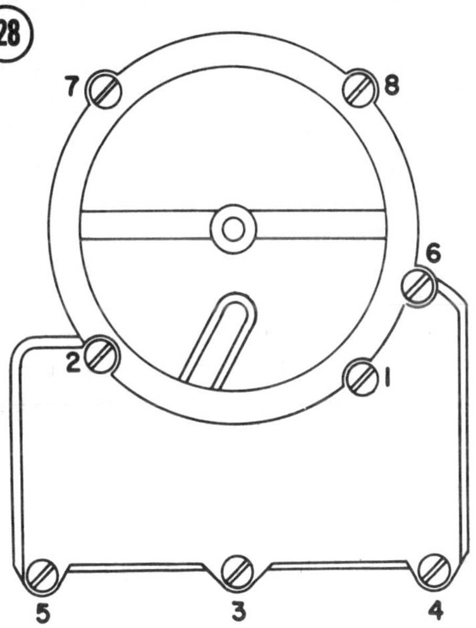

5. Install vacuum hose between vacuum break diaphragm and fitting on throttle body.

Pump Rod

Refer to **Figure 29**.

1. Back out idle speed screw and hold throttle valve completely closed.

2. Measure from top of air horn ring to top of pump rod, as shown in **Figure 29**. See **Tables 2 and 3** for specifications.

3. If adjustment is required, bend rod as shown in **Figure 29**. Reset idle using the procedure given above.

Fast Idle Cam

Refer to **Figure 30**

1. Turn idle speed screw in until it contacts low step of fast idle cam, and then turn screw in one complete turn.

(29)

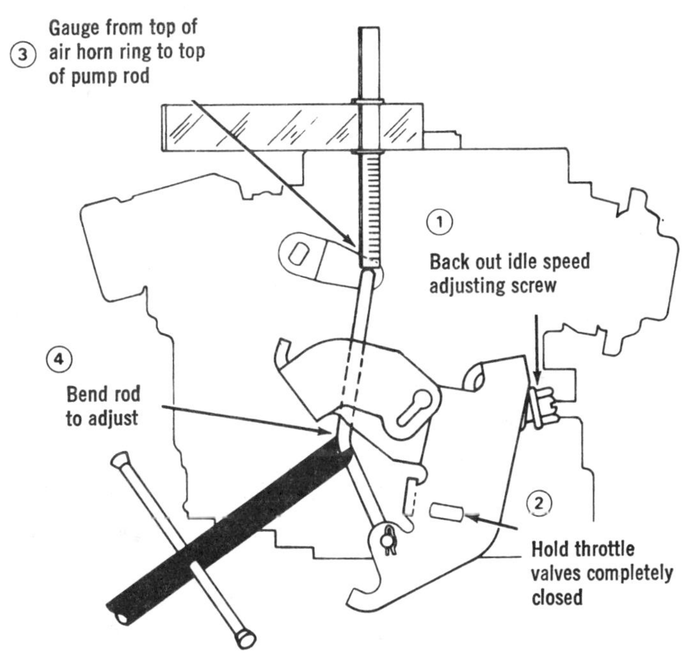

2. Place idle speed screw on step 2 of fast idle cam, with screw resting against high step as shown in inset No. 1, **Figure 30**.
3. Place gauge between upper edge of choke valve and inside air horn wall. See **Tables 2 and 3** for specifications.
4. If required, adjust by bending choke lever tang as shown in inset No. 2 or No. 3, **Figure 30**.

Choke Unloader

Refer to **Figure 31**.

1. Place choke valve toward closed position while holding throttle valve wide open.
2. Place specified gauge (**Tables 2 and 3**) between the upper edge of the choke valve and air horn wall.
3. If clearance adjustment is required, bend tang on throttle lever.

Intermediate Choke Rod

1. Remove thermostatic cover, coil assembly, gasket, and inside baffle plate (**Figure 32**) by removing 3 attaching screws.
2. Position idle speed screw on high step of fast idle cam.
3. Close choke by pushing up on intermediate choke lever.
4. Verify that edge of coil lever inside choke housing lines up with edge of 0.120 in. plug gauge.
5. If adjustment is required, bend intermediate choke rod as shown in inset, **Figure 32**.
6. Replace thermostatic cover, coil assembly, gasket, and inside baffle plate and secure with 3 ataching screws. Adjust coil using procedure given below.

Automatic Choke Coil

Refer to **Figure 33**.

1. Place idle speed screw on high step of fast idle cam.
2. Loosen thermostatic choke coil cover retaining screws.
3. With choke fully open, rotate choke cover against coil tension until choke valve begins to close. Continue rotating until index mark lines up with index point on choke housing.
4. Tighten 3 retaining screws.

30

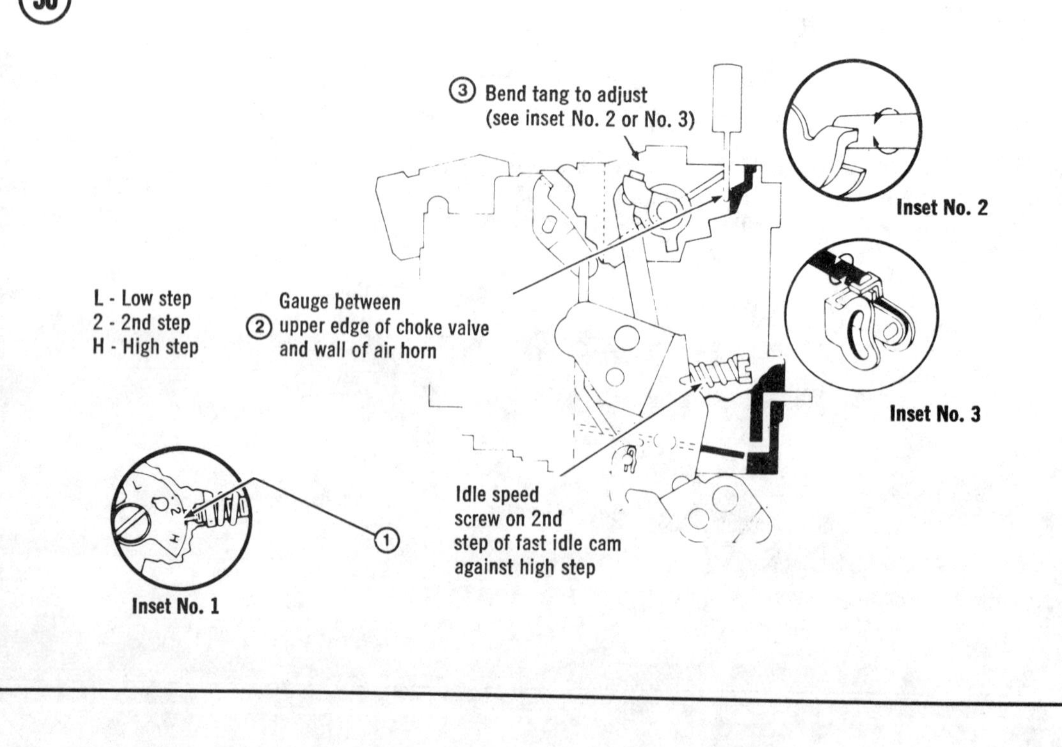

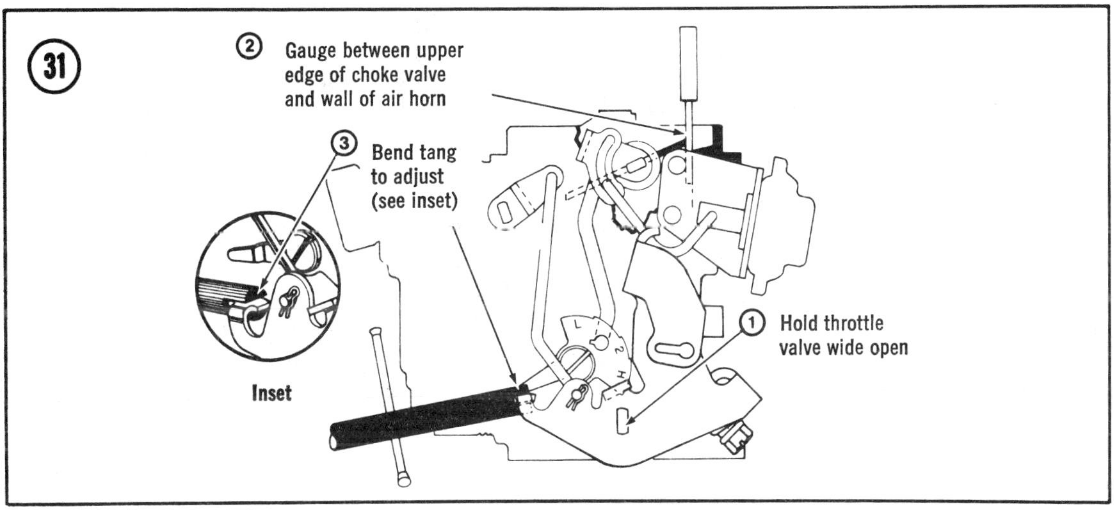
31
② Gauge between upper edge of choke valve and wall of air horn
③ Bend tang to adjust (see inset)
Inset
① Hold throttle valve wide open

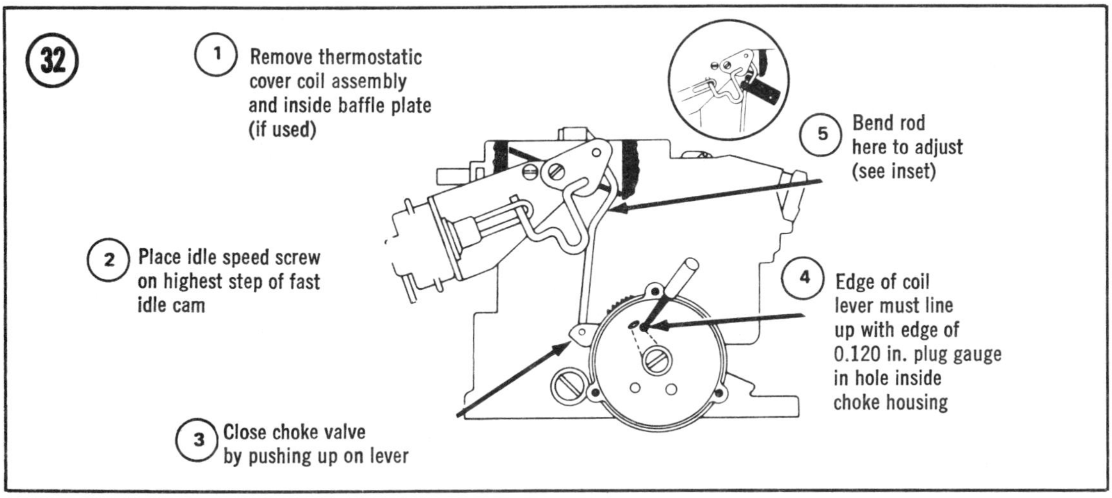
32
1 Remove thermostatic cover coil assembly and inside baffle plate (if used)
5 Bend rod here to adjust (see inset)
2 Place idle speed screw on highest step of fast idle cam
4 Edge of coil lever must line up with edge of 0.120 in. plug gauge in hole inside choke housing
3 Close choke valve by pushing up on lever

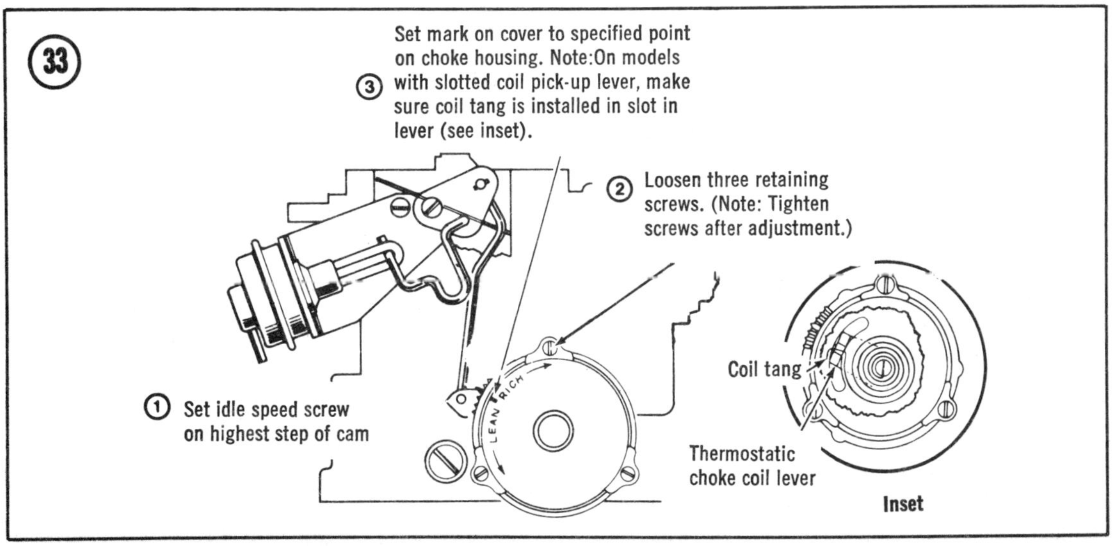
33
③ Set mark on cover to specified point on choke housing. Note:On models with slotted coil pick-up lever, make sure coil tang is installed in slot in lever (see inset).
② Loosen three retaining screws. (Note: Tighten screws after adjustment.)
Coil tang
① Set idle speed screw on highest step of cam
Thermostatic choke coil lever
Inset

Vacuum Break

Refer to **Figure 34**.

1. Use an outside vacuum source to seat vacuum break diaphragm.
2. Use a piece of tape to cover vacuum break bleed hole so diaphragm will not bleed down.
3. Locate idle speed screw on fast idle cam high step.
4. Hold choke coil lever (inside choke housing) toward closed choke position.
5. Gauge between upper of choke valve and air horn wall. See **Tables 2** and **3** for specifications.
6. Bend vacuum break rod to adjust. Remove tape and reconnect vacuum hose.

Idle Mixture Adjustment (1970-1977)

1. Start engine and allow it to reach normal operating temperature. Air cleaner should be on, air conditioner should be off, and choke should be fully open. Connect tachometer to engine, using manufacturer's instructions.
2. Set parking brake and block rear wheels.
3. Disconnct fuel tank hose from vapor canister. Disconnect vacuum hose from distributor and plug hose. Disconnect and plug other hoses as directed by VECI decal.
4. Check engine timing and adjust if required, using the procedures given above. Reconnect vacuum hose to distributor.
5. Position the transmission selector in DRIVE (automatic transmission) or NEUTRAL (manual transmission).

NOTE
Obtain idle speeds from Vehicle Emission Control Information decal located in the engine compartment.

NOTE
Idle mixture screws are equipped with factory-installed caps. These caps allow acceptable adjustment and must not be removed or altered, unless the carburetor has been overhauled.

6. Adjust idle speed screw to obtain the higher of 2 idle speeds given.
7. Turn idle mixture screws out equally to obtain highest possible idle speed. Reset idle speed screw to higher specified rpm if required.
8. Turn idle mixture screws in equally until the lower specified idle speed is achieved.
9. Reconnect fuel tank hose to vapor canister.

Idle Mixture Adjustment (1978)

When the idle mixture must be adjusted following major overhaul on 1978 2GC and 2GE carburetors, refer to *Propane Enrichment Idle Mixture Adjustment*, elsewhere in this chapter.

ROCHESTER M2M-M2MC-E2ME CARBURETOR

The Rochester M2M-M2MC-E2ME carburetors used on many late models are closely interrelated to other emission control systems. They can seldom be accurately set without the help of expensive test equipment, including a flow meter. Service should be limited to the procedures given in Chapter Three. Other repairs and adjustments should be entrusted to your Oldsmobile dealer or a reputable repair shop.

Removal/Installation

NOTE
When removing the carburetor, take care not to spill the fuel in the fuel bowl. Contents of the bowl then can be examined to provide possible clues as to carburetor trouble.

1. Remove the air cleaner and gasket and then remove the electrical wire from the solenoid, if so equipped.
2. Identify and mark all fuel and vacuum hoses so they may be returned to their original positions and then remove the hoses from the carburetor.
3. Identify and mark all electrical connections so they may be reconnected in their original positions and then disconnect the connectors.
4. Disconnect the accelerator linkage.
5. On models with automatic transmissions, disconnect the downshift cable at the carburetor.
6. Disconnect the cruise control linkage, if so equipped.
7. Remove the attaching bolts and then remove the carburetor and insulator.

34

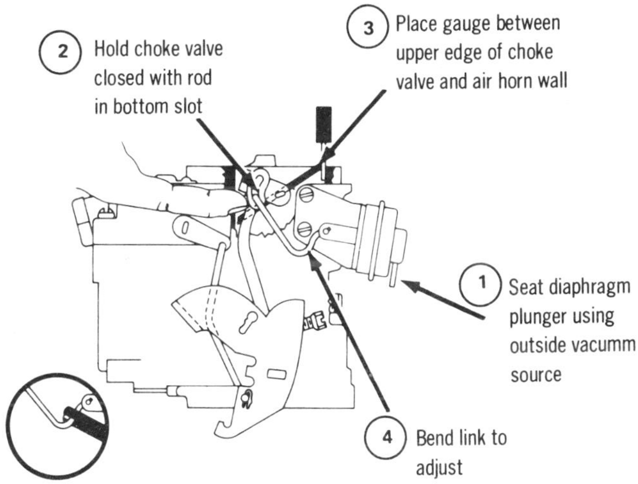

35

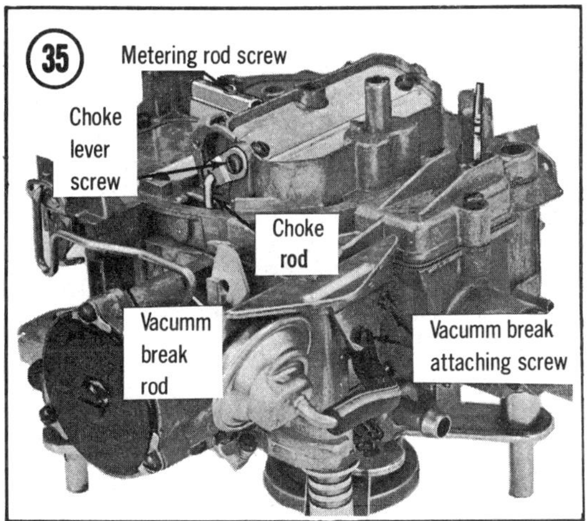

8. Installation is the reverse of these steps. Tighten attaching bolts to 145 in.-lb. Check and adjust idle speed as required, using the procedures given in Chapter Three.

ROCHESTER 4MC

Removal/Installation

This is a 4-barrel, 2-stage carburetor which is easily adapted for small to large engines without design changes. See **Table 4** for specifications.

The fuel bowl is centrally located to avoid fuel slosh. The float needle valve is pressure balanced to permit use of a small single float.

The primary side has 2 small bores, each with a triple venturi for fine fuel control in the idle and economy ranges. The secondary side has large bores and an air valve for high air capacity.

1. Remove air cleaner, gasket, and stud.
2. Disconnect vacuum, fuel, and choke pipes at carburetor.
3. On automatic models, disconnect transmission control rod from throttle lever.
4. On models with positive crankcase ventilation (PCV), disconnect ventilation hose at valve on carburetor base, if so equipped.
5. Disconnect accelerator rod and throttle return spring at carburetor.
6. Remove mounting nuts and lift carburetor from manifold.
7. Reverse procedures to install. Be sure carburetor base and manifold flange are clean. Tighten attaching bolts to 15 ft.-lb.

Disassembly

Air Horn

1. Remove the upper choke lever (see **Figure 35**) from the end of the choke shaft (one retaining screw) and remove the choke rod.
2. Remove the vacuum break hose and then remove the vacuum break (2 retaining screws). See **Figure 35**.

3. Remove the secondary metering rods by lifting on the secondary rod hanger (**Figure 36**) and remove them from the air horn.
4. Remove the attaching screws (**Figure 37**) and separate the air horn from the float bowl assembly. See **Figure 38**. Turn the air horn sideways while lifting to disengage the pump rod from the upper pump lever. Take care not to bend the small bleed tubes and accelerating tubes in the air horn casting.

Float Bowl and Throttle Body

Refer to **Figure 39**.

1. Remove pump plunger from pump well.
2. Lift air horn gasket from dowels on

38

36

39

37

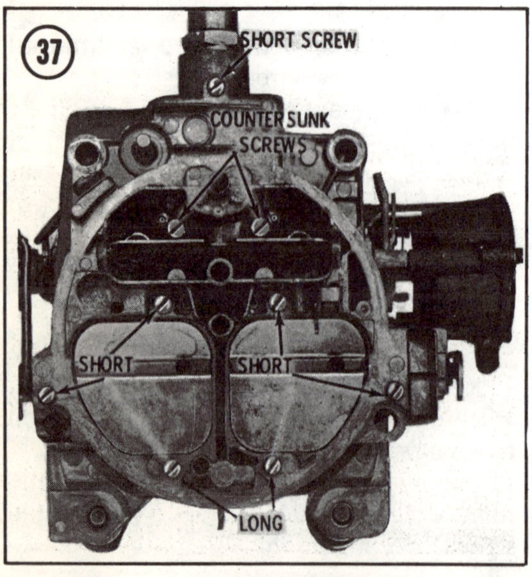

40

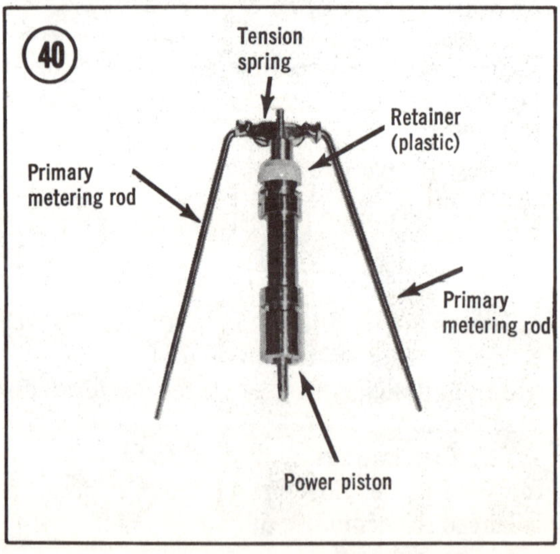

secondary side of bowl, then remove gasket from around power piston and primary metering rods.
3. Remove pump return spring.
4. Remove plastic filler over float valve.
5. Remove power piston and primary metering rods. See **Figure 40**.
6. Disconnect tension spring at top of each primary metering rod. Rotate each rod and remove from hanger on power piston.
7. Remove float assembly. See **Figure 41**.
8. Remove pull clip and fuel inlet needle.
9. Remove fuel inlet needle seat with *wide* blade screwdriver.
10. Remove primary metering jets. See **Figure 42**.

CAUTION
Do not remove secondary metering discs.

11. Remove pump discharge check ball retainer and check ball.
12. Remove baffle from secondary side of bowl.
13. Remove vacuum hose.

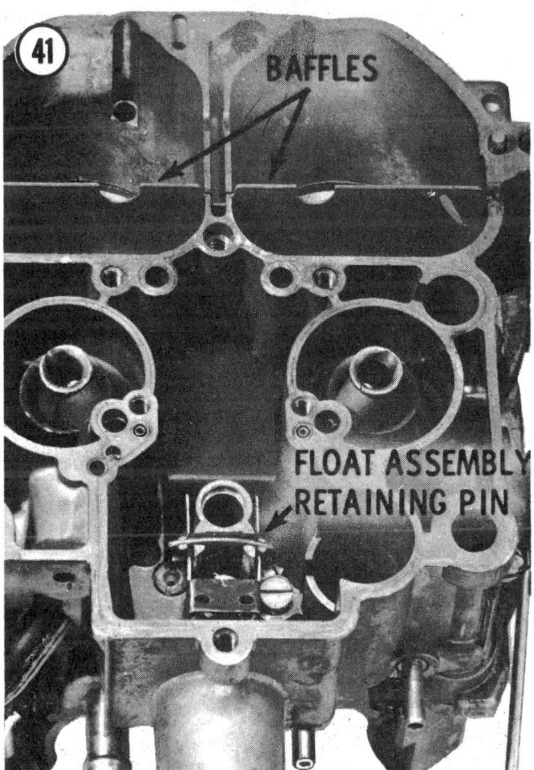

14. Remove choke assembly from bowl.
15. Remove secondary lockout link from bowl.
16. Remove vacuum break rod and vacuum break.
17. Remove fast idle cam.
18. Remove lower choke rod and lever from inside float bowl well.
19. Remove fuel inlet filter nut, gasket, filter, and spring.
20. Remove throttle body from bowl.
21. Remove pump rod from throttle lever.
22. Remove idle mixture screws and springs.

Cleaning and Inspection

The most frequent causes of carburetor trouble are dirt, gum, and water in the carburetor. Carefully clean and inspect all parts before assembling carburetor.

1. Clean all carburetor castings and metal parts in carburetor cleaner.

CAUTION
Do not immerse any rubber parts, plastic parts, diaphragms, or pump plungers in carburetor cleaner. The Delrin cam on the air valve shaft, however, will withstand normal cleaning carburetor cleaner.

2. Check bearing surfaces of all operating levers, shafts, and castings for excessive wear.
3. Check floats for dents. Immerse floats in hot water. If bubbles appear, float is leaking and the float *must* be replaced. Do not attempt to solder the hole. This increases float weight and causes high fuel level.

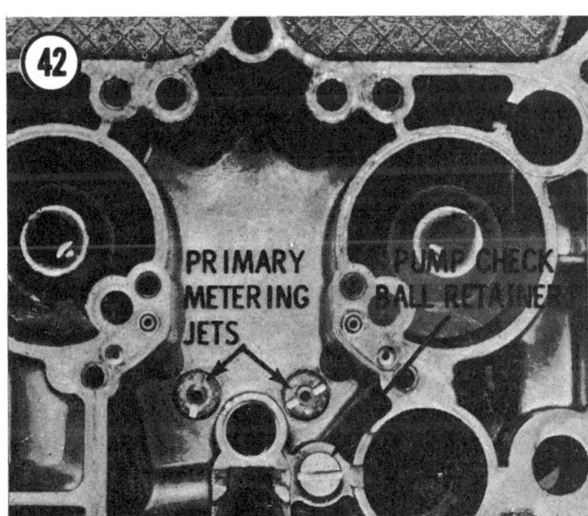

5

4. Check pump plunger leather for cracks and creases.
5. Check float needles and seats for burrs and ridges. If present, replace needle and seat; never replace either alone.
6. Check metering rods and jets for bends, burrs, ridges, or other distortion. Replace rod jet if either is damaged.
7. Inspect edges of throttle valves for nicks or gouges. Replace if necessary.
8. Check all springs for obvious distortion or bends.
9. Make certain all gasket mating surfaces of choke housing, top cover, and carburetor body are free of burrs, gouges, deep scratches, or other irregularities which could prevent a good seal.

Assembly

Float Bowl and Throttle Body

See **Figure 43**.

1. Install idle mixture needles and springs. Screw in until they seat *lightly*, then back them out 6 turns.
2. Install pump rod in throttle lever.
3. Install throttle body with new gasket. See **Figure 43**.
4. Install fuel inlet filter spring, filter, new gasket, and nut. See **Figure 44**.
5. Install vacuum break and rod.
6. Install fast idle cam on vacuum break assembly.
7. Connect choke rod to choke rod actuating lever. Hold choke rod with grooved end pointing inward. Position lever in well of float bowl and install choke assembly so that shaft engages with hole in actuating lever. See **Figure 45**. Install choke housing retaining screw.
8. Install vacuum hose to bowl and vacuum break.
9. Install baffle in secondary side of bowl with notches toward top.
10. Install pump discharge check ball and retainer in passage next to pump well.
11. Install primary main metering jets.
12. Install fuel inlet needle seat and gasket. Use *wide* blade screwdriver.
13. Install fuel inlet needle. Secure with the pull clip.

43

44

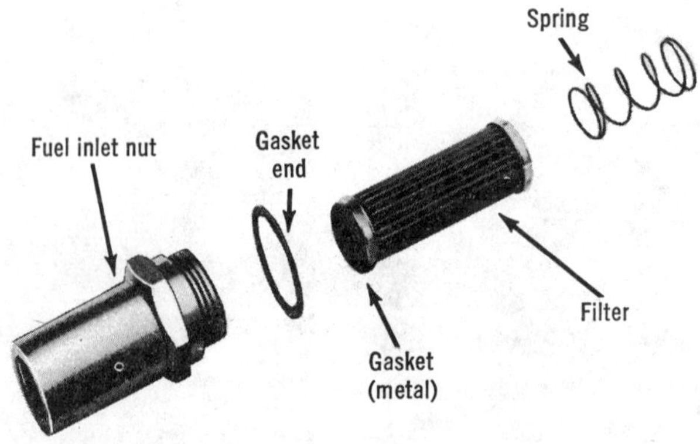

14. Install float.

15. Measure from top of float bowl gasket surface (gasket removed) to top of float at toe. See **Figure 46**. Proper measurement is listed in **Table 4**.

16. Bend float as necessary to adjust.

17. Install power piston spring in well.

18. Install main metering rods to power piston assembly.

19. Install power piston assembly in well with main metering rods positioned in jets. Secure with retainer.

20. Install plastic filler over float needle.

21. Install pump return spring in pump well.

22. Install air horn gasket over bowl.

23. Press power piston down firmly to ensure correct alignment.

24. Install pump plunger in pump well.

Air Horn

1. Install secondary metering rods. Refer to **Figure 36**.

2. Place air horn on float bowl carefully. Secure with 7 screws. Tighten screws is sequence shown in **Figure 47**.

3. Install idle vent rod in pump lever.

4. Connect pump rod in pump lever and secure with spring clip.

5. Connect choke rod in lower choke lever. Secure upper end with spring clip.

6. Install idle vent valve.

45

47

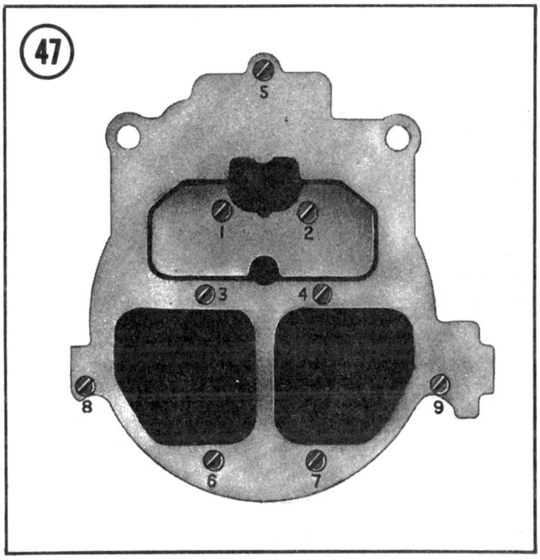

46

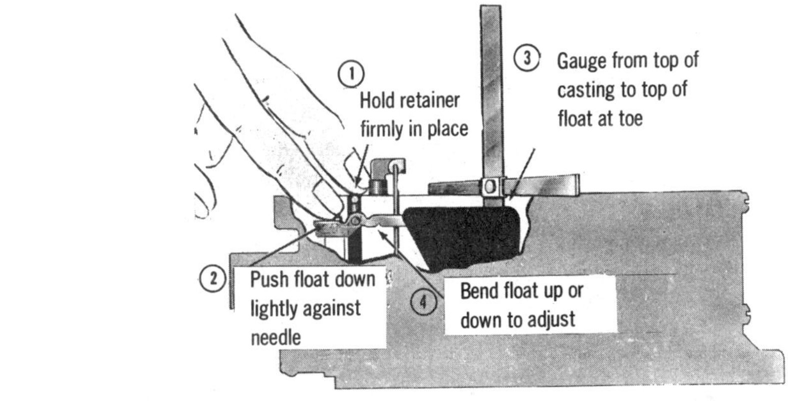

48

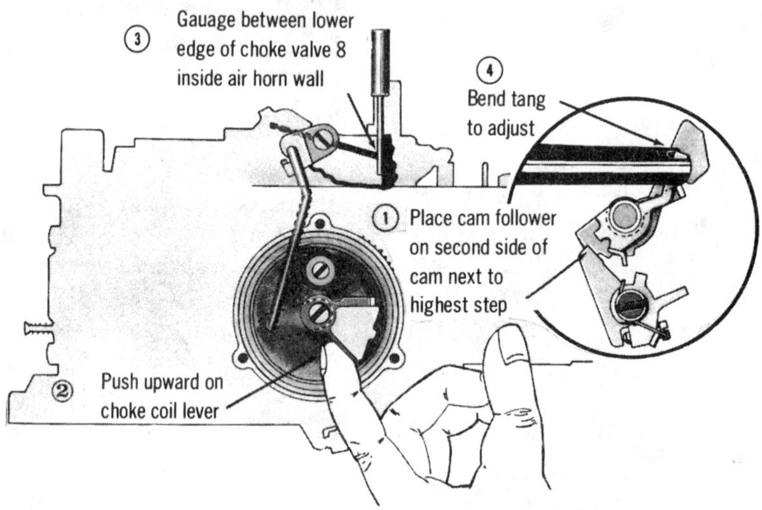

Fast Idle Cam Adjustment

Refer to **Figure 48**.

1. Place cam follower on second step of fast idle cam.
2. Rotate choke valve toward closed position by pushing down on vacuum break tang.
3. Measure between lower edge of choke valve and wall. See **Table 4** for dimensions.
4. Bend choke rod if necessary to adjust.

Vacuum Break

1. Hold choke valve closed with a rubber band. Cam follower must be on highest step of fast idle cam.
2. Hold vacuum break diaphragm stem against its seat so vacuum link is at end of slot, using an outside vacuum source of at least 10 in.Hg.
3. Measure clearance between lower edge of choke valve and air horn wall. Check against specifications in **Table 4** and bend lower end of rod as shown in **Figure 49** if adjustment is required.

Air Valve Dashpot

Refer to **Figure 50**.

1. Make sure no vacuum is applied to the vacuum diaphragm of the dashpot.
2. Make sure the air valves are completely open, then insert the specified gauge between the rod and the end of the slot as shown in **Figure 50**. See **Table 4**.
3. If the gauge is not a snug fit, bend the rod at the indicated point to obtain the specified clearance.

Unloader

Refer to **Figure 51**.

1. Close the choke by pushing lightly on the vacuum break lever and use a rubber band to hold the choke in this position.
2. Hold the primary throttle valve wide open and insert a gauge between the air horn wall and the lower edge of the choke valve. See **Table 4** for gauge size.
3. If the gauge does not fit snugly, bend the tang on the lever at the point shown in **Figure 51** to adjust as required.

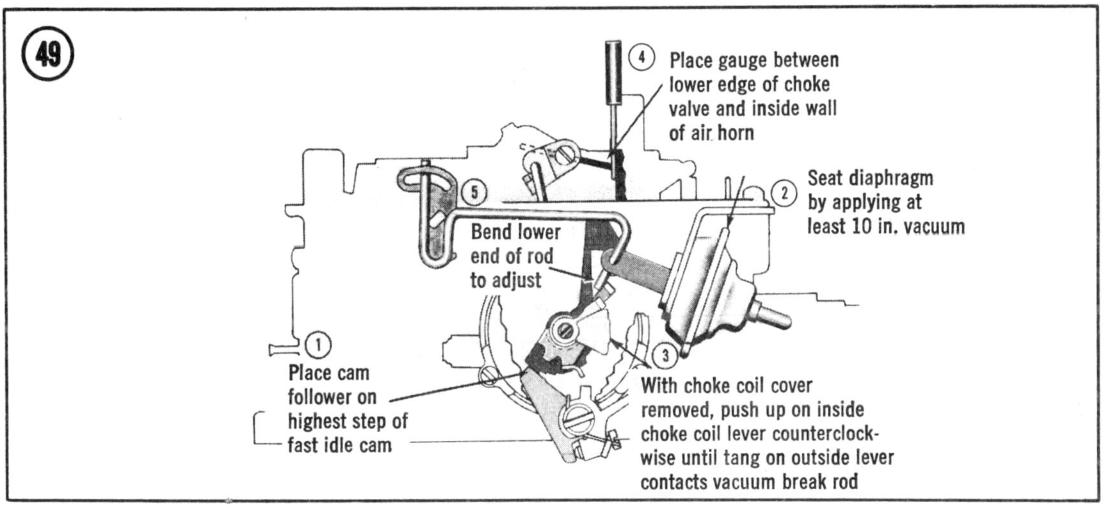
49
4 Place gauge between lower edge of choke valve and inside wall of air horn
2 Seat diaphragm by applying at least 10 in. vacuum
5 Bend lower end of rod to adjust
1 Place cam follower on highest step of fast idle cam
3 With choke coil cover removed, push up on inside choke coil lever counterclockwise until tang on outside lever contacts vacuum break rod

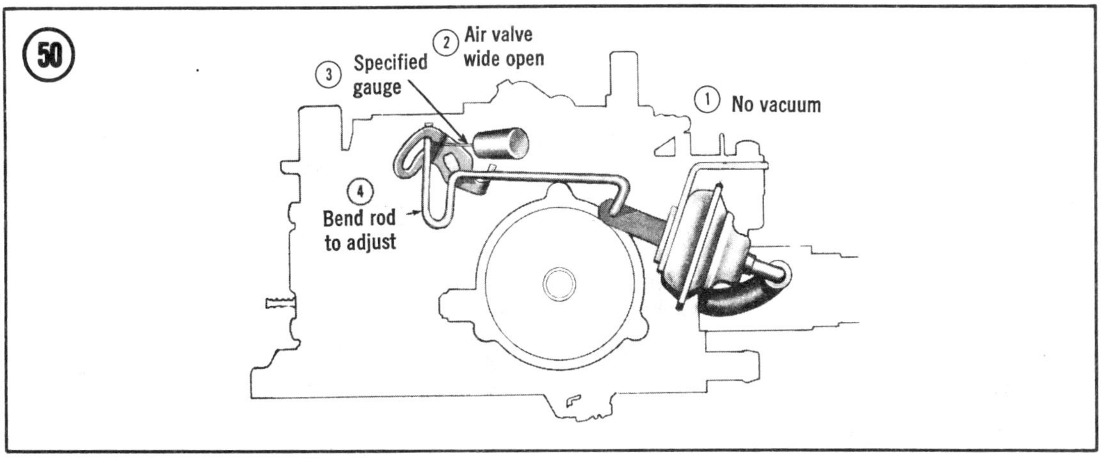
50
2 Air valve wide open
3 Specified gauge
1 No vacuum
4 Bend rod to adjust

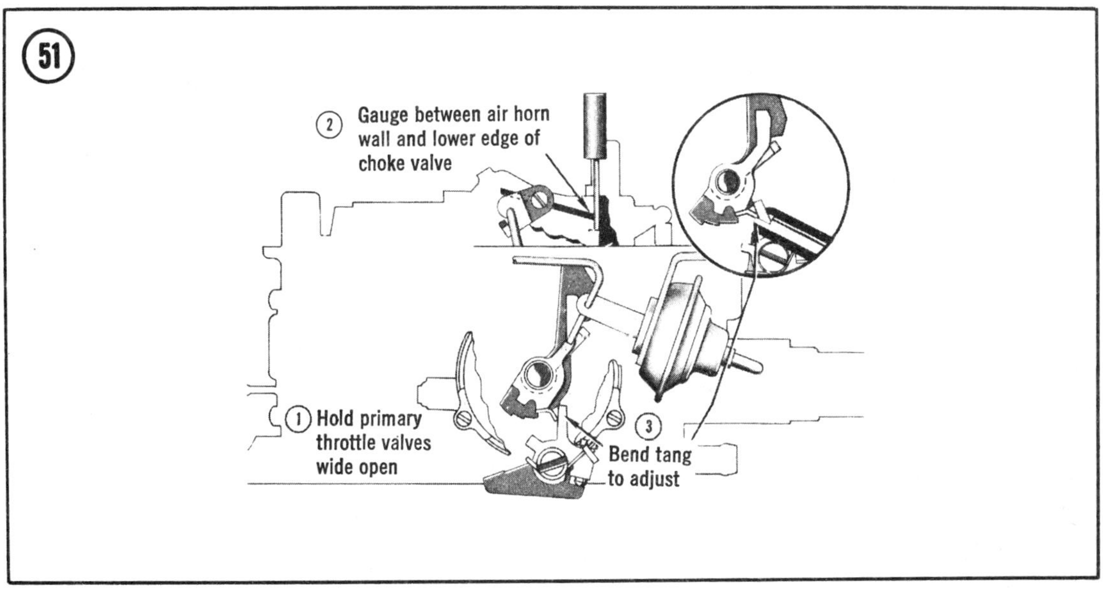
51
2 Gauge between air horn wall and lower edge of choke valve
1 Hold primary throttle valves wide open
3 Bend tang to adjust

52

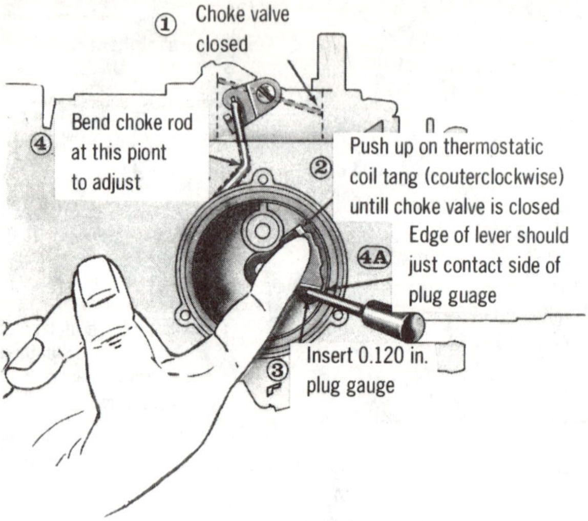

Choke Coil Lever Adjustment

1. With the choke valve closed, push up on the thermostatic coil tang (counterclockwise) as shown in **Figure 52** to hold the choke fully closed.

2. Insert a 0.120 in. plug gauge as shown in **Figure 52**. The edge of the lever should just touch the side of the gauge. Bend rod at the point shown to adjust, if required.

Idle Mixture Adjustment

The *Idle Mixture Adjustment* procedure for the Rochester 2GC-2GE carburetor, elsewhere in this chapter, may be used for adjusting the idle mixture on the 4MC carburetor. Use the specifications listed on the VECI sticker in the engine compartment of your car.

ROCHESTER M4MC-E4ME

The M4MC-E4ME model Quadrajet carburetors (**Figure 53**) are used on some 1975 and later model V8 engines. These carburetors were developed from the 4MV series.

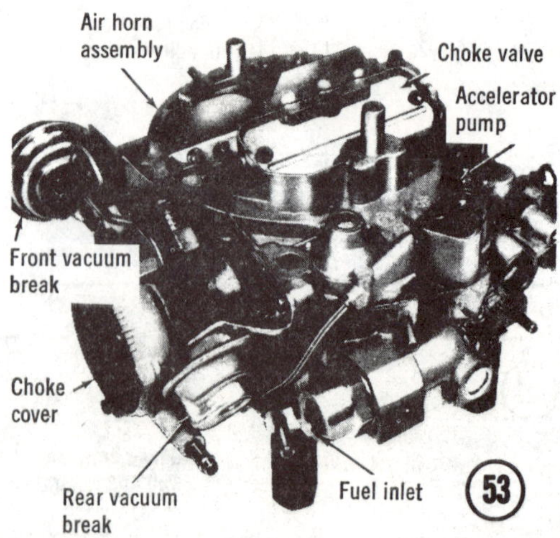

53

All these carburetors are an integral part of the emission control system. Adjustment and repair by the home mechanic, other than adjustment of idle speed and idle mixture as described in Chapter Three, is not recommended. This is a job for your dealer or a carburetor shop equipped with the precision equipment and skills necessary for working with this complicated device. Removal and installation procedures are essentially the same as those for the 4MV, this chapter.

PROPANE ENRICHMENT IDLE MIXTURE ADJUSTMENT (1978-ON, EXCEPT 1980 CALIFORNIA)

This procedure must be used on all 1978 and later vehicles except 1980 cars produced for sale in California that are equipped with the electronic carburetors (E2ME and E4ME) when idle mixture adjustment is required to meet air pollution control requirements. Idle mixture is preset at the factory to meet these requirements, and limiter caps or plugs have been placed on the adjustment screw to prevent tampering. These caps or plugs should not be removed, and no attempt should be made to adjust the idle mixture, unless the carburetor has received a major overhaul.

In addition, most General Motors carburetors have been modified so that backing out the idle mixture adjustment screw(s) does not necessarily enrich the mixture. In view of this, the time-honored "lean drop" system (or any other system, for that matter) cannot be substituted satisfactorily for the propane enrichment method.

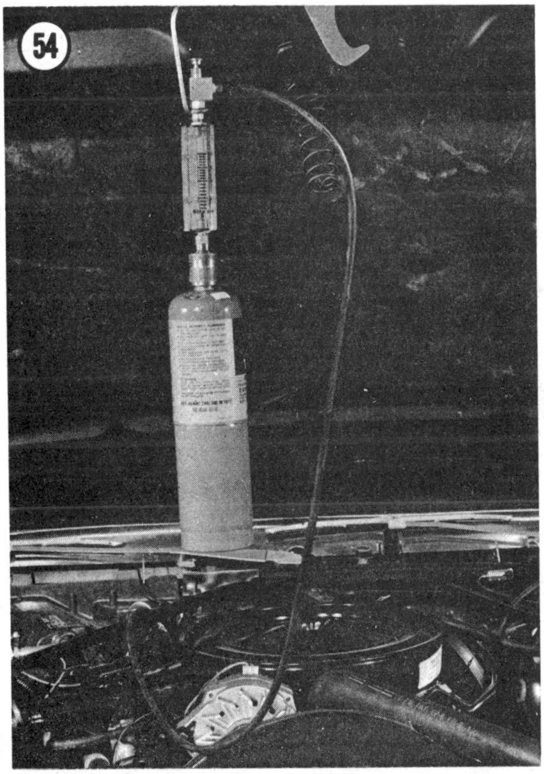

A special propane tool (General Motors J-26911) is required for this procedure. See **Figure 54**. You may be able to find a suitable substitute at an auto supply or parts store. If the tool is not available, do not try to adjust the idle speed mixture.

1. Set the parking brake and block the rear wheels. If the car is equipped with a vacuum brake release system, disconnect the hose at the brake and plug the hose.
2. Refer to the Vehicle Emissions Control Information (VECI) decal in the engine compartment and disconnect and plug the hoses listed.
3. Connect a tachometer to the engine, following the manufacturer's instructions.
4. Disconnect and plug the vacuum advance hose at the distributor and check and adjust the timing to the specification given on the VECI decal. See *Ignition Timing Adjustment*, this chapter.
5. If the engine is cold, allow it to run until it reaches normal operating temperature. Make sure the choke is open and the air conditioner (if so equipped) is off.
6. Set the idle speed to the rpm shown on the VECI decal.
7. Remove the crankcase ventilation hose from the air cleaner and insert the propane tool hose and plug into the hose opening in the air cleaner. See **Figure 55**. The propane cartridge must be in the vertical position as shown in **Figure 54**.

5

56

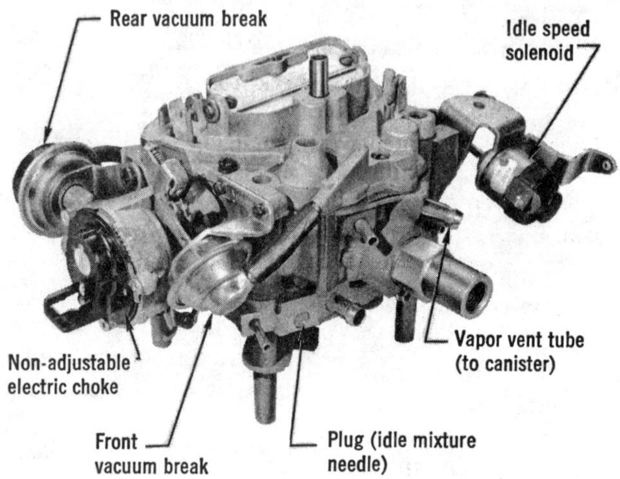
Rear vacuum break
Idle speed solenoid
Non-adjustable electric choke
Front vacuum break
Plug (idle mixture needle)
Vapor vent tube (to canister)

57

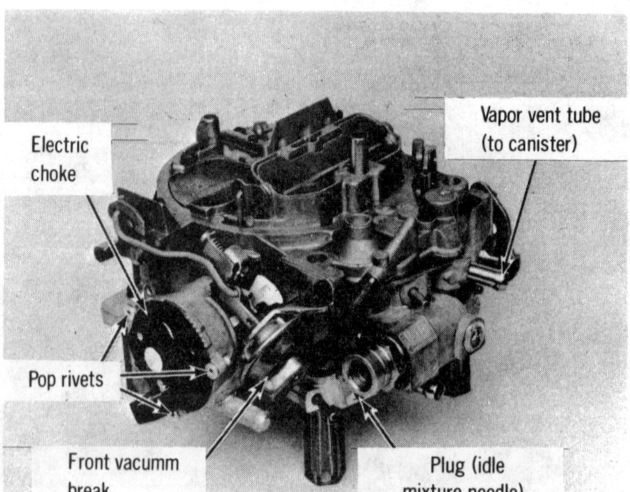
Electric choke
Vapor vent tube (to canister)
Pop rivets
Front vacumm break
Plug (idle mixture needle)

8. With the engine running at idle speed, slowly open the propane control valve until maximum speed is reached with the transmission in NEUTRAL.
9. If the speed is within the enriched idle speed range given on the VECI label, no further adjustment is required. If the speed is incorrect, remove the limiter caps or plugs from the idle mixture screws. See **Figure 56** (2bbl) or **Figure 57** (4bbl).
10. Turn the mixture screw(s) in (clockwise) until they are lightly seated, then back them out until the best idle point (maximum speed) is reached. Make sure propane is flowing during this adjustment.
11. Turn off propane and, if necessary, turn idle mixture screw(s) clockwise until the specified curb idle speed is reached.
12. Recheck the idle speed with the propane turned on. If not within the enriched idle speed specification, repeat Steps 10 and 11.
13. Turn the engine off and remove the propane tool. Reconnect the crankcase ventilation hose to the air cleaner.
14. Recheck the idle speed and reset to specification, if necessary, using the procedure on the VECI decal.

FUEL PUMP (DIESEL AND GASOLINE)

All engines use a non-serviceable mechanical fuel pump which cannot be repaired or rebuilt; the pump is replaced as a unit. Three tests can be performed to check fuel pump operation without removing the pump from the engine.

Pretest

1. Make sure there is fuel in the fuel tank.
2. Inspect for loose fuel line connections at the fuel pump and tighten if necessary.
3. Check fuel lines/hoses for kinks or bends which might reduce flow.

Flow Test

1. Disconnect fuel line at carburetor or filter inlet and connect an 8-10 inch length of rubber hose to line.
2. Place end of hose in quart-size container.
3. Start engine and run at idle for 30 seconds. If pump is satisfactory, it should pump at least one pint of fuel in 30 seconds.

Pressure Test

1. Disconnect fuel line at carburetor or filter inlet and connect an 8-10 inch length of rubber hose to line.
2. Connect low-reading pressure gauge to line and position about 16 inches above pump. Clamp off fuel return line on gasoline engines.
3. Start engine and run at slow idle. Pressure reading on gauge should be between 5 1/2 to 6 1/2 psi. If pressure is above or below this range or varies considerably at different engine speeds, replace pump.

5

Vacuum Test

1. Disconnect fuel tank-to-pump line at pump. Tie line upright to prevent fuel from running out.
2. Connect a vacuum gauge to fuel pump inlet with a short length of hose.
3. Start engine and note gauge reading. If less than 15 in. of vacuum, replace pump.

Removal/Installation

1. Disconnect negative battery cable.
2. Disconnect fuel inlet hose at pump.
3. Disconnect vapor return hose, if used.
4. Loosen fuel line at carburetor/injection pump and disconnect outlet line at fuel pump.
5. Remove hex-head mounting bolts and remove fuel pump and gasket.
6. Clean gasket residue from engine block mounting pad.
7. Position new gasket on mounting pad and install new fuel pump. Make sure pump rocker arm rides on crankshaft eccentric.
8. Install and tighten mounting bolts alternately and snugly.
9. Install fuel inlet and outlet hoses. Connect vapor return hose, if used.
10. Connect negative battery cable, start engine and check for leaks.

FUEL TANK

Removal/Installation

1. Remove ground cable from battery.

2. Disconnect meter wire at rear harness connector. Push out grommet and thread meter wire through trunk floor.
3. Raise vehicle and drain fuel tank.
4. Remove gauge ground wire screw from floor pan.
5. Disconnect fuel line at gauge pickup line.
6. Remove vent hoses.
7. Remove tank strap bolts and lower tank.
8. Reverse procedure to install tank.
9. Lower vehicle.

EXHAUST SYSTEMS

The exhaust system used in practically all vehicles consists of an exhaust pipe (or pipes in dual systems), mufflers, and tailpipes. Resonators, located between the muffler and the rear of the vehicle, are used on some models. In a typical installation, the exhaust pipes are constructed in one piece and are clamped to the muffler inlet. Tailpipes, except in installations having resonators, are also one piece and are, as a rule, welded to the muffler outlets. The following procedures are for typical installations and should be modified as required if the installation does not agree with the description given above.

MUFFLER AND TAILPIPE

Removal

1. Raise vehicle.
2. Remove clamp(s) at muffler inlet(s).
3. Remove tailpipe-to-hanger screws.
4. Disengage muffler(s) from exhaust pipe(s).
5. On single exhaust system, remove muffler and tailpipe. On dual exhaust system, cut tailpipe(s) from muffler(s), then remove muffler(s).

Installation

1. Connect new muffler(s) to exhaust pipe(s).
2. Install tailpipe(s) to new muffler and hanger.

NOTE

Do not tighten or fasten screws or clamps at this time.

3. Check complete length of exhaust system for clearance and adjustment as required.
4. Tighten all clamps, screws, and attachments, and weld tailpipe to muffler, if required.
5. Lower vehicle.

EXHAUST PIPE

Removal

1. Raise vehicle.
2. Remove nuts attaching exhaust pipe to exhaust manifold.
3. Remove clamp attaching exhaust pipe to muffler.
4. Remove exhaust pipe from muffler and then from vehicle.

Installation

1. Insert exhaust pipe into muffler and loosely install clamp.
2. Using new gaskets, loosely attach exhaust pipe to exhaust manifold with nuts.
3. Check all clearances between exhaust manifold and muffler.
4. Tighten clamp and nuts.
5. Lower vehicle.

CATALYTIC CONVERTER

Exhaust systems on 1975 and later gasoline engine models have a catalytic converter between the front exhaust pipe and the tailpipe. This is an emission control device which reduces hydrocarbon and carbon monoxide pollutants in the exhaust gases. The converter contains beads which are coated with a catalytic material containing platinum and palladium. The converter does not require periodic maintenance. However, it may become necessary to replace either the converter or catalytic beads. Procedures for replacement are given below. Replacement of the catalytic beads requires expensive special tools, making it more economical to have the job done by a dealer or qualified garage.

Removal/Installation

1. Raise the vehicle.
2. Disconnect converter at front and rear.
3. Remove converter.
4. Reverse procedure for installation.

Table 1 ROCHESTER MV SPECIFICATIONS

	1970	1971	1975	1976[1]
Float level	1/4 in.	1/4 in.	11/32 in.	11/32 in.
Choke rod (automatic trans.)	0.170 in.	0.160 in.	0.160 in.	0.100 in. Fed. 0.135 in. Cal.
Choke rod (manual trans.)	0.190 in.	0.180 in.	0.275 in.	0.140 in.
Vacuum break (automatic trans.)	0.200 in.	0.200 in.	0.200 in.	0.140 in. Fed. 0.165 in. Cal.
Vacuum break (manual trans.)	0.225 in.	0.225 in.	0.350 in.	0.165 in.
Choke unloader	0.350 in.	0.350 in.	0.275 in.	0.265 in.
Thermostatic choke rod	1 rod dia.	1 rod dia.	—	—
Metering rod	0.070 in.	0.080 in.	0.080 in.	0.080 in.
Auxiliary vacuum break (automatic transmission)	—	—	0.215 in.	0.260 in. Fed. 0.320 in. Cal.
Auxiliary vacuum break (manual transmission)	—	—	0.312 in.	0.320 in.

1. Fed. - cars sold new in any state except California.
   Cal. - cars sold new in California.

Table 2 ROCHESTER 2GC SPECIFICATIONS (1970-1973)

	1970	1971	1972	1973
Float level	9/16 in.	9/16 in.	17/32 in.	15/32 in.
Float drop	1 3/8 in.	1 3/8 in.	1 3/8 in.	1 9/32 in.
Choke rod	0.140 in.	0.140 in.	0.160 in.	0.160 in.
Vacuum break (manual trans.)	0.160 in.	0.200 in.	0.215 in.	—
Vacuum break (auto. trans.)	0.160 in.	0.200 in.[1]	0.200 in.	0.200 in.
Unloader	0.170 in.	0.170 in.	0.170 in.	0.250 in.
Pump rod	1 11/32 in.	1 3/8 in.	1 11/32 in.	1 11/32 in.

1. 0.215 on 455 CID engines.

Table 3 ROCHESTER 2GE/2GC SPECIFICATIONS (1977-1978)

	1977	1978
Float level	7/16 in.	7/16 in.
Float drop	1 5/32 in.	1 5/32 in.
Pump rod	1 19/32 in.	1 19/32 in.[1]
Choke rod (fast idle cam)	0.080 in.	0.080 in.
Vacuum break (throttle side)	0.110 in.[2]	0.110 in.[3]
Unloader	0.140 in.	0.140 in.

1. 1 9/16 in. on California models.
2. 0.130 in. for California auto. trans. and high altitude models;
   0.140 in. for California manual trans. models.
3. 0.140 in. for all California and high altitude models.

Table 4 ROCHESTER 4MC SPECIFICATIONS

	1970		1971	
Engine	350	455	350	455
Float level	¼ in.	¼ in.	¼ in.	¼ in.
Accelerator pump rod	⅜ in.	⅜ in.	⅜ in.	⅜ in.
Fast idle cam step	—	—	second	second
Choke rod	0.120 in.	0.120 in.	0.120 in.	0.120 in.
Vacuum break (manual trans.)	0.275 in.	0.200 in.	0.200 in.	0.200 in.
Vacuum break (automatic trans.)	0.200 in.	0.200 in.	0.200 in.	0.200 in.
Unloader	0.200 in.	0.200 in.	0.200 in.	0.200 in.
Air valve dashpot	0.030 in.	0.030 in.	0.050 in.	0.050 in.
	1972		**1973**	
Engine	**350**	**455**	**350**	**455**
Float level	¼ in.	¼ in.	¼ in.	¼ in.
Accelerator pump rod	⅜ in.	⅜ in.	—	—
Fast idle cam step	second	second	second	second
Choke rod	0.120 in.	0.120 in.	—	—
Vacuum break (manual trans.)	0.275 in.	0.230 in.	0.120 in.	0.120 in.
Vacuum break (automatic trans.)	0.230	0.215 in.	0.275 in.	0.215 in.
Unloader	0.200 in.	0.200 in.	0.200 in.	0.200 in.
Air valve dashpot	0.050 in.	0.050 in.	0.300 in.	0.300 in.
	1974			
Engine	**350**	**455**		
Float level	¼ in.	¼ in.		
Accelerator pump rod	⅜ in.	⅜ in.		
Fast idle cam step	second	second		
Fast idle cam gauge	0.230 in.	0.230 in.		
Choke rod	0.120 in.	0.120 in.		
Vacuum break (manual trans.)	0.200 in.	0.200 in.		
Vacuum break (automatic trans.)	0.200 in.	0.275 in.[1]		
Unloader	0.300 in.	0.300 in.		
Air valve dashpot	0.030 in.	0.030 in.		

1. 0.200 in. on L-75 engine option.

NOTE: If you own a 1981 and later model, first check the Supplement at the back of the book for any new service information.

CHAPTER SIX

EMISSION CONTROL SYSTEMS

All models covered by this book have a number of systems added to their engines to minimize harmful emissions. The numbers and types of systems vary from model to model and from year to year. Cars first sold in California have some systems that may not be found on cars that are first sold elsewhere, and vice versa.

Service to emission systems that can be performed by car owners is described in detail in Chapter Three. Other service, especially on late models, is not recommended. Most of the components are non-adjustable and non-repairable. In many cases, expensive tools and test equipment are required.

The principal emission control systems include those in the following list.

a. Positive crankcase ventilation (PCV)
b. Controlled combustion system (CCS)
c. Air injection reaction (AIR)
d. Fuel evaporation control system (ECS)
e. Exhaust gas recirculation system (EGR)
f. Carburetor calibration
g. Distributor calibration
h. Catalytic converter
i. Early fuel evaporation system (EFE)
j. Computer controlled catalytic converter system (C-4)

POSITIVE CRANKCASE VENTILATION

All vehicles have positive crankcase ventilation. Clean air is drawn from the air cleaner; the oil filler cap is not vented. The fresh air scavenges emissions (e.g., piston blowby) from the crankcase and manifold vacuum draws the emissions into the carburetor. Eventually they can be reburned in the normal combustion process. **Figure 1** shows the closed system.

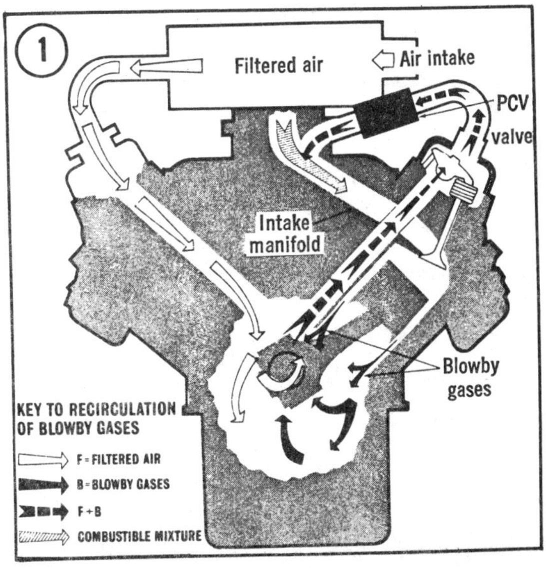

6

Either a PCV valve or mixed metered orifice mounted on the carburetor controls the volume of flow from crankcase to manifold. The valve or orifice tends to clog and must be removed and cleaned or replaced periodically. See Chapter Three. **Figure 2** shows a PCV valve.

CONTROLLED COMBUSTION SYSTEM

The controlled combustion system on 1970 and later models includes a special thermostatically controlled air cleaner, special calibrated carburetor and distributor, and a higher temperature thermostat.

The thermostatically controlled air cleaner maintains air to the carburetor at 100 degrees F or more. The air cleaner includes a temperature sensor, vacuum motor, and control damper assembly.

When the engine is off, absence of manifold vacuum permits the vacuum motor to close off the hot air pipe. When engine is running and underhood temperatures are below 85 degrees F, the temperature sensor bleed valve is closed and manifold vacuum operates the vacuum motor. The vacuum motor closes off the underhood air supply and the carburetor draws the much hotter air from the hot air pipe. See **Figure 3**. Between 85 degrees F and 128 degrees F, the temperature sensor air bleed is partially open. See **Figure 4**. The vacuum motor opens both the underhood air inlet and the hot air tube inlet. The resulting blend is maintained around 100 degrees F or more. Finally, if the underhood temperature is above 128 degrees F, the temperature sensor air bleed is fully open, the vacuum motor cannot operate and carburetor air is drawn from underhood air inlet only. See **Figure 6**.

Inspection

1. Check all heat pipe and hose connections.
2. Check for kinked or deteriorated hoses.
3. Remove air cleaner cover and install thermometer as close as possible to sensor. Install cover without wing nut. Lift cover after temperature stabilizes and read thermometer; temperature must be below 85 degrees F before proceeding. Put cover back in place.

2

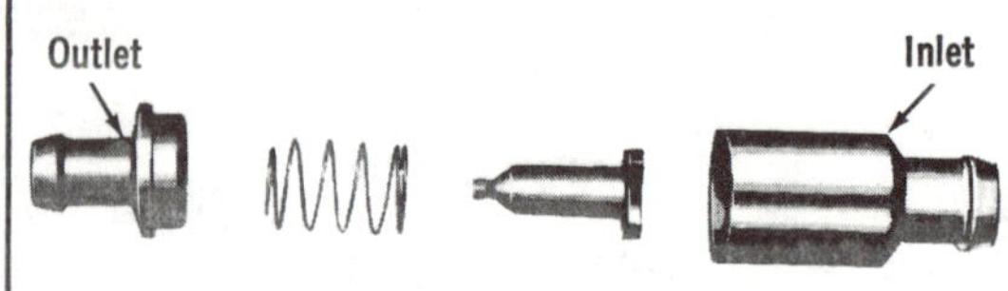

NOTE
Use a relatively fast acting thermometer such as a photographic darkroom thermometer. These are available for about $8.

4. With engine off, observe damper door position through snorkel opening. Snorkel passage should be completely open (**Figure 5**). If not, check for binds in linkage.
5. Start the engine. With the air temperature below 85 degrees F, snorkel passage should close. When damper door begins to open passage, remove air cleaner cover and observe temperature, which should be between 85-115 degrees F.
6. If damper door does not close completely or open at the right temperature, check the vacuum motor as described below.

Vacuum Motor Checking

1. Turn engine off. Disconnect vacuum hose between sensor and vacuum motor at sensor.
2. Suck on vacuum hose (or otherwise apply at least 9 in. Hg of vacuum). Damper door should completely close snorkel passage. If not, check for vacuum leak at other end of hose.

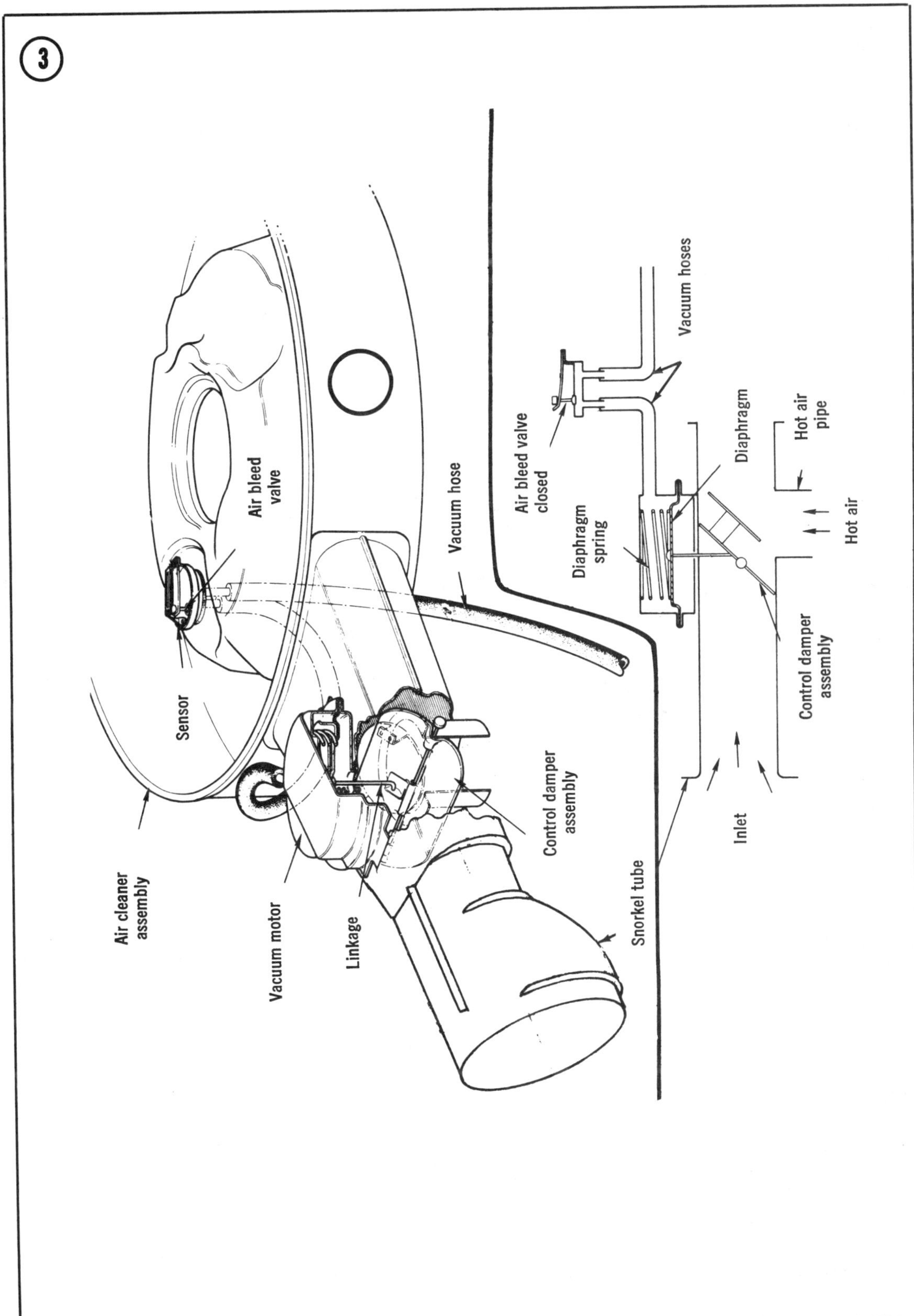
3
Air bleed valve
Sensor
Air cleaner assembly
Vacuum motor
Linkage
Vacuum hose
Control damper assembly
Snorkel tube
Air bleed valve closed
Diaphragm spring
Vacuum hoses
Diaphragm
Inlet
Control damper assembly
Hot air pipe
Hot air

4

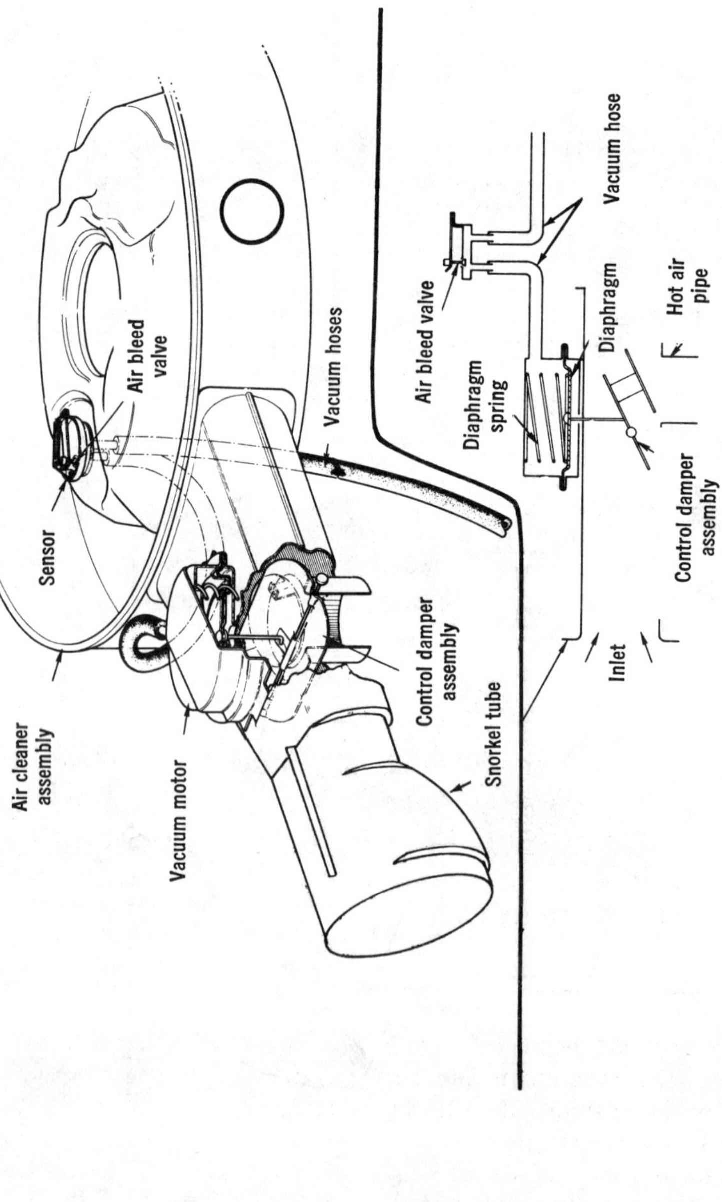
Air cleaner assembly
Vacuum motor
Sensor
Air bleed valve
Vacuum hoses
Control damper assembly
Snorkel tube
Inlet
Air bleed valve
Diaphragm spring
Diaphragm
Vacuum hose
Hot air pipe
Control damper assembly

5

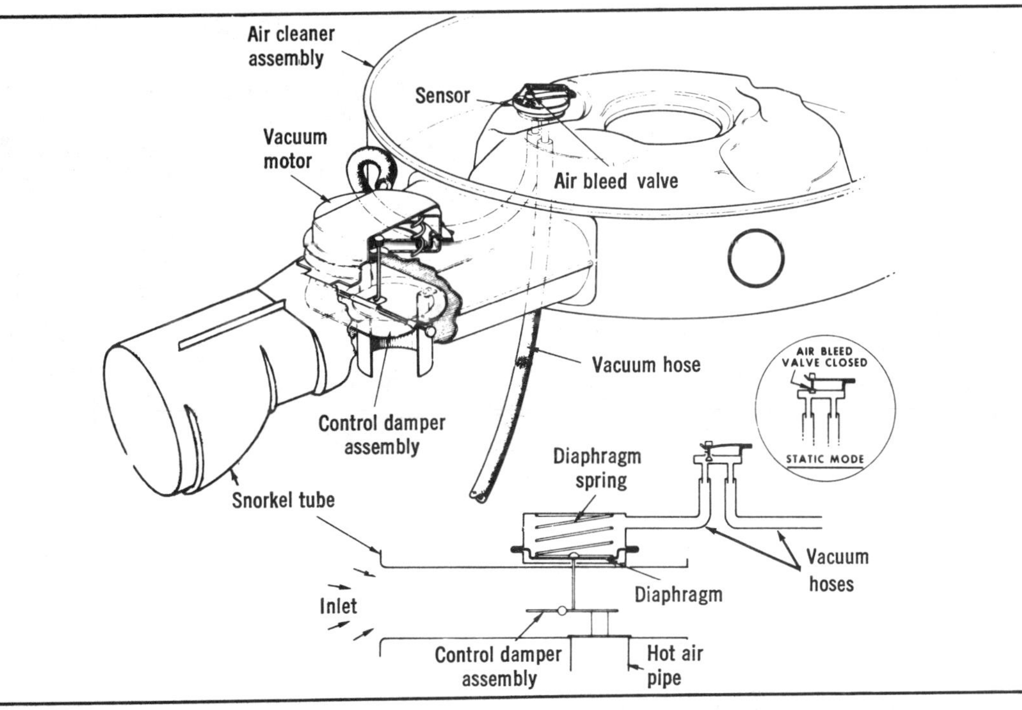

6

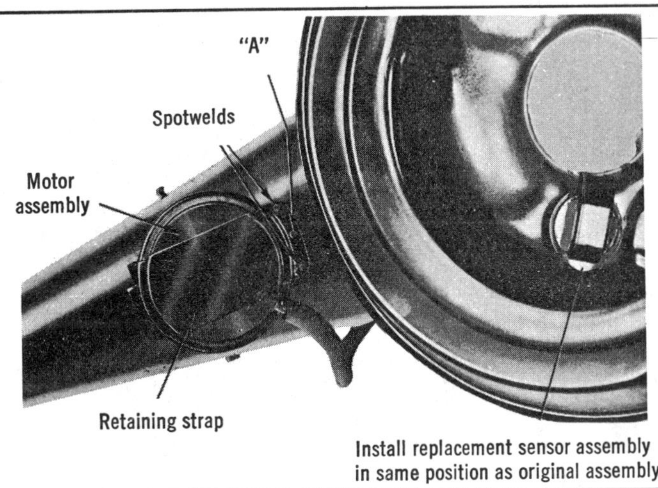

3. With vacuum applied, bend or tightly clamp hose to trap vacuum in motor. Damper door must remain in position. If not, the vacuum motor leaks and must be replaced.
4. If vacuum motor is good, yet system does not work properly, replace temperature sensor.

Vacuum Motor Replacement

Refer to **Figure 6** for the following procedure.

1. Remove air cleaner from engine.
2. Drill out spot welds fastening motor retaining strap to snorkel tube.
3. Unhook vacuum motor from damper door.
4. Drill 7/64 in. hole in snorkel tube at center of vacuum motor retaining strap.
5. Connect vacuum motor to damper door.
6. Fasten retaining strap to air cleaner with sheet metal screw.
7. Install air cleaner and check operation of vacuum motor.

Temperature Sensor Replacement

1. Remove air cleaner from engine and disconnect vacuum hose at sensor.
2. Pry up sensor clip tubes. See **Figure 7**.

NOTE
Observe position of sensor. New sensor must be installed in the same position.

3. Remove clip and sensor from air cleaner.
4. Install sensor and gasket in air cleaner in exactly the same position as the old one.
5. Press clip on sensor. Hold sensor by its sides only; do not touch control mechanism.
6. Install air cleaner and connect vacuum hoses.

7

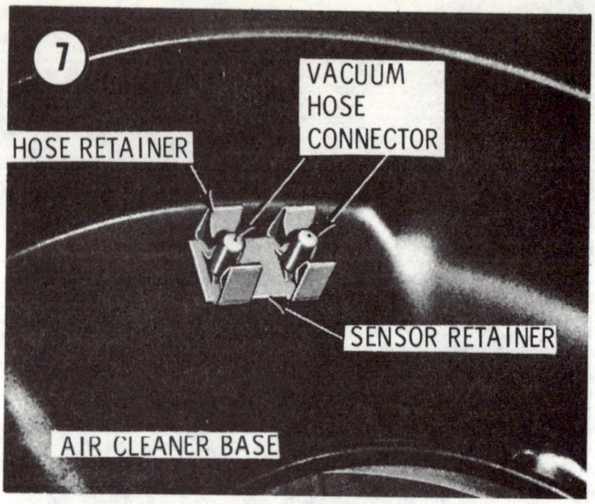

AIR INJECTION REACTOR SYSTEM

The air injection reactor system reduces air pollution by oxidizing hydrocarbons and carbon monoxide as they leave the combustion chamber. See **Figure 8**.

The air injection pump (**Figure 9**), driven by the engine, compresses filtered air and injects it at the exhaust port of each cylinder. The fresh air mixes with the unburned gases in the exhaust and promotes further burning.

Backfire is prevented by cutting the fresh air to the exhaust system and diverting the pump output to the atmosphere. See **Figure 10**.

The check valves prevent exhaust gases from entering and damaging the air pump if the pump becomes inoperative, e.g., from a drive belt failure. Under normal conditions, the pump delivers sufficient air pressure to prevent exhaust gases from entering the pump.

The air injection reactor system also depends on a special calibrated carburetor, distributor, and other related components.

8

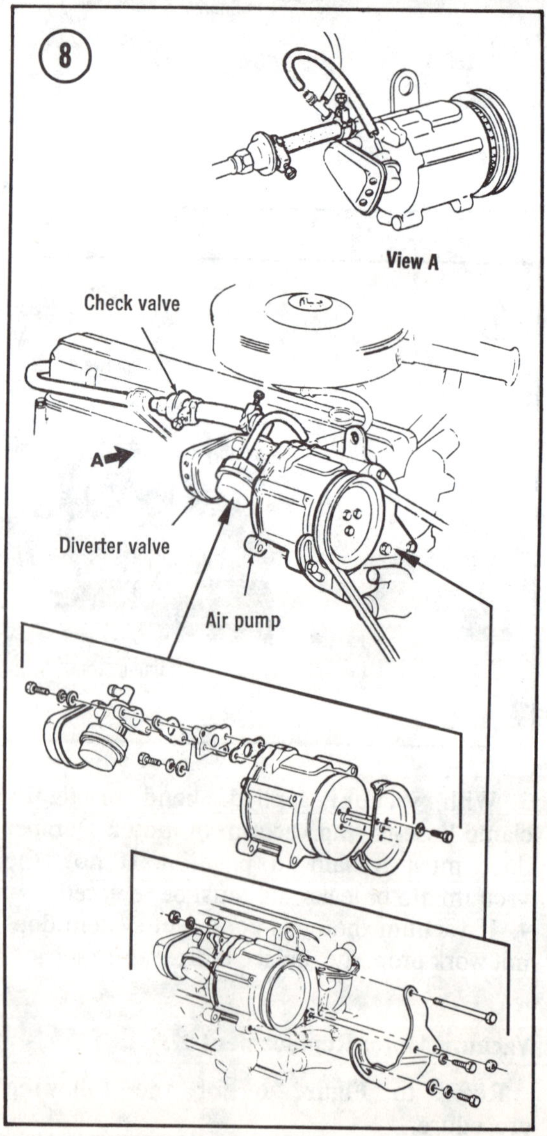

Air Injection Pump Removal/Installation

1. Disconnect hoses at pump.
2. Compress sides of pump belt tightly together to hold pump pulley bolts.
3. Loosen pump mounting bolt and pump adjustment bracket bolt. Swing pump until drive belt can be removed.

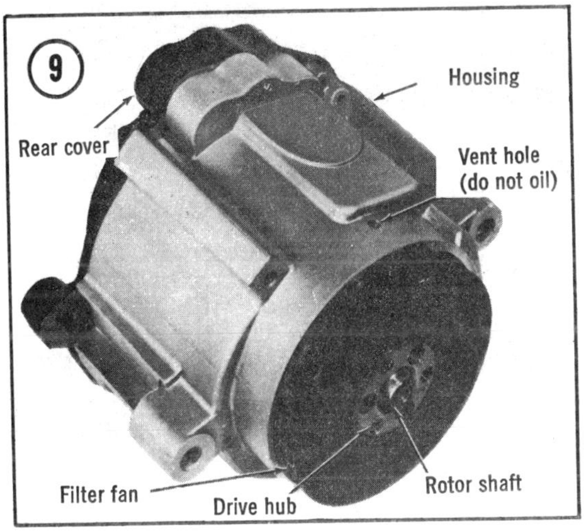

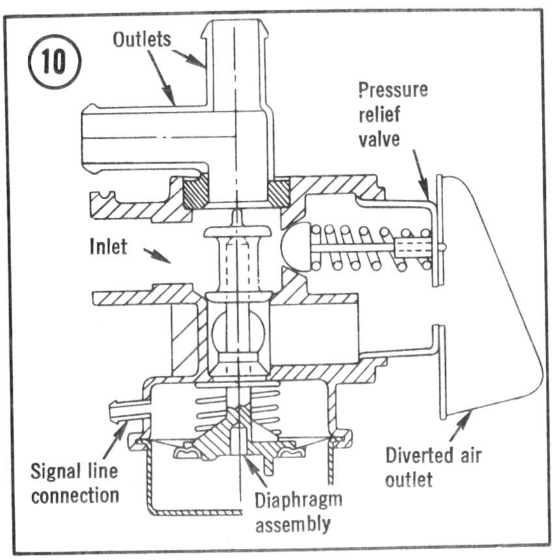

4. Remove pump pulley.
5. Remove pump mounting bolts and remove the pump.
6. Installation is the reverse of these steps.
7. Adjust belt tension (Chapter Three).

Air Injection Tube Removal/Installation

Refer to **Figure 11**.

1. Remove carbon from tubes.
2. Work tubes out of exhaust manifold. Use a penetrating oil, such as Liquid Wrench, liberally.
3. Install tubes in manifold. Use anti-seize compound on threads.

CAUTION
Straight pipe threads are used at the exhaust manifold. Do not use 1/4 in. tapered pipe threads.

Check Valve Replacement

1. Disconnect pump outlet hoses from valve.
2. Unscrew check valve from air manifold. Be careful not to bend the manifold.
3. Installation is the reverse of these steps.

Air Hose Replacement

To remove any hose or tube, note routing, then remove hose or tube. Install new hose or tube and tighten connection.

6

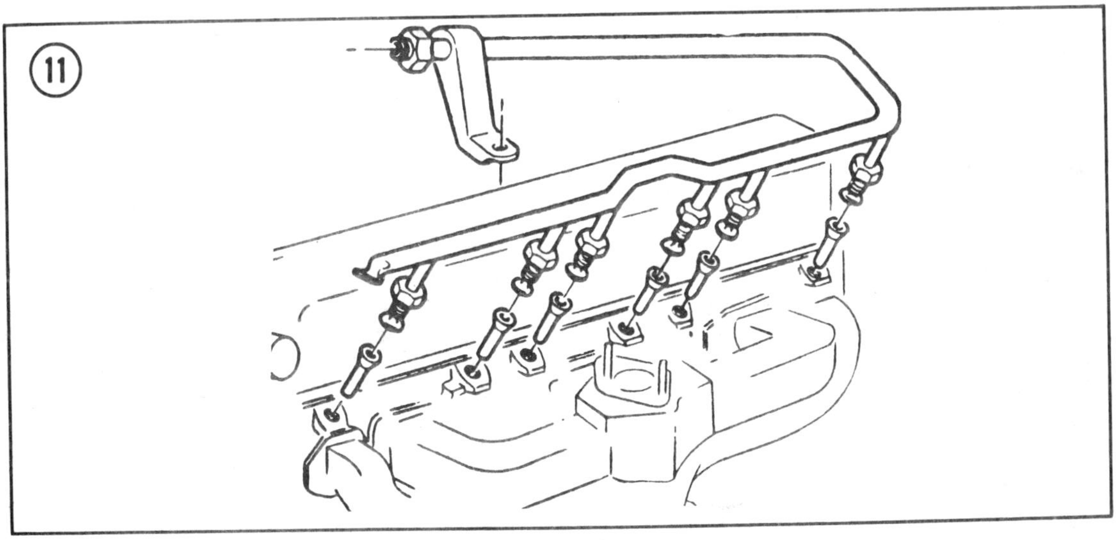

12

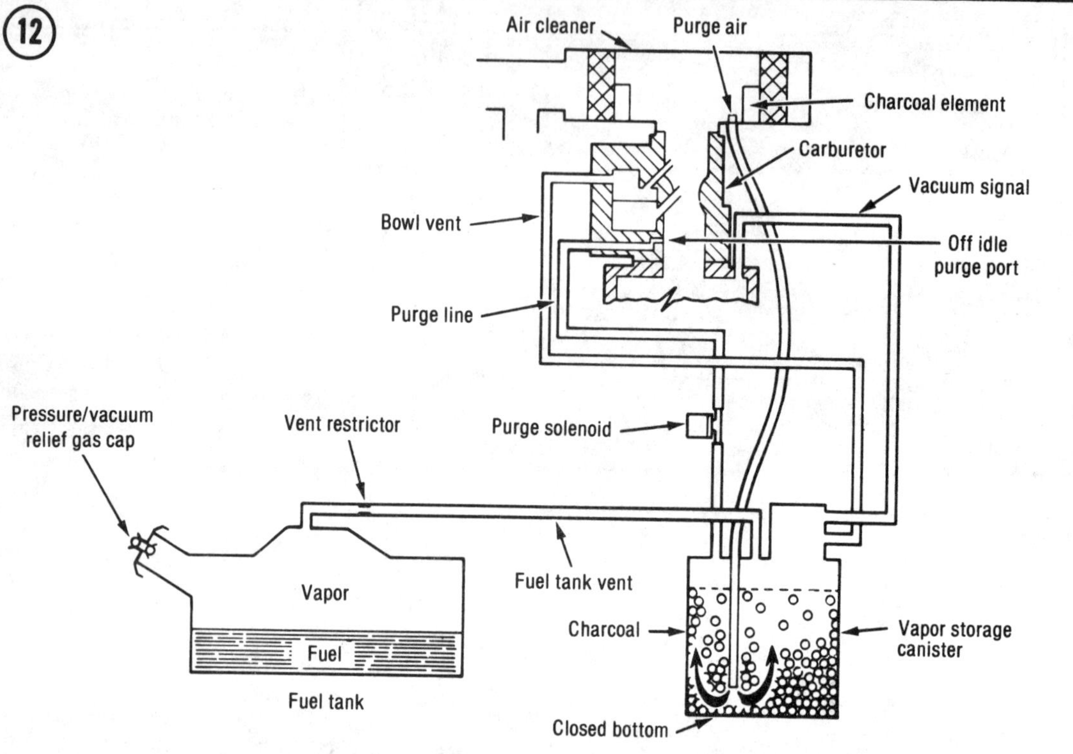

CAUTION
Air reactor system hoses are made from special materials to withstand high temperatures. Do not use substitutes.

FUEL EVAPORATION CONTROL SYSTEM

All 1972 and later vehicles (1970 and later in California) are equipped with a fuel evaporation control system which prevents release of fuel vapor into the atmosphere.

Refer to **Figure 12**. Fuel vapor from the fuel tank passes through the liquid/vapor separator to the carbon canister. The carbon absorbs and stores the vapor when the engine is stopped. When the engine runs, manifold vacuum draws the vapor from the canister. Instead of being released into the atmosphere, the fuel vapor takes part in the normal combustion process.

There is no preventive maintenance other than replacing the filter on the bottom of the carbon canister (if present) every 12,000 miles and checking tightness and condition of all lines connecting the parts of the system.

EXHAUST GAS RECIRCULATION (GASOLINE ENGINE, 1973-ON)

The Exhaust Gas Recirculation (EGR) system is used to reduce the emission of nitrogen oxides (NOX). Relatively inert exhaust gases are introduced into the combustion process to slightly reduce peak temperatures. This reduction in temperature reduces the formation of NOX.

The exhaust gases are introduced into the intake manifold by way of an EGR valve. See **Figure 13** (L6). This shut-off and metering valve operates on vacuum from the intake manifold via a signal port in the carburetor. On 1974 and later models, a thermal vacuum switch cuts off vacuum to the EGR valve until water temperature reaches 100-130 degrees F. At idle speed, recirculation is not required. Thus, the carburetor signal port is located above the throttle valve and vacuum to the EGR valve diaphragm is cut off at idle speeds. This causes the EGR to close, halting the introduction of exhaust gas to the intake manifold. As the throttle valve is opened, the

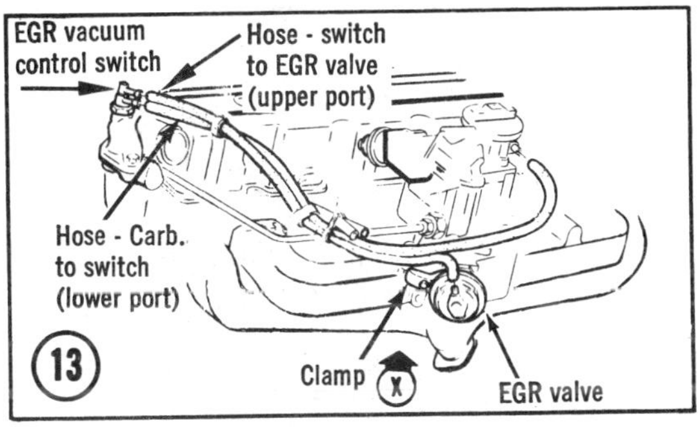

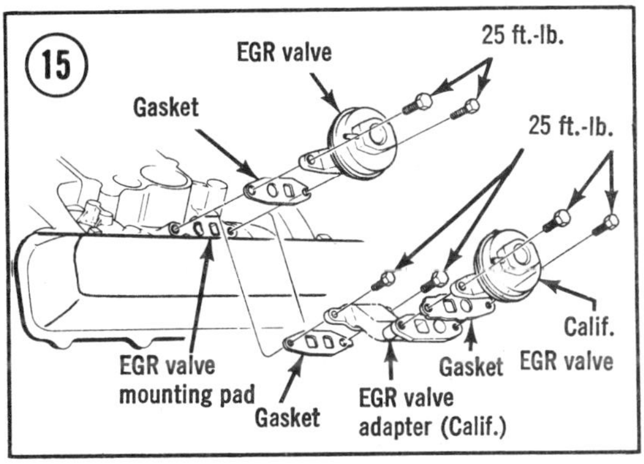

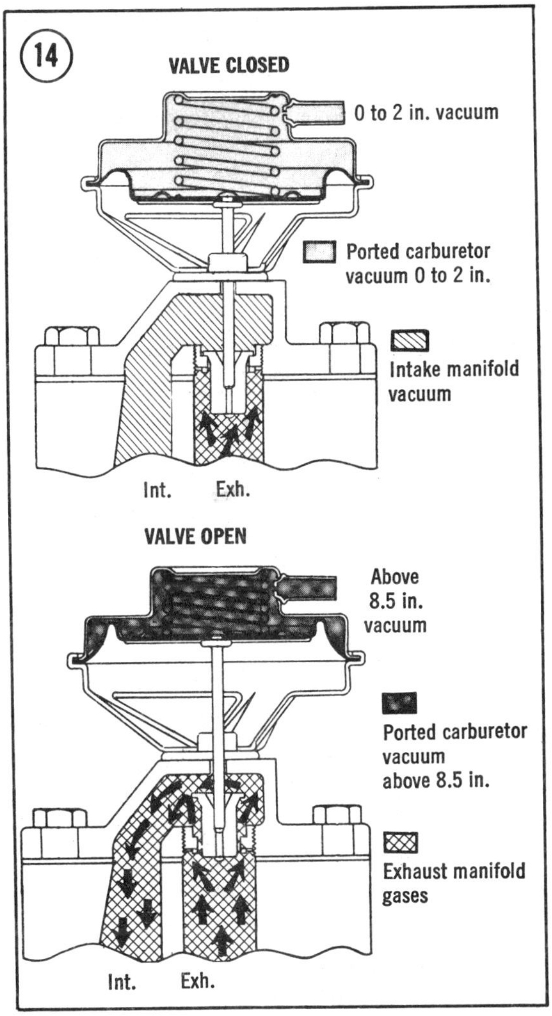

signal port is again exposed to manifold vacuum. This actuates the EGR valve diaphragm, which opens the valve and allows exhaust gas to be metered (through an orifice) into the intake manifold. See **Figure 14**.

On L6 engines, the EGR valve is located on the intake manifold next to the carburetor. See **Figure 13**. On V6 and V8 engines, the valve is located externally on the right side of the intake manifold next to the rocker arm cover. See **Figure 15**, typical.

EGR Valve Replacement

1. Disconnect vacuum line at top of valve.
2. Remove bolt and clamp securing valve to manifold, then remove valve from manifold.
3. Reassemble valve to manifold, using new gasket, and torque bolt to 25 ft.-lb. Bend lock tab up over bolt head. Replace vacuum line.

EGR Valve Cleaning

CAUTION

Do not wash valve assembly in solvents or degreaser. Permanent damage could result.

1. Use wire brush or wire wheel to clean valve base and remove exhaust deposits from mounting surface. Tapping with a soft-faced hammer may help.
2. A spark plug cleaner (sandblaster) can be used to clean valve seat and pintle. Insert valve and pintle into machine and blast for 30 seconds.

NOTE

Many service stations have this type of spark plug cleaner and will clean the valve for a small fee.

3. Compress the diaphragm spring so the valve is fully unseated and repeat sandblasting for 30 seconds.
4. Make sure all exhaust deposits have been removed. Repeat cleaning if required.
5. Use compressed air to remove all abrasive material from the valve.

Thermal Vacuum Switch Replacement

Refer to **Figure 16**.

1. Disconnect vacuum lines and remove switch from thermostat housing.
2. Apply a sealer to threads and install switch in thermostat housing. Torque to 15 ft.-lb.
3. Rotate head of switch as required for proper hose routing and install vacuum hoses.

NOTE

Thermal vacuum switch is non-repairable. If defective, replace.

EXHAUST GAS RECIRCULATION (DIESEL ENGINE)

Diesel engines used in 1980 models have an Exhaust Gas Recirculation (EGR) system to control NOx emission levels. The diesel EGR system is somewhat different from that used on gasoline engines. It is a throttle position-modified orifice system using a flow orifice and EGR valve (**Figure 17**).

16

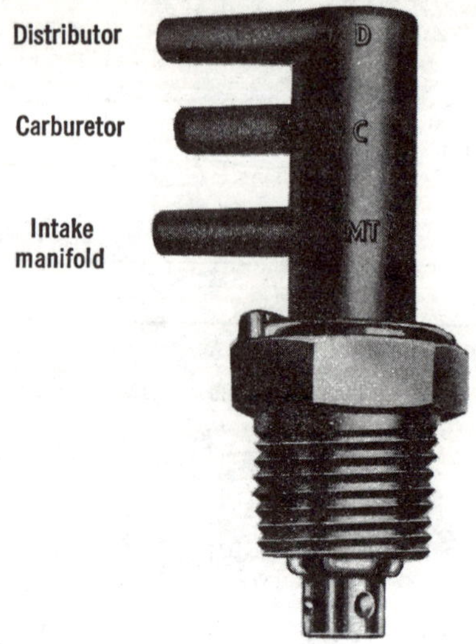

17

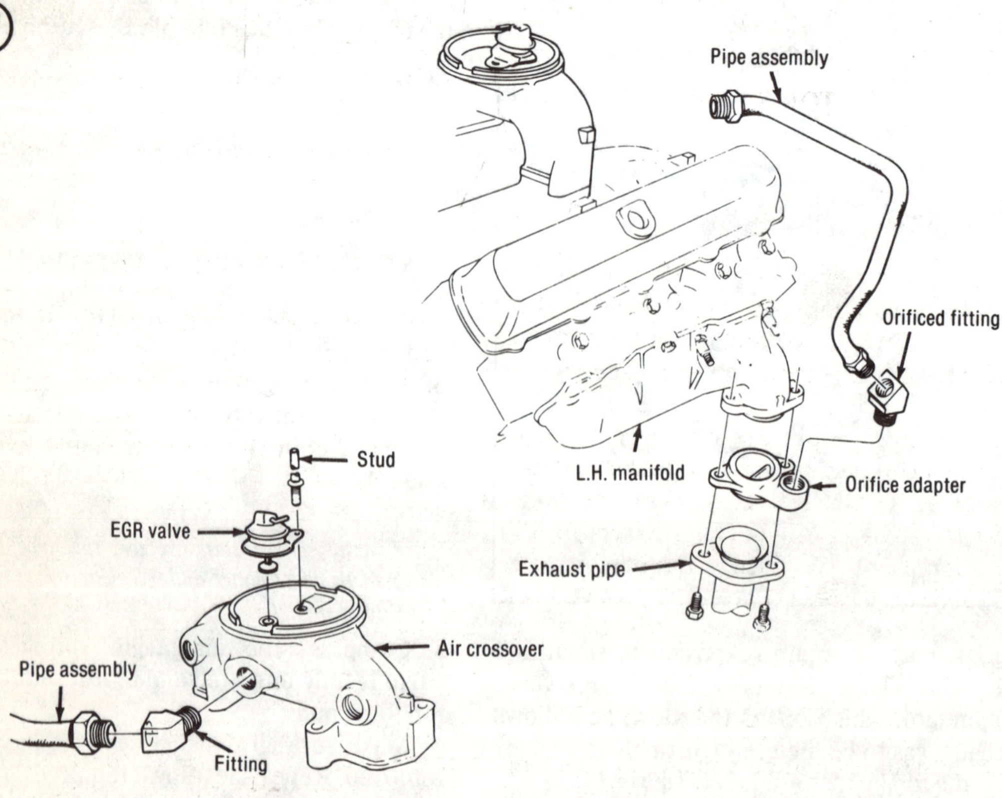

The orifice allows exhaust gas to flow into the intake manifold, except during wide-open throttle conditions. A vacuum regulator valve on the injection pump's throttle shaft senses throttle position. At wide-open throttle, it shuts off the vacuum signal to the EGR valve, closing the valve.

EGR System Diagnosis

Since the diesel EGR system is a constant-flow type, a system failure will be noticed only if the EGR valve does not close. If this happens, heavy smoke from the exhaust pipe will be seen during acceleration. This indicates that the valve is either stuck, binding, or receiving a vacuum signal when it shouldn't.

If heavy smoke is noted on medium but not heavy acceleration, the vacuum switch or regulator valve is at fault. If you notice these symptoms, check to make sure the vacuum line under the air cleaner is properly connected to the EGR valve and have your Olds dealer check the EGR system operation. No other service of the EGR system is necessary.

CARBURETOR CALIBRATION

In addition to providing the engine with a combustible mixture of air and fuel, the carburetor is also calibrated to maintain proper emission levels. The idle, off-idle, power enrichments, main metering, and accelerator pump systems all are calibrated to provide the best possible combination of performance, economy, and exhaust emission control.

Calibration is especially critical on 1975 and later models, and the tasks involved require special skills and test equipment the home mechanic is not likely to have. Except for the adjustments of idle speed and idle mixture described in Chapter Three, carburetor work on late models should be entrusted to an expert.

DISTRIBUTOR CALIBRATION

Distributor calibration consists of adjusting initial timing, centrifugal advance, and vacuum advance to obtain the best combination of engine performance, fuel economy, and exhaust emission level control. Timing specifications for each engine are given on the Vehicle Emission Control Information (VECI) decal located in the engine compartment. See *Ignition Timing Adjustment*, Chapter Three, for instructions on how to set the initial timing. Adjustment or repair of the centrifugal and vacuum spark advance systems on 1975 and later cars should be left to an expert.

CATALYTIC CONVERTER

Catalytic converters, used since 1975, reduce air pollutants by promoting further burning of the exhaust gases. The converter is located in the exhaust line ahead of the muffler and contains a material coated with platinum and palladium. Both of these materials are catalysts, and cause the reduction of hydrocarbons and carbon monoxide by burning as the exhaust gases pass over them. The catalytic converter should be checked for general condition at the same time the remainder of the exhaust system is checked (see *Periodic Maintenance*, Chapter Three). See *Catalytic Converter*, Chapter Five, for replacement procedure.

6

EARLY FUEL EVAPORATION SYSTEM

The early fuel evaporation (EFE) system, used on 1975 and later models, provides heating to the intake manifold while the engine is cold, to promote vaporization of the fuel. This helps cut down on choke time and also promotes more thorough burning of the fuel. This, in turn, reduces the amount of pollutants released into the air. The heart of the system is the EFE valve, which controls the amount of heat (exhaust gas) directed under the intake manifold. The valve is controlled by a thermostatic switch in the cooling system that applies or removes vacuum to the EFE valve according to the temperature of the engine coolant. Use the following procedure to check operation of the EFE system.

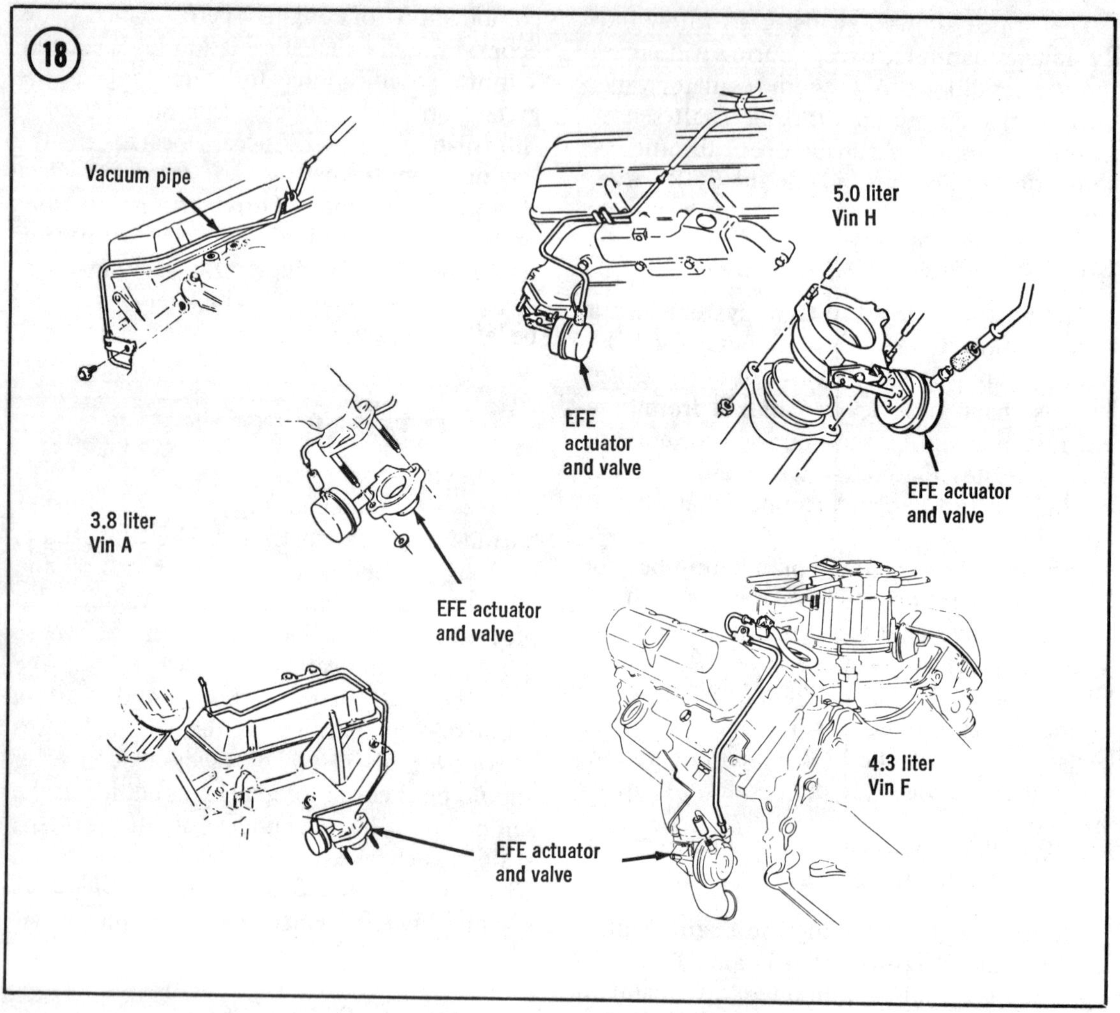

EFE System Check

Refer to **Figure 18** (V6 and V8 engines) or **Figure 19** (L6 engines).

1. With engine cold, place transmission in NEUTRAL or PARK, apply handbrake, and start engine. Observe movement of EFE actuator rod and heat valve. Valve should move to closed position.
2. If valve does not close, remove vacuum hose from actuator and check for presence of vacuum. If vacuum is present, replace the actuator. If no vacuum is present, remove vacuum input hose at thermovacuum switch (TVS) and check for vacuum. If vacuum is present, replace TVS. If no vacuum is present, check for damaged vacuum hose and replace as required.

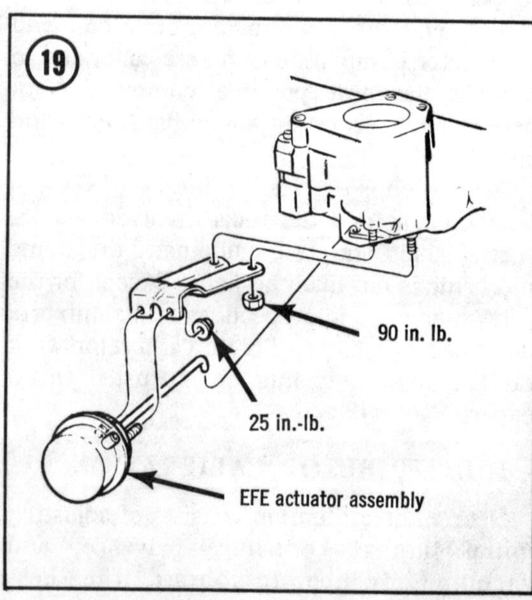

20

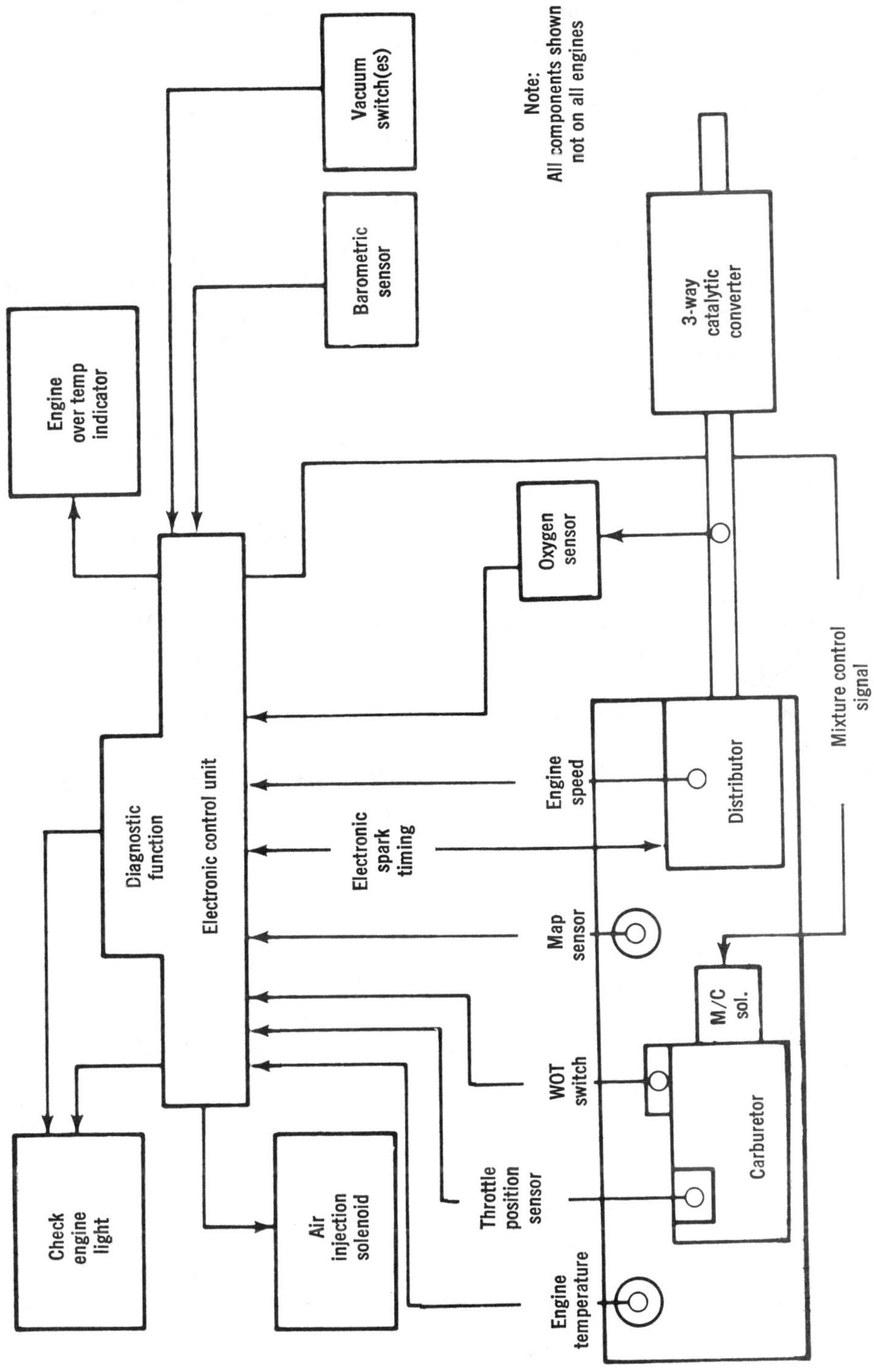
Vacuum switch(es)
Barometric sensor
Engine over temp indicator
Note:
All components shown not on all engines
3-way catalytic converter
Oxygen sensor
Diagnostic function
Electronic control unit
Engine speed
Distributor
Electronic spark timing
Mixture control signal
Map sensor
M/C sol.
WOT switch
Carburetor
Throttle position sensor
Check engine light
Air injection solenoid
Engine temperature

3. If valve closes, allow engine to warm up until coolant temperature reaches 180 degrees F (V6 and V8) or 150 degrees F (L6). Exhaust heat valve should move to open position. If not, remove hose from actuator and check for vacuum. If no vacuum is present, replace actuator. If vacuum is present, replace TVS.

NOTE
When replacing TVS on V8 engines, drain coolant below level of coolant outlet housing. Apply a soft setting sealant to replacement switch threads before installation, and torque to 120 in.-lb.

COMPUTER CONTROLLED CATALYTIC CONVERTER SYSTEM (C-4)

The computer controlled catalytic converter system used on some 1980 models (those with electronic carburetors) reduces harmful emissions by constantly controlling the air/fuel ratio and by using a three-way catalytic converter. A schematic diagram of the system is shown in **Figure 20**.

The essential components of the system are an oxygen sensor in the exhaust system (OS), a three-way catalytic converter (ORC), an electronically-controlled carburetor, and an electronic control module (ECM).

Other elements of the system are a barometric pressure sensor (BARO), an engine temperature sensor, a manifold absolute pressure sensor (MAP), a lean authority limit switch, a throttle position sensor, and a vacuum control switch.

The ECM receives and processes signals from the various sensors, and uses the processed data to control an air injection solenoid and a mixture control solenoid on the carburetor. This provides the correct air/fuel mixture for every operating condition.

In addition, the catalytic converter used with the C-4 system contains 3 catalysts: platinum, palladium, and rhodium. These elements promote further burning of the exhaust gases and help reduce the levels of oxides of nitrogen (NOX), hydrocarbons (HC), and carbon monoxide (CO). Most conventional catalytic converters use only platinum and palladium as catalysts.

All service to the C-4 system and its components should be entrusted to an expert. This includes the carburetor.

CHAPTER SEVEN

COOLING, HEATING, AND AIR CONDITIONING SYSTEMS

All models use pressurized cooling systems, sealed with a pressure type radiator cap. The higher operating pressure of the system raises the boiling point of the coolant. This increases the efficiency of the radiator.

COMPONENTS

Cooling system components are the radiator, pressure cap, water pump, thermostat, fan, thermostatic fan clutch (on 1972 and later models so equipped), and associated hoses and water passages.

Radiator

The cross-flow type radiator, used on the Cutlass, cools the coolant fluid by allowing it to flow from side to side through the radiator and transferring heat to the air passing through the radiator.

Radiator Cap

The pressure-type radiator cap (**Figure 1**) allows the cooling system to operate at higher than atmospheric pressure, thus raising the boiling point of the coolant. This permits operation of the vehicle with coolant temperatures between 245 degrees F and 260 degrees F without vaporization of the coolant through boiling. The pressure cap has a pressure relief valve which allows excessive pressure to be vented. It also has a vacuum valve which opens to relieve the vacuum created when the system cools.

Fan/Fan Clutch

The cooling fan increases cooling system efficiency by drawing air through the radiator. Two types of fans are used. One is a fixed drive fan which rotates at water pump speed.

1

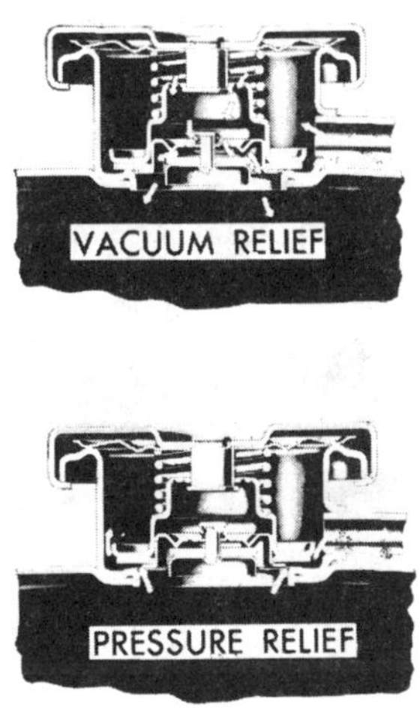

7

2

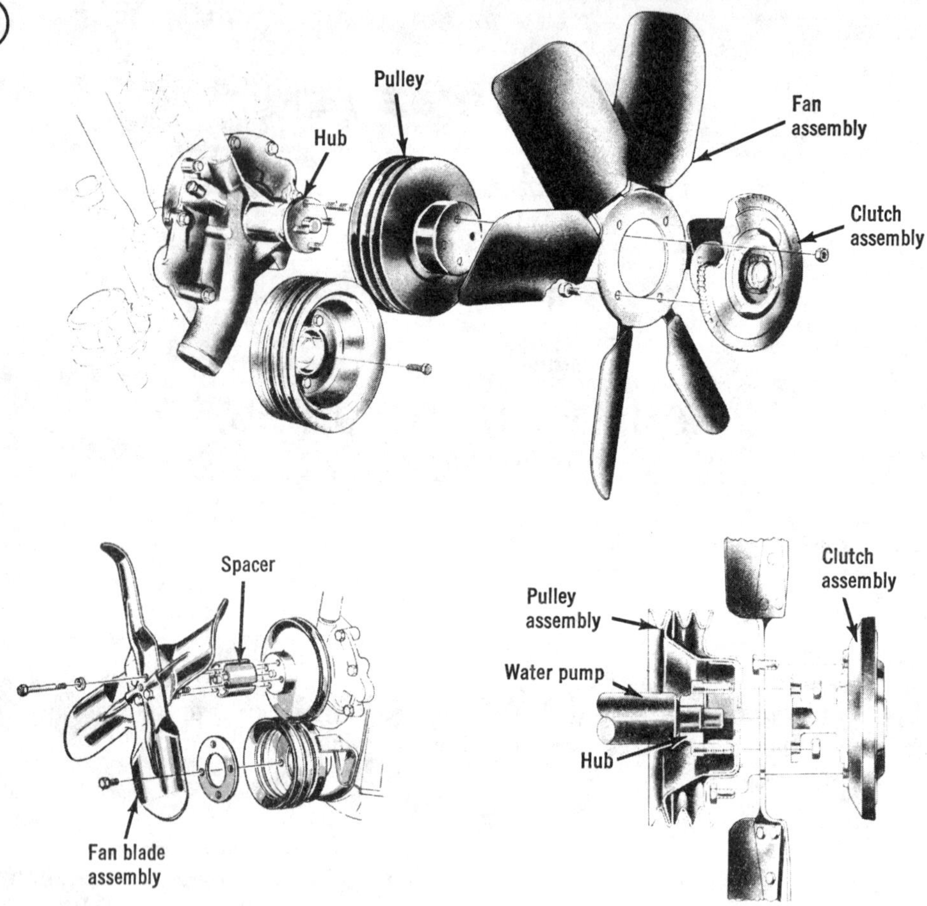
Pulley
Hub
Fan
assembly
Clutch
assembly
Spacer
Fan blade
assembly
Clutch
assembly
Pulley
assembly
Water pump
Hub

3

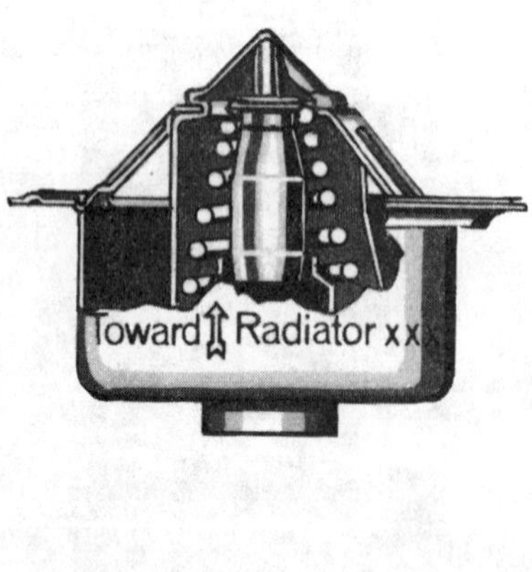
Toward Radiator xx

4

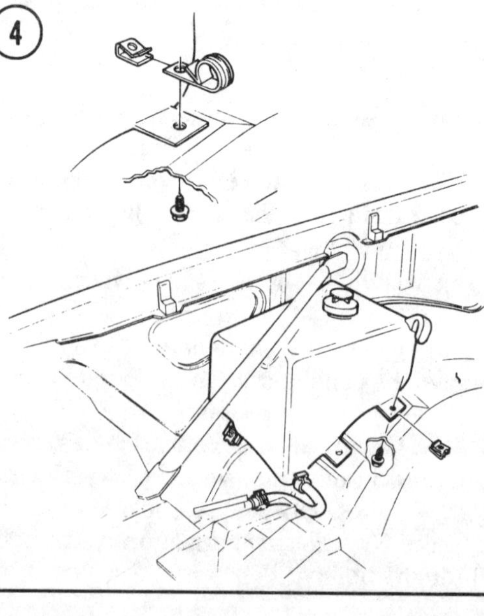

The other type has a thermostatically-controlled clutch which ensures adequate cooling at reduced engine speeds and eliminates overcooling, excessive noise, and power loss at high speeds. See **Figure 2**. The clutch is used on cars equipped with air conditioning or the heavy duty cooling option.

NOTE
Oldsmobile has indicated that fan/blade assemblies on 1978 models may break apart during operation. 1978 Cutlass models should be returned to an Olds dealer for fan inspection and replacement, if not already done.

Thermostat

The thermostat (**Figure 3**) is basically a heat-controlled valve in the cooling system, located in the engine water passage outlet to the radiator. When the engine and coolant are cold, the valve remains closed, preventing the circulation of coolant through the radiator. When the engine and coolant approach normal operating temperature, the thermostat opens to allow coolant to flow to the radiator for cooling.

Coolant

Only ethylene glycol-based coolant meeting the requirements of GM Specification 1899-M should be used.

Coolant Recovery System

This system, which has a translucent reservoir containing a supplemental supply of coolant, catches the coolant that escapes through the radiator cap pressure relief valve. When the engine cools, fluid is drawn back into the radiator by vacuum. See **Figure 4**.

Water Pump

The centrifugal vane impeller water pump in the cooling system has sealed bearings and requires no periodic maintenance. The pump inlet is connected to the bottom of the radiator by a hose. The pump causes water to circulate through the water passages in the engine block and cylinder heads, where engine heat is transferred to the water. The engine and head passages are connected and terminate at the water outlet of the thermostat housing, which is connected to the top of the radiator by a hose.

MAINTENANCE

Coolant level

NOTE
On vehicles equipped with coolant recovery systems the coolant is checked by observing the liquid in the recovery system reservoir. The radiator cap should not be removed. If additional coolant is required, it should be added to the recovery reservoir. Level should be at COLD FULL mark when engine is cool or at HOT FULL mark when engine is hot.

If vehicle does not have a coolant recovery system, coolant level should be checked only when engine is cool.

CAUTION
Removal of radiator cap from a hot engine, especially if an air conditioner has been in use, can result in coolant "blow out." This results in coolant loss and could cause injuries to bystanders.

Coolant level in cross-flow radiators should be maintained at 3 inches below the bottom of the filler neck when the system is cold.

Cooling System Checks

1. Check the radiator by warming up engine. Turn engine off and feel radiator. Cross-flow radiator should be hot along left side and warm along right side with an even temperature rise from right to left. Cold spots indicate obstructed radiator sections.
2. Check water pump operation by running the warmed up engine while squeezing upper radiator hose. If a pressure surge is felt, the water pump is functioning. If not, check for plugged vent hole in pump.
3. Check for exhaust leaks into cooling system by draining coolant until level is just above top of cylinder head. Disconnect upper radiator hose and remove thermostat and fan belt. Start engine and accelerate engine several times while observing coolant. If level rises or bubbles appear, chances are that exhaust gases are leaking into the cooling system.

Periodic Maintenance

The concentration of antifreeze in the coolant should be maintained to provide protection to -20 degrees, regardless of expected temperatures. This is required for protection from corrosion and for proper temperature indicator light operation.

Every 2 years the cooling system should be drained and back flushed with clear water. If required, the system should be cleaned with a good cleaning solution (follow manufacturer's instructions). Back flush the system by removing radiator upper and lower hoses and replacing radiator cap. Attach a lead-away hose to upper radiator opening and a length of garden hoses to lower opening. If an air/water flushing gun is available, connect it to garden hose and connect gun water hose to water supply and air hose to compressed air supply. Allow water to fill radiator, then apply air in short bursts. Repeat process until water coming from lead-away hose runs clear. If gun is not available, connect garden hose to pressurized water supply and back flush until water from lead-away hose runs clear. Then the system should be refilled with ethylene glycol-based coolant/water solution sufficient to provide protection to at least -20 degrees F. GM Cooling System Inhibitor and Sealer, or equivalent, should also be added at this time to retard the formation of rust or scale. This inhibitor should also be added every fall thereafter.

CAUTION
Alcohol or methanol coolants or plain water are not recommended for use at any time.

Fan Belt Adjustment

Loosen bolts at alternator mounting and pull unit away from engine until desired tension (75 +/- 5 lb. used or 125 +/- 5 lb. new) is reached (use strand tension gauge). Tighten all mounting bolts securely and repeat tension check. Readjust if required.

Thermostat Check and Replacement

1. Drain coolant until level is slightly below thermostat housing base.

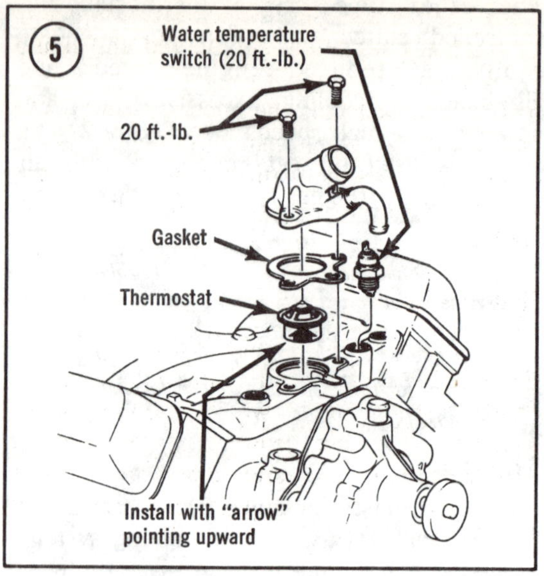

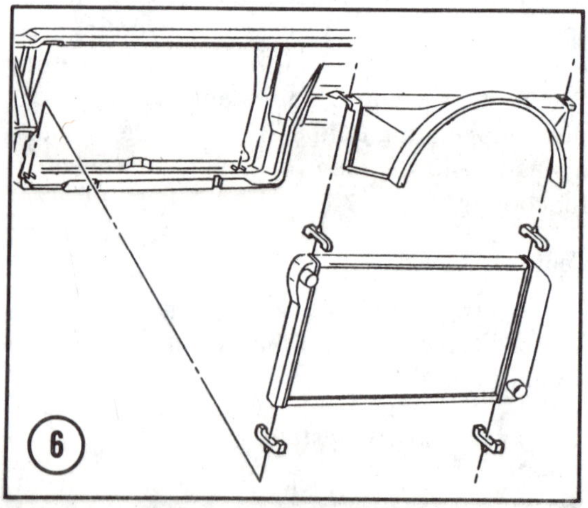

2. Remove upper radiator hose.
3. Remove thermostat housing bolts and then remove water outlet and gasket from thermostat housing. See **Figure 5**.
4. Remove and inspect thermostat valve for condition.
5. Test thermostat valve as follows:
   a. Place thermostat in a 33% solution of glycol heated to 25 degrees F above the temperature stamped on the thermostat.
   b. Submerge thermostat and agitate liquid. Valve should open fully.
   c. Remove valve and place in another 33% glycol solution heated to 10 degrees F under the temperature stamped on the thermostat.

d. Thermostat should close completely when completely submerged and liquid is agitated.

6. If thermostat fails the above test it should be replaced. If OK, reinstall in housing, using a new gasket. Tighten bolts securely.

7. Replace radiator hose and refill with coolant.

Belt and Hoses

All engine belts and hoses should be inspected for wear and damage at each oil change period. Belts which show signs of wear, cracking, etc., and hoses with bulges, soft spots, leaks, etc., should be replaced. The inspection should include heater hoses as well as radiator hoses.

Thermostatic Fan Clutch

Check with engine cool. Fan blade should rotate freely, with only a slight drag. If fan does not move, or drag is either excessive (or a rough grating is felt) or not present (fan revolves over 5 times when spun by hand) the clutch should be replaced.

WATER PUMP

Removal

1. Siphon or drain coolant from radiator. Remove fan shroud if equipped, and break loose fan pulley bolts.

WARNING

If coolant is syphoned from radiator, do not use mouth to start syphon. Coolant solution is poisonous and can cause death or serious illness if swallowed.

2. Disconnect heater hose, lower radiator hose and bypass hose (if so equipped) from water pump.

3. Remove alternator upper brace (V6 and V8 engines), loosen swivel bolt and remove fan belt. If so equipped, disconnect power steering and air conditioner belts and swivel power steering pump unit to one side.

4. Remove fan blade attaching bolts, fan, and pulley.

CAUTION

Bent or damaged fans should not be reused, as any distortion will affect fan balance and operation. Damaged fans cannot be properly repaired and should be replaced.

NOTE

When a thermostatic fan clutch is removed from the car it should be supported so that clutch disc remains vertical to prevent silicone fluid leakage.

5. Remove pump-to-cylinder block and power steering-to-pump bolts. Remove water pump and old gasket from engine.

CAUTION

On L6 cylinder engines, pull pump straight forward out of block first to avoid impeller damage.

Installation

1. Install water pump assembly on engine block, using a new sealer-coated gasket. Tighten bolts to 15 ft.-lb. for L6 engines, 22 ft.-lb. for V6 and V8's (13 ft.-lb, if self-tapping screws are used).

2. Install pump pulley and fan on pump hub and tighten to 15 ft.-lb.

NOTE

For ease in aligning pulley and fan, install a 5/16 in.-24 guide stud (bolt with head removed) in one hole of the fan hub. Remove stud and install bolt after other 3 bolts have been started.

3. If disconnected, install power steering and air conditioner belts.

4. Connect hoses and fill cooling system with solution of ethylene glycol antifreeze sufficient to withstand -20 degrees.

5. Install alternator upper brace (V6 and V8 only) and power steering pump bolt. Install fan belts and tension to 75 +/- 5 lb. (used belt) or 125 +/- 5 lb. (new belt).

6. Start engine and inspect installation for leaks.

RADIATOR

Removal

Refer to **Figure 6** for this procedure.

1. Disconnect the battery negative cable at

the battery. Then drain the radiator, remove the fan shroud (if so equipped), and remove the fan. See **Figure** 7 (typical).

NOTE

If the vehicle is equipped with a fan clutch, store the fan clutch in an upright position to prevent leakage of the clutch solution through the seal.

2. Remove the upper and lower radiator hoses from the radiator.
3. If the vehicle has an automatic transmission, disconnect the transmission fluid cooler lines from the radiator. Plug the lines to prevent fluid loss and contamination.
4. Remove the radiator attaching bolts and remove the radiator and fan shroud assembly by lifting it straight up.

INSTALLATION

1. If a new radiator is being installed, remove the necessary fittings from the old radiator and install them on the new one.
2. Install the radiator in the vehicle and securely tighten the attaching bolts.
3. Reconnect the automatic transmission fluid cooler lines to the radiator.
4. Connect the upper and lower radiator hoses to the radiator. If the hoses are more than 2 years old or are in poor condition, it is a good idea to replace them at this time.
5. Install the fan and fan shroud and fill the radiator with the proper solution of ethylene glycol antifreeze and water. Follow the antifreeze manufacturer's instructions to arrive at the suitable solution for your climate.
6. Connect the battery negative cable and start the engine. Check the water hoses and transmission cooler line connections for leaks. Also check the automatic transmission fluid level, if so equipped.

HEATER

Blower Motor Replacement (1970-1977)

Refer to **Figure 8**.

1. Disconnect the battery negative cable at the battery and then disconnect the electrical lead wire from the heater blower motor.

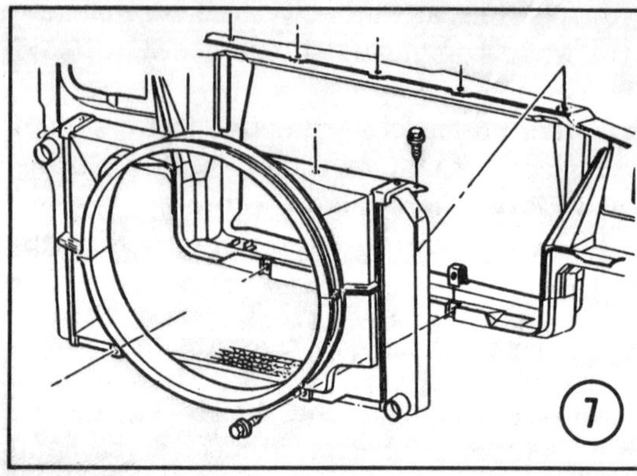

7

8

2. Remove the blower-to-case attaching screws and remove the blower assembly from the vehicle. Pry gently on the flange, if necessary, if the sealer does not allow easy removal.

NOTE

On some early models, it may be necessary to shift the fender skirt to remove the blower assembly. In this case, remove all the skirt attaching screws except those holding the skirt to the radiator support. Then insert a 2 in. x 4 in. wood block between the skirt and the fender to provide the necessary clearance.

3. Remove the blower wheel retaining nut and separate the wheel and motor.
4. Installation is the reverse of these steps. Make sure the open end of the blower wheel is away from the blower motor.

Blower Motor Replacement (1978-On)

Refer to **Figure 9**.

1. Disconnect the battery negative cable and then disconnect the electrical connector at the

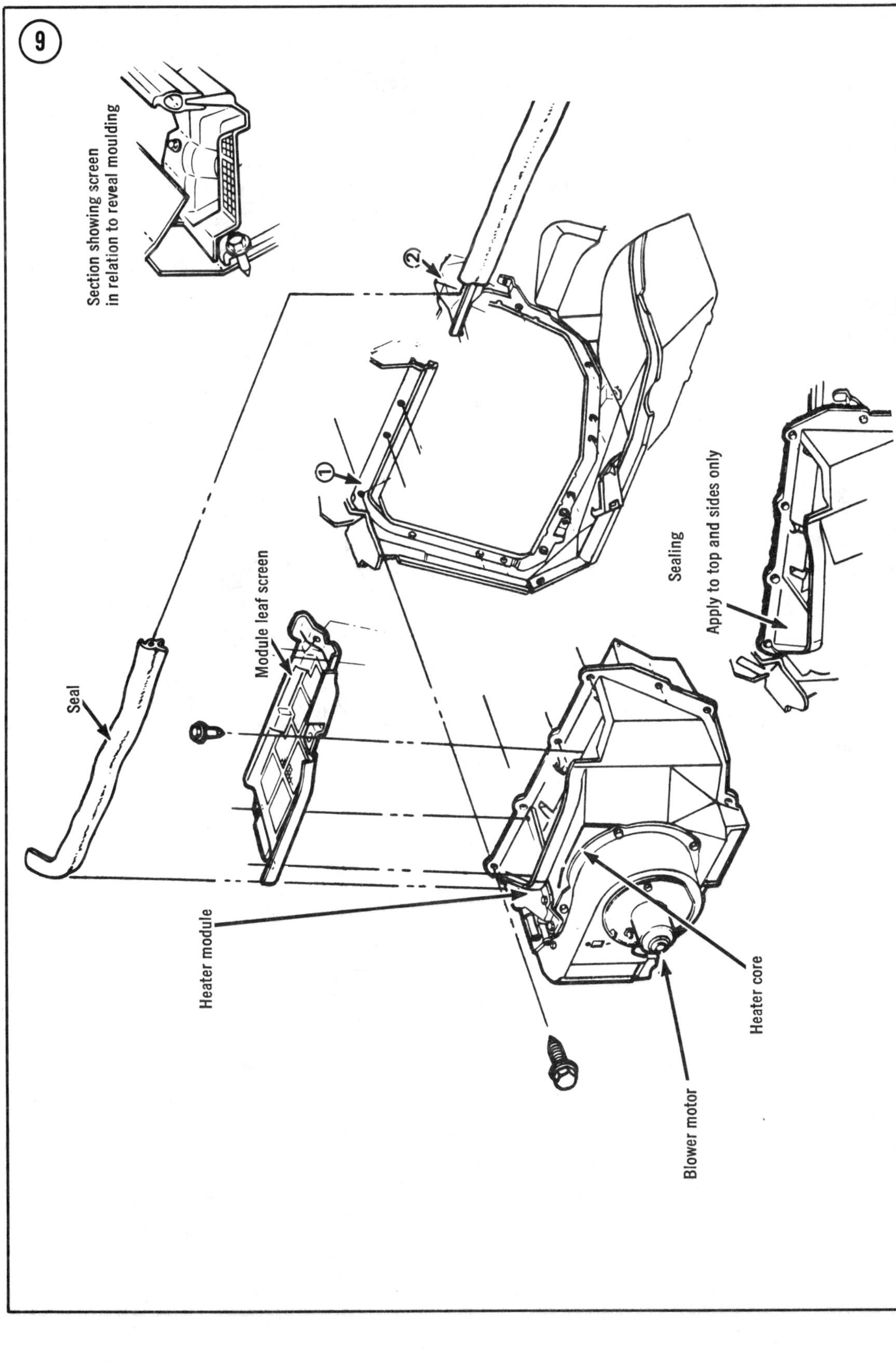
9
Section showing screen
in relation to reveal moulding
1
2
Seal
Module leaf screen
Sealing
Apply to top and sides only
Heater module
Heater core
Blower motor

blower assembly. Disconnect heater and radio ground straps.

2. Remove the blower-to-case attaching screws and remove the blower motor assembly.

3. Installation is the reverse of these steps. Make sure the ground straps are reinstalled in their original locations.

Control Head Assembly Replacement (All Models)

Refer to **Figure 10** (typical).

1. Disconnect battery negative cable and disconnect heater control wiring connectors.
2. Remove the instrument cluster bezel. See **Figure 11** (typical).
3. Remove the radio center speaker, if so equipped and if necessary for clearance.
4. Remove the attaching screws from the control assembly and pull the assembly far enough to the rear to disconnect the cables. Also disconnect the wiring harness from the blower switch. Then remove the control assembly. Take care not to kink the cables.
5. Installation is the reverse of these steps.

AIR CONDITIONING

Major service and repair to air conditioning systems requires specialized training and tools, and the difficulty of the work is compounded in the late heating/air conditioning systems. However, most air conditioning problems do not involve major repair; they are well within the ability of an experienced hobbyist mechanic, armed with an understanding of how the system works.

10

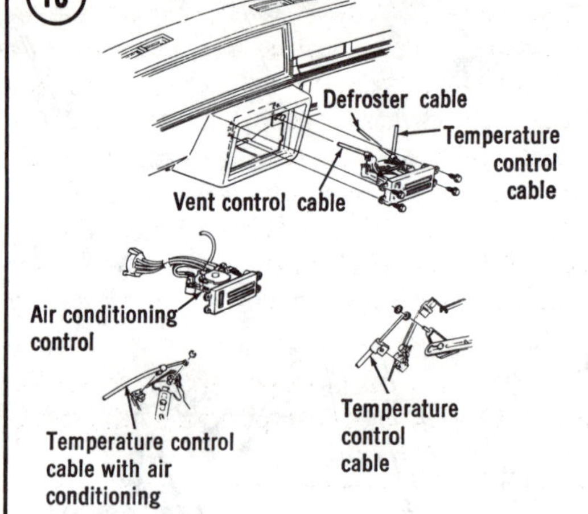

11

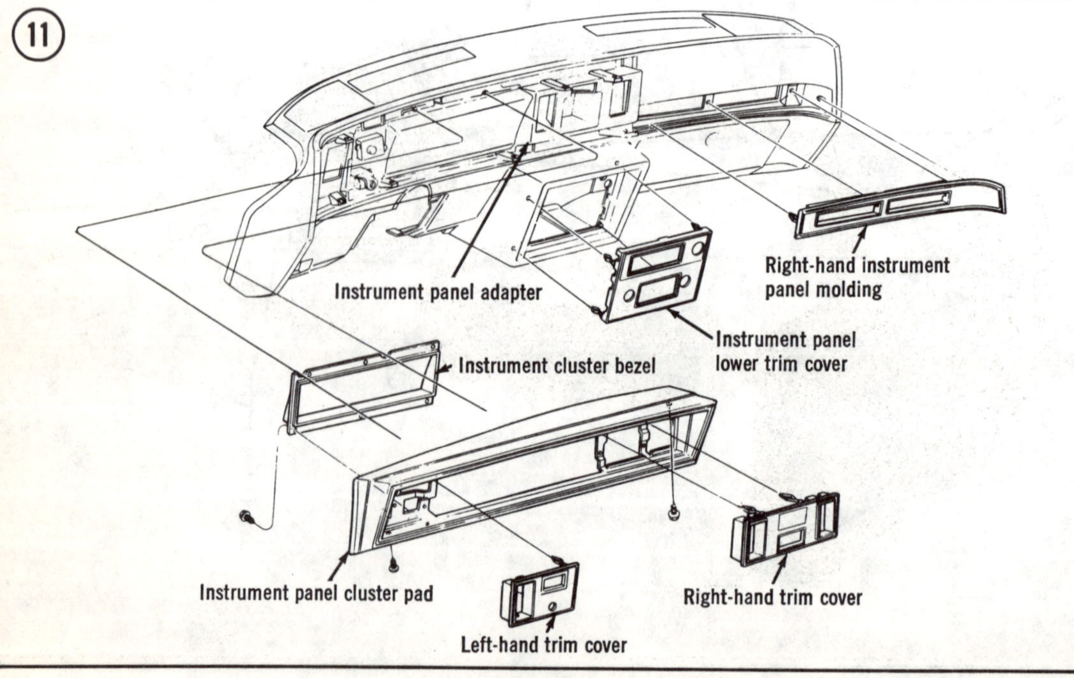

SYSTEM OPERATION

A typical air conditioning system is shown in **Figure 12**. Actual component locations may differ, depending on model.

The five basic components are common to all air conditioning systems:

1. Compressor
2. Condenser
3. Receiver/drier
4. Expansion valve
5. Evaporator

The components, connected with high-pressure hoses and tubes, form a closed loop. A refrigerant, dichlorodifluoromethane—more commonly referred to as R-12, circulates through the system under high-pressure—as much as 300 psi. As a result, work on the air conditioning system is potentially hazardous if certain precautions are ignored. For safety's sake *read this entire section* before attempting any troubleshooting, checks, or work on the system.

A typical system is shown schematically in **Figure 13**. For practical purposes, the cycle begins at the compressor. The refrigerant, in a warm, low-pressure vapor state, enters the low-pressure side of the compressor. It is compressed to a high-pressure hot vapor and pumped out of the high-pressure side to the condenser.

Air flow through the condenser removes heat from the refrigerant and transfers the heat to the outside air. As the heat is removed, the refrigerant condenses to a warm, high-pressure liquid.

The refrigerant then flows to the receiver/drier where moisture is removed and impurities are filtered out. The refrigerant is stored in the receiver/drier until it is needed. Generally, the receiver/drier incorporates a sight glass that permits visual monitoring of the condition of the refrigerant as it flows. This is discussed later.

From the receiver/drier, the refrigerant flows to the expansion valve. The expansion valve is thermostatically controlled and meters refrigerant to the evaporator. As the refrigerant leaves the expansion valve it changes from a warm, high-pressure liquid to a cold, low-pressure liquid.

7

12

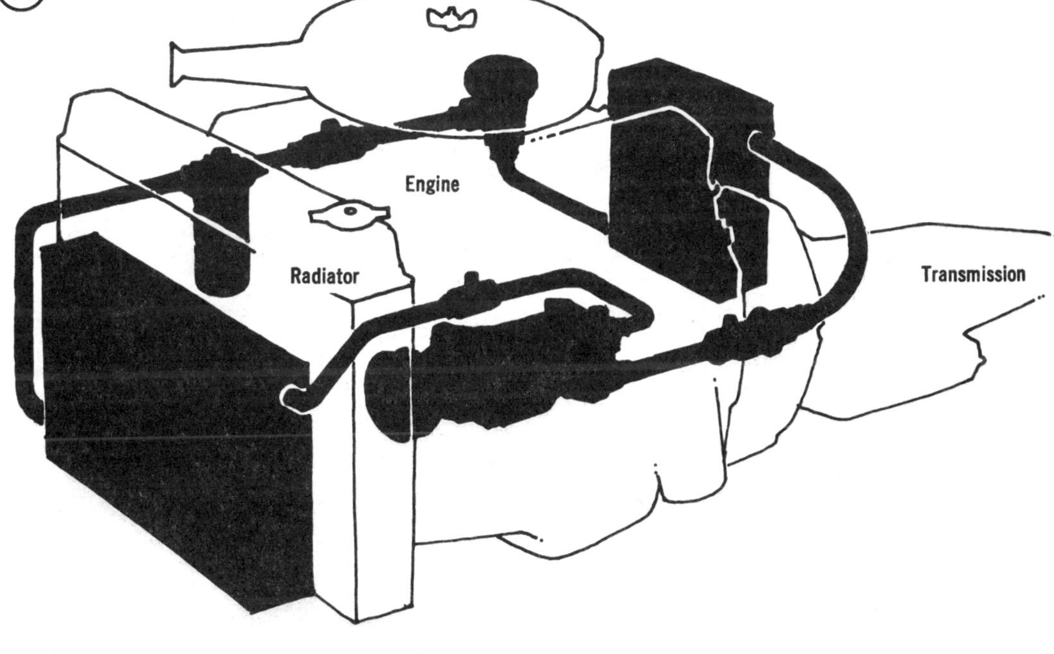

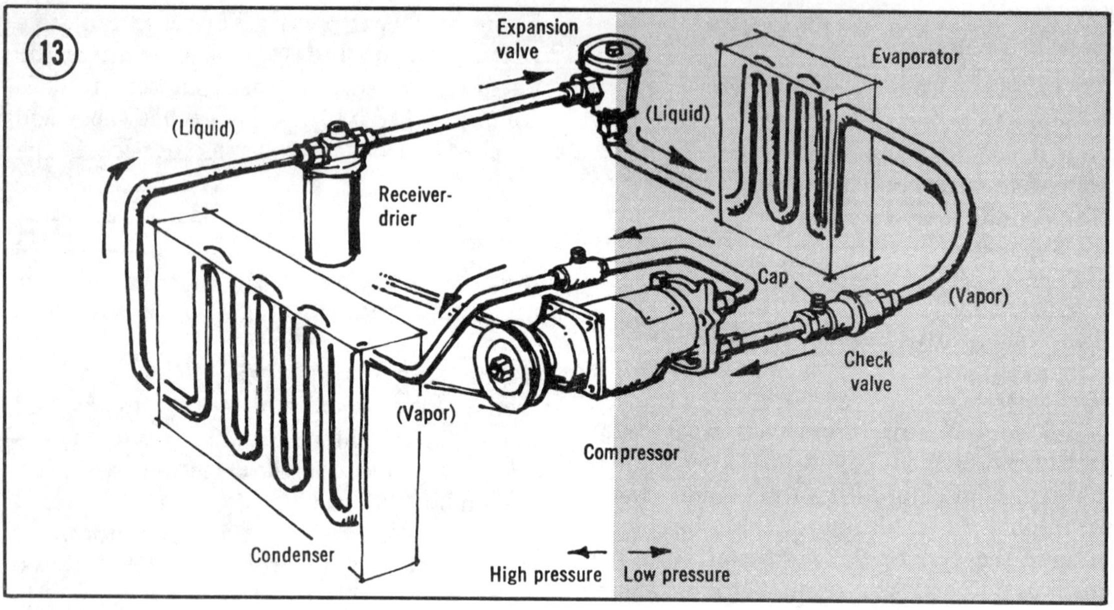

In the evaporator, the refrigerant removes heat from the cockpit air that is blown across the evaporator's fins and tubes. In the process, the refrigerant changes from a cold, low-pressure liquid to a warm, high-pressure vapor which flows back to the compressor where the refrigeration cycle began.

GET TO KNOW YOUR VEHICLE'S SYSTEM

With **Figure 13** as a guide, begin with the compressor and locate each of the following components in turn:

1. Compressor
2. Condenser
3. Receiver/drier
4. Expansion valve
5. Evaporator

Compressor

The compressor is located on the front of the engine, like an alternator, and is driven by one or two drive belts (**Figure 14**). The large pulley on the front contains an electromagnetic clutch that is activated and operates the compressor when the air conditioning controls are switched on. There are 2 compressor types—piston-and-crank (**Figure 15**), and swashplate (axial plate). See **Figure 16**.

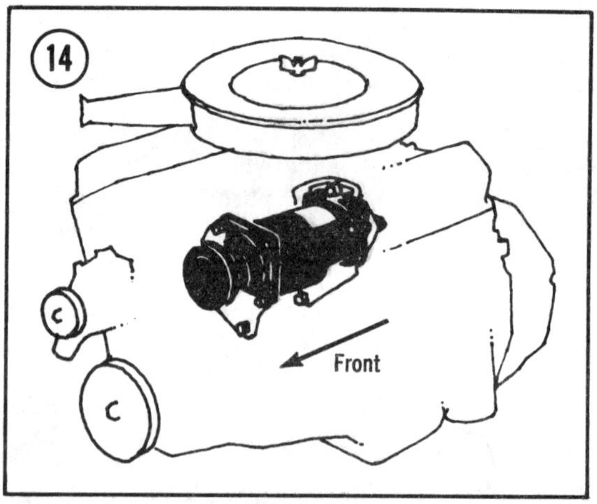

Condenser

In most cases, the condenser is mounted in the front of the radiator (**Figure 17**). Air passing through the fins and tubes removes heat from the refrigerant in the same manner it removes heat from the engine coolant as it passes through the radiator.

Receiver/Drier

The receiver/drier is a small tank-like unit (**Figure 18**), usually found mounted to one of the wheel wells. Many receiver/driers incorporate a sight glass through which refrigerant flow can be seen when the system is operating (**Figure 19**). Some systems have

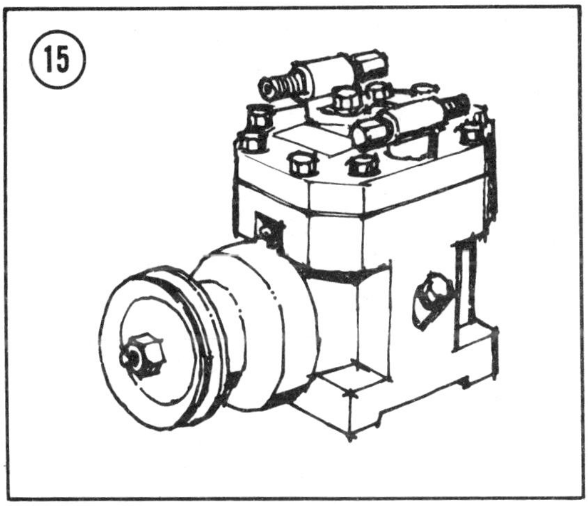

15

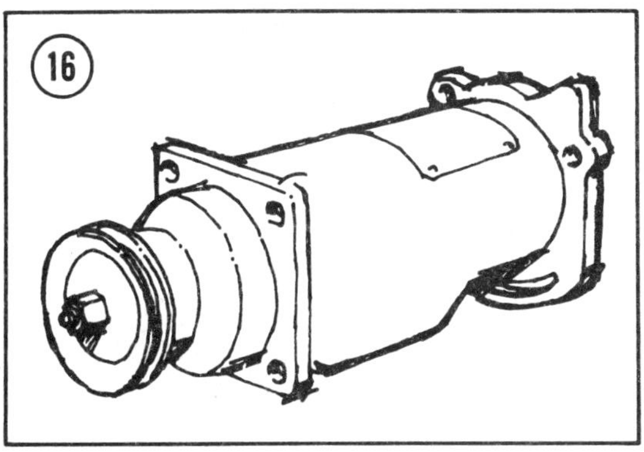

16

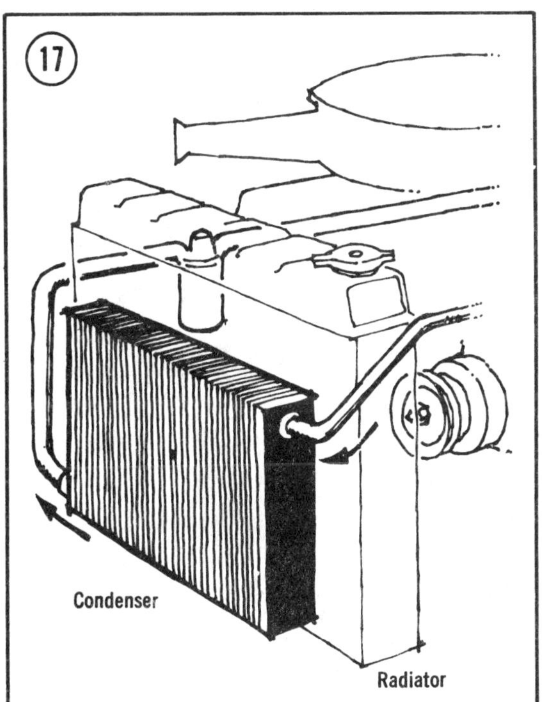

17

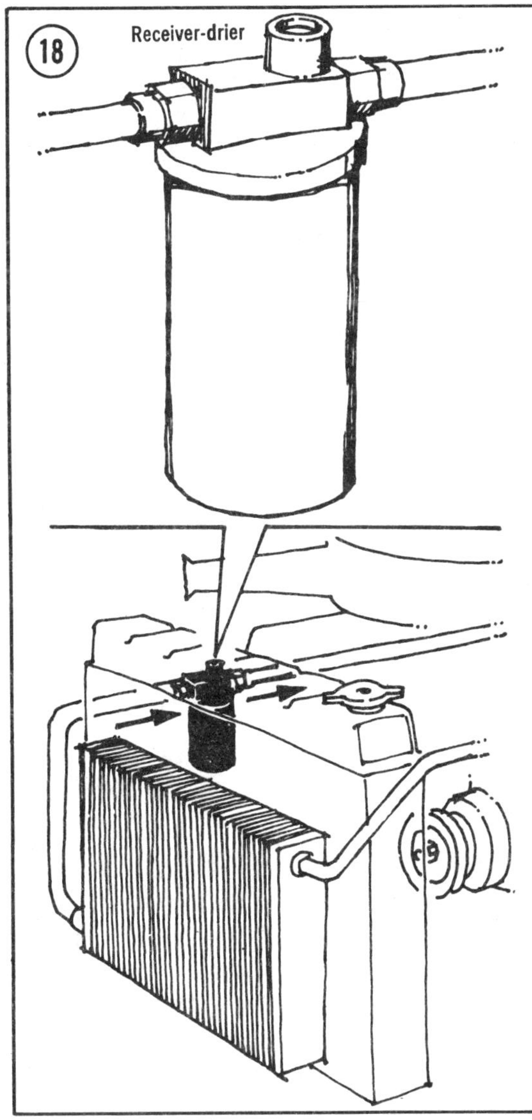

18

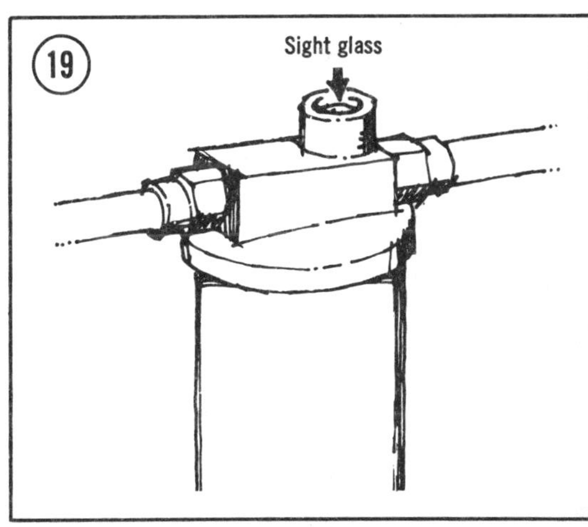

19

an inline sight glass (**Figure 20**). Some early systems do not have a sight glass, but it's not essential to system operation—just handy to help diagnose air conditioning troubles.

Expansion Valve

The expansion valve (**Figure 21**) is located between the receiver/drier and the evaporator. It is usually mounted on or near the firewall, in the engine compartment. In some very late systems, the valve is concealed in a housing on the firewall.

Evaporator

The evaporator is located in the passenger compartment, beneath the dashboard, and is hidden from view by the fan shrouding and ducting (**Figure 22**). Warm air from the passenger compartment is blown across the fins and tubes in the evaporator where it is cooled and dried and then ducted back into the compartment through the air outlets.

ROUTINE MAINTENANCE

Preventive maintenance for your air conditioning system couldn't be simpler; at least once a month, even in cold weather, start your engine and turn on the air conditioner and operate it at each of the switch and control settings. Allow it to operate for about 5 minutes. This will ensure that the compressor seal will not deform from sitting in the same position for a long period of time. If this occurs, the seal is likely to leak.

The efficiency of your air conditioning system depends in great part on the efficiency of your engine cooling system. Periodically check the coolant for level and cleanliness. If it is dirty, drain and flush the system and fill it with fresh coolant and water, following the coolant manufacturer's instructions for coolant/water ratio. Have your radiator cap pressure tested and replace it if it will not maintain 13 psi pressure. If the system requires repeated topping up and the radiator cap is in good condition, it is likely that there is a leak in the system. Pressure test it as described earlier in this chapter.

With an air hose and a soft brush, clean the radiator fins and tubes to remove bugs , leaves and any other imbedded debris.

Check and correct drive belt tension as described earlier.

If the condition of the cooling system thermostat is in doubt, check it as described earlier and replace it if it is faulty.

When you're confident that the engine cooling system is working correctly, you are ready to inspect and test the air conditioning system.

20

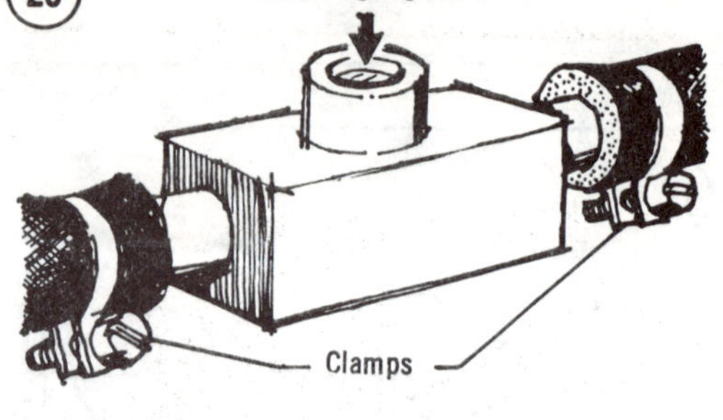

21

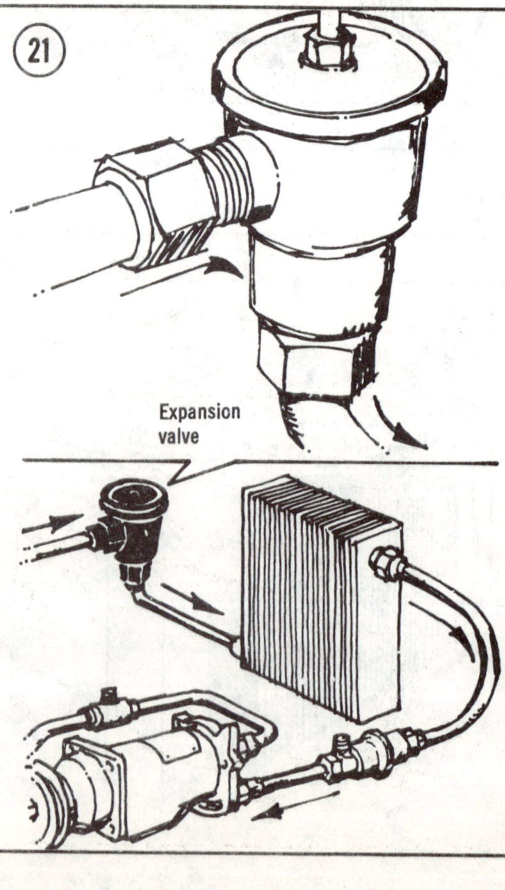

Inspection

1. Clean all lines, fittings, and system components with solvent and a clean rag. Pay particular attention to the fittings; oily dirt around connections almost certainly indicates a leak. Oil from the compressor will migrate through the system to the leak. Carefully tighten the connection, taking care not to overtighten and risk stripping the threads. If the leak persists it will soon be apparent once again as oily dirt accumulates. Clean the sight glass with a clean, dry cloth.

(22)

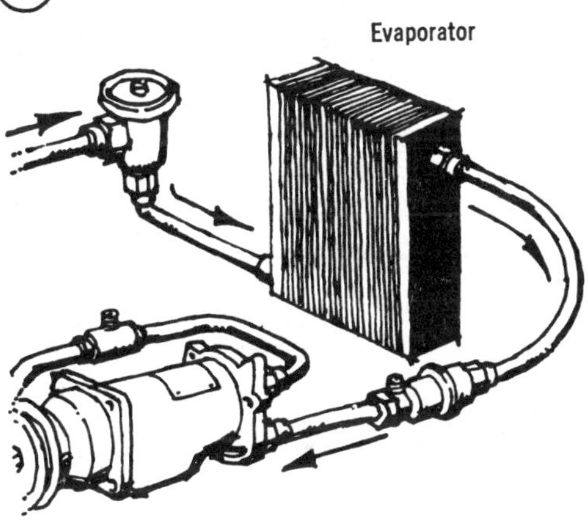

(23)

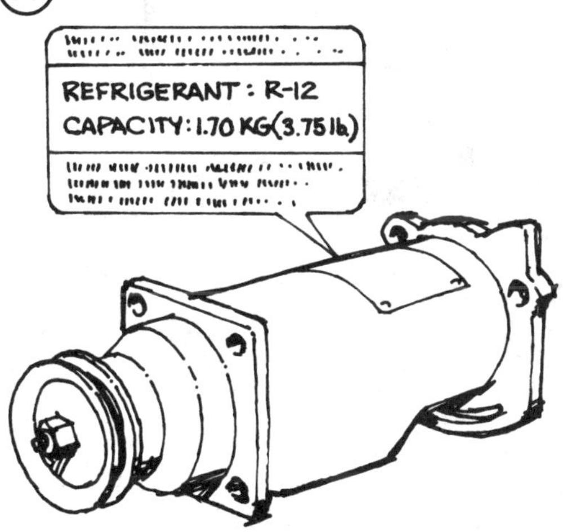

2. Clean the condenser fins and tubes with a soft brush and an air hose, or with a high-pressure stream of water from a garden hose. Remove bugs, leaves and other imbedded debris. Carefully straighten any bent fins with a screwdriver, taking care not to dent or puncture the tubes.
3. Check the condition and tension of the drive belts and replace or correct as necessary.
4. Start the engine and check the operation of the blower motor and the compressor clutch by turning the controls on and off. If either the blower or the clutch fails to operate, shut off the engine and check the condition of the fuses. If they are blown, replace them: if not, remove them and clean the fuse holder contacts. Then, recheck to be sure the blower and clutch operate.

Testing

1. With the transmission in PARK (automatic) or NEUTRAL (manual) and the handbrake set, start the engine and run it at a fast idle.
2. Set the temperature control to its coldest setting and turn the blower to high. Allow the system to operate for 10 minutes with the doors and windows open. Then close them and set the blower on its lowest setting.
3. Place a thermometer in a cold-air outlet. Within a few minutes, the temperature should be 35-45 degrees F. If it is not, it's likely that the refrigerant level in the system is low. Check the appearance of the refrigerant flow through the sight glass. If it is bubbly refrigerant should be added.

REFRIGERANT

The majority of automotive air conditioning systems use a refrigerant designated R-12. However, a commercial grade, designated R-20, is used in heavy-duty systems. The two are not compatible. Look for an information sticker, usually mounted near the compressor to determine which refrigerant your system uses (**Figure 23**). Also, check the system capacity indicated on the sticker. Capacity can range from 2 to 5 pounds, depending on the system.

24

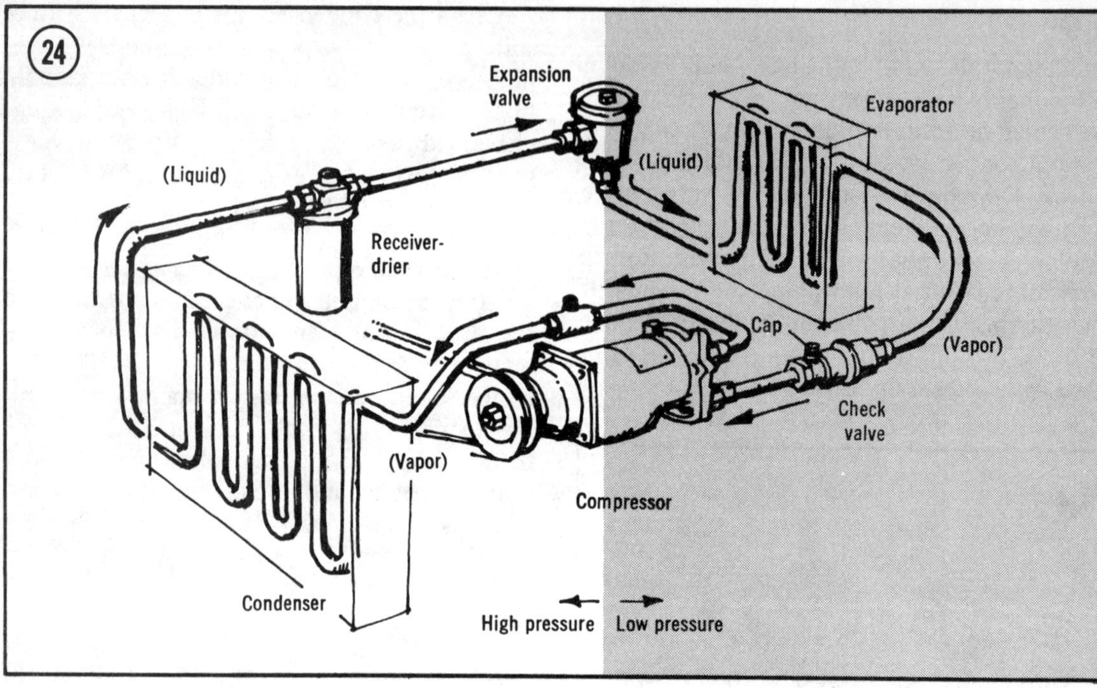
Expansion valve
Evaporator
(Liquid)
(Liquid)
Receiver-drier
Cap
(Vapor)
Check valve
(Vapor)
Compressor
Condenser
High pressure
Low pressure

25

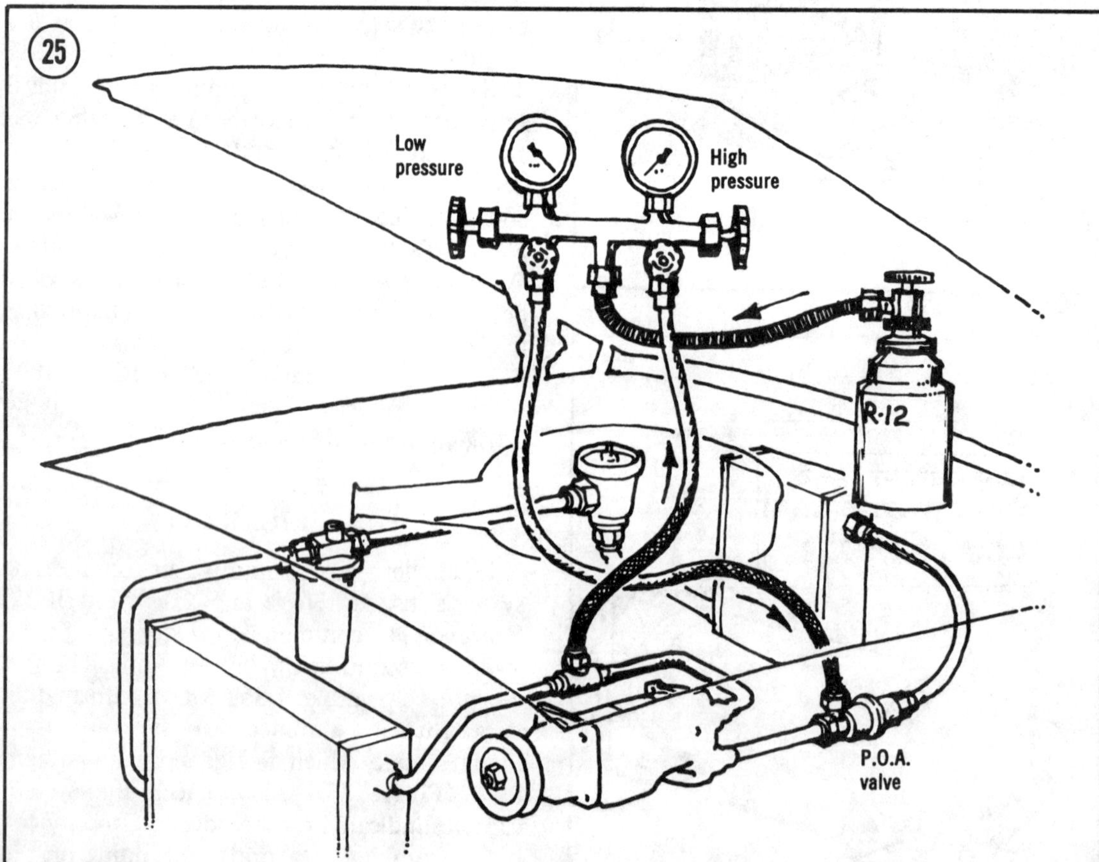
Low pressure
High pressure
R-12
P.O.A. valve

(26)

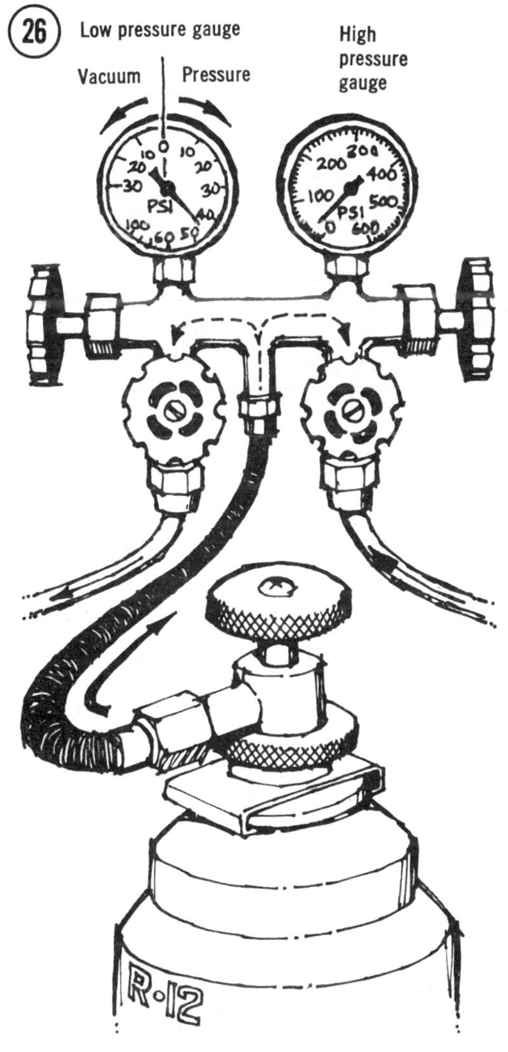

That harmless-looking little can of refrigerant is potentially hazardous. If it is hooked up to the high-performance side of the compressor, or is hooked up without a gauge set, it becomes a hand grenade.

Charging

WARNING
Do not attempt to add refrigerant to the system without using a gauge set; it's essential that the system pressure during charging not exceed 50 psi.

1. Carefully read and understand the gauge manufacturer's instructions before charging the system.
2. Remove the cap from the Schrader valve on the low-pressure side of the compressor (**Figure 24**). The low-pressure side is labelled SUCTION, SUCT., or SUC.
3. Connect the gauge set to the low-pressure Schrader valve. Connect the refrigerant can to the gauge set and hang the gauge set on the hood (**Figure 25**).
4. Start the engine and run it at a fast idle (about 1,000 rpm).
5. Set the temperature control at its coldest setting. Set the blower at its lowest setting.
6. Slowly open the refrigerant feed valve on the gauge set (**Figure 26**). Do not allow the refrigerant pressure to exceed 50 psi.
7. Watch the refrigerant as it flows through the sight glass (**Figure 27**). When it's free of bubbles, the system is charged. Shut off the refrigerant feed valve on the gauge set.

7

(27)

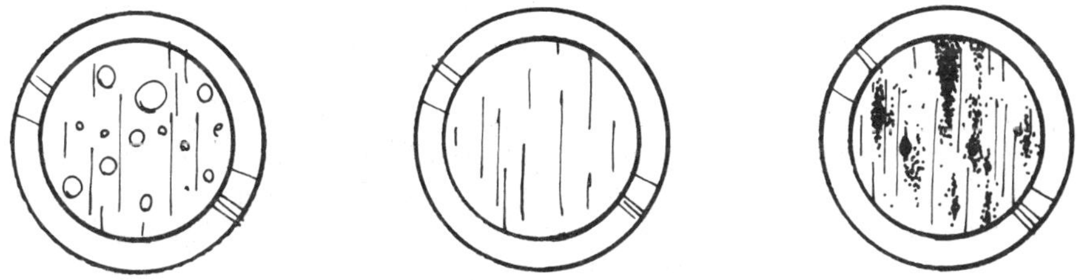

Bubbles — low charge

Clear — correct charge

Cloudy and oily — contaminated

TROUBLESHOOTING

Preventative maintenance like that just described will help to ensure that your system is working efficiently. Still, trouble can develop and while most of it will invaribly be simple and easy to correct, you must first locate it. The following sequence will help to diagnose system troubles when your air conditioning ceases to cool the passenger compartment.

1. First, stop the vehicle and look at the control settings. One of the most common sources of air conditioning trouble occurs when the temperature control is set for maximum cold and the blower is set on low. This arrangement promotes ice buildup on the fins and tubes of the evaporator, and particularly so in humid weather. Eventually, the evaporator will ice over completely, and restrict air flow. Turn the blower on high and place a hand over an air outlet. If the blower is running but there is little or no air flowing through the outlet, the evaporator is probably iced up. Leave the blower on high and turn off the temperature control or turn it down to its lowest setting—and wait; it will take 10 or 15 minutes before the ice begins to melt.
2. If the blower is not running, the motor may be burned out, there may be a loose connection, or the fuse may be blown. First check the fuse panel for a blown or incorrectly seated fuse. Then, check the wiring for loose connections.
3. Shut off the engine and check the condition and tension of the compressor drive belt. If it is loose or badly worn, tighten or replace it.
4. Start the engine and check the condition of the compressor clutch by turning the air conditioner on and off. If the clutch does not energize, it may be defective; if its fuse is defective, replace it. If the clutch still does not energize, refer the problem to an air conditioner specialist.
5. If all components checked so far are OK, start the engine, turn on the air conditioner, and watch the refrigerant through the sight glass; remember, if it's filled with bubbles after the system has been operating for a few seconds, the refrigerant level is low. If the sight glass is oily or cloudy, the system is contaminated and should be serviced by an expert as soon as possible. Corrosion and deterioration occur rapidly and if it's not taken care of at once, it will result in a very expensive repair job.
6. If the system still appears to be operating satisfactorily, but the air flow into the passenger compartment is not cold, check the condenser and cooling system radiator for debris that could block the air flow, and recheck the cooling system as described earlier under *Inspection.*
7. If the above steps do not uncover the difficulty, have the system checked and corrected by a specialist as soon as possible.

DISCHARGING THE SYSTEM

The pressure in the system must be relieved before any fittings are disconnected. To do this, connect the gauge set in the same manner as though the system were to be charged; however, do not attach a refrigerant can to the center hose. Slowly open both the high- and low-pressure valves on the gauge set. Then, slowly open the valve for the center hose.

WARNING

Route the center hose down through the engine compartment so it will discharge on the ground. Wear safety goggles and take care not to let the refrigerant touch your skin.

When the system pressure has been relieved, disconnect the gauge set. Slowly loosen any fittings that are involved until you are sure the system is not pressurized. *Wear safety goggles.*

Immediately plug any open fittings to keep moisture out of the system; corrosion will begin almost instantly if the system is left open to the atmosphere, and it will quickly make some very expensive components unserviceable.

When all of the system components and lines have been reconnected, charge the system as described earlier.

CHAPTER EIGHT

ELECTRICAL SYSTEM

Making repairs to electrical components such as the alternator or starter motor is usually beyond the capability of both the inexperienced mechanic and his tool box. Such repairs are best left the specialized mechanic who is equipped with specialized tools.

By using the troubleshooting procedures given in Chapter Two it is possible, however, to isolate problems in many cases to a specific component.

This chapter provides some additional tests and procedures for removing, inspecting (in some cases), and replacing electrical components.

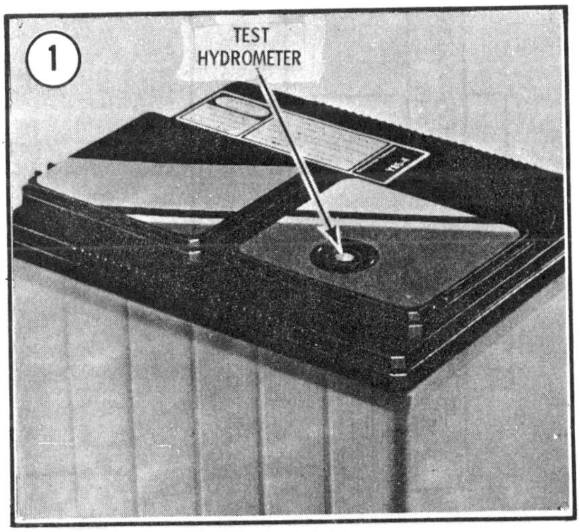

In many cases it may be faster and more economical to obtain new or rebuilt components instead of making repairs. Make certain, however, that the new or rebuilt part is an exact replacement. Also, make sure that the cause of the failure has been isolated and corrected before installing a replacement. For instance, an uncorrected short in an alternator circuit will in all probability burn out a new alternator as quickly as it damaged the old one. If in doubt, always consult an expert.

Wiring diagrams are located at the end of the book, and **Tables 1-3** are at the end of the chapter.

BATTERY MAINTENANCE

NOTE

*Some models are equipped with sealed "Freedom" batteries which do not require routine maintenance. See **Figure 1**.*

The level of the electrolyte in each cell of the battery should be checked regularly, especially in hot weather. The proper level is even with the split vent located at the bottom of the vent well in each cell. Distilled water should be added as required to maintain this level in each cell. Never add electrolyte (acid).

8

If water must be added frequently, chances are that the battery is being overcharged (normal usage is 1 or 2 ounces per month per battery—not each cell). Common causes of overcharge are high battery operating temperature, voltage regulator set too high, and a poor ground connection from the regulator.

The battery terminals, top, carrier, and hold-down should be kept free of corrosion, oil, and dirt. The carrier should be in sound mechanical condition and the hold-down bolts should be tight to keep the battery level and prevent excessive shaking of the battery. However, the bolts should not be so tight that a severe strain is placed on the battery case.

Cleaning

To clean the battery, make certain the cell vent plugs are fully seated. Cover the vent holes, if accessible, with small pieces of tape. Wash the top and sides of the battery case and carrier, and the hold-down with a mild solution of ammonia or baking soda to neutralize any acid or corrosion present, then flush with clean water. Pay special attention to terminals and the areas around them. Be sure to remove the tape from the holes in the vent caps.

Most models use a battery with "sealed" terminals on the side. See **Figure 2**. This positions them out of the "wet" area on the top of the battery surrounding the vent holes. Normal spillage, spewage, and other sources of moisture are far less likely to collect around the terminal. These terminals, as a rule, require no maintenance.

Some older models use the "conventional" type battery with the terminals on the top. These terminals should be cleaned whenever they become corroded. See **Figure 3**.

Following cleaning they should receive a light coat of petroleum jelly to help control future corrosion.

Common Causes of Battery Failure

All batteries eventually fail. Their life can be prolonged, however, with a good maintenance program. Some of the reasons for premature failure are listed below.

2

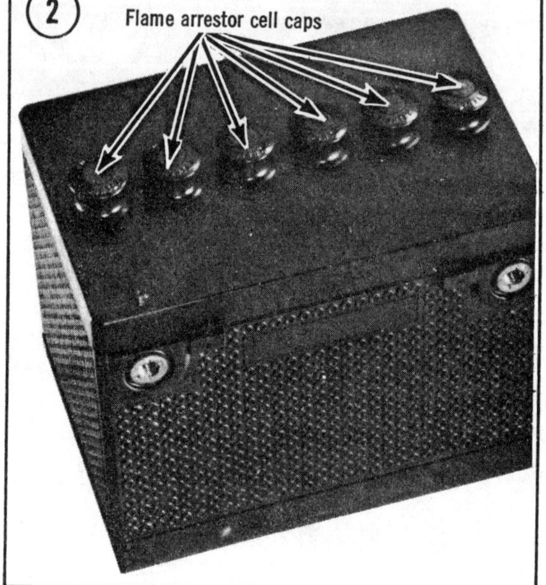

3

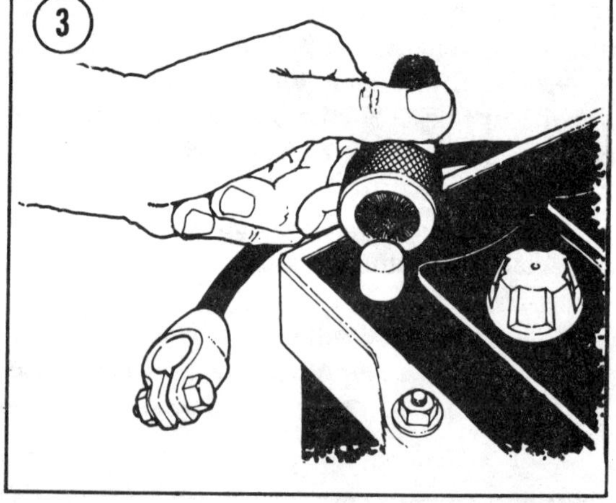

1. Vehicle accessories left on overnight or longer, causing a discharged condition.
2. Slow driving speeds on short driving trips causing an undercharged condition.
3. Vehicle electrical load exceeding the alternator capacity.
4. Charging system defects, such as high resistance, slipping alternator belt, or faulty alternator or regulator.
5. Abuse of the battery, including failure to keep battery top and terminals clean, failure to keep cable attaching bolts clean and tight, and failure to add water when needed or habitually adding too much water.

BATTERY TESTING

Use of the following test procedures will provide a basis for deciding whether a battery is good and usable, requires recharging, or should be replaced. A complete analysis of battery condition requires a visual inspection, an instrument test, and, if required, a full charge hydrometer test.

Visual Inspection

This test is accomplished as follows.

1. Check outside of battery for broken or cracked case or cover. If any such damage is present, replace battery.
2. Check electrolyte level (except Freedom battery). Levels too high or too low may cause poor performance.
3. Check for loose and/or corroded cable connections. Correct as required before proceeding with instrument tests.

Instrument Tests

If an instrument with battery testing capabilities is available, follow the manufacturer's instructions. If such an instrument is not available, use a hydrometer to make a specific gravity cell comparison test (except Freedom battery) as follows:

1. Measure specific gravity of each cell.
2. Compare readings of all cells. If readings between highest and lowest cell show a difference of 0.050 (50 points) or more, the battery is defective and should be replaced.

Full Charge Hydrometer Test

This test should only be made on batteries which check out OK after the tests in Step 2 above but which subsequently fail in service.

1. Remove battery from vehicle and add distilled water, if required, to adjust electrolyte to proper level.
2. Fully charge battery at the slow charging rate, using the procedure given below.
3. When battery is fully charged, use a hydrometer to measure the specific gravity of the electrolyte in each cell. Interpret results as follows:
   a. Full charge hydrometer reading of less than 1.230, corrected for temperature, battery is defective and should be replaced.

NOTE

For every 10° above 80° F electrolyte temperature, add 0.004 to specific gravity reading. For every 10° below 80° F, subract 0.004 from the reading.

   b. Hydrometer readings above 1.310 corrected for electrolyte temperature indicate that cells have been improperly filled (activation) or improperly serviced. Poor service and short battery life will result.
   c. Readings between 1.250 and 1.290 in all cells indicate that battery condition is good. The problem is probably in the charging system or battery cables.

CHARGING SYSTEM

The charging system consists of the battery (discussed above), the alternator, the voltage regulator, the "tell-tale" lamp or ammeter and the wiring necessary to connect these components. Two basic types of alternators are used. The Series ID Delcotron (**Figure 4**) used on older models has an external mechanical or transistorized voltage regulator. The Series 10- and 15-SI and 10 SE Delcotrons (**Figure 5**) have internal solid state voltage regulators which employ integrated circuits.

All models have an internal bridge which changes the stator AC voltage to DC voltage. The blocking action of the diodes in the bridge prevents battery discharge back through the alternator.

Neither type of alternator requires periodic maintenance, other than a check for loose mounting bolts and belt tension. Certain precautions should be observed, however, since the alternator and regulator are designed for use only on negative polarized systems. These precautions are as follows:

1. Do not attempt to "polarize" the alternator.
2. Do not short across or ground any of the terminals in the charging system except as specifically instructed in these procedures.
3. Never operate the alternator with the output terminal open-circuited.
4. Make sure the alternator, external voltage regulator (if so equipped), and battery are of the same ground polarity.

8

4

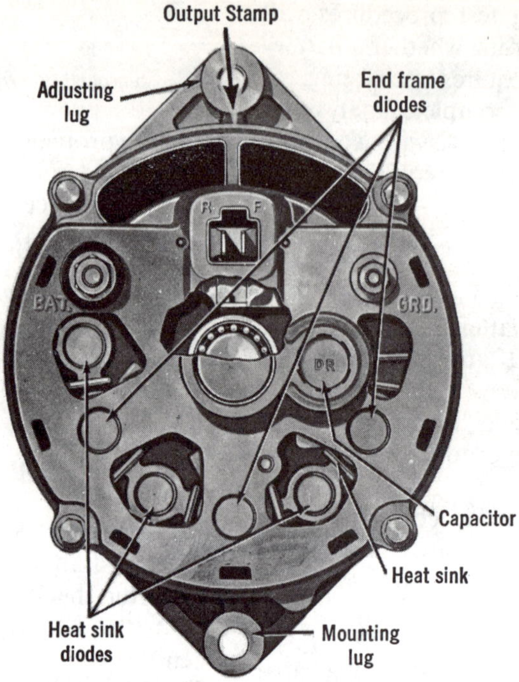

SERIES 1D DELCOTRON ALTERNATOR

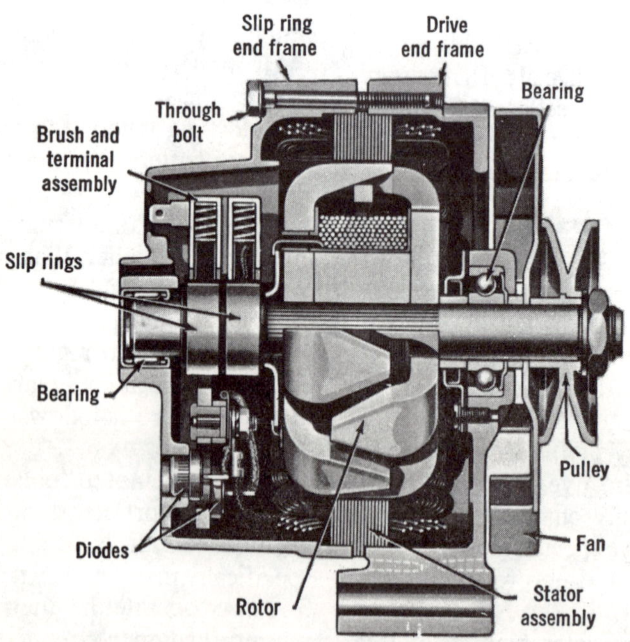

5

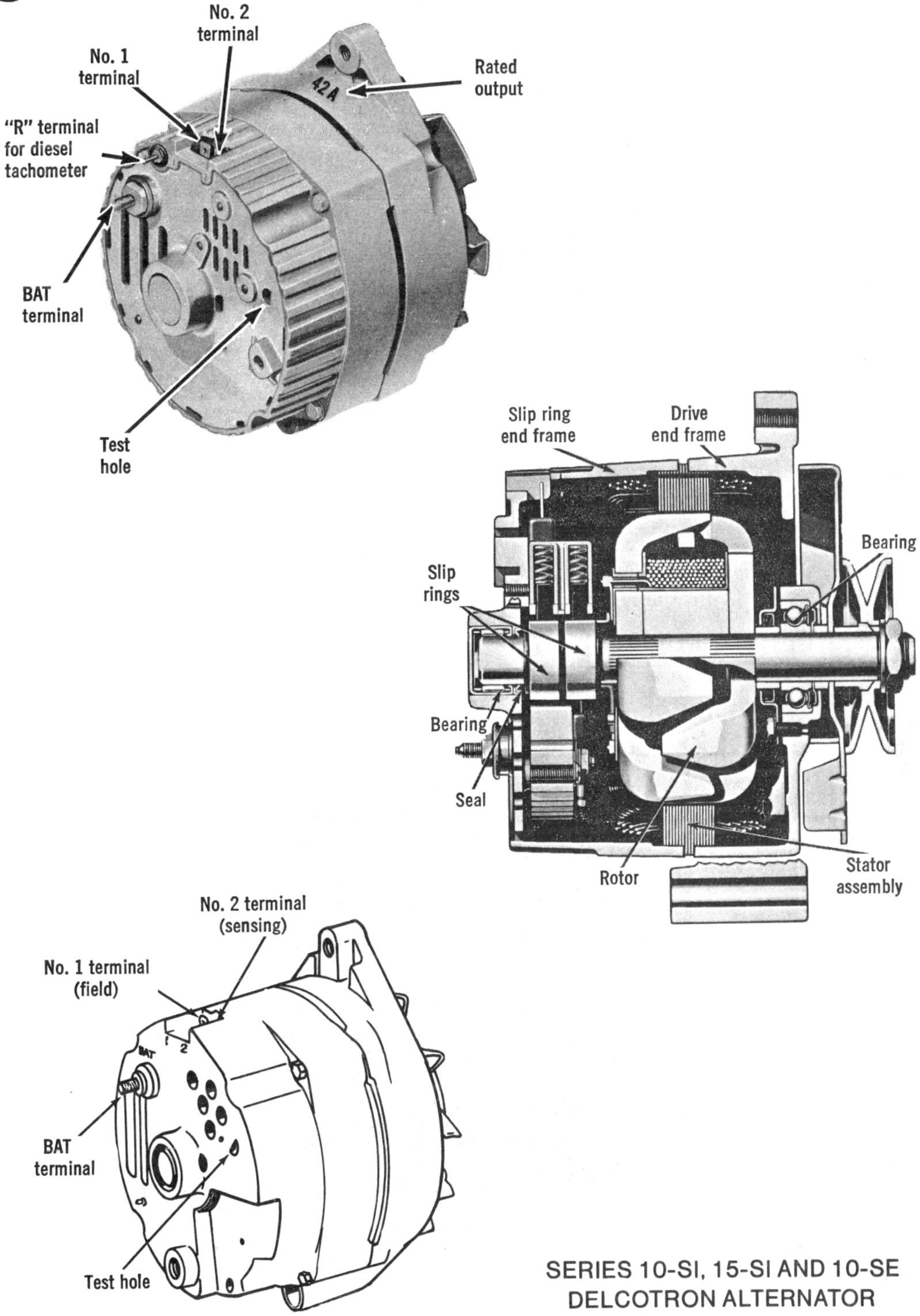

SERIES 10-SI, 15-SI AND 10-SE
DELCOTRON ALTERNATOR

5. When connecting a charger or booster battery to the vehicle battery, connect negative terminal to negative terminal and positive terminal to positive terminal.
6. Whenever a lead is disconnected from the alternator the battery ground cable should be disconnected.

Static Checks (All Series)

Before making any electrical checks on the charging system, visually inspect all connections to make sure that they are clean and tight. Verify that battery is serviceable and charged. Inspect all wiring for frayed, broken, or cracked insulation. Check for loose mounting bolts and proper belt tension.

SI DELCOTRON TROUBLESHOOTING

NOTE
If vehicle is equipped with ammeter instead of indicator lamp, omit indicator lamp circuit check.

Indicator Lamp Circuit Check

Check indicator lamp for normal operation as follows:

Switch	Lamp	Engine
Off	Off	Stopped
On	Off	Stopped
On	Off	Running

If indicator lamp operates normally, proceed to *Undercharged Battery* or *Overcharged Battery*. Otherwise, make the appropriate abnormal condition checks below:

1. *Switch off, lamp on*—Disconnect leads from alternator No. 1 and No. 2 terminals. If lamp remains on, there is a short circuit between the 2 leads. If the lamp goes out, the rectifier bridge is faulty and must be replaced, as this condition will result in an undercharged battery.
2. *Switch on, lamp off, engine stopped*—This defect can be caused by the defects listed in Step 1 above, or by reversal of the No. 1 and No. 2 leads at these 2 terminals, or by an open circuit. To determine where the open exists, proceed as follows:

   a. Connect voltmeter between alternator No. 2 terminal and ground. If reading is zero, repair open circuit between No. 2 terminal and battery. If lamp comes on, no further check is required.
   b. Disconnect leads from No. 1 and No. 2 terminals on alternator and turn ignition switch on. Momentarily ground No. 1 terminal lead.

   CAUTION
   Do not ground No. 2 lead.

   c. If lamp does not come on, check for blown fuse or fusible link, burned out lamp bulb, defective bulb socket or open in No. 1 lead circuit between alternator and ignition switch.
   d. If lamp lights, remove ground on No. 1 terminal and reconnect No. 1 and No. 2 wires to alternator. Insert a screwdriver in test hole (see **Figure 5**) to ground wiring. If lamp does not go on, check connection between wiring harness and alternator No. 1 terminal. If wiring is okay, have alternator tested and repaired or replaced. If lamp lights, repeat voltmeter check in Step a. If reading is obtained, have the regulator replaced.

3. *Switch on, lamp on, engine running*—See *Undercharged Battery Condition Check*, below, for possible causes.

Undercharged Battery Condition Check

Symptoms of this condition are slow cranking and low specific gravity readings (battery cells). The condition can be caused by one or more of the following, even though the ammeter may be operating properly.
1. Accessories left on for extended periods.
2. Improper belt tension.

3. Defective or discharged battery (see battery tests above).

4. Wiring defects. Check all connectors for tightness and cleanliness, including connectors at alternator and firewall, cable clamps, and battery posts.

5. Open circuit between alternator and battery. Check by connecting voltmeter between BAT terminal on alternator and ground, No. 1 terminal and ground, and No. 2 terminal and ground. A zero reading indicates an open between the voltmeter connection point and the battery.

CAUTION

An open in the alternator No. 2 lead circuit will cause uncontrolled voltage, battery overcharge and possible damage to the battery and accessories. The 10-SI Series alternator has a built-in feature to prevent overcharge and accessory damage by preventing the alternator from turning on if there is an open in the No. 2 lead circuit. Such an open could occur between terminals, at the crimp between the harness wire or terminal or in the wire itself.

Overcharged Battery Condition Check

1. Verify that battery is in serviceable condition and fully charged.

2. Connect a voltmeter from the alternator No. 2 terminal to ground. If reading is zero, an open exists in No. 2 lead circuit.

3. If battery and No. 2 lead circuit are OK, but an obvious overcharged condition, such as excessive battery water usage, exists, have the following items checked:

   a. Field windings (for short).
   b. Brushes and brush leads (for grounding or defective regulator).

1D DELCOTRON TROUBLESHOOTING

Charging System Condition Test

This test may be used to indicate overall charging system condition and to isolate a malfunctioning component (if present).

1. Turn ignition off and make static tests outlined above. If tests fail to isolate trouble, set handbrake and place transmission in NEUTRAL.

2. Connect voltmeter from junction block on horn relay to ground at voltage regulator base.

CAUTION

Make certain voltmeter lead does not touch a resistor or terminal extension under the regulator, as this could cause damage.

3. Connect a tachometer to engine, using the tachometer manufacturer's instructions.

4. Start engine and run at 1,500 rpm a few minutes to develop battery surface charge. Turn on headlights and heater blower motor (high speed), operate at 1,500 rpm and read voltmeter. If reading is 12.5 volts or more, turn off headlights and heater blower and stop engine. Have voltage regulator adjusted (see procedure below). If reading was below 12.5 volts, check alternator output using the procedure given below.

   a. *Alternator tests bad*—Have alternator repaired or replaced.

   b. *Alternator tests okay*—Disconnect regulator connector, remove cover and reconnect connector. Repeat load test at beginning of Step 4. Turn adjusting screw (**Figure 6**) to raise voltmeter reading to 12.5 volts.

Regulator Voltage Adjustment

Operate engine at 1,200 rpm for at least 15 minutes (longer in cold weather), then cycle regulator voltage control (by stopping and

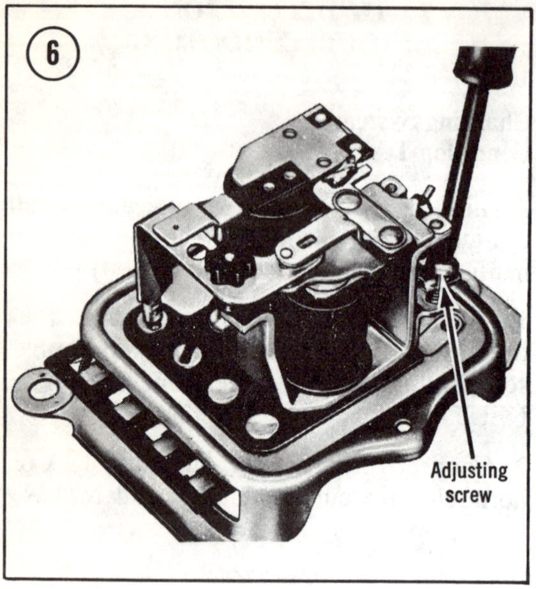

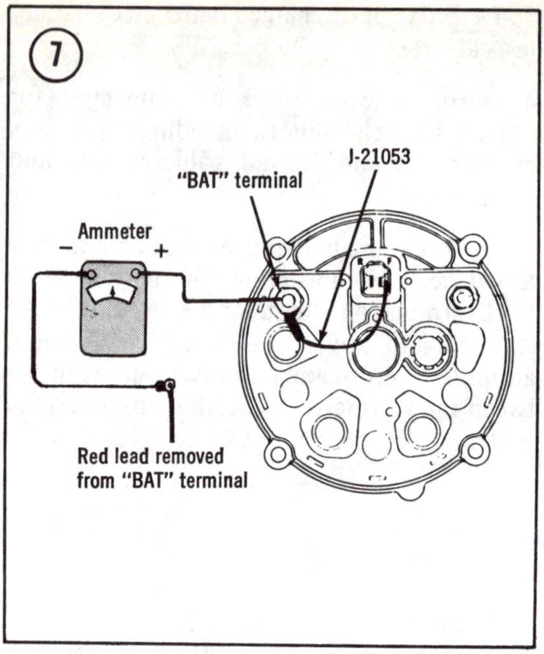

starting the engine) and read voltage. If reading is between 13.5-15.2, regulator is OK. If not within these limits, leave engine running at 1,200 rpm and perform the following:

a. Disconnect regulator connector and remove regulator cover. Reconnect connector and set voltage to 14.2-14.6 volts. See Step 4 under *Charging System Condition Test* and **Figure 6**.
b. Disconnect connector, reinstall cover, and reconnect connector.
c. Allow engine to operate another 5-10 minutes at 1,200 rpm to reestablish regulator internal operating temperature.
d. Cycle regulator voltage by stopping and starting engine, then increase speed to 2,200 rpm and read voltage. If between 13.5 and 14.6 volts, regulator is OK. Stop engine and remove voltmeter.

Alternator Output Test (Voltmeter Method)

Refer to **Figure** 7.

1. Disconnect 2-terminal connector from alternator "F" and "R" terminals.
2. Connect a jumper between BAT terminal to F terminal to provide field excitation.
3. Connect voltmeter between BAT terminal and GRD terminal.
4. Start engine and turn on high beam headlights and high speed heater blower. Slowly increase engine speed to 1,500 rpm and read voltage. A reading of 12.5 volts or higher indicates output is okay. Stop engine, remove voltmeter and reconnect wiring. If reading was less than 12.5 volts, have alternator repaired or replaced.

ALTERNATOR REMOVAL/INSTALLATION (ALL MODELS)

1. Disconnect battery ground cable at battery.
2. Identify and disconnect wiring leads at alternator.
3. Remove alternator brace bolt (if equipped with power steering), loosen pump brace and mount nuts and then remove drive belt(s).
4. Support the alternator and remove bolt(s). Remove alternator from vehicle.
5. Reverse the procedure to install alternator. Adjust drive belt(s) tension.

STARTER MOTOR

The procedure below is general in nature and applies, with slight variations, to all starters used on all vehicles. **Figure 8** is a cross-sectional view of a typical starter motor/solenoid assembly. **Figure 9** (Olds V8), **Figure 10** (L6), **Figure 11** (Chevy V8) and **Figure 12** (V6) show typical starter installations.

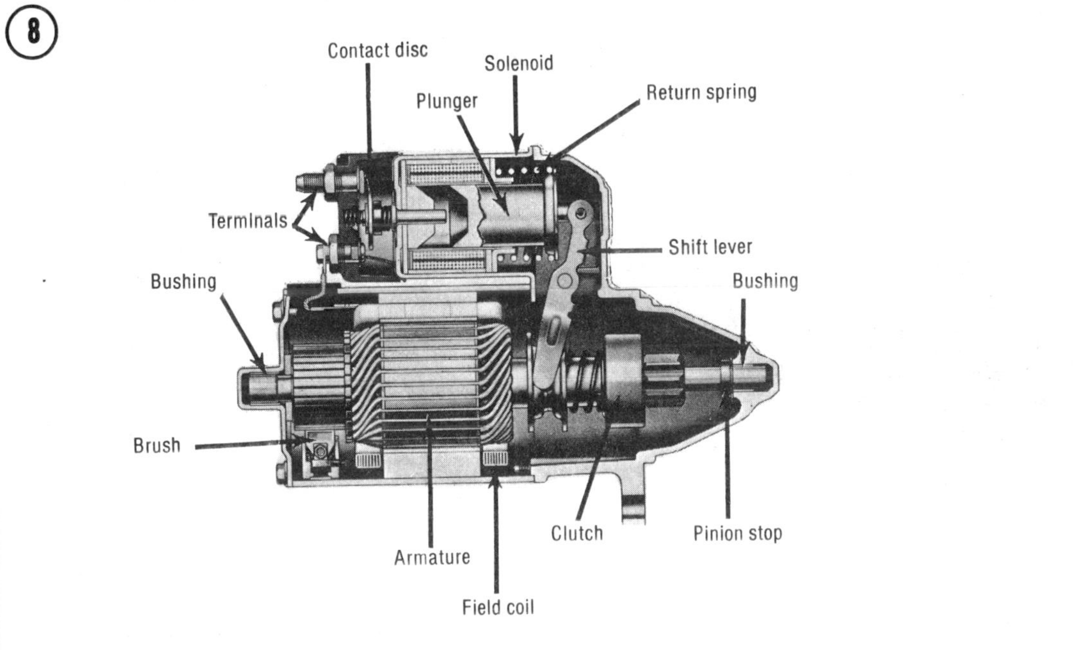

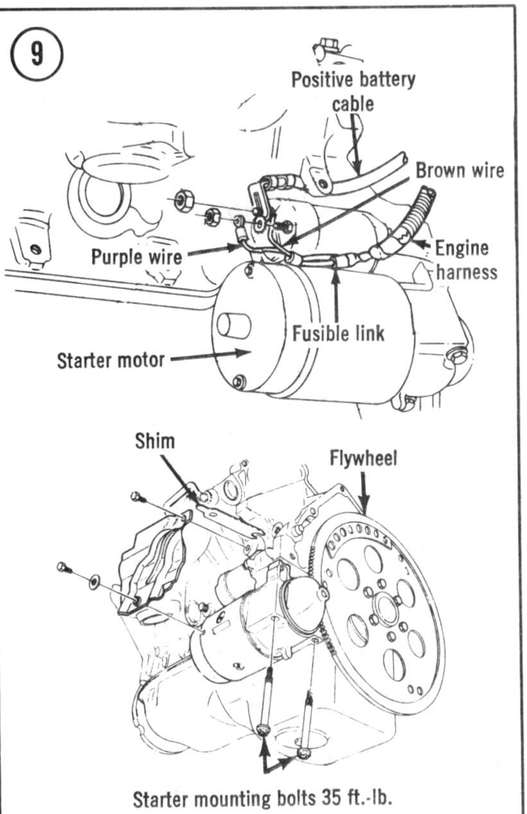

OLDS V8 STARTER MOUNTING

Periodic lubrication or other maintenance is not required for either the starter motor or the solenoid.

Diagnostic procedures for troubleshooting the starter motor are given in Chapter Two. If these procedures are used and it is determined that the starter motor and/or solenoid are defective, the starter/solenoid assembly should be removed from the engine and repaired or replaced. Starter repairs are beyond the scope of this book. See a qualified mechanic.

Starter Motor Replacement

Refer to **Figures 9-12**.

1. Disconnect battery ground cable at the battery and, if possible, raise vehicle to a good working height.
2. Disconnect all wires at solenoid terminals.

NOTE
Reinstall nut as each wire is removed. Thread sizes are different and stripped threads could result if nuts become mixed.

3. Automatic transmission only—If oil cooler lines interfere with starter removal, disconnect

8

and plug. Remove upper support mounting bolts and flywheel housing cover.

4. Manual transmission only—Loosen crossmember braces and remove crossmember if it interferes with access to starter motor.

NOTE

On V8 engines using a solenoid heat shield, remove the front bracket upper bolt and detach bracket from starter motor.

5. Remove starter mounting bolts, loosen and rotate any bracket clear of work area, then lower starter from vehicle front end first.

6. Reverse the removal procedure to install starter motor. Tighten mounting bolts first (25-35 ft.-lb.), then tighten bracket nut or bolt. Reconnect all wires and make sure connections are tight.

7. Install crossmember and braces on manual transmission models.

8. Install flywheel housing cover and support bolts and reconnect oil cooler lines on automatic transmission models.

9. Check starter operation.

NOTE

*If any shims were removed (see **Figure 9** and **Figure 12**), make sure they are reinstalled in original position for proper drive pinion-to-flywheel alignment.*

BODY ELECTRICAL SYSTEM

Instructions and pointers for troubleshooting the electrical wiring system are given in Chapter Two. Service which the owner/mechanic can be expected to perform includes isolation and repair of short and open circuits, wire, connector, and fuse replacement, and sealed beam and other bulb replacement.

Maintenance of the lighting and wiring systems consists of an occasional check to see that wiring connections are tight and clean, that lighting units are tightly mounted and that headlights are properly adjusted. The latter task should be performed by a qualified mechanic having access to the specialized equipment required.

10

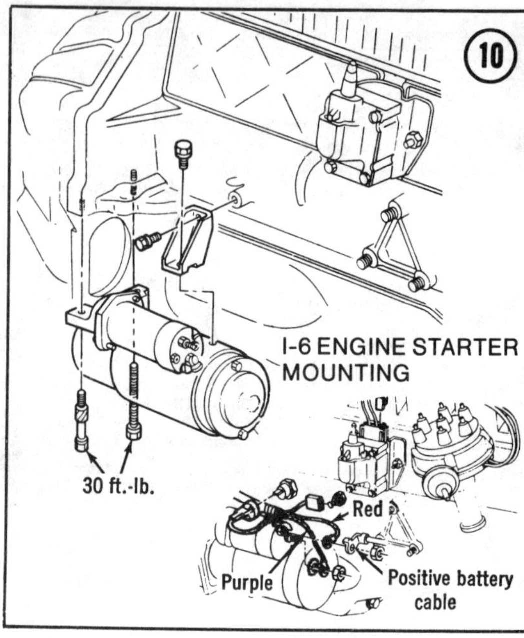

I-6 ENGINE STARTER MOUNTING

11

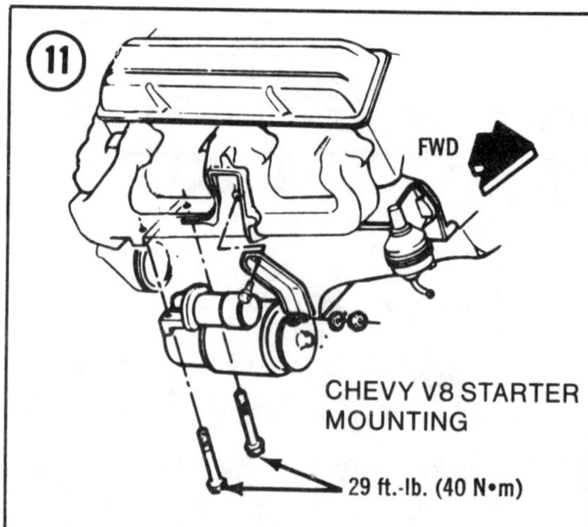

CHEVY V8 STARTER MOUNTING

12

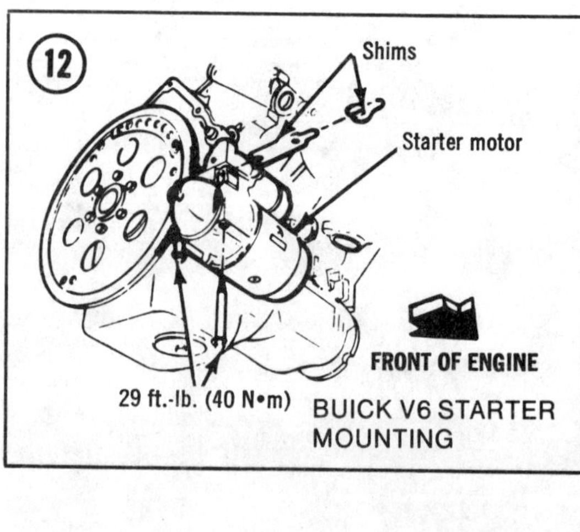

BUICK V6 STARTER MOUNTING

Loose or corroded connectors can result in a discharged battery, dim lights, and possible damage to the alternator or voltage regulator. Should insulation become burned, cracked, abraded, etc., the affected wire or wiring harness should be replaced.

Rosin core solder—never use acid core solder on electrical connections—must always be used with splicing wires. Splices should be covered with insulating tape.

Replacement wires must be of the same gauge as the replaced wire—never use a smaller gauge. All harnesses and wires should be held in place with clips, cable ties, or other holding devices so that chafing and abrasion can be avoided.

Headlight Replacement

NOTE

On models so equipped, open headlamp doors.

1. Remove headlamp bezel retaining screws and then remove bezel.
2. Disengage spring from retaining ring and remove 2 attaching screws.
3. Remove retaining ring. Disconnect sealed beam unit from connector.
4. Install replacement sealed beam unit in connector and position in place with number molded into lens face at top of unit.
5. Position and install retaining ring, using screws and spring (as applicable).
6. Check operation of the sealed beam unit, then install bezel.

Parking and Rear Lamp Replacement

Because of the large number of models, it would be impossible to provide individual procedures covering all types of lamp installations. Close examination will usually reveal the approach to be taken for bulb replacement. Many installations use twist-out sockets which are removed from the rear of the lamp housings. Other installations require the removal of the lamp lens (held in place by 2 or more screws through the lens face) for access to the bulb and socket. In other installations, nuts must be removed from the rear of the lamp unit to release the lens. See **Table 2** for bulb usage.

FUSES/CIRCUIT BREAKERS

Each electrical circuit is protected from overload by either a circuit breaker, a fuse, or a fusible link.

Circuit Breaker

The circuit breaker is a protective device that causes an open circuit when the current load exceeds the rated breaker capacity. The circuit breaker will remain open until the problem causing the overload is found and corrected. The Cutlass uses one circuit breaker, which is located on the headlight switch and protects the headlamp circuit. The headlamp switch circuit breaker will cause a flickering condition whenever an overload is present in the circuit.

Fuses

Most major circuits are protected by fuses (**Table 3**). The fuse contains an element that melts when the protected circuit is overloaded. When this occurs, the fuse must be replaced. The cause of the overload should be located and corrected before replacing a burned out fuse. If this does not happen, chances are that the replacement fuse will burn out too. Replacement fuses must be of the same amperage capacity as the fuses they are replacing. Fuses of higher ratings could cause damage to wiring and insulation, or even cause fires. A fuse with a lower than specified rating could cause unnecessary interruption of the circuit at normal loading.

Fusible Links

Fusible links provide protection to circuits which are not normally fused, such as the ignition system circuit. The links are usually 4 gauges smaller than the wiring in the circuit they are protecting, and are covered with heavy red insulation. The gauge size is plainly marked on the insulation. If a link burns out, the cause must be isolated and corrected. A new fusible link must then be spliced into the circuit. The following procedure should be used to replace fusible links.

1. Disconnect battery ground cable at battery.

13

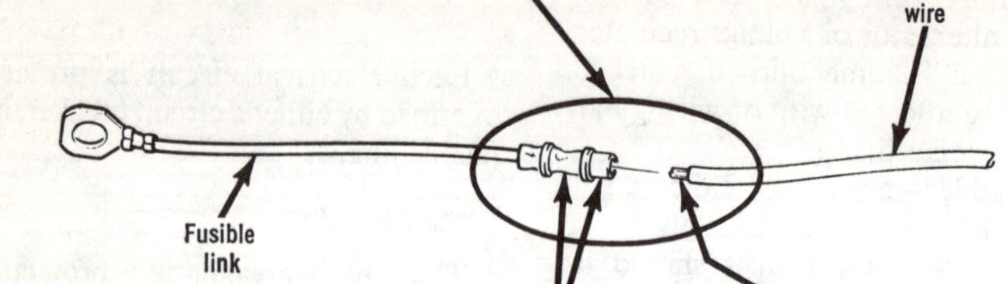

2. Disconnect old fusible link from components to which it is attached.
3. Cut harness behind connector (**Figure 13**) to remove damaged fusible link.
4. Strip harness wire insulation approximately to 1/2 in.
5. Position clip around ends of new fusible link (**Figure 13**) and harness wire and crimp so that both wires are securely fastened.
6. Solder the connection, using rosin core solder. Use sufficient heat to obtain a good solder joint, but do not overheat.
7. Tape all exposed wires with insulating tape.
8. Connect fusible link to component from which it was removed.
9. Reconnect ground cable to battery.

14

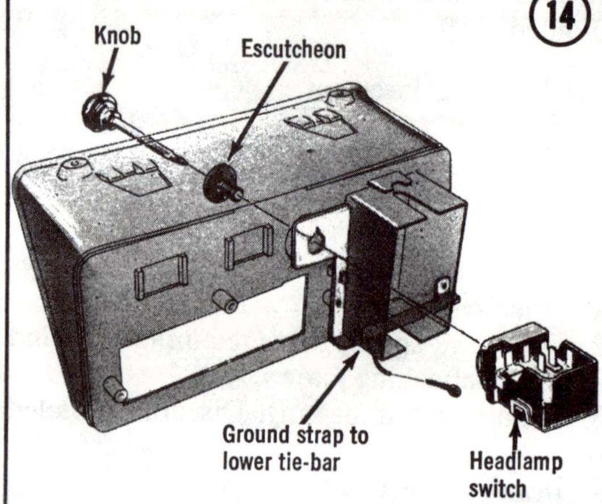

LIGHTING SWITCH REPLACEMENT

Refer to **Figure 14** (1970-1972), **Figure 15** (1973-1977), or **Figure 16** (1978-on).

1. Disconnect the battery negative cable.
2. On 1970-1977 models, remove the lower left hand trim panel. On 1978 and later models, remove instrument cluster pad assembly.
3. Pull headlight switch knob to ON position.
4. On 1978 and later models, remove the screws (2) attaching the switch mounting plate to the cluster and pull the assembly rearward.
5. On 1970-1977 models, reach under the panel and depress switch shaft retaining button and pull out shaft and knob.
6. Remove nut attaching switch to plate or panel and remove switch.
7. Disconnect wiring connector from switch.
8. Installation is the reverse of these steps.

15

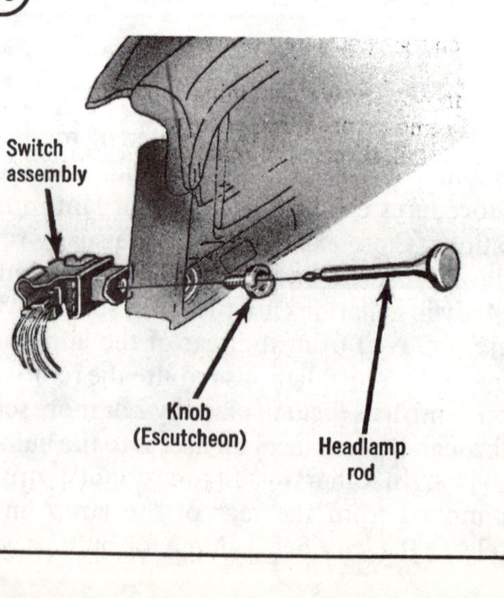

16

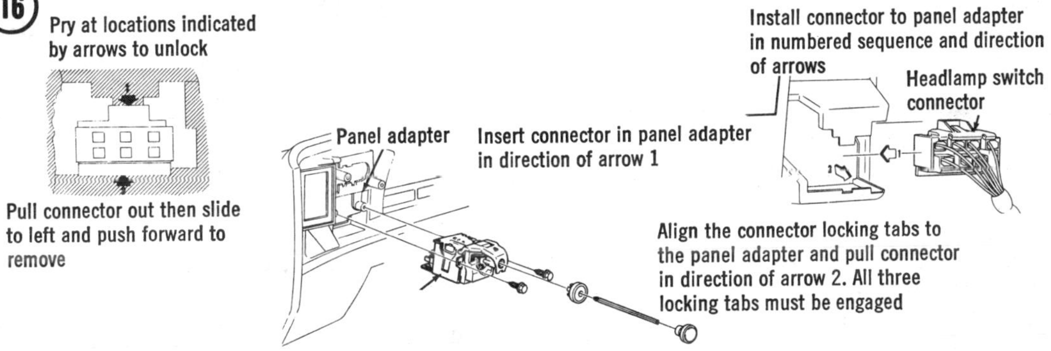

17

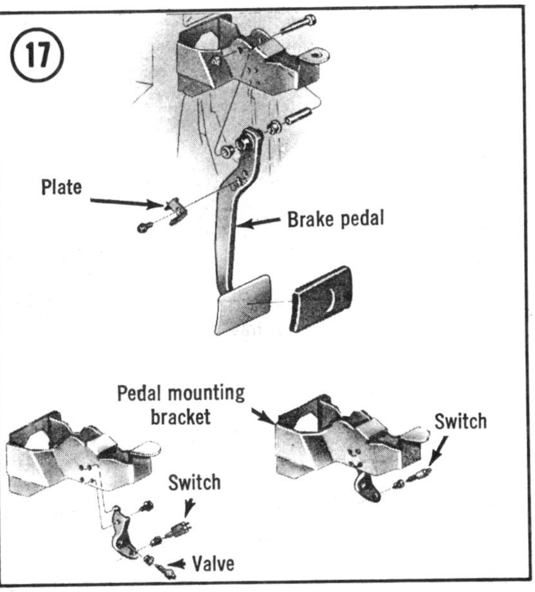

DIMMER SWITCH REPLACEMENT

1. Fold back floor mat and disconnect connector from dimmer switch.
2. Remove dimmer switch by removing attaching screws.
3. Install new switch by connecting it to connector and checking operation. Then attach switch to floor pan with attaching screws and replace floor mat.

STOPLIGHT SWITCH REPLACEMENT

See **Figure 17** (typical) for this procedure.

1. Disconnect wiring harness connectors from switch.
2. Remove retainer (nut, clip, screw, etc.) and unscrew the switch from the bracket.
3. Press brake pedal in and install new switch in bracket until shoulders bottom out. Plug in connector.
4. Check switch position to see that electrical contact is made (observe stoplights) when brake pedal is depressed 3/8-5/8 in.

DISTRIBUTOR REMOVAL

1. Remove distributor cap and position it out of the way.

NOTE

If necessary, remove secondary wires (spark plug and coil high tension wires) from distributor cap, after first marking cap tower for No. 1 cylinder spark plug wire.

2. Disconnect distributor primary wire from coil terminal.
3. Crank engine until No. 1 cylinder is in the firing position and the timing mark on the harmonic balancer is aligned with the proper mark on the timing tab on the engine front cover.
4. Make alignment marks on the distributor housing and the engine block in line with the rotor pointer.
5. Disconnect vacuum line to distributor.
6. Remove the distributor hold-down bolt and clamp. Remove distributor from engine.

8

Note that rotor moved during removal, and mark new position of rotor pointer on distributor housing.

NOTE

If possible, avoid any further cranking of engine until distributor has been replaced.

DISTRIBUTOR INSTALLATION (ENGINE NOT DISTURBED)

1. Turn rotor pointer to align with second mark made on the distributor housing (about 1/8 turn clockwise from first mark) during the removal procedure.
2. Insert distributor into its opening in engine block and align first mark on housing with mark made on block during removal. Push down into position. Rotor should now be aligned with first mark on distributor housing and mark on engine block.

NOTE

During installation it may be necessary to move rotor slightly so drive gear will match with camshaft gear.

3. Install distributor hold-down clamp and screw. Connect vacuum line to vaccum advance mechanism. Connect primary lead wire to coil terminal and install distributor cap. Replace spark plug wires and coil high tension lead wire, if removed.

NOTE

Verify that spark plug wires are connected to proper spark plugs.

4. Time ignition, using the procedure given above.

DISTRIBUTOR INSTALLATION (ENGINE DISTURBED)

1. Locate No. 1 piston in firing position, using either of the following methods:
   a. Remove No. 1 spark plug (the one closest to front of engine) and place finger over spark plug hole. Crank engine until compression is felt, then continue cranking until timing mark on crankshaft pulley lines up with proper mark on timing tab attached to front cover of engine.
   b. Remove rocker cover (left side on V6 and V8 engines) and crank engine until No. 1 intake valve closes. Continue to slowly crank until timing mark on pulley lines up with mark on timing tab.

2. Place distributor in opening in block in normal installed position. Hold distributor in this position and position rotor to point toward front of engine.

3. Turn rotor approximately 1/8 turn more toward left of engine and push distributor down to mesh drive gear with camshaft gear. It may be necessary to turn rotor slightly to engage the gears.

4. Press down firmly on distributor housing and crank engine a few times to verify that oil pump shaft is engaged. Install hold-down clamp and bolt.

5. Turn distributor body slightly until points just open, then tighten hold-down clamp bolt.

6. Hold distributor cap in place and verify that rotor lines up with terminal for No. 1 spark plug.

7. Install distributor cap. If spark plug wires were removed, replace them in their proper towers.

8. Connect vacuum hose to vacuum advance unit, and distributor primary lead wire to coil terminal.

9. Set dwell angle and time ignition using the procedures given above.

HIGH ENERGY IGNITION SYSTEM

All 1975 and later and some 1974 engines are equipped with High Energy Ignition (HEI) systems. All V6 and V8 engines are equipped with a distributor that contains all ignition components in one unit. The system used on L6 engines is identical in operating theory and uses similar system components, but has an externally mounted ignition coil.

The HEI is a pulse-triggered transistor controlled, inductive discharge system. Conventional breaker points are not used. Principal system elements are the ignition coil, electronic module, pick-up assembly, and the centrifugal and vacuum advance mechanisms.

Ignition Coil

The HEI coil operates in basically the same way as a conventional coil, but is smaller in size and generates higher secondary voltage when the primary circuit is broken. The coil is built into the cap of the V6 and V8 systems and is externally mounted in the L6 system.

Electronic Module

Circuits within the electronic module perform 5 functions. These are spark triggering, switching, current limiting, dwell control, and distributor pickup.

Pick-up Assembly

The pick-up assembly consists of a rotating timer core with external teeth (turned by the distributor shaft), a stationary pole piece with internal teeth, and a pick-up coil and magnet located between the pole piece and a bottom place.

Centrifugal and Vacuum Advance

The centrifugal and vacuum advance mechanisms are basically identical to units used in breaker point ignition systems.

Operation

As the distributor shaft turns the timer core teeth out of alignment with the pole piece teeth, a voltage is created in the magnetic field of the pick-up coil. The pick-up coil sends this voltage to the electronic module, which determines from the rotational speed of the distributor shaft when to start building current in the ignition coil primary windings. When the timer core teeth are again aligned with the pole piece teeth, the magnetic field is changed, creating a different voltage. This signal is sent to the electronic module by the pick-up coil and causes the module to shut off the ignition coil primary circuit. This collapses the coil magnetic field and induces a high secondary voltage to fire one spark plug.

The electronic module limits the 12-volt current to the ignition coil to 5-6 amperes. The module also triggers the opening and closing of the coil primary circuit with zero energy loss.

Corrective Maintenance

If a part in the HEI systems fails, it must be replaced. See *Troubleshooting*, this chapter, for diagnostic procedures. If the replacement of a part is necessary (other than the ignition coil), it may be necessary or more convenient to remove the distributor from the engine.

In a breaker point ignition system some energy can be lost due to point arcing and/or capacitor charging-time lag. Although a capacitor is present in the HEI system, it is used only as a radio noise suppressor. The efficiency of the triggering system allows up to approximately 35,000 volts to be delivered through the secondary wiring system to the spark plugs.

The module circuit controlling dwell angle causes the angle to increase as engine speed increases.

Troubleshooting

The following procedures are for the diagnosis of problems in High Energy Ignition (HEI) system distributors.

1. *Engine cranks but will not start*—Turn ignition on and place automatic transmission selector in PARK or manual transmission in NEUTRAL. Connect a test light to the BAT lead terminal on side of distributor. Connect other test light lead to ground. If light goes on, remove a spark plug and hold it 1/4 inch from engine block and crank engine. If spark is present, problem is not in the ignition system. Check spark plugs, fuel system, or for flooded condition. If test light did not go on, check ignition switch and repair or replace as necessary. If engine still does not start perform spark test described above. If not spark resulted from spark test, connect test light to distributor B terminal by inserting test lead into red B wire. If test light does not go on, repair or replace B wire or connector. If a test light goes on, or if engine fails to start after repairs, electronic module should be checked. Replace module if necessary.

NOTE

Checking of the electronic module requires special equipment. This task should be referred to your Oldsmobile dealer or a qualified mechanic who has the necessary equipment.

8

If the electronic module is OK, remove the distributor cap assembly, disconnect the 3-wire connector (HEI system), and inspect cap and distributor for evidence of moisture, dust, cracks, burns, etc., and replace or repair as required. If no defects were noted, remove green and white leads from module and connect an ohmmeter from either lead to ground. If any reading less than infinity is obtained (on the X1000 scale), pick up coil should be replaced.

If the engine does not start, connect an ohmmeter between the 2 outside terminals in the distributor cap side connector. If a reading above one ohm (on the X1 scale) is obtained, replace the ignition coil.

If the reading from the green or white module lead to ground was infinite, connect the ohmmeter between the 2 leads (green and white). If the reading is between 500 and 1500 ohms, repeat the reading while moving the vacuum advance with a screwdriver. If the reading is still within the 500-1500 ohms limits, check the ignition coil as outlined in the paragraph above.

If a reading of less than 500 or more than 1500 ohms was obtained, replace the pick up coil. If engine still does not start, check the ignition coil as outlined above.

If the ohmmeter reads less than one ohm (on the X1 scale) when the ignition coil is checked, connect the ohmmeter between the center terminal in the distributor cap side connector and the ignition coil case. If the reading is above 30,000 ohms or less than 6000 ohms, replace the ignition coil. If the reading is between 6000 and 30,000 ohms, repeat the procedure above until the problem is isolated.

2. *Engine runs rough or cuts out*—First check spark plugs and spark plug wires and repair or replace as required. If engine still runs rough, have the electronic module checked and replace if indicated. If this is not the problem, remove the distributor cap and inspect cap and distributor for evidence of moisture, dust, cracks, burns, etc. and repair and replace as necessary. If electronic module was OK, remove green and white leads from module and connect an ohmmeter between either and ground. If the reading is less than infinity (on the X1000 scale), replace the pick-up coil. If the reading was infinity, connect the ohmmeter between green and white leads. If the reading is between 500 and 1500 ohms, retake the reading while moving the vacuum advance with a screwdriver. If the reading is still within the 500-1500 ohm limits, check the ignition coil as outlined below. If the reading was not within the 500-1500 ohm limits, replace the pick-up coil. If engine still does not function properly, check ignition coil as outlined below.

To check ignition coil, connect an ohmmeter between the center terminal in the distributor cap side connector and the ignition coil case. If the reading is more than 30,000 ohms or less than 6,000 ohms, replace the ignition coil. If the reading is between 6000 and 30,000 ohms, repeat the procedure above until the problem is isolated.

Distributor Removal

1. Unplug wiring harness connectors at side of distributor cap.
2. Remove distributor cap and position to one side out of the way.
3. Disconnect hose from the vacuum advance mechanism.
4. Note the alignment of the rotor and make marks on the engine and the distributor housing so alignment can be duplicated when distributor is reinstalled.
5. Remove distributor hold-down clamp.
6. Lift distributor straight up and away from engine. Note and mark housing with final position of rotor. Also note position of vacuum advance unit in relation to engine.

Distributor Installation (Engine Undisturbed)

If engine has not been disturbed since distributor was removed, proceed as follows:

1. Position the rotor to align with the last mark made on distributor body (Step 6 above).
2. Insert the distributor shaft into its hole and push down into position in the engine block. When fully inserted, the rotor should be aligned with the marks made in Step 4 above.

NOTE

It may be necessary to turn rotor slightly to mesh distributor gear with camshaft

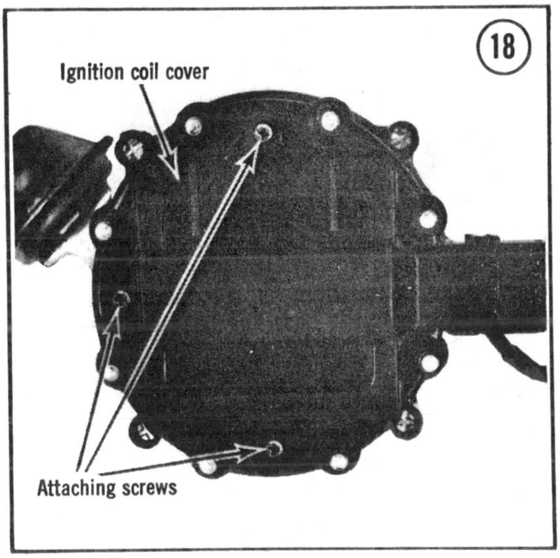

gear. However, rotor should line up with mark when fully inserted. If not, remove and repeat Steps 1 and 2.

3. Install hold-down clamp and nut. Tighten nut snugly after making certain distributor body and rotor are in same relative position to engine as noted during removal.
4. Reinstall distributor cap with 4 latches after verifying that tab in base of cap is aligned with notch in distributor housing.
5. Plug in wiring harness connector to terminal on side of distributor cap.
6. Time engine in accordance with procedure given in Chapter Three.

Distributor Installation (Engine Disturbed)

If the engine has been disturbed, reinstall the distributor as follows:

1. Position the No. 1 piston in firing position by one of the following methods.

NOTE

The No. 1 cylinder piston or spark plug is the one closest to the front of the vehicle. In V6 and V8 engines, it is located on the left side (note that the left bank of cylinders is offset slightly further foward than the right bank).

a. Remove spark plug from No. 1 cylinder and place finger over plug hole. Crank engine until compression is felt. Continue cranking until timing mark on crankshaft pulley lines up with proper timing mark on tab attached to engine front cover.
b. Remove rocket cover (left cover on V8 engines) and crank engine until intake valve for No. 1 cylinder closes. Continue cranking until timing marks on pulley and timing tab are aligned.

2. Place the distributor in its hole in the normally installed position, as noted in the removal procedure.
3. While holding the distributor in the position described in Step 2, turn the rotor to point to the front of the engine, then turn rotor about 1/8 turn counterclockwise. Push down on distributor to engage gear with camshaft gear, rotating the rotor slightly to assist in engagement, if necessary.
4. Verify that rotor lines up with No. 1 spark plug terminal on distributor cap when cap is in proper position.
5. Install distributor hold-down clamp and nut.
6. Install distributor cap and secure with 4 latches, making certain tab in base of cap is aligned with notch in distributor housing.
7. Plug wiring harness connectors into terminals on side of distributor cap.
8. Time engine per the procedure given in Chapter Three.

Ignition Coil Removal (V8 and V6 Engines)

1. Disconnect battery wire and harness connector from distributor cap.
2. Remove 3 coil cover attaching screws and remove coil cover. See **Figure 18**.
3. Remove 4 ignition coil attaching screws. See **Figure 19**.
4. Remove ground wire from coil.
5. Remove coil leads from connectors, then remove coil. Remove arc seal. See **Figure 20**.

Ignition Coil Installation (V8 and V6 Engines)

1. Install arc seal and place coil into distributor cap with terminals over connector at side of cap.

19

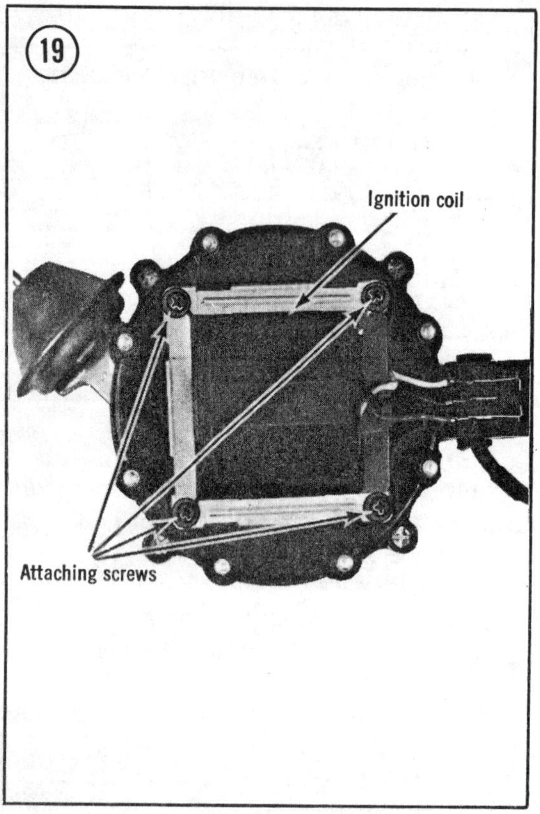

20

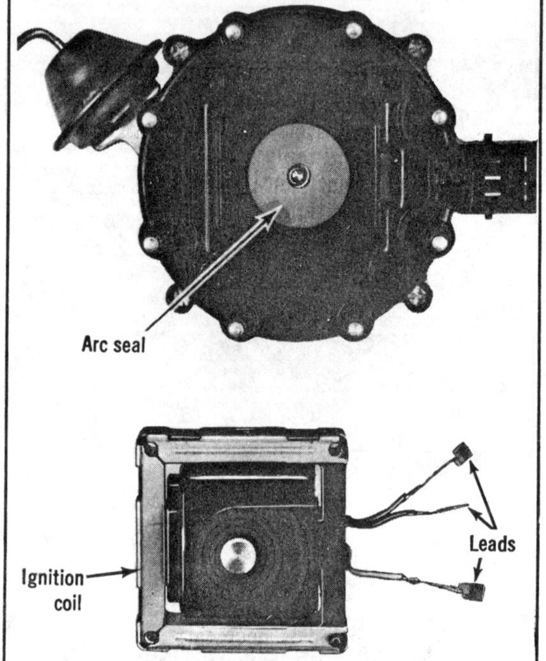

2. Push coil lead wires into connector on side of cap as follows: brown wire next to vacuum advance unit, black wire in center, and pink wire opposite vacuum advance unit.

3. Secure coil to distributor cap with 4 attaching screws, placing ground wire under one of the screws.

4. Install coil cover and secure with 3 attaching screws.

Ignition Coil Removal/Installation (L6 Engines)

Refer to **Figure 21**.

1. Disconnect ignition switch and distributor lead wires from coil.

2. Remove attaching screws, then remove coil from side of engine.

3. Installation is the reverse of these steps.

21

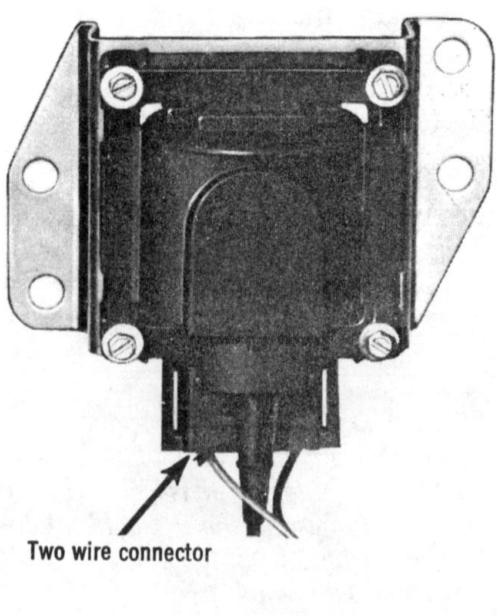

Table 1 ALTERNATOR SPECIFICATIONS

	Model						
	1100693 1100628	1100694	1100857 1100859 1100833 1100696	1100697	1100825 1100750	1100544	1102353 1100950 1100543
Voltage	14	14	14	14	14	14	14
Output current (hot)	37	55	42	60	61	61	42
Field current	1.9-2.3	2.2-2.6	2.2-2.6	2.8-3.2	2.2-2.6	4.0-4.5	4.0-4.5

Table 2 LAMP USAGE

Usage	Bulb Identification Number			
	1970-1972	1973-1975	1976-1977	1978-On
Back-up	1156	1156	1156	1156
Headlamp (outer or single)	L-4002	6014	4652	6052
Headlamp (inner)	L-4001	—	4651	—
Headlamp (Halogen)	—	—	—	H6052
License	97	194	194	194
Parking and turn signal	1157NA	1157NA	1157NA	1157NA
Side markers (front and rear)	194	194	194	194
Tail and stop light	1157	1157	1157	1157
A/C or heater control lamp	161	—	—	194
Ash tray light	1445	—	—	1445
Brake warning light	161	161	161	161
Brake warning (w/fuel economy gauge)	—	74	74	—
Ceiling reading lights	—	1004	1004	1004
Check engine (Calif. only)	—	—	—	161
Clock	168	1816	1816	194
Clock (digital)	—	—	—	168
Console compartment light	1895	1893	1893	1891
Console shift indicator light	1445	1895	1895	194
Courtesy light, left front	90	89	906*	906
Courtesy light, right front	90	89	906*	89
Console courtesy lamp	97	97	97	906
Cruise control switch light	—	—	—	1893
Dome light	211	211	211	211
Dome light (between reading lights)	—	—	—	212
Engine oil warning light	168	168	168	161
Fasten belts light	—	1895	1895	161
Fuel gauge light (gauge cluster)	194	194	194	194
Fuel gauge cluster, top bulbs	—	161	161	161
Fuel gauge cluster, bottom lights	—	194	194	194

(continued)

8

Table 2 LAMP USAGE (continued)

Usage	Bulb Identification Number			
	1970-1972	1973-1975	1976-1977	1978-On
Fuel gauge cluster, salon	—	168	168	168
Generator warning light	161	168	168	168
Glove box lights	563	563	563	1891
High beam indicator	194	161	161	194
Engine temperature indicator	168	557	168	161
Lights on indicator	—	—	—	161
Low fuel indicator	—	—	—	161
Radio dial (exc. tapes & CB)	1445	1893	1893	216
Radio dial (tapes & CB)	—	563	216	1893
Shift indicator	—	—	—	161
Side pillar lights	212	212	212	212
Station wagon rear lamp	212	212	212	212
Trunk lamp	631	1003	1003	1003
Turn signal indicator	194	194	194	194
Underhood lamp	631	631	631	631
Visor vanity mirror light	—	562	562	562
Instrument panel floodlights	161	563	563	—
Wait light (diesel only)	—	—	—	168
Windshield wiper switch light	—	—	—	168

*On 1976 models, use 89.

Table 3 FUSES

Circuit	Fuse Rating (amps) 1970	1971	1972	1973	1974	1975
Wiper	25	25	25	25	25	25
Back-up lamp and gauges	25	20	20	20	20	20
Heater/AC	25	25	25	20	20	20
Radio	10	10	10	10	10	10
Instrument lamps	4	4	4	4	4	4
Stop and taillamp	20	20	20	20	20	20
Clock, lighter and courtesy	25	25	25	25	25	25
In-line AC	30	30	30	30	30	30
Gauges and telltales	10	10	10	20	20	20
Tail, license, luggage & park	20	20	20	20	20	20

Circuit	Fuse Rating (amps) 1976	1977	1978	1979	1980
Wiper	25	25	25	25	25
Back-up lamp and gauges	20	20	20	20	20
Heater/AC	20	20	25	25	25
Radio	10	10	10	10	10
Instrument lamps	4	4	5	5	5
Stop and taillamp	20	20	20	20	20
Clock, lighter and courtesy	25	25	20	20	20
In-line AC	30	30	30	—	—
Gauges and telltales	20	20	20	20	20
Tail, license, luggage & park	20	20	20	20	20

8

CHAPTER NINE

CLUTCH AND TRANSMISSION

This chapter provides procedures for replacement of manual and automatic transmissions and for clutch removal, replacement, and adjustment.

A number of manual transmissions were offered during the period covered by this book. Repairs requiring disassembly are not recommended for home mechanics or garage mechanics without special skills and a large assortment of special tools. In fact, the cost of the necessary tools far exceeds the price of a professionally rebuilt transmission.

Considerable money can be saved by removing the transmission and installing a new or rebuilt one yourself. See Chapter Two, *Troubleshooting*, for procedures which will help isolate the fault.

Table 1, at the end of the chapter, gives tightening torques.

CLUTCH

There is one linkage adjustment (clutch fork pushrod or pedal pushrod) to compensate for all normal clutch wear. The clutch should have a specified amount of free travel before the throw-out bearing engages clutch diaphragm spring levers (see individual procedures below for amounts specified). Lash is required to prevent clutch slippage which would occur if the bearing was held against the fingers or to prevent the bearing from running continually until failure.

A clutch that was slipping before adjustment may still slip after adjustment because of prior heat damage. Allow clutch to cool for at least 12 hours, then check for slippage as follows.

1. Drive in high gear at 20-25 mph.
2. Depress clutch pedal to floor and accelerate engine speed to about 3,000 rpm.
3. Snap foot off clutch pedal and at the same time press accelerator to floor board. Engine speed should drop and then accelerate. If clutch is bad, engine speed will increase immediately (before vehicle accelerates).

NOTE
Clutch will overheat if this test is repeated before allowing clutch to cool for at least 12 hours.

Clutch Linkage Adjustment

Refer to **Figure 1** (1970-1973) or **Figure 2** (1975-on).

1. Disconnect return spring (at clutch fork).
2. Raise clutch lever and shaft assembly until clutch pedal rests against rubber bumper and dash brace.
3. Move outer end of clutch fork rearward until throwout bearing just contacts pressure plate fingers.
4. Install pushrod in gauge hold and adjust length until all lash is removed from linkage system.
5. Remove rod from gauge hole and reinstall in lower hole in lever. Install retainer and tighten locknut. Be careful not to alter length of rod.
6. Reinstall return spring and check free travel of pedal. Travel should be 3/4-1 in. (1970-1972); 3/4 - 1-1/4 in. (1973-1977); or 11/16-5/8 in. (1978 and later).

1

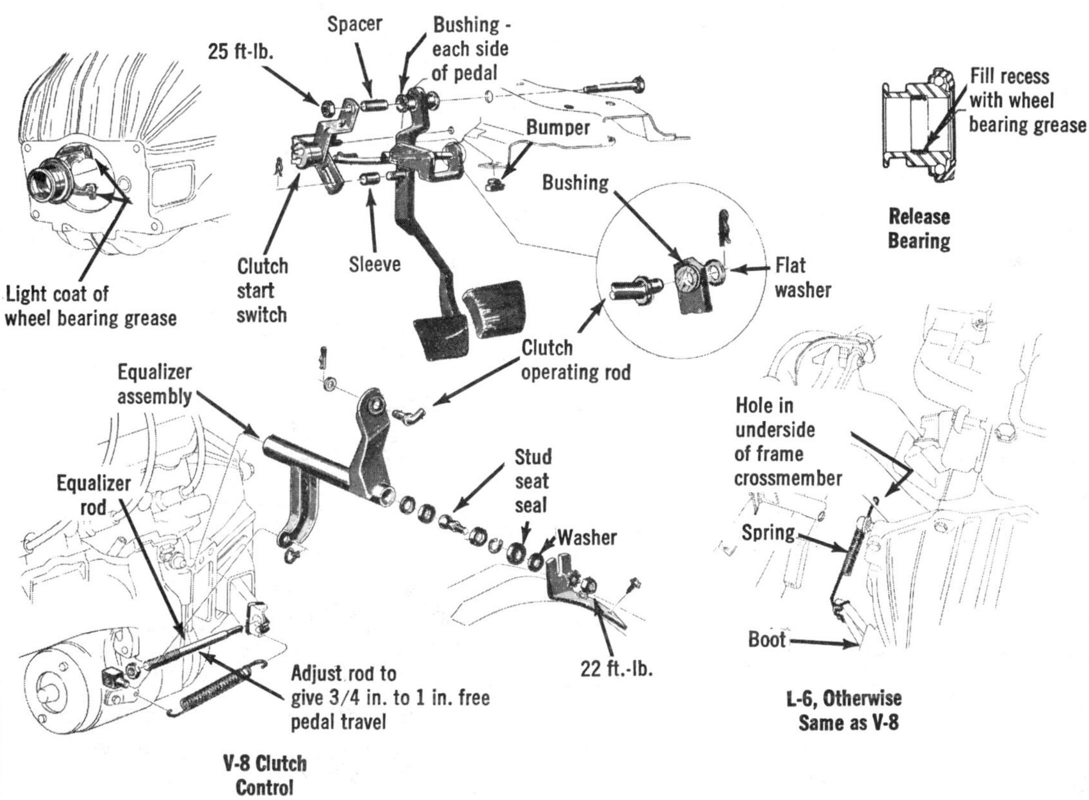

Clutch Mechanism Removal

Refer to **Figure 1** or **Figure 2**.

1. Support engine and remove transmission as described later.
2. Disconnect clutch fork pushrod and spring.
3. Remove flywheel housing.
4. Slide clutch fork from ball stud and remove fork from dust boot.

NOTE

Ball stud is threaded into clutch housing and is easily replaced.

5. Install tool J-5824 or equivalent clutch pilot to support clutch.
6. Look for "X" mark on flywheel and clutch cover. If not visible, make small punch marks on these parts to aid reassembly.
7. Loosen clutch-to-flywheel bolts evenly, one turn at a time, until spring pressure is released.
8. Remove bolts and clutch assembly.

Inspection

Never replace clutch parts without giving thought to the reason for failure. To do so only invites repeated troubles.

1. Clean the flywheel face and pressure plate assembly in a non-petroleum base cleaner.
2. Check the friction surface of the flywheel for cracks and grooves. Attach a dial indicator and check runout. Compare with specifications for your engine. If necessary, have the flywheel reground; replace it in cases of severe damage.
3. Check the pressure plate for cracked or broken springs, evidence of heat, cracked or scored friction surface, and looseness. Check release lever ends for wear. On diaphragm spring clutches, check the spring fingers for wear. If there is any damage, replace with a professionally rebuilt pressure plate assembly.
4. Check the clutch disc (drive plate) lining for wear, cracks, oil, and burns. The assembled thickness of the disc should be at least 0.36 in. See **Figure 3**. Check for loose rivets and cracks in the spring leaves or carrier plate. Ensure that the disc slides freely on the transmission spline without excessive radial play. If the disc is defective, replace it with a new one.

2

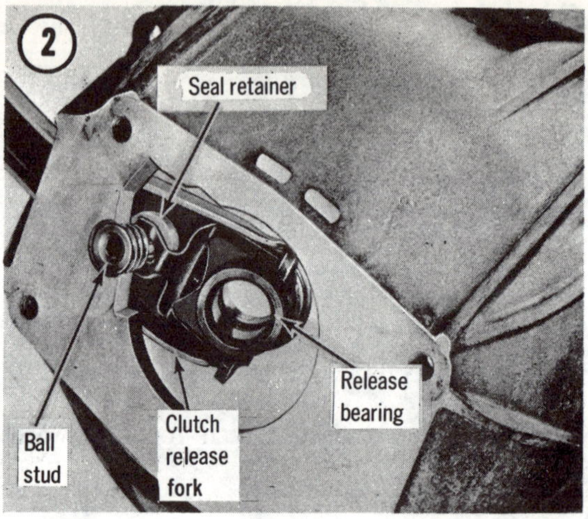

3

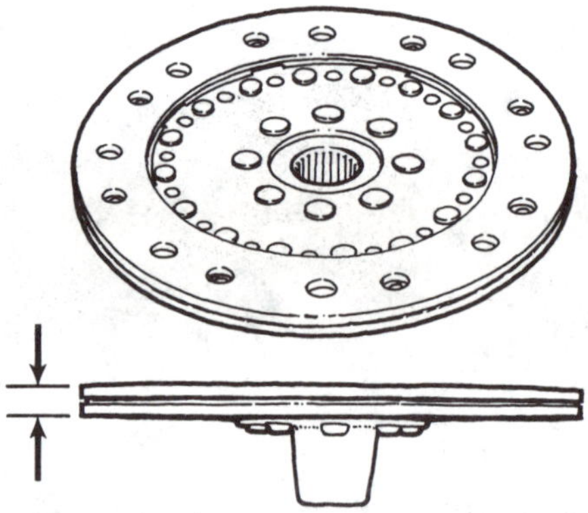

4

All models except 2 bbl. 350 cu. in. and L-6

350 cu. in. 2 bbl. and L-6

5. Check the release bearing for wear to determine if it caused the original trouble. Never reuse a release bearing unless necessary. When other clutch parts are worn, the bearing is probably worn. If it is necessary to reinstall the old bearing, do not wash it in solvent; wipe it with a clean cloth.

Installation

1. Make sure your hands are *clean* before proceeding.

5

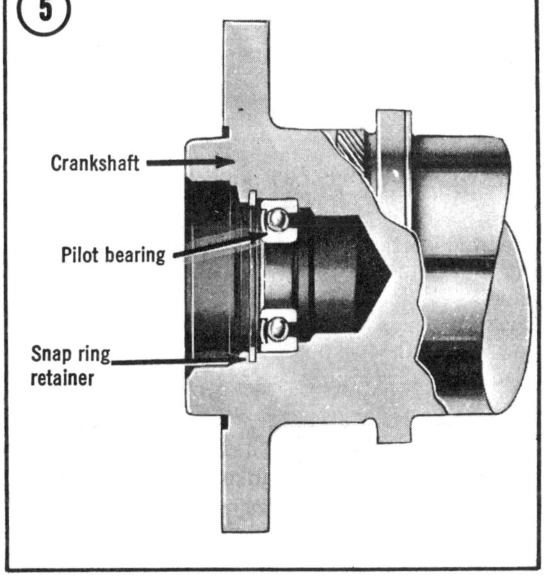

6

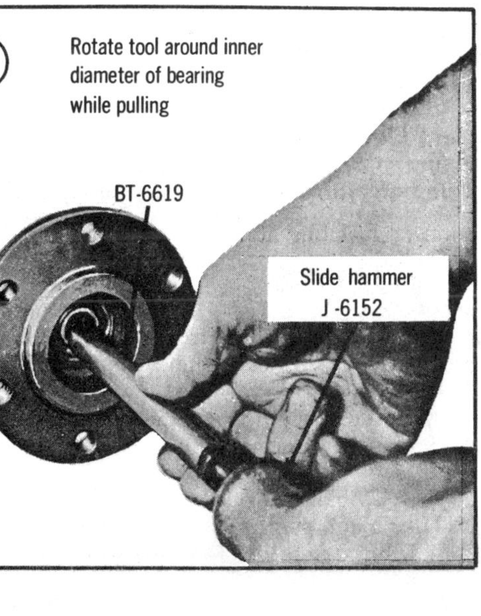

2. Sand the friction surface of the flywheel and pressure plate with a medium-fine emery cloth. Sand lightly across the surfaces (not around) until they are covered with fine scratches. This breaks the glaze and aids seating a new clutch disc.
3. Clean the flywheel and pressure plate with non-petroleum solvent.
4. Position clutch disc and pressure plate on engine and support with tool J-23720 or equivalent pilot tool. Clutch disc damper springs face the pressure plate.

NOTE

An excellent pilot tool can be made by cutting off about one foot from the forward end of an old transmission main shaft. Other tools made from wooden dowellings are available from most auto parts suppliers.

5. Align "X" marks or punch marks on clutch cover and flywheel. Install bolts finger-tight.
6. Tighten diagonally opposite bolts a few turns at a time until all are tight. Torque to 30 ft.-lb.
7. Remove pilot tool.
8. Unhook clutch fork. Lubricate ball socket with high melting point grease, e.g., graphite, and reinstall fork on ball stud.
9. Lubricate recess on inside of throwout bearing with molybdenum grease. See **Figure 4**, typical.
10. Install clutch fork and dust boot into clutch housing.
11. Install throw-out bearing to fork.
12. Install flywheel housing.
13. Install transmission.
14. Connect fork pushrod and spring.
15. Adjust clutch pedal travel and free play.

Clutch Pilot Bearing Removal/Installation

The clutch pilot bearing should be removed for inspection whenever the clutch is removed.
1. Remove retaining ring, if used, and pull bearing from crankshaft with tool BT-6619 or equivalent puller and a slide hammer. See **Figure 5** and **Figure 6**.
2. Clean bearing with a clean cloth dipped in solvent.

3. Check bearing for excessive wear or other damage. Replace if necessary.
4. Drive bearing into the crankshaft with tool 3T6618 or equivalent driver. See **Figure 7**.

TRANSMISSION REPLACEMENT

NOTE

This procedure is general and covers all transmissions, manual or automatic.

1. On manual transmissions remove shifter knob and on 4-speed models, remove spring and "T" handle.
2. Raise vehicle.
3. Disconnect speedometer cable and all electrical leads from transmission.
4. On automatic transmissions, disconnect shift control linkage, oil cooler pipes (if present), and vacuum line modulator (if so equipped).
5. Disconnect and remove the drive shaft (see Chapter Twelve).
6A. Remove manual transmission:
   a. Remove the transmission mount-to-crossmember bolts and crossmember-to-frame bolts. Raise and support engine and remove crossmember.
   b. Remove shift levers from transmission and disconnect back drive rod at bell housing (floor shift models).
   c. Remove the shift control assembly-to-transmission support attaching bolts (floor shift models) and pull unit down until shift lever clears boot. Remove assembly.
   d. Remove the upper bolts holding transmission to clutch housing and install guide pins in holes. Then remove lower bolts.
   e. Support transmission while sliding it to the rear to remove it from vehicle.

7

8

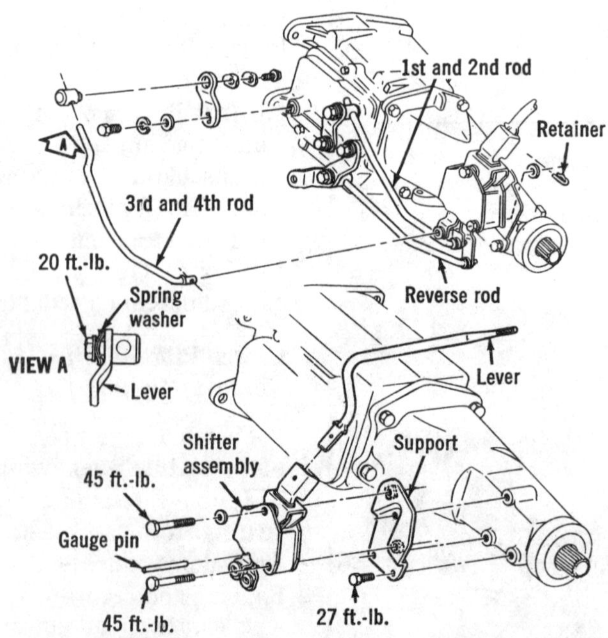

6B. Remove automatic transmission:
   a. Support transmission with a suitable jack.
   b. Disconnect rear mount from crossmember. Then remove attaching bolts and remove crossmember from frame.
   c. Remove converter under pan.
   d. Remove converter-to-flywheel bolts.
   e. Lower transmission until jack is barely supporting it and then remove transmission-to-engine mounting bolts. Remove oil filter tube at transmission.
   f. Raise transmission to normal position, support engine with a jack, and remove transmission by sliding it rearward. Lower it away from vehicle.

7. Reverse the above procedures to install transmission.

NOTE
*Tighten all nuts and bolts to specifications. See **Table 1**.*

NOTE
On automatic transmissions, before installing flexplate-to-converter bolts, make certain weld nuts on converter are flush with flexplate and the converter freely rotates in this position. Then hand start all bolts and tighten finger-tight before torquing to specifications. This will ensure proper alignment of converter.

SHIFT LINKAGE ADJUSTMENT

Manual Transmission Floor Shift (4-Speed)

Refer to **Figure 8**.

1. Raise vehicle. Ignition should be off.
2. Loosen locknuts on shift rod swivels so that rods move freely through swivels.
3. Place all transmission levers in NEUTRAL.
4. Move shift control lever to neutral detent and install locating gauge (1/4 in. diameter) into lever alignment slot.
5. Tighten shift rod swivel locknuts to 10 ft.-lb., and remove gauge.

NOTE
*The following steps apply to 1973 and earlier models with back drive linkage. See **Figure 9**.*

9

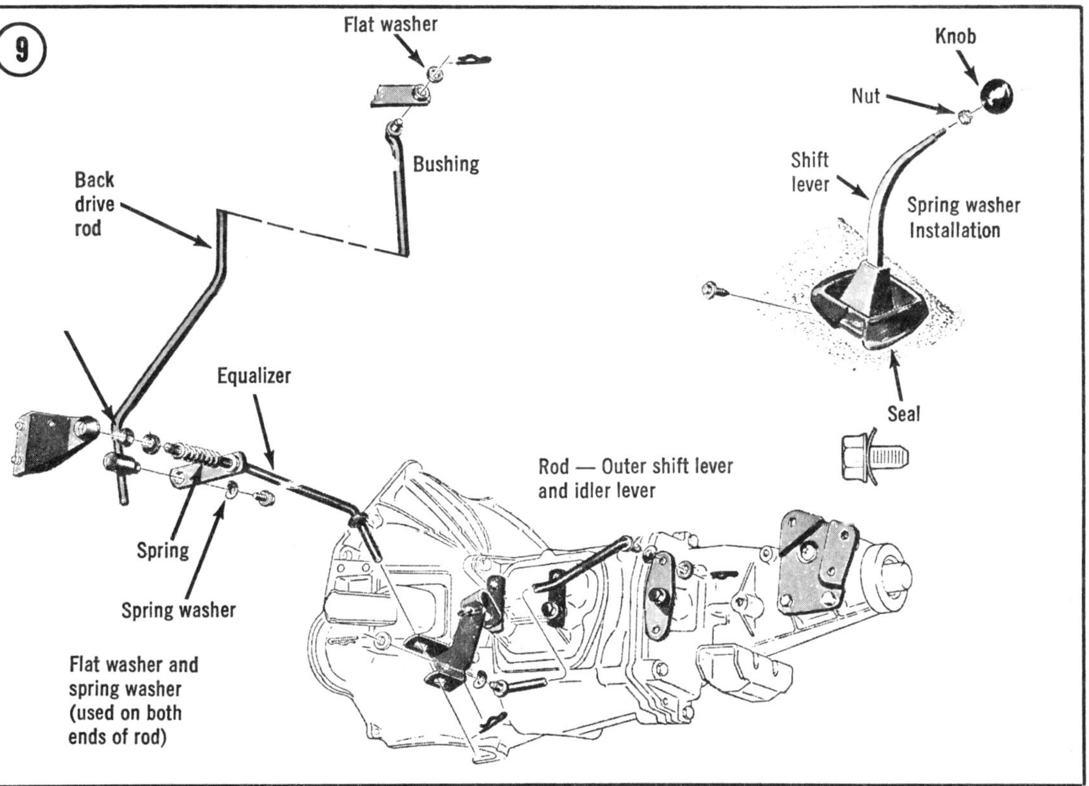

9

6. Shift control lever into reverse and plate ignition switch in LOCK position. Loosen back drive control rod swivel locknut. Pull down slightly on rod to remove all slack and tighten jam nut.
7. Verify that ignition key moves freely to and from LOCK position. Readjust if required.
8. Check shift operation through all gears. Readjust rods if necessary. Lower vehicle.

Manual Transmission Floor Shift (1978-On 3-Speed)

Refer to **Figure 10**.

1. Turn ignition switch to OFF and raise the vehicle.

2. Remove the shift rod retainers and place the transmission shift levers in NEUTRAL position.

10

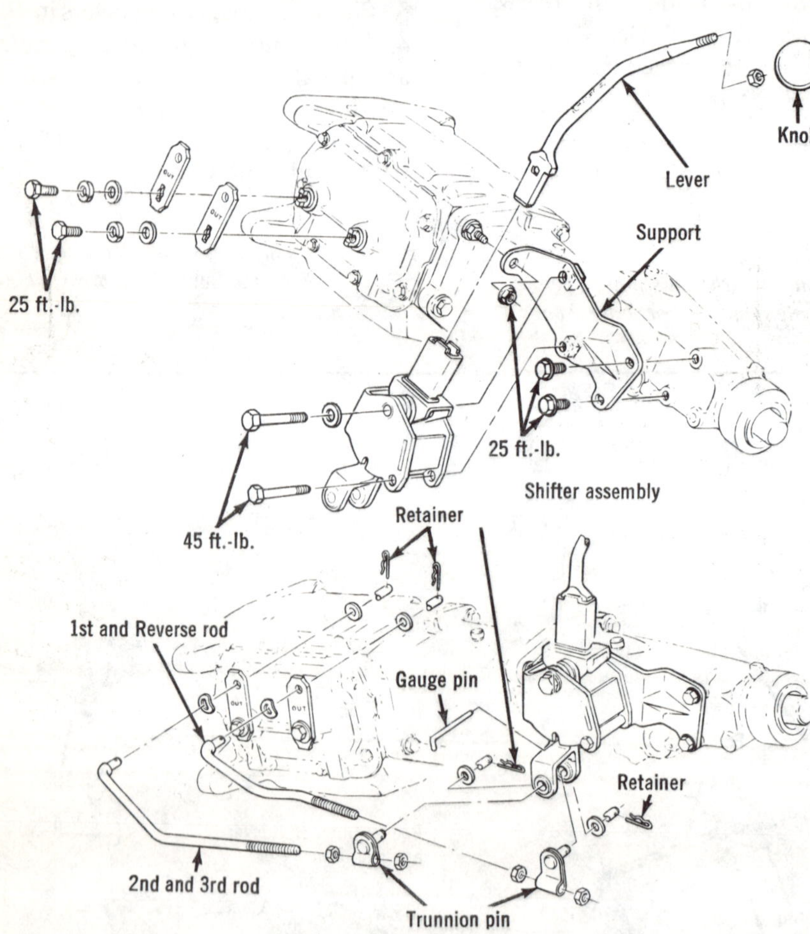

3. Place the shift control lever in NEUTRAL position and align the shift levers. Insert a gauge pin through the levers and the bracket.
4. Loosen the adjusting nuts on the ends of the shift rods and adjust the trunion and pin assemblies, then install retainers and tighten nuts.
5. Remove the gauge pin and check the linkage to make sure that it operates properly and smoothly through all gear changes. Readjust, if required.

Manual Transmission Floor Shift (5-Speed)

The shift lever and cover of the 5-speed transmission mount directly on the transmission extension housing. See **Figure 11**. As no linkage is involved, adjustment is not required.

Manual Transmission Column Shift

Refer to **Figure 12**.

(11)

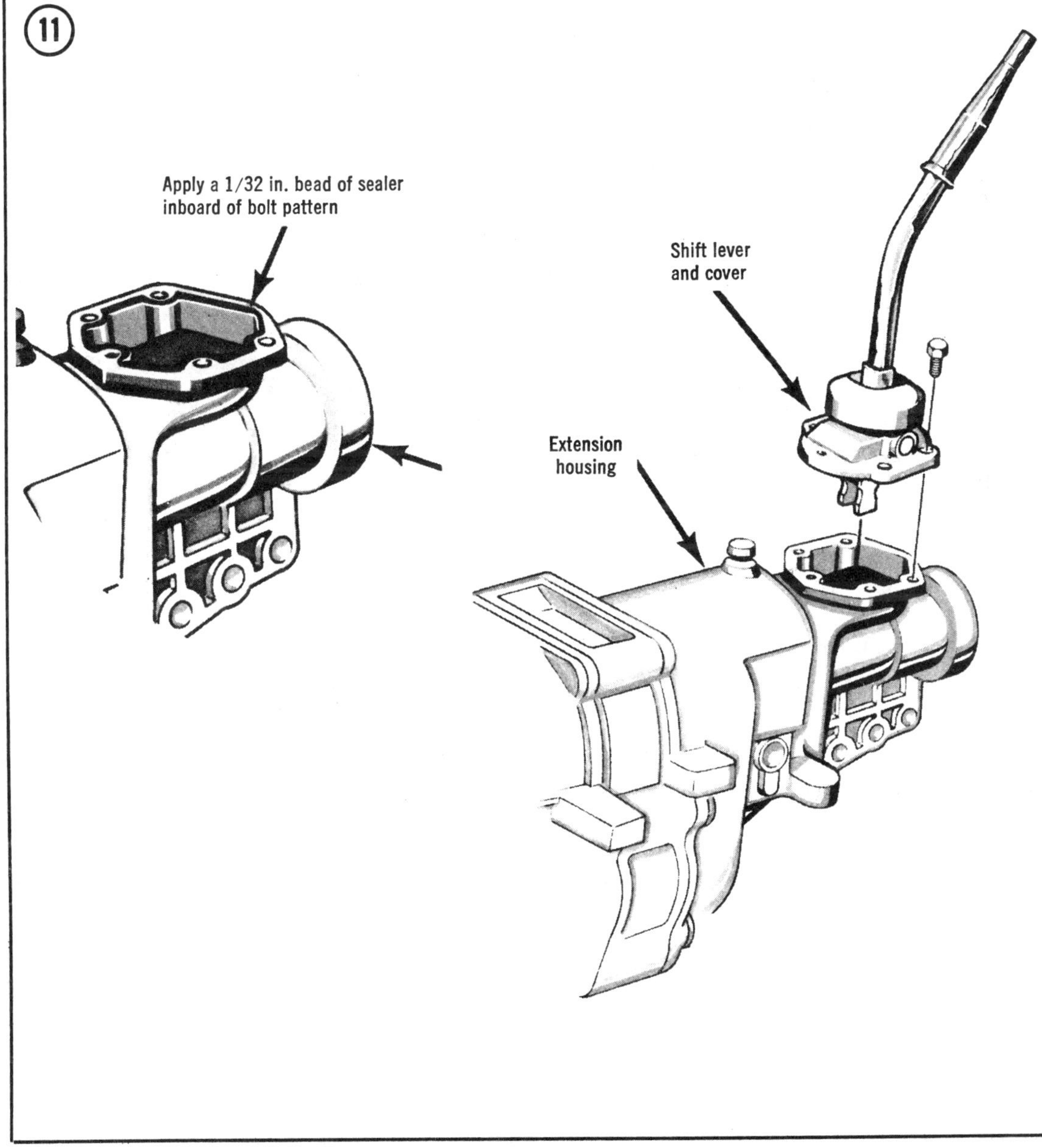

12

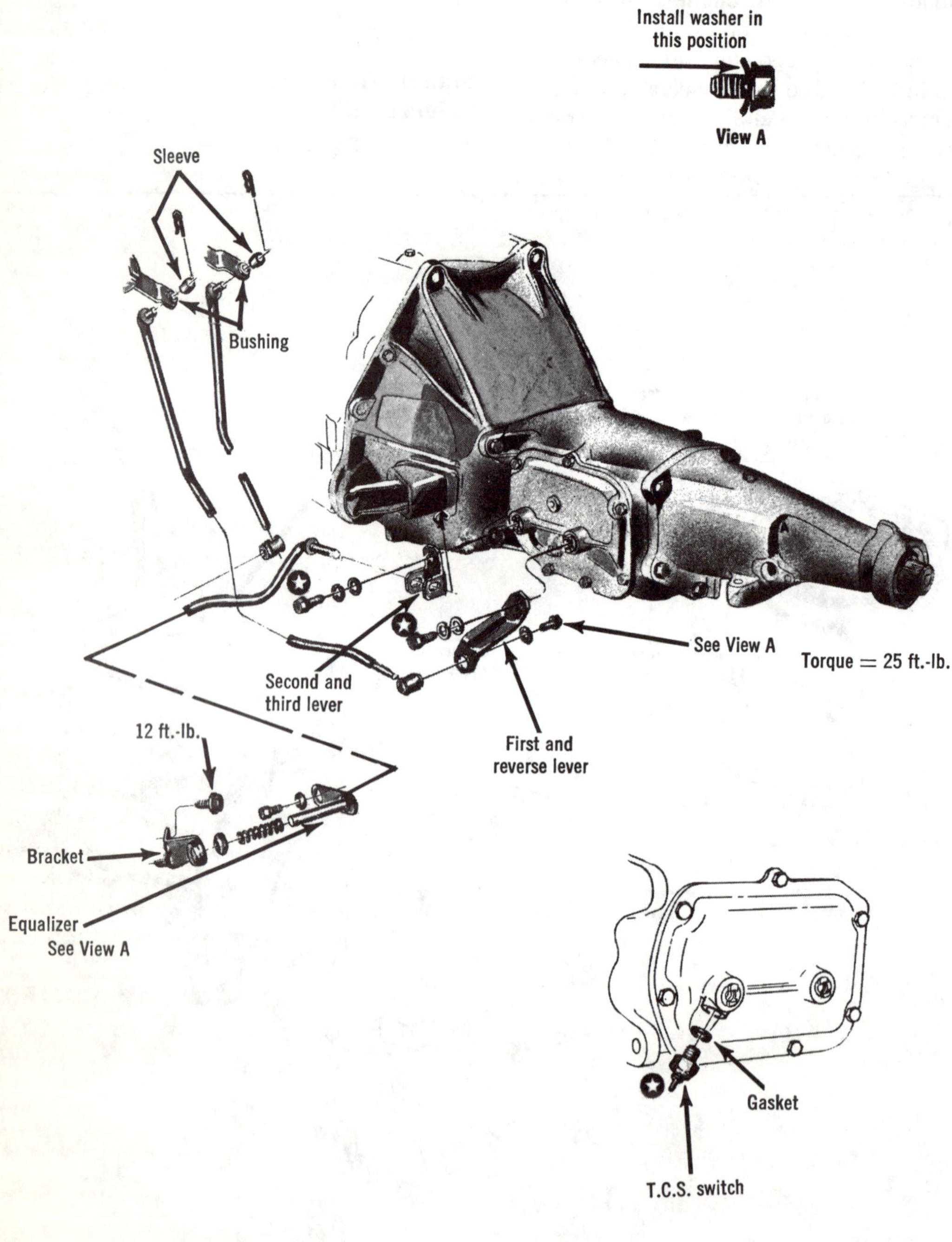
Install washer in
this position
View A
Sleeve
Bushing
See View A
Torque = 25 ft.-lb.
Second and
third lever
First and
reverse lever
12 ft.-lb.
Bracket
Equalizer
See View A
Gasket
T.C.S. switch

1. With the transmission in the REVERSE position, raise the vehicle.
2. Loosen all swivel bolts at the transmission shift rods, and make sure the rods move freely in the swivels.
3. Hold the 1st-REVERSE column relay lever in position, and then push up on the shift rod until you can feel it slip into the column detent. Then tighten the swivel bolt to 26 ft.-lb.
4. Shift the transmission into NEUTRAL and insert a 3/16 in. rod through the column relay lever and the alignment hole. Then tighten the 2nd-3rd swivel bolt to 26 ft.-lb.
5. Lower the vehicle and check the linkage to make sure that it operates properly and smoothly through all gear changes.
6. Shift the transmission to reverse, turn the ignition switch to LOCK position. Make sure that the key can be removed, that the transmission cannot be shifted out of REVERSE, and that the steering wheel will not turn.
7. Turn the ignition key to RUN position and shift into 2nd gear. Make sure that the steering wheel will turn and that the key cannot be removed from the switch.
8. If the linkage fails any of the checks in Steps 5 through 7, readjust as required.

Turbo-Hydramatic Shift Linkage Adjustment (1970-1977 Console Shift)

Refer to **Figure 13**.

1. Loosen shift rod clamp screw and pin in transmission manual lever.
2. Place both shift lever and transmission manual lever in P positions. Make sure ignition key is in LOCK position.
3. Pull the shift rod tightly against the lock stop and tighten the clamp screw.
4. Move the pin in the transmission manual lever until it fits freely, and then tighten the nut to 26 ft.-lb.
5. Lower the vehicle and check the linkage to make sure that it operates properly and smoothly through all gear changes.
6. Shift the transmission to reverse, and turn the ignition switch to LOCK position. Make sure that the key can be removed, that the transmission cannot be shifted out of REVERSE and that the steering wheel will not turn.
7. Turn the ignition key to RUN position and shift into 2nd gear. Make sure that the steering wheel will turn and that the key cannot be removed from the switch.
8. If the linkage fails any of the checks in Steps 5 through 7, readjust as required.

Turbo-Hydramatic Shift Linkage Adjustment (1970-1977 Console Shift)

Refer to **Figure 13**.

1. Loosen shift rod clamp screw and pin in transmission manual lever.
2. Place both shift lever and transmission manual lever in P positions. Make sure ignition key is in LOCK position.
3. Pull the shift rod tightly against the lock stop and tighten the clamp screw.
4. Move the pin in the transmission manual lever until it fits freely, and then tighten the nut to 26 ft.-lb.
5. Check operation of the linkage by shifting the selector through all positions and making sure the transmission manual lever is properly seated in the proper detent. Turn key to RUN with transmission in REVERSE and make sure key cannot be removed from ignition switch. Then turn key to LOCK and transmission to PARK and make sure that key can be removed and that steering wheel locks. Then make sure engine will start when transmission is in PARK and NEUTRAL. Readjust if required.

Turbo-Hydramatic Shift Linkage Adjustment (1978-On Console Shift)

Refer to **Figure 14**.

1. Loosen the shift rod clamp screw and then loosen the nut on the pin in the transmission manual lever.

13

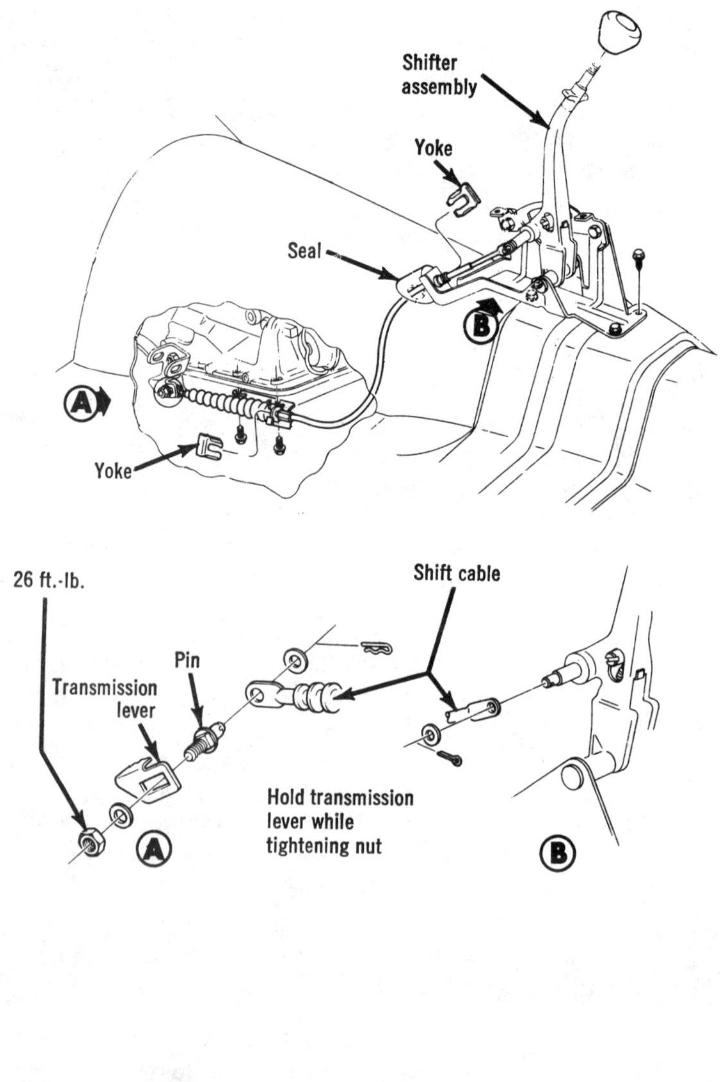

Shifter assembly
Yoke
Seal
B
A
Yoke
26 ft.-lb.
Shift cable
Pin
Transmission lever
Hold transmission lever while tightening nut
A
B

14

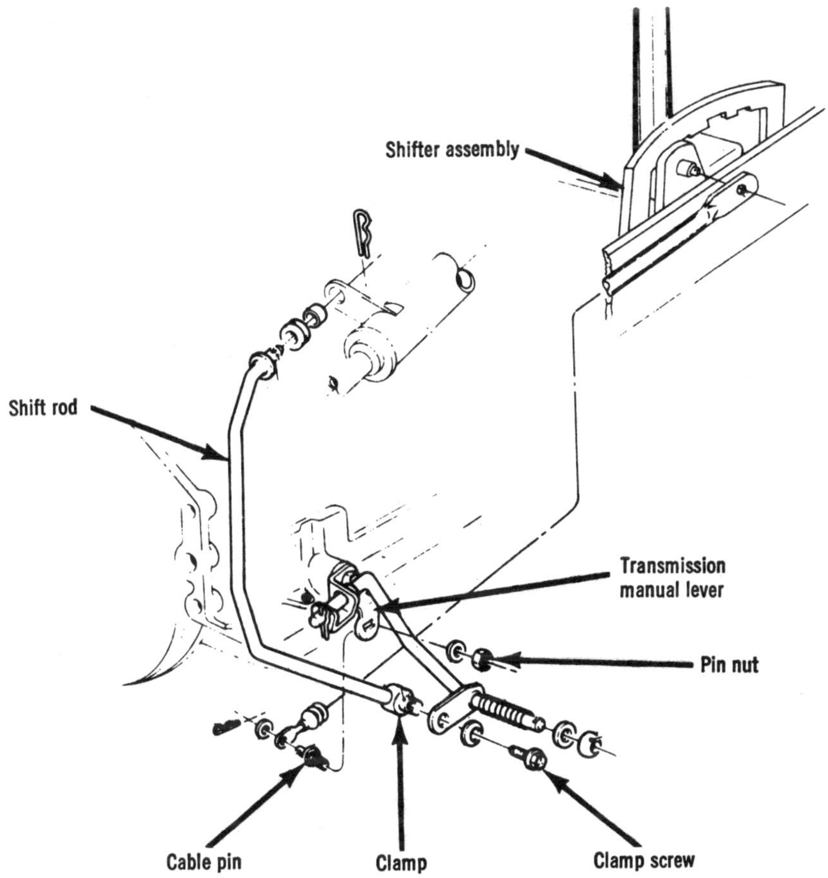
Shifter assembly
Shift rod
Transmission manual lever
Pin nut
Cable pin
Clamp
Clamp screw

9

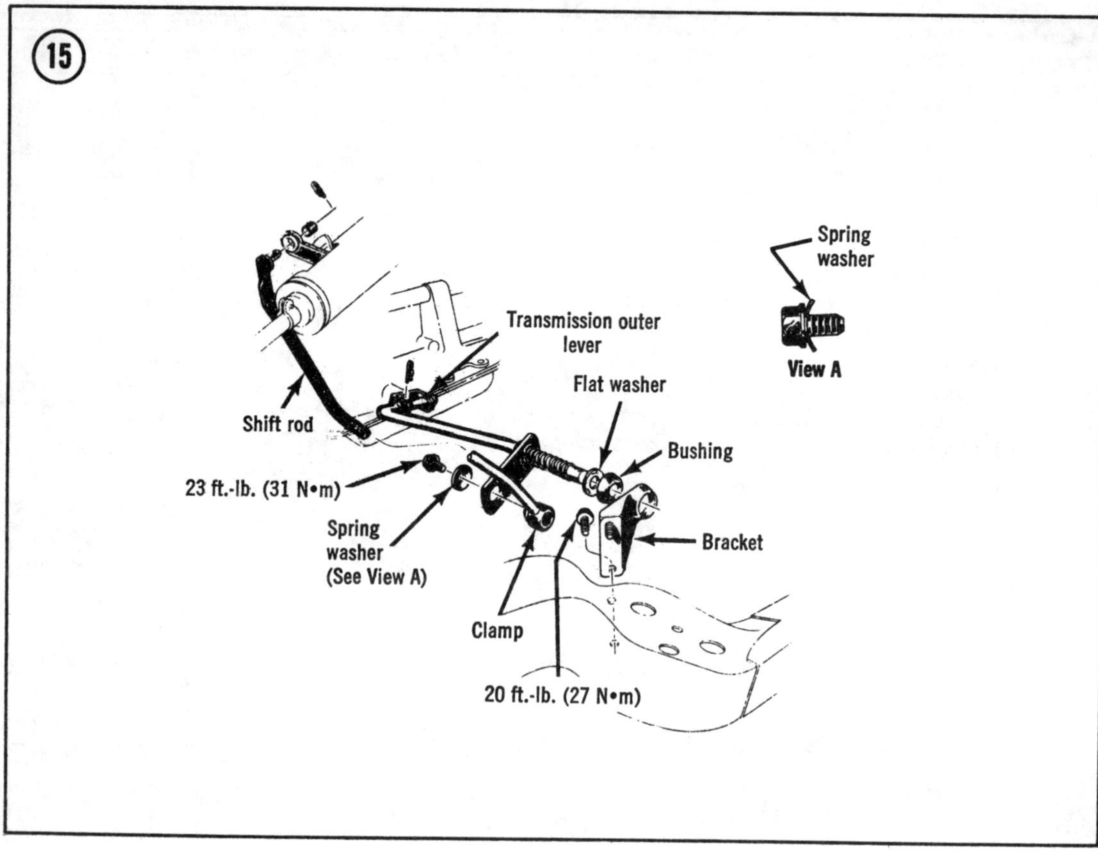

2. Place the shift selector lever in PARK and then place the transmission manual lever in the PARK detent. Make sure the ignition key is in the LOCK position.
3. Tighten the cable pin nut to 20 ft.-lb.
4. Push the transmission manual lever fully against the PARK stop and release it. Then pull the shift rod down against the stop to eliminate lash. Tighten the clamp screw to 20 ft.-lb.
5. Shift the selector lever through each gear position and make sure the transmission manual lever is located in the proper detent. Turn ignition key to RUN, and place transmission in REVERSE, and make sure key cannot be removed and steering wheel is not locked. Turn key to LOCK, place transmission in PARK, and make sure key can be removed and that steering wheel locks. Make sure engine will start in both PARK and NEUTRAL. Apply brakes and make sure starter will operate in any position except PARK and NEUTRAL.

Turbo Hydramatic Column Shift Linkage Adjustment

Refer to **Figure 15**.

1. Loosen shift rod clamp screw and place transmission outer lever in NEUTRAL position.
2. Hold the shift selector against the NEUTRAL stop in the upper steering column, making sure the the lever is not raised.
3. Tighten the clamp screw to 23 ft.-lb.
4. Check operation by turning ignition key to RUN, placing transmission in REVERSE, and making sure key cannot be removed and that steering wheel is not locked. Turn key to LOCK, place transmission in PARK, and make sure that key can be removed and that steering wheel locks. Make sure engine can be started in NEUTRAL and PARK. Apply brakes firmly and make sure that engine cannot be started in any position except NEUTRAL and PARK.

Table 1 TIGHTENING TORQUES

Manual Transmission	Ft.-lb.
Side cover bolts	20
Transmission-to-clutch housing bolts	53
Cross support bar-to-frame bolts	25
Transmission mount-to-transmission bolts	35
Transmission mount-to-cross support bar	35
Propeller shaft strap bolts	16
Bearing retainer-to-transmission case bolts	25
Rear bearing retainer-to-transmission bolts	50
T.C.S. plug	30
Transmission filler plug	18
Automatic Transmissions	**Ft.-lb.**
Transmission mount-to-transmission	35
Speedometer sleeve retainer on extension housing	13
Converter-to-flywheel	35
Oil pan-to-transmission case	13
Transmission case-to-cylinder block bolts	40
Propeller shaft-to-companion flange	20
Crossmember-to-frame	40
Transmission-to-crossmember	60

NOTE: If you own a 1981 and later model, first check the Supplement at the back of the book for any new service information.

CHAPTER TEN

BRAKES

Front disc brakes are standard on all 1975 and later vehicles and have been optional equipment from 1970 to 1974. Rear drum brakes are standard on all models. All covered models use a split brake system, with a dual master hydraulic cylinder, providing essentially 2 separate braking systems, one for the front wheels and one for the rear wheels. When a failure is encountered in either system, the other can be used to stop the car. Failure of either system will cause a dash warning lamp to light. All brakes are self-adjusting (except parking brake).

Torque specifications (**Table 1**) are at the end of this chapter.

BRAKE INSPECTION

At the intervals recommended in Chapter Three, brake linings and pads and other internal brake components should be inspected (drums, rotors, wheel cylinders, springs, hydraulic lines, etc.). More frequent checks should be made if brakes are used frequently. Parking brake should also be checked and adjusted, if required, when linings are checked. Linings and pads should be replaced when worn to within 1/32 in. of shoe table or rivet head, whichever applies. Brake shoes and pads should be replaced in axle sets (right and left sides). Disc brakes are checked by inspecting the outboard shoe, or pad at both ends of the caliper (inboard shoe should also be checked in case premature wear has occurred).

NOTE

Oldsmobile has indicated that the combination stoplight/cruise control switch on 1979 models may stick. This will prevent stoplight operation when brake pedal is depressed. The switch should be checked by an Olds dealer, if not already done.

BLEEDING

The hydraulic brake system must be bled when brakes are overhauled, when any hydraulic line has been disconnected, or whenever air enters the hydraulic system in any way.

CAUTION

Do not attempt to bleed the system while any brake drum is off the car or any caliper is removed from a disc, as damage to wheel cylinders and linings could occur.

NOTE

Bleed one valve at a time to prevent master cylinder fluid level from becoming low enough to introduce more air into system. Proper sequence is to bleed valve nearest master cylinder (either front or back) first, then bleed the opposite valve.

NOTE

Do not mix brake fluid types. Use only Delco Supreme 11 or other DOT-3 type fluid.

1. Clean all dirt, etc., from top of master cylinder and remove cover and rubber diaphragm.
2. If fluid level is low, fill and reinstall diaphragm and cover.
3. Install bleeder wrench (GM Part No. J-2472 or equivalent) on bleeder valve nearest master cylinder. See **Figure 1** (disc brakes) or **Figure 2** (drum brakes). Install a bleeder hose on the valve. Place other end of hose in a transparent container with enough brake fluid to ensure that end of hose will remain submerged.

NOTE

Carefully monitor fluid level in master cylinder and add fluid as required during bleeding to prevent draining of reservoirs.

4. Open bleeder valve 3/4 turn and have an assistant depress brake pedal. As pedal nears the end of its travel, close bleeder valve tightly and allow brake pedal to return slowly to released position. Repeat this step until expelled brake fluid is completely free of air bubbles, then close bleeder valve tightly.

CAUTION

Do not re-use brake fluid which is bled from brakes. To do so might introduce contamination into the hydraulic system and result in damage.

5. Remove bleeder hose and wrench.
6. Repeat Steps 1 through 5 for opposite bleeder valve, then for valves on other end of car, starting with valve closest to master cylinder.
7. Fill master cylinder to within 1/4 inch of top, then reinstall diaphragm and cover.

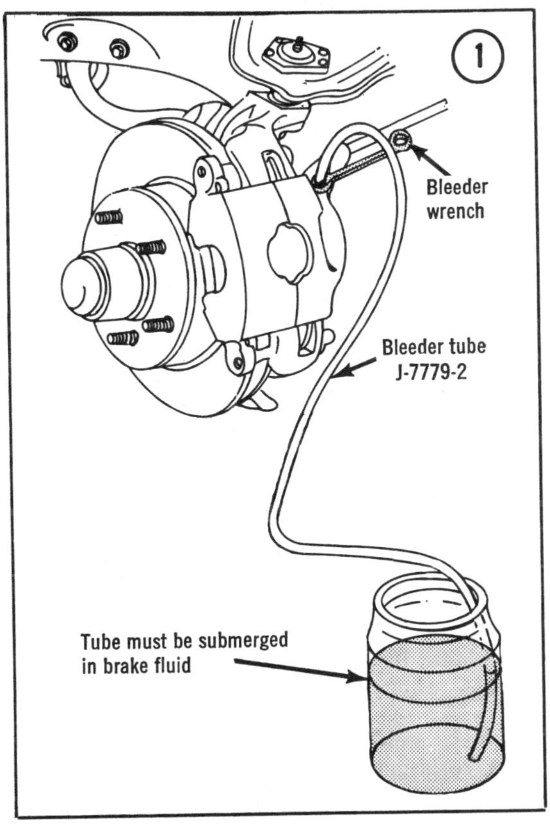

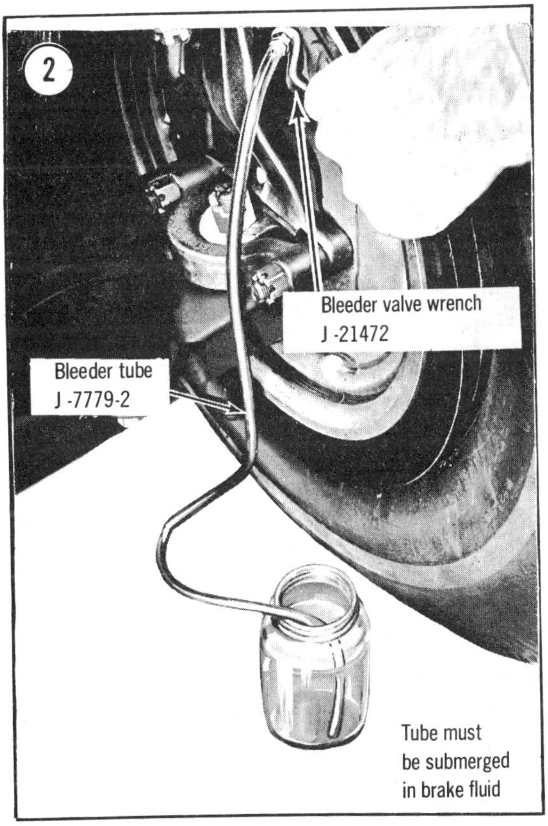

10

3

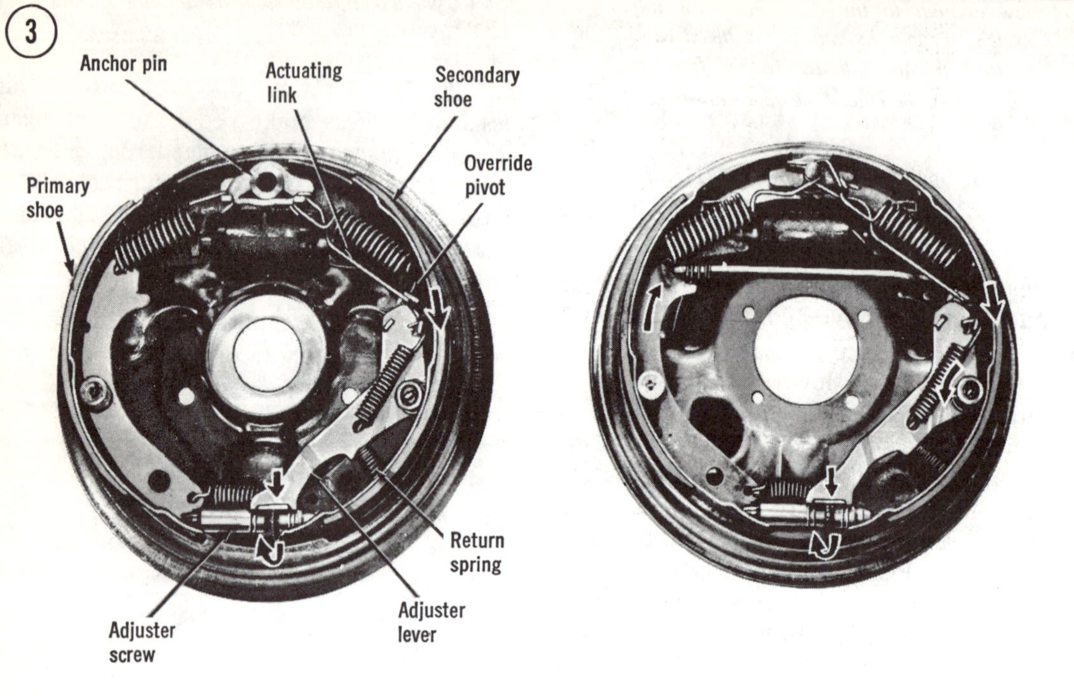

Actuating parts

Operation

4

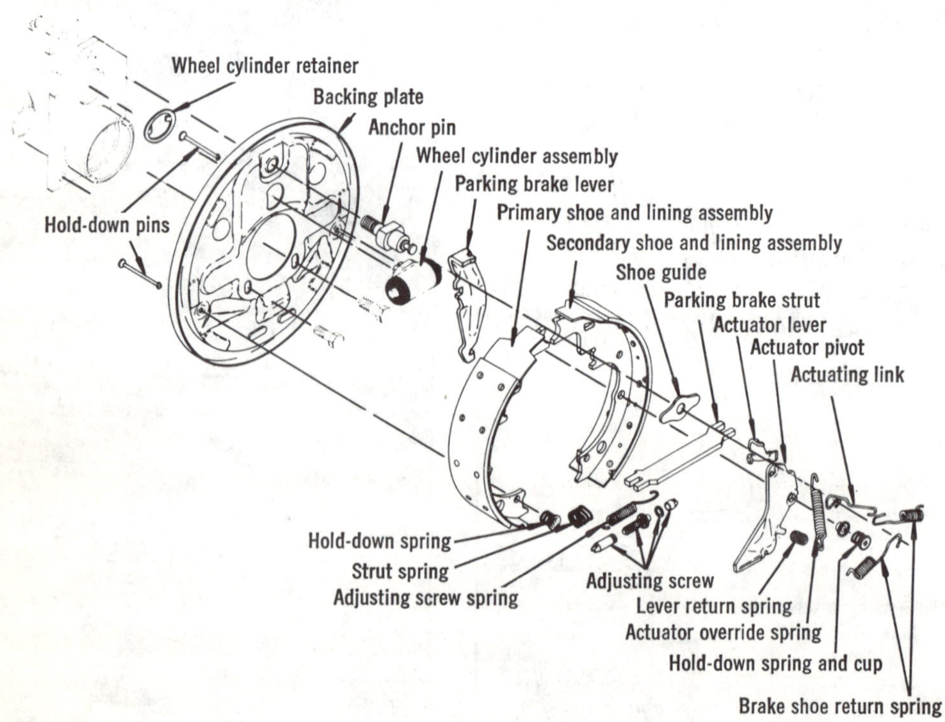

5

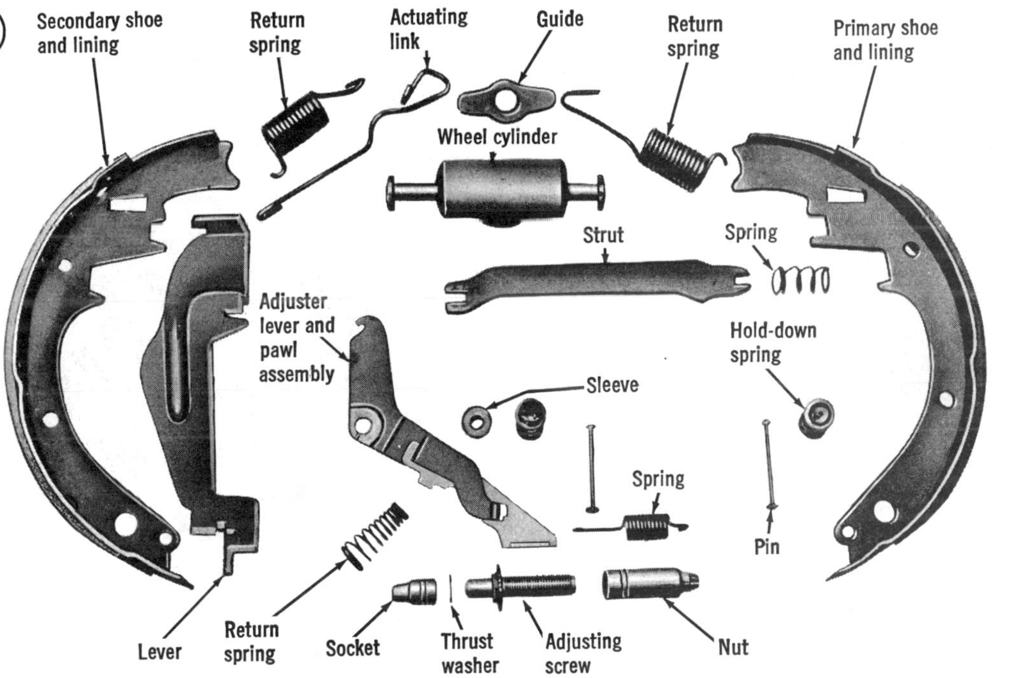

BRAKE LININGS

Refer to **Figures 3, 4, and 5**.

It may be necessary to retract adjusting screw if brake drums are severely worn. If drum does not slide off easily, remove access slot plug in backing plate. See **Figure 6**.

Removal

1. Mark drum position so it can be reinstalled in its original position.
2. Loosen check nuts at parking brake equalizer enough to remove all tension from cables.
3. Remove brake drums.

CAUTION

Do not depress brake pedal while drums are removed from brakes.

4. Unhook brake shoe return springs. See **Figures 4 and 5**.

NOTE

A spring removal tool (GM J-8049 or equivalent) will save time and possibly help prevent damage to springs and brake parts.

6

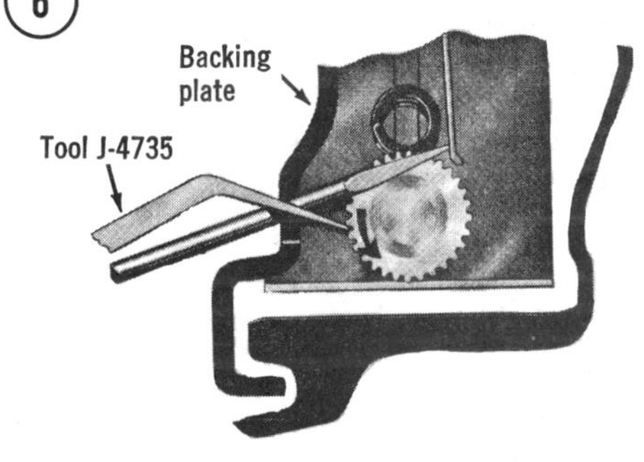

5. Remove actuator return spring and link.
6. Remove brake shoe hold-down pins and springs and remove actuator assembly (**Figures 4 and 5**).

NOTE

Mark positions of brake shoes if they are to be replaced.

7. Separate brake shoes by removing spring and adjusting screw.
8. Remove parking brake lever from secondary brake shoe (**Figure 4**) on rear brakes only.

10

Inspection

1. Clean all dirt out of brake drum (avoid getting dirt into wheel bearings) and inspect them for roughness, scoring, or out-of-round condition. See **Figure 7**.

NOTE

If measuring equipment is not available, take all drums to an automotive machine shop or brake repair shop and have them checked, and turned on a lathe, if required. Brake drums that are more than 0.006 in. out-of-round must be reconditioned or replaced to ensure effective brake action.

CAUTION

Brake drums must not be refinished more than 0.060 in. over maximum standard diameter.

WARNING

Under no circumstances should a cracked brake drum be reused. Do not attempt to weld a cracked drum, as welding heat will cause distortion. Use of a cracked or deformed brake drum could result in brake failure, which could result in injury or loss of life.

2. Remove and inspect wheel bearings and grease seals and replace parts as required. Repack and replace bearings, and replace seals.
3. On wheel cylinders with external rubber boots, carefully pull lower edges of boot away from cylinders and check to see if interior is wet with brake fluid. If excessive fluid is noted, cylinder must be overhauled or replaced.

NOTE

A slight amount of fluid is almost always present and is needed to lubricate the piston.

4. On wheel cylinders having internal boots, carefully pull a small part of boot out of cylinder and check as in Step 3, observing note above.
5. Check all flange plate attaching bolts to verify that they are tight. Clean all dirt and rust from contact faces on flange plate, using fine emery or crocus cloth, if required.

7

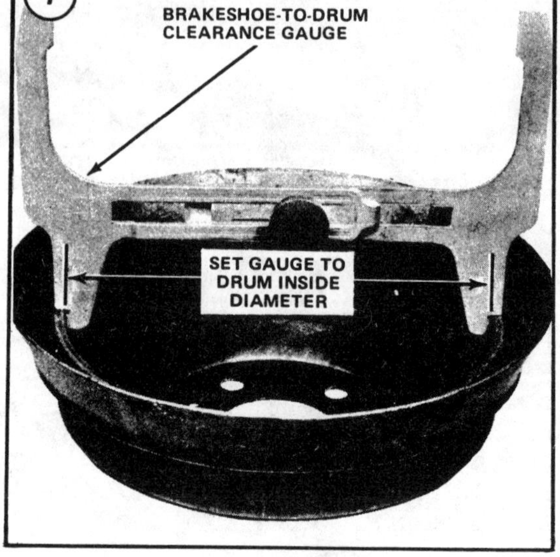

Installation

CAUTION

Make certain only exact replacement shoe and lining assemblies are used. If old assemblies are to be reinstalled, they must be installed in original positions (as marked when removed).

NOTE

Use only Delco Brake Lubricant (No. 5450032) or equivalent in the following procedure when lubrication is required.

1. Inspect new shoe and lining assemblies and remove all nicks and burrs on bonding material, shoe edges, or on any contact surface.

CAUTION

Do not permit oil or grease to come into contact with linings. Keep hands clean.

2. Lubricate parking brake cable (if working on rear brakes). Also lubricate fulcrum end of parking brake lever, then attach lever to secondary shoe. Verify that lever moves freely.
3. Lightly lubricate pads on flange plate and threads of adjusting screw (make sure adjusting screw is clean before lubricating).
4. Connect primary and secondary brake shoes together with adjusting screw spring. Then position adjusting screw, socket and nut in position.

CAUTION

Verify that proper adjusting screw is being used (marked "L" for left side, "R" for right side of vehicle). Star wheel should be installed closer to secondary shoe. Adjusting screw spring must not intefere with starwheel operation. Verify that star wheel lines up with adjusting slot in flange plate.

5. On rear wheels, connect parking brake cable to lever.
6. Attach primary brake shoe (shorter lining faces forward) using hold-down spring and pin. Make certain shoes engage with wheel cylinder connecting links.
7. Install and secure actuator assembly and secondary brake shoe with hold-down spring and pin. Position parking brake strut and strut spring (rear wheels).
8. Install guide pin over anchor pin and install wire link.

CAUTION

To avoid damage, fasten wire link to actuator assembly first and place other end over anchor pin stud by hand (do not use tool) while holding adjuster assembly in full down position.

9. To install actuator return spring by easing it into place with a screwdriver or other suitable flat tool. Do not pry on actuator lever to install it.

NOTE

If old brake return (pull back) springs are nicked, distorted or if strength or condition is doubtful, new spring should be installed.

10. Hook brake return in shoes and install other ends of primary spring over anchor pin. Then hook other end of secondary shoe spring over the wire link end (see **Figure 3**).
11. After installation is complete, make certain actuator lever works easily by hand operating the self-adjusting feature.
12. Repeat the above procedure for all wheels equipped with drum brakes.
13. If wheel cylinders were removed and replaced, bleed the hydraulic lines using procedure given above.
14. Adjust service and parking brakes, using procedures given below, and install drums and wheels.
15. Lower vehicle to floor.

SERVICE BRAKE ADJUSTMENT

1. Disengage actuator from star wheel and rotate star wheel by hand.
2. Using brake drum as a fixture, turn star wheel until drum slides over brake linings with slight drag.
3. Use star wheel to retract lining 1 1/4 turns.
4. Install drum and wheel, making certain drum is in same position as when removed.
5. Make final adjustment after all drums and wheels are replaced, using Steps 1 through 4. This is accomplished by making numerous forward and reverse stops, using firm pressure on brake pedal, until satisfactory pedal height and braking action are achieved.

PARKING BRAKE ADJUSTMENT

1. Raise vehicle (if not already raised).
2. Apply parking brake 2 notches from fully released position.
3. Loosen the brake equalizer check nut until a light to moderate drag is felt when rear wheels are rotated frontward.
4. Hold front nut while tightening jam nut securely.
5. Release parking brake and rotate rear wheels. No drag should be felt.
6. Lower vehicle to floor.

WHEEL CYLINDER SERVICE (DRUM BRAKES)

Removal

1. Raise vehicle and remove wheel and brake drum, and brake shoe pull back springs, using procedure given above.
2. Carefully clean all dirt away from line to wheel cylinder connection. Disconnect hydraulic line from cylinder and cover end of line with clean, lint-free cloth to prevent contamination from entering hydraulic system.

3. On 1977 and earlier models, remove screws securing cylinder to flange plate. Disengage cylinder pushrods from brake shoes and remove cylinder.
4. On 1978 and later models, insert small (1/8 in. diameter or less) awls or similar tools into access slots between cylinder pilot and retainer locking tabs (see **Figure 8**). Bend both tabs at the same time to release cylinder from backing plate. Discard retaining clip.

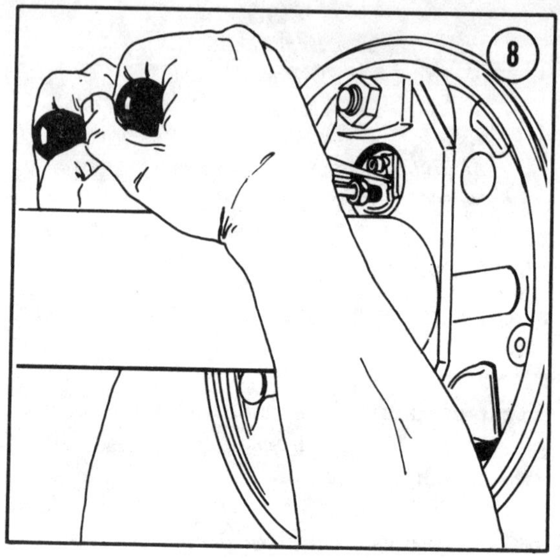

Disassembly

Refer to **Figure 9** (1970-1977) or **Figure 10** (1978-on).

1. Remove boots from cylinder ends and discard.
2. Remove and discard pistons and caps.

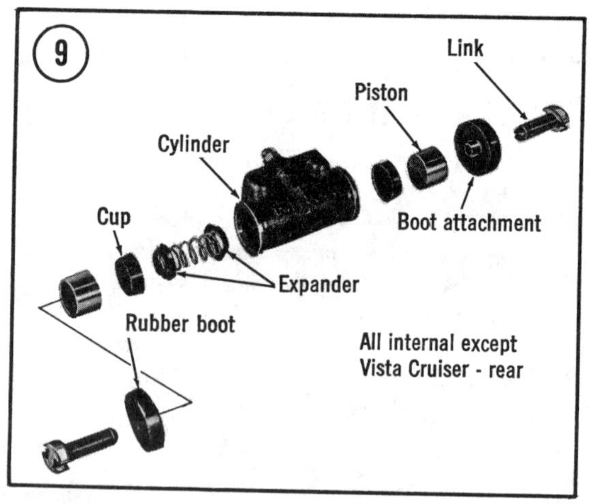

Cleaning and Inspection

NOTE

Most brake fluids are colored to assist in detecting hydraulic leaks. Staining from this coloring should not be confused with corrosion, which can be identified by pits or roughness in the cylinder bore.

1. Check cylinder bore for staining and corrosion. Discard if corroded (see note above) and obtain new cylinder.
2. Remove stains by polishing cylinder bore with crocus cloth by revolving the cylinder while supporting the cloth in the bore with a finger. Do not polish by sliding the cloth lengthwise in the bore under pressure.
3. Clean hands thoroughly with soap and water (do not use gasoline or any other petroleum based solvent) and then clean the cylinder and all metal parts in clean brake fluid.
4. Shake excess fluid from cylinder. Never use a rag to dry cylinder bore as even the smallest fragment of lint could cause problems.

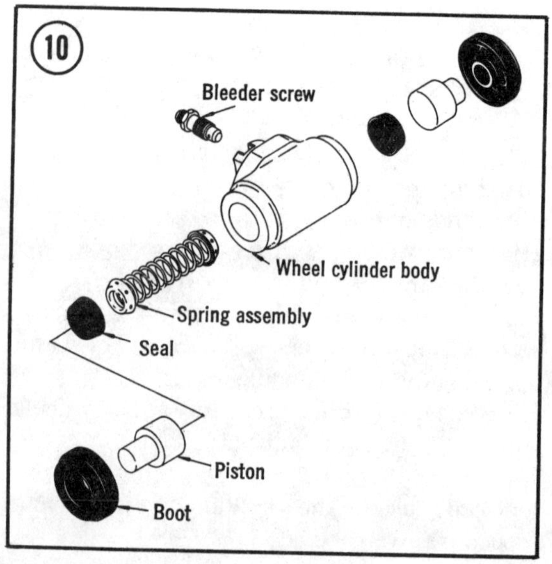

Assembly

1. Lubricate cylinder bore and counterbore with clean brake fluid and insert the spring-expander assembly.
2. Install new cups with flat surfaces toward end of cylinder. Do not lubricate cups prior to assembly.

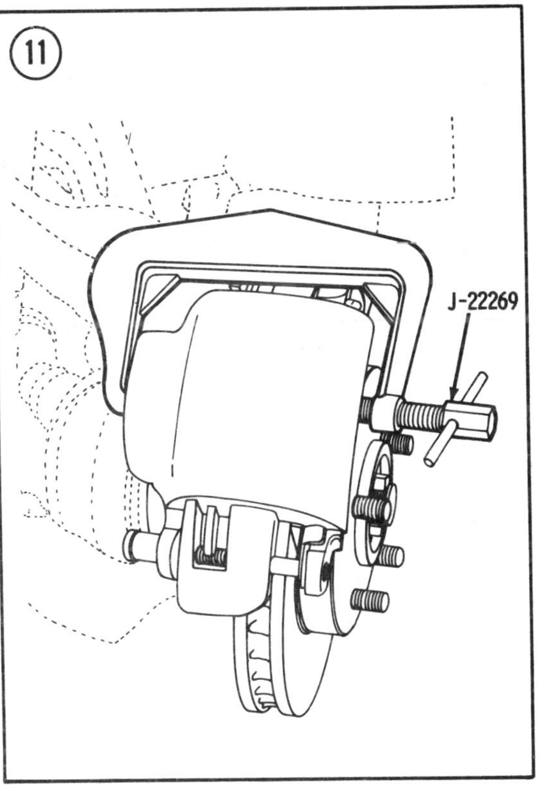

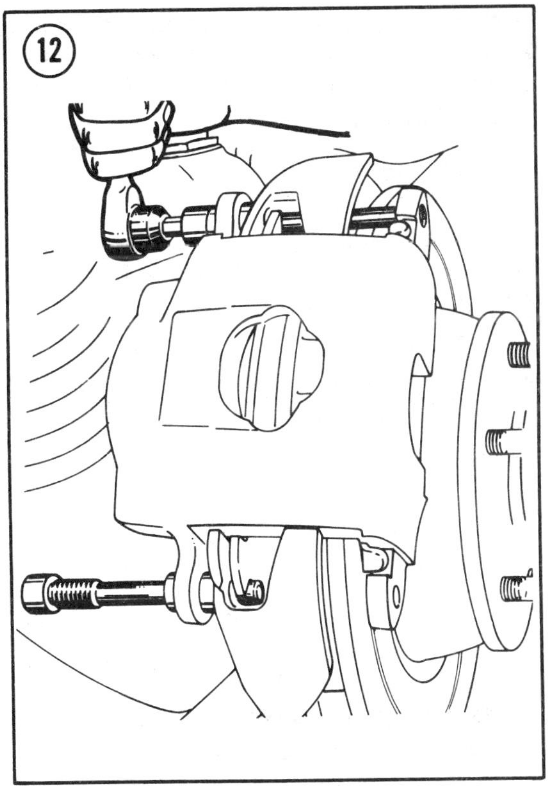

3. Install new pistons with flat surfaces toward center of cylinder. Do not lubricate pistons before assembly.
4. Press new boots into cylinder counterbore by hand. Do not lubricate boots prior to installation.

Installation

1. On 1970-1977 models, position wheel cylinder to flange plate and install and securely tighten screws.
2. On 1978 and later models, hold cylinder against backing plate with a block installed between the cylinder and the wheel flange. Then install new retaining spring over cylinder, lining tabs up with the cylinder grooves. Drive retainer into position with a 1-1/8 in. socket and a 10 in. extension. Make sure both retainer tabs are properly engaged.
3. Replace all pushrods and pull back screws.
4. Reconnect hydraulic line to wheel cylinder.
5. Install brake drum and wheel.
6. Bleed the hydraulic system using the procedure given above.
7. Lower vehicle to floor.

PAD REPLACEMENT (FRONT DISC BRAKES)

Removal

1. If master cylinder reservoir is more than 1/3 full, syphon out enough fluid to reduce level to 1/3 full (to avoid overflow when caliper piston is pushed back into bore). Discard fluid removed as brake fluid should never be reused.
2. Raise vehicle and remove front wheels.
3. Using a C-clamp as shown in **Figure 11**, push caliper piston back into its bore.
4. Remove the 2 mounting bolts attaching caliper to support (see **Figure 12**).
5. Lift caliper from disc.
6. Remove inboard shoe and dislodge outboard shoe. Position caliper on front suspension arm so brake hose does not support caliper weight.

NOTE

Mark shoe positions if shoes are to be reinstalled.

10

7. Remove shoe support spring from piston.
8. Remove 2 sleeves from inboard ears of caliper; then remove 4 rubber bushings from the grooves in caliper ears.

Cleaning and Inspection

NOTE

If shoes have 1/32 in. or less lining left above rivet heads they should be replaced. Always replace shoes in axle (2 wheel) sets.

1. Thoroughly clean mounting bolts and holes and bushing grooves in caliper ears.

CAUTION

If bolts are damaged or corroded they should be replaced to prevent later brake failure.

2. Examine inside of calipers for leakage evidence. If present, caliper should be overhauled (see procedure below).
3. Clean inside of calipers, including exterior of dust boot check boot for damage.

CAUTION

Do not use compressed air for cleaning as it could unseat boot.

4. If disc is badly worn or scored it should be refinished or replaced. Remove hub and take it to an automotive machine shop or brake repair shop for checking and refinishing.

Installation

CAUTION

If old shoes are to be reinstalled they must be installed in original positions.

1. Lubricate new sleeves, new rubber bushings, bushing grooves and ends of mounting bolts with Delco Silicone Lube No. 5459912 or equivalent. See **Figure 13**.

CAUTION

New sleeves and rubber bushings must be used and lubrication instructions followed to ensure that calipers function properly.

2. Install new rubber bushings in grooves in caliper ears and install new sleeves to inboard ears.

NOTE

Position sleeves with ends toward shoe and lining assemblies flush with machined surfaces of ears.

3. Install shoe support spring (see **Figure 14**) and inboard shoe in center of piston cavity as shown in **Figure 15**. Push down until shoe lays flat against caliper.
4. Position the outboard shoe in the caliper with ears at top of shoe over caliper ears and tab at bottom of shoe engaged in caliper cutout. See **Figure 16**.
5. Lift caliper (with shoes installed) up and rest bottom edge of outboard lining on outer edge of brake disc. Verify that there is no clearance between tab at bottom of outboard shoe and caliper abutment.
6. Position caliper over brake disc, lining up holes in caliper ears with holes in mounting bracket. See **Figure 17**.

NOTE

Verify that brake hose is not kinked or twisted.

7. Insert mounting bolts through sleeves in inboard caliper ears and mounting bracket, making sure that ends of bolts pass under retaining ears on inboard shoe.
8. Engage holes in outboard shoes and outboard caliper ears with mounting bolts, and thread bolts into mounting bracket. Torque to 35 ft.-lb.
9. Pump brake pedal to seat linings against the discs.
10. Using arc-joint pliers, bend upper ears of outboard shoe until no radial clearance exists between shoe and caliper housing (see **Figure 18**). Locate pliers on small notch of caliper housing during bending procedure.

NOTE

Repeat Step 10 if radial clearance still exists after initial clinching.

NOTE

Outboard shoes (with formed ears) are designed for original installation only and are fitted to calipers. The shoes should never be relined or reconditioned for reinstallation.

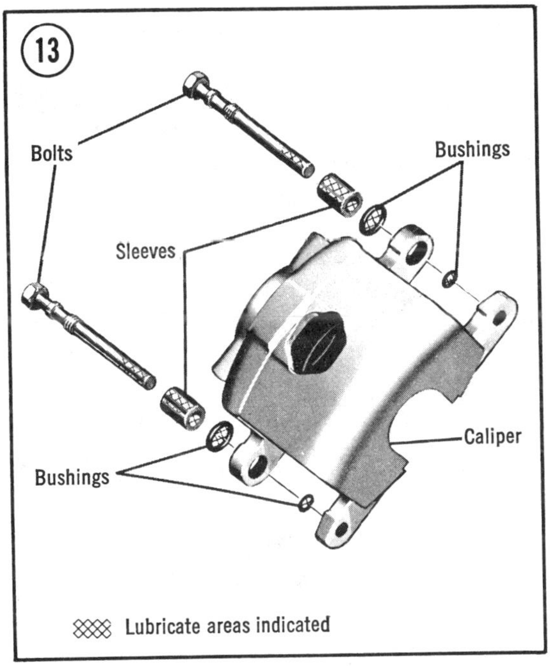
13
Bolts
Bushings
Sleeves
Caliper
Bushings
Lubricate areas indicated

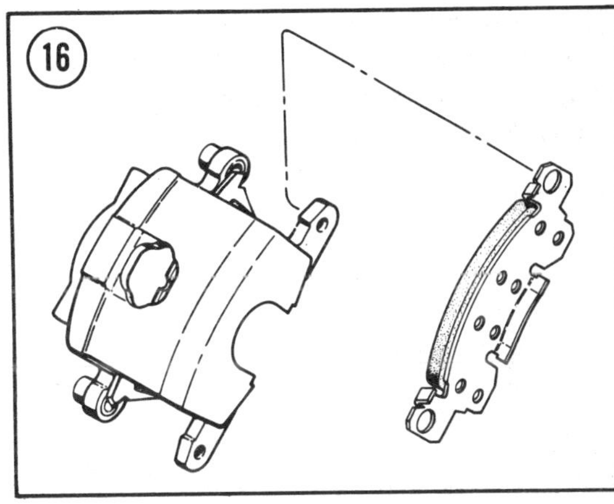
16

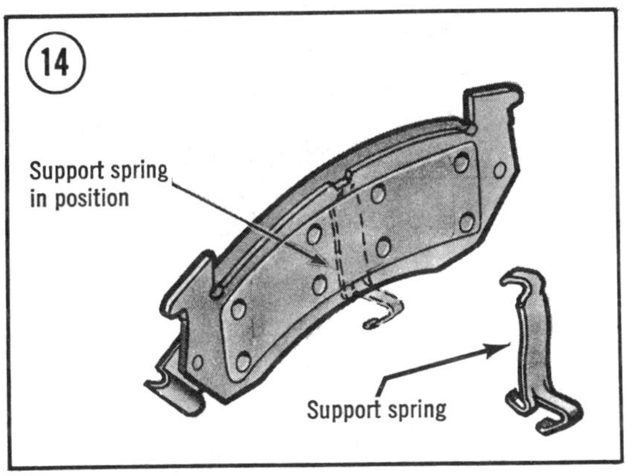
14
Support spring
in position
Support spring

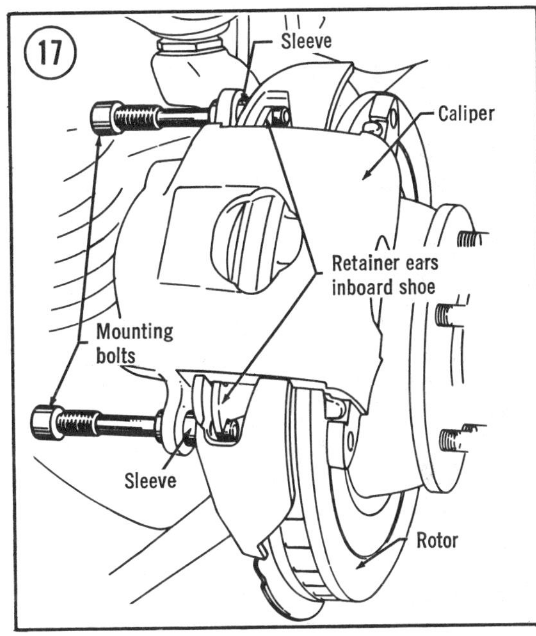
17
Sleeve
Caliper
Retainer ears
inboard shoe
Mounting
bolts
Sleeve
Rotor

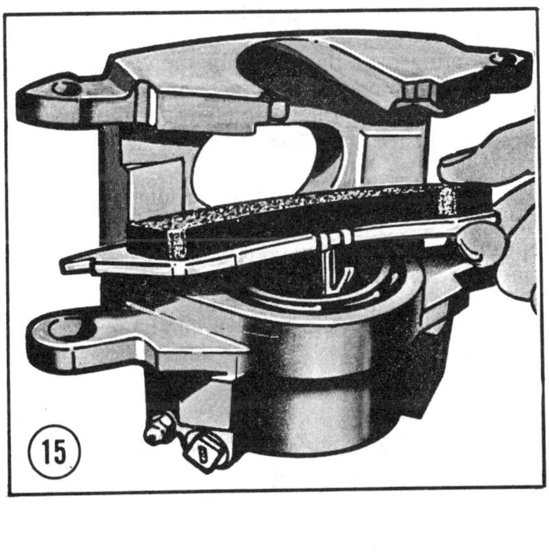
15

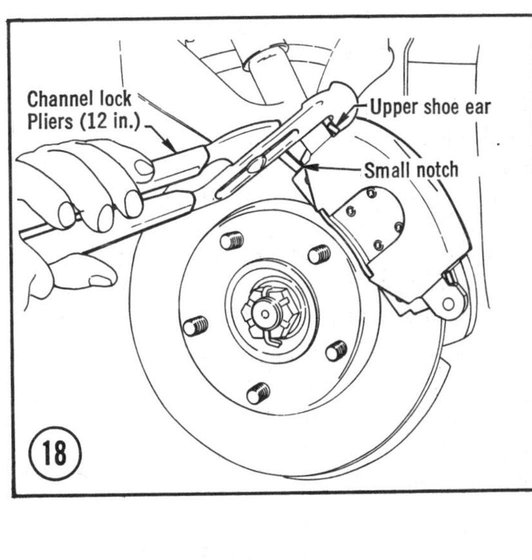
Channel lock
Pliers (12 in.)
Upper shoe ear
Small notch
18

11. Reinstall front wheel and lower vehicle.
12. Add brake fluid to master cylinder reservoir to bring level to within 1/4 in. of top.
13. Before moving vehicle, pump brake pedal several times to make certain it is firm. Recheck master cylinder after firm pedal is achieved.

CALIPER

Removal/Installation

1. Remove caliper assembly from disc, using procedure given above.
2. Remove hose to caliper bolt and cap or tape open connections to prevent introduction of contamination.
3. Remove caliper assembly from vehicle; then remove brake shoes from caliper.
4. Installation is the reverse of these steps.
5. Connect hydraulic lines to calipers, using new copper gaskets.

CAUTION
Hose must be positioned in caliper locating gate (between locating beads) to assure proper positioning to caliper.

6. Bleed calipers, using procedure given elsewhere in this chapter.

Disassembly

1. Clean caliper exterior with clean brake fluid and place caliper on a clean working surface.
2. Drain brake fluid from caliper.

WARNING
Follow Step 3 and do not attempt to catch or protect piston with fingers, as injury could result.

3. Pad interior of caliper with clean shop towels and remove piston by application of compressed air as shown in **Figure 19**.

CAUTION
Do not blow piston out of bore. Use only enough air to ease the piston out.

NOTE
If compressed air is not available, remove piston by applying brake pedal after caliper has been removed from disc, but before disconnecting hydraulic line.

19
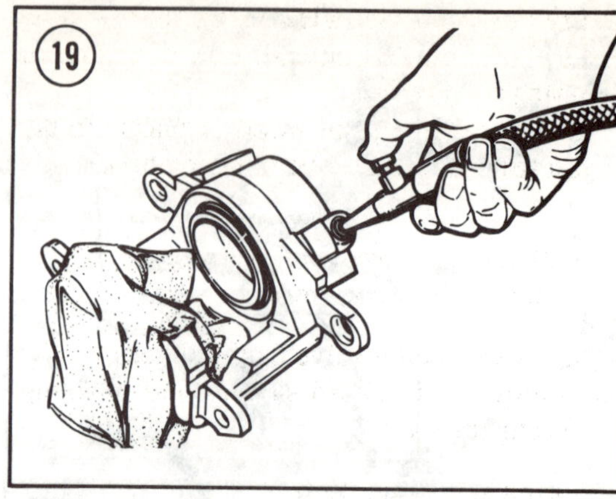

20
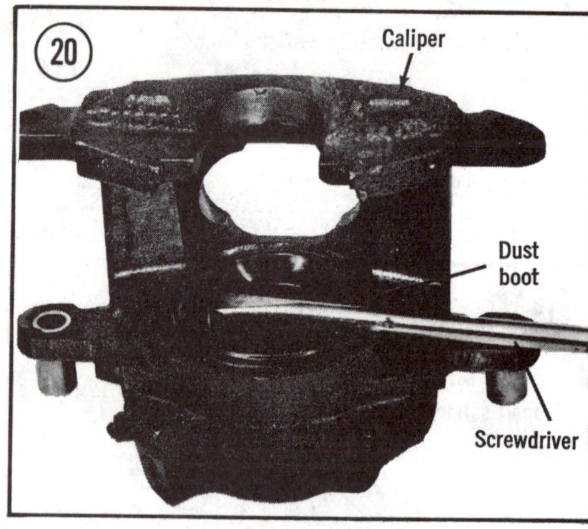

21
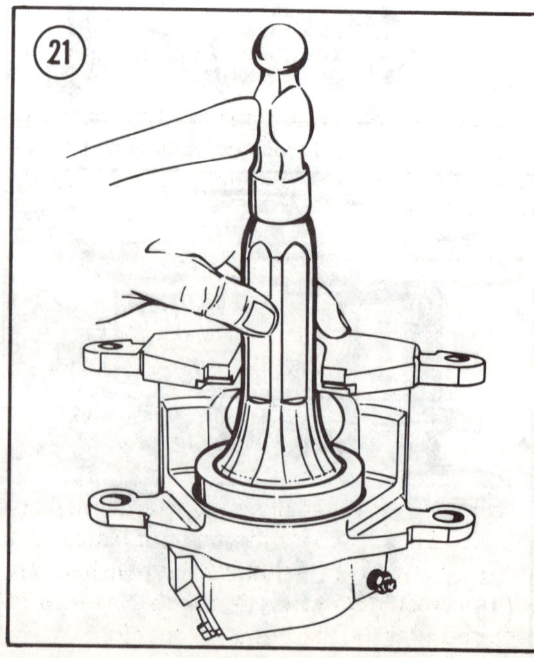

4. Pry dust boot (see **Figure 20**) out of caliper piston bore, exercising care not to scratch or damage piston bore.
5. Use a small piece of wood or plastic to pry piston seal from its groove in piston bore.

CAUTION
Do not use metal tool of any kind, as damage to bore may result.

6. Remove bleeder valve from caliper.
7. Remove and discard sleeves and bushings from caliper ears.

Cleaning and Inspection

CAUTION
Dust boot, piston seal, rubber bushings, and sleeves must be discarded when the caliper is overhauled. Replace them with new parts.

1. Clean all parts (except those discarded) in clean brake fluid. Use only dry, filtered, compressed air to blow out all passages in caliper and bleeder valve.

CAUTION
Use of unfiltered shop air may leave a film of mineral oil on internal passages. Mineral oil may damage rubber parts. If filtered air is not available, allow parts to drain completely dry before attempting reassembly.

2. Examine all mounting bolts and replace any that appear corroded or damaged.
3. Carefully examine piston for scoring, nicks, corrosion and worn or damaged chrome plating. If any surface defects are noted, replace piston.

CAUTION
Piston surfaces and plating are manufactured to close tolerances. Do not attempt to resurface the piston by any means. The use of any abrasive, including crocus cloth, is not acceptable.

4. Check bore of caliper for same defects listed in Step 3. Piston bore is not plated and stains or minor corrosion can be polished out with a crocus cloth. A thorough clean-up is required after the use of the crocus cloth.

CAUTION
Do not use emery cloth or any other type of abrasive. If crocus cloth does not remove the stain or corrosion, the caliper must be replaced.

Assembly

1. Lubricate caliper piston bore and new piston seal with clean brake fluid and place seal in caliper bore groove.
2. Lubricate piston with clean brake fluid and assemble new dust boot into groove in piston. Fold must face open end of piston.
3. Insert piston into caliper bore, using care not to disturb the seal seating. Force piston to bottom of bore (50-100 lb. of force required).
4. Position dust boot in caliper counterbore and seat, using a boot installer tool (GM No. J-26267) as shown in **Figure 21**.

CAUTION
Check boot installation carefully to ensure that retaining ring molded into boot is not bent and that boot is installed below caliper face and evenly all around.

NOTE
If boot installer tool is not available, the inventive mechanic can improvise a substitute. Care should be taken, however, not to damage the boot, piston, or caliper. A better alternative is to take the assembly to a brake shop and have the boot installed there.

5. Install bleeder valve and torque to 65 in.-lb. (If valve will not seal at 65 in.-lb., torque to 100 in.-lb. maximum to obtain seal. If valve will not seal at 100 in.-lb., replace valve.).

POWER BRAKE UNIT

Inspection

1. Check vacuum line and connections, including check valve, for vacuum loss.
2. Inspect hydraulic lines and connections for leaks.
3. Check for scored brake drums, worn or contaminated linings.
4. Check and fill fluid reservoirs, as required.
5. Inspect for loose master cylinder and power unit mounting bolts.

6. Check power piston air filter and replace if indicated.
7. Check for misalignment and/or binding of brake pedal linkage.

Bleeding

Bleeding of brake hydraulic system should be done in accordance with procedure given elsewhere in this chapter.

CAUTION
Do not use power assist when bleeding brake lines. Engine must be off. Pump brake pedal several times to expend vacuum reserve before starting.

Removal

1. Disconnect vacuum line from power unit check valve.
2. Remove nuts holding power unit to master cylinder (in models so attached). Pull forward on master cylinder until cylinder clears power unit mounting studs. Carefully move cylinder to one side with all hydraulic lines attached.

NOTE
On older models, the power unit and master cylinder must be removed as a unit. In this case, disconnect hydraulic lines from master cylinder.

3. Remove nuts holding power cylinder to dash.
4. Remove retainer holding pushrod to brake pedal. Remove power unit from engine compartment.

NOTE
Repair of the power unit requires special tools and knowledge. Instead of attempting to make repairs, the home mechanic should take the unit to a dealer or competent power brake mechanic for repairs or replacement. Always replace vacuum check valve and grommet when power unit is overhauled.

Installation

1. Install all components in reverse order of the removal procedure. See **Table 1** at the end of the chapter for torque specifications.
2. Check brake operation. If hydraulic lines were removed, the system must be bled to remove all air from lines and components. Check stoplight switch adjustment.

CAUTION
Start engine and allow time for vacuum to build up in power unit before operating brake pedal.

MASTER CYLINDER

Removal

1. Remove the brake fluid from the master cylinder reservoir. A suction-type turkey baster makes an excellent tool for fluid removal.
2. Disconnect the brake lines from the master cylinder. Plug the lines or place tape over their ends to prevent contamination from entering the system.
3. On non-power brake systems, disconnect the brake pedal from the master cylinder pushrod.
4. Remove the nuts attaching the master cylinder to the fire wall or the power brake booster. Remove the master cylinder from the car, taking care not to drip brake fluid on the paint.

Installation

1. Position the master cylinder in the engine compartment and install the nuts to attach it to the fire wall or power brake booster. Tighten nuts to 24 ft.-lb.
2. On non-power brake models, connect the master cylinder pushrod to the brake pedal pin and install the retainer.

NOTE
It may be easier to connect the pushrod before installing the attaching nuts (as described in Step 1).

3. Install plugs in the brake line outlets and fill the master cylinder reservoir with fresh brake fluid. Bleed the master cylinder by placing a suitable container under it, then opening the plugs one at a time. With the plug open, push the brake pedal in and hold it. Then tighten the plug and release the pedal. Continue until all air is expelled, and then repeat the procedure with the other plug.

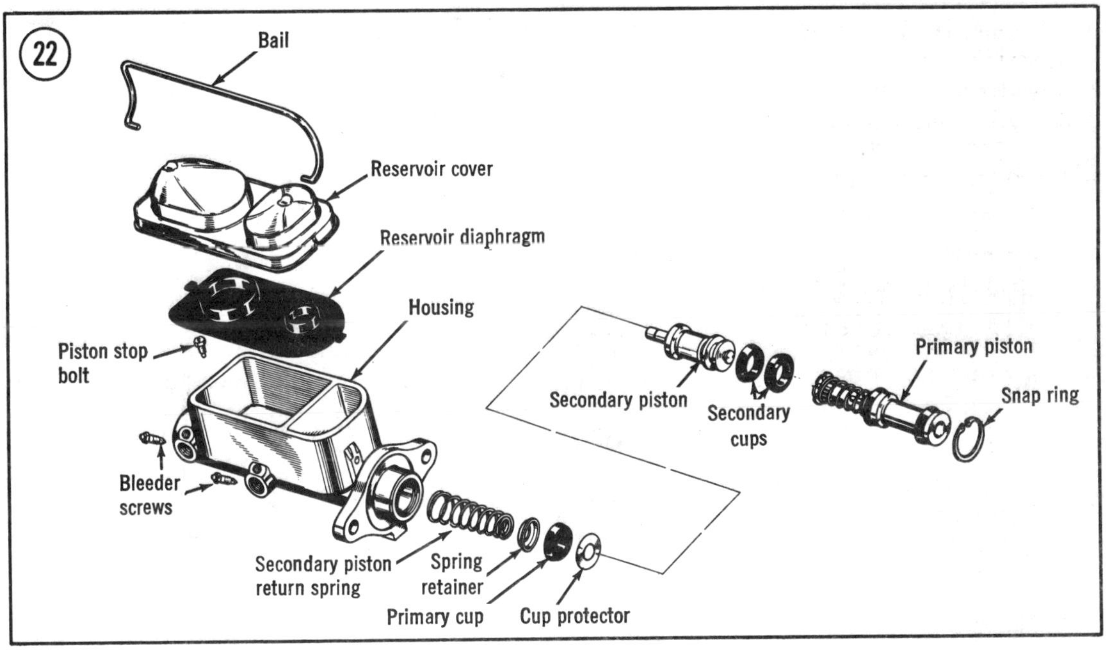

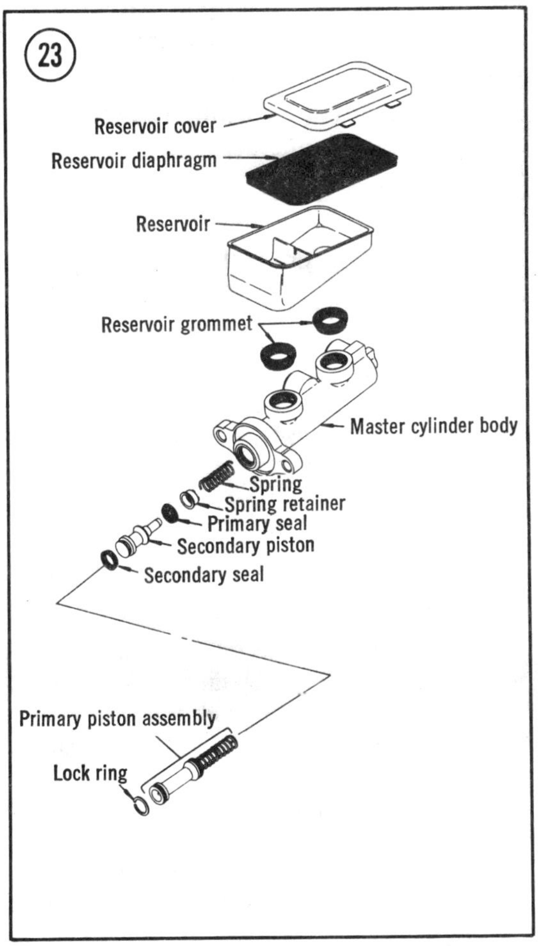

4. Remove the plugs one at a time and install the brake lines. Then remove the container that was placed under the master cylinder in Step 3.
5. Replenish the brake fluid and then bleed the entire hydraulic system as described in *Bleeding*, this chapter.
6. Road test the vehicle to ensure that the brake system is working properly.

Disassembly

NOTE

Although master cylinders can be rebuilt, and kits containing all of the necessary replacement parts are available, it is recommended that a defective master cylinder be replaced with a new or professionally rebuilt unit. If you are sure of your mechanical skills and you decide to rebuild the cylinder yourself, bear in mind that a number of master cylinder configurations were used in the models covered by this book, and that the following disassembly, inspection, and assembly procedures are general in nature. During disassembly, carefully note the arrangement of parts so the cylinder can be correctly reassembled.

Refer to **Figures 22 and 23** for typical master cylinder configurations.

1. Thoroughly clean the outside of the master cylinder. Then remove the cover and pour out any residual brake fluid.
2. On non-power brake models, pull the boot away from the master cylinder and pry up on the retainer tab to release the pushrod retainer.
3. Depress the piston and remove the secondary piston stop screw (if so equipped) from the front master cylinder reservoir.
4. Install the master cylinder in a vise and remove the lock or retaining ring from the groove in the bore. Then remove the primary piston assembly. On 1978 and later models, use a pry bar or similar tool (see **Figure 24**) to separate the reservoir from the cylinder. Remove and discard the grommets.
5. Remove the secondary piston, spring retainer, and spring. This is best done by applying compressed air through the stop screw hole. If air is not available, the parts can be removed by hooking a bent piece of wire under the edge of the secondary piston and pulling it out. Take care not to damage the cylinder surface.
6. If replacement of the brake line outlet inserts is required, use a 13/64 in. drill to enlarge the holes in the inserts. Then install a heavy washer over an outlet and thread a 1/4-20x3/4 in. screw into the enlarged insert hole. Turn the screw as required to remove the insert, then repeat for the other insert.
7. Remove the primary and secondary seals from the secondary piston.
8. Clean all metal parts in clean brake fluid. Blow the parts dry with compressed air and store them in a clean place on clean paper or a clean lint-free cloth.

24

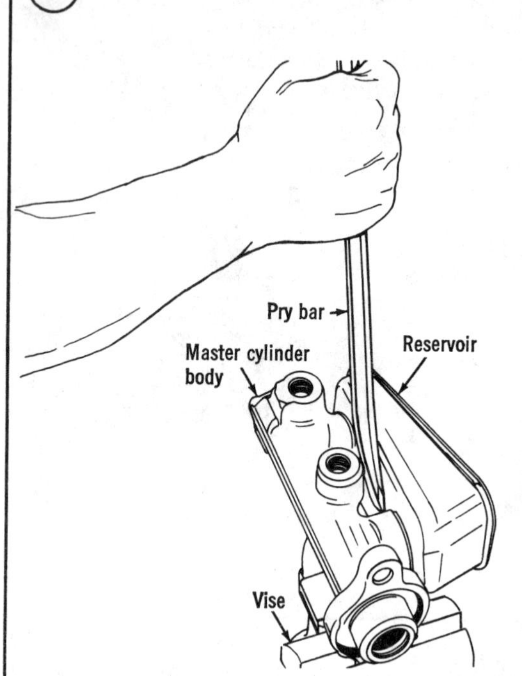

Inspection

1. Check the cylinder bore for scoring and/or corrosion (pits or excessive roughness). If damage is present, replace the cylinder.
2. Remove stains or discoloration from the cylinder bore with crocus cloth. Do not use emery cloth or other abrasive material that might remove metal or leave harmful residue.
3. Clean the master cylinder bore thoroughly with fresh brake fluid. Do not use any other solvent. Shake excessive fluid from the cylinder and blow dry with compressed air. If air is not available, allow the cylinder to drain thoroughly.

CAUTION

Do not dry the cylinder with a cloth, as even a small amount of lint could lead to brake failure.

4. Make sure the compensating port in the cylinder is clear.

Assembly

1. Install new inserts in the brake line outlets and then seat them by installing a spare brake line nut in the outlet and tightening it as far as it will go. Remove the nut and repeat for the other outlet.
2. Install a new secondary seal in the groove in the end of the secondary piston. Then install a new primary seal over the secondary piston so that the flat side of the seal seats against the piston flange.

25

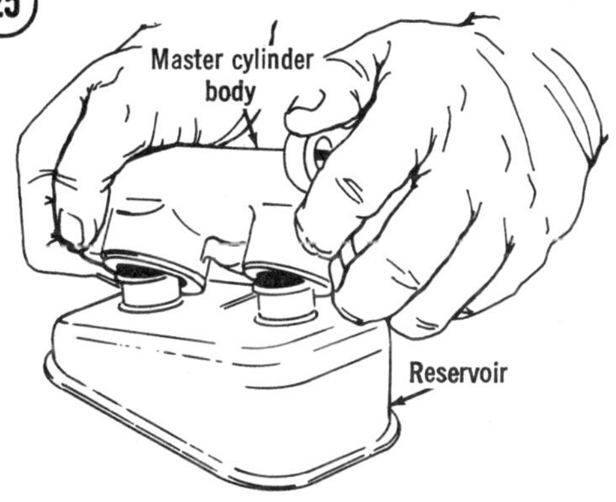

3. Install a second new secondary seal in the groove in the pushrod end of the secondary piston.
4. Lubricate the master cylinder bore and the secondary and primary seals on the secondary piston with clean brake fluid.
5. Install the spring retainer into the secondary piston spring and then install the retainer and spring assembly onto the end of the secondary piston. The retainer should be located inside the lip of the primary seal.
6. Install the secondary piston and spring assembly in the master cylinder bore so that the spring seats in the end of the bore.
7. Install the master cylinder in a vise with the open end up and lubricate the seals of the primary piston with clean brake fluid. Install the primary piston assembly, spring end first, into the bore. Hold the primary piston down and install the lock ring.
8. Continue holding the primary piston down and install the secondary piston stop screw. On 1978 and later models, lubricate new grommets in clean brake fluid and install them in the cylinder casting. After grommets are properly seated, remove the casting from the vise and install the reservoir as shown in **Figure 25**, using a rocking motion. The shoulder of the grommet should be in continuous contact with both the casting and the reservoir.
9. Install a new diaphragm in the cover, if required, and install the cover on the master cylinder.
10. On non-power brake models, install the pushrod through the pushrod retainer (if disassembled) and push the retainer over the end of the master cylinder. Install a new boot over the pushrod and the retainer. Thread the thread nut down to the shoulder on the pushrod and thread the clevis down to the jam nut. Tighten the nut against the clevis to 14 ft.-lb.

10

Table 1 TORQUE SPECIFICATIONS

Component	Torque
Master cylinder to dash	15 ft.-lb.
Master cylinder to power unit	24 ft.-lb.
Power unit to dash	15 ft.-lb.
Brake line nuts (all)	150 in.-lb.
Brake bleeder valves	65-100 in.-lb.
Brake shoe to anchor pin	120 ft.-lb.
Wheel cylinder to flange plate	12 ft.-lb.
Caliper mounting bolt	35 ft.-lb.
Flex hose-to-caliper	25 ft.-lb.
Pedal mounting pivot bolt (nut)	30 ft.-lb.
Combination valve mounting	15 ft.-lb.

NOTE: If you own a 1981 and later model, first check the Supplement at the back of the book for any new service information.

CHAPTER ELEVEN

FRONT SUSPENSION AND STEERING

The Cutlass independent front suspension uses unequal upper and lower control arms and coil springs. The shock absorbers mount inside the coil springs. A stabilizer bar attaches to the lower control arms and the frame. **Figure 1** shows the complete front suspension.

Figure 2 shows the steering system. Pitman arm motion is transmitted to a center relay rod supported at one side by an idler arm. Center relay rod movement in turn moves the steering knuckles through a tie rod. A steering damper incorporated in the steering linkage dampens road shock.

The power steering system uses the same recirculating ball steering gear, and linkage, with an added hydraulic power assist. An engine driven vane-type pump delivers hydraulic pressure to a control valve. The control valve senses the requirement for power assistance and supplies the power cylinder which operates the linkage.

WHEEL ALIGNMENT

Several front suspension dimensions affect running and steering of the front wheels. These variables must be properly aligned to maintain directional stability, ease of steering, and proper tire wear.

The dimensions involved define:

a. Caster
b. Camber
c. Toe-in
d. Steering axis inclination

All except steering axis inclination are adjustable. Since these adjustments are critical, they must be done by a competent front end alignment shop or your dealer.

Pre-alignment Check

Several factors influence suspension angles, or steering. Before any adjustments are attempted, perform the 10 following checks.

1. Check tire pressure and wear.
2. Check play in front wheel bearings. Adjust, if necessary.
3. Check play in ball-joints.
4. Check for broken springs.
5. Remove any excessive load.
6. Check shock absorbers.
7. Check steering gear adjustments.
8. Check play in pitman arm and tie rod parts.
9. Check wheel balance.
10. Check *rear* suspension for looseness.

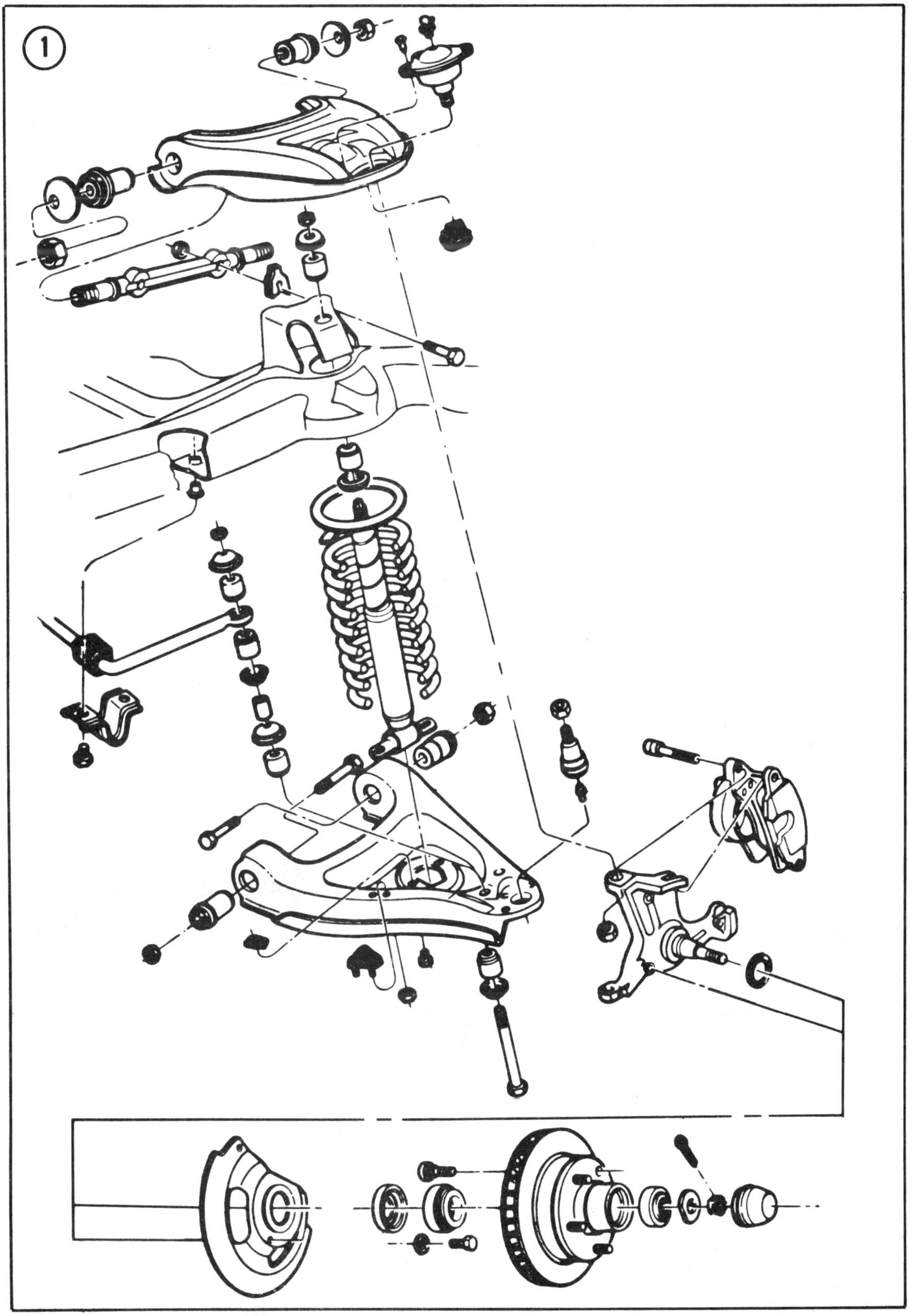
1

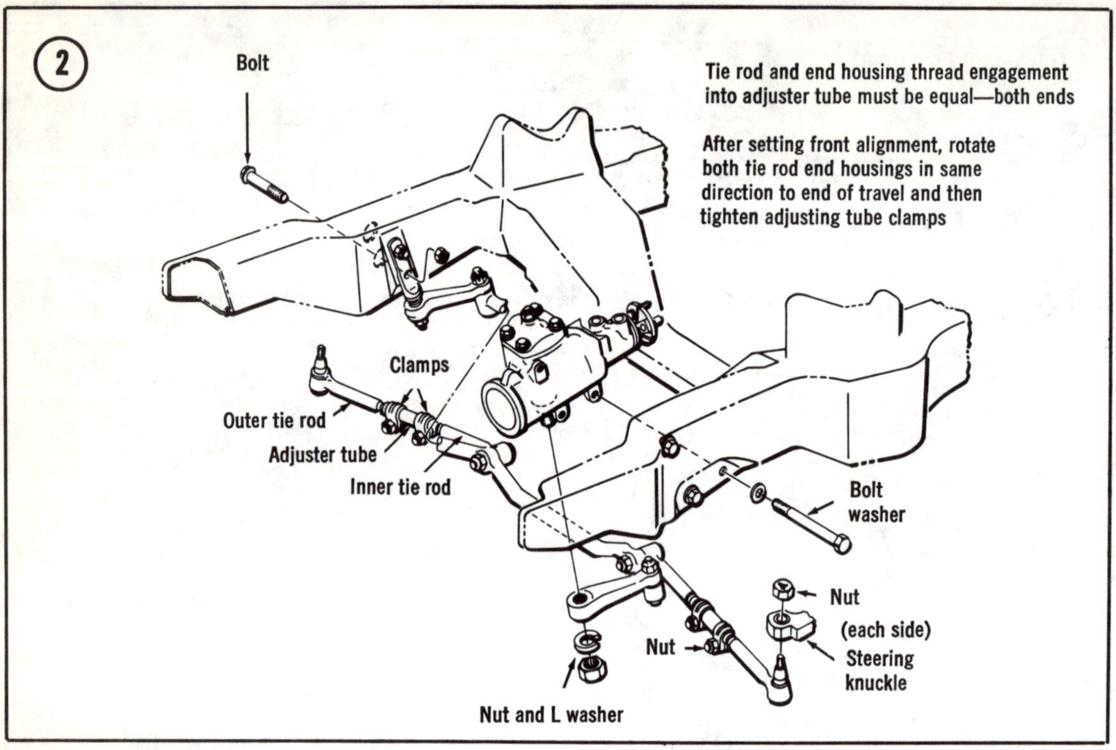
2
Bolt
Tie rod and end housing thread engagement into adjuster tube must be equal—both ends
After setting front alignment, rotate both tie rod end housings in same direction to end of travel and then tighten adjusting tube clamps
Clamps
Outer tie rod
Adjuster tube
Inner tie rod
Bolt
washer
Nut
(each side)
Steering knuckle
Nut
Nut and L washer

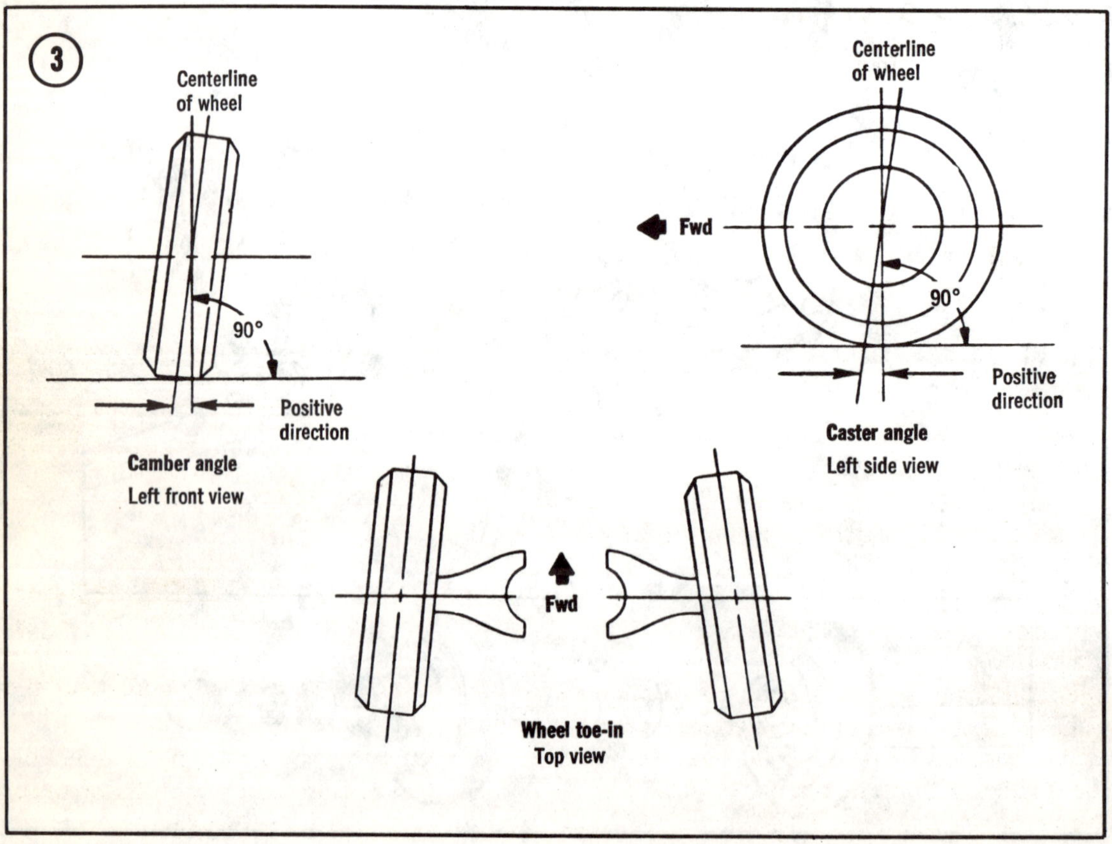
3
Centerline of wheel
90°
Positive direction
Camber angle
Left front view
Centerline of wheel
Fwd
90°
Positive direction
Caster angle
Left side view
Fwd
Wheel toe-in
Top view

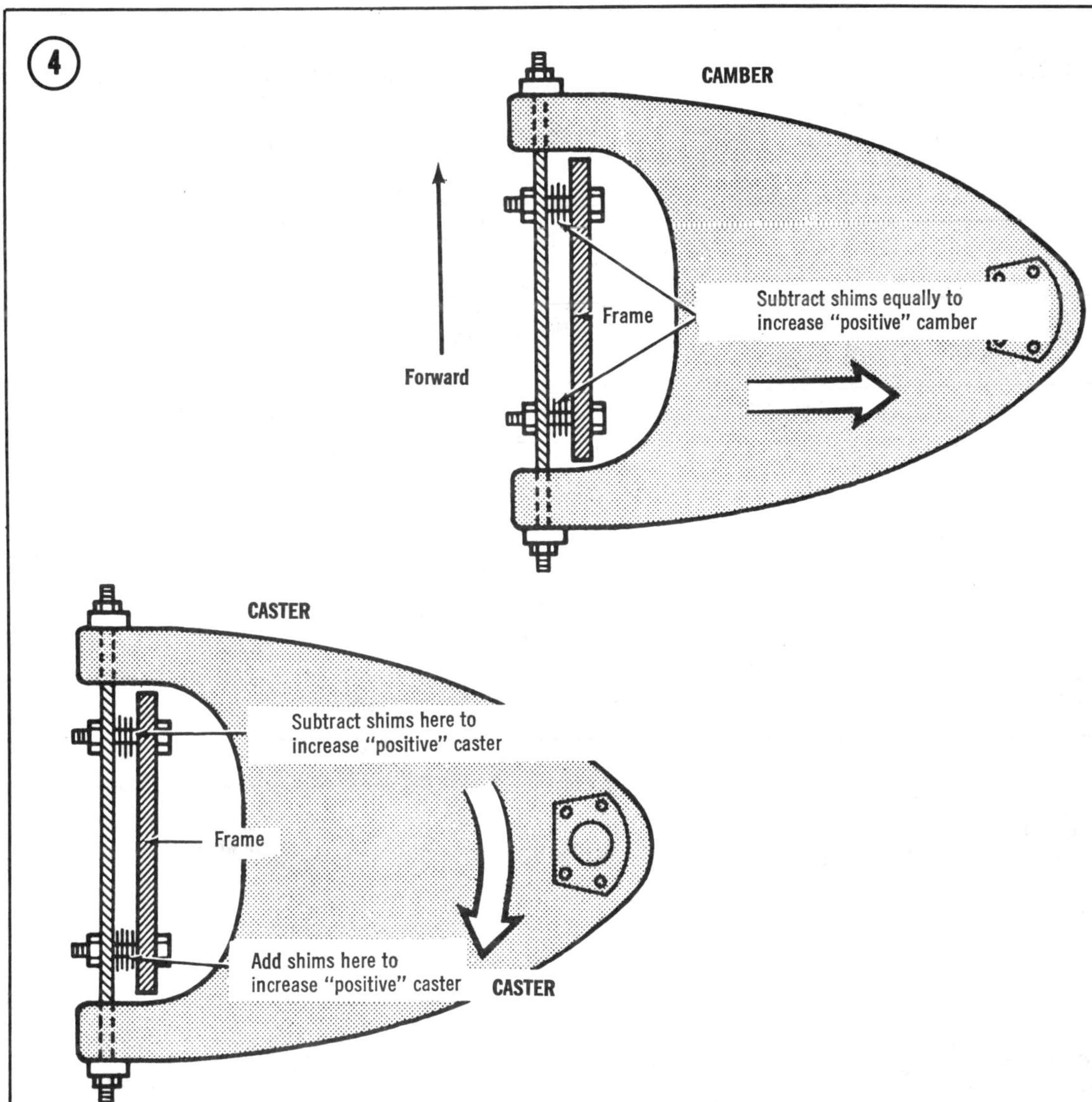

A proper inspection of front tire wear can point to several alignment problems. Tires worn primarily on one side show problems with toe-in. If toe-in is incorrect on one wheel, the car probably pulls to one side or the other. If toe-in is incorrect on both wheels, the car probably is hard to steer in either direction.

Incorrect camber may also cause tire wear on one side. Tire cupping (scalloped wear pattern) can result from worn shock absorbers, one wheel out of alignment, a bent spindle, or a combination of all. Tires worn in the middle, but not the edges, or worn nearly even on both edges, but not in the middle, are probably over-inflated or under-inflated, respectively. These conditions are not caused by suspension misalignment.

Camber

Camber is the inclination of the wheel from vertical as shown in **Figure 3**. Note that camber angle is positive camber, i.e., the top of the tire inclines outward more than the bottom.

Camber is adjusted by adding or removing shims at both front and rear bolts of upper control arm shaft. See **Figure 4**.

5

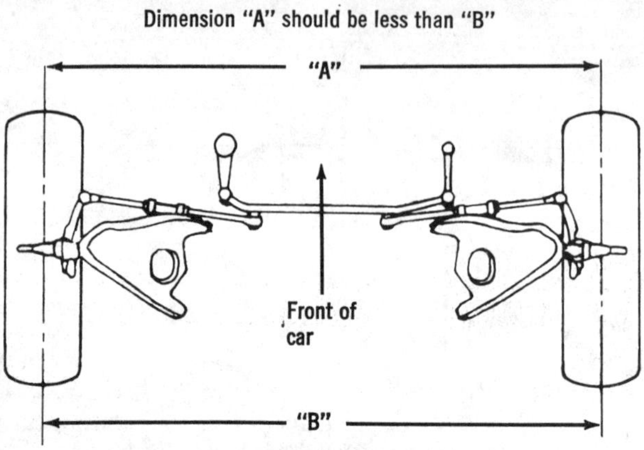

Caster

Caster is the fore and aft inclination of the steering knuckle centerline from vertical. See **Figure 3**. The Cutlass has positive caster, i.e., the bottom of the wheel is shifted forward. Caster causes the wheel to return to a position straight ahead after a turn. It also prevents the car from wandering due to wind, potholes, or uneven road surfaces.

Caster is adjusted by adding or subtracting shims to the front bolt or the rear bolt of the upper control arm shaft. See **Figure 4**.

Steering Axis Inclination

Steering axis inclination, shown in **Figure 3**, is the inward inclination of the steering knuckle centerline from vertical. This angle is not adjustable, but can be checked with proper front end racks to find bent suspension parts.

Toe-in

Camber and rolling resistance tend to force the front wheels outward at their forward edge. To compensate for this tendency, the front edges are turned slightly inward when the car is at rest. This is toe-in. See **Figure 3**.

Toe-in is adjusted by lengthening or shortening the tie rods. Each tie rod is threaded so that the center section can be rotated to make the adjustment. See **Figure 5**.

SHOCK ABSORBERS

Replacement

Refer to **Figure 6**.

6

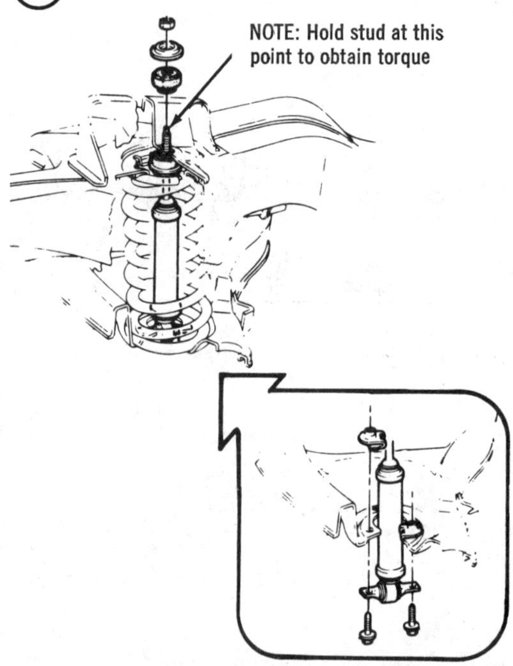

1. Raise front of vehicle on hoist or jackstands.
2. Hold upper stem of shock absorber with a 1/4 in. wrench. Remove upper nut, retainer, and grommet.
3. Disconnect lower shock mount from control arm. Pull shock out of arm.
4. Installation is the reverse of these steps. Tighten upper nut to 15-25 ft.-lb. Tighten lower bolts to 8-11 ft.-lb.

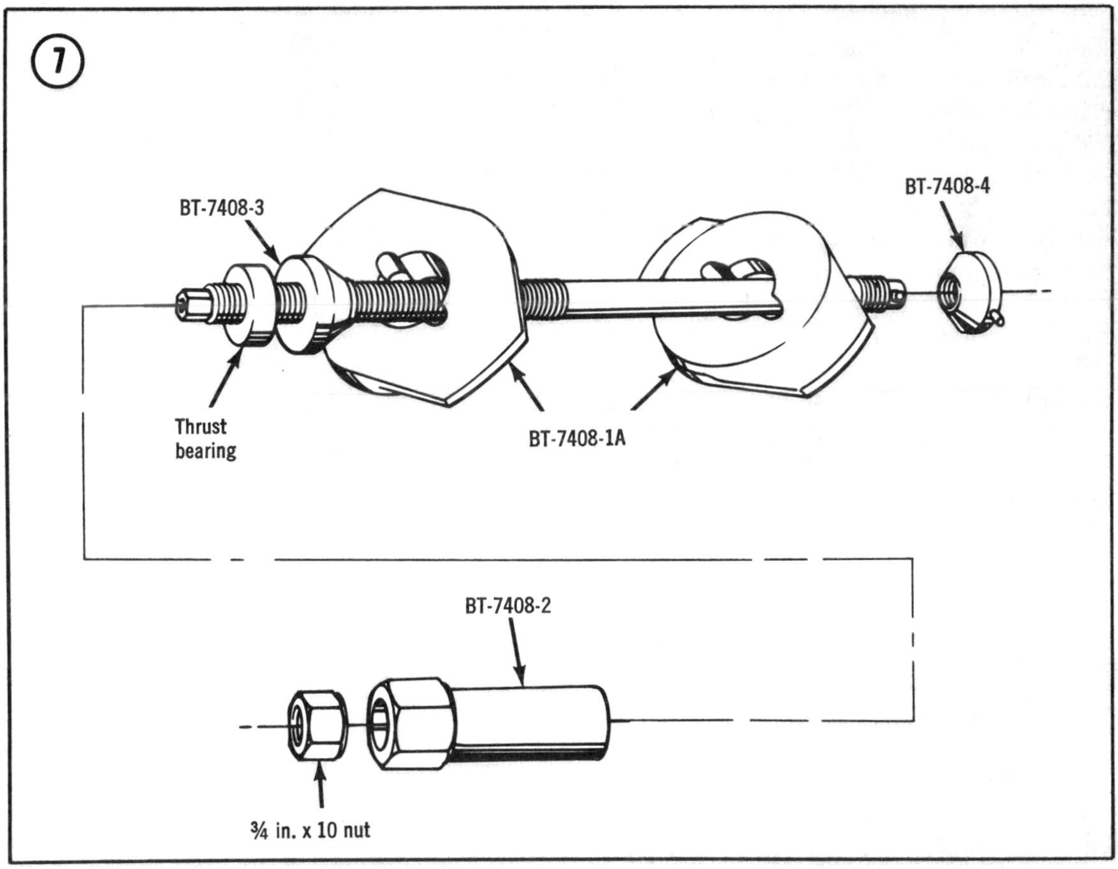

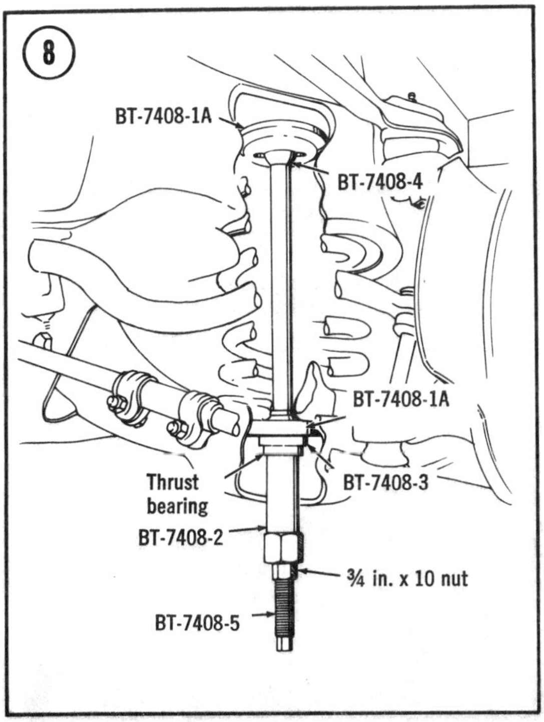

FRONT SPRINGS

WARNING

Due to the danger of working with a compressed spring, special tools must be used. Do not improvise.

Removal

1. Raise car on jackstands.
2. Remove wheel and tire.
3. Remove stabilizer bar and shock absorber.
4. Install special compression tool (**Figure 7**) in the spring as shown in **Figure 8**. Install the plate, pivot seat down, into the spring coils and rotate it downward until it is fully seated in the lower control arm spring seat.
5. Install the other plate, pivot ball seat up, in the upper coils of the spring. Then install the ball nut (BT-7408-4) in the ball seat of the upper plate.
6. Make sure the compression rod is clean and lubricated with high pressure grease, and that the threads are in good condition.

11

7. Insert the compression rod through the shock absorber opening in the lower control arm, and through the upper and lower tool plates. Depress the rod lock pin and thread the rod into the upper ball nut until the rod extends above the ball nut. Make sure the tang on the ball nut is engaged with the slot in the upper plate, then turn the rod until the upper plate contacts the upper spring seat.
8. Install the lower ball nut, thrust bearing, and nut on the lower end of the compression rod.
9. Turn the nut until the spring is compressed enough to be free in its seats, then remove the lower control arm pivot bolts. Disengage the lower control arm and swing it rearward and remove the spring.

NOTE

*If the spring is to be replaced with another one, a special holding tool (BT-7408-9) is required to prevent accidents while the old spring is being decompressed. See **Figure 9**. The same tool can be used while compressing the new spring.*

10. Installation is the reverse of the above steps. See **Figure 10** for correct spring positioning.

BALL-JOINTS

Lower Ball-Joint Removal/Installation

1. Raise the vehicle and support it on jackstands, then remove the front wheels.
2. Place a jack under the lower control arm spring seat. Raise the jack just enough to support the control arm, and leave the jack in this position until this procedure has been completed.
3. Remove the ball-joint stud nut and install a special tool (J-8806 or equivalent) as shown in **Figure 11** to break the stud loose from the steering knuckle.
4. Guide the lower control arm past the splash shield with a putty knife or screwdriver. See **Figure 12**. Use a wood block between the frame and the upper control arm to keep the knuckle out of the way.

9

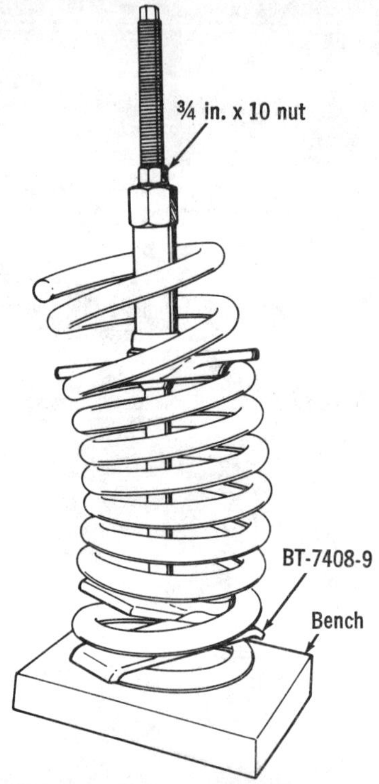

10

Spring to be installed with tape at lowest position. Bottom of spring is coiled helical, and the top is coiled flat with a gripper notch near end of wire.

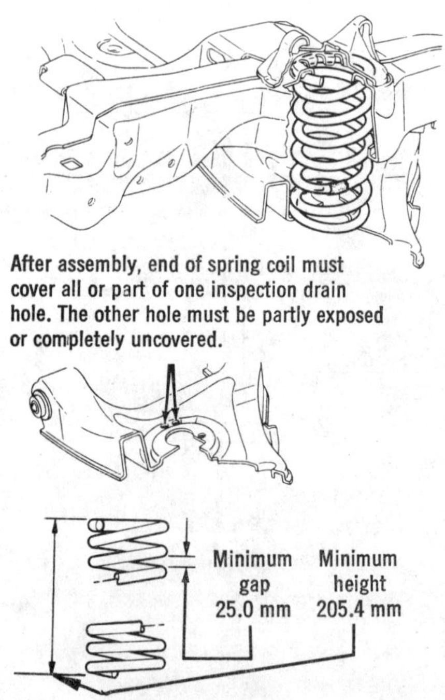

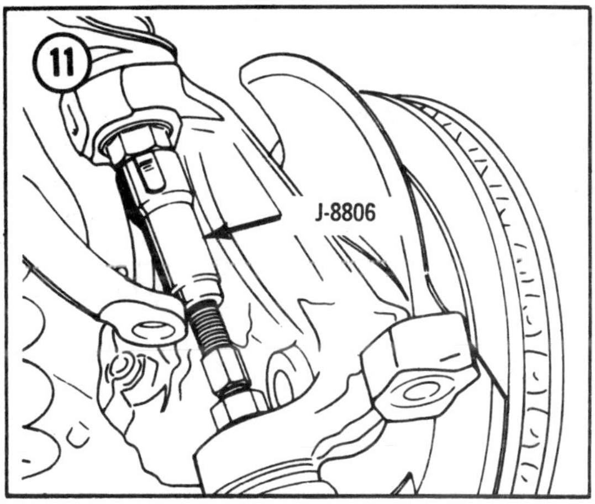

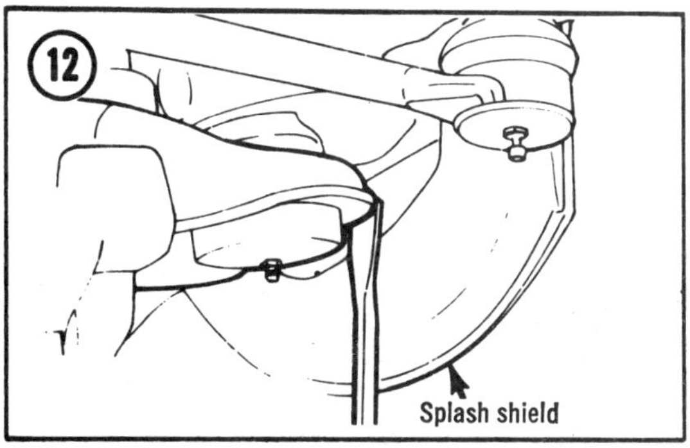

5. Remove retainer and remove ball-joint seal. Use special tools as shown in **Figure 13** to remove the ball-joint from the lower control arm. Remove grease fittings, if present, before installing tools.
6. Place new ball-joint in lower control arm, making sure seal grease purge is facing inboard, and use special tools as shown in **Figure 14** to press into place. Make sure ball-joint bottoms on lower control arm.
7. Install ball-joint stud in steering knuckle, using tool J-29194, and tighten tool to 14 ft.-lb. Remove tool, install nut, and tighten nut to 40 ft.-lb.
8. Install grease fitting and lubricate ball-joint until grease appears at the seal.
9. Install the front wheels, remove the jack, and lower the vehicle.

Upper Ball-Joint Removal/Installation

1. Raise vehicle on jackstands and remove the wheels.
2. Remove cotter pin; loosen ball-joint stud nut about one turn.

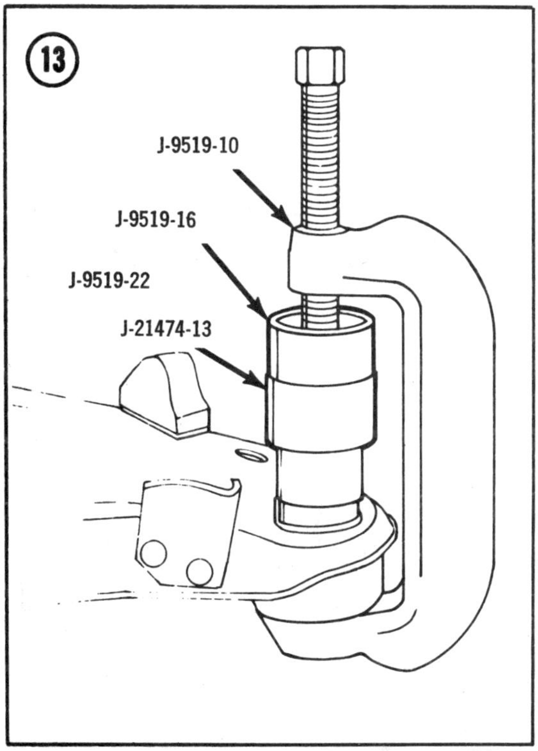

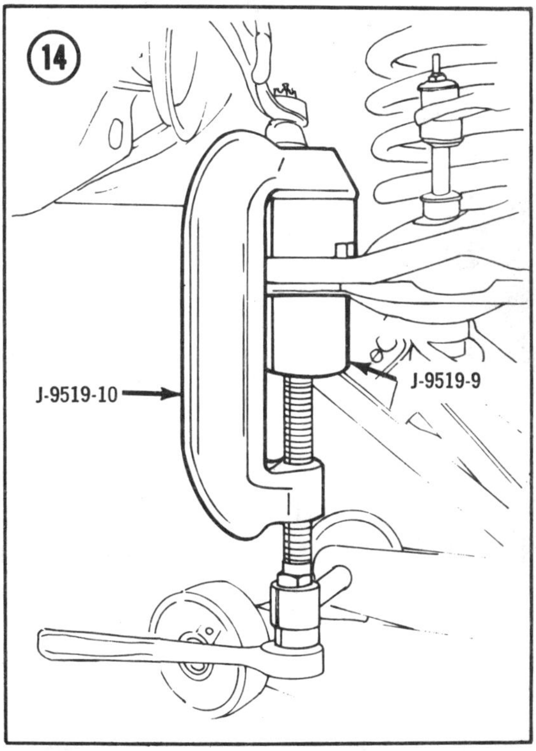

3. Place jack under lower control arm (spring seat). Jack the arm up just enough to support control arm.
4. Remove ball-joint stud nut and install tool J-8806 or equivalent as shown in **Figure 15**. Turn tool to break stud free of steering knuckle.
5. Remove tool and disconnect steering knuckle from ball-joint.
6. Drill rivet heads off ball-joint assembly as shown in **Figures 16 and 17**. Then use a punch (**Figure 18**) to remove the ball-joint.
7. Check ball-joint mounting area on control arm for cracks or metal fatigue.
8. Drill rivet holes in control arm to diameter specified in ball-joint service kit (if required).
9. Install new ball-joint with special bolts supplied in service kit. Insert bolts from bottom with nut on top. Torque bolts to 8 ft.-lb.
10. Turn ball stud cotter pin hole so it faces fore and aft.
11. Raise control arm and insert upper ball stud into steering knuckle. Install ball stud nut and torque to 65 ft.-lb. Secure with new cotter pin. If cotter pin hole does not line up, *tighten* nut additionally; do not *loosen* nut.
12. Connect stabilizer link to the control arm bracket. Torque to 13 ft.-lb.

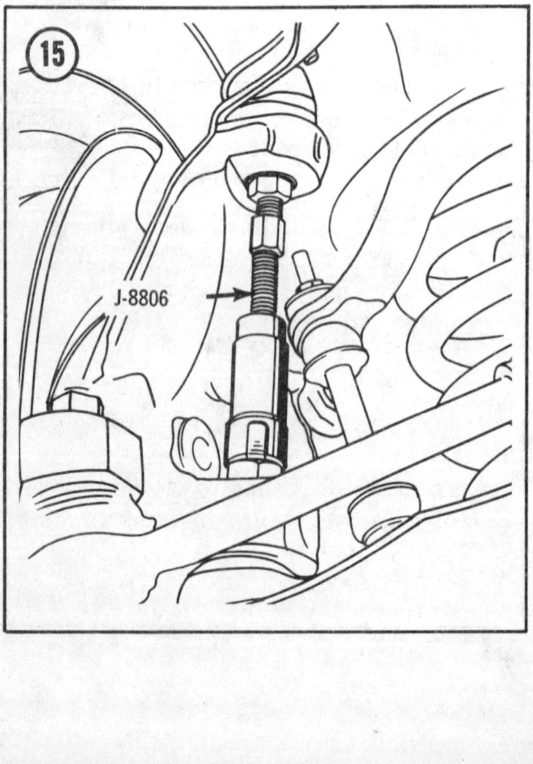

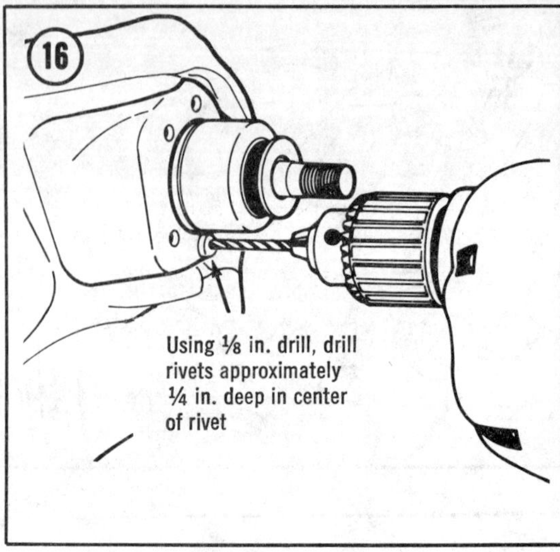

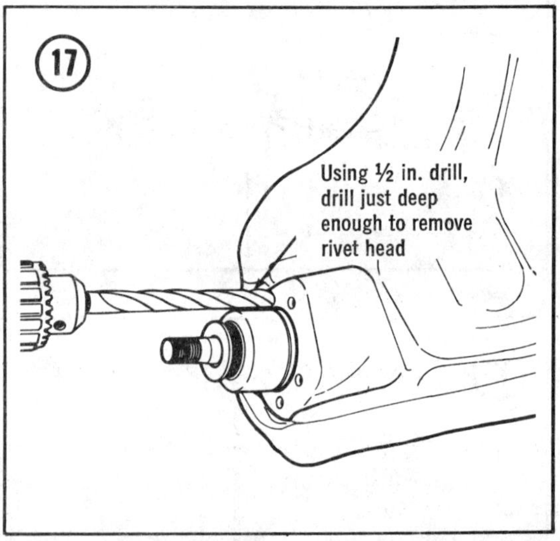

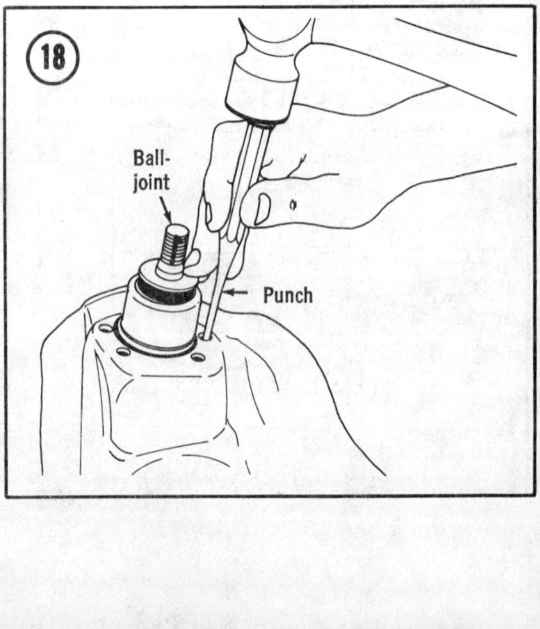

19

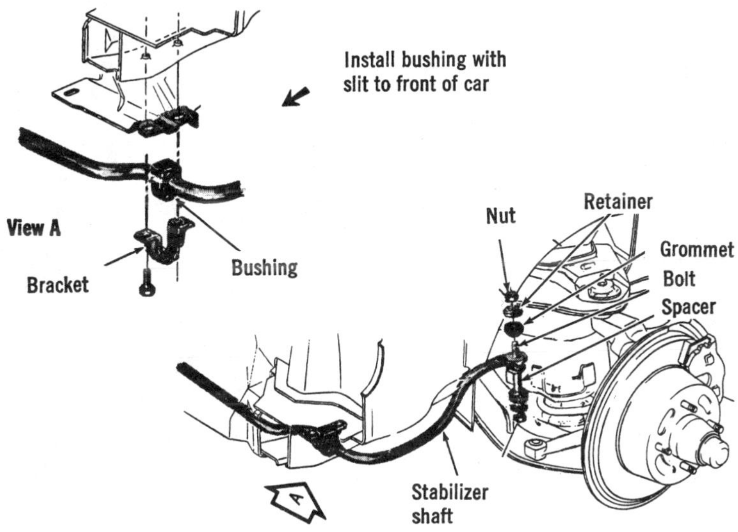

STABILIZER BAR

Removal

Refer to **Figure 19** (typical) for these procedures.

1. Raise car on jackstands.
2. Disconnect both stabilizer links from control arms. Remove link bolt bushings and retainers.
3. Remove bolts securing stabilizer supports to frame. Bolt heads are accessible through holes in frames.
4. Remove stabilizer bar.

Installation

1. Secure stabilizer bar with brackets.
2. Assemble link bolts, bushings, and retainers. See **Figure 19**. Secure to control arms.
3. Lower car. Bounce front several times to settle rubber bushings.
4. Retighten link nuts to 13 ft.-lb. and bracket nuts to 24 ft.-lb.

STEERING KNUCKLE

Removal

1. Raise vehicle on jackstands. Support lower control arms with jacks so that spring stays compressed as if car were on floor.
2. Remove front wheels.

3a. On models with front drum brakes, remove brake drum, wheel hub, and brake shoes. Disconnect backing plate and steering knuckle. See Chapter Ten.

3b. On models with disc brakes, remove brake caliper and disc. See Chapter Ten.

4. Separate upper and lower ball-joints from knuckle as described in *Ball-joints*, this chapter.
5. Installation is the reverse of these steps. Tighten ball-joint nuts to 65 ft.-lb. (upper) and to 90 ft.-lb. (lower).
6. Adjust wheel bearings as described earlier.
7. Check wheel alignment.

11

FRONT WHEEL HUB (DRUM BRAKES)

Removal/Installation

1. Perform Steps 1 and 2 of *Wheel Bearing Replacement* (this chapter) procedure to remove hub.
2. Remove, clean, and repack bearings as described in Steps 3-5, 11, and 12, *Wheel Bearing Replacement*.
3. Reinstall wheel hub by following Steps 13-18, *Wheel Bearing Replacement*.

FRONT WHEEL BEARINGS

Adjustment

1. Raise front of vehicle until tires clear the ground.
2. Remove hub cap and dust cap. Remove cotter pin from spindle.
3. Tighten spindle nut to 15 ft.-lb. while rotating wheel.

NOTE

Oldsmobile has indicated that some 1979 models may experience bearing failure due to high temperatures in the outer bearing area. This will reduce bearing lubrication and result in noticeable front end vibration and noise. This condition should be checked by an Olds dealer and modified grease retainer caps installed, if not already done.

20

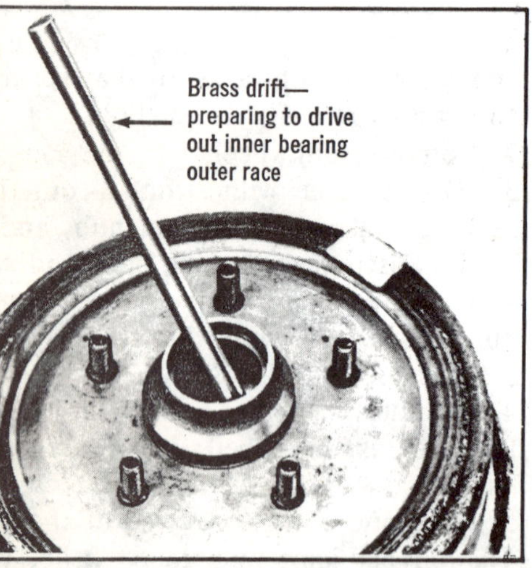

21

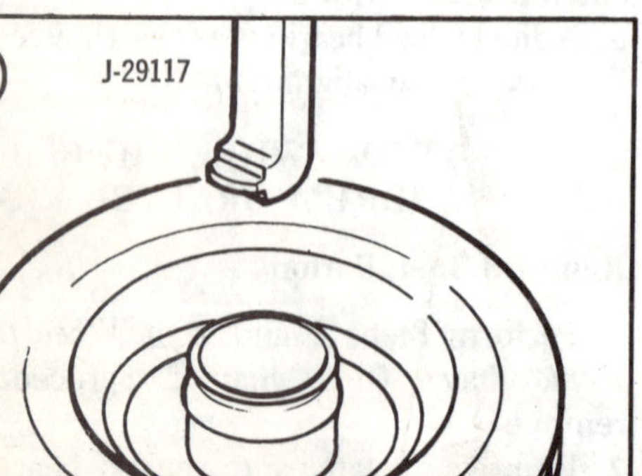

22

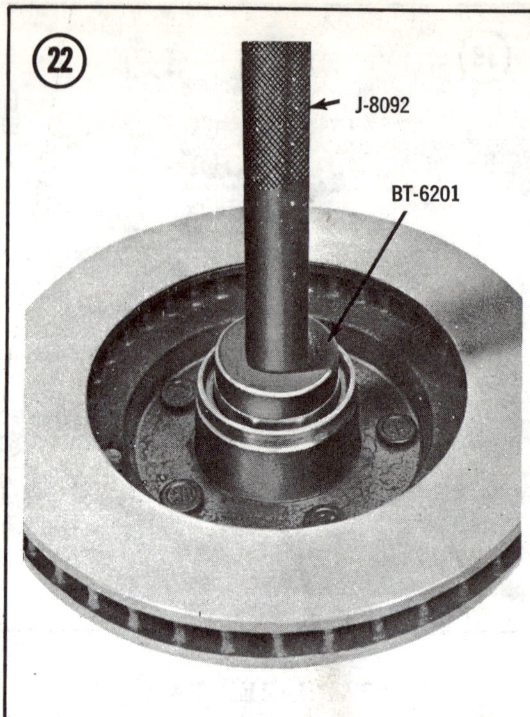

23

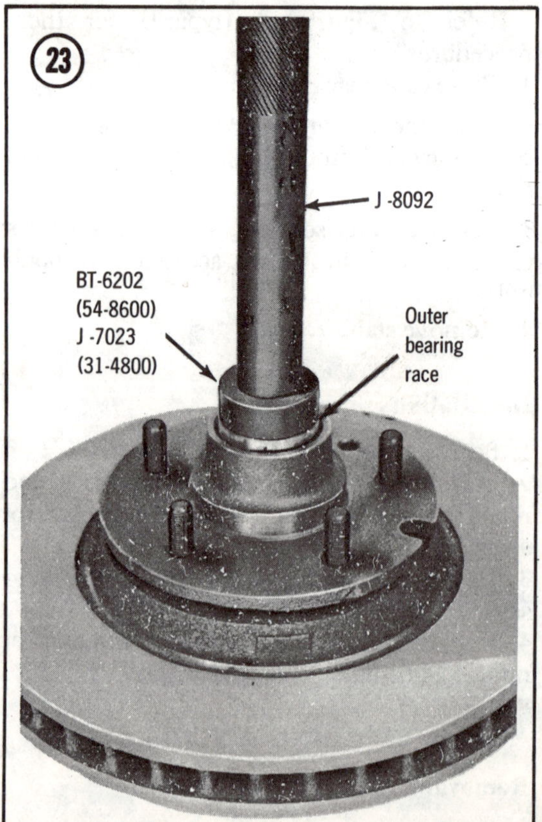

4. Back off nut one flat (1/6 turn) and insert cotter pin. If slot and hole do not line up, back off an additional 1/2 flat or less, as necessary.
5. Make sure wheel turns freely. Then measure hub end play with a dial indicator. Readjust if not 0.001-0.005 in.
6. Install dust and hub caps and lower vehicle.

Replacement

1a. On models so equipped, remove brake drums as described in Chapter Ten. Remove hub dust cap, cotter pin, spindle nut, and washer. Pull hub off.
1b. On disc brake models, remove brake caliper and disc. See Chapter Ten.
2. Remove outer bearing from hub on the brake disc.
3. Pry out inner grease seal from hub on brake disc and remove inner bearing race.
4. Clean wheel bearings in hub or brake disc in solvent. Blow dry. Check bearing rollers for scores, wear, and evidence of overheating (bluish tint). Check bearing races in hub or brake disc. Do not mix bearings up if they are good; they must be replaced on the same wheel assembly.

CAUTION
If a bearing race is damaged, the bearing and race must be replaced.

5. If bearings and races are good, reinstall by following Steps 10-17. Otherwise, proceed to Step 6.

24

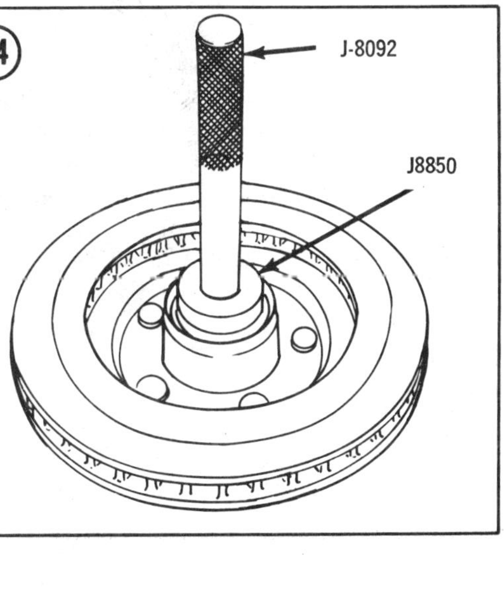

6. On drum brake models, use a brass drift (**Figure 20**) to remove the outer bearing races.
7. On disc brake models, use a special tool (**Figure 21**) to remove the outer races.
8. On drum brake models, use special tool BT-6201 to press in new inner bearing outer race, and tool BT-6202 to press in outer race of outer bearing. See **Figures 22 and 23**.
9. On disc brake models, use special tool J-8850 (**Figure 24**) for inner bearing and J-8457 (**Figure 25**) for outer bearing to press in outer races.
10. Pack inner and outer bearings with high temperature wheel bearing grease. Do not pack cavity in hub with grease.
11. Fit inner bearing in hub or brake disc. Install new grease seal with flange facing bearing race.
12a. On drum brake models, install wheel hub over steering spindle.
12b. On disc brake models, install brake disc.
13. Install outer bearing. Press it in firmly by hand.
14. Install spindle washer and adjusting nut.
15. Adjust wheel bearings as described earlier.
16a. On drum brake models, install brake drum and wheel.
16b. On disc brake models, install brake caliper and wheel.
17. Lower vehicle.

25

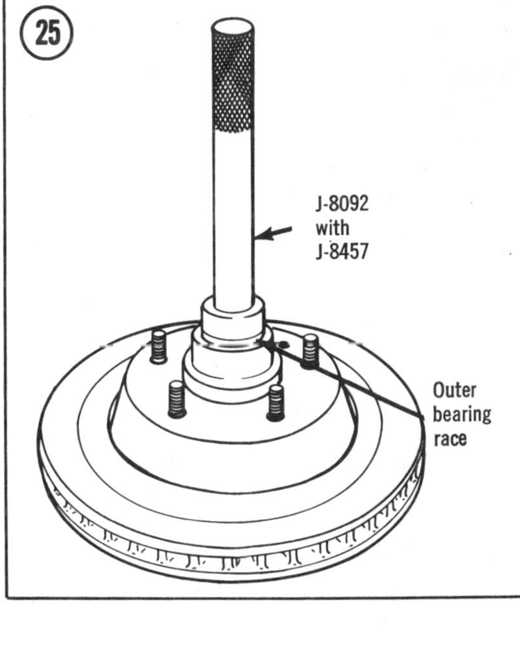

11

26

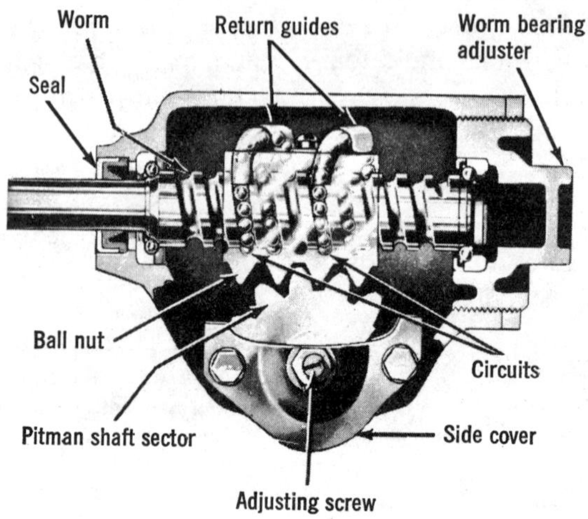

STEERING ADJUSTMENTS

Before attempting to correct steering deficiencies with these adjustments, carefully check:

a. Front end alignment
b. Shock absorber condition
c. Wheel balance
d. Tire pressure

Manual Steering Gear Adjustment

1. Remove pitman arm nut. Mark relation of pitman arm on sector shaft.
2. Remove pitman arm using tool J-5504 or similar puller.
3. Loosen lash adjuster locknut (**Figure 26**) and turn adjuster a few turns counterclockwise.
4. Turn steering wheel very gently and slowly in one direction until stopped by gear. Back away about one turn.

CAUTION
Do not turn wheel hard against stops when relay rod is disconnected or ball guides in steering gear will be damaged.

5. Using a torque wrench, measure pull required to keep wheel in motion. See **Figure 27**. Scale centerline must remain at right angles to wheel spoke. From 4-8 in.-lb. should be required; adjust worm bearings if force is higher or lower.

NOTE
Use a torque wrench with a maximum reading of 50 in.-lb.

6. To adjust worm bearings, loosen worm adjuster locknut. See **Figure 26**. Turn adjuster until there is no appreciable end play in worm. Tighten locknut.
7. Repeat Steps 5 and 6 as necessary.
8. If gear feels rough after adjustment, bearings are probably damaged.
9. Count number of turns required to turn steering wheel from one stop to the other. Turn wheel back from stop exactly 1/2 the number of turns. Mark on top of worm shaft just below clamp should be at top of shaft and in line with the sawcut at the lower coupling clamp.
10. Turn lash adjuster screw clockwise to take out all lash in gear teeth. Tighten locknut.
11. Check force required to pull wheel through center position with spring scale. From 7/8 -1 1/2 lb. force should be required. Repeat Steps 9-11 as necessary to obtain correct force.

12. Reassemble pitman arm to sector shaft. Line up marks made during disassembly. Torque nut to 200 ft.-lb.

Steering Wheel Removal/Installation

1. Disconnect the column harness at wiring connector.
2. Disconnect battery ground cable.

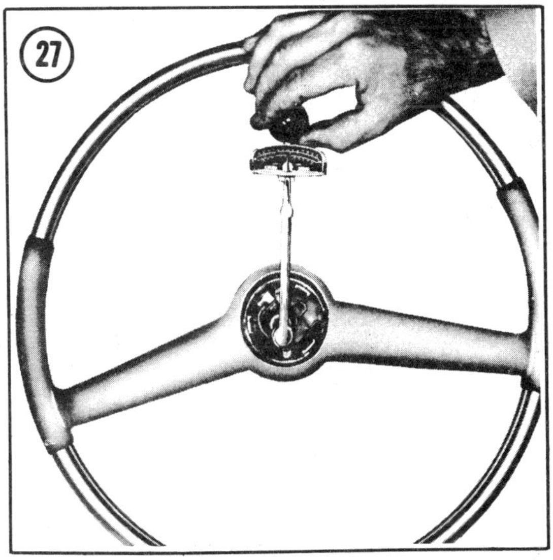

3. Pry off horn button cap or remove pad attaching screws. See **Figure 28**.
4. Remove screws securing horn contact to spacer and hub.
5. Remove screws securing lock screw to lock knob and remove lock screw, lock knob and spacer.
6. Remove screws securing steering wheel to hub and remove wheel. If necessary use a puller tool. See **Figure 29**.
7. Installation is the reverse of these steps.

STEERING GEAR

Removal/Installation

1. Remove lower clamp bolt on steering coupling to worm shaft.

2. Raise vehicle on hoist or jackstands.

3. Remove pitman arm with suitable puller.

4. Remove steering gear mounting bolts and remove steering gear.

5. Installation is the reverse of these steps. Be sure that wheels are straight ahead, steering wheel is straight, and steering gear is set to the middle of its travel.

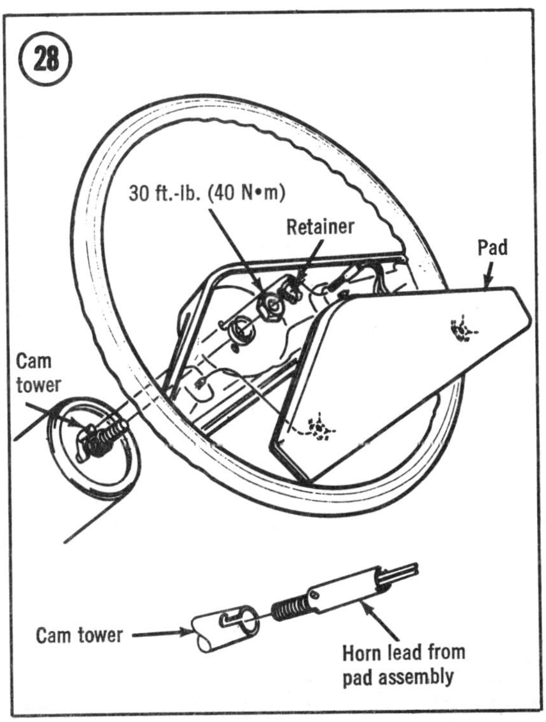

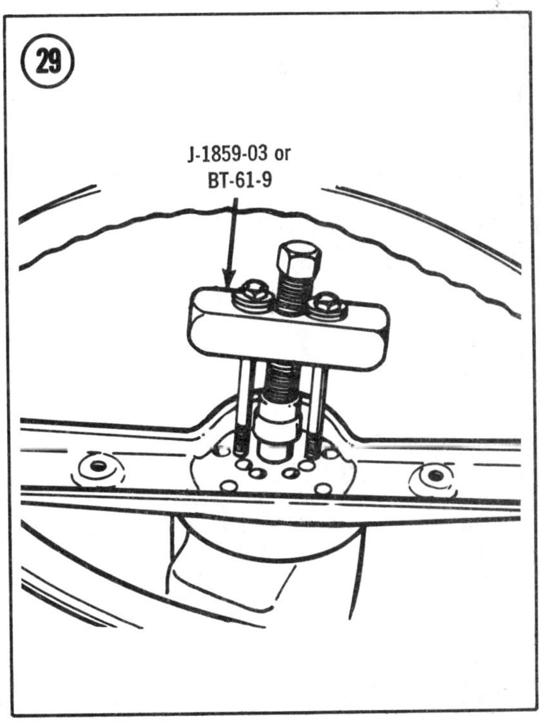

11

30

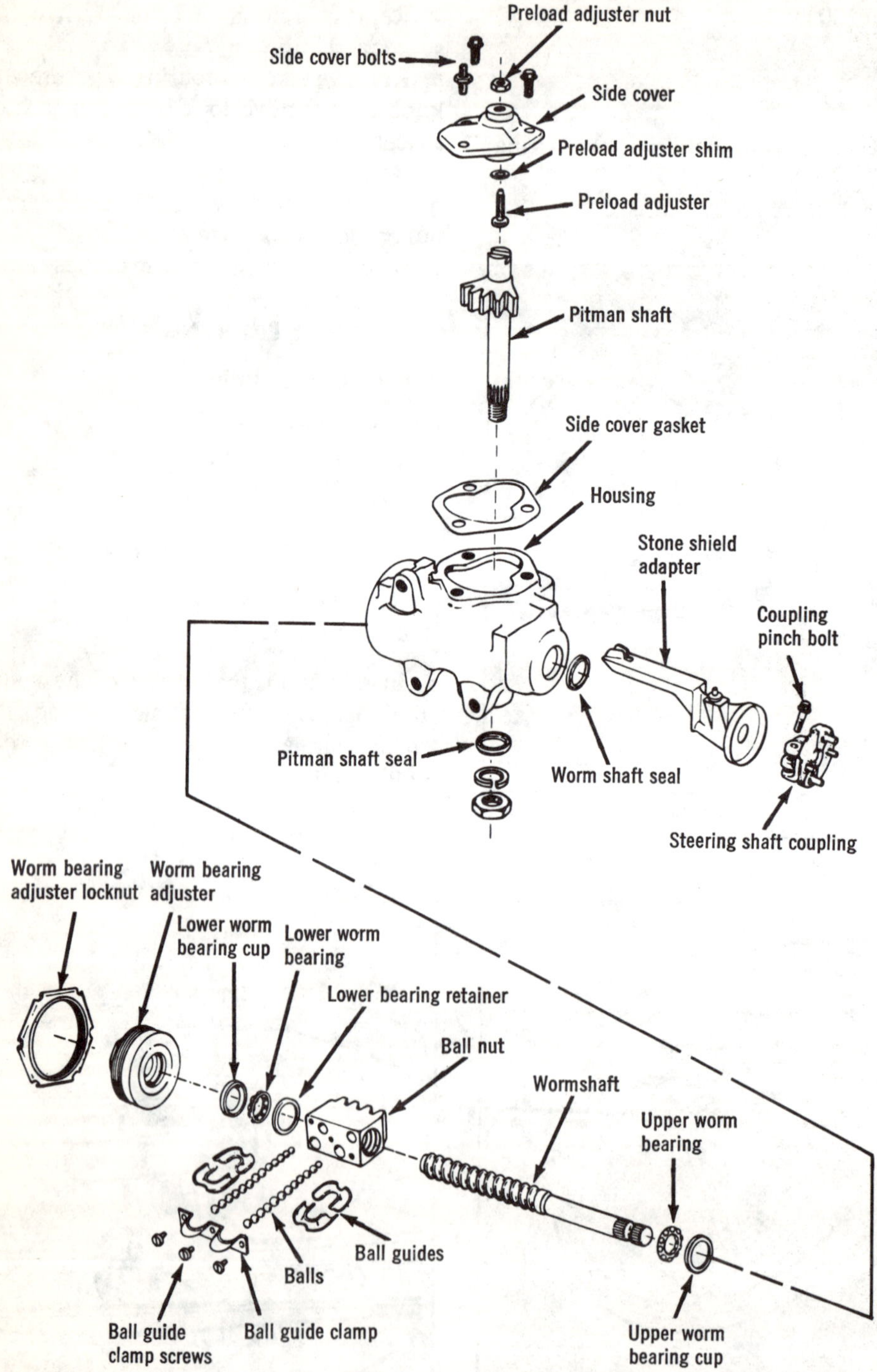
Preload adjuster nut
Side cover bolts
Side cover
Preload adjuster shim
Preload adjuster
Pitman shaft
Side cover gasket
Housing
Stone shield adapter
Coupling pinch bolt
Pitman shaft seal
Worm shaft seal
Steering shaft coupling
Worm bearing adjuster locknut
Worm bearing adjuster
Lower worm bearing cup
Lower worm bearing
Lower bearing retainer
Ball nut
Wormshaft
Upper worm bearing
Ball guides
Balls
Ball guide clamp screws
Ball guide clamp
Upper worm bearing cup

31

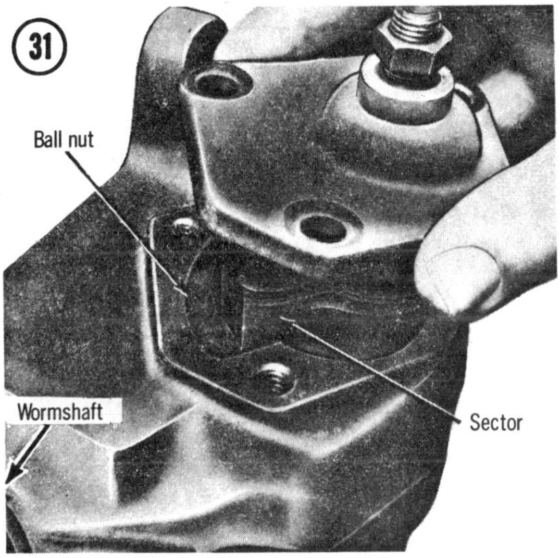

Disassembly

Refer to **Figure 30** for this procedure.

1. Clean exterior thoroughly in solvent to remove road dirt and grease.
2. Cover *clean* work surface with clean paper or rags.
3. Hold gear in vise by lower mounting ear.
4. Loosen lash adjuster locknut and turn lash adjuster several turns counterclockwise.
5. Remove 3 cap screws holding side cover. Pull side cover/sector assembly out. See **Figure 31**.

NOTE

Turn worm shaft if necessary until sector clears opening.

32

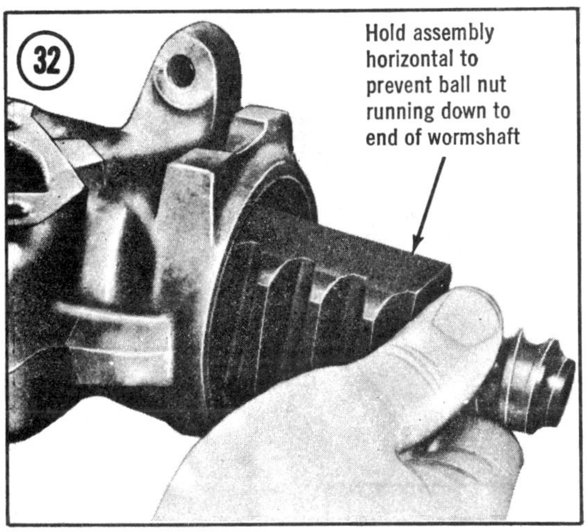

6. Remove worm bearing adjuster and locknut from housing.
7. Remove worm shaft and ball nut assembly from housing. See **Figure 32**. Remove upper ball bearing from worm shaft. Remove lower bearing retainer and bearing from adjuster plug housing (**Figure 33**).
8. Remove locknut from lash adjuster and unscrew adjuster clockwise from side cover. Slide adjuster and shim out of slot in end of sector shaft.
9. Remove sector shaft seal.
10. Remove worm shaft seal.
11. Remove clamp retaining ball guides in nut. Draw guides out of nut.
12. Turn nut upside down and rotate worm shaft back and forth until all the balls have dropped out into a clean pan.
13. Pull nut off worm.

33

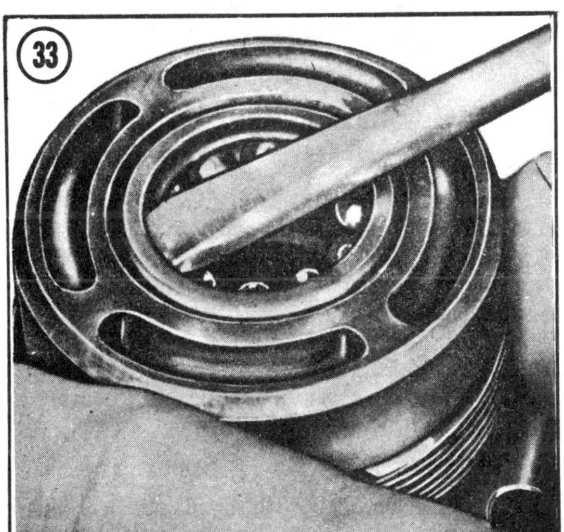

Inspection

1. Wash all parts in solvent. Dry them with clean rags.
2. Check ball bearings, races, worm and nut grooves for signs of wear and scoring.
3. Inspect sector shaft and bushings for wear and check the fit of the shaft in the housing bushings.
4. Check ball guides for damaged ends. Replace if necessary.
5. Check worm shaft assembly for damage.

Rebuilding and Assembly

Refer to **Figure 30**.

> NOTE
>
> *Bushings require replacement only if worn or damaged. Consider dealer replacement with proper tools.*

1. Press the pitman shaft bushing out of housing. See **Figure 34**.
2. Coat new bushing with steering gear lube.
3. Press new pitman shaft bushing into end of bore so that bushing is flush with, or slightly lower than, end of bore. See **Figure 35**.

> NOTE
>
> *Worm shaft seal and sector shaft seal must be replaced anytime gear is assembled.*

4. Pry out old seals.
5. Coat new seals with steering gear lube inside and out.
6. Position new worm shaft seal over center hole in worm bearing adjuster and press in using socket of suitable outside diameter. Seal edge should be flush with inner edge of small bore adjuster.
7. Position new pitman shaft seal over pitman shaft bore. Press in with the same method used for worm shaft seal.

> NOTE
>
> *Worm shaft bearing races require replacement only if worm is damaged. Consider dealer replacement with proper tools.*

8. Place steering gear housing or worm adjuster in vise with bore horizontal.
9. Remove worm bearing races, using a brass drift.
10. Install new races in housing, using tool J-8811 or equivalent.

> NOTE
>
> *Steps 11 through 20 assemble ball nut on worm shaft.*

11. Lay worm shaft flat on bench. Slip ball nut over worm with ball guide holes up and the shallow end of the rack teeth to the left as viewed from the steering wheel position. See **Figure 36**. Insert ball guides as shown.

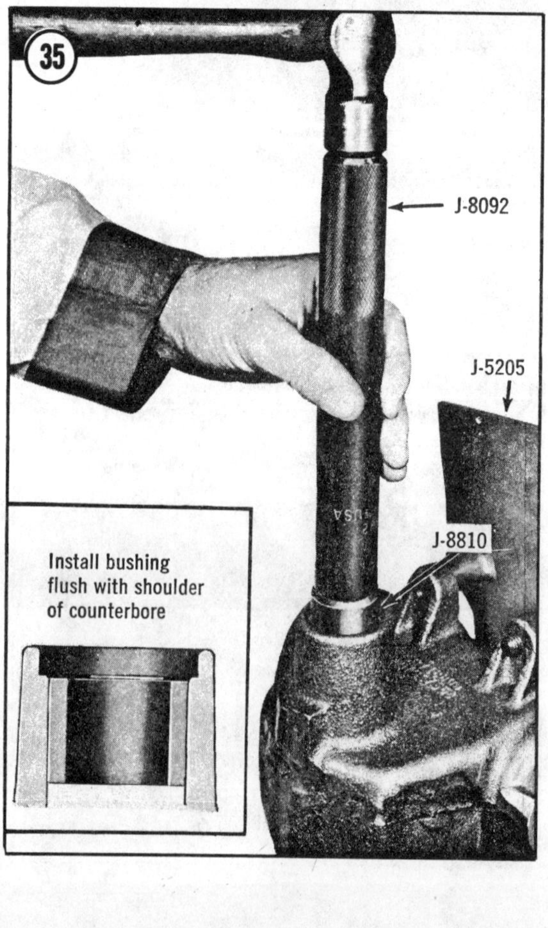

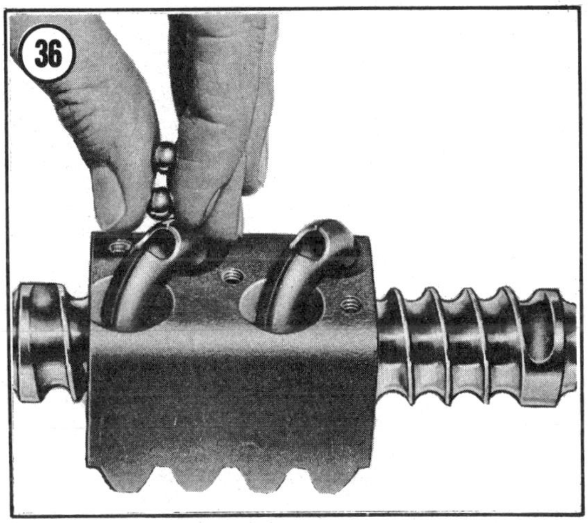

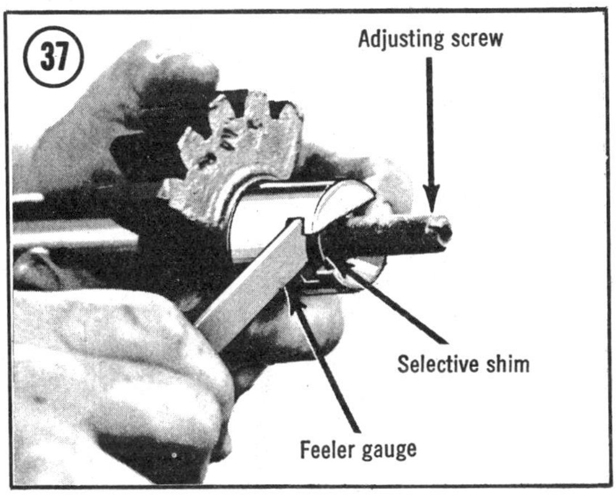

12. Count 24 balls into a clean shallow container like a pie tin. Each circuit requires 24 steel balls.
13. Align grooves in the worm and nut by sighting through the ball guide holes.
14. Insert balls for one circuit into one of the guide holes one at a time while gradually turning the worm away from that hole. Continue until all 24 balls have been inserted.
15. Fill second circuit by repeating Steps 12 through 14.
16. Mount ball guide clamp on nut. Use lockwasher under clamp screw.
17. Rotate nut on worm to be sure it rotates freely. Do not bottom nut on either end of worm. If nut binds slightly, ball guide ends are probably damaged and require replacement.
18. Hold steering gear housing in vise with the worm shaft bore vertical and the side cover opening up.
19. Slip upper ball bearing over worm shaft and install worm shaft and nut assembly into housing. Feed shaft end through ball bearing race and seal.
20. Place a bearing in adjuster race and press the retainer into place with a suitable socket.
21. Install locknut and adjuster into lower end of housing until nearly all end play is removed from worm shaft. Take care while guiding end of worm shaft into bearing.
22. Assemble lash adjuster with shim in slot in end of sector shaft.
23. Check end clearance as shown in **Figure 37**. End clearance should not exceed 0.002in. Change shim installed in Step 22 if necessary. Shims are available 0.063-0.069 in. thick only, in 0.002 in. increments.
24. Install gasket on side cover and insert sector shaft into housing. Do not damage seal. Be sure to index center tooth of sector in center space of ball nut rack.
25. Align holes in side cover with holes in housing. Install upper cap screw.
26. Hold steering gear housing in vise in its approximate position when installed in car.
27. Inject steering gear lubricant into lowest side cover cap screw opening until lubricant appears in another opening. Install 2 remaining capscrews and lockwashers.

Preliminary Adjustment

Worm bearing and lash adjustments may be made before gear is installed in car.

1. Hold steering gear in vise in its approximate position when installed in car.
2. Temporarily install steering wheel on worm shaft.
3. Perform Steps 5-7, *Steering Gear Adjustment* procedure, described earlier.

STEERING LINKAGE

Tie Rod Replacement

Refer to **Figure 2**.

1. Raise car on jackstands and remove front wheels.
2. Remove nuts from ball studs.
3. Hold heavy hammer as backing. Tap steering arm with another hammer to loosen ball stud.

4. Remove inner ball stud in a similar manner.
5. Loosen clamp bolts and unscrew tie rod ends.
6. Install tie rod ends on tie rod. Make sure both ends are threaded on an equal amount.
7. Make sure that ball stud threads and nut threads are perfectly clean and smooth.
8. Install neoprene seals on ball studs.
9. Install ball studs in the steering arm and the relay rod.
10. Install ball stud nuts. Torque to 45 ft.-lb. Tighten additionally if necessary to align cotter pin holes; do not loosen nut.
11. Install new cotter pin.
12. Lubricate tie rod ends. See Chapter Three.
13. Adjust toe-in.

CAUTION
*Before locking clamp bolts, make sure that tie rod ends are aligned with their ball studs. Otherwise, binding will result. See **Figure 38** for alignment.*

Idler Arm Removal/Installation

Refer to **Figure 39**.

1. Raise car on jackstands.
2. Remove cotter pin and nut from idler arm.
3. Disconnect relay rod from idler arm.

38

Bolts must be installed in direction shown, rotate both inner and outer tie rod housings rearward to the limit of ball joint travel before tightening clamps. With this same rearward rotation all bolt centerlines must be between angles shown after tightening clamps.

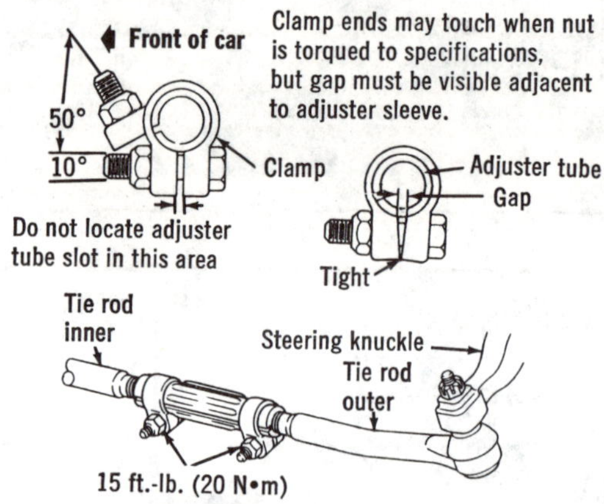

Clamp must be between and clear of dimples. Torque nuts to specification.

39

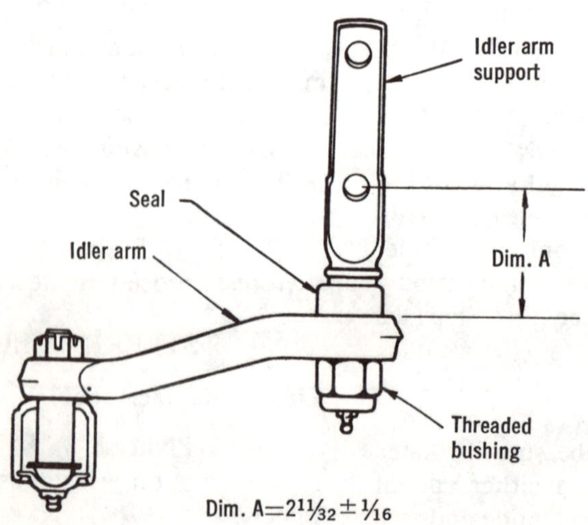

Dim. A=$2\frac{11}{32} \pm \frac{1}{16}$

4. Remove idler arm from frame.
5. Mount idler arm on frame. Torque the bolts to 40 ft.-lb.
6. Connect relay rod to idler arm. Use new seal and tighten nut to 35-45 ft.-lb.
7. Install cotter pin. Tighten nut additionally if necessary to align cotter pin holes.

POWER STEERING PUMP

Belt Tension Adjustment

1. Loosen the power steering bracket-to-power steering pump attaching bolts.
2. Move pump, with belt in plce, until belt tension is correct. This occurs when a 15 lb. force applied at the midpoint between the power steering pump pulley and the drive pulley causes 1/2-3/4 in. belt deflection.
3. Tighten pump mounting bolts.

Removal

1. Disconnect hoses at pump. Secure ends to prevent drainage of oil. Cap or tape the ends of the hose to prevent entry of dirt.
2. Install 2 caps at pump fittings to prevent drainage of oil from pump.
3. Loosen bracket-to-pump mounting nuts.
4. Remove pump belt.
5. Remove bracket-to-pump bolts and remove pump from vehicle.
6. Remove drive pulley attaching nut.
7. Slide pulley from shaft.

CAUTION

Do not hammer pulley off shaft as this will damage the pump.

Installation

1. Slide pulley on shaft.

CAUTION

Do not hammer pulley on shaft.

2. Install pulley nut and torque to 35-45 ft.-lb. Use a new pulley nut.
3. Position pump assembly on vehicle and install nuts loosely.
4. Connect and tighten hose fittings.
5. Fill reservoir. Bleed pump by turning pulley backward (counterclockwise as viewed from front) until air bubbles cease to appear.
6. Install pump belt over pulley.
7. Adjust belt tension as described earlier.
8. Bleed the hydraulic system by raising the front wheels, starting the engine, and turning the steering wheel from stop to stop several times. Lower car, test operation, and repeat the procedure if the unit is not operating properly.

NOTE: If you own a 1981 and later model, first check the Supplement at the back of the book for any new service information.

CHAPTER TWELVE

REAR SUSPENSION AND DRIVE SHAFT

In this rear suspension system the rear axle assembly is connected to the frame by a lower control arm and an upper control arm on each side. Individual coil springs at each wheel are located between brackets on the axle tube and spring seats in the frame. Double acting shock absorbers provide ride control.

The rubber-bushed lower control arms maintain fore and aft relationships between the axle and the chassis, while the upper control arms, which are shorter than the lower ones, control side sway and maintain the pinion nose angle. A typical rear suspension system is shown in **Figure 1**.

A one-piece drive shaft with a splined slip yoke and universal joint in front and another universal joint at the rear is used on all models. The slip yoke fits into splines in the transmission housing, while the rear universal joint is connected to a companion flange at the front of the differential. See **Figure 2**. Universal joint bearing caps are retained by plastic injected during manufacture.

In this type of suspension system, the relationship of the angles of the front and rear universal joints is very important. If these angles are not almost identical, rough operation and vibration can result. The front angle is the angle between the engine-transmission center line and the drive shaft, and can be adjusted by the addition or removal of shims between the transmission rear bearing retainer and the transmission mount. The rear angle is determined by the length of the upper control arms, and the angle is adjusted by installing different length upper control arms. Special tools and skills are required to make these adjustments properly. If the procedures provided in this chapter do not clear up vibration and rough operation problems, the car should be taken to your dealer.

REAR SUSPENSION

Coil Spring Removal/Installation

NOTE

Springs must be removed and installed one side at a time.

1. Raise the rear of the vehicle and support at the rear lifting points (see **Figure 3**) with jackstands. Make sure the rear wheels clear the floor by several inches.
2. Support the rear axle with a jack or other adjustable lifting device.

1

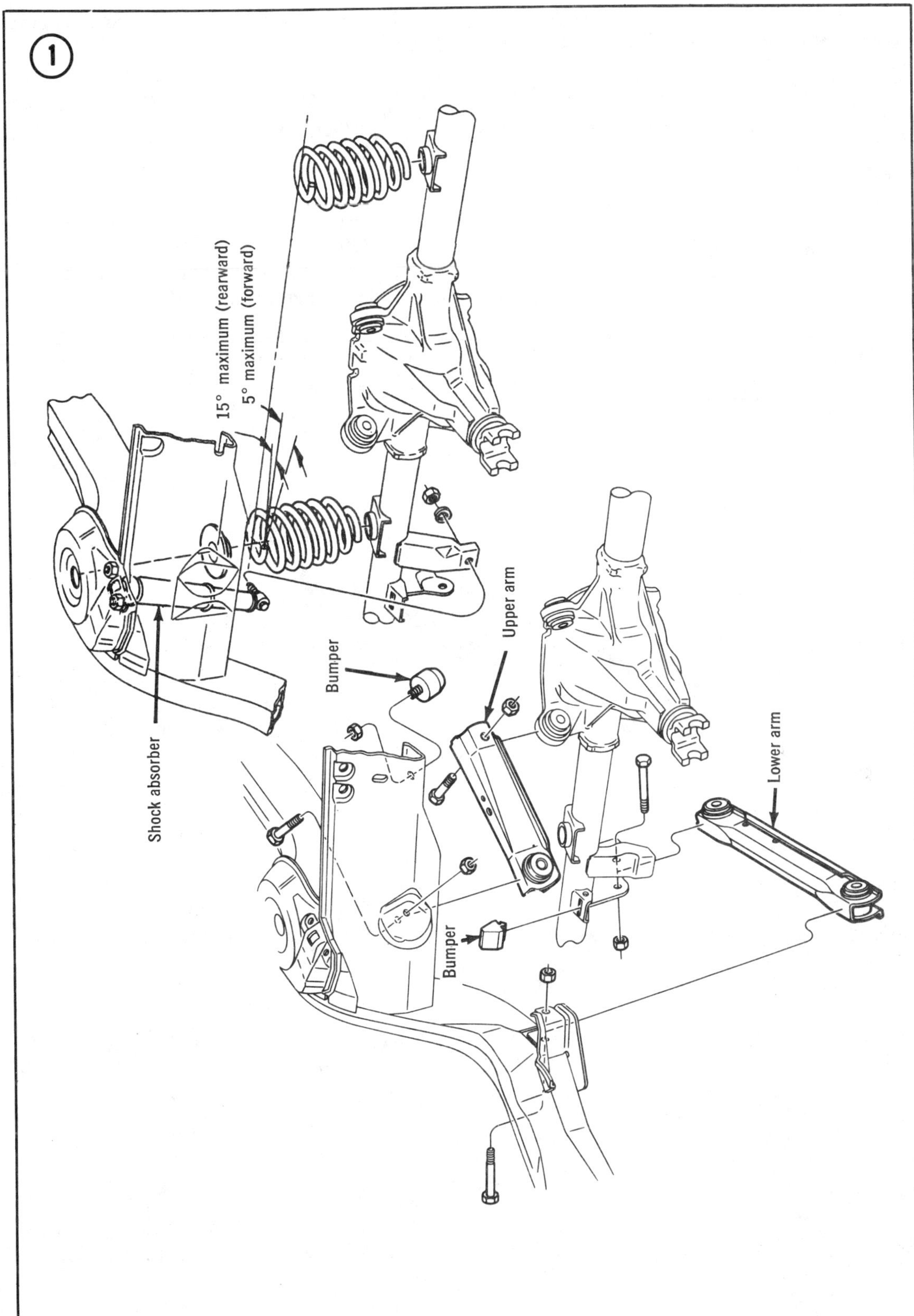
15° maximum (rearward)
5° maximum (forward)
Shock absorber
Bumper
Upper arm
Lower arm
Bumper

3. Disconnect both upper control arms at the axle. If the car has a stabilizer bar, disconnect it at the side of the control arm. See **Figure 4**.

4. Remove the brake hose support, if necessary, to allow for the drop when the spring is removed. It is not necessary to disconnect the brake hose.

5. Remove the shock absorber lower attachments, then slowly lower the jack under the axle until the spring can be removed. Remove the spring and insulator.

6. Install the insulator on the spring and position the spring and insulator on the axle. See **Figure 5** (typical). Upper end of coil must face as shown (within the limits listed). Left side is shown.

7. Raise the axle and reinstall the lower attachments of the shock absorber. Tighten the nut to 62-65 ft.-lb.

8. Connect the brake hose support and securely tighten the bolt.

9. Connect the stabilizer bar, if so equipped, and tighten the nut to 35 ft.-lb.

10. Connect the upper control arms to the axle and tighten the bolts to 70 ft.-lb.

11. Remove the jack from the axle and lower the rear of the vehicle.

2

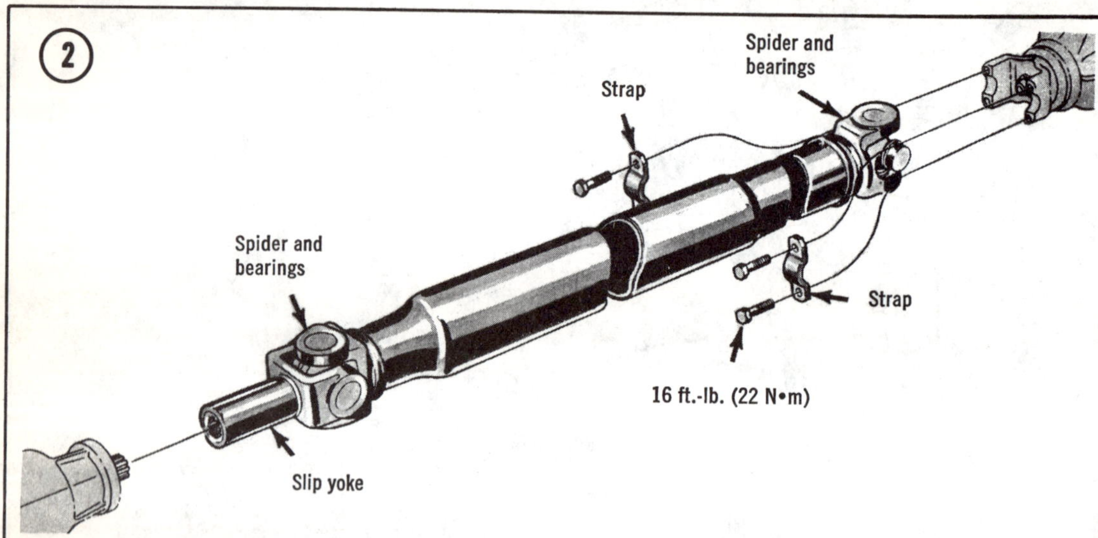

3

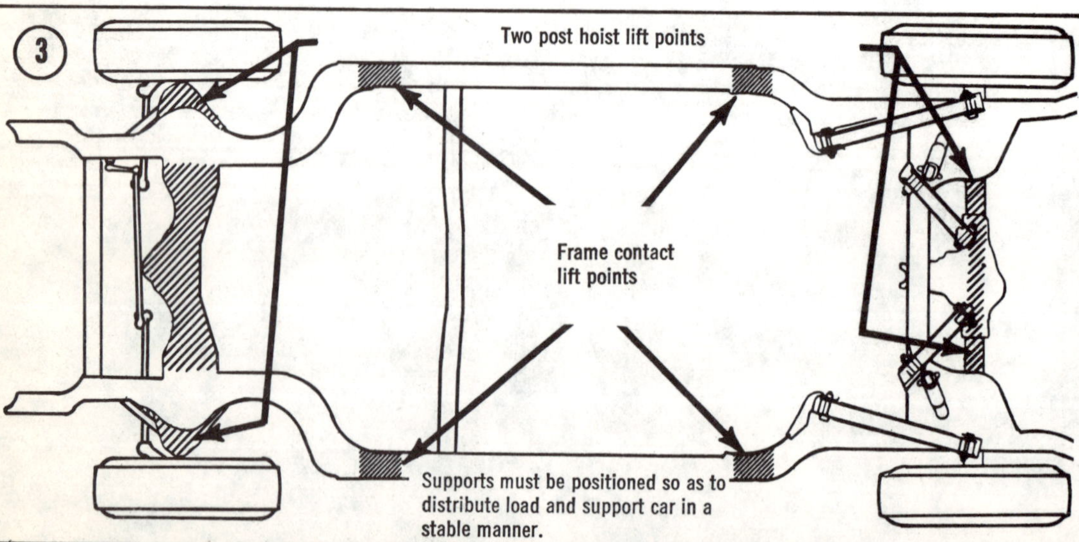

Shock Absorber Removal/Installation

1. Raise the vehicle and place jackstands under the rear lifting points (see **Figure 3**). Make sure the rear wheels clear the floor by several inches.
2. If the vehicle is equipped with superlift shock absorbers, disconnect the air line from the shock absorber. A snap-on connector is used.
3. Support the rear axle with a jack and remove the 2 bolts attaching the shock absorber to the upper mounting bracket. Removal of the rear wheels, while not necessary, will make this task easier.
4. Disconnect the shock absorber at the lower mounting point by holding the hex head on the end of the mounting stud and removing the nut.

CAUTION

Failure to hold the hex head of the shock absorber lower mounting stud during removal or installation of the shock absorber can result in damage to the bond between the stud and the shock absorber bushing.

5. Loosely install the shock absorber to the upper mounting bracket with the attaching bolts (see **Figure 6**) and then position the lower mounting stud in the lower bracket and loosely install the washer and nut.
6. Tighten the upper attaching bolts to 20 ft.-lb. or upper nuts to 12 ft.-lb. (20 ft.-lb. on superlift shocks).

4

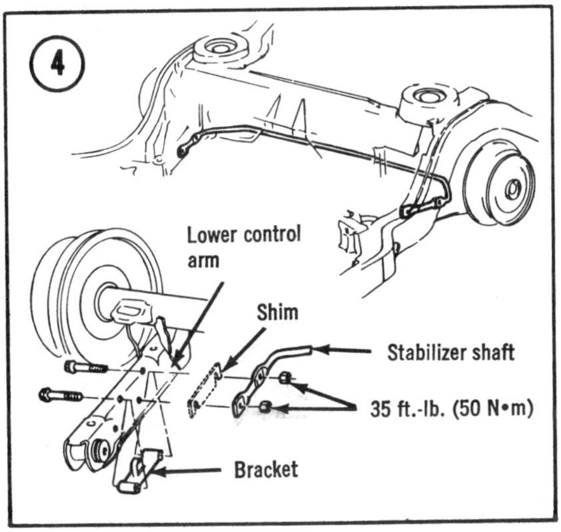

5

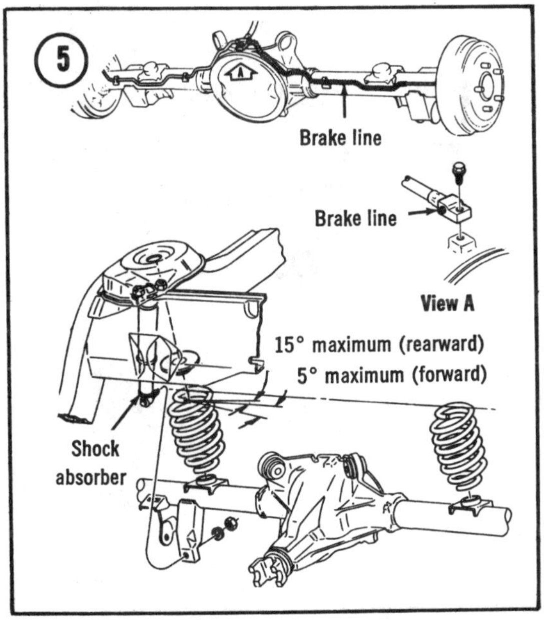

6

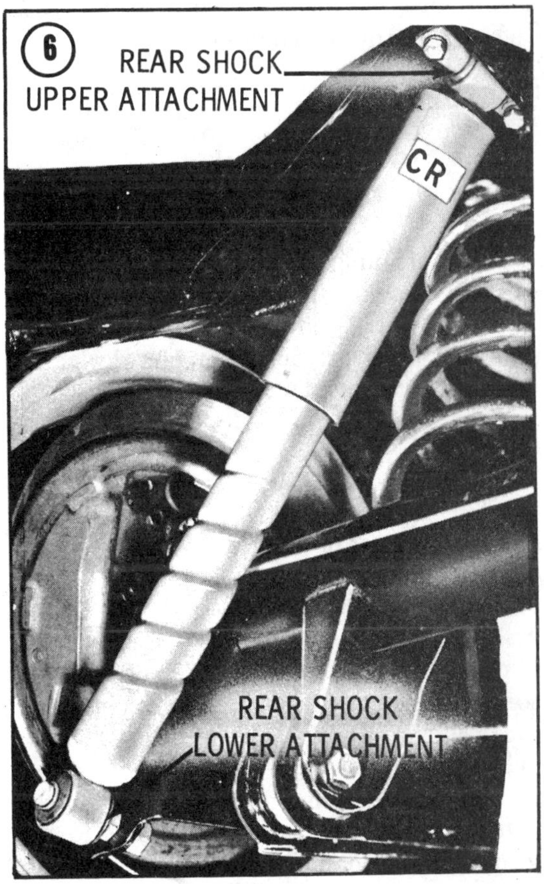

12

7. Hold the hex on the end of the lower mounting stud (see CAUTION above) and tighten the nut to 62-65 ft.-lb.
8. If superlift shocks are being installed, connect the air line to the shock absorber. Then add air to obtain a minimum of 10 psi to avoid damage to the shock absorber.
9. Lower the vehicle.

Upper or Lower Control Arm Removal/Installation

NOTE

*If both the upper **and** lower control arms are to be removed, first remove the coil spring as described earlier in this chapter.*

1. Raise the rear of the vehicle and support it with jackstands under the rear body lifting points. See **Figure 3**.
2. Use a jack to raise the axle until tension is removed from the control arm being removed.
3. Remove the attaching bolts and remove the control arm. See **Figure 7**.
4. Installation is the reverse of these steps. Tighten nuts to 70-73 ft.-lb.

Control Arm Bushing Replacement

Removal and installation of the control arm bushings requires the use of special tools and an arbor press. Remove the control arm(s) and take them to your Chevrolet dealer for bushing replacement when required.

Drive Shaft Removal/Installation

1. Raise the rear of the vehicle and support it on jackstands at the rear body lifting points. See **Figure 3**.
2. Make matching marks to show the relationship of the drive shaft to the companion flange at the rear universal joint. This will permit installation of the drive shaft in its original position.
3. Remove the trunion bearing straps (see **Figure 2**) and tape the bearing cups to the trunion to prevent their loss.

(7)

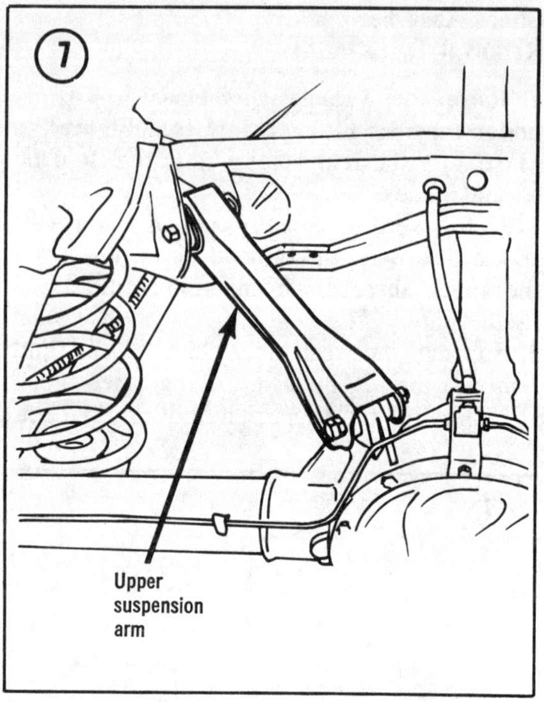

(8)

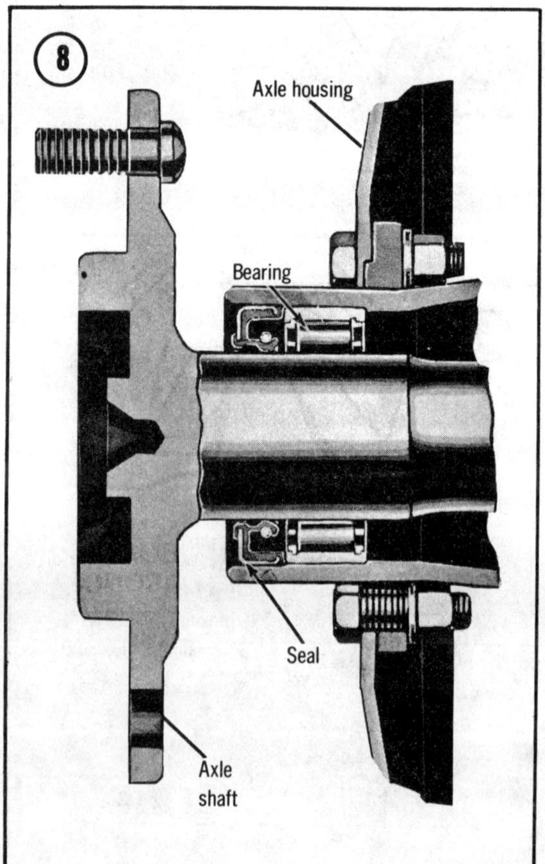

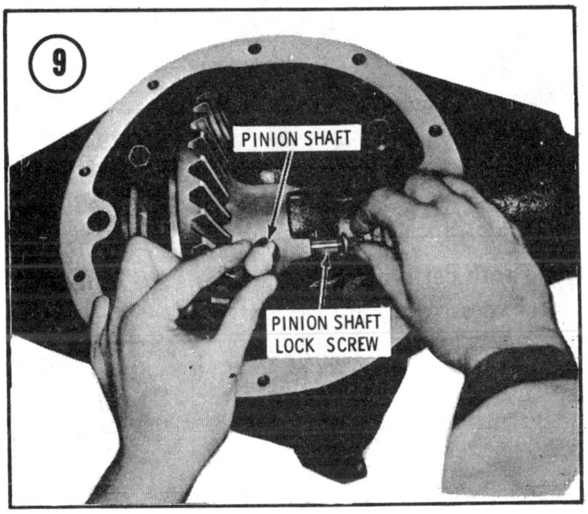

4. Withdraw the drive shaft front yoke carefully from the transmission by pulling the shaft to the rear. Check for oil leakage at the transmission output shaft housing. Make sure the yoke seal is in good condition.
5. Installation is the reverse of these steps. Lubricate the yoke splines with a light coat of transmission oil before inserting the yoke. Make sure the alignment marks made during disassembly are lined up, then (after removing tape from the bearing cups) install the retaining straps and tighten the attaching bolts to 70 ft.-lb.

Universal Joint Repair

Special tools, including an arbor press, are required to shear the plastic bearing cap retainers and disassemble and assemble this type of universal joint. If repairs are required, remove the drive shaft and take it to your Oldsmobile dealer for rebuilding.

REAR AXLE

Most rear axle maintenance and repair involves axle shafts, axle shaft bearings, seals, or the pinion shaft seal. If the correct ring and pinion gear adjustment and bearing preloads were made in production, and if there is adequate lubricant, there should be no need for differential adjustment or repair. Except for replacement of the housing or complete assembly, all rear axle repairs and adjustments can be accomplished without removal from the vehicle.

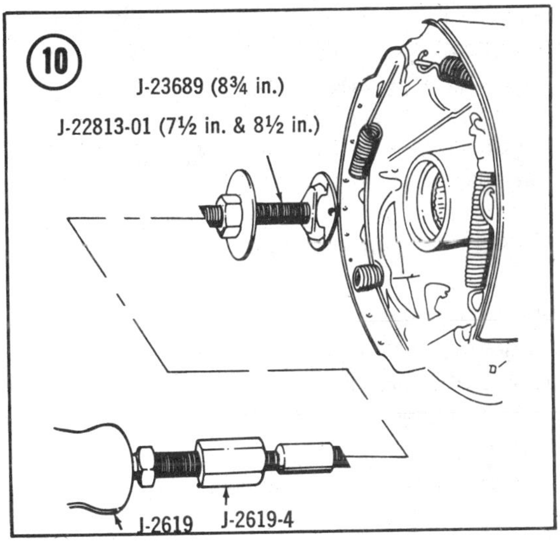

NOTE

Oldsmobile has indicated that some 1977 rear axle shafts are flawed and may break, causing an accident. If you have not already done so, have your Olds dealer inspect the rear axle and replace it, if defective.

Axle Shaft, Seal, or Bearing Replacement

Refer to **Figure 8**.
1. Raise the vehicle by the frame, until wheels clear the ground.

CAUTION

Use safety stands when working under the car. Do not rely solely on hydraulic or mechanical jacks.

2. Remove the wheels and the brake drums (Chapter Ten).
3. Place a container under the differential to catch lubricant, and remove cover.
4. Unscrew the pinion shaft lockscrew and remove the pinion shaft. See **Figure 9**.
5. Without turning differential or axles, tap each axle toward the differential, and remove C-locks. Replace the pinion shaft and lockscrew. Axle shafts can now be removed.
6. With a suitable puller, remove oil seal and bearing (**Figure 10**). If replacing seal only, it can be removed by inserting the lock end of the axle shaft behind the seal and prying it out of its bore. Use caution not to damage the bore with the shaft.

12

7. Lubricate the new bearing and seal lip cavity with wheel bearing grease. Install, referring to **Figures 11 and 12**.
8. Now is a good time to replace any damaged lug bolts in the axle flange; this can be done with the axle installed, however. Studs can be driven out (or pressed out with tool J-6627 or J-5504). A new stud can be drawn into place by starting it with finger pressure, then threading the lug nut on backwards and tightening until bolt head is seated.
9. When reinstalling the axle shafts, use care not to damage the seals.
10. Remove the pinion shaft, insert the C-locks, and pull the shafts outward to seat the C-locks in the recessed end of the side gears before reinstalling the pinion shaft. Install cover on differential, using a new gasket, and tighten bolts to 20 ft.-lb.

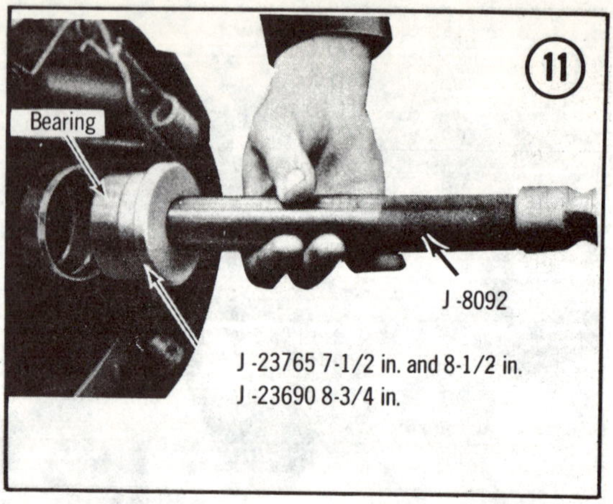

Pinion Seal Replacement

1. With rear universal disconnected and rear brake drums removed, measure pinion bearing preload with an in.-lb. torque wrench on the pinion flange nut. Record the torque required to keep the pinion turning (for reference during reassembly).
2. Mark the pinion shaft-to-flange relationship for correct reassembly.
3. Holding the pinion flange, remove the self-locking nut. See **Figure 13**.
4. With a suitable puller, remove pinion flange (see **Figure 14**). Inspect oil seal surface, drive splines, and bearing cup surfaces. If deflector is damaged, replace it by tapping it from the flange, cleaning up the stake points and staking a new deflector into place at 3 new equally spaced locations. Be careful not to damage the sealing surface when staking.
5. Measure the old seal location with respect to the rim of the seal bore, and top out seal. See **Figure 13**. Install the new seal to that same location after applying grease to the cavity between the seal lips. **Figure 16** shows the position of the pinion oil seal. Note that the seal will be seated at the inner shoulder before its outside flange reaches the rim of the differential carrier housing.
6. Start flange onto pinion shaft, located according to marks previously made, and tap lightly until 2 or 3 threads of the pinion shaft are showing.
7. Using the old nut, draw flange onto pinion shaft *only until end play is removed.* Remove the old nut, apply a non-hardening sealer between the washer and the flange, and start the new nut onto the shaft. Tighten in small increments until previously recorded (Step 1) bearing preload is reached. See **Figure 17**.

Differential Removal

1. With axle shafts removed, remove the differential pinion gear shaft. Identify the differential pinions and thrust washers, also the side gears and thrust washers. Then remove, keeping gears and thrust washers together.
2. Loosen the bearing cap bolts, after marking the caps and housing, and tap each cap lightly to unseat it; do not pry caps. A pry bar may be used to help release the caps by inserting one end into the differential case web—where the spider gears are located—then prying against the axle housing.
3. Identify cups, caps, and shims, and keep them together for each side.

Cleaning and Inspection

Wash all parts and inspect for scored, chipped, nicked, or worn bearing surfaces. Examine differential case for cracks. Inspect thrust washers for wear. Check fit of all splines.

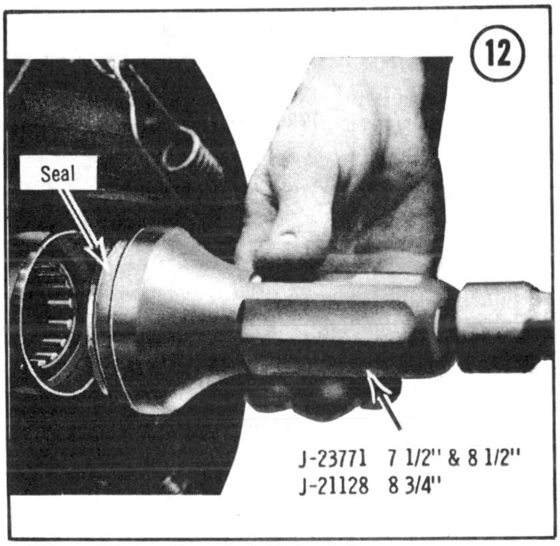
12
Seal
J-23771 7 1/2" & 8 1/2"
J-21128 8 3/4"

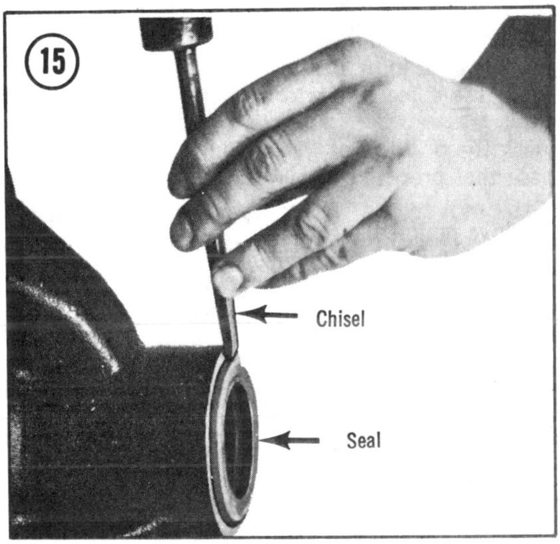
15
Chisel
Seal

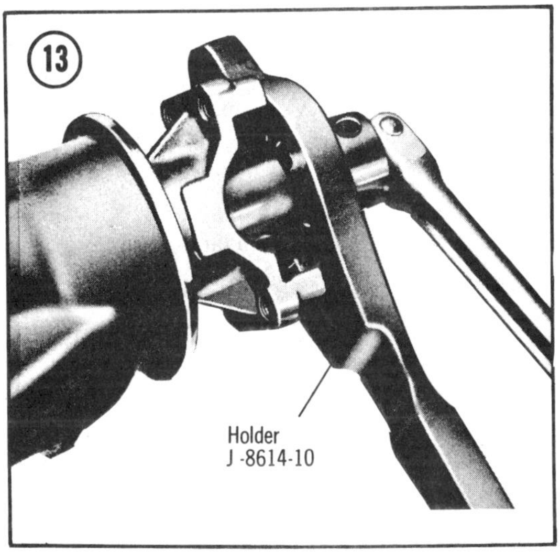
13
Holder
J-8614-10

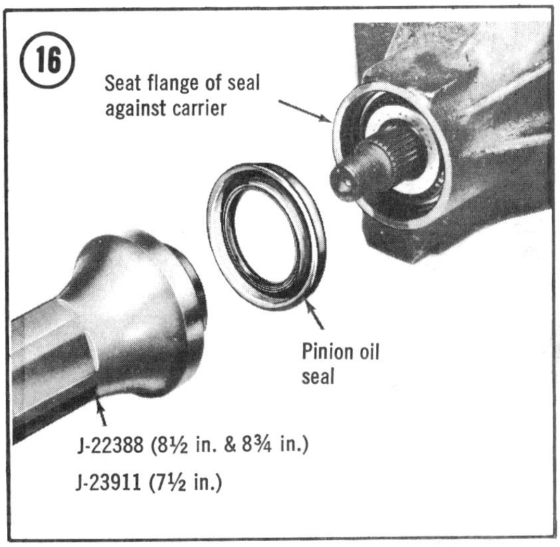
16
Seat flange of seal
against carrier
Pinion oil
seal
J-22388 (8½ in. & 8¾ in.)
J-23911 (7½ in.)

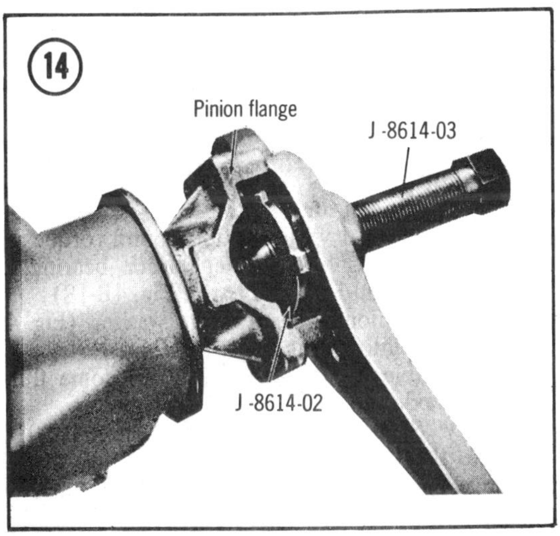
14
Pinion flange
J-8614-03
J-8614-02

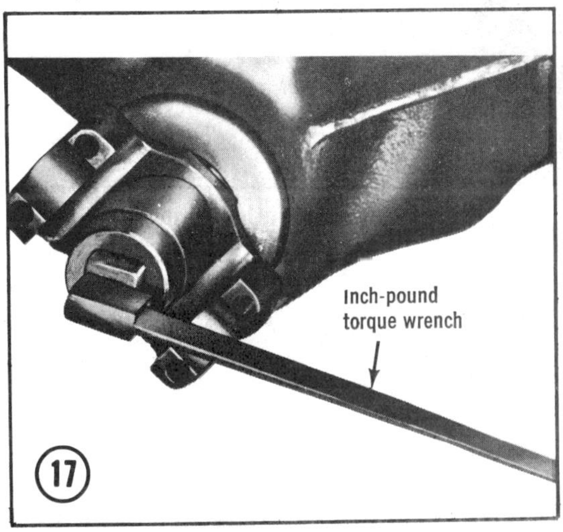
Inch-pound
torque wrench
17

12

Rear Axle Assembly Removal/Installation

1. Raise the rear of the car and support it on jackstands placed under the rear body lifting points. See **Figure 3**.
2. Remove the drive shaft as described elsewhere in this chapter.
3. Disconnect the parking brake cable from the equalizer by removing the adjusting nut. Then remove the cable from the rear connectors and remove it from the car.
4. Disconnect the rear brake hose at the floor pan and cover the hose and pipe connectors to prevent the entrance of dirt.
5. Disconnect the shock absorbers at their lower ends. See *Shock Absorber Removal/Installation*, this chapter, for instructions and CAUTION. Push shocks up out of the way.
6. Remove the springs as described in *Coil Spring Removal/Installation*, this chapter.
7. Remove the upper and lower control arms as described in *Upper* or *Lower Control Arm Removal/Installation*, this chapter.
8. Roll the axle and wheel assembly out from under the car.
9. Installation is the reverse of these steps. Tighten all bolts and nuts to the specifications given in the individual procedures.

SUPPLEMENT

1981 AND LATER SERVICE INFORMATION

This supplement contains service and maintenance information for the 1981-1982 rear-wheel drive Oldsmobile Cutlass. The information supplements the procedures in the main body (Chapters One through Twelve) of the book, referred to in this supplement as the "basic book."

The chapter headings and titles in this supplement correspond to those in the basic book. If a chapter is not included in the supplement, there are no changes affecting 1981-1982 rear-wheel drive models.

If your vehicle is covered by this supplement, carefully read the supplement and then read the appropriate chapters in the basic book before beginning any work.

CHAPTER THREE

LUBRICATION, MAINTENANCE, AND TUNE-UP

PERIODIC MAINTENANCE

Tables 1-4 summarize all changes in lubricants, capacity and maintenance periods, while **Table 5** provides spark plug data for 1981-1982 models. **Table 6** provides tune-up data for 1981 models. For 1982 tune-up specifications, see the Vehicle Emission Control Information (VECI) label under the hood.

Ignition timing for 1982 models (unless specified otherwise on VECI label) is as follows:

a. 321 cid V6—15° BTDC
b. All other gasoline engines—20° BTDC

The following periodic maintenance services reflect changes or additions to those specified in Chapter Three of the basic book:

1. *Tires*—Radial tires on diesel cars should be rotated at first 5,000 miles and every 15,000 miles thereafter. Bias-belted tires should be rotated every 5,000 miles.

2. *Positive Crankcase Ventilation System (PCV)*—Check system operation and replace PCV valve every 30,000 miles. Replace air cleaner PCV filter (if so equipped) every 30,000 miles.

3. *Exhaust Pressure Regulator Valve (Diesel)*—Valve is closed at idle and opens above idle. Check for proper operation and correct any binding. Check for damaged hoses and replace as required.

CARBURETOR ADJUSTMENTS

Idle Speed Adjustment

Many 1981 carburetors used on Cutlass models have an Idle Speed Control (ISC) device (**Figure 1**) mounted on the float bowl. This is part of the Computer Command Control (CCC) system and is controlled by the electronic control module. The normal curb idle speed is programmed into the module and no adjustments are possible. The module monitors idle speed, increasing or decreasing it as required by engine operating condition.

Some V8 engines may be equipped with an Idle Load Compensator (ILC) mounted on the carburetor float bowl instead of an Idle Speed Control device. This factory-adjusted compensator controls idle speed according to engine manifold vacuum and you should not attempt to change the adjustment.

If a carburetor equipped with an ISC or ILC device is not working properly, take the car to

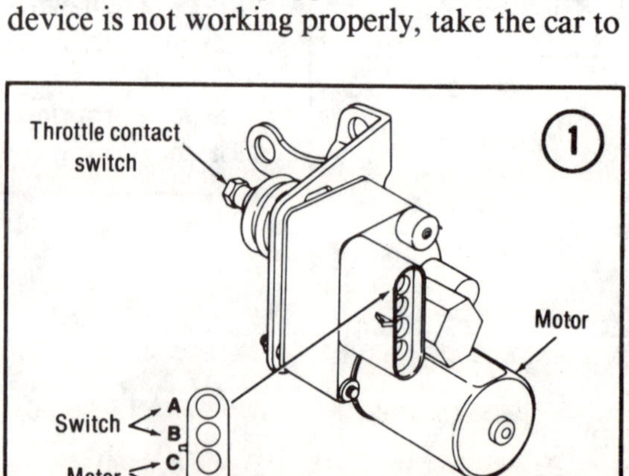

an Olds dealer where a factory-trained technician can diagnose the problem and make the proper adjustments according to factory procedures.

Some 1981 carburetors are not equipped with an ISC or ILC device. Idle speed adjustment on these carburetors should be made according to the procedure and specifications provided on the Vehicle Emission Control Information (VECI) sticker in the engine compartment of the car.

All 1982 carburetors are equipped with an ISC or ILC device. Do not attempt to change the factory-set adjustment.

DIESEL ENGINE FUEL FILTER REPLACEMENT

A Housing Pressure Cold Advance feature is used on 1981-1982 diesel engines. This is a temperature-operated circuit designed to advance injection timing 3° during cold operation. When installing a fuel filter, this device should be activated manually to reduce cranking time. To do this, disconnect the wire connector at the engine temperature switch as shown in **Figure 2** and use a jumper wire to bridge the connection. Once the engine is running, remove the jumper wire and reconnect the wire connector to the temperature switch.

2

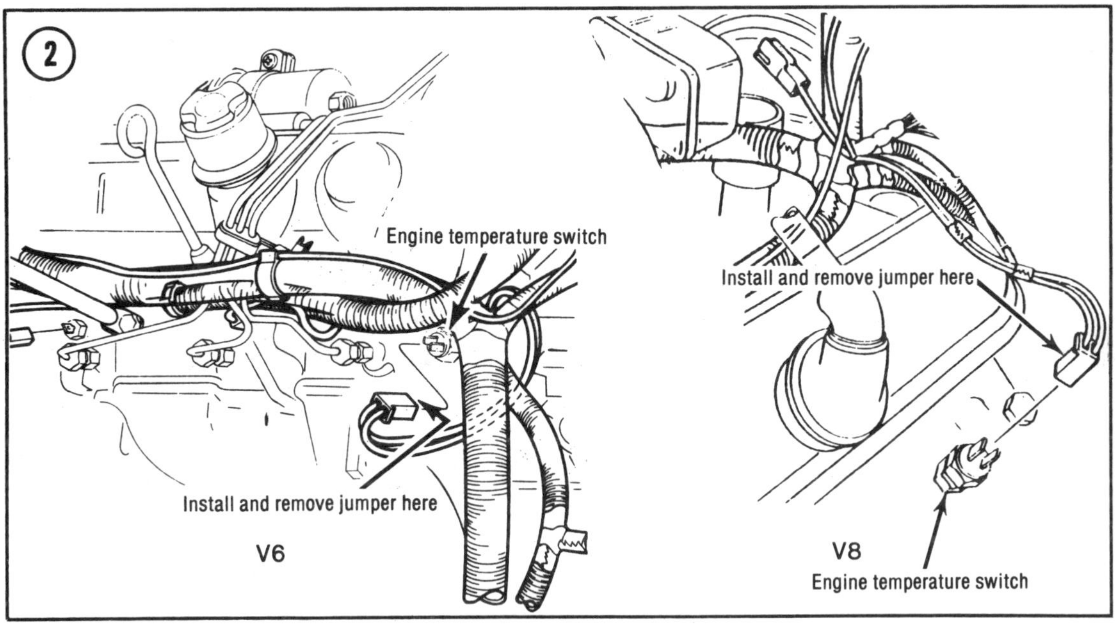

Table 1 RECOMMENDED LUBRICANTS AND FLUIDS

Use	Recommendation
Engine oil	Product bearing letter designation SF, SF/CC or SF/CD (gasoline engine); SF/CC, SF/CD or SE/CD (diesel)
Differential—Positraction	Add 4 oz. GM lubricant additive part No. 1052358, then fill with GM lubricant part No. 1052271, 1052272 or eqivalent
Steering gear—manual	GM lubricant part No. 1052182 or equivalent
Front wheel bearings	GM lubricant part No. 1051344 or equivalent
Windshield washer	GM washer solvent part No. 1051515 or equivalent

13

Table 2 MAINTENANCE SCHEDULE (DIESEL ENGINE)

Frequency	Item No.	Service
Section A—Lubrication and General Maintenance		
Every 5,000 miles	1	Chassis lubrication
	2	Fluid levels check
	3	Engine oil change
	4	Oil filter change
See explanation in text	5	Tire rotation
	6	Rear axle lubrication check
Every 12 months or 15,000 miles	7	Air conditioner charge and hose condition check
	8	Cooling system service
	47	Crankcase ventilation system service
Every 30,000 miles	9	Front wheel bearing lubrication
Every 100,000 miles	12	Automatic transmission fluid change
Section B—Safety Maintenance		
Every 12 months or 9,000 miles	13	Owner safety checks
	14	Tires and wheels condition check
	15	Exhaust system condition check
	17	Suspension and steering systems check
	18	Brake and power steering hoses check
	19	Disc brake condition check
Every 12 months or 15,000 miles	16	Drive belt condition and adjustment check
	20	Drum and parking brake condition check
	21	Throttle linkage operation and condition check
	24	Bumper check
Section C—Emission Control Maintenance		
First 5,000 miles, then every 15,000 miles	50	Exhaust pressure regulator valve
First 5,000 miles, then every 30,000 miles	48	Engine idle speed adjustment
Every 30,000 miles	42	Air cleaner filter element replacement
	49	Fuel filter replacement

Table 3 OIL FILTER USAGE

Engine	Filter Brand AC	Lee	Fram	Purolator
231 cid V6	PF-47	LF-111	PH-3387A	PER-111
All gasoline V8	PF-45	LF-45	PH-3429	PER-123
All diesel (except 260 cid)	PF-45	LF-45	PH-3429	PER-123
260 cid diesel	PF-51	——	PH-3535	——

Table 4 AUTOMATIC TRANSMISSION FLUID CAPACITY

Transmission	Capacity
Turbo Hydramatic 350,350C	6 1/2 U.S. pints
Turbo Hydramatic 250,250C	8 U.S. pints
Turbo Hydramatic 200, 200C	7 U.S. pints

Table 5 SPARK PLUG SPECIFICATIONS (1981-ON)

Engine	Spark Plug Type (AC)	Gap Setting (in.)
231 cid V6	R45TS8	0.080
260 cid 8	R46SX	0.080
307 cid 8	R46SX	0.080

Table 6 TUNE-UP SPECIFICATIONS (1981)

	Timing (BTDC @ rpm)	
Engine Description and Usage[2]	Manual Transmission	Automatic Transmission
231 cid 6 (Fed)	15 @ 800	15 @ Idle
231 cid 6 (Cal)	—	15 @ Idle
260 cid 8 (Fed)	—	See VECI sticker
260 cid 8 (Cal)	—	See VECI sticker
307 cid 8 (Fed)	—	See VECI sticker
307 cid 8 (Cal)	—	15 @ 1100
350 cid 8 (Diesel)	—	—

	Slow Idle (rpm)[1]		Fast Idle (rpm)	
Engine Description and Usage[2]	Manual Transmission (in Neutral)	Automatic Transmission (in Drive)	Manual Transmission (in Neutral)	Automatic Transmission
231 cid 6 (Fed)	800	See Note 3	2,200	1,800(P)
231 cid 6 (Cal)	—	See Note 3	—	1,800(P)
260 cid 8 (Fed)	—	See Note 3	—	700(D)
260 cid 8 (Cal)	—	See Note 3	—	700(D)
307 cid 8 (Fed)	—	See Note 3	—	650(D)
307 cid 8 (Cal)	—	See Note 3	—	650(D)
350 cid 8 (Diesel)	—	750/600	—	—

1. If two speeds are listed, higher speed is with the solenoid energized and lower speed is with the solenoid de-energized.
2. All=all models; Fed=cars first sold in any state except California; Cal=Cars first sold in California.
3. Controlled by Idle Speed Control Motor, no adjustment possible.

(D)=Drive

(P)=Park

CHAPTER FOUR

ENGINE

The 1981 Cutlass is available with the following engines: 231 cid V6, 260 cid V8, 307 cid V8 and 350 cid V8 diesel. The 1982 engine line-up is the same, with the addition of a 260 cid V6 diesel.

Engine specifications for the 307 V8 are identical to those for the 260 cid V8 listed in Chapter Four of the basic book, except as noted in **Table 7** of this supplement. Tightening torques for the 307 cid V8 are the same as those found in Table 2, Chapter Four of the basic book under Olds V8 heading.

Engine specification changes for the 231 cid V6 for 1982 are given in **Table 8**. Specifications for the 260 cid V6 diesel are provided in **Table 9** and tightening torques in **Table 10**. For cylinder head bolt torque pattern on the 260 cid diesel engine, see **Figure 3**. All other 1981-1982 engine specifications are the same as 1980 models.

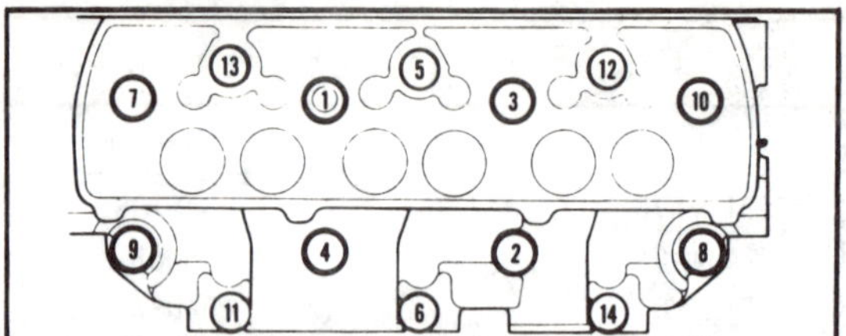

3 Torque all bolts (except 5, 6, 11, 12, 13 & 14) to 193 N•m (142 ft.-lb.). Numbers 5, 6, 11, 12, 13 & 14: torque to 80 N•m (59 ft.-lb.).

Table 7 ENGINE SPECIFICATIONS
OLDS 307 CID V8 (5.0 LITER)

General	
Bore	3.800 in.
Cylinder bore	
Diameter	3.800 in.
Piston rings	
Ring end gap	
Top and 2nd	0.009-0.019 in.
Main bearing journals	
Diameter, No. 1	2.4998-2.4993 in.
Diameter, No. 2, 3, 4 & 5	2.4995-2.4990 in.

Table 8 ENGINE SPECIFICATIONS
231 CID V6 (3.8 LITER)

Ring end gap (top and second)	0.013-0.023 in.
Piston pin	
Clearance in rod	0.00075-0.00125 in.
Crankshaft end play	0.003-0.009 in.
Valve springs (outer)	
Load at length (lb. @ in.)	
Closed	88-98 @ 1.727
Open	210-230 @ 1.340

Table 9 ENGINE SPECIFICATIONS
260 CID V6 DIESEL (4.3 LITER)

General	
Bore	4.057 in.
Stroke	3.385 in.
Firing order	1-6-5-4-3-2 (No. 1, left front)
Cylinder bore	
Diameter	4.057 in.
Maximum taper	0.001 in.
Piston	
Clearance in bore	0.603-0.004 in.
Piston rings	
Number per cylinder	3
Ring end gap	
Top	0.015-0.025 in.
2nd	0.015-0.025 in.
Oil control	0.015-0.055 in.
Ring side clearance	
Top	0.005-0.007 in.
2nd	0.003-0.005 in.
Oil control	0.001-0.005 in.
Piston pins	
Diameter	1.0949-1.0952 in.
Clearance	
In piston	0.0003-0.0005 in.
In rod	0.0003-0.0013 in.
Crankshaft	
End play	0.0035-0.0135 in.
Main bearing journals	
Diameter	2.9993-3.0003 in.
Out-of-round	0.0015 in.
Main bearing clearance	
No. 1, 2, & 3	0.0005-0.0021 in.
No. 4	0.0020-0.0034 in.
Connecting rods	
Side clearance	0.0082-0.0214 in.
Bearing clearance	0.0003-0.0025 in.
Camshaft	
Journal diameter*	
No. 2	2.185-2.224 in.
No. 3	2.1653-2.20472 in.
No. 4	2.145-2.1850 in.
Journal clearance in bearing (all)	0.0020-0.0059 in.
End clearance	0.0008-0.0228 in.
Valve system	
Lifter type	Hydraulic
Intake valve	
Head diameter	1.875 in.
Stem diameter	0.3425-0.3432 in.
Face angle	44°
Seat angle	45°
Seat width	0.037-0.075 in.
Clearance in guide	0.0010-0.0027 in.
Exhaust valve	
Head diameter	1.625 in.
Stem diameter	0.3420-0.3427 in.
Face angle	30°
Seat angle	31°
Seat width	0.046-0.084 in.
Clearance in guide	0.0015-0.0032 in.
Valve springs	
Free length	2.09 in.
Load @ length (lb. @ in.)	
Closed	85-95 @ 1.670
Open	203-217 @ 1.220

* No. 1 cam bearing is mounted in the camshaft retainer. No bearing journal diameter is provided.

13

Table 10 TIGHTENING TORQUES
260 CID V6 DIESEL (4.3 LITER)

Fastener	Ft.-lb.
Air crossover to intake manifold	28
Camshaft bearing retainer bolts	48
Camshaft sprocket to camshaft bolt	64
Cylinder head to block	
Except 5, 6, 11, 12, 13 & 14	142
5, 6, 11, 12, 13 & 14	59
Connecting rod cap bolts	42
Crankshaft bearing cap bolts	107
Crankshaft balancer bolt	160-350
Crankshaft pulley to balancer bolts	30
Crossover and exhaust pipe to exhaust manifold bolts	23
Exhaust manifold to cylinder head	29
Flywheel to crankshaft bolts	48
Flywheel to torque converter bolts	35
Front cover bolts	41
Fuel pump mounting nut	18
Injection pump adapter to block	22
Injection pump intermediate adapter to adapter	35
Intake manifold to cylinder head	41
Engine mount	
To block	37
To frame	38
To bracket	55
Oil cooler lines	
To oil filter base	29
To radiator	25
Oil filter adapter	29
Oil pan bolts	10
Oil pan drain plug	26
Oil pump cover to pump body screws	7
Oil pump drive assembly clamp bolt	26
Oil pump to bearing cap bolts	18
Rocker arm pivot to cylinder head	28
Valve cover to cylinder head screws	*
Water outlet to intake manifold	18
Water pump/front cover bolts **	21
Water pump to front cover screws **	13
Water pump pulley to water pump	20

* Fully driven, seated and not stripped.
** Dip in sealer.

CHAPTER FIVE

FUEL AND EXHAUST SYSTEMS

All 1981-on engines use either Rochester E2MC/E2ME or E4MC/E4ME carburetors. These carburetors are an integral part of the Computer Command Control (CCC) system. Adjustment and repair by the home mechanic is not recommended. Idle mixture needles are covered by a metal plug to prevent such adjustment and idle speed on most is controlled by the CCC electronic control module or by a factory-preset Idle Load Compensator. Adjustment and overhaul is a job for your dealer or a carburetor shop equipped with the precision equipment and skills necessary for working with this complex device. Removal and installation procedures remain the same as those given in Chapter Five of the basic book.

CHAPTER SIX

EMISSION CONTROL SYSTEM

DIESEL ENGINE SYSTEMS

Diesel EGR System

The modified orifice EGR system used on 1980 diesel engines remains on 1981 diesels sold in California, but a modulated EGR system replaces it on all other 1981 diesel engines. Instead of providing a constant flow of exhaust gas into the intake manifold, the modulated system uses a series of vacuum controls (**Figure 4**) to regulate the EGR valve opening.

For 1982, 2 simplified EGR systems are used on diesel engines. The system used with all V8 engines is shown in **Figure 5**, while the V6 system is shown in **Figure 6**.

A system malfunction will cause the same symptoms as described in Chapter Six of the basic book. Locating the exact cause is a more complicated procedure due to the increased number of vacuum controls and should be left to your Oldsmobile dealer.

Diesel Engine Crankcase Ventilation System

The ventilation regulator valve system (**Figure 7**) used on previous diesel engines is carried over to 1981 diesels sold in California. A new depression regulator valve system is used on all other 1981 and all 1982 Cutlass diesel engines. **Figure 8** shows the V8 application; the V6 is similar. The new valve limits crankcase vacuum as crankcase gases are

4

MODULATED EGR SYSTEM (1981 DIESEL EXCEPT CALIFORNIA)

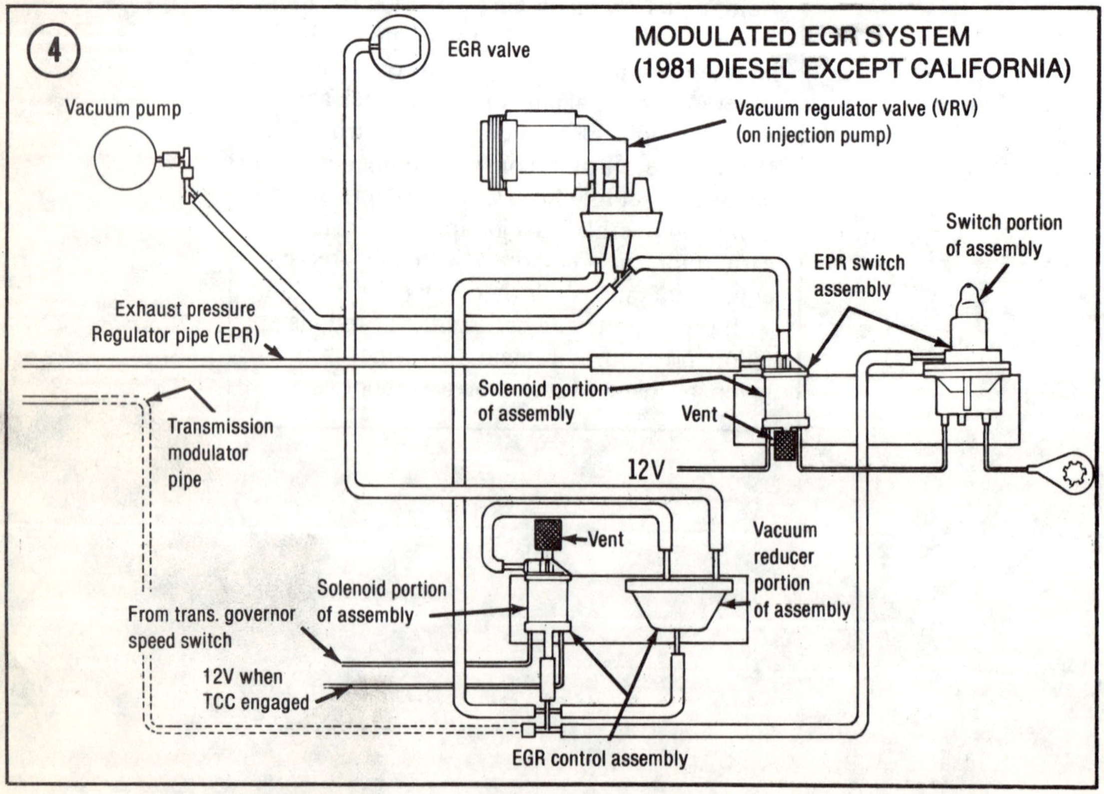

5

EGR SYSTEM (1982 V8 DIESEL)

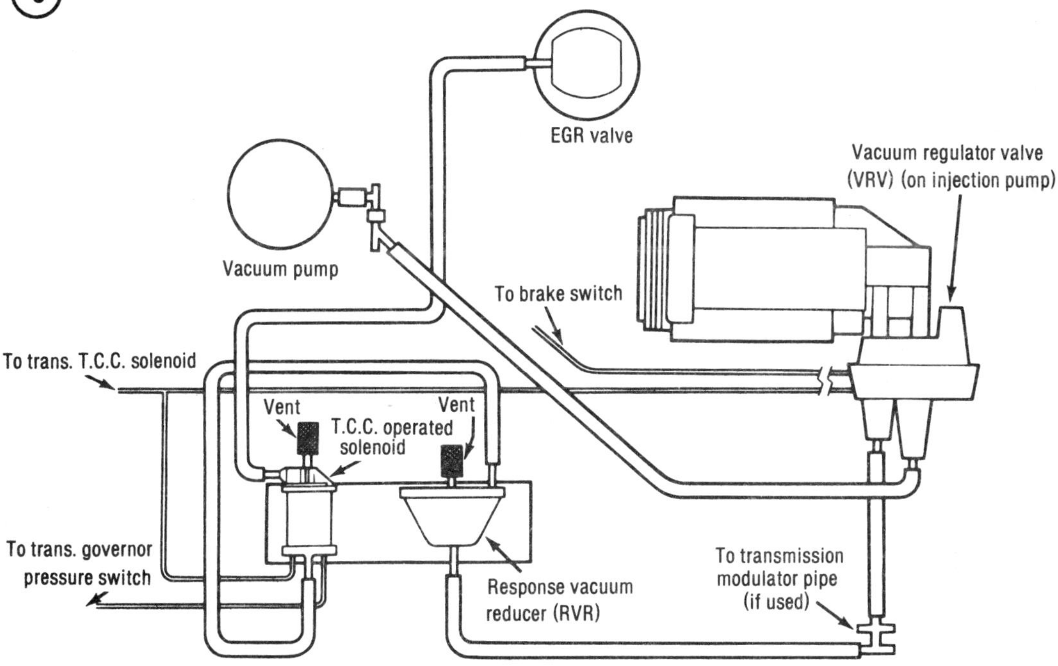

6

EGR SYSTEM (1982 V6 DIESEL)

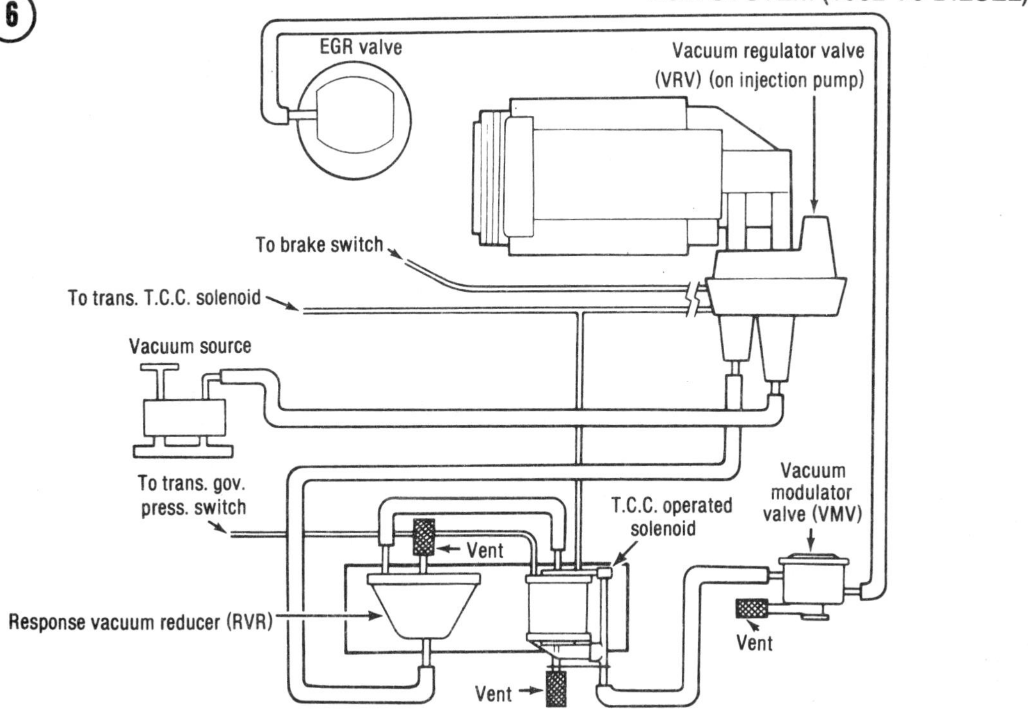

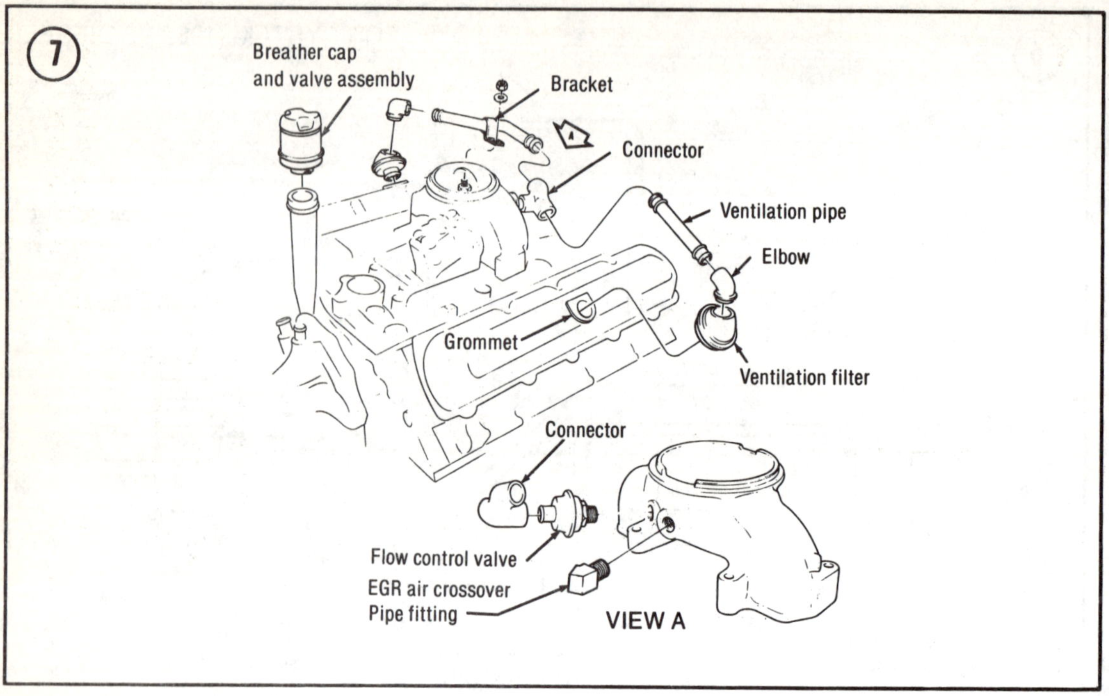

drawn into the air crossover. Service procedures remain the same for both systems, but Oldsmobile now recommends replacement of the breather cap and ventilation regulator valve every 30,000 miles instead of cleaning.

COMPUTER COMMAND CONTROL (CCC) SYSTEM

This electronic fuel management system is a far more sophisticated version of the 1980 C-4 system described in Chapter Six of the basic book. It monitors up to 15 different engine/vehicle operating conditions to control as many as 9 related engine systems. A schematic diagram of the system is shown in **Figure 9**. It should be noted that the CCC system design and components used differs according to engine and application.

The Electronic Control Module (ECM) contains a removable calibration unit called a PROM. This unit is programmed specifically for a given engine/vehicle combination. If defective, the PROM must be replaced by one carrying the same part number. All service to the CCC system and its components should be referred to your dealer.

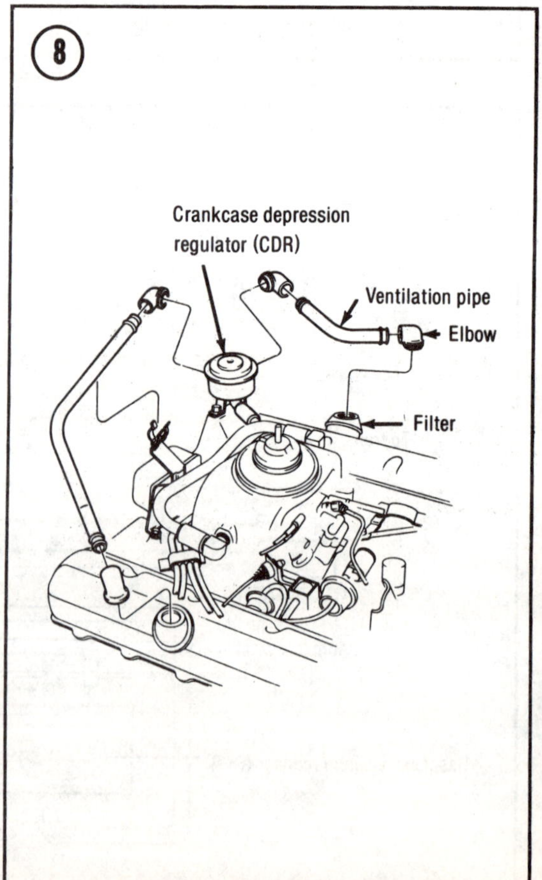

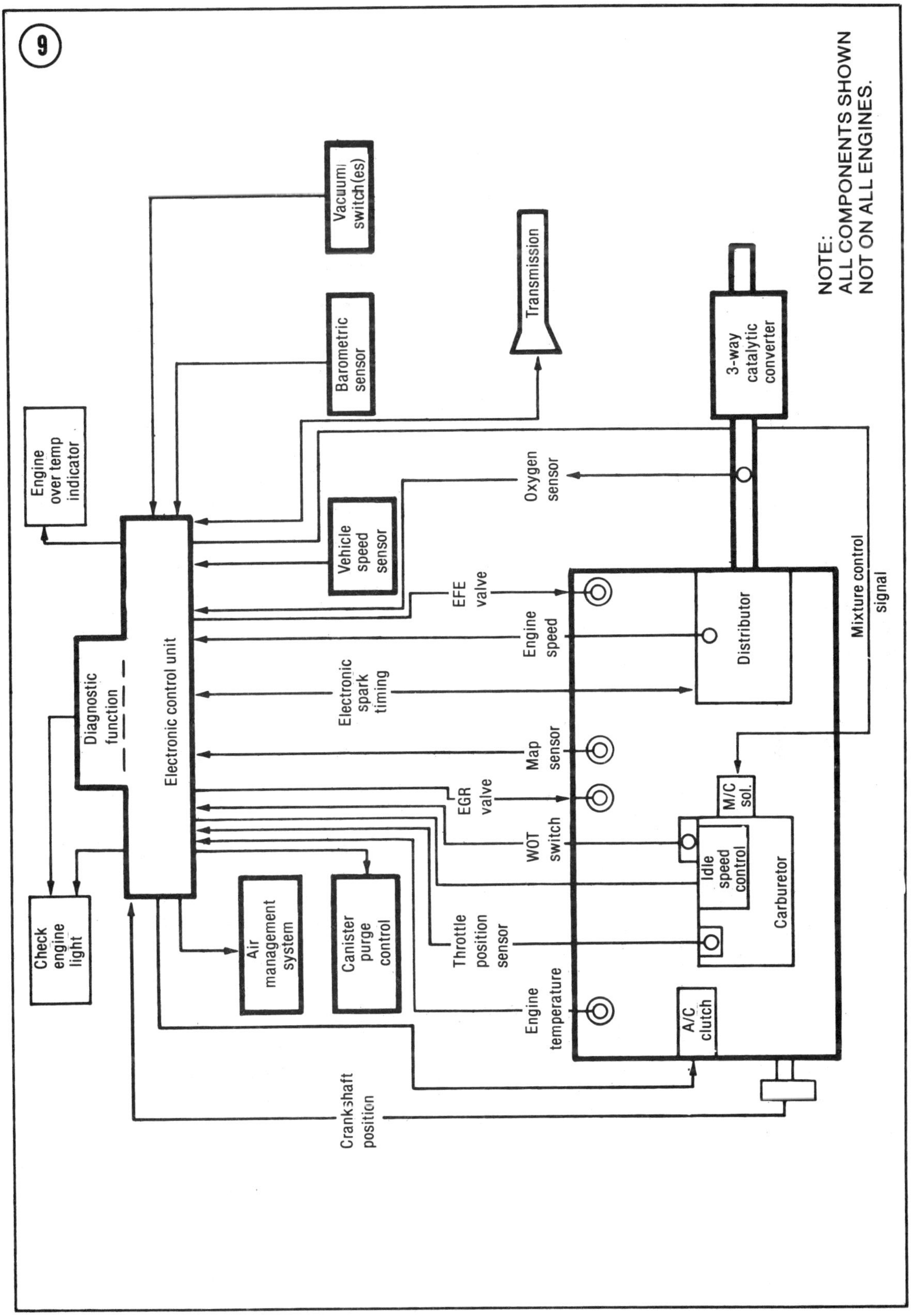
9
Vacuum switch(es)
Barometric sensor
Transmission
Engine over temp indicator
Vehicle speed sensor
Oxygen sensor
3-way catalytic converter
NOTE: ALL COMPONENTS SHOWN NOT ON ALL ENGINES.
Diagnostic function
Electronic control unit
EFE valve
Engine speed
Distributor
Electronic spark timing
Mixture control signal
Map sensor
M/C sol.
EGR valve
WOT switch
Idle speed control
Carburetor
Check engine light
Air management system
Canister purge control
Throttle position sensor
Engine temperature
A/C clutch
Crankshaft position

13

CHAPTER TEN

BRAKES

Maintenance and repair procedures for 1981-1982 models are the same as for earlier models. Some specifications differ. See **Table 11**.

Table 11 TIGHTENING TORQUES

Master cylinder to power unit	20 ft.-lb.
Caliper mounting bolt	28 ft.-lb.
Flex hose to caliper	32 ft.-lb.

CHAPTER ELEVEN

FRONT SUSPENSION AND STEERING

Maintenance and repair procedures for 1981-1982 models are the same as for earlier models. Some specifications differ. See **Table 12**.

Table 12 TIGHTENING TORQUES

Shock absorber upper nut	8 ft.-lb.
Shock absorber lower bolts	20 ft.-lb.
Lower ball-joint stud nut	70 ft.-lb.
Upper ball-joint stud nut	50 ft.-lb.

CHAPTER TWELVE

REAR SUSPENSION AND DRIVE SHAFT

Maintenance and repair procedures for 1981-1982 models are the same as for earlier models. Some specifications differ. See **Table 13**.

Table 13 TIGHTENING TORQUES

Upper arm to rear axle bolts	80 ft.-lb.
Lower arm to rear axle nuts	70 ft.-lb.
Stabilizer shaft to lower control arm bolt	35 ft.-lb.

INDEX

A

B

C

Front Wheel Brg. Adj.
Page 260-Chp. 11

WIRING DIAGRAMS

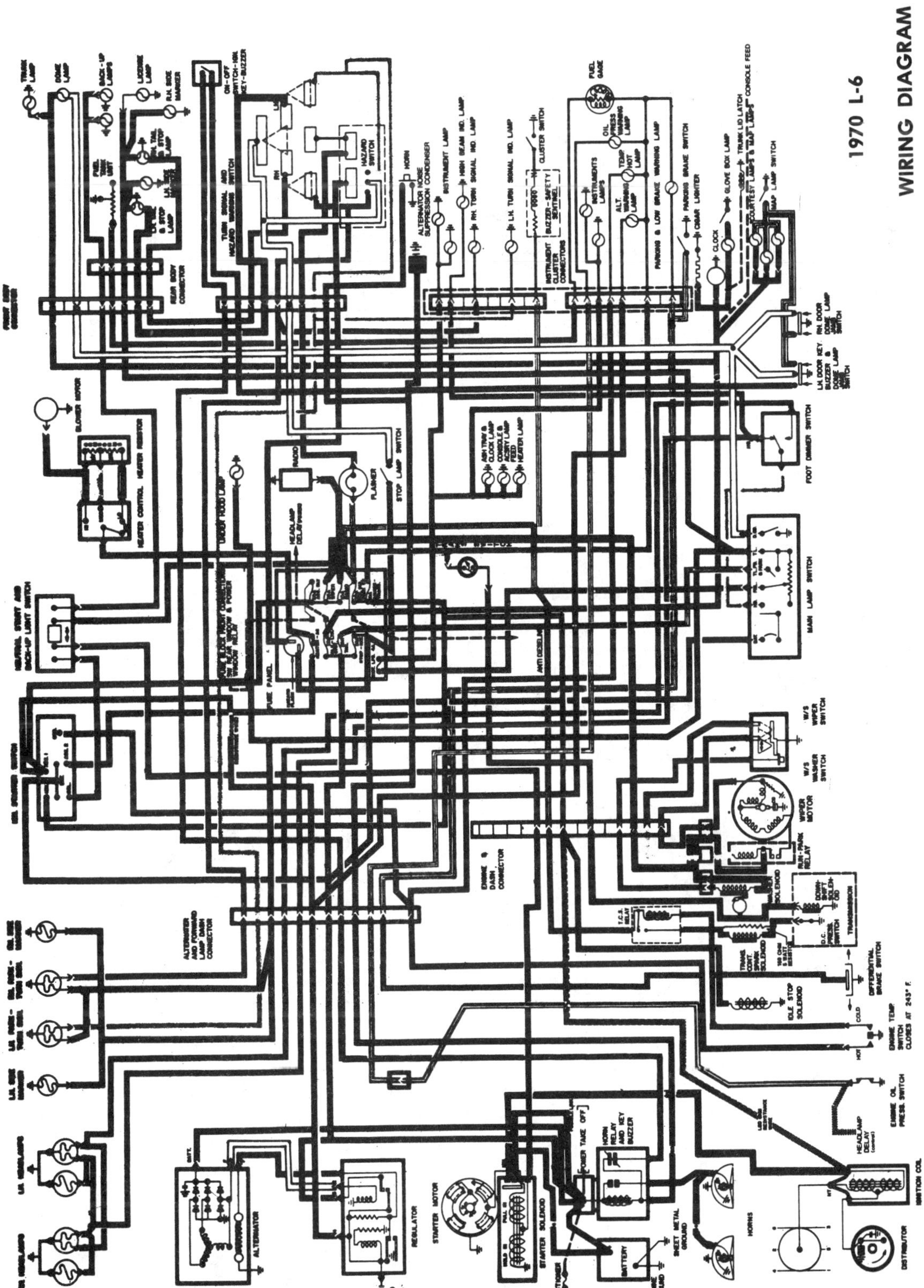
1970 L-6
WIRING DIAGRAM

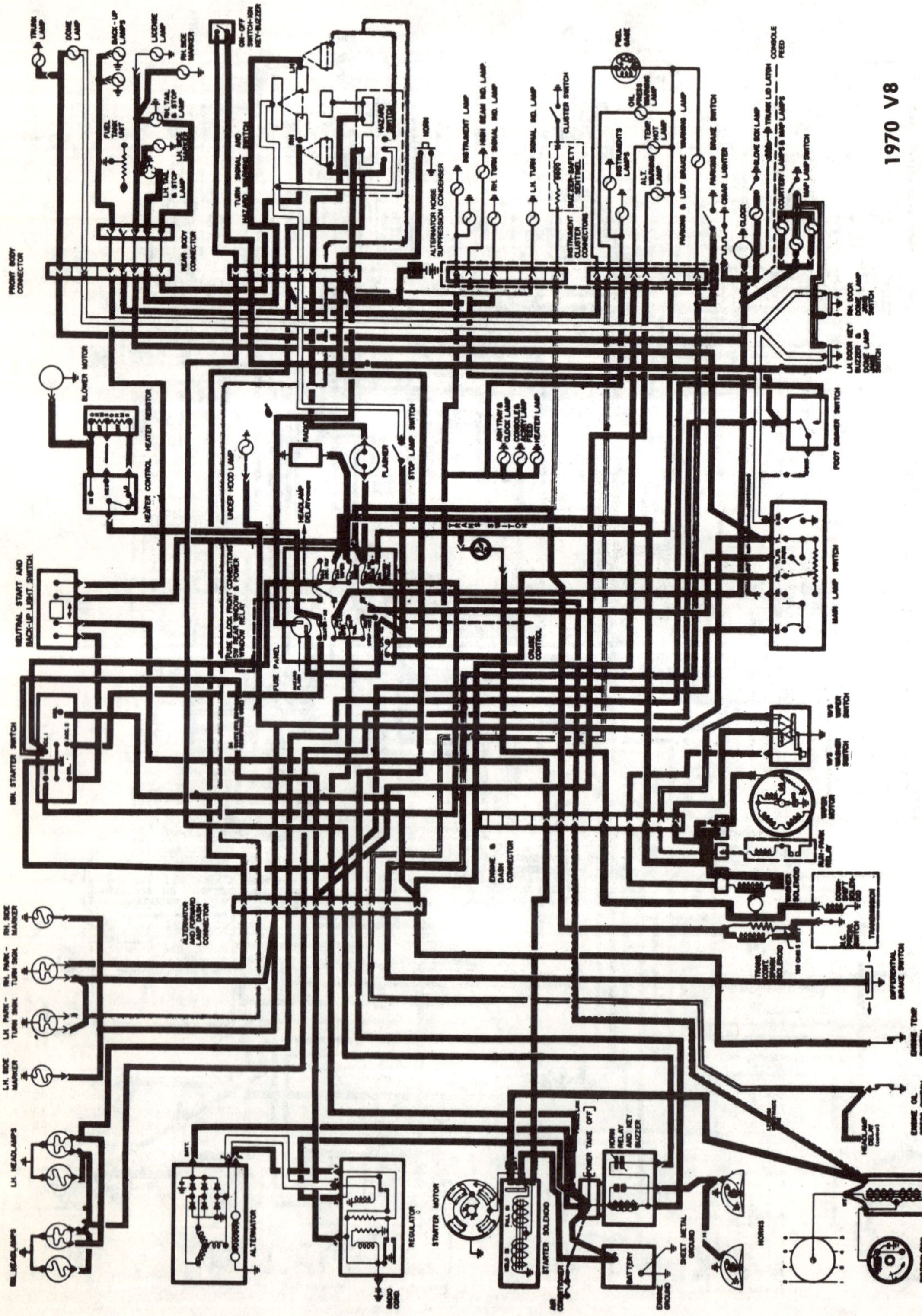

1970 V8
WIRING DIAGRAM

1971 L-6 WIRING DIAGRAM

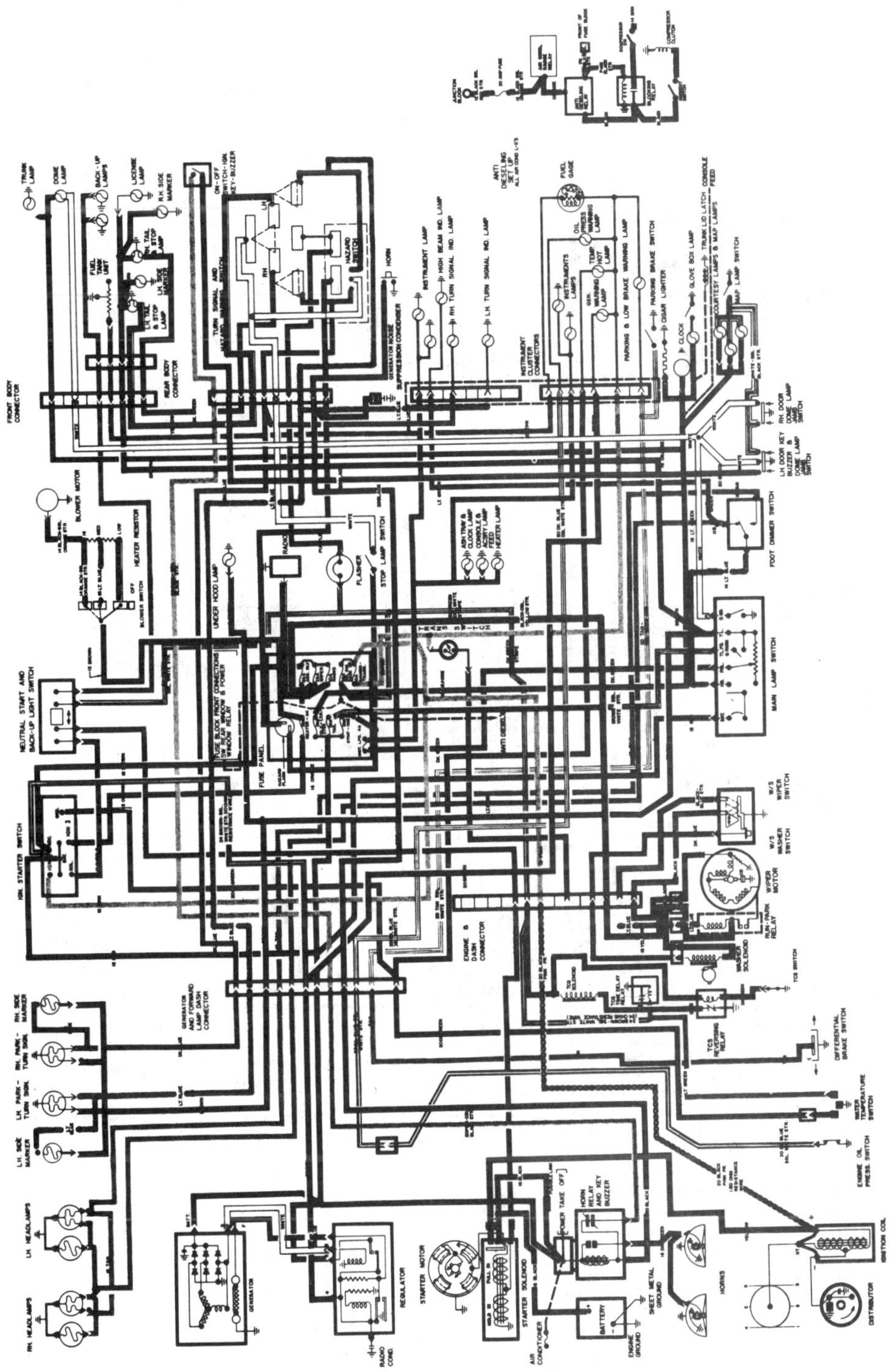

1971 V8 WIRING DIAGRAM

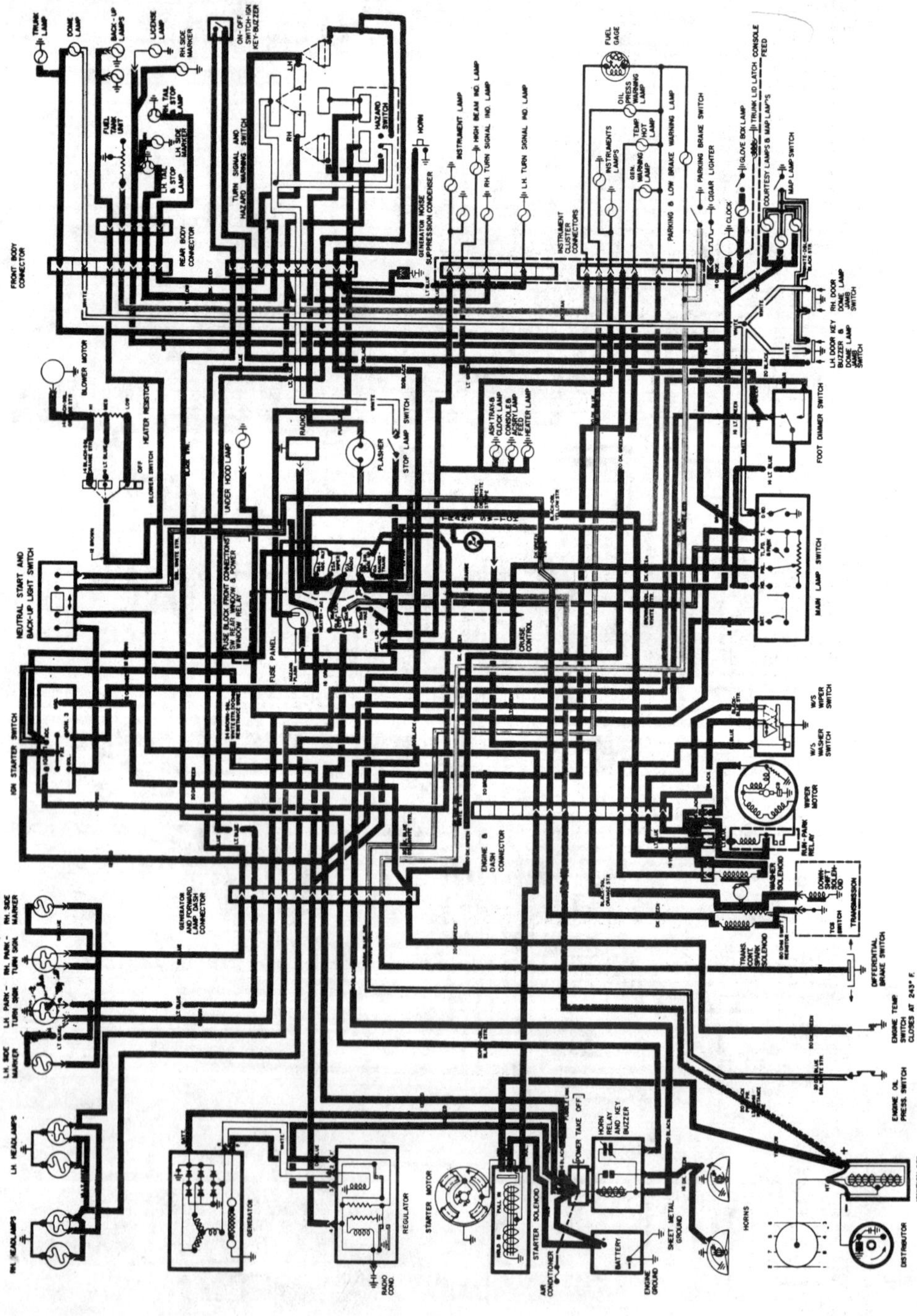

1971 442 WIRING DIAGRAM

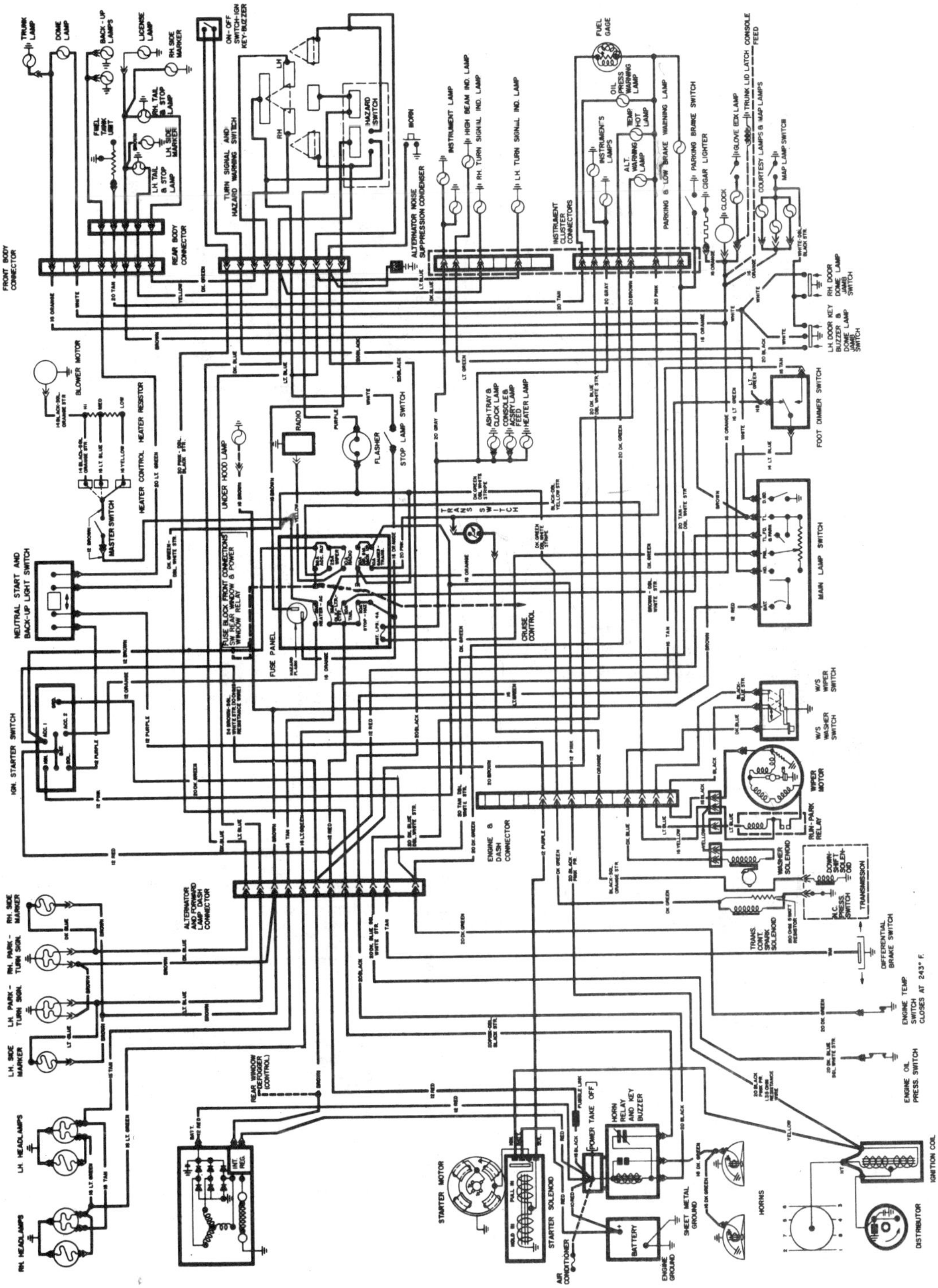

1972 WIRING DIAGRAM

IGNITION SWITCH OPERATION	
POSITION	CIRCUIT
ACCESSORY	ACC. & BAT CONNECTED
OFF-LOCKED	OPEN
OFF	OPEN
RUN	IGN1, IGN3, ACC & BAT CONNECTED
START	IGN1, SOL & BAT CONNECTED GRD TERMINAL GROUNDED

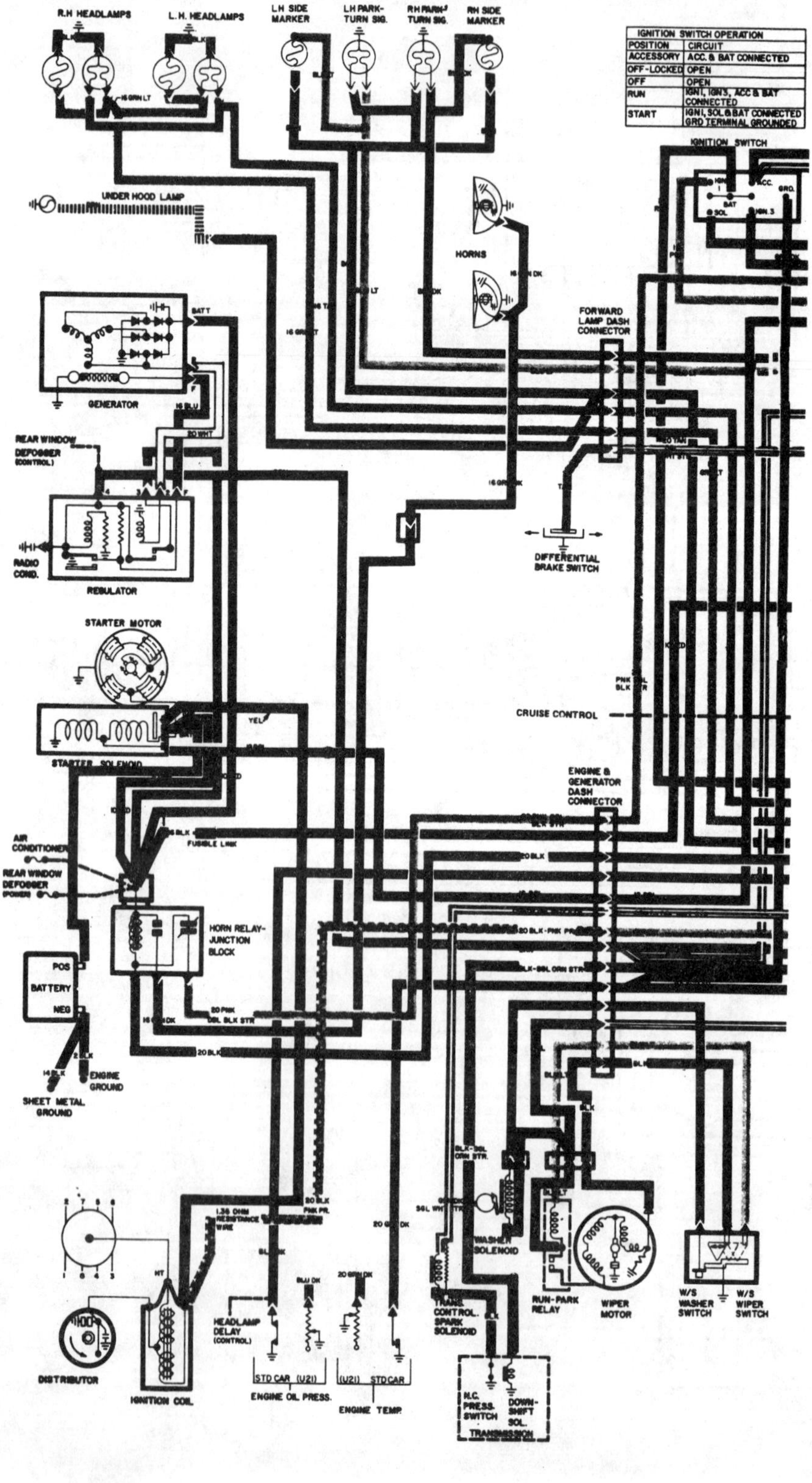

1972 WIRING DIAGRAM

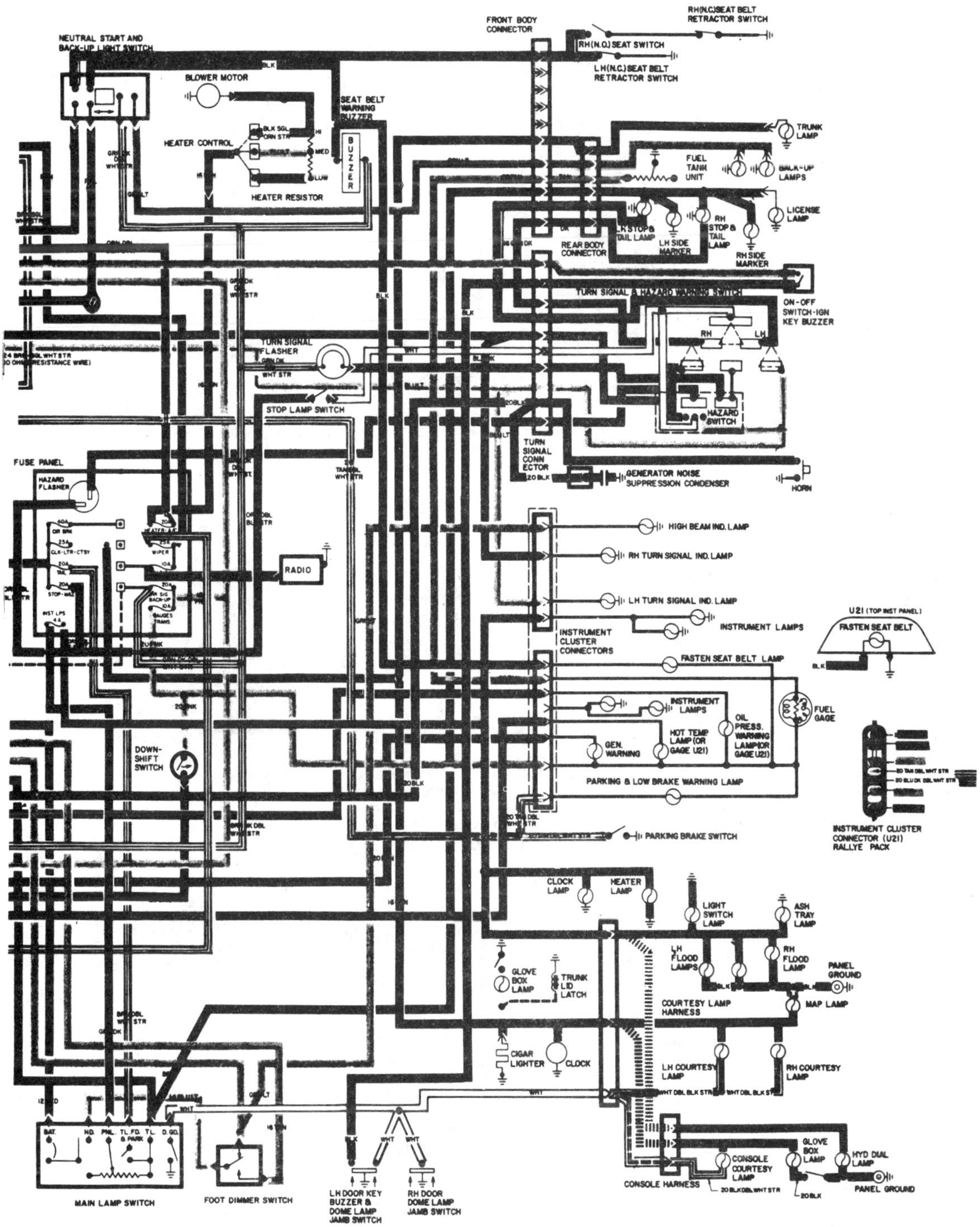

NOTE
ALL WIRES ARE 18 GAGE UNLESS OTHERWISE SPECIFIED

ALL CONNECTIONS ON MAIN HARNESS OPEN END VIEW

15

1973 WIRING DIAGRAM

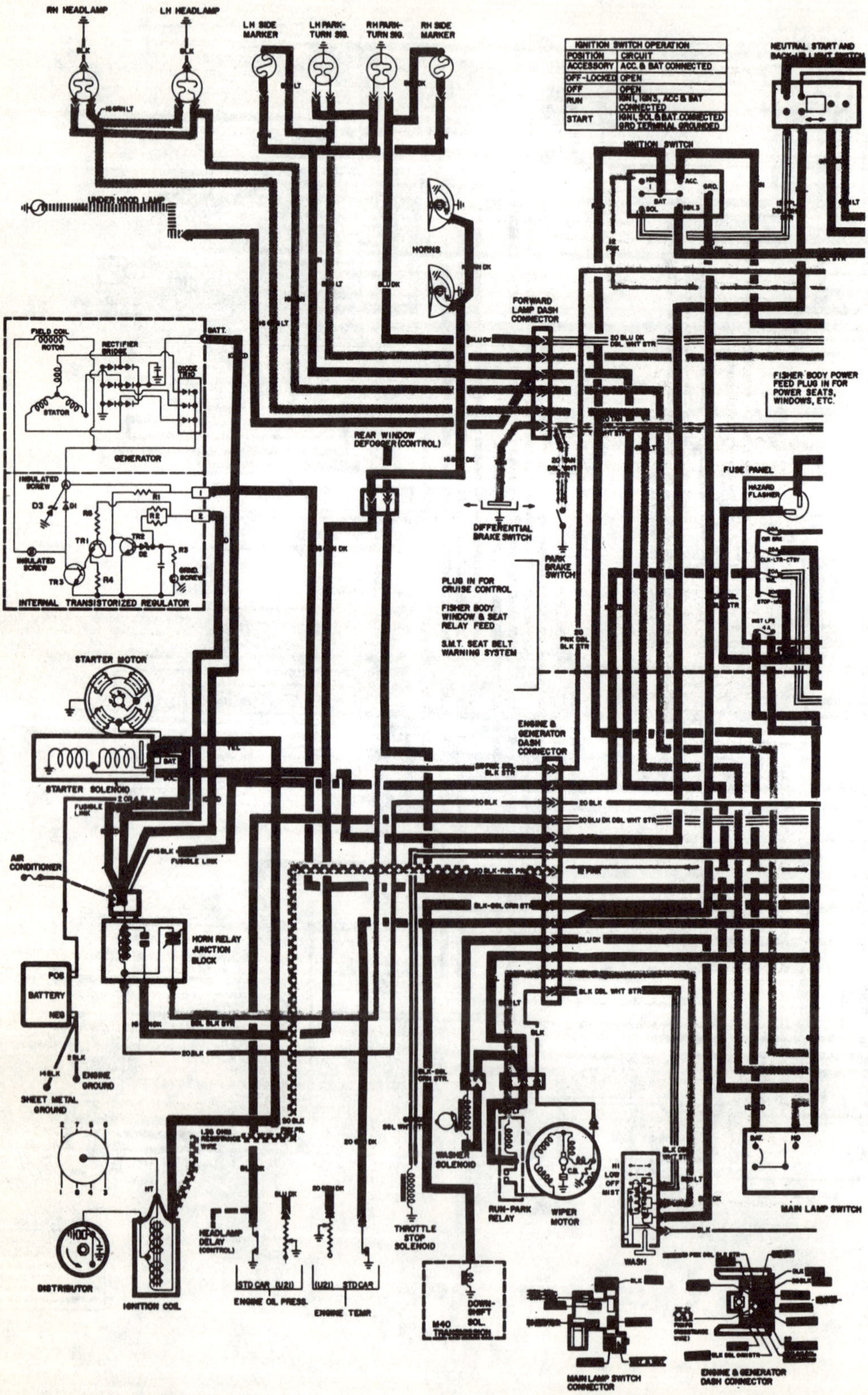

1973 WIRING DIAGRAM

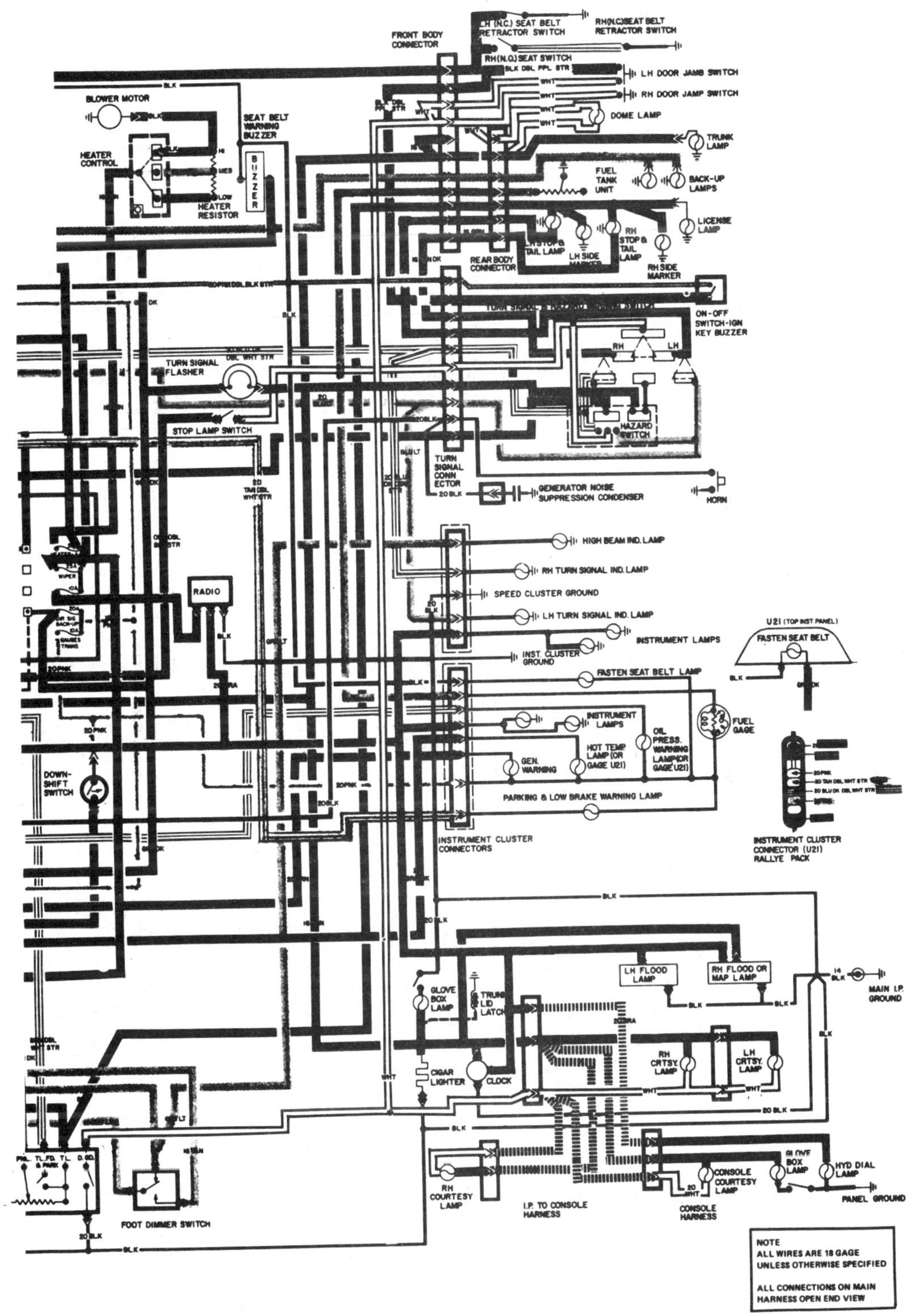

NOTE
ALL WIRES ARE 18 GAGE UNLESS OTHERWISE SPECIFIED

ALL CONNECTIONS ON MAIN HARNESS OPEN END VIEW

1974 WIRING DIAGRAM

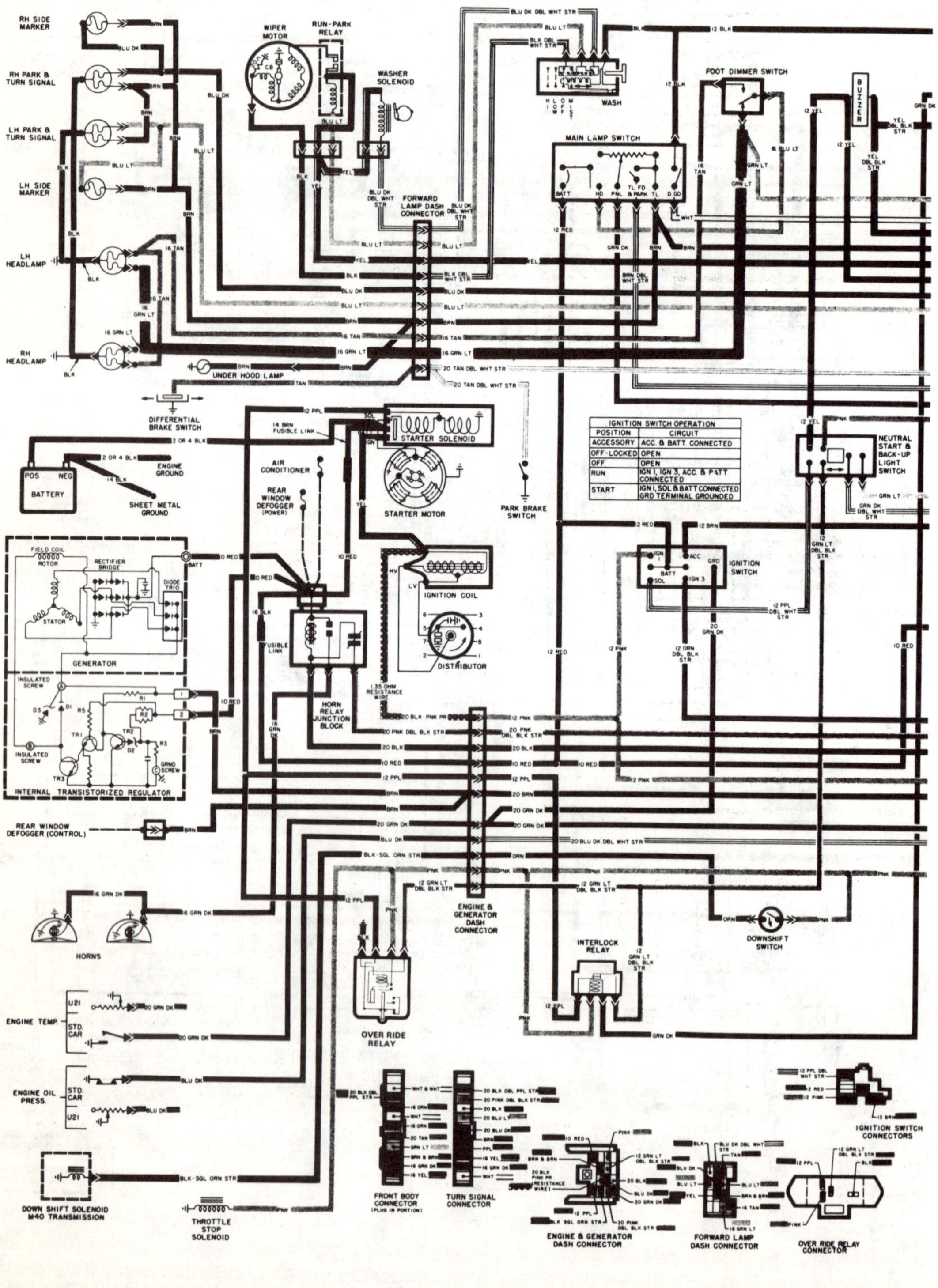

1974 WIRING DIAGRAM

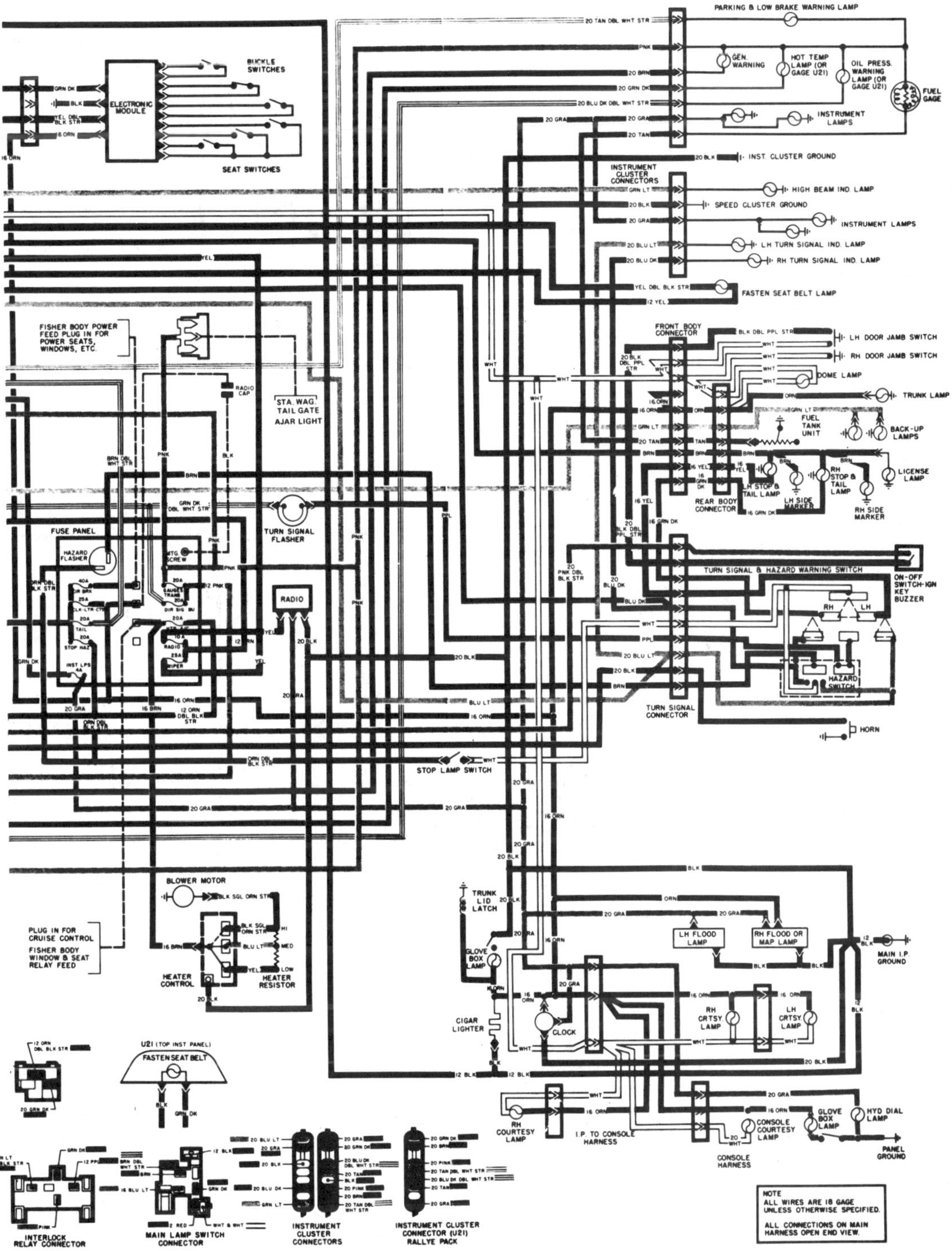

NOTE
ALL WIRES ARE 18 GAGE UNLESS OTHERWISE SPECIFIED.
ALL CONNECTIONS ON MAIN HARNESS OPEN END VIEW.

1975 WIRING DIAGRAM

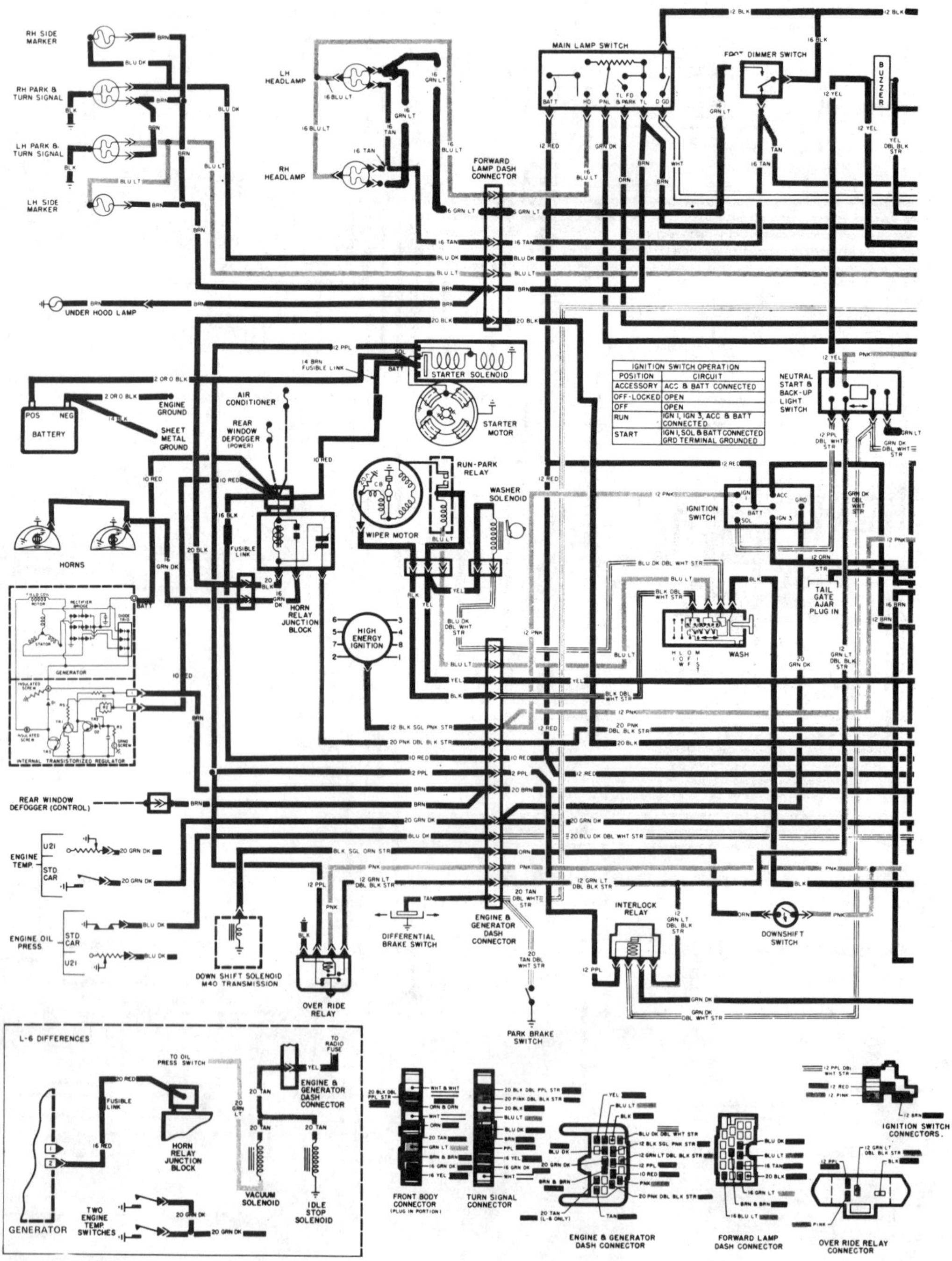

IGNITION SWITCH OPERATION	
POSITION	CIRCUIT
ACCESSORY	ACC & BATT CONNECTED
OFF-LOCKED	OPEN
OFF	OPEN
RUN	IGN I, IGN 3, ACC & BATT CONNECTED
START	IGN I, SOL & BATT CONNECTED GRD TERMINAL GROUNDED

1975 WIRING DIAGRAM

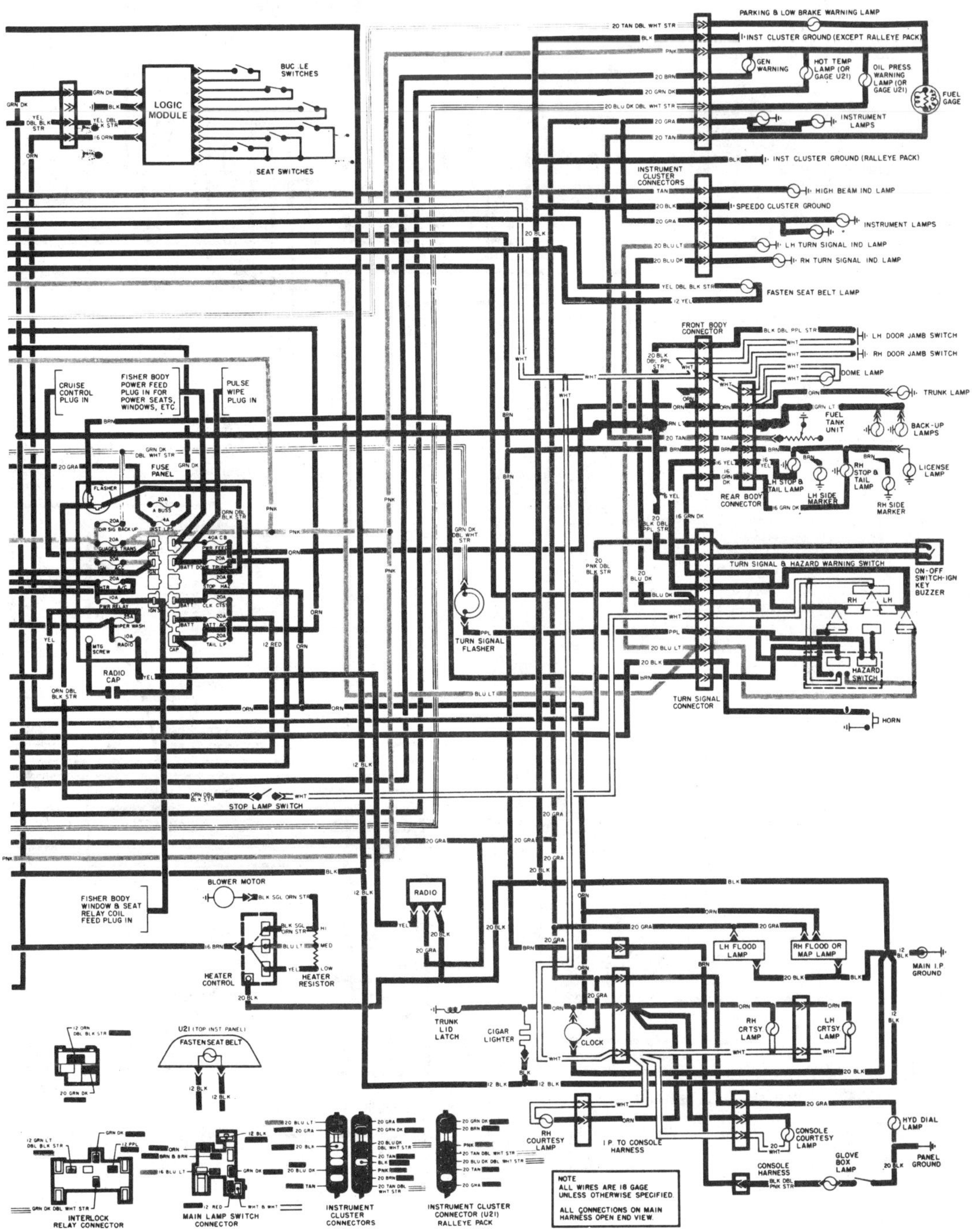

1976 WIRING DIAGRAM

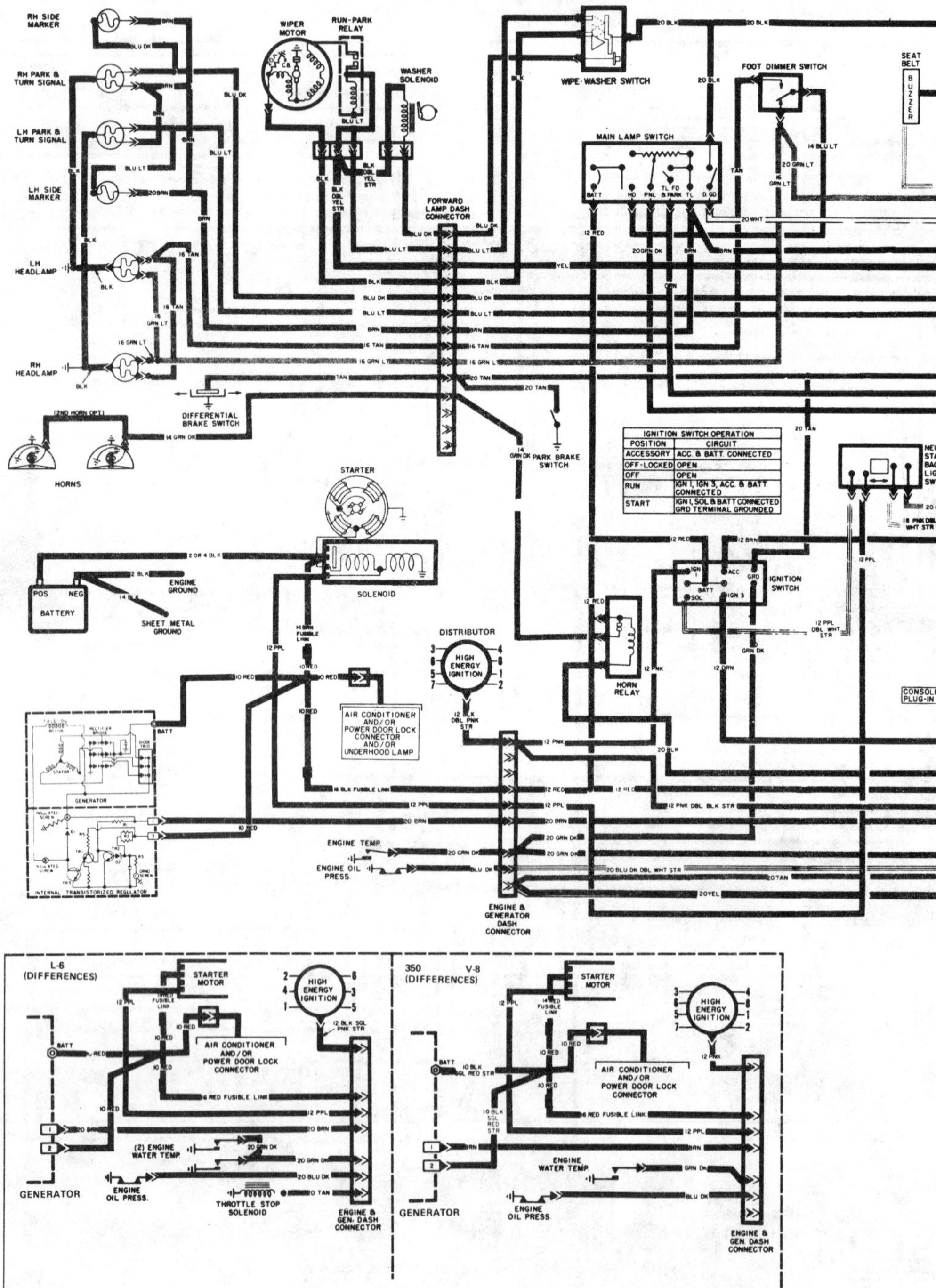

IGNITION SWITCH OPERATION	
POSITION	CIRCUIT
ACCESSORY	ACC. & BATT. CONNECTED
OFF-LOCKED	OPEN
OFF	OPEN
RUN	IGN I, IGN 3, ACC. & BATT CONNECTED
START	IGN I, SOL & BATT CONNECTED GRD TERMINAL GROUNDED

1976 WIRING DIAGRAM

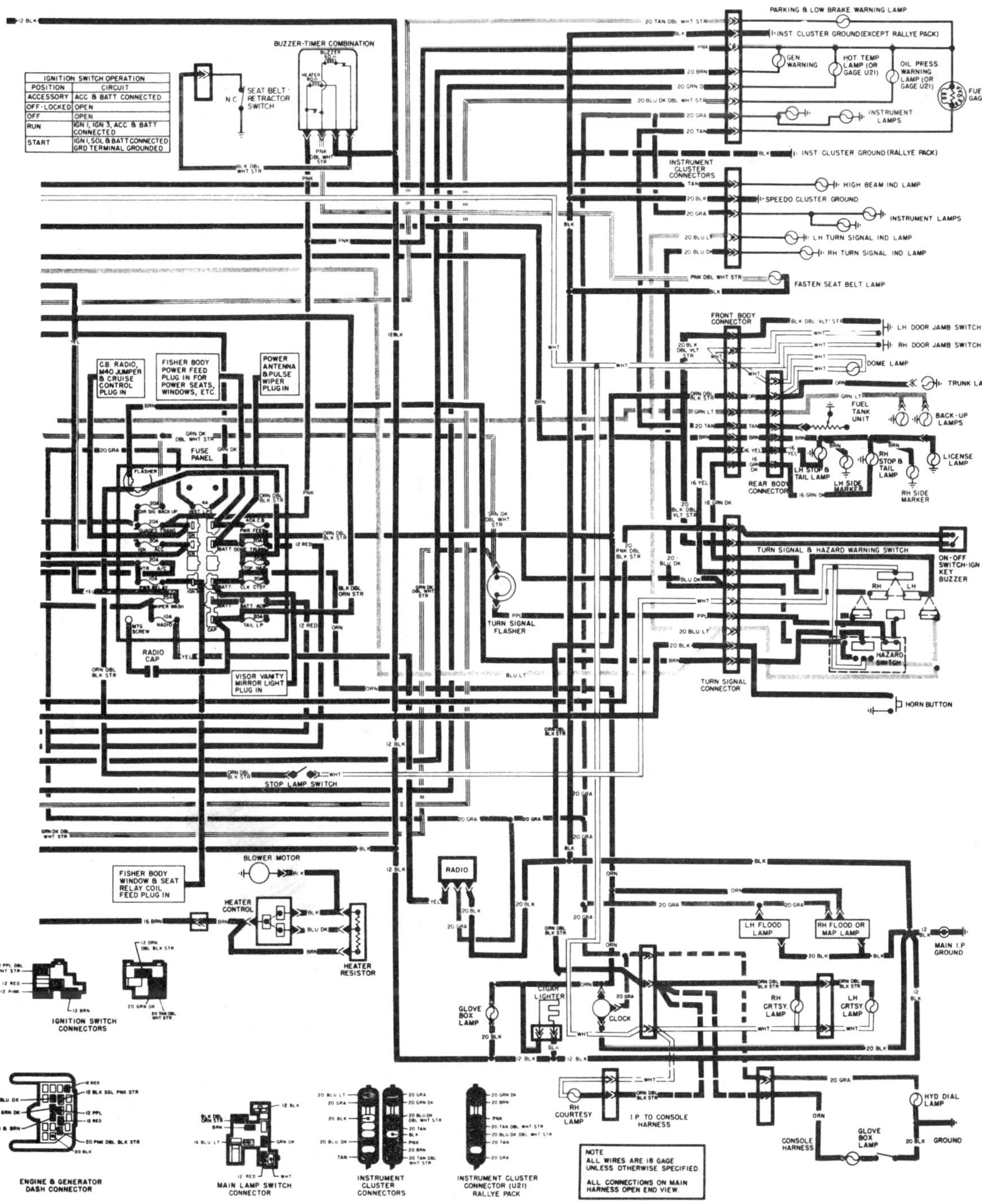

1977 WIRING DIAGRAM

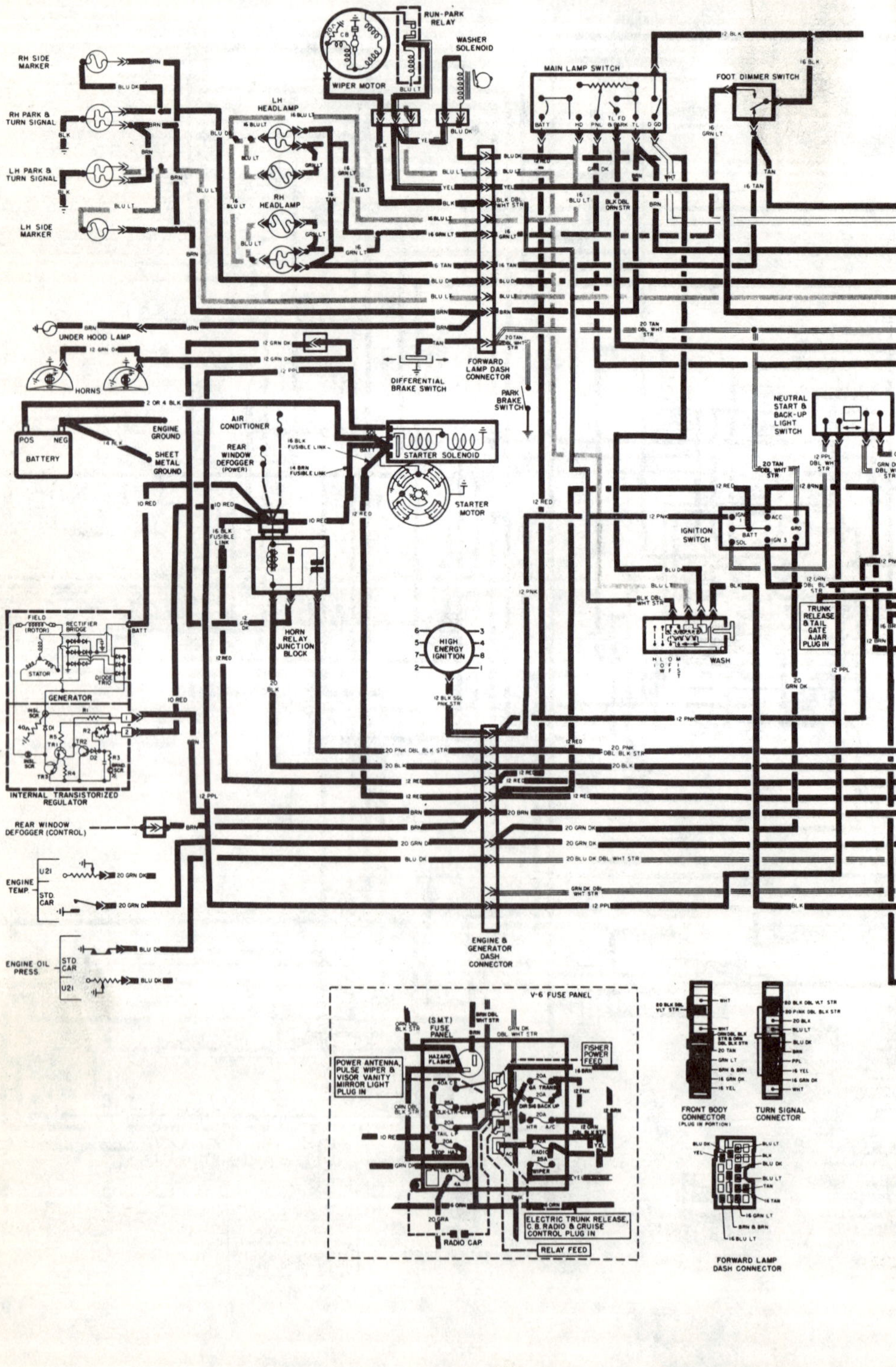

1977 WIRING DIAGRAM

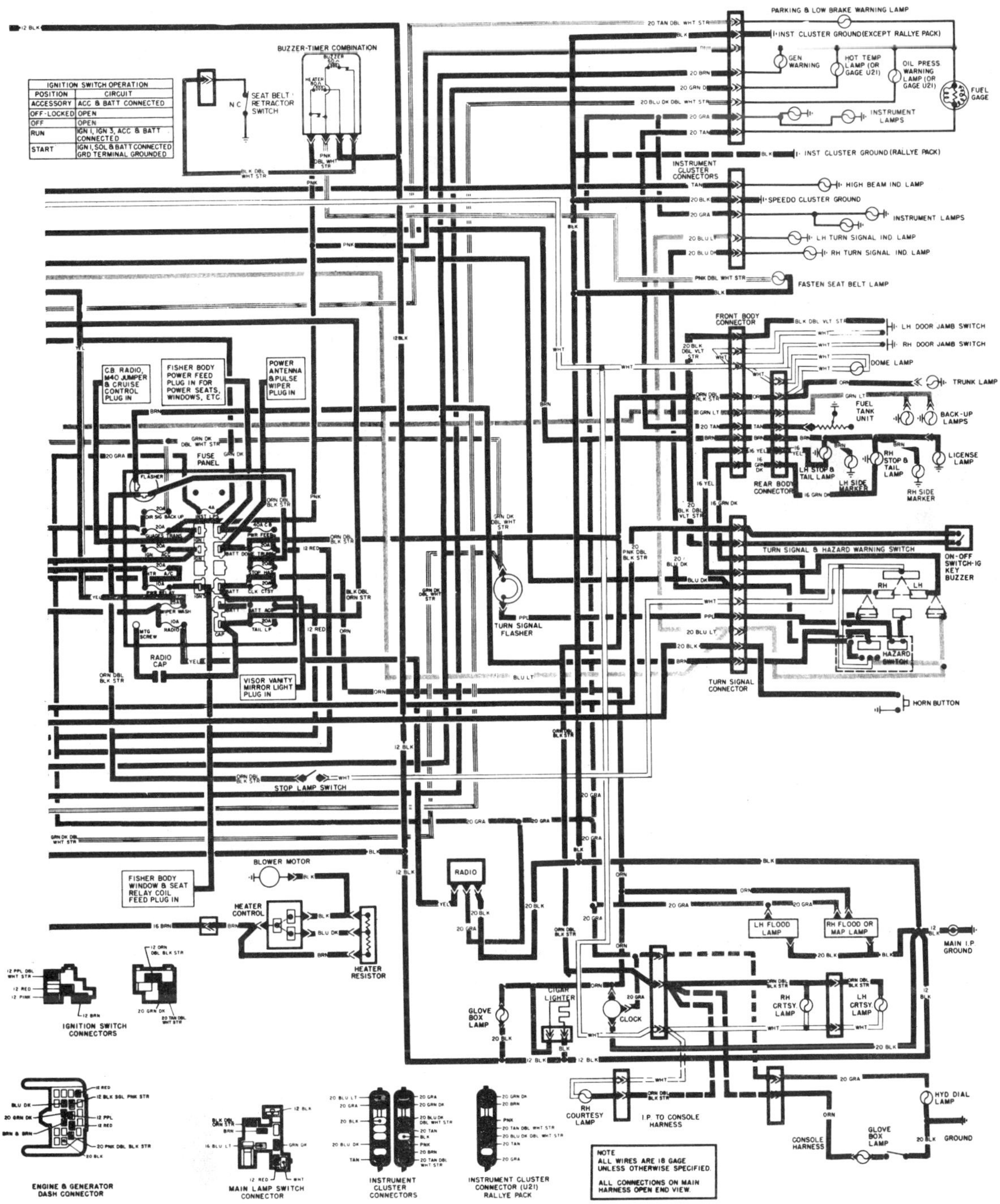

POSITION	CIRCUIT
ACCESSORY	ACC & BATT CONNECTED
OFF-LOCKED	OPEN
OFF	OPEN
RUN	IGN I, IGN 3, ACC & BATT CONNECTED
START	IGN I, SOL & BATT CONNECTED GRD TERMINAL GROUNDED

1978 (PART 1) WIRING DIAGRAM

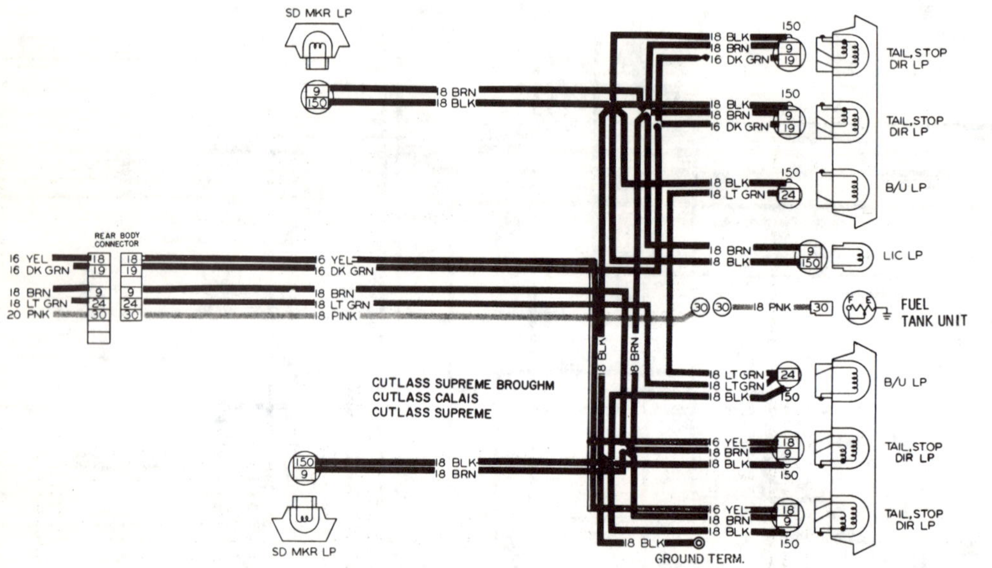

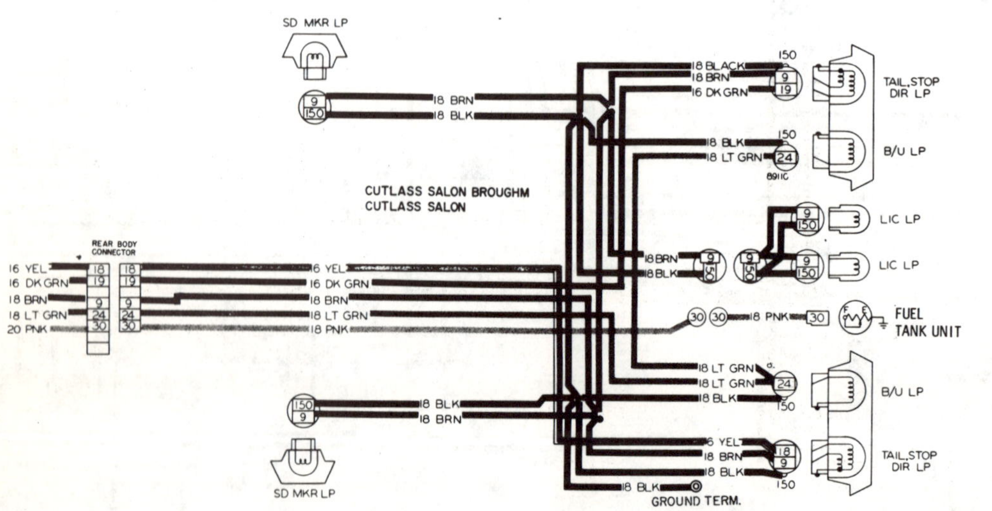

1978 (PART II) WIRING DIAGRAM

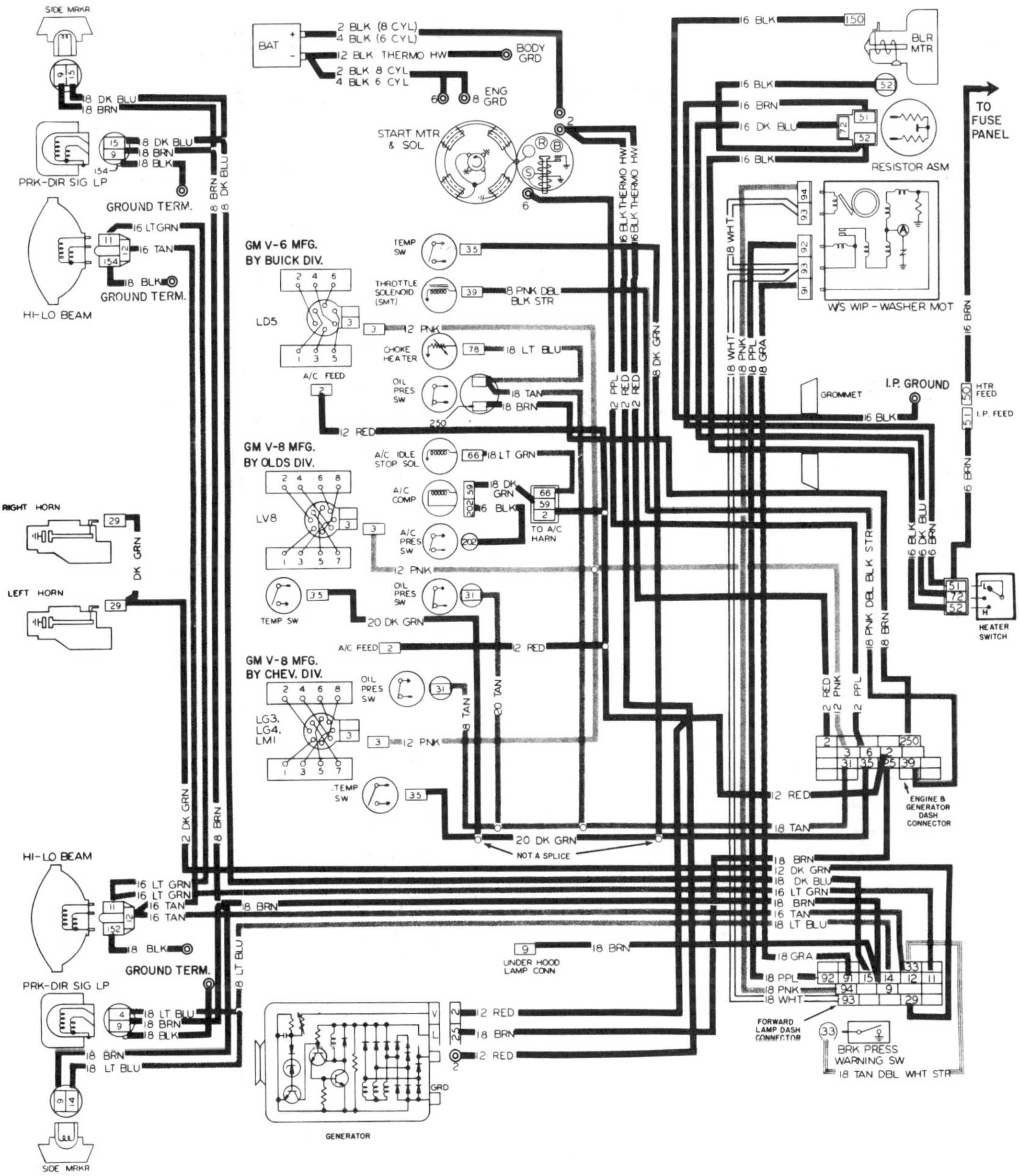

1978 (PART III) WIRING DIAGRAM

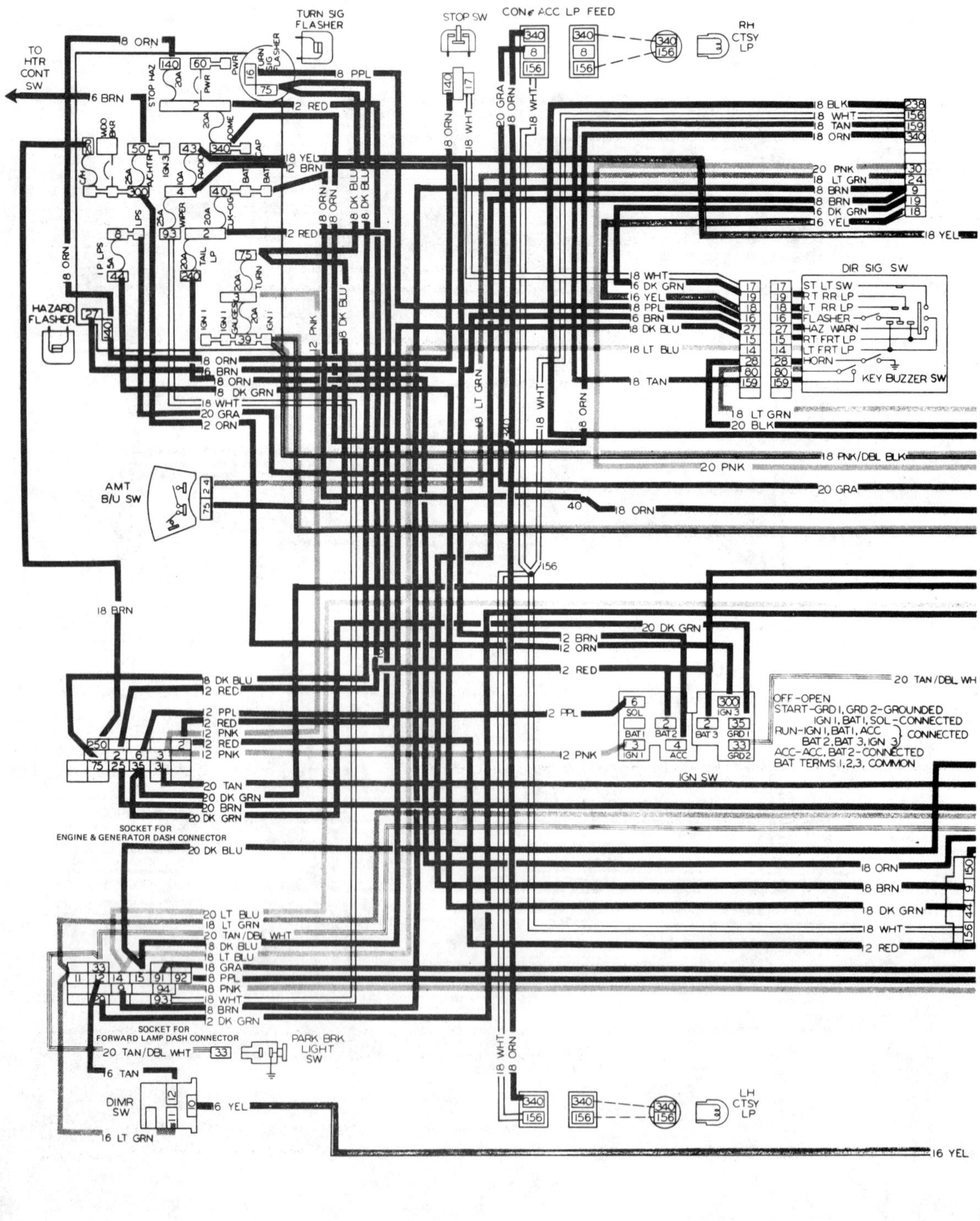

1978 (PART IV) WIRING DIAGRAM

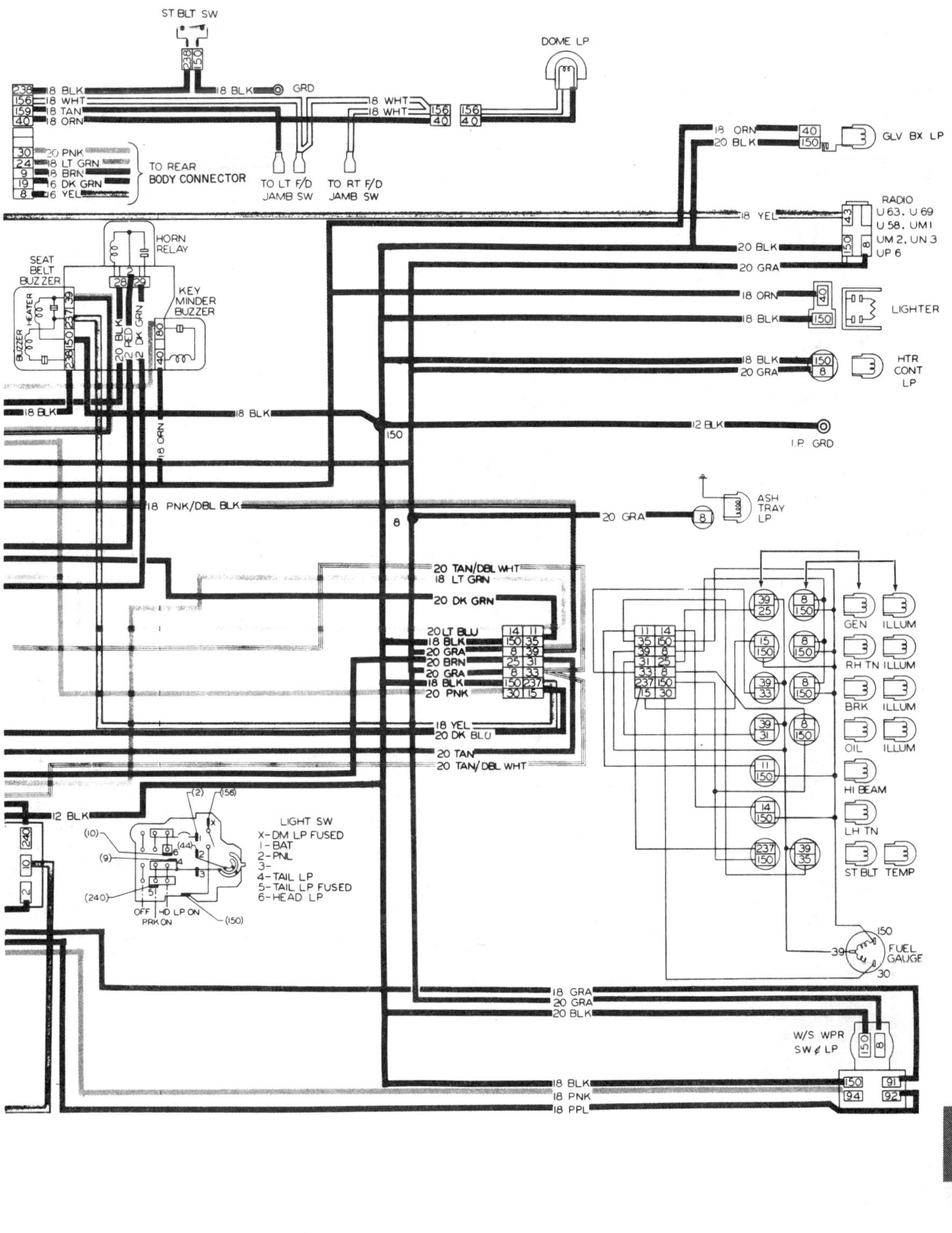

1979 (PART I) WIRING DIAGRAM

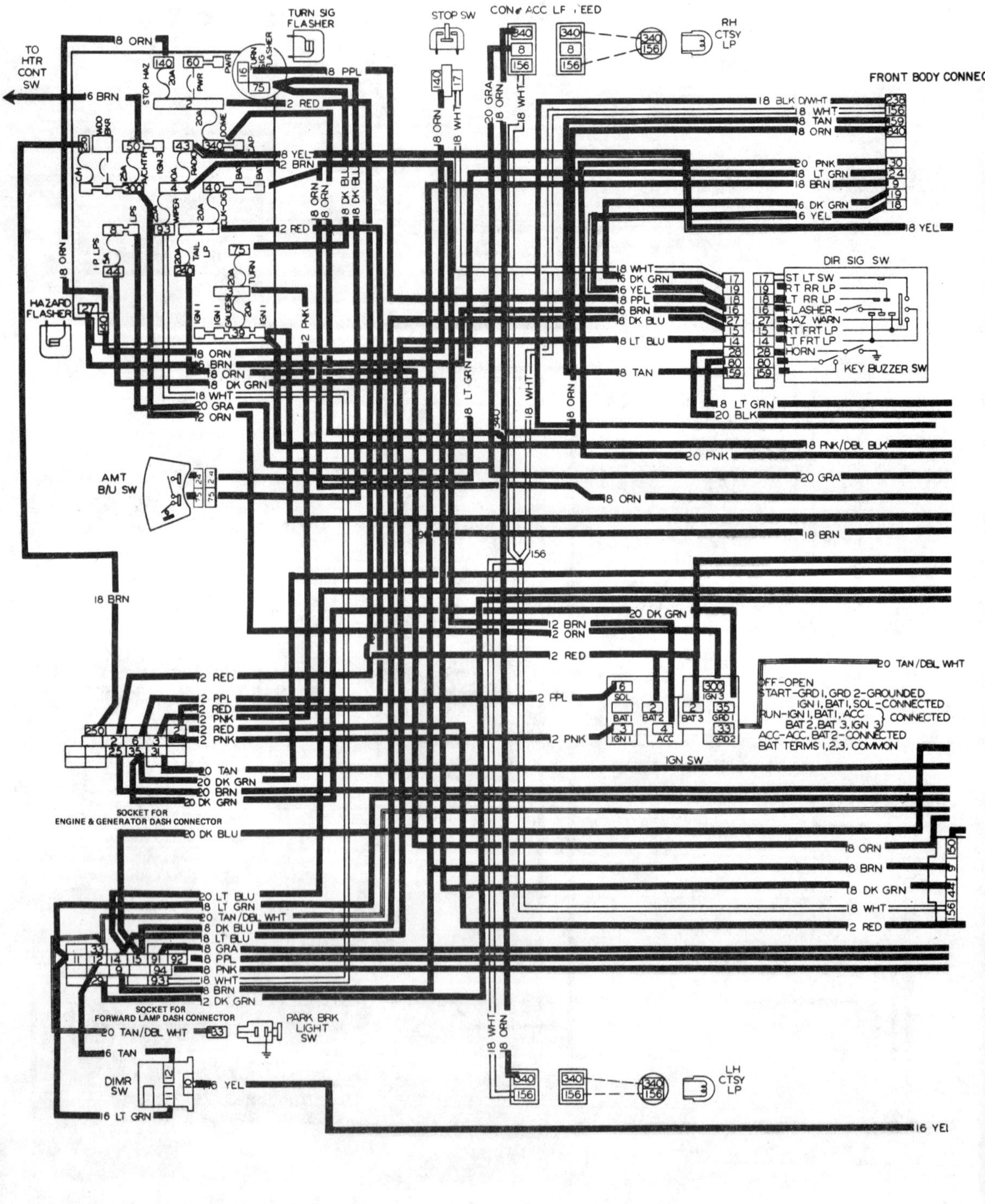

1979 (PART II) WIRING DIAGRAM

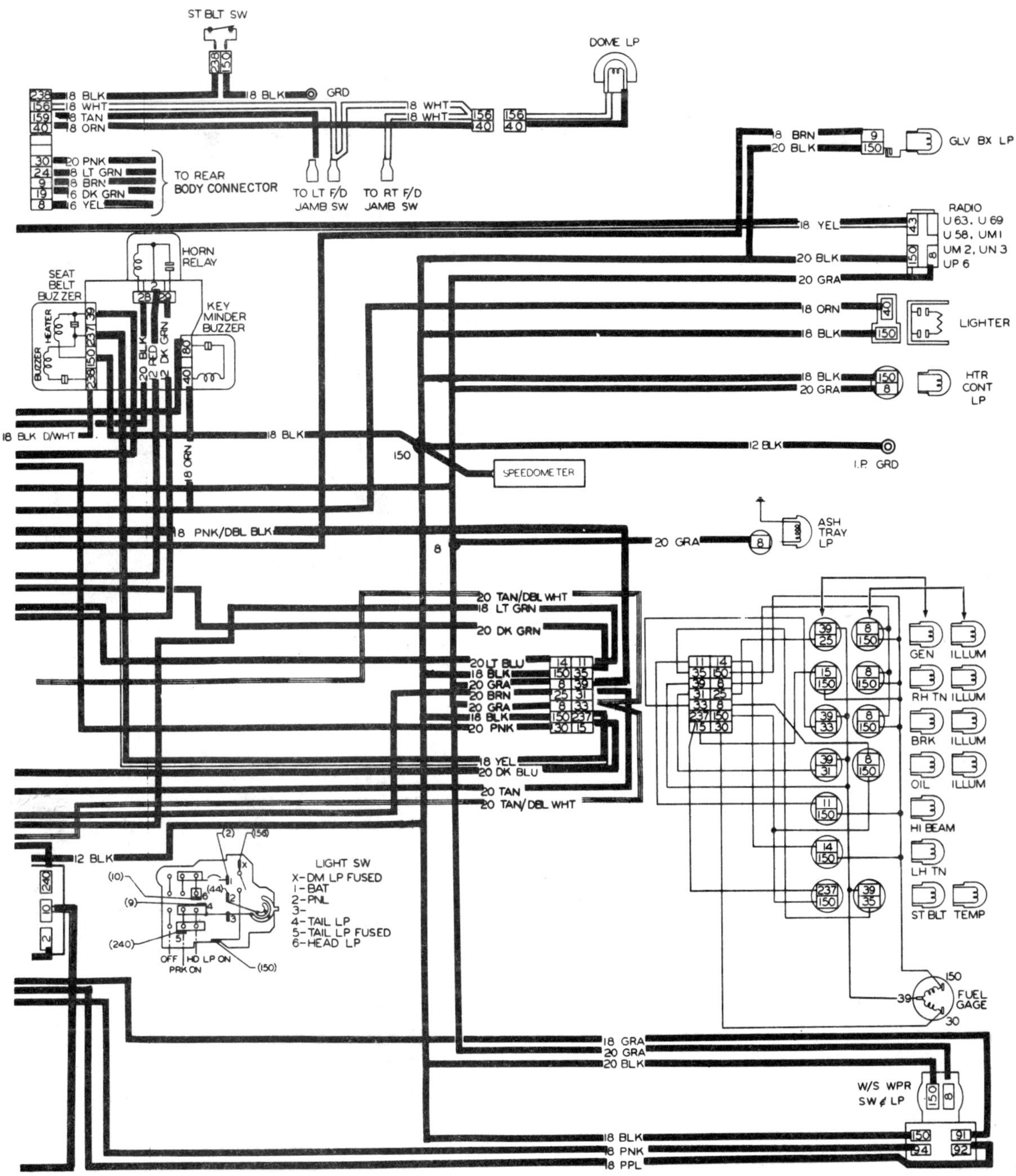

1979 (PART III) WIRING DIAGRAM

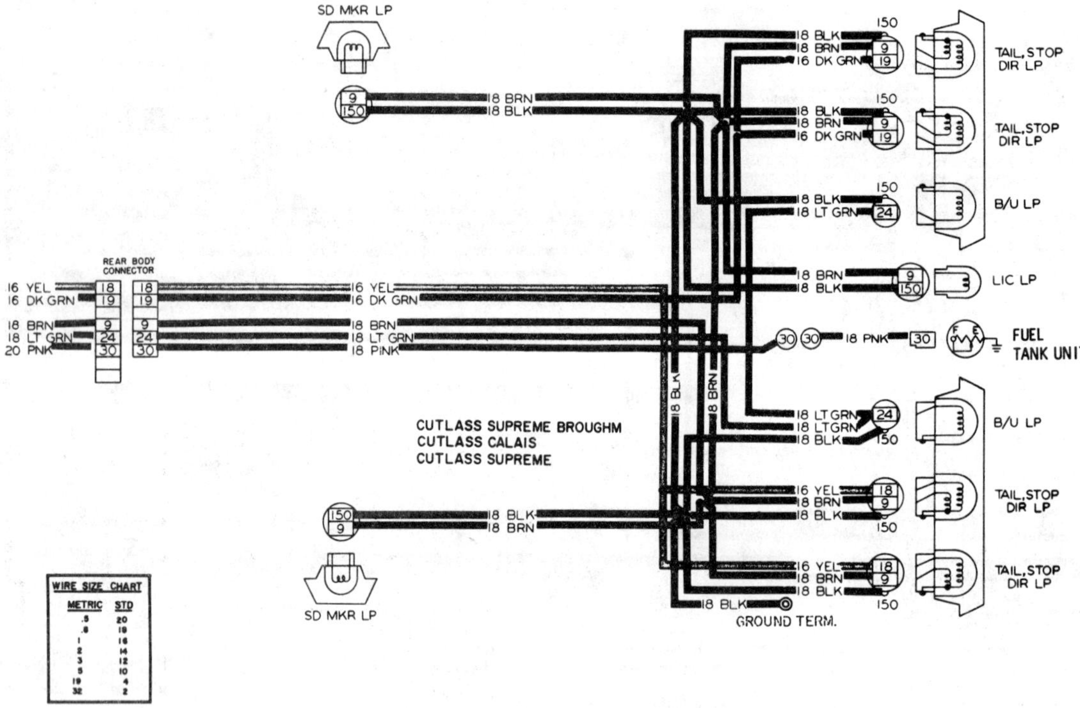

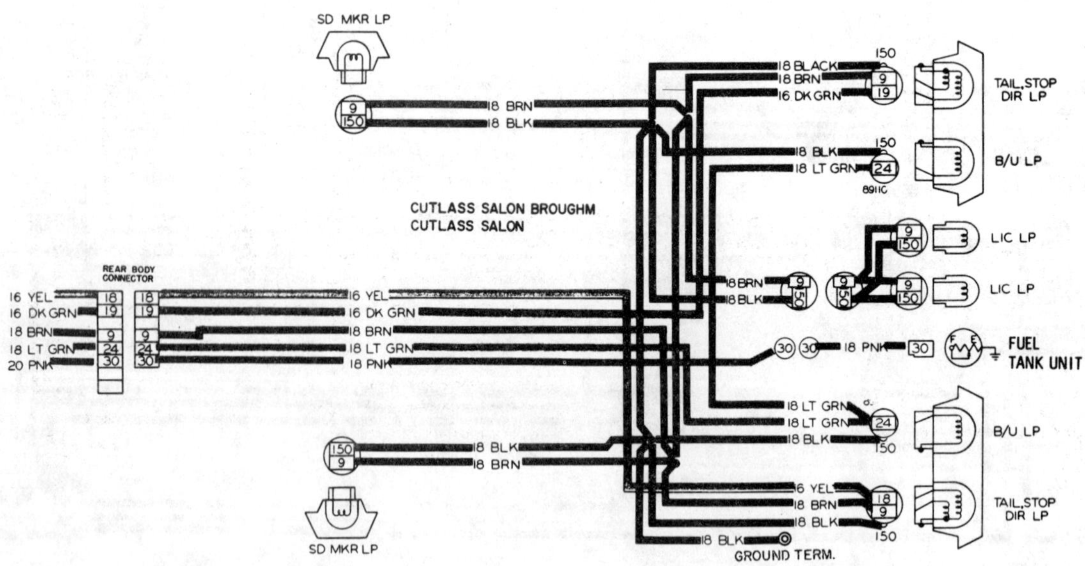

1979 (PART IV) WIRING DIAGRAM

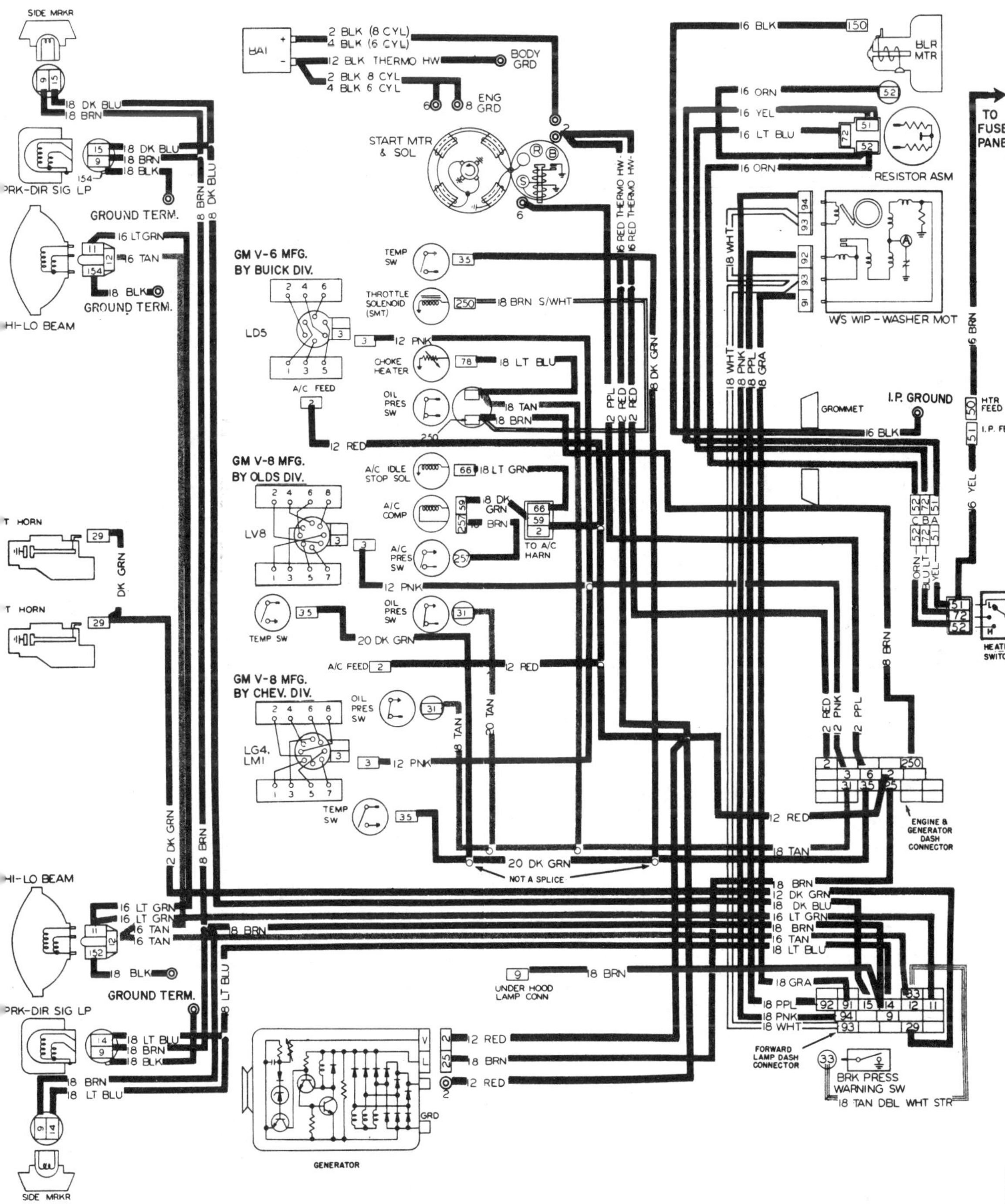

1980 WIRING DIAGRAM SYMBOLS

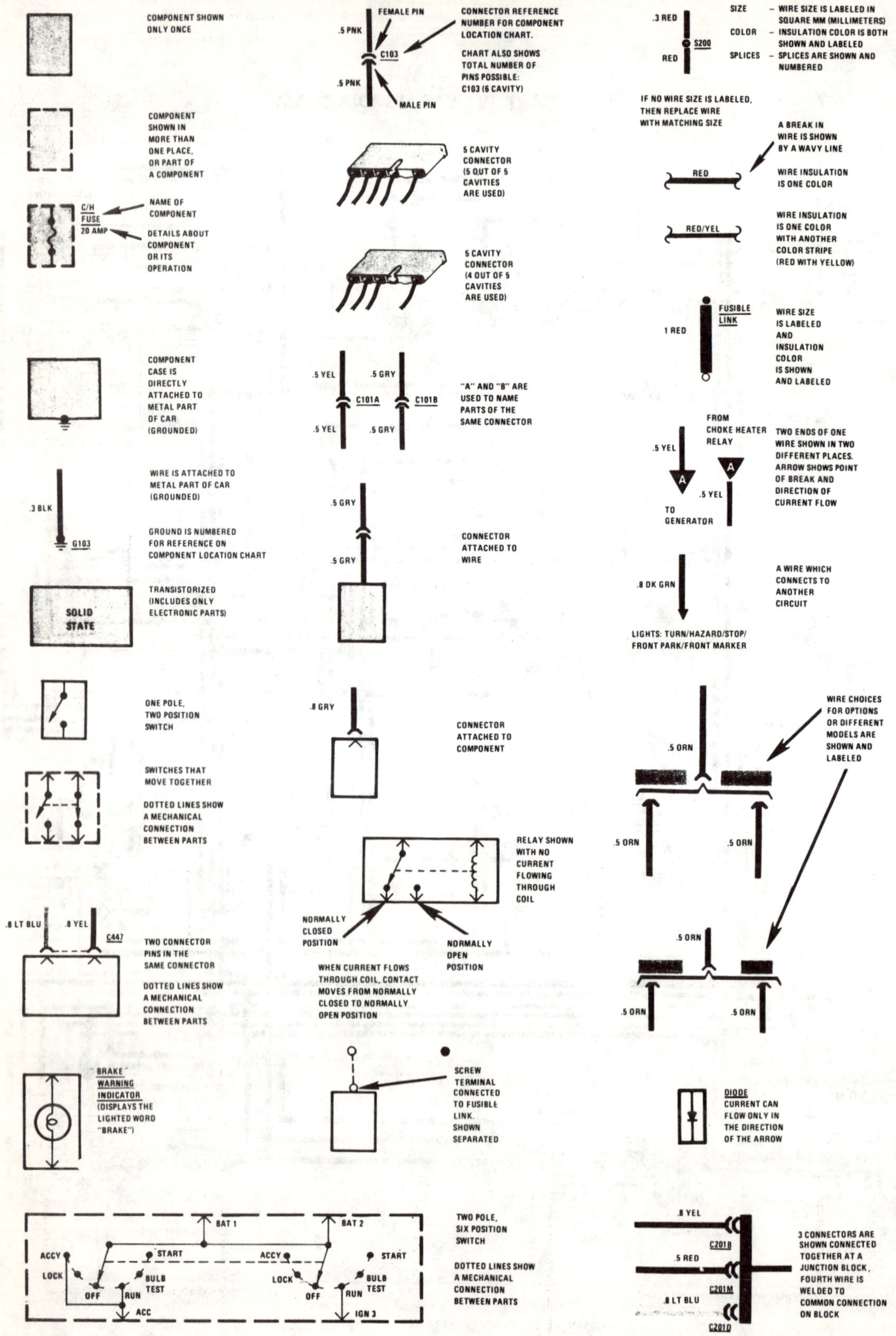

1980 GROUND DISTRIBUTION

FRONT LIGHTS GROUNDS — EXCEPT: CUTLASS CALAIS COUPE, CUTLASS SUPREME COUPE, CUTLASS SUPREME BROUGHAM COUPE

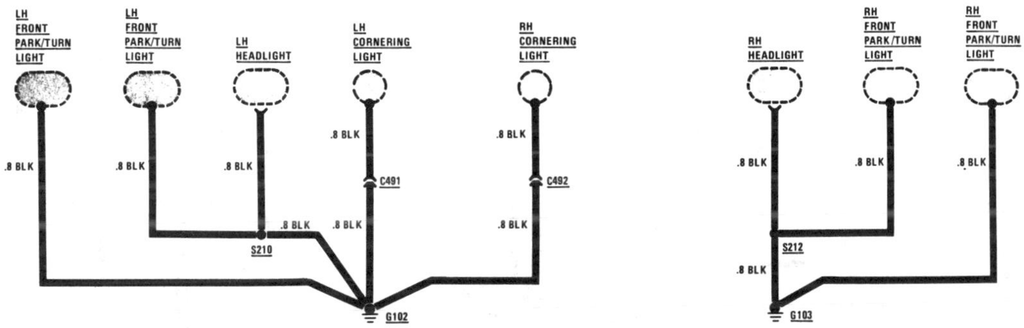

REAR LIGHTS GROUND — STATION WAGON

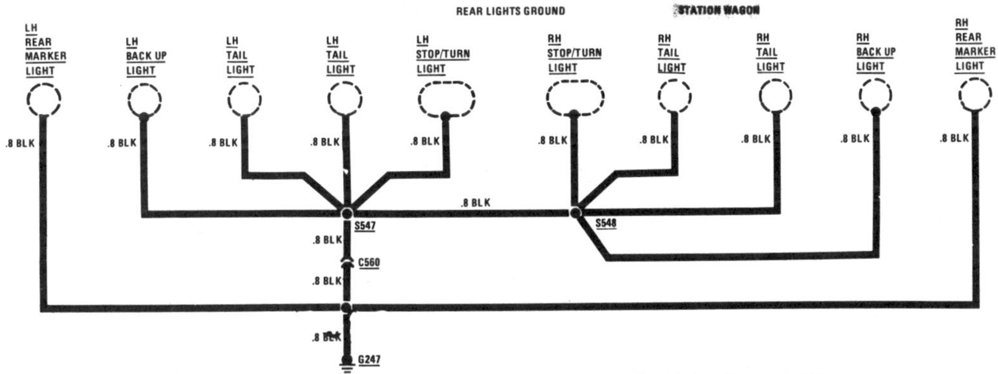

REAR LIGHTS GROUND — CUTLASS CALAIS COUPE, CUTLASS SUPREME COUPE, CUTLASS SUPREME BROUGHAM COUPE ONLY

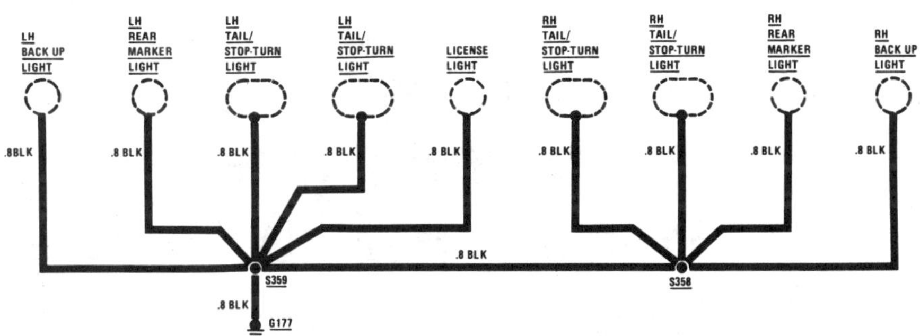

FRONT LIGHTS GROUNDS — CUTLASS CALAIS COUPE, CUTLASS SUPREME COUPE, CUTLASS SUPREME BROUGHAM COUPE

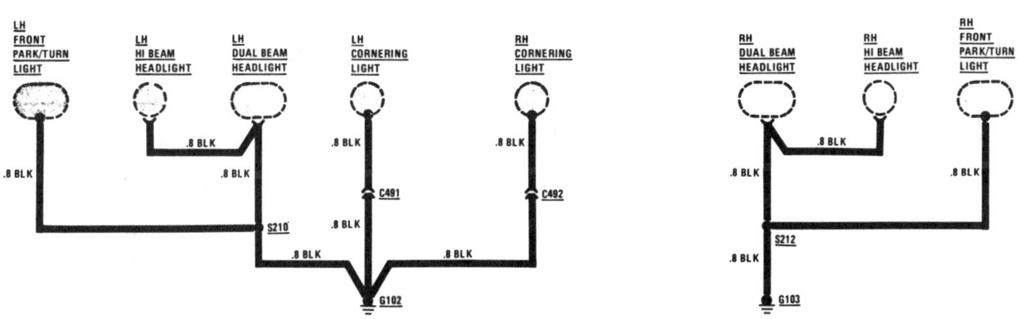

15

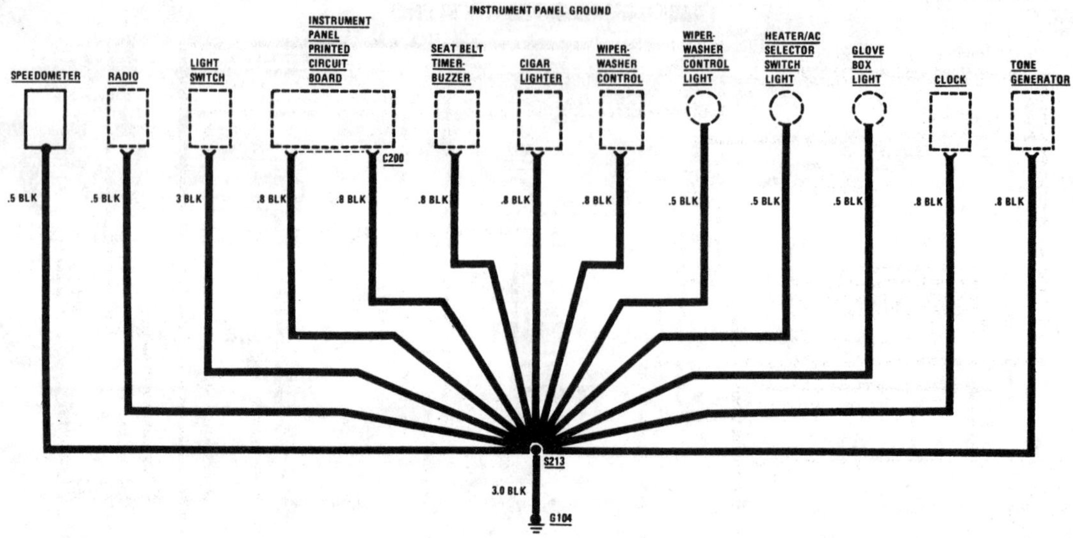

1980 FUSE BLOCK

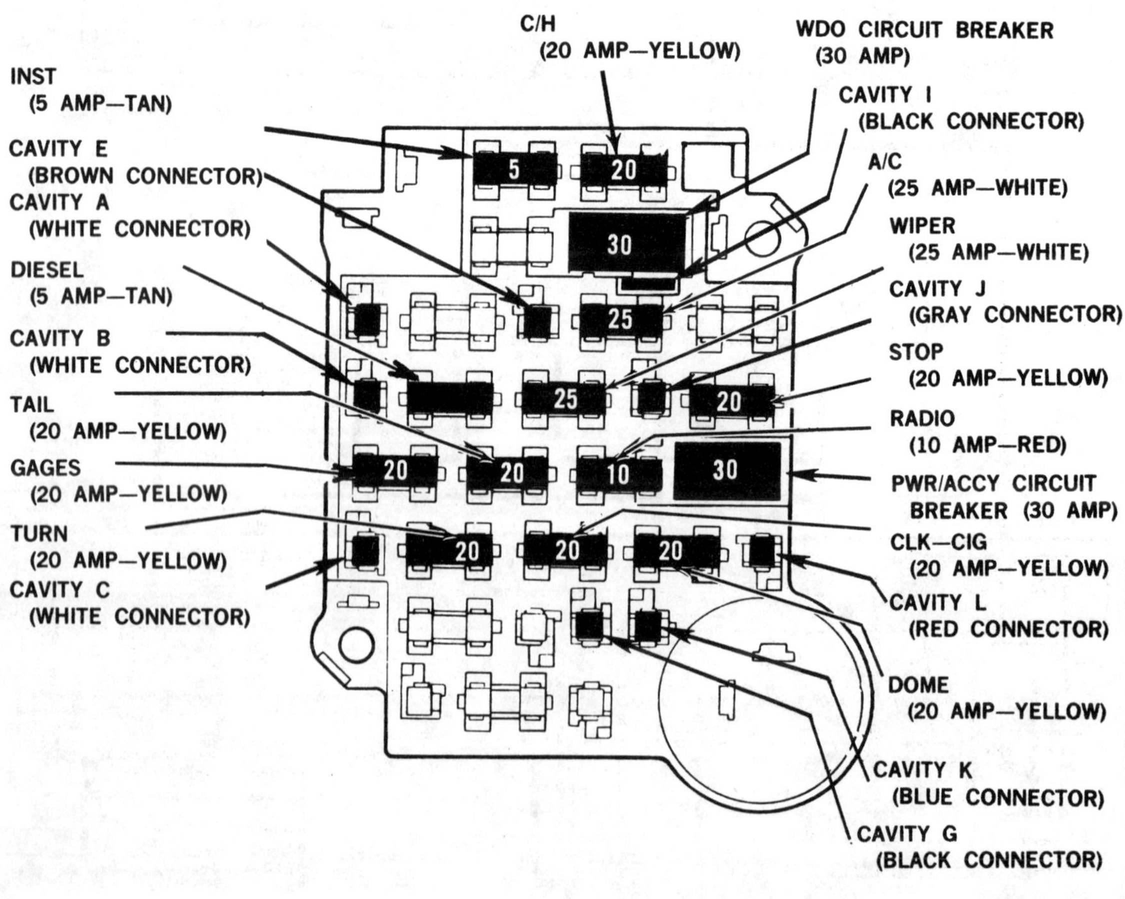

1980 CHARGING SYSTEM

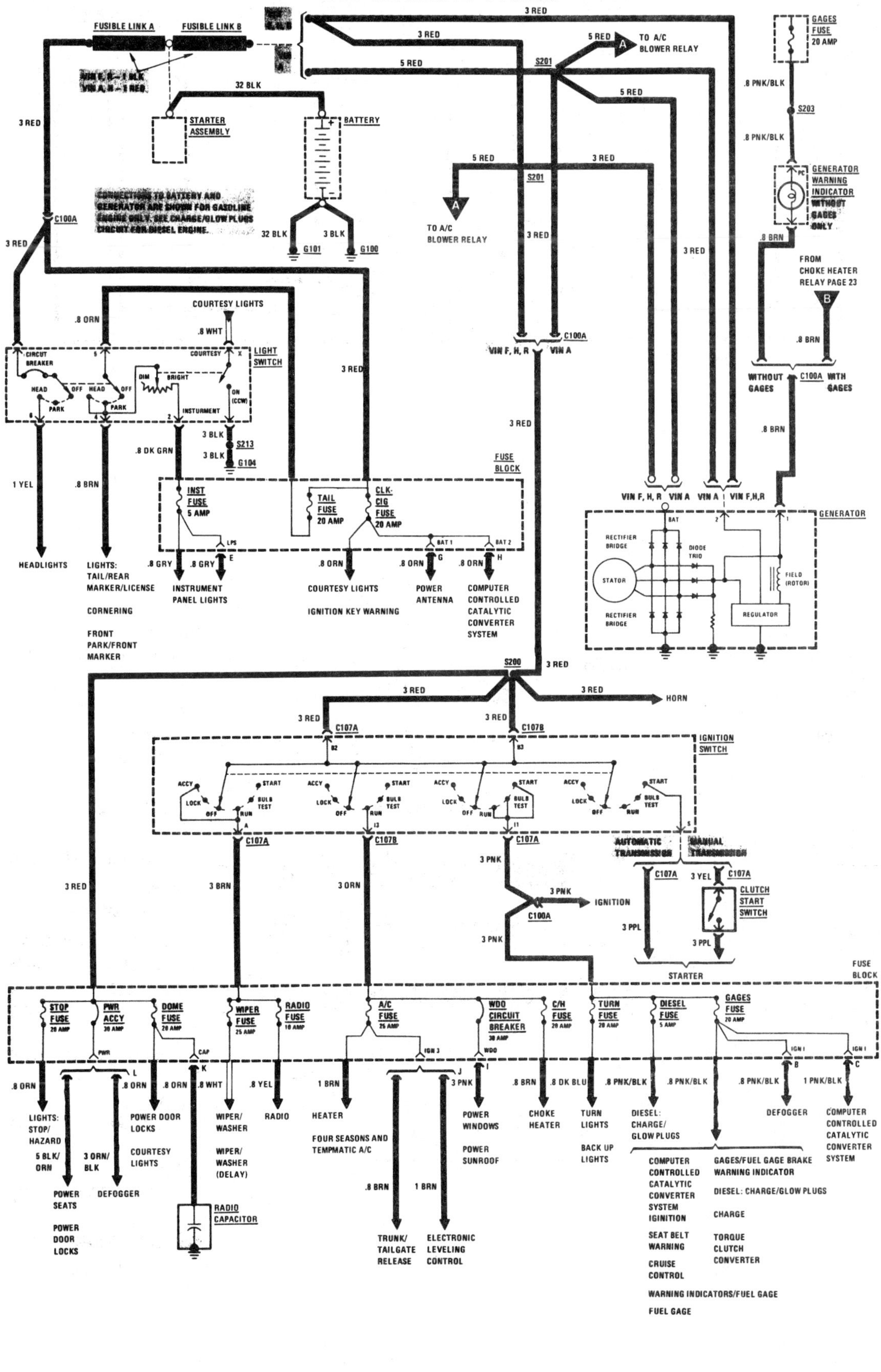

15

1980 IGNITION/TACH

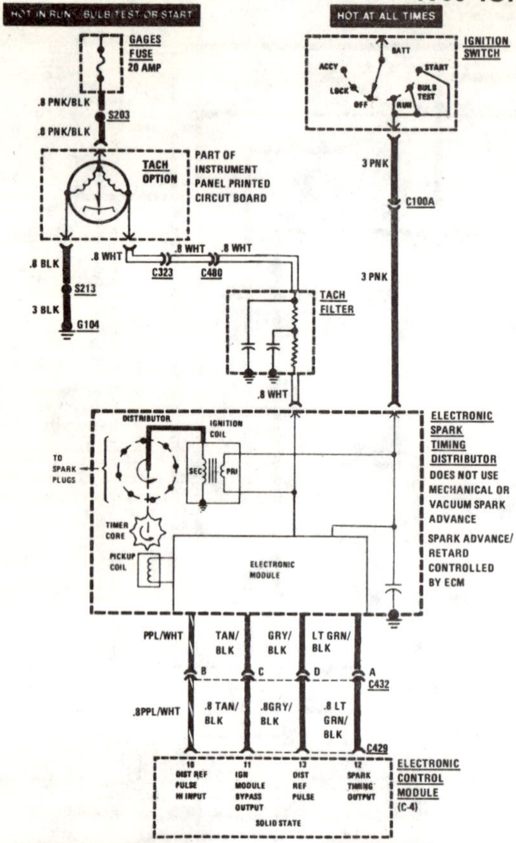

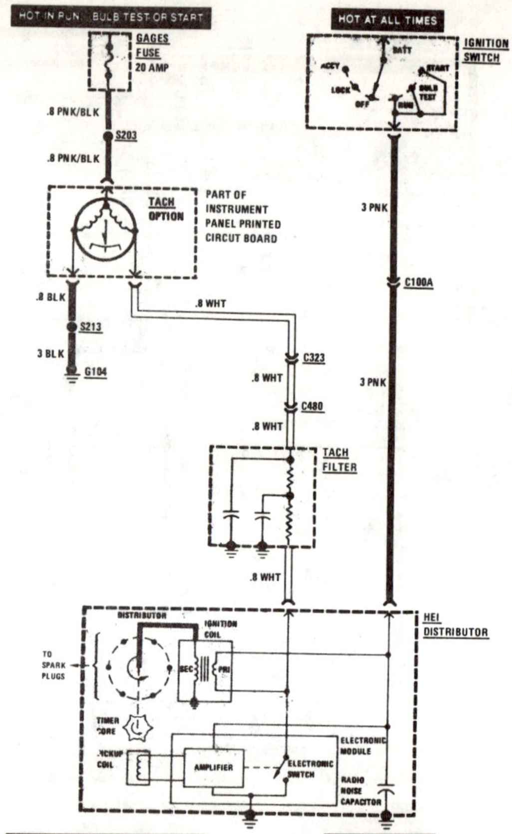

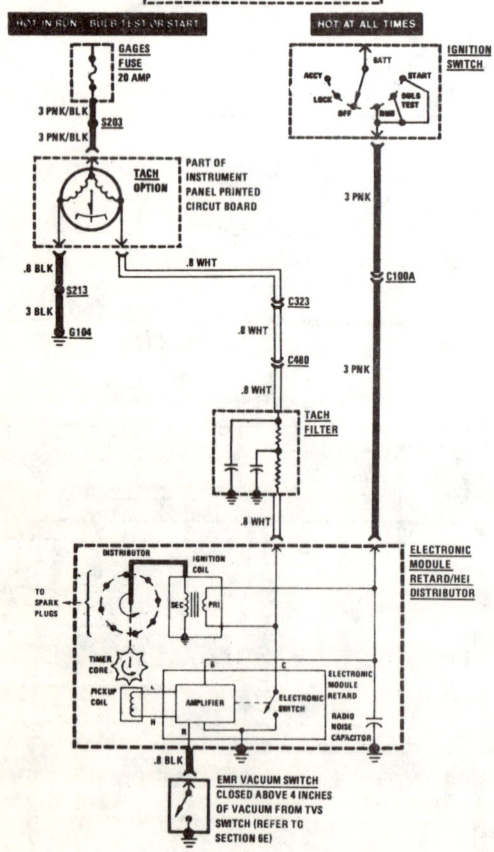

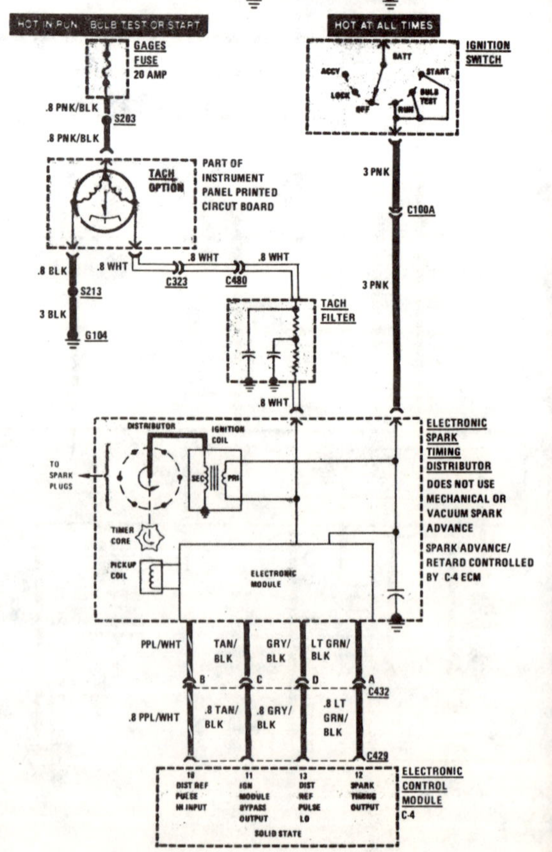

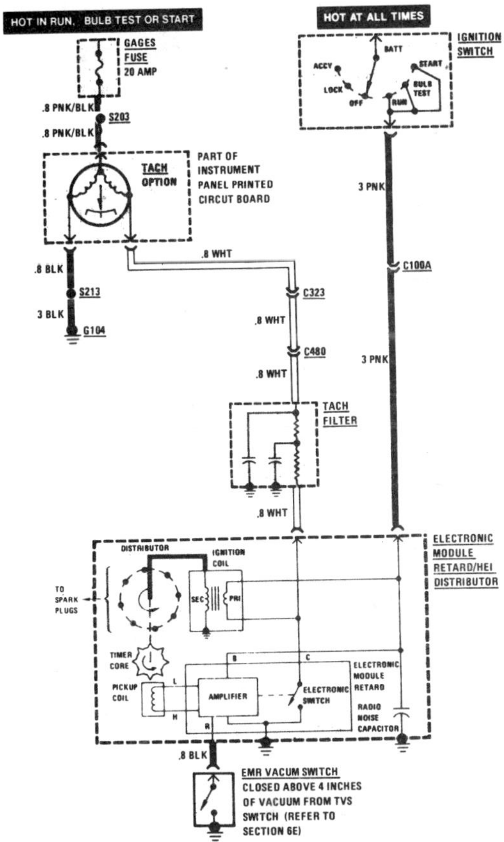

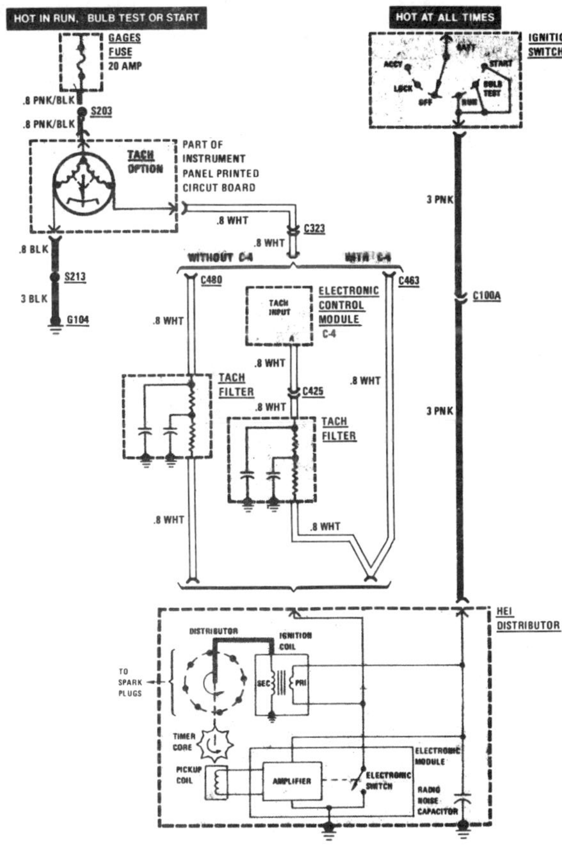

1980 CHOKE HEATER

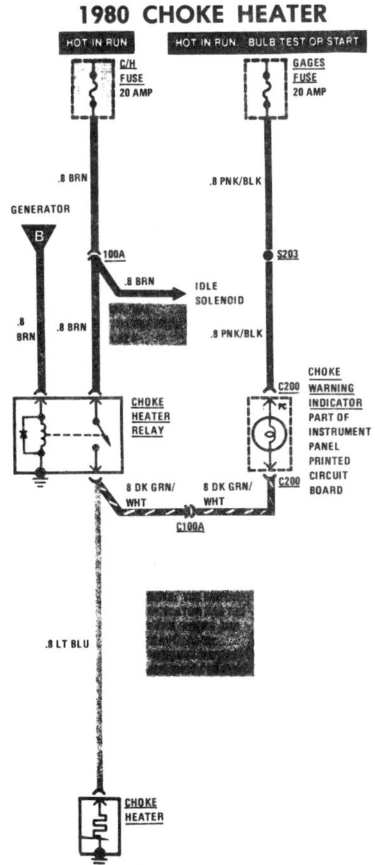

1980 STARTER

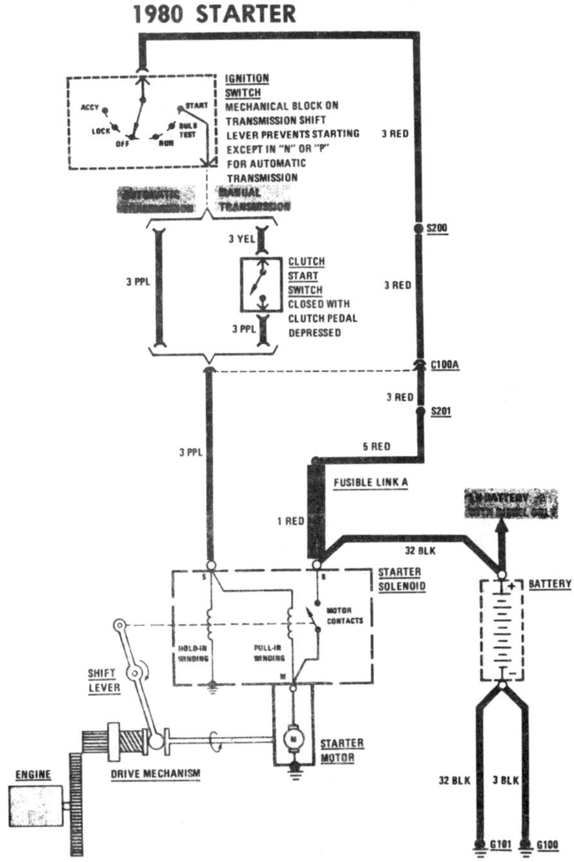

15

1980 CATALYTIC CONVERTER

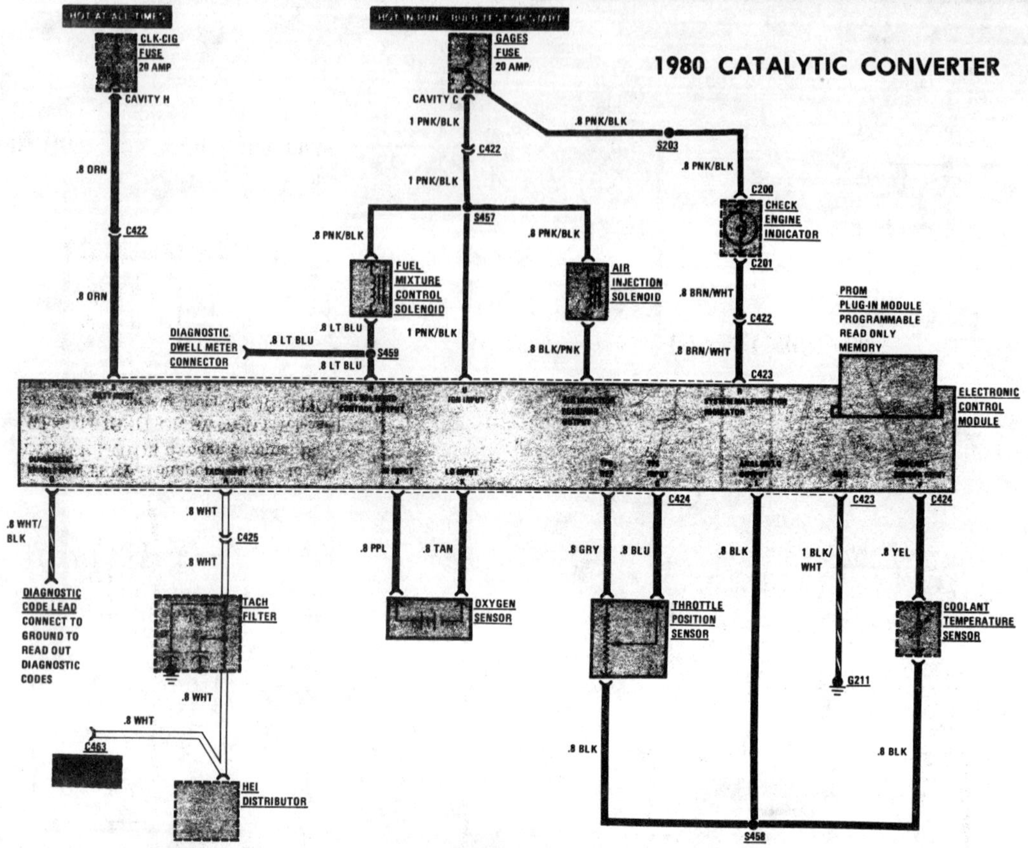

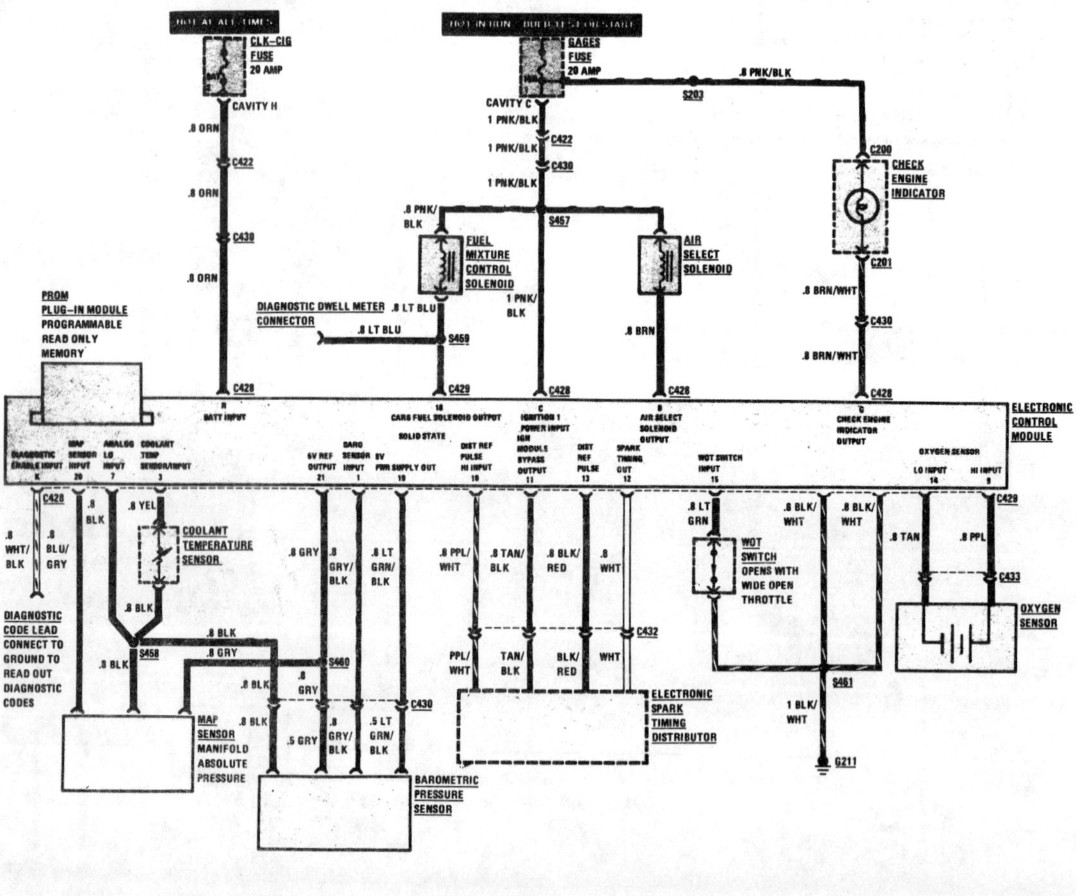

1980 TORQUE CONVERTER CLUTCH

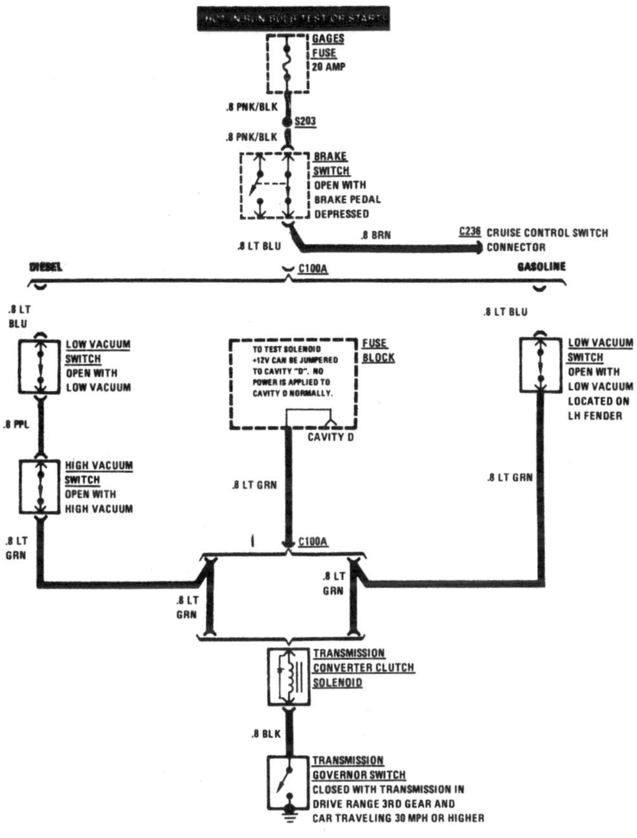

1980 IGNITION KEY WARNING

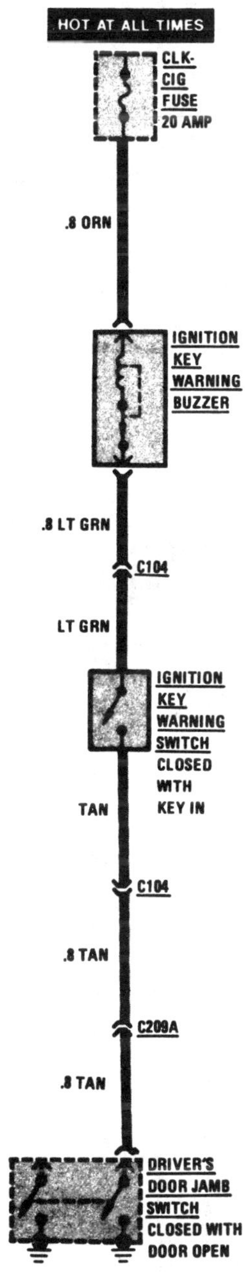

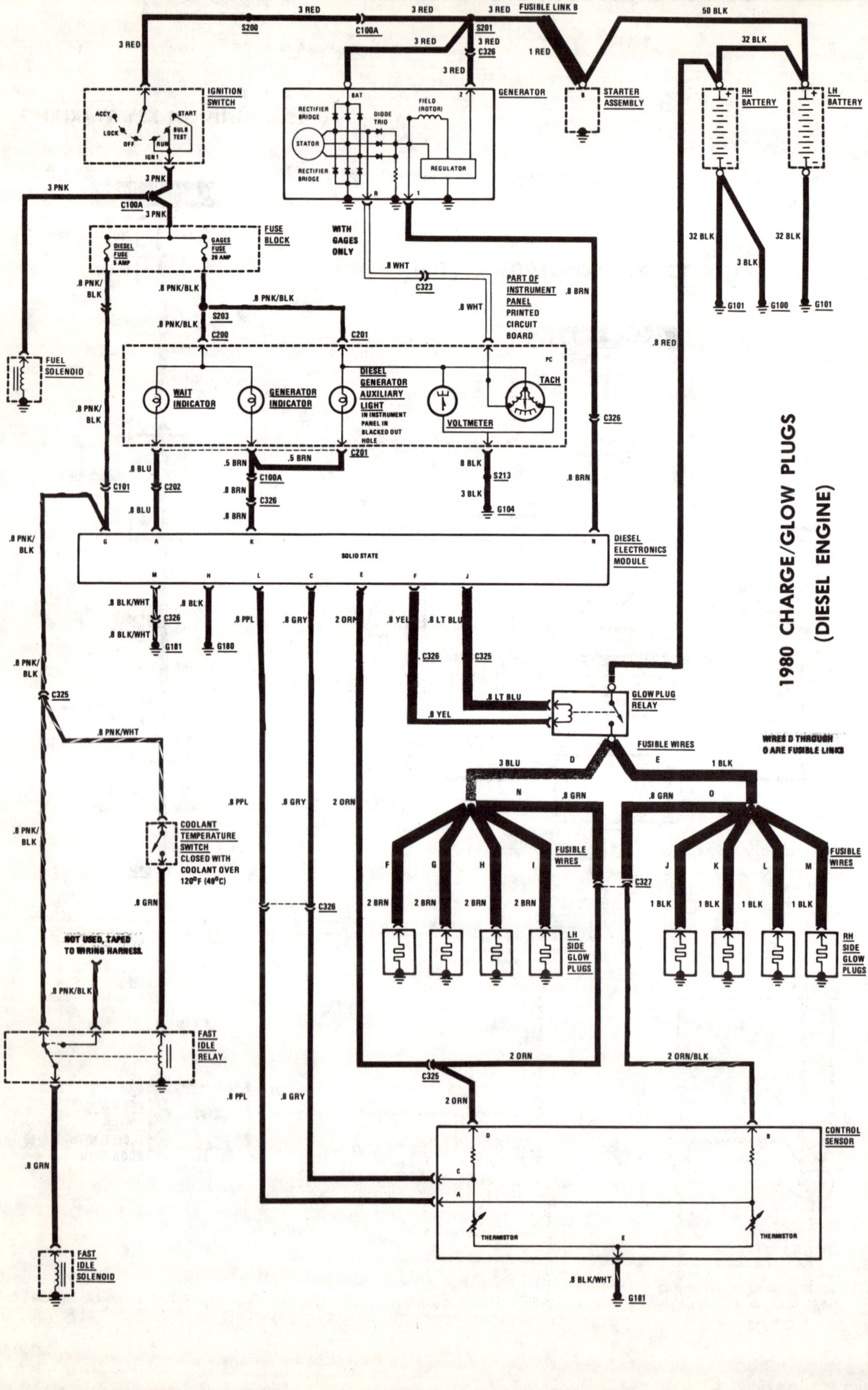

1980 CHARGE/GLOW PLUGS (DIESEL ENGINE)
FUSIBLE LINK B
IGNITION SWITCH
GENERATOR
STARTER ASSEMBLY
RH BATTERY
LH BATTERY
FUSE BLOCK
WITH GAGES ONLY
PART OF INSTRUMENT PANEL PRINTED CIRCUIT BOARD
FUEL SOLENOID
WAIT INDICATOR
GENERATOR INDICATOR
DIESEL GENERATOR AUXILIARY LIGHT IN INSTRUMENT PANEL IN BLACKED OUT HOLE
VOLTMETER
TACH
DIESEL ELECTRONICS MODULE
SOLID STATE
GLOW PLUG RELAY
FUSIBLE WIRES
WIRES D THROUGH O ARE FUSIBLE LINKS
COOLANT TEMPERATURE SWITCH CLOSED WITH COOLANT OVER 120°F (49°C)
LH SIDE GLOW PLUGS
RH SIDE GLOW PLUGS
NOT USED, TAPED TO WIRING HARNESS.
FAST IDLE RELAY
CONTROL SENSOR
THERMISTOR
FAST IDLE SOLENOID

1980 CORNERING LIGHTS

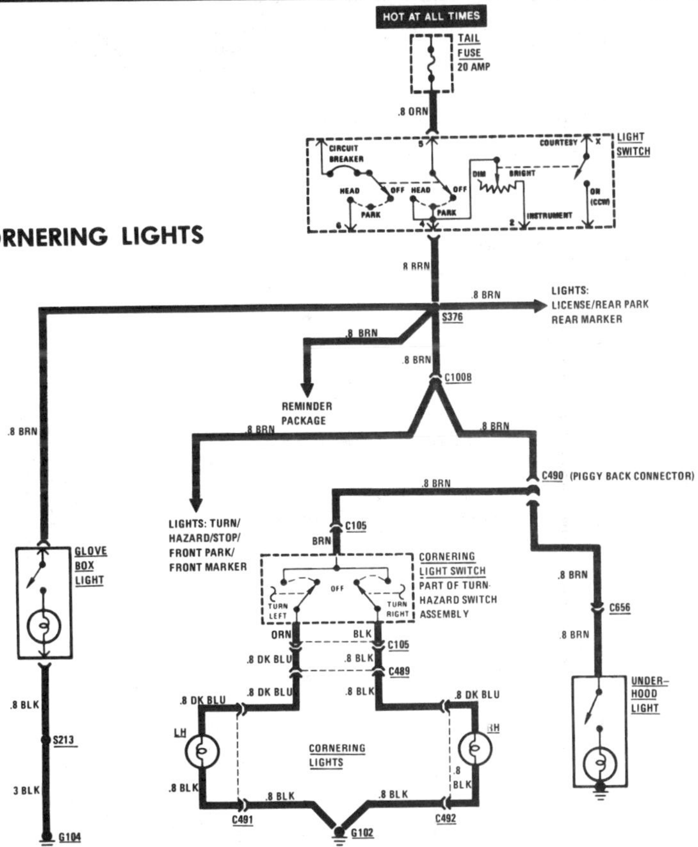

1980 HEADLIGHTS

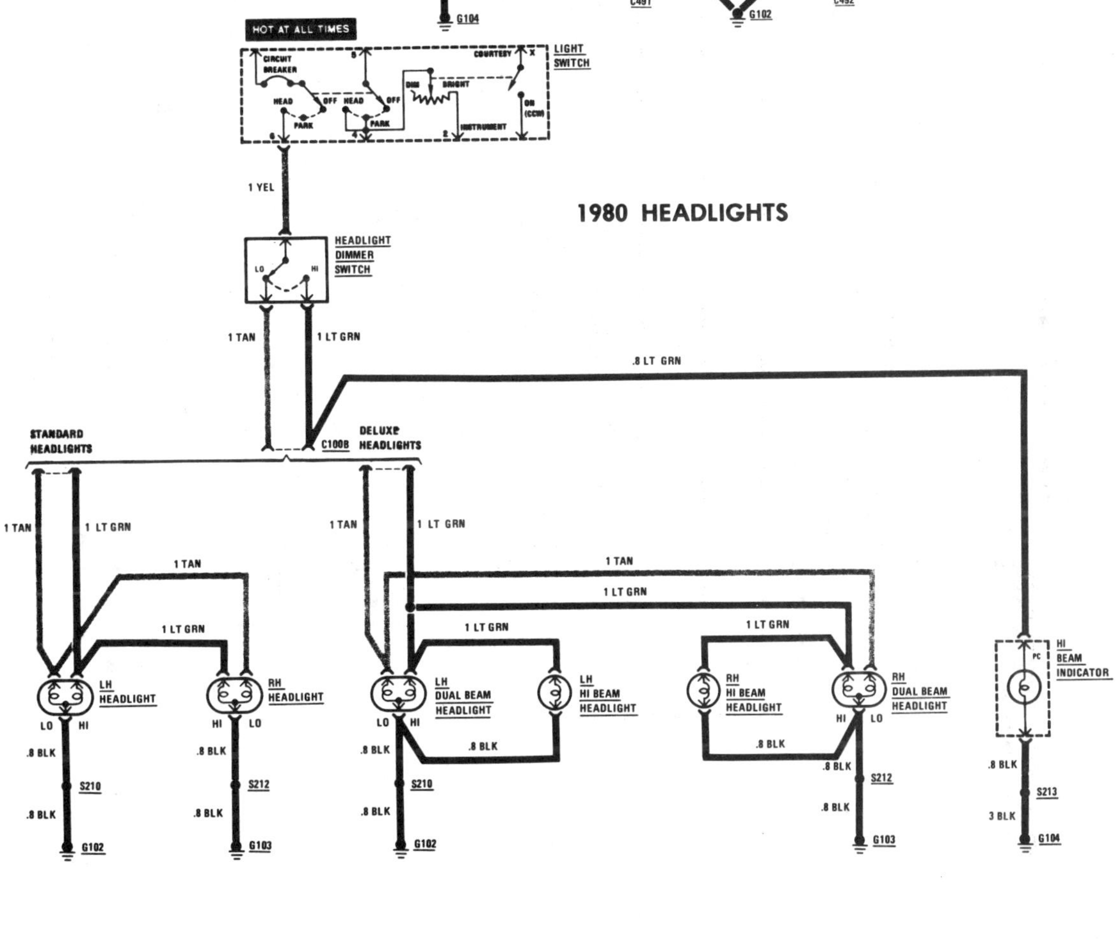

1980 TURN/HAZARD/STOP/FRONT PARK/FRONT MARKER

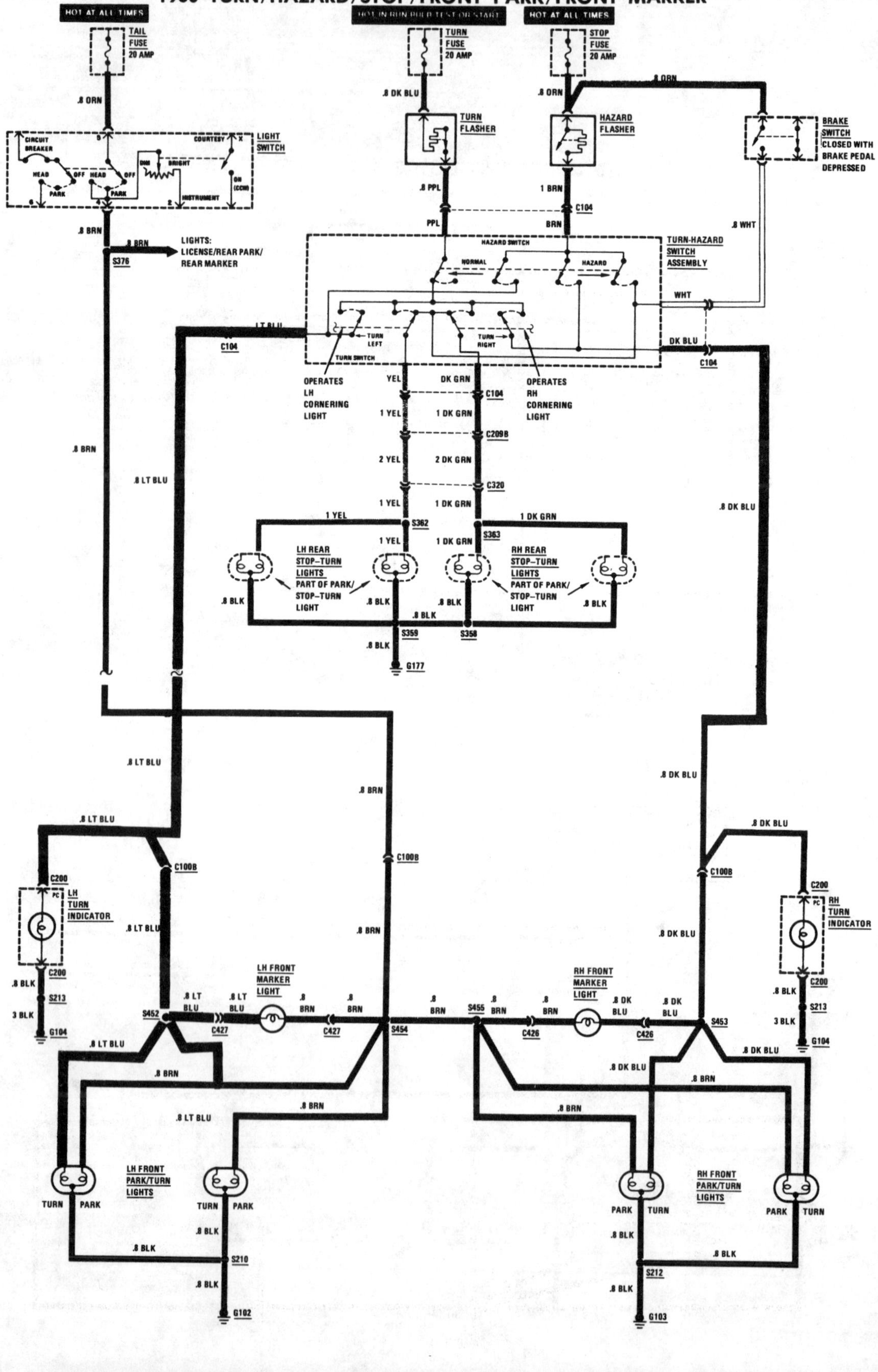

1980 TAIL/REAR MARKER/LICENSE (STATION WAGON)

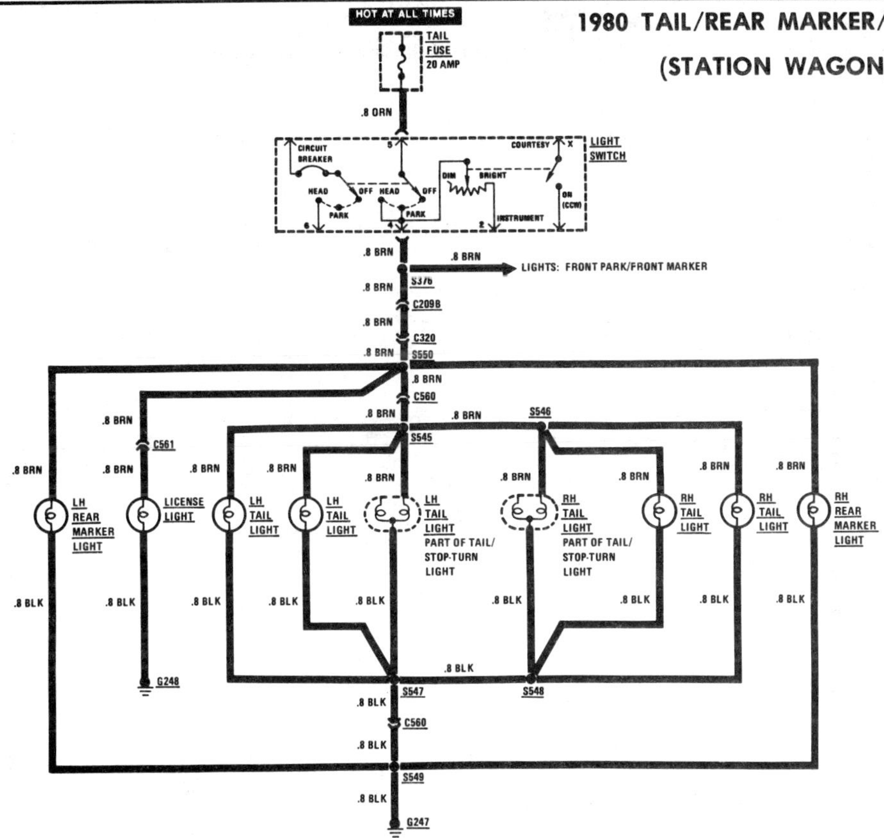

1980 TAIL/REAR MARKER/LICENSE (SEDAN)

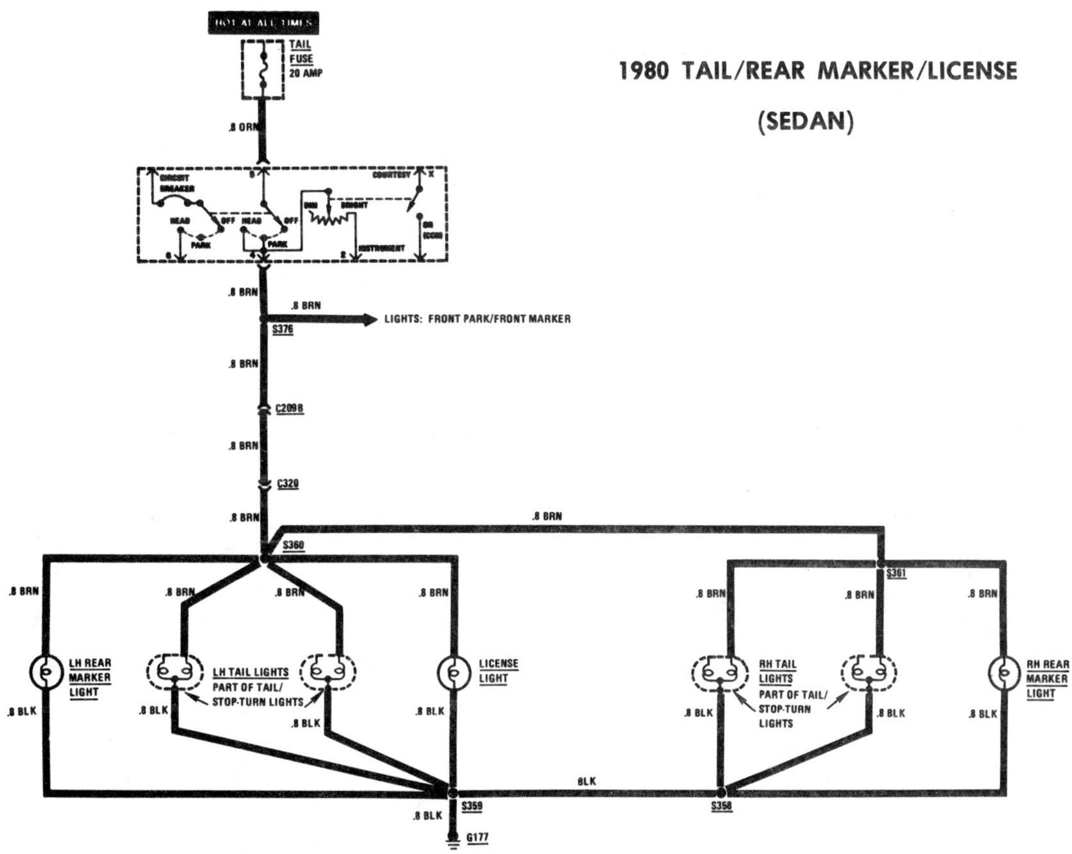

1980 COURTESY LAMPS

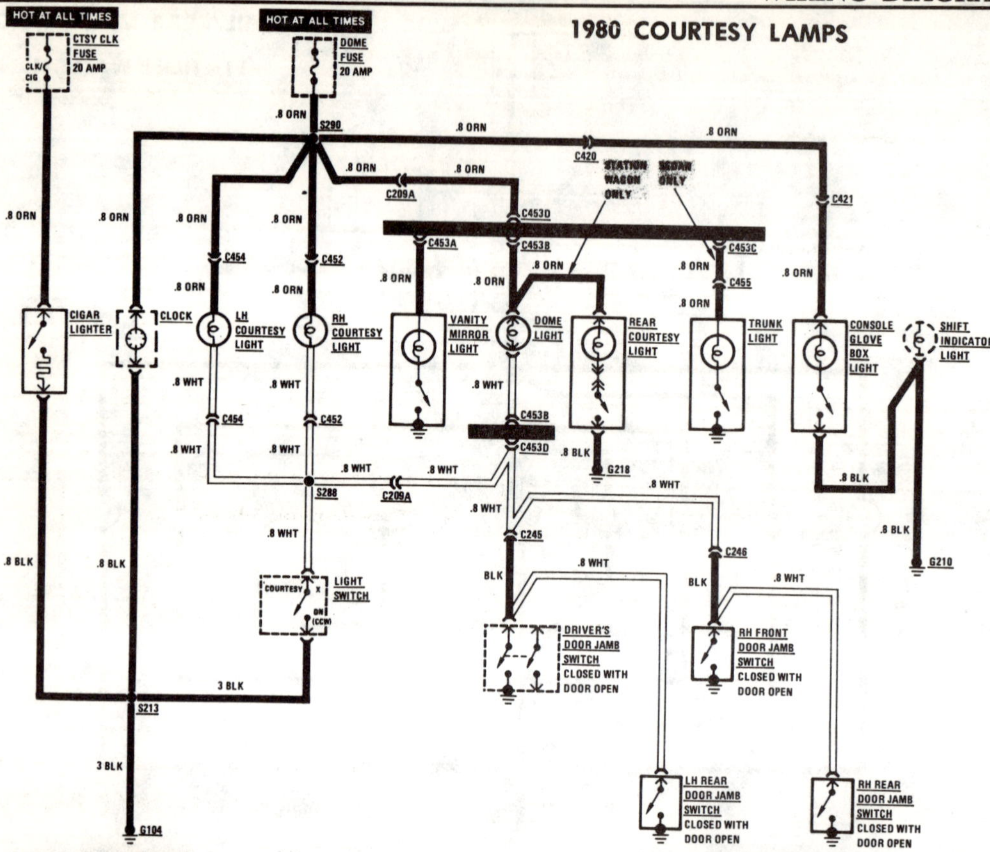

1980 BACK-UP LIGHTS

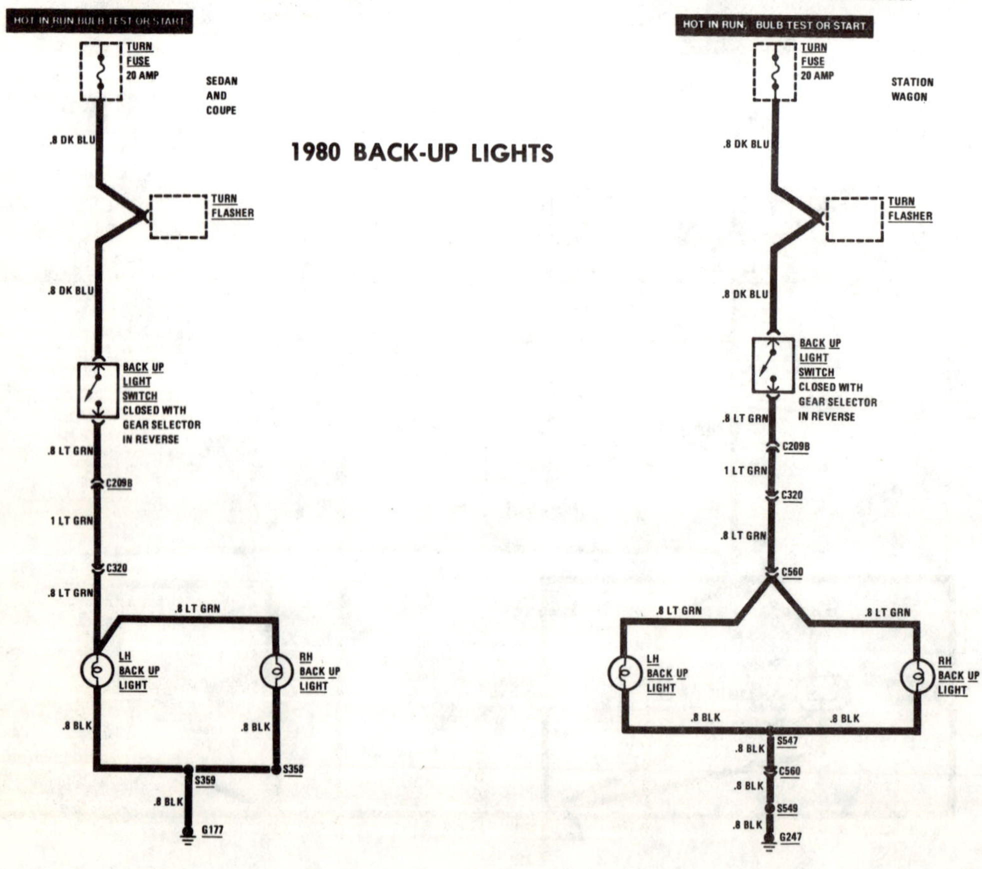

1980 INSTRUMENT PANEL

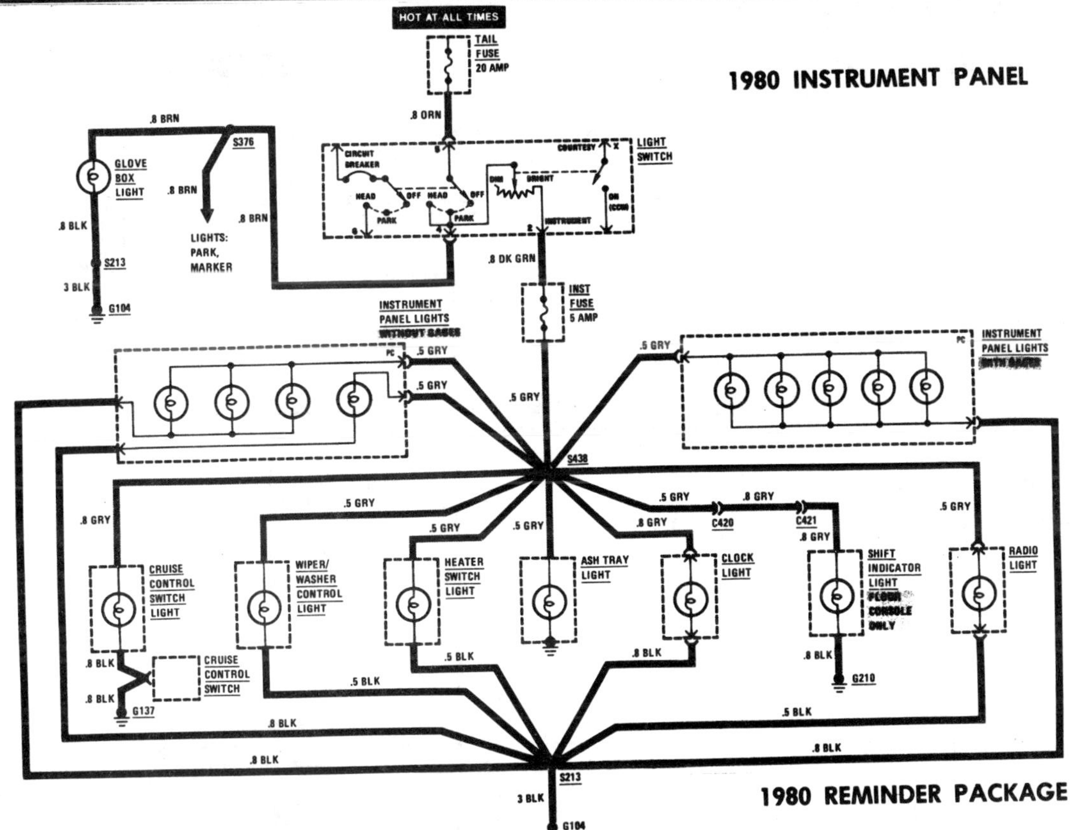

1980 REMINDER PACKAGE

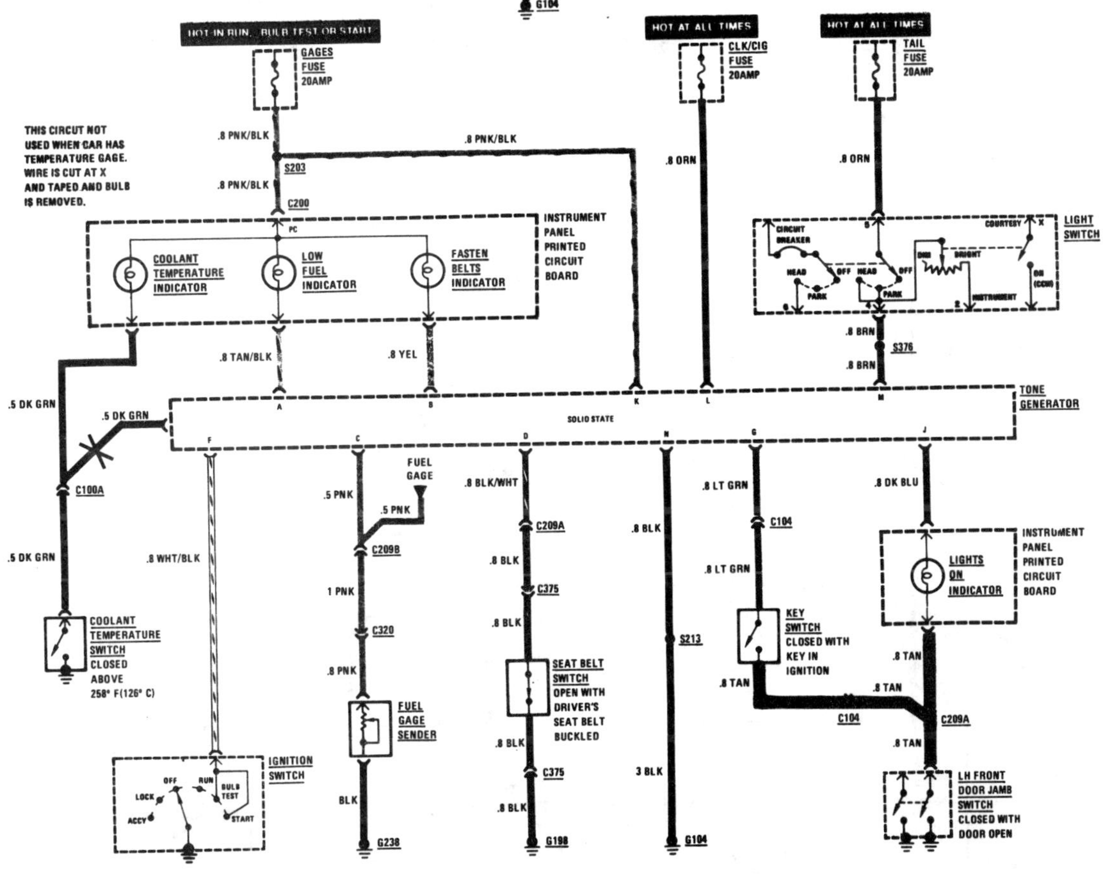

15

1980 WIPER/WASHER

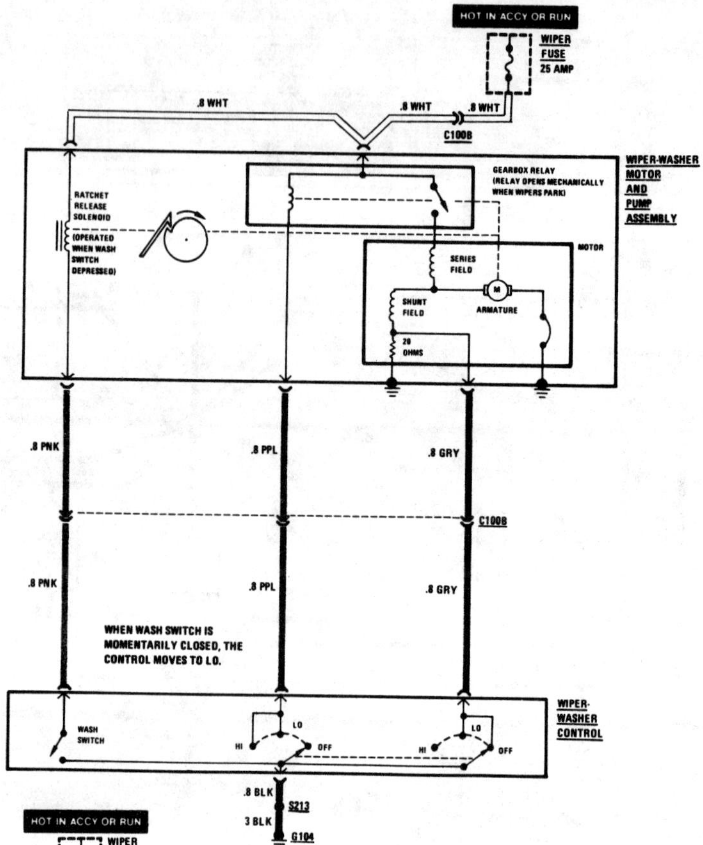

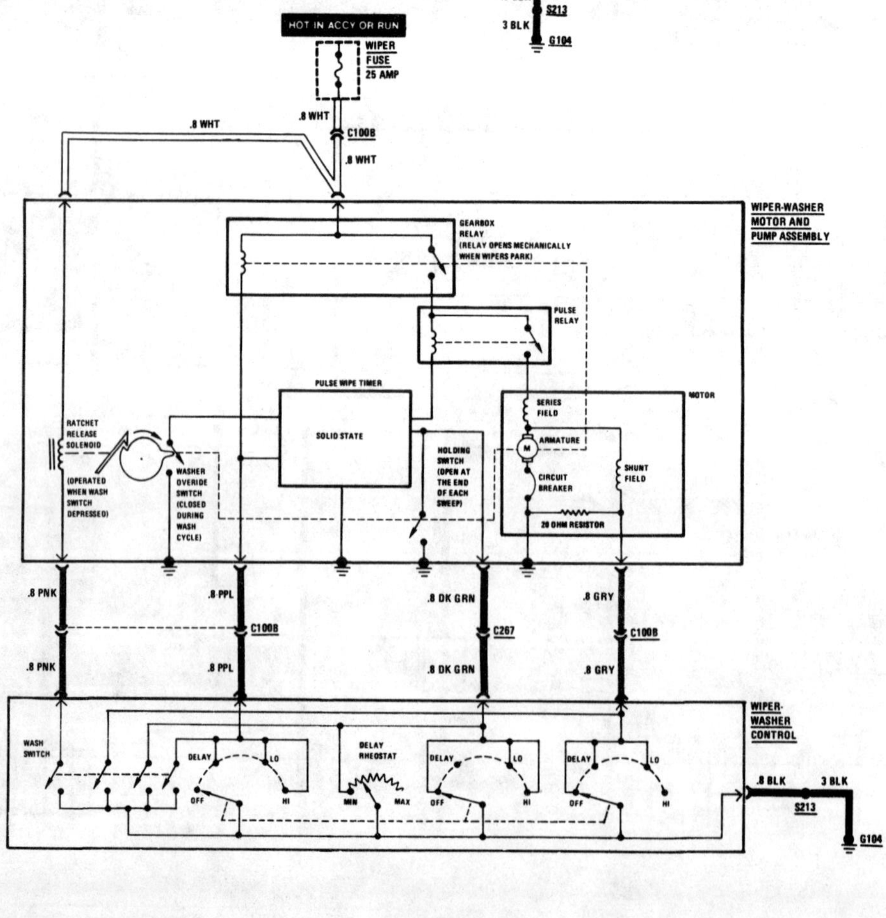

1980 HORN

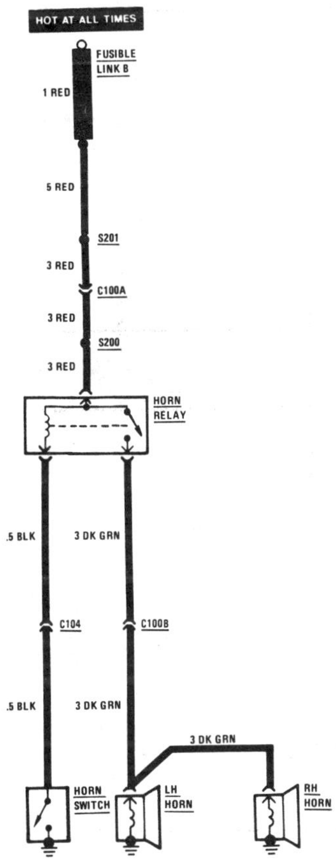

1980 SEAT BELT WARNING

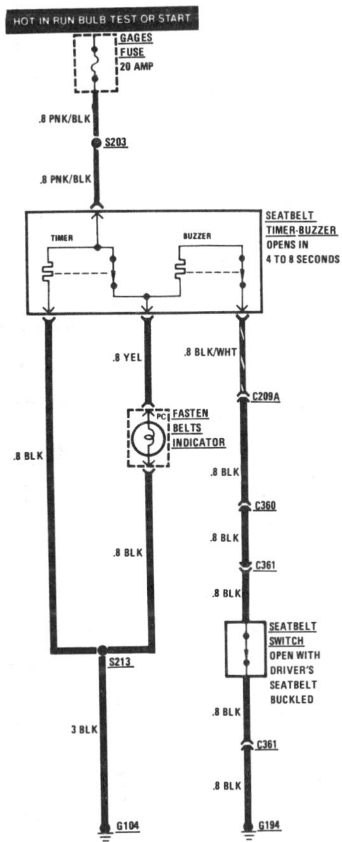

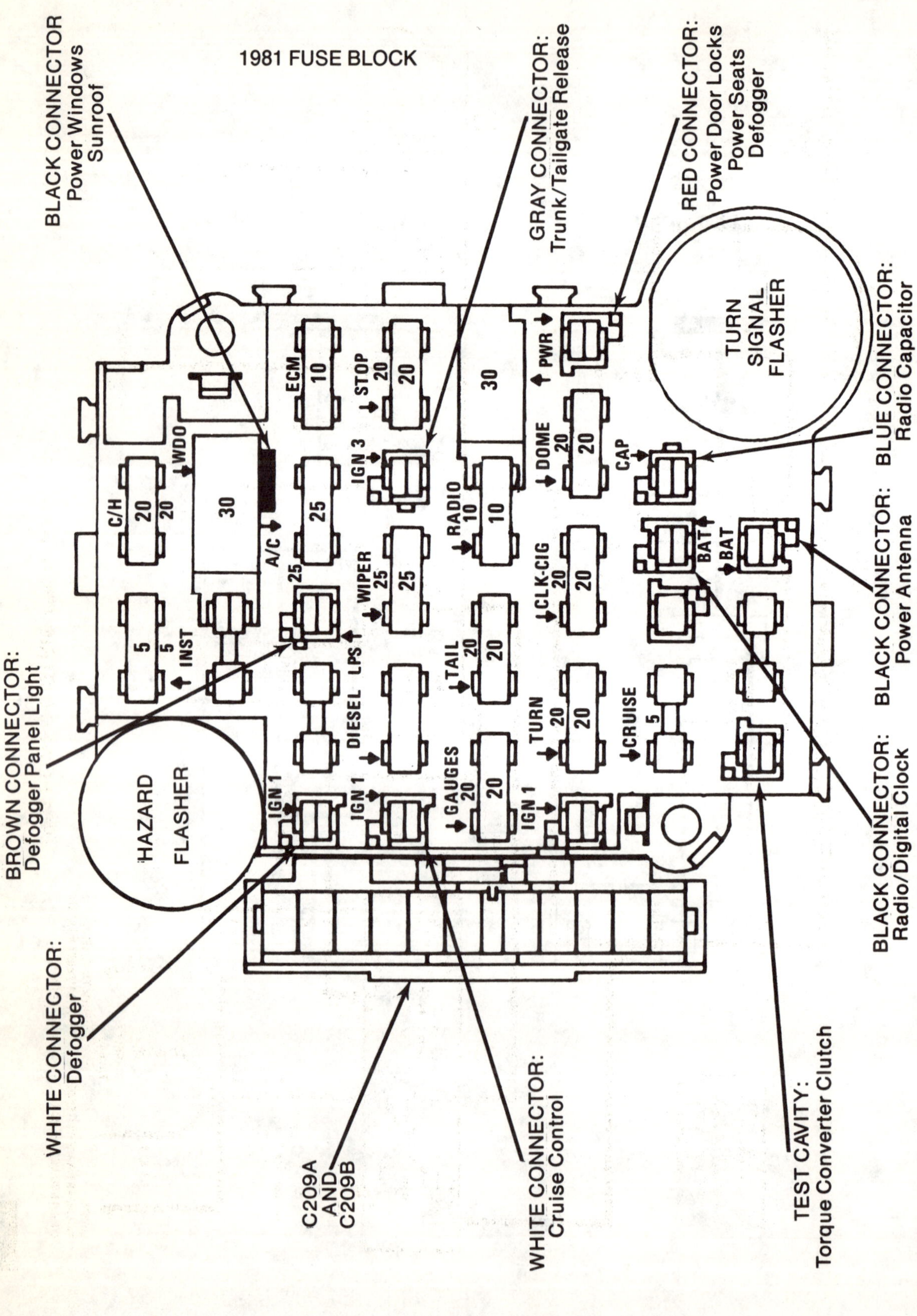
1981 FUSE BLOCK
BLACK CONNECTOR: Power Windows Sunroof
GRAY CONNECTOR: Trunk/Tailgate Release
RED CONNECTOR: Power Door Locks Power Seats Defogger
TURN SIGNAL FLASHER
BLUE CONNECTOR: Radio Capacitor
BLACK CONNECTOR: Power Antenna
BLACK CONNECTOR: Radio/Digital Clock
TEST CAVITY: Torque Converter Clutch
WHITE CONNECTOR: Cruise Control
C209A AND C209B
WHITE CONNECTOR: Defogger
BROWN CONNECTOR: Defogger Panel Light
HAZARD FLASHER
C/H 20
WDO 20
30
A/C
INST 5
ECM 10
25
LPS 25
DIESEL
STOP 20
IGN 3
WIPER 25
IGN 1
GAUGES 20
RADIO 10
TAIL 20
PWR
DOME 20
CLK-CIG 20
TURN 20
CAP
BAT
CRUISE 5

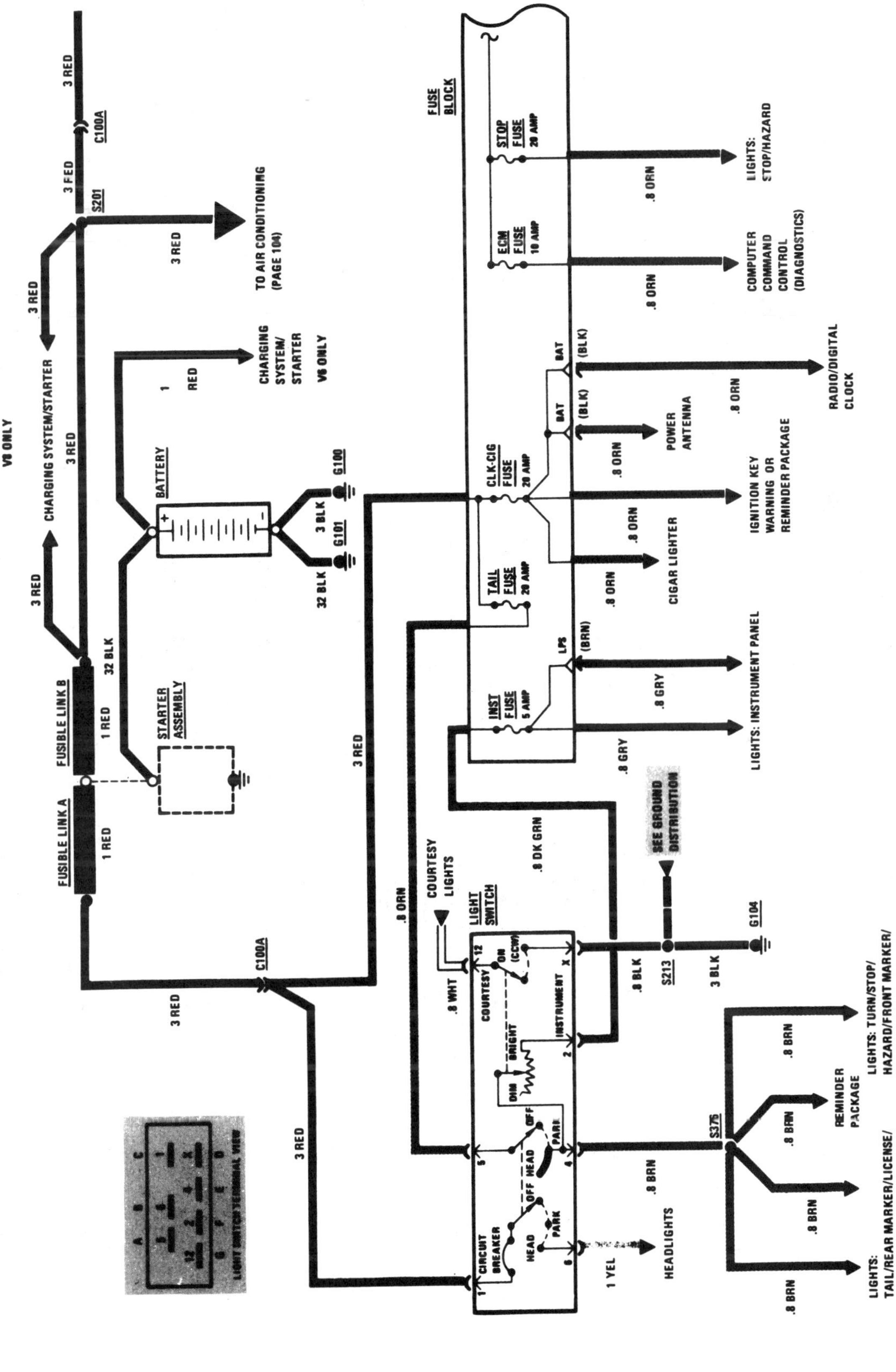

1981 POWER DISTRIBUTION (GASOLINE)

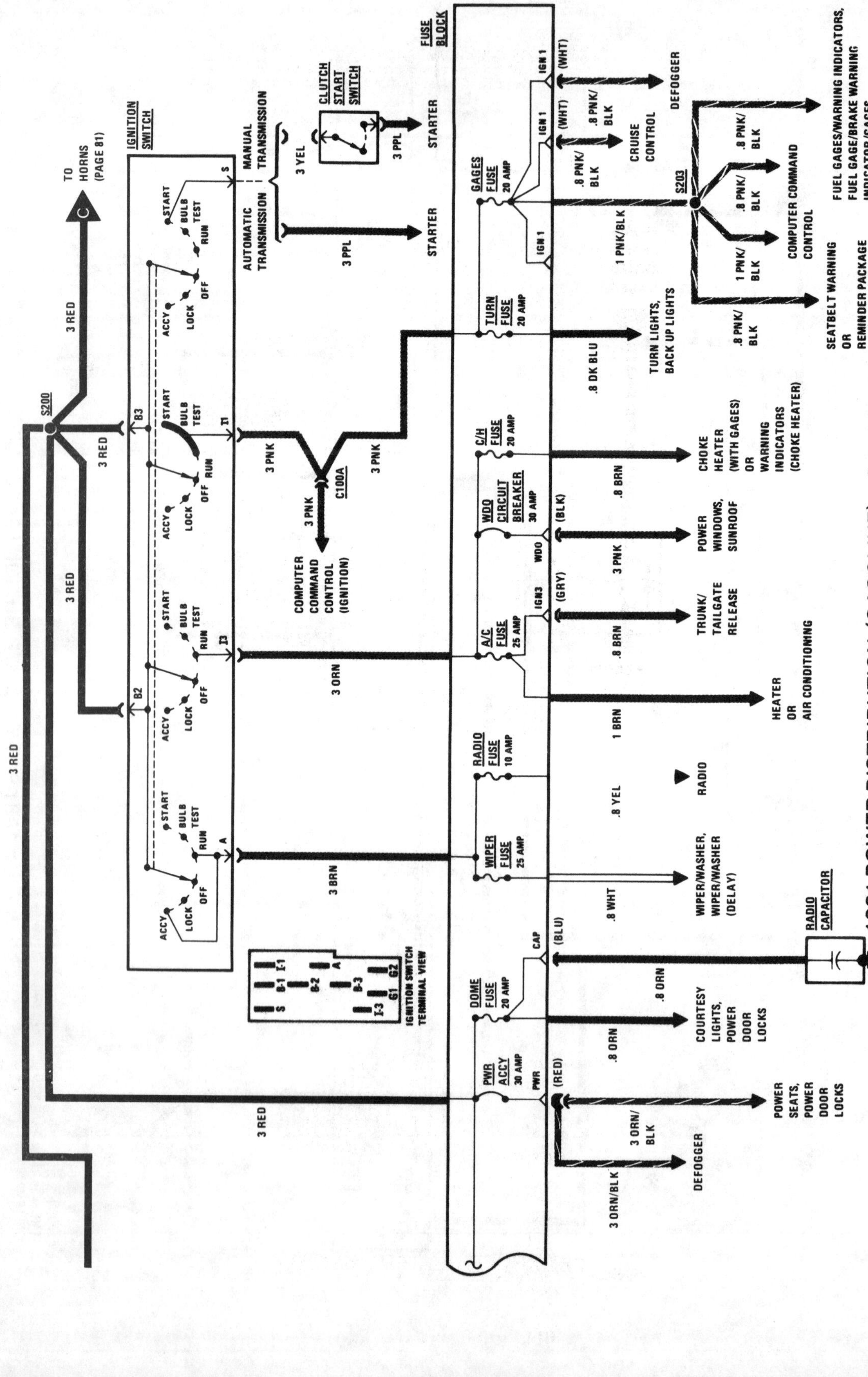

1981 POWER DISTRIBUTION (GASOLINE)—Continued

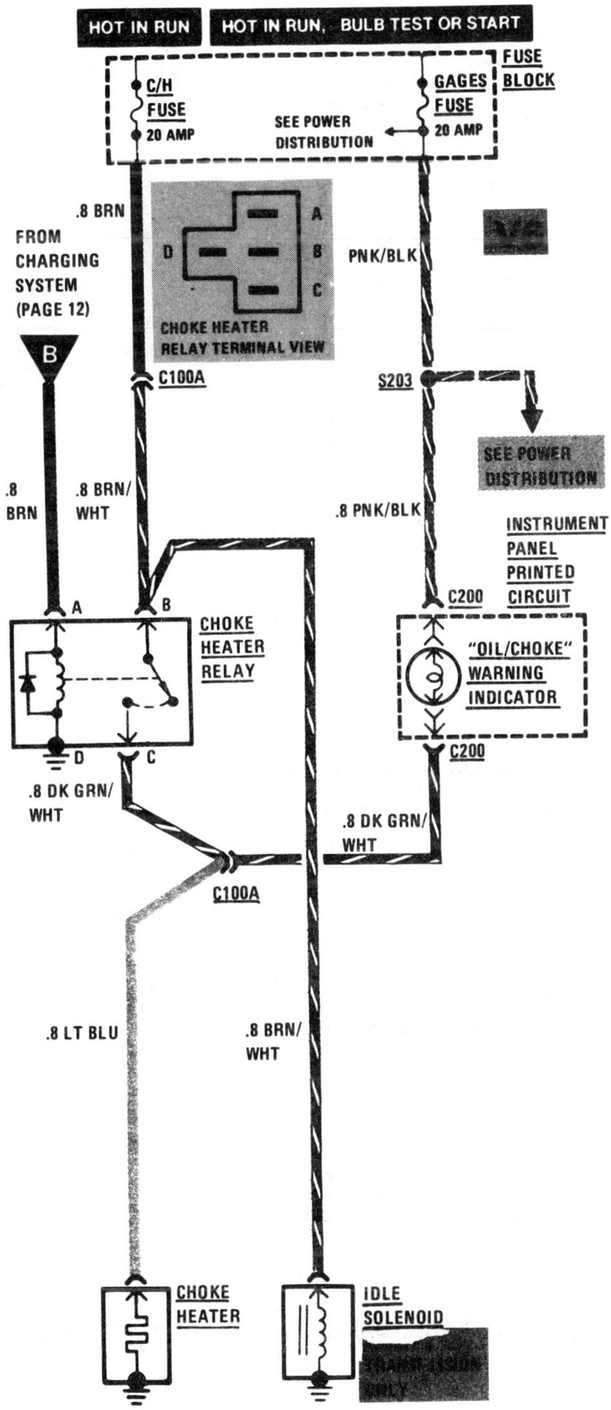

1981 CHOKE HEATER—V6 ONLY

15

1981 CHARGING SYSTEM/STARTER (GASOLINE)

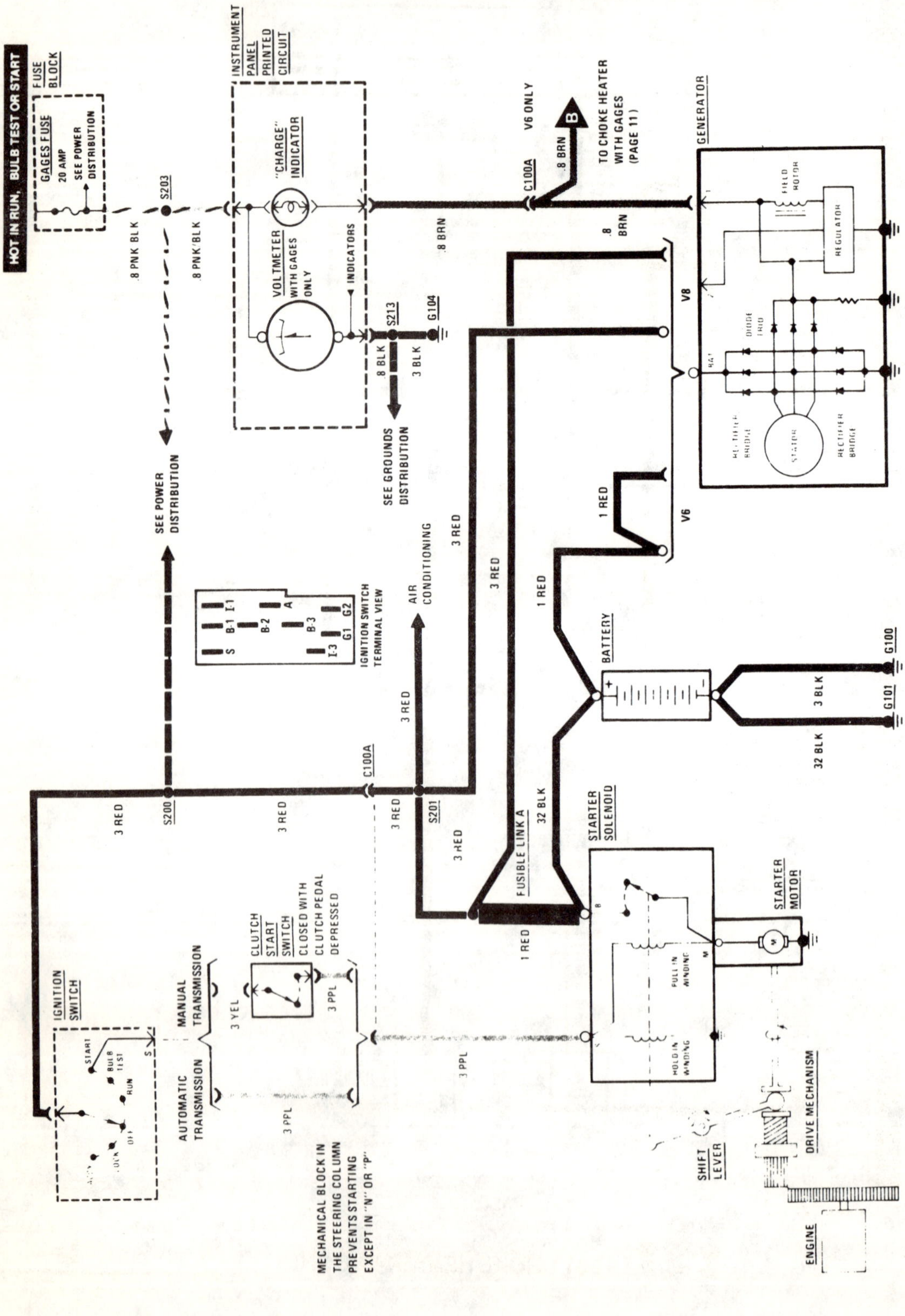

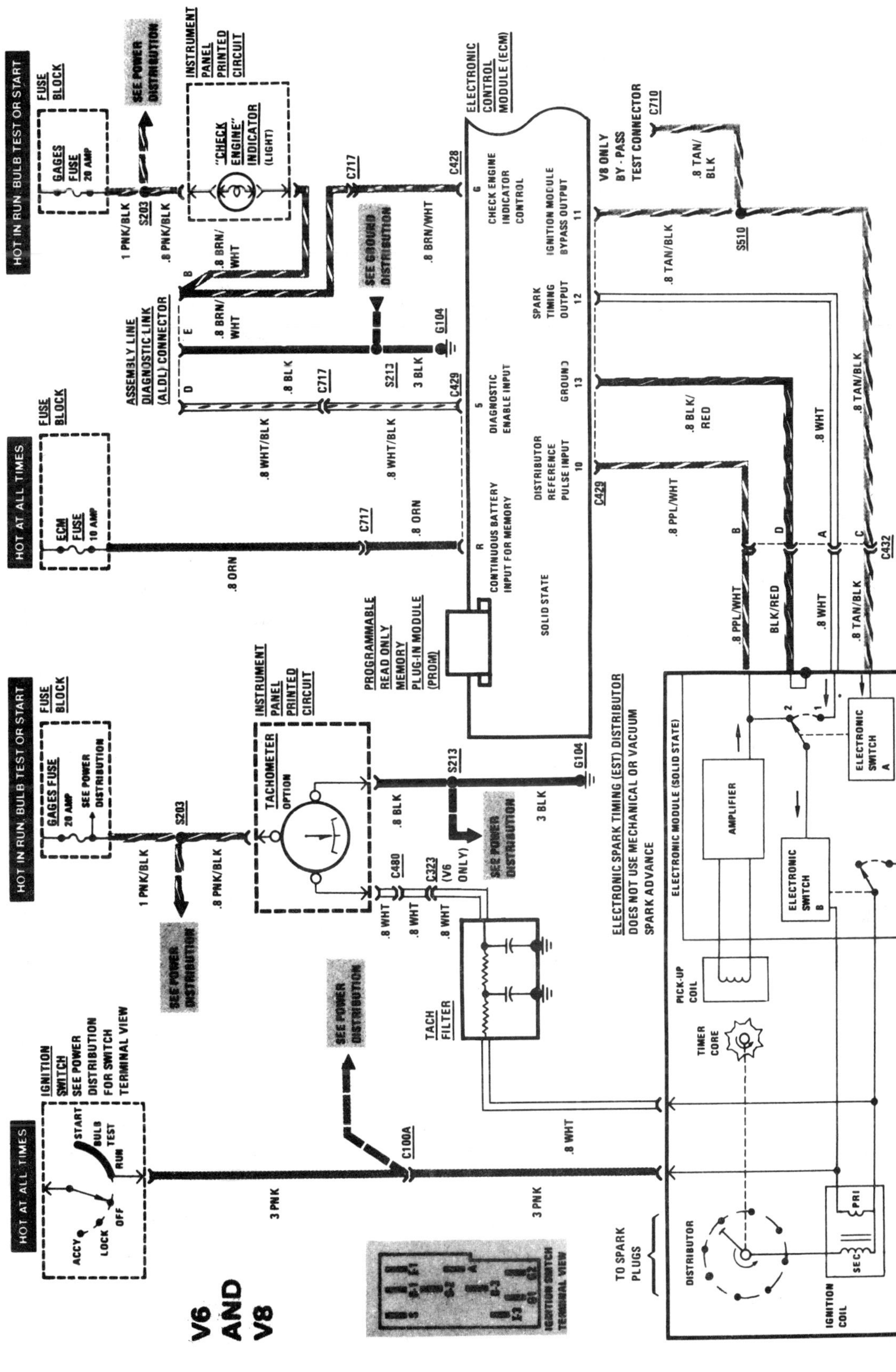

1981 COMPUTER COMMAND CONTROL (DIAGNOSTICS, IGNITION)

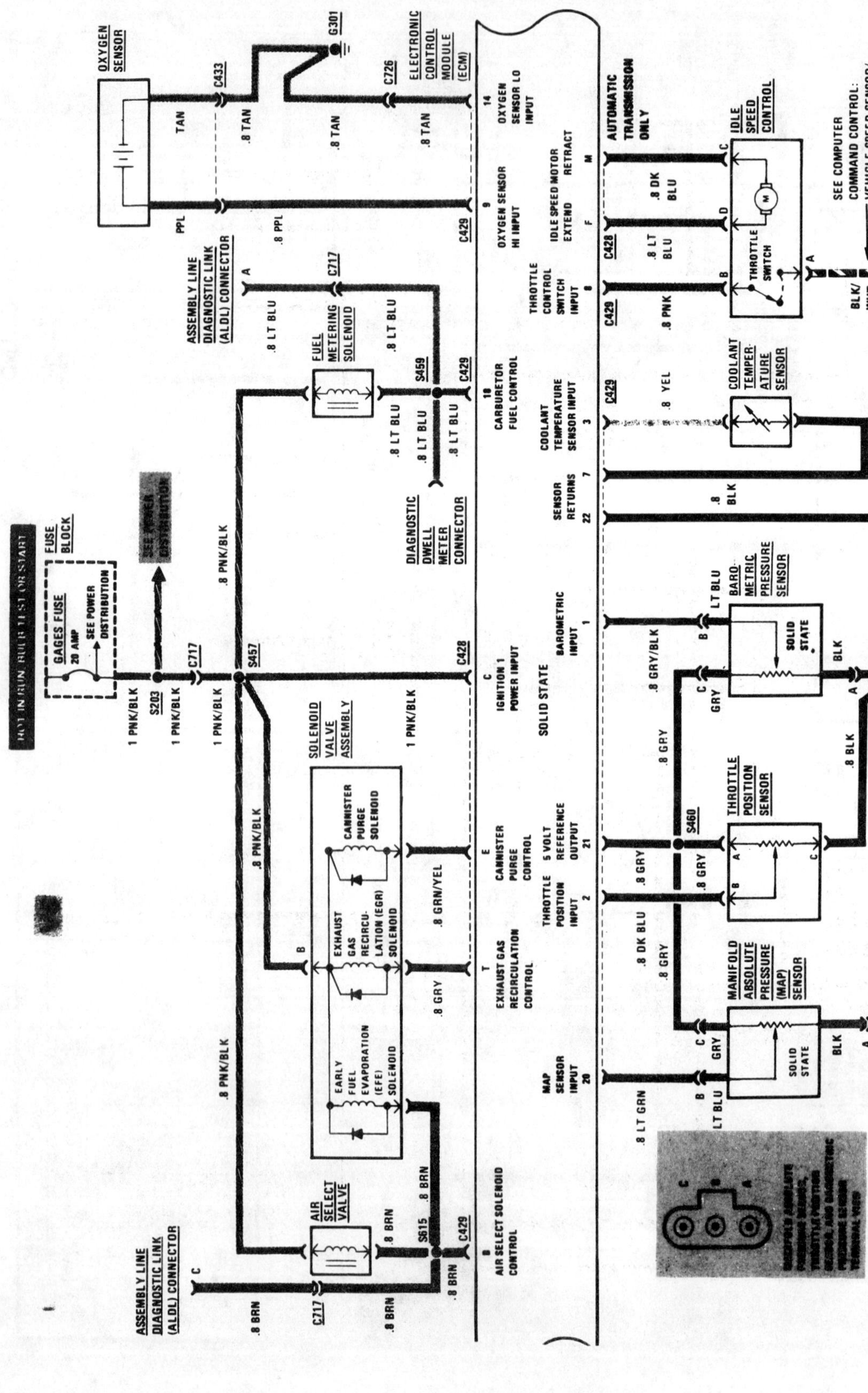

1981 COMPUTER COMMAND CONTROL—V6
(ENGINE DATA SENSORS, EMISSION CONTROLS)

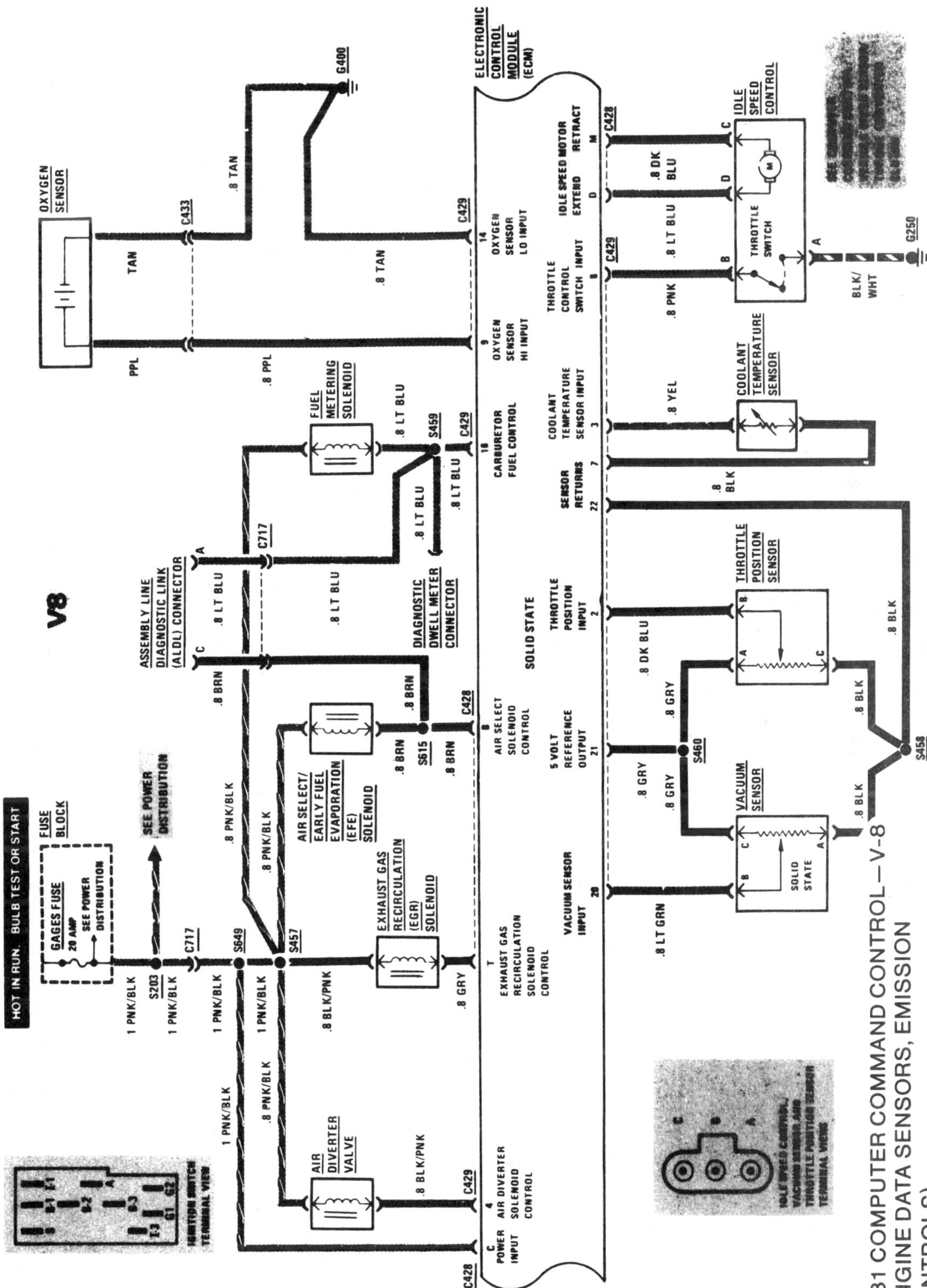

1981 COMPUTER COMMAND CONTROL—V-8 (ENGINE DATA SENSORS, EMISSION CONTROLS)

V6 AND V8

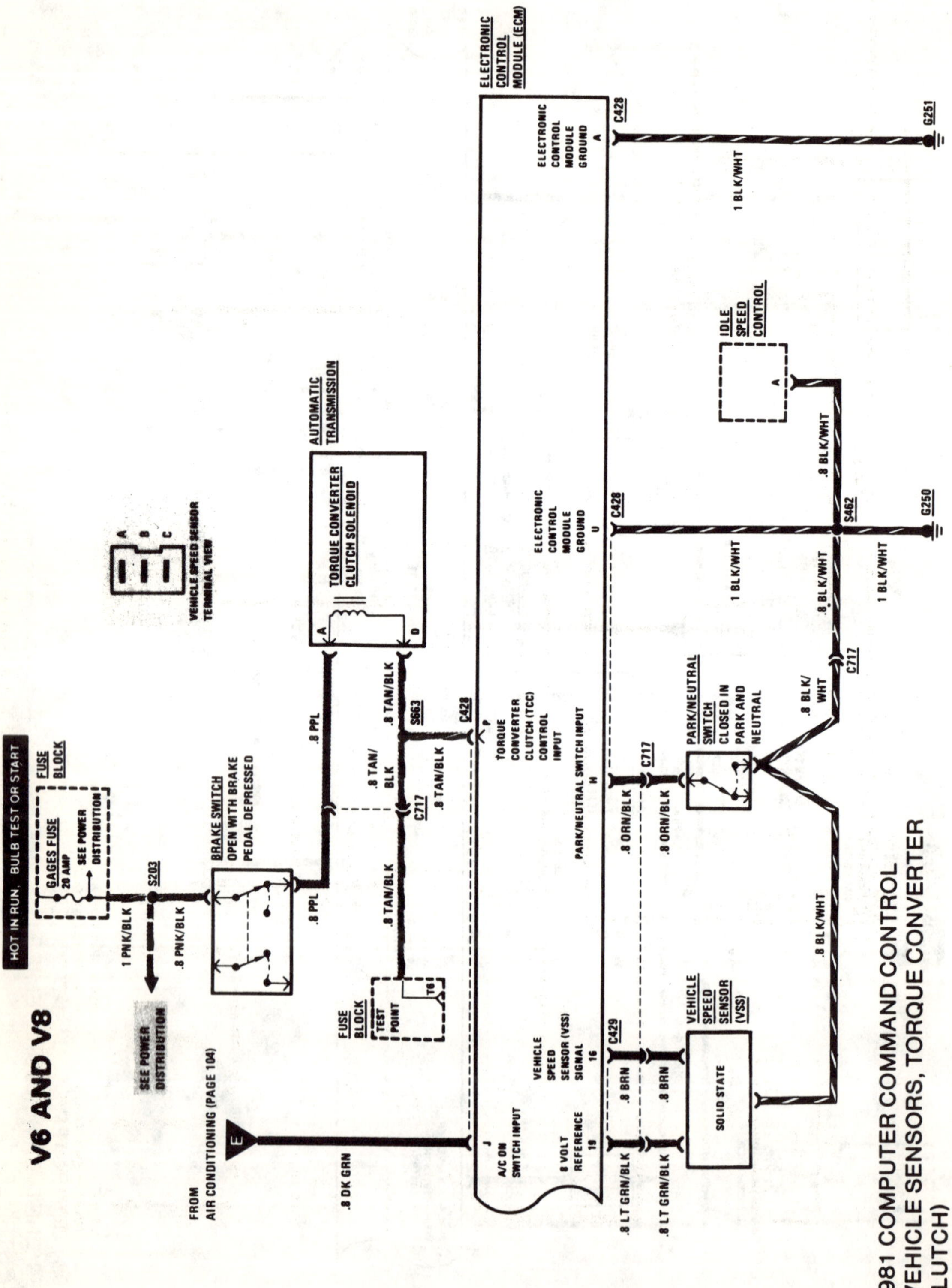

1981 COMPUTER COMMAND CONTROL
(VEHICLE SENSORS, TORQUE CONVERTER CLUTCH)

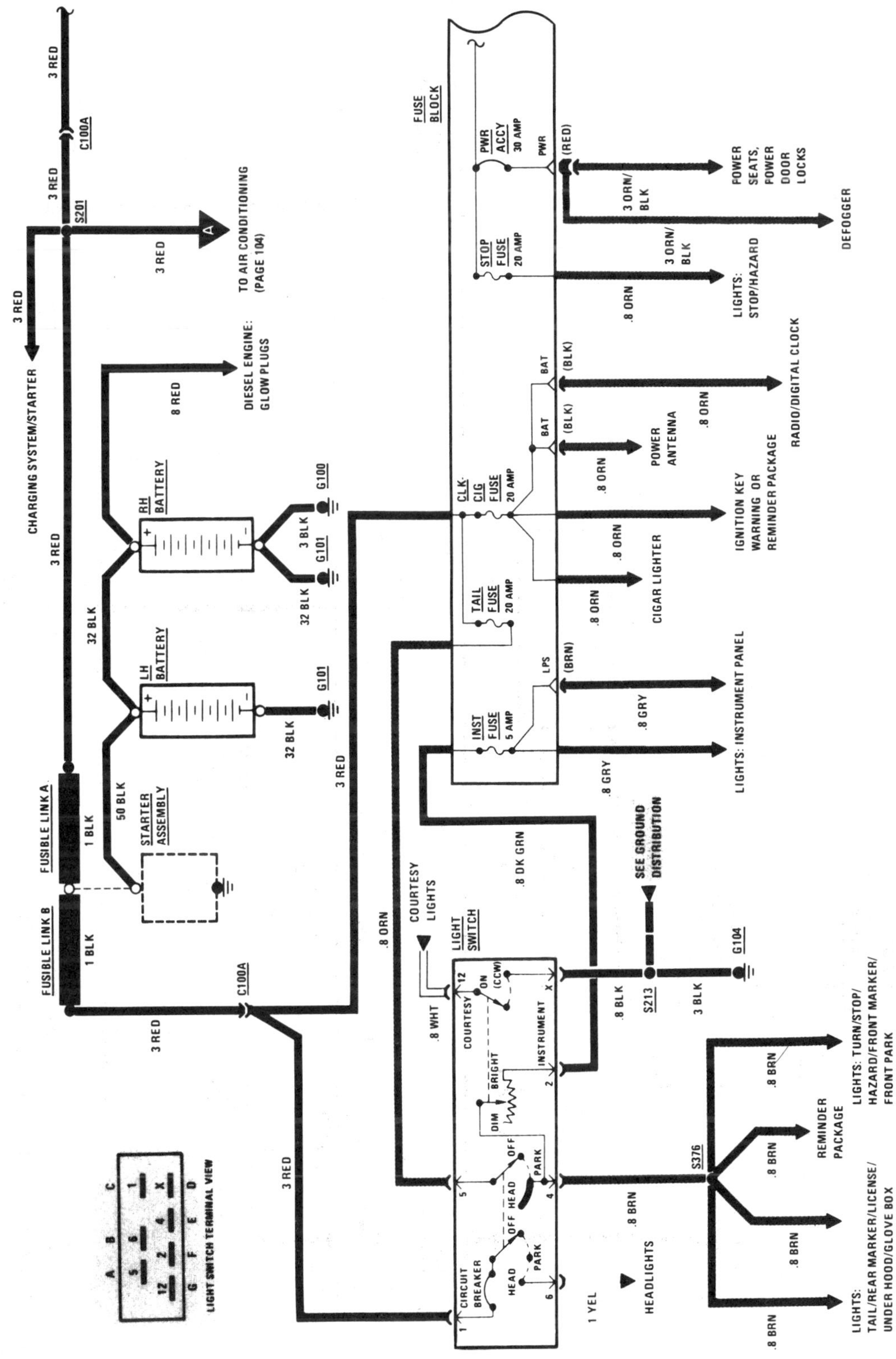

1981 POWER DISTRIBUTION (DIESEL)

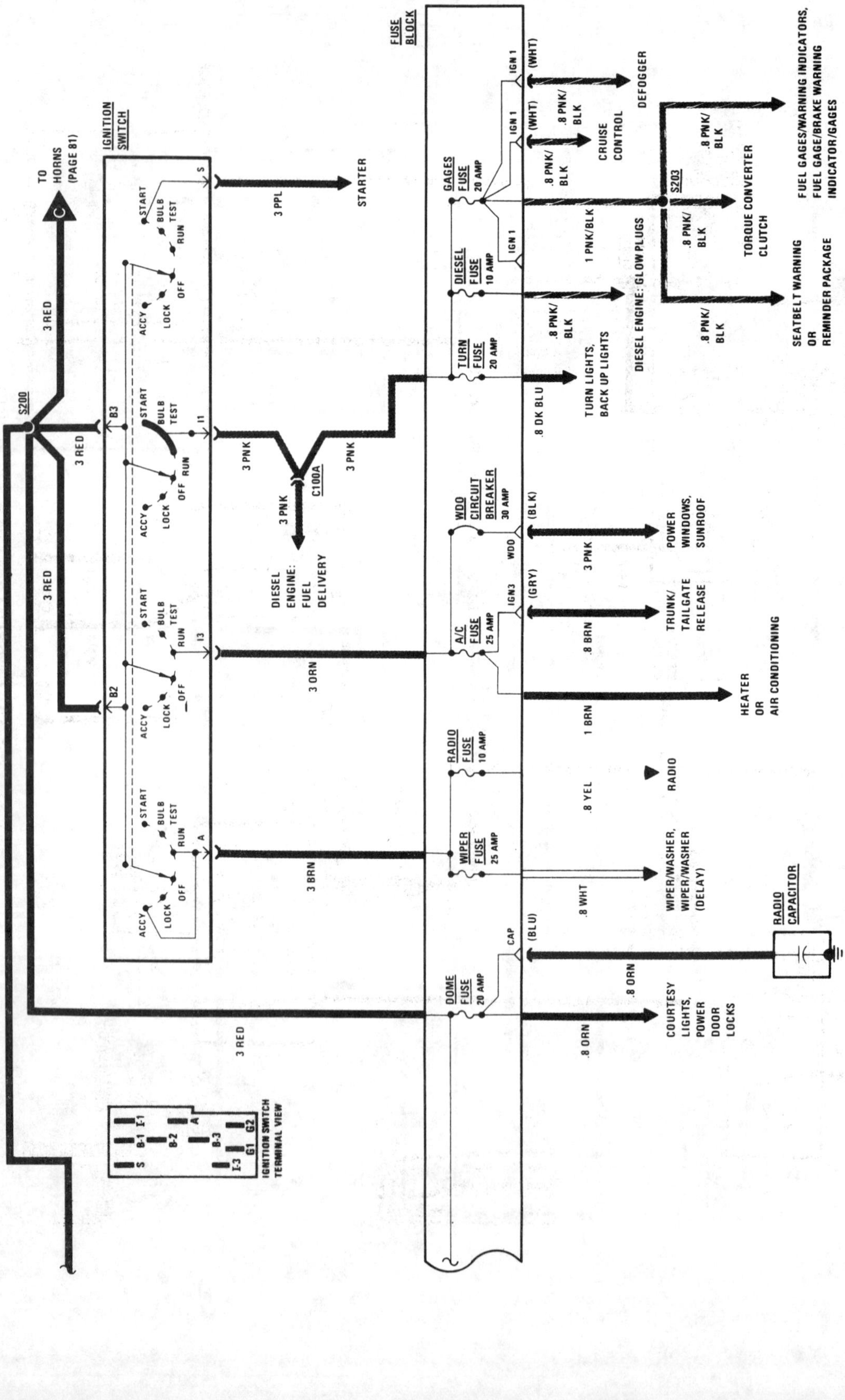

1981 POWER DISTRIBUTION (DIESEL)—Continued

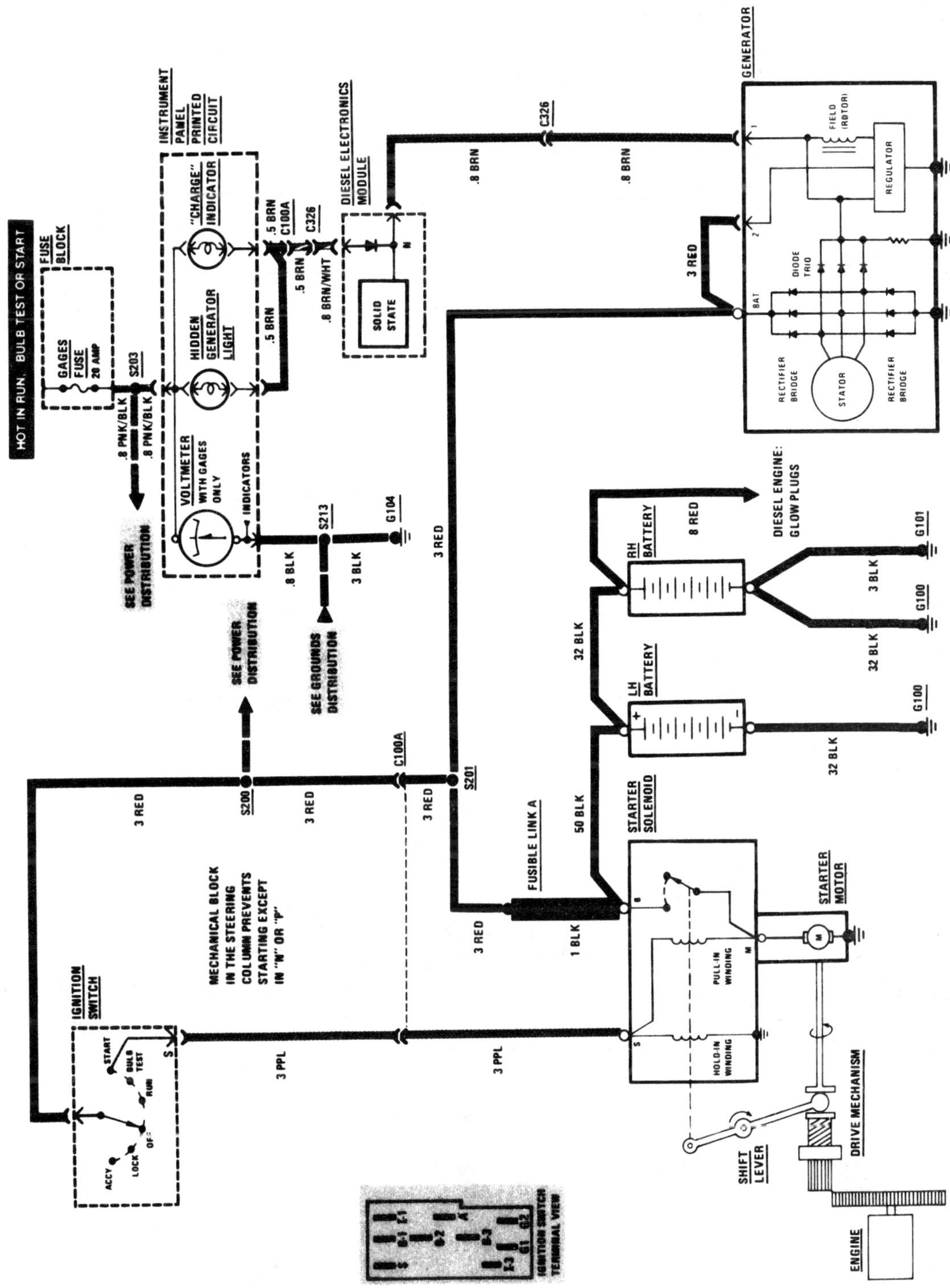

1981 CHARGING SYSTEM/STARTER (DIESEL)

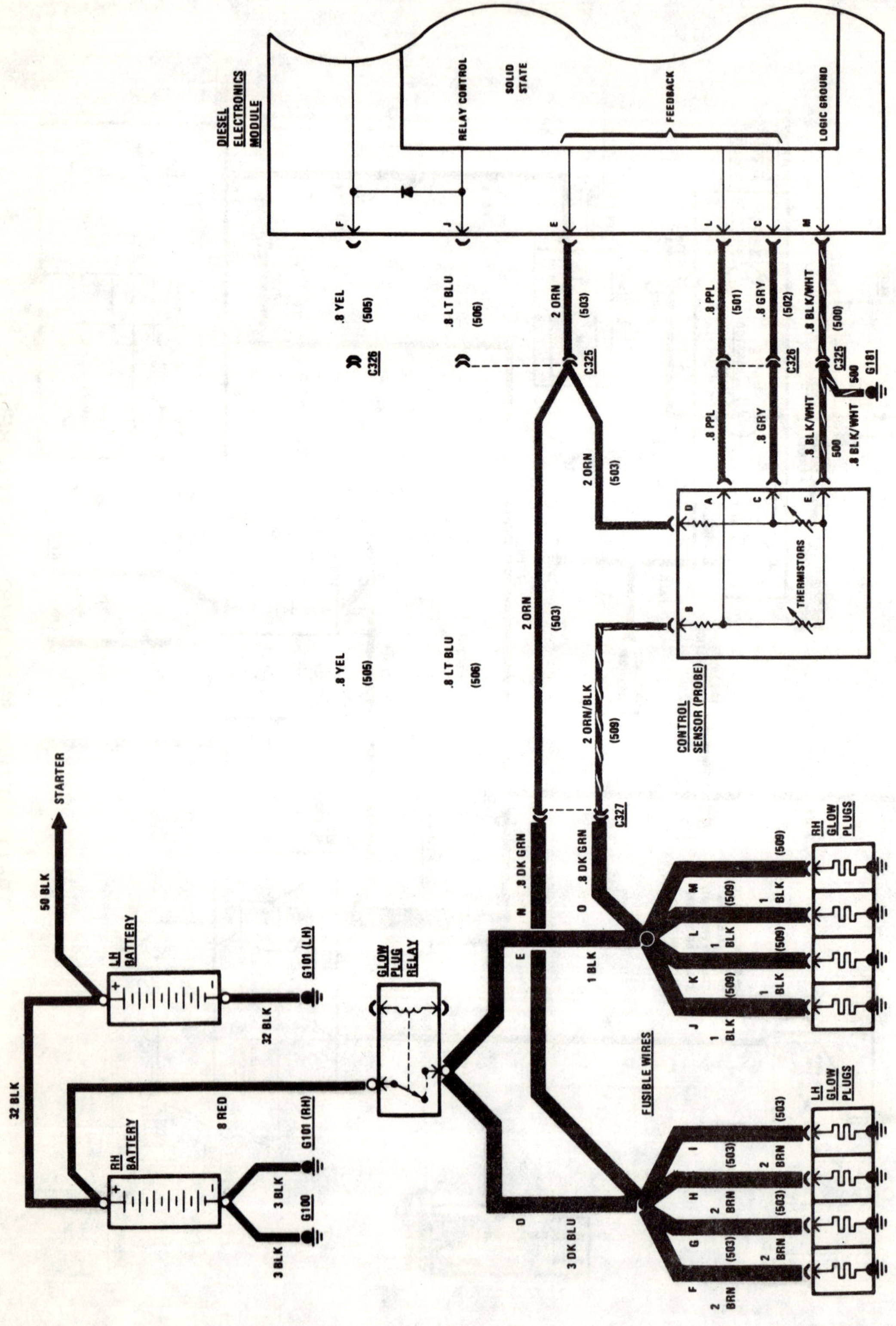

1981 GLOW PLUGS (DIESEL)

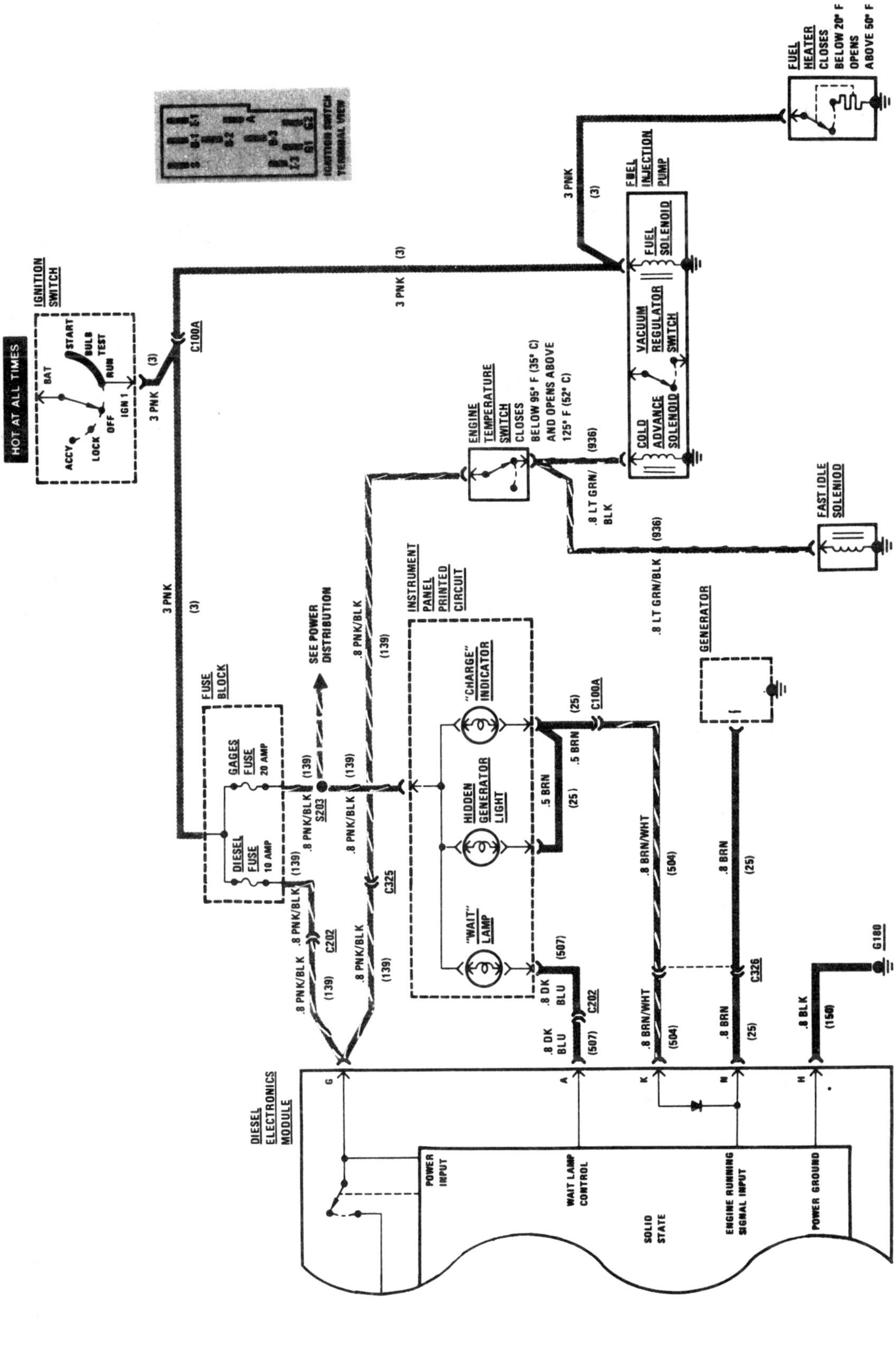

1981 GLOW PLUGS (DIESEL)—Continued

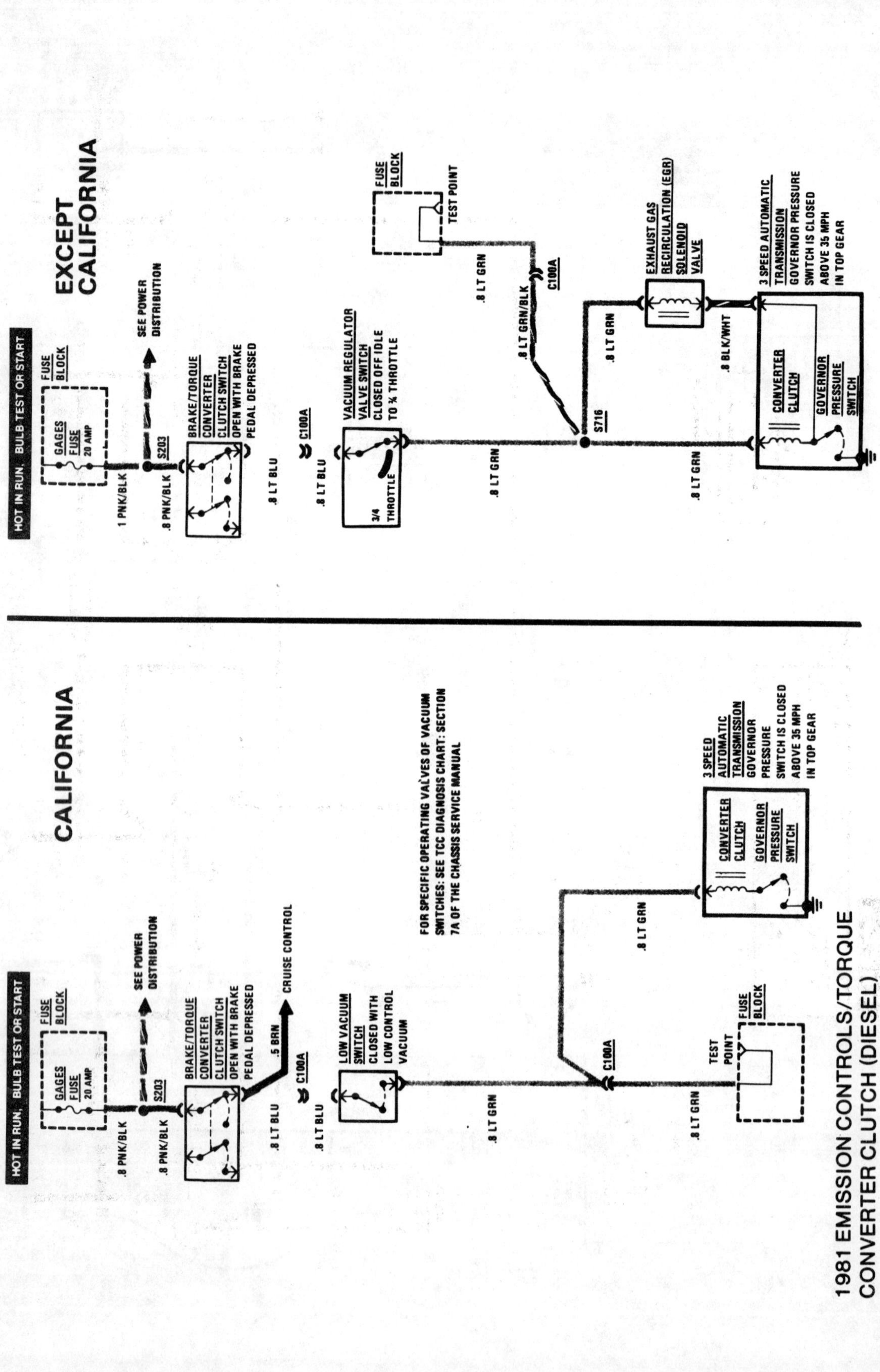

1981 EMISSION CONTROLS/TORQUE CONVERTER CLUTCH (DIESEL)

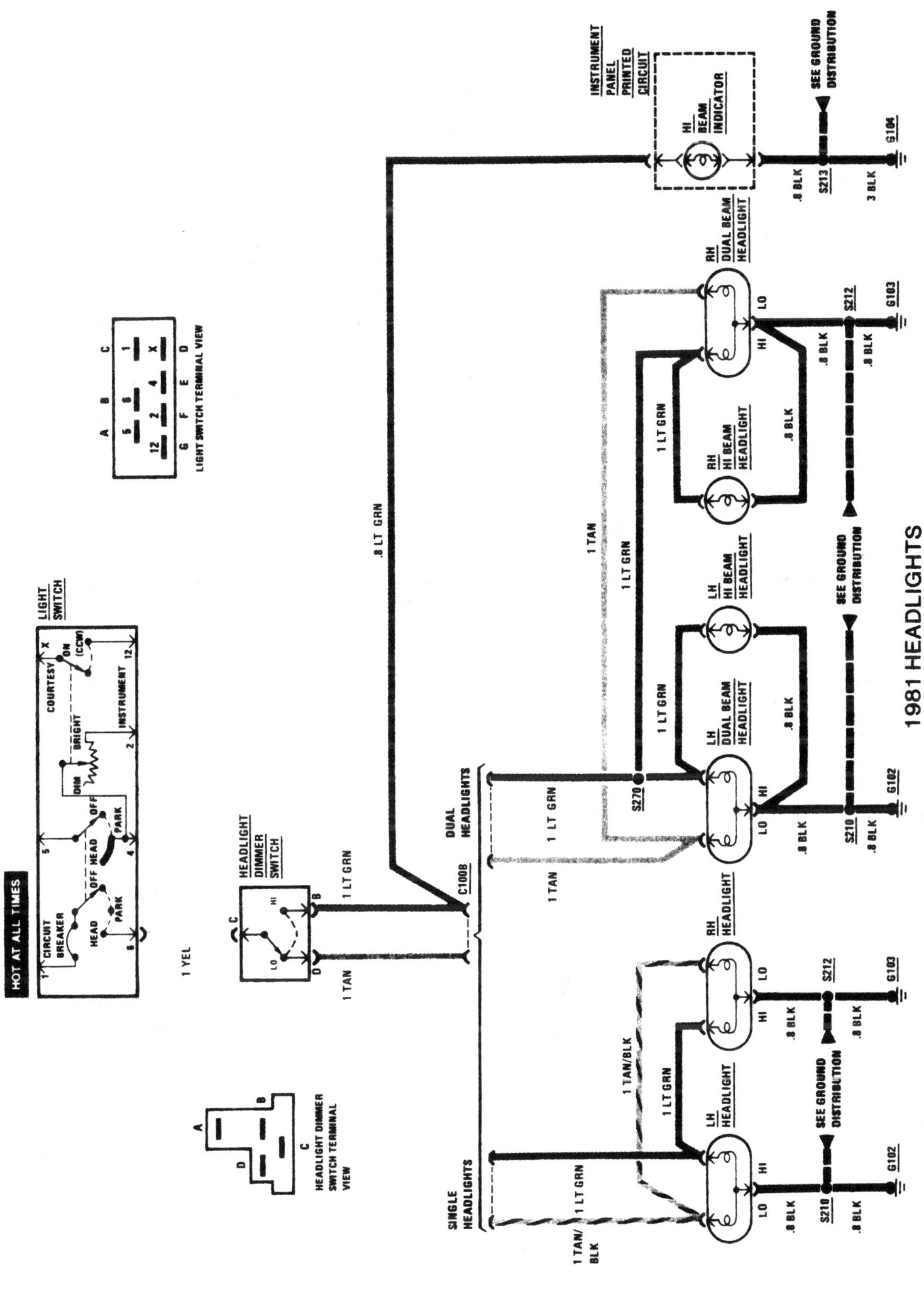

1981 HEADLIGHTS

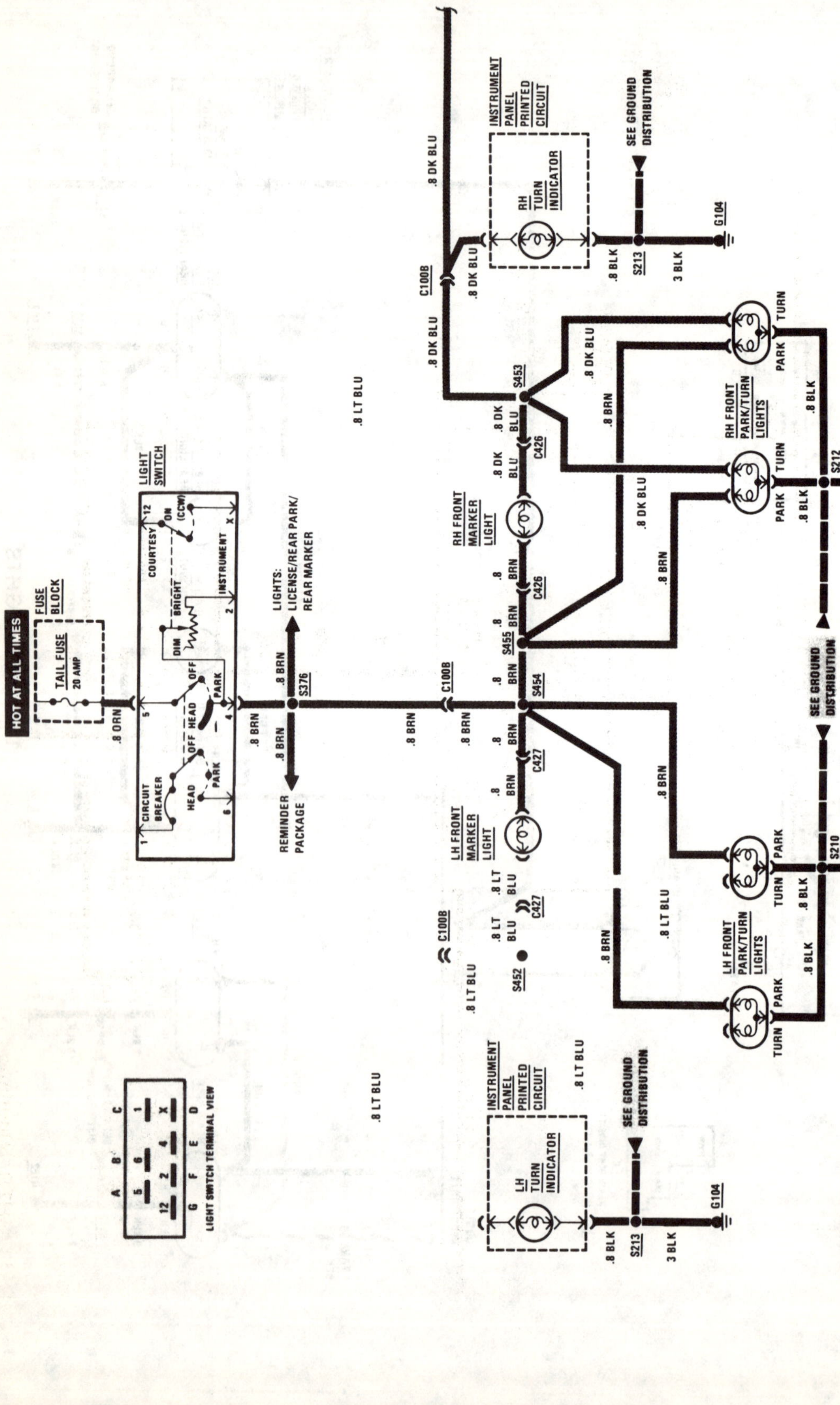

1981 LIGHTS:
TURN/HAZARD/STOP/FRONT
PARK/FRONT MARKER

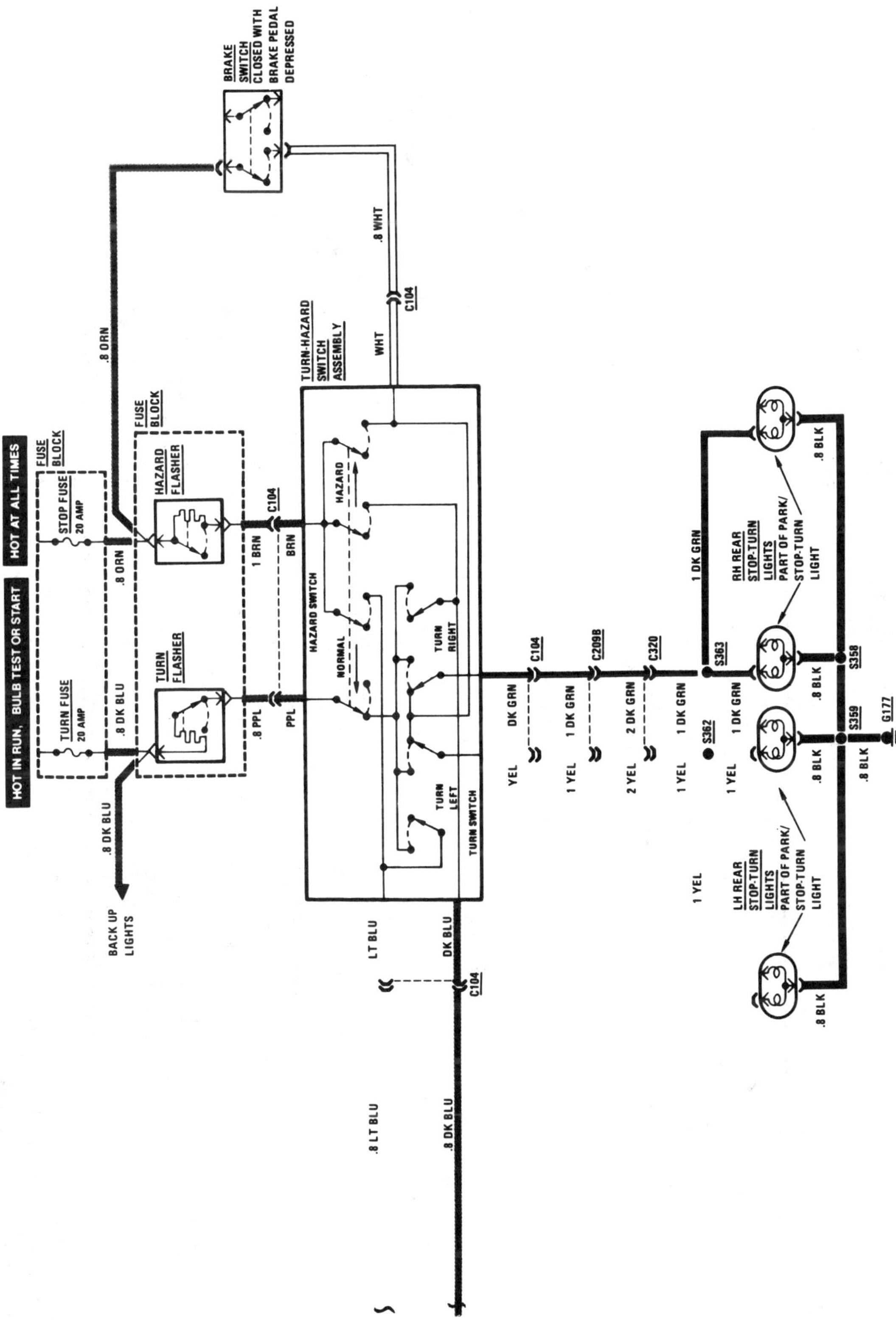

1981 LIGHTS—Continued

15

EXCEPT STATION WAGON

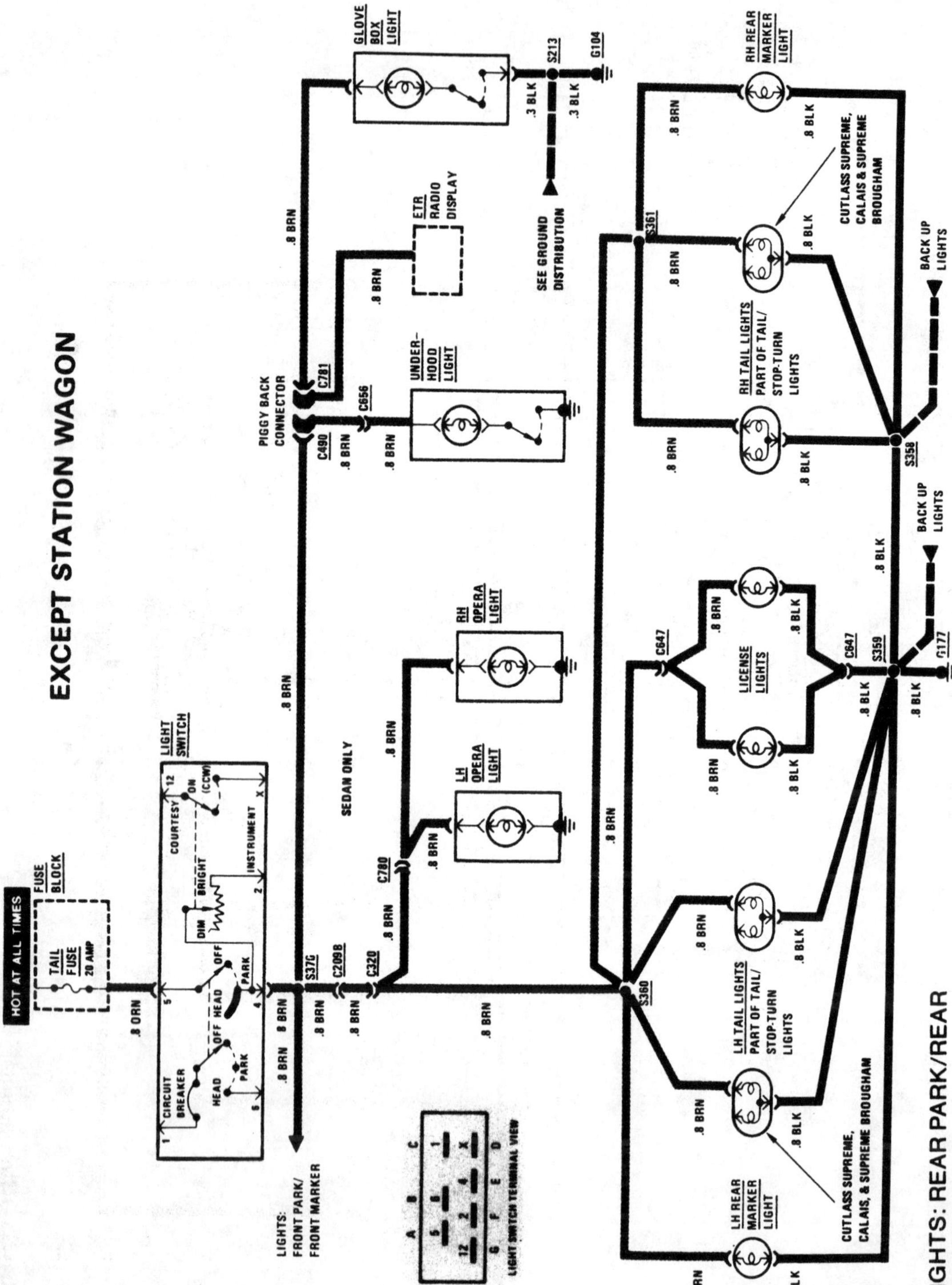

1981 LIGHTS: REAR PARK/REAR MARKER/LICENSE

STATION WAGON

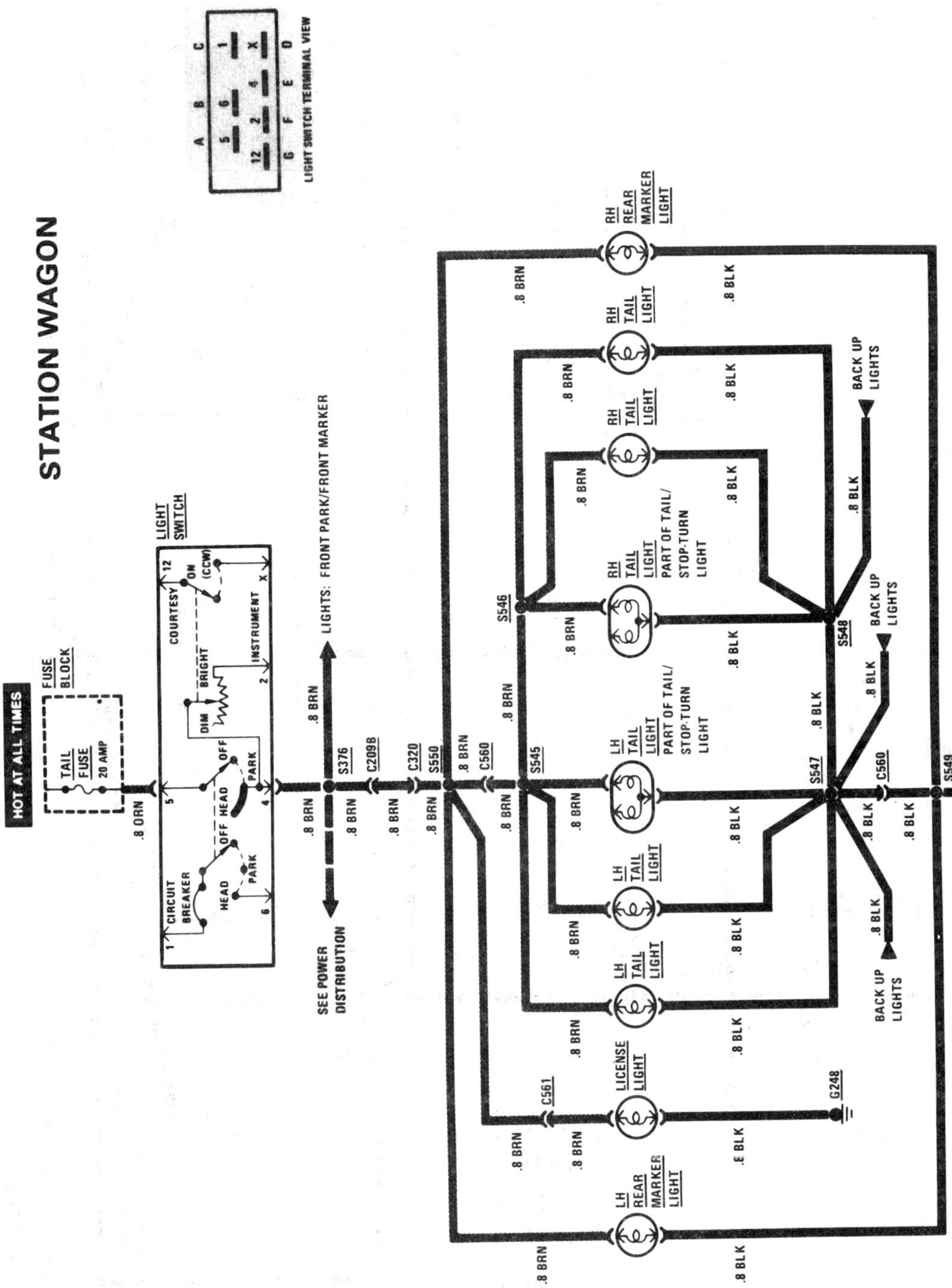

1981 LIGHTS—Continued

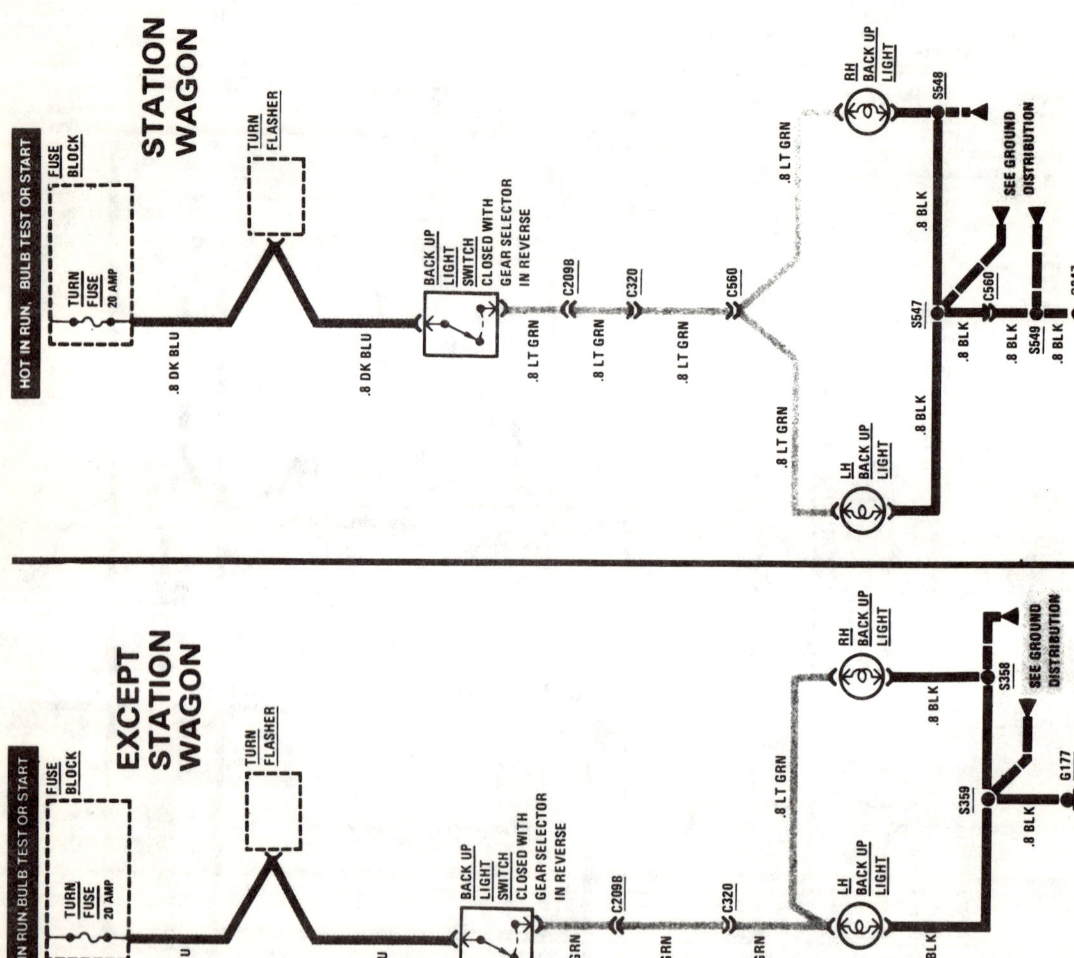

1981 BACK-UP LIGHTS

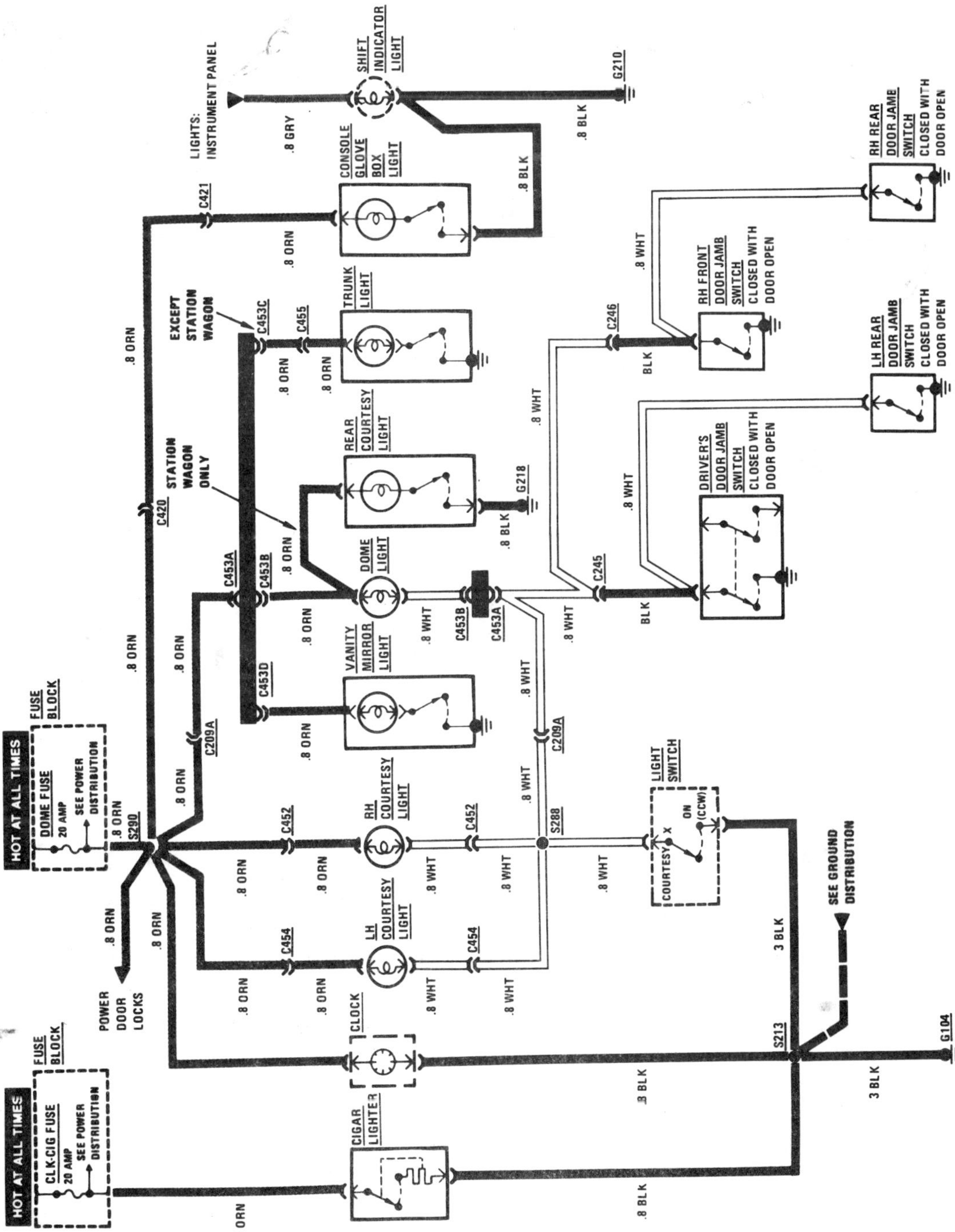

1981 COURTESY LIGHTS

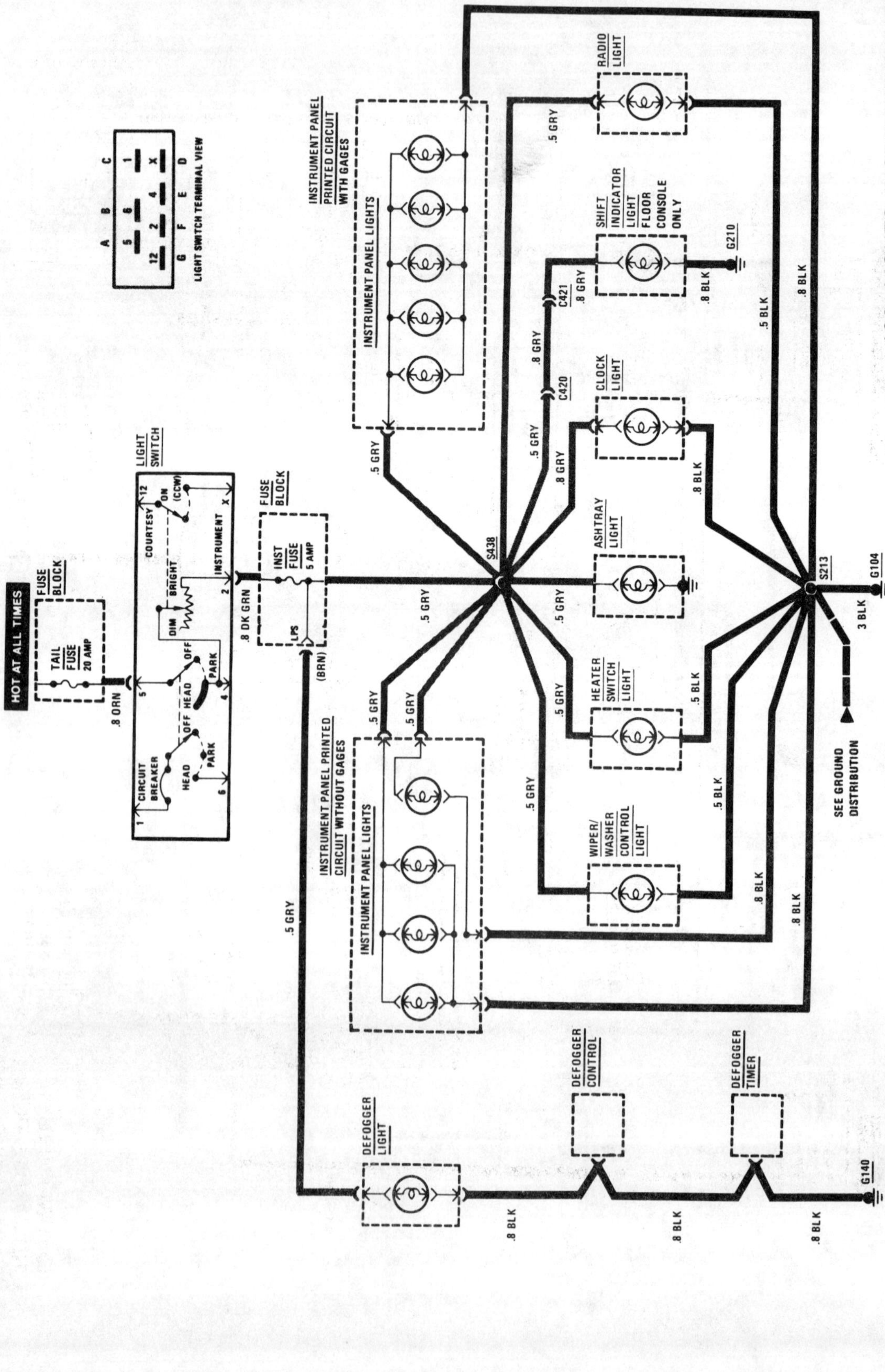

1981 LIGHTS: INSTRUMENT PANEL

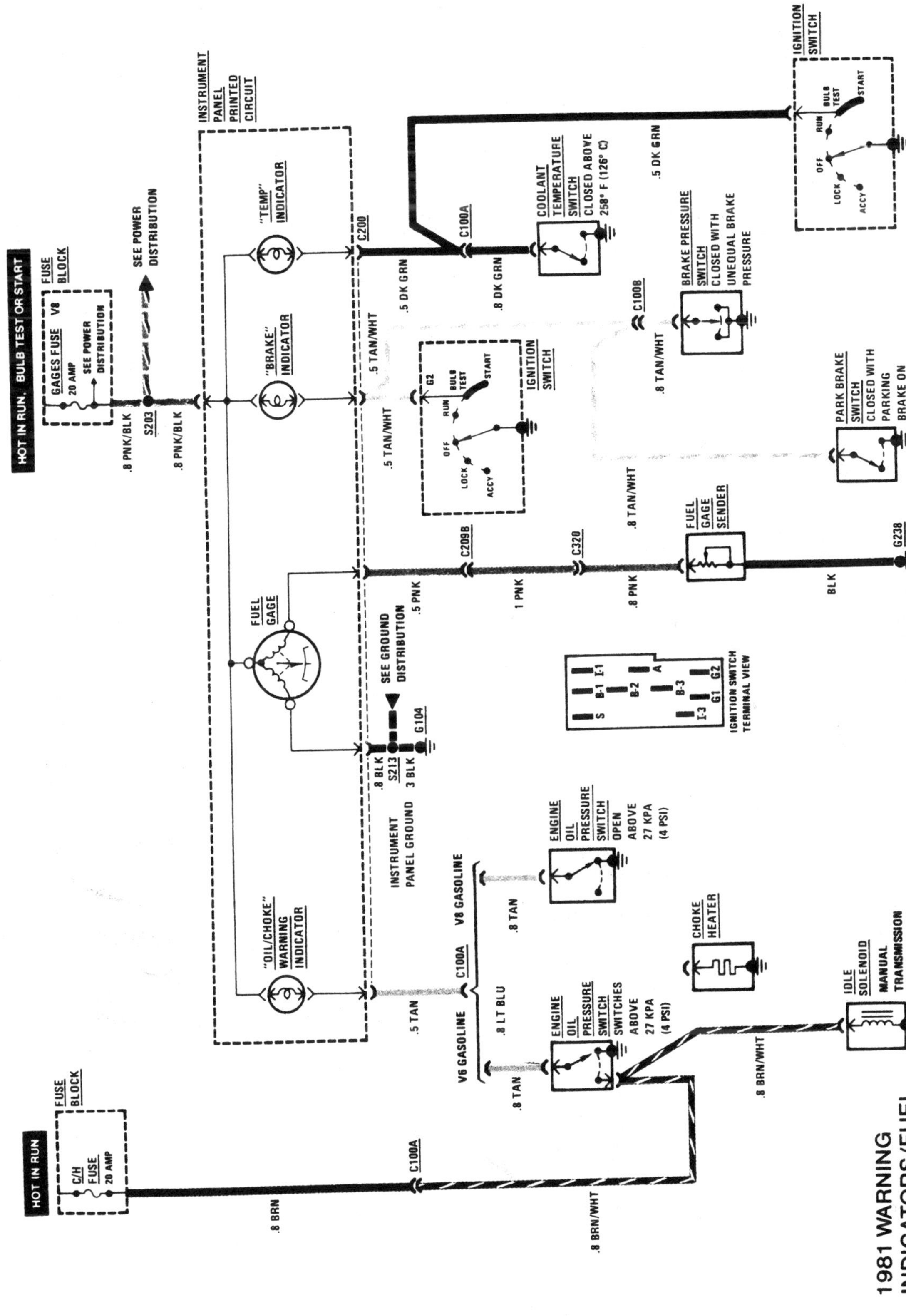

1981 WARNING INDICATORS/FUEL GAGE/CHOKE HEATER

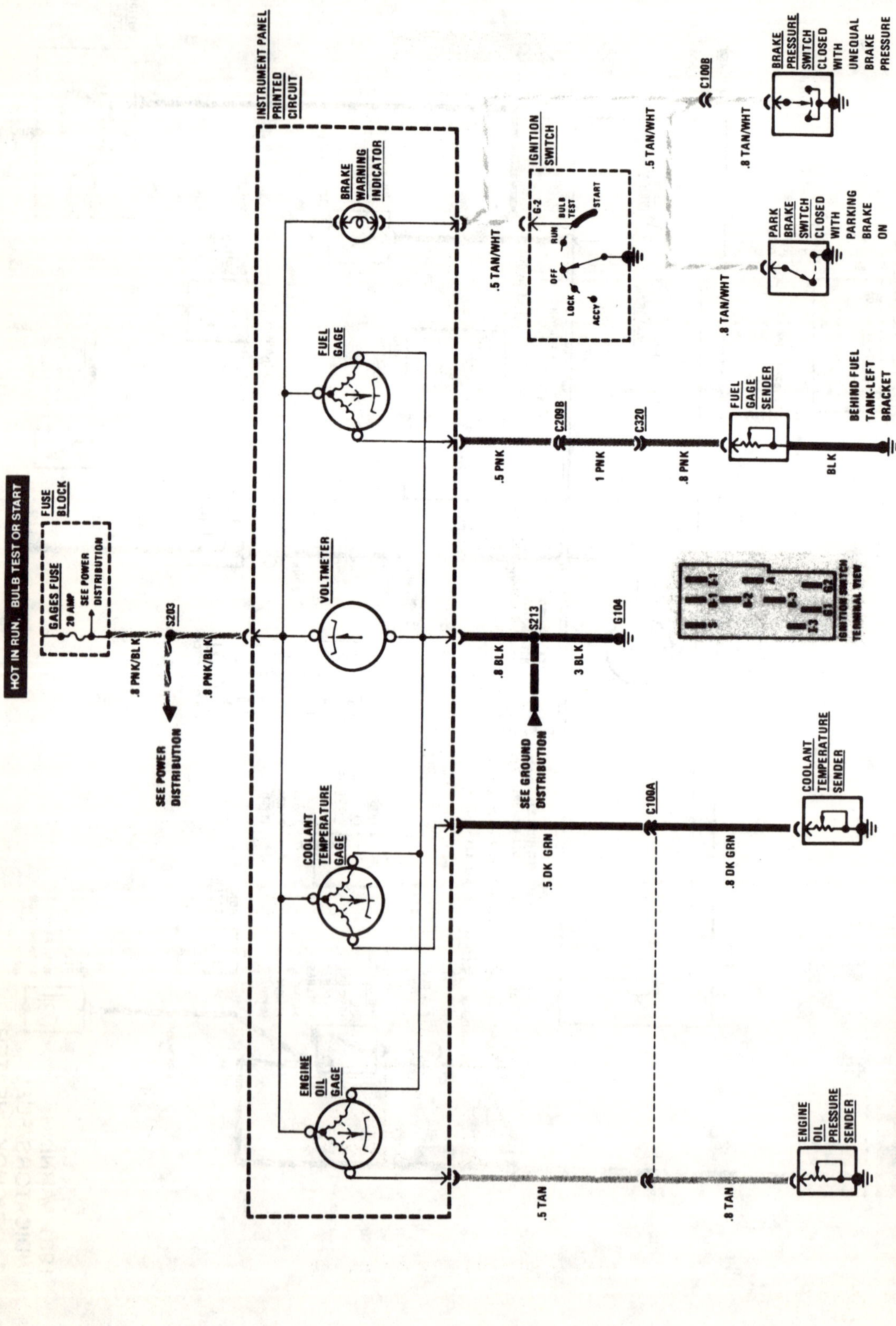

1981 GAGES/BRAKE WARNING INDICATOR

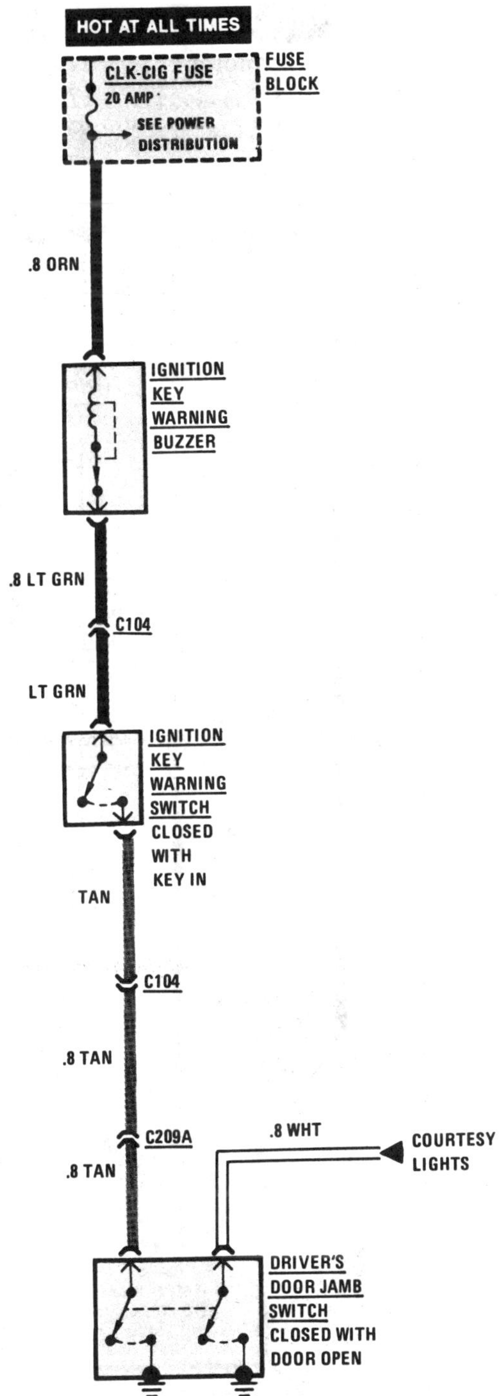

1981 IGNITION KEY WARNING

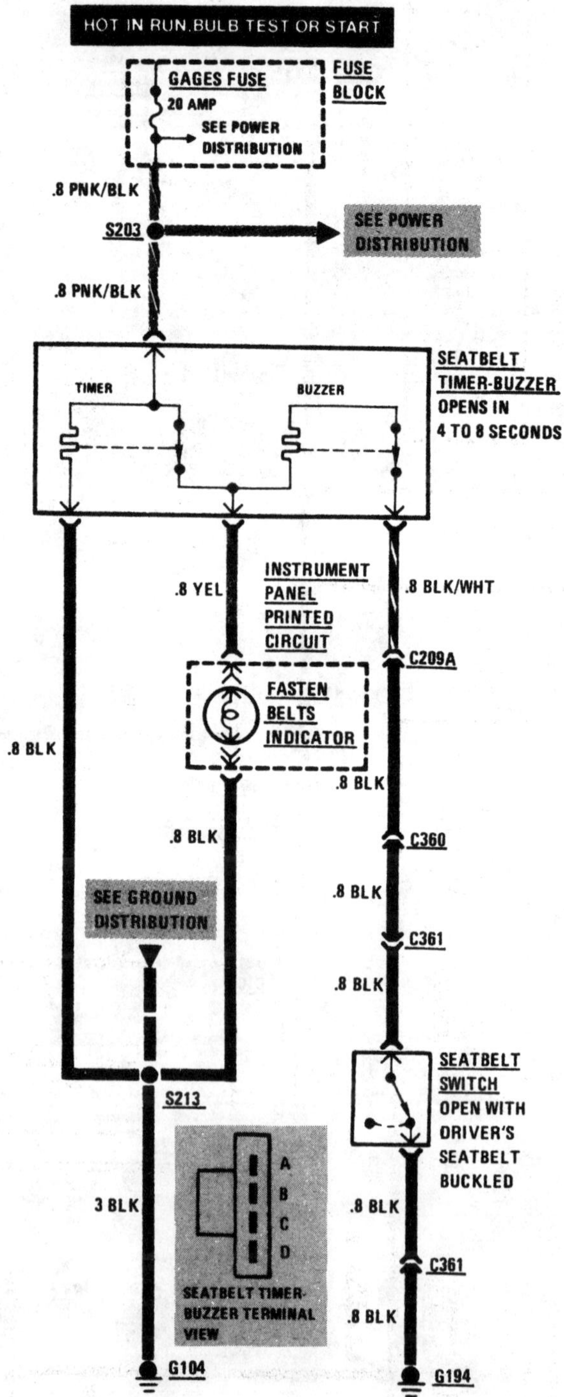

1981 SEATBELT WARNING

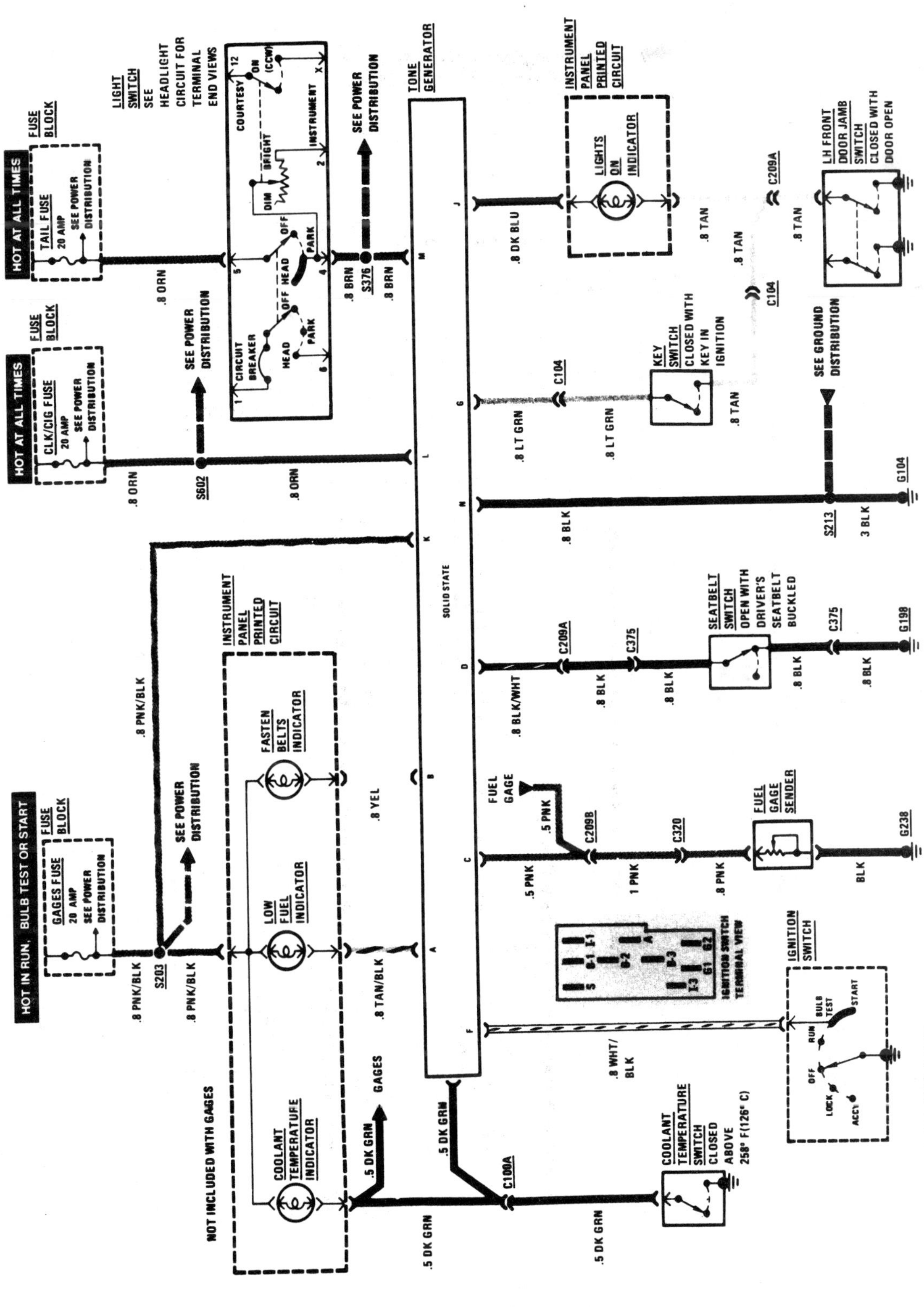

1981 REMINDER PACKAGE

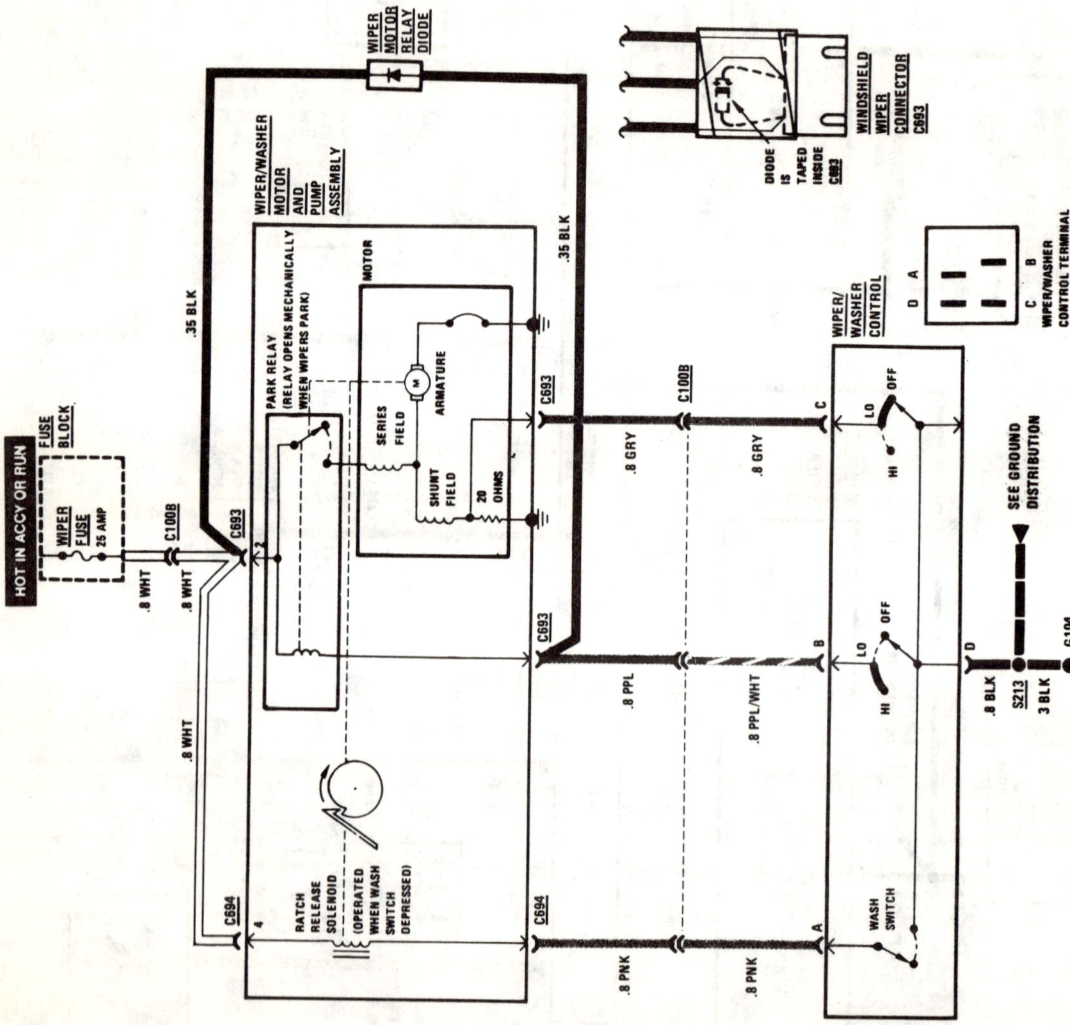

1981 WIPER/WASHER

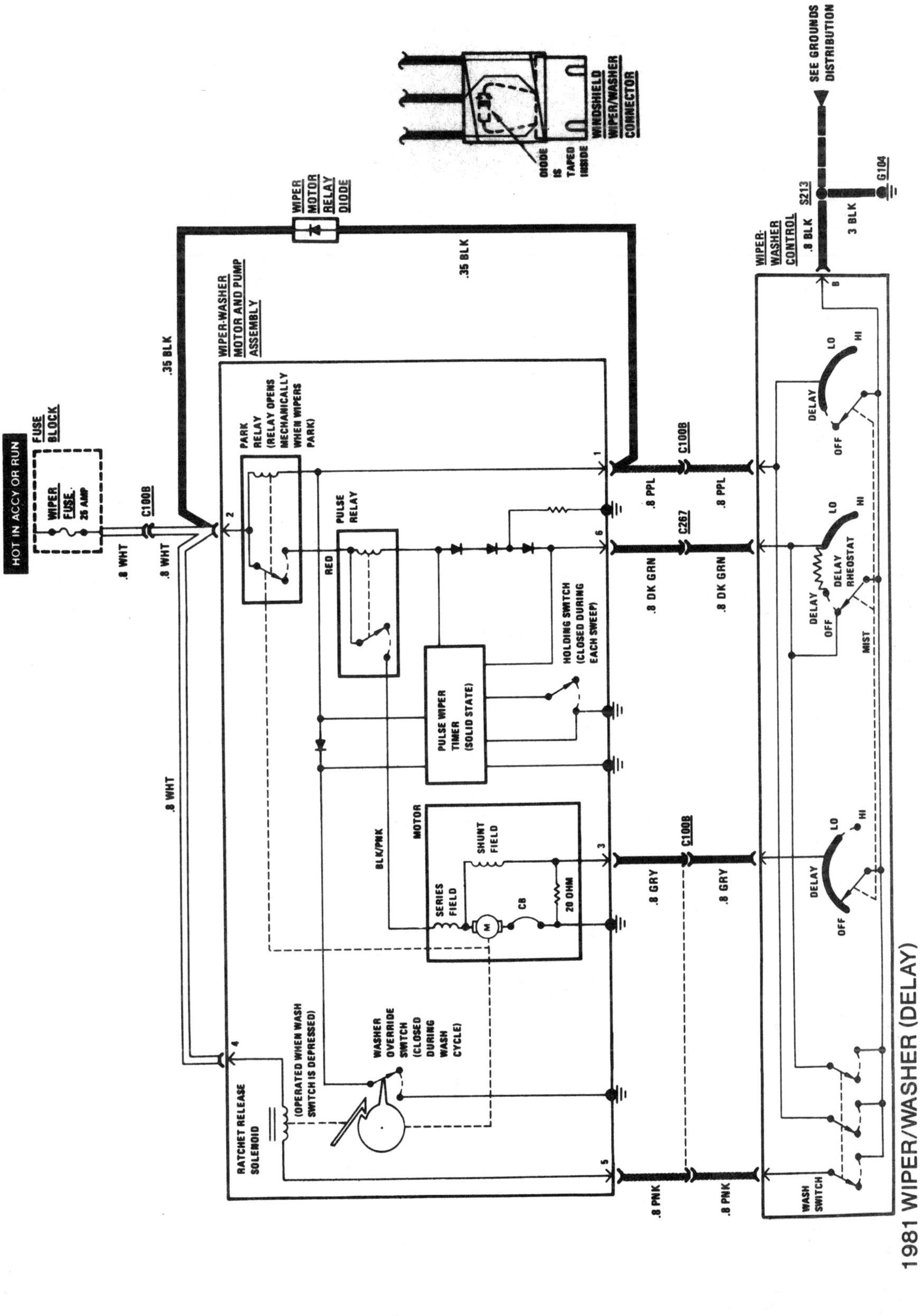

1981 WIPER/WASHER (DELAY)

15

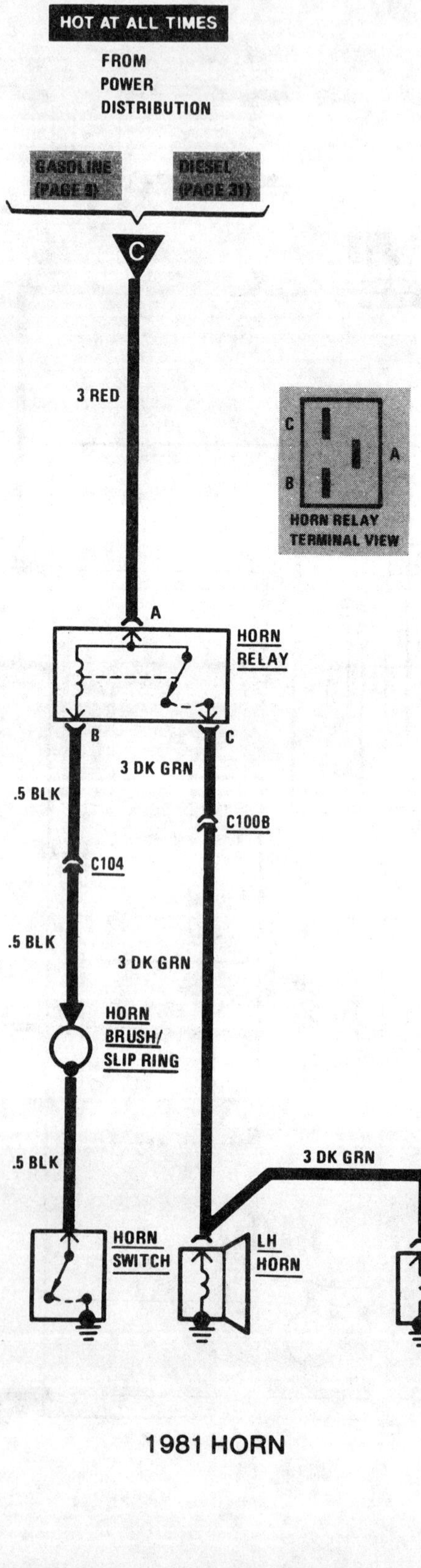
HOT AT ALL TIMES
FROM
POWER
DISTRIBUTION
GASOLINE
(PAGE 8)
DIESEL
(PAGE 31)
C
3 RED
C
A
B
HORN RELAY
TERMINAL VIEW
A
HORN
RELAY
B
C
3 DK GRN
.5 BLK
C100B
C104
.5 BLK
3 DK GRN
HORN
BRUSH/
SLIP RING
.5 BLK
3 DK GRN
HORN
SWITCH
LH
HORN
RH
HORN

1981 HORN

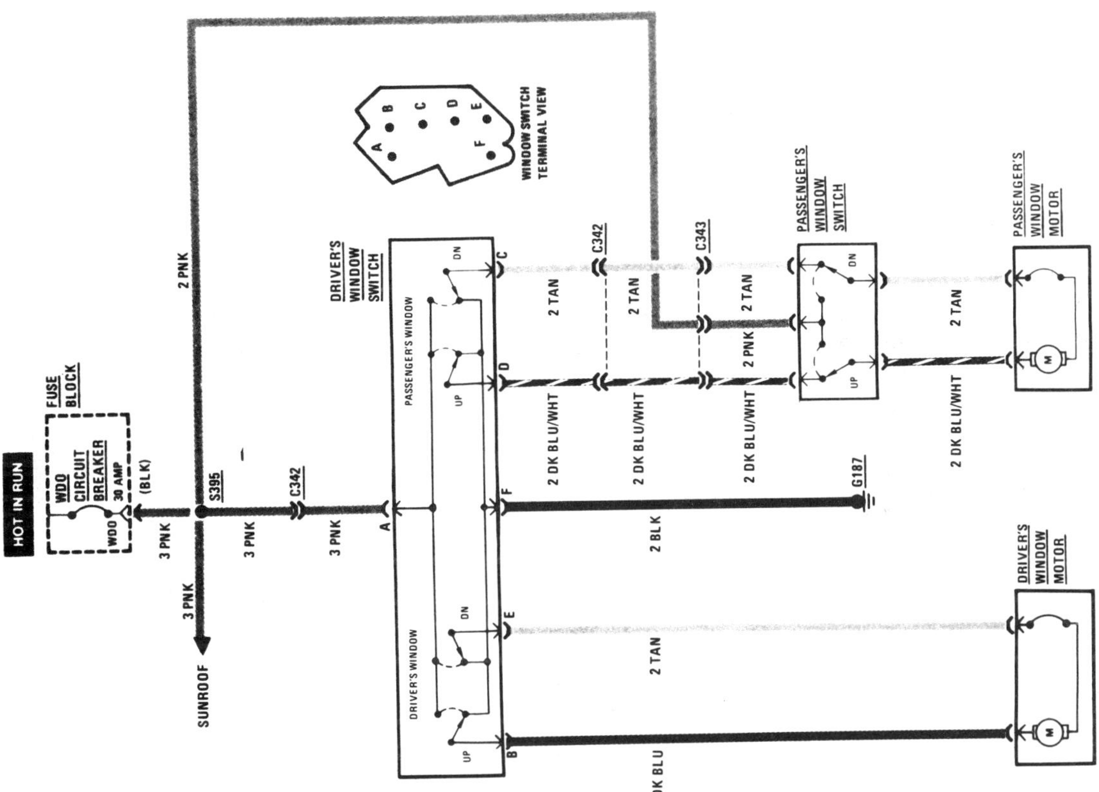

1981
POWER WINDOWS
(2-DOOR)

15

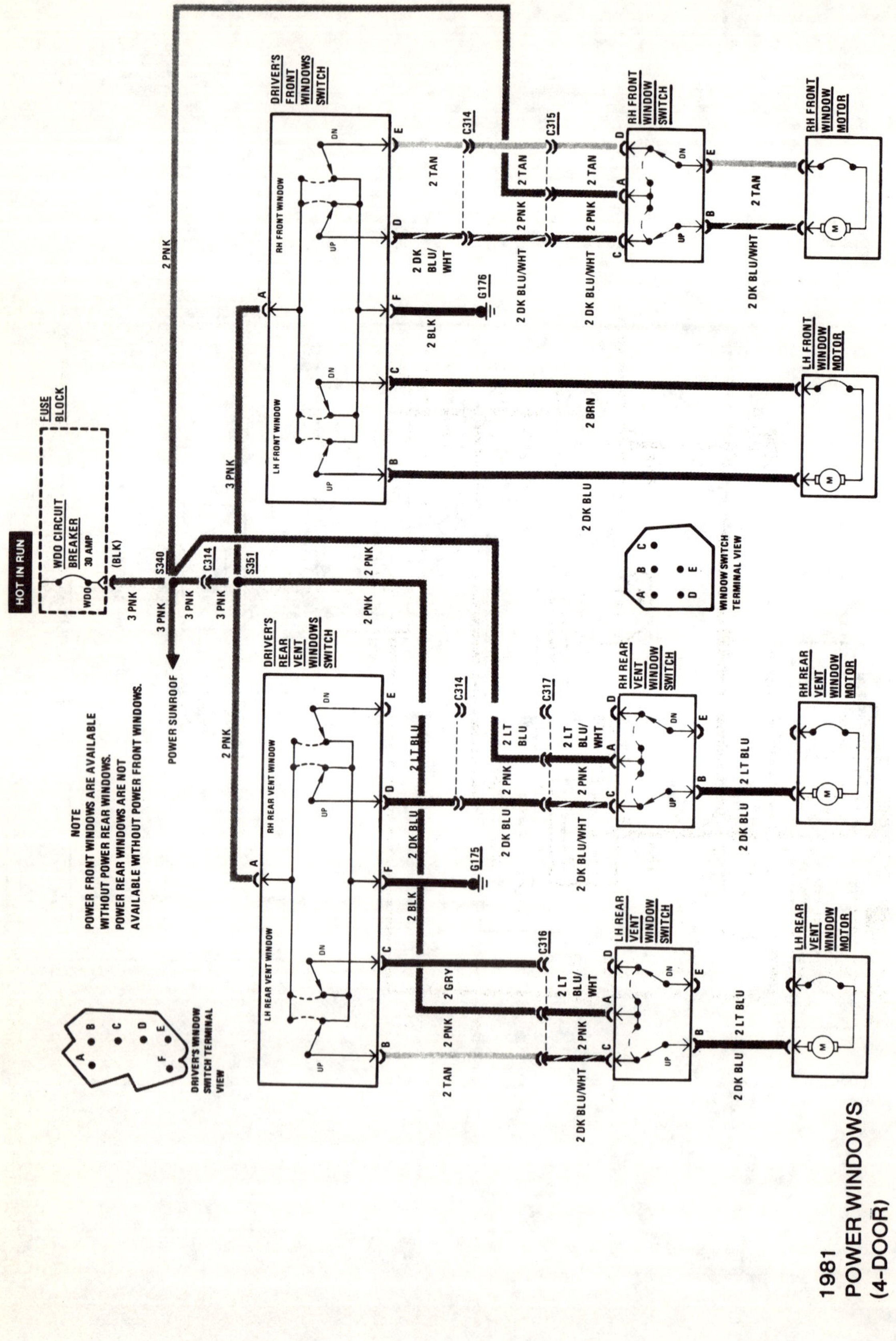

1981 POWER WINDOWS (4-DOOR)

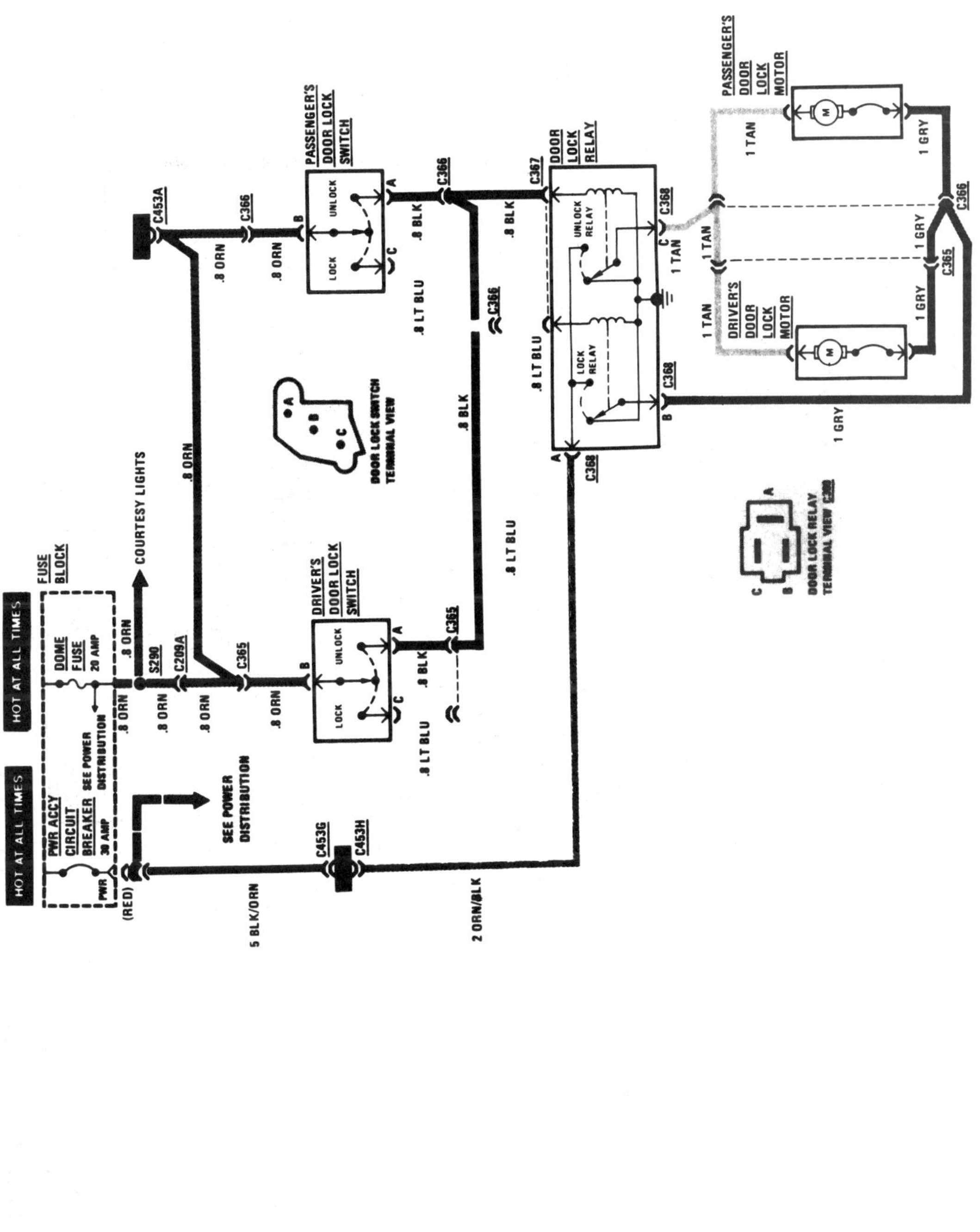

1981 POWER DOOR LOCKS (2-DOOR)

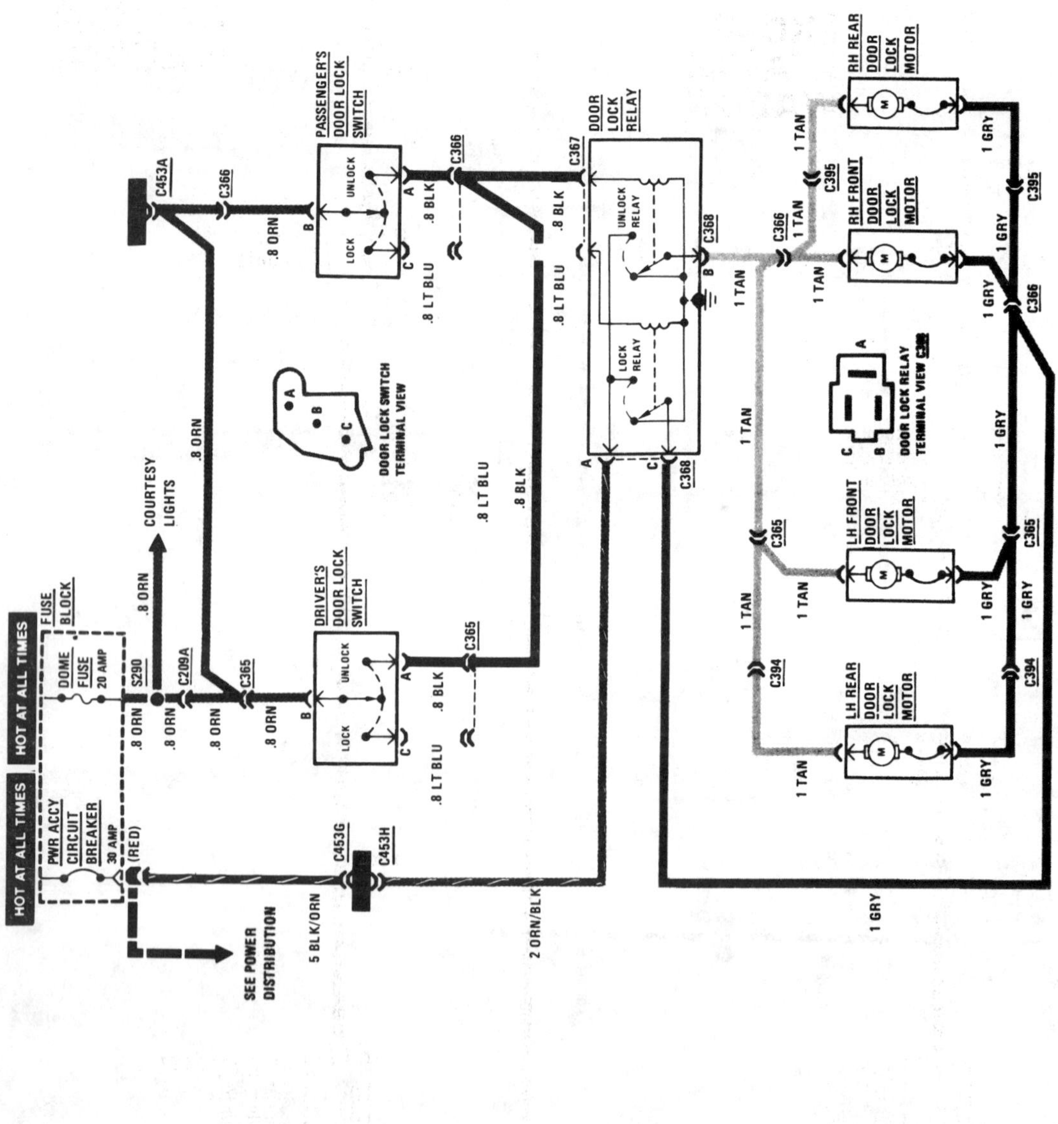

1981 POWER DOOR LOCKS (4-DOOR)

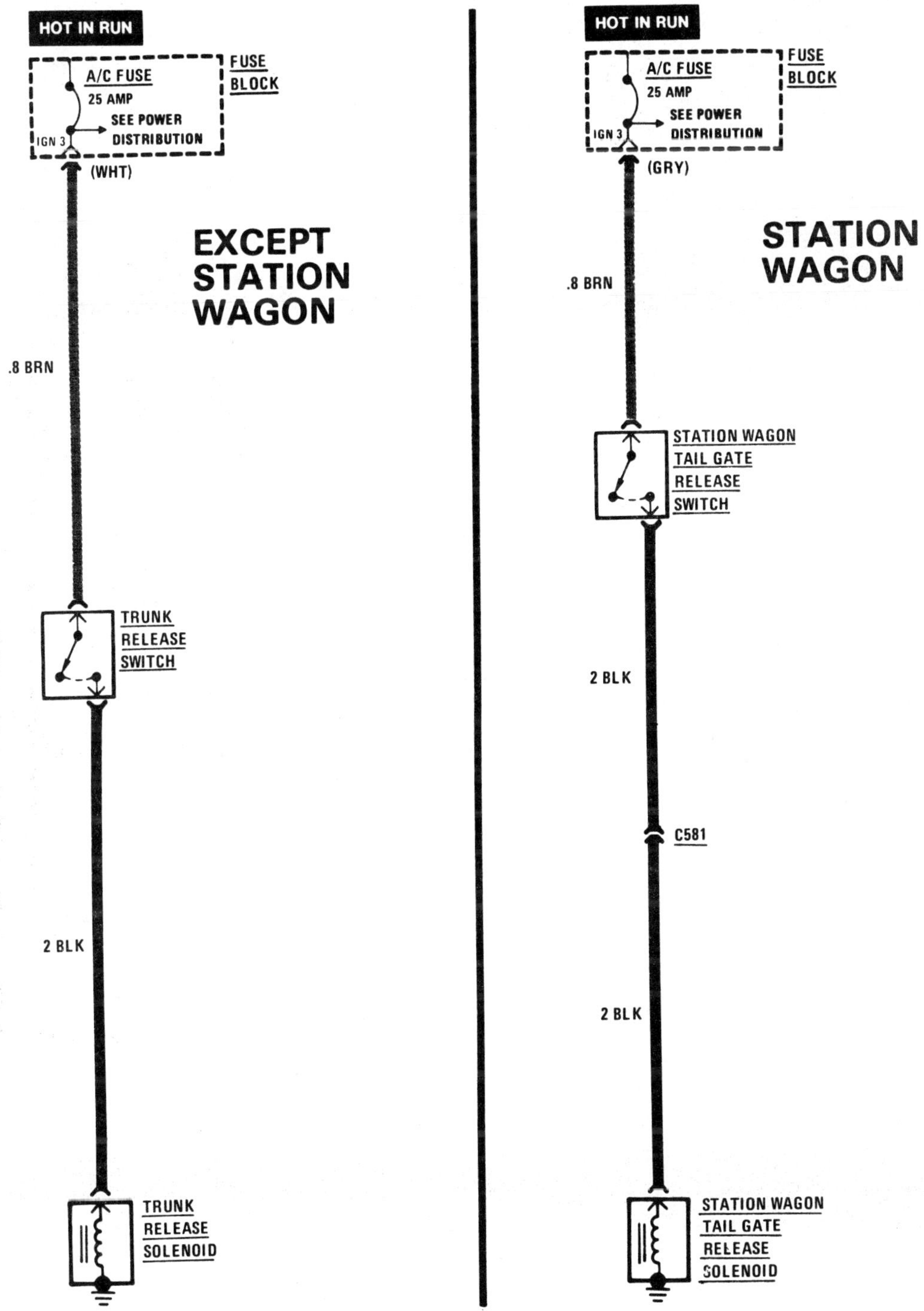

1981 TRUNK/TAILGATE RELEASE

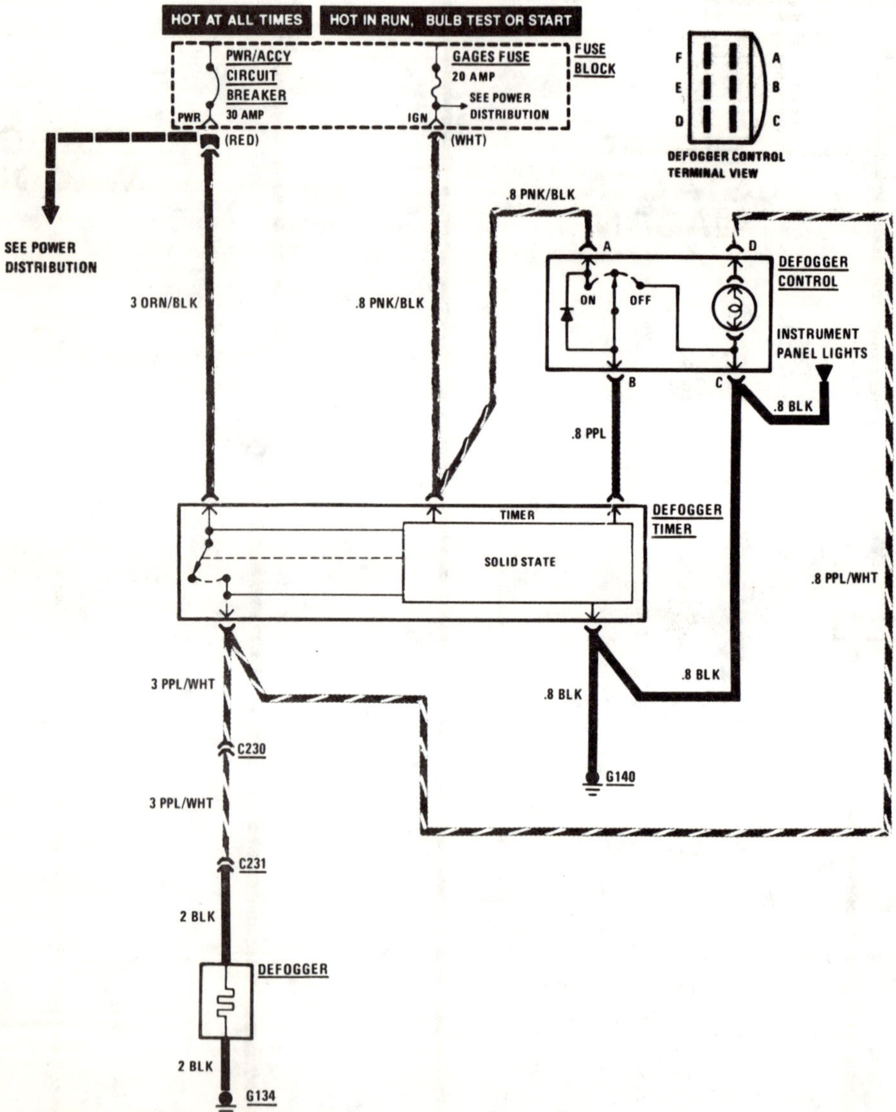
HOT AT ALL TIMES
HOT IN RUN, BULB TEST OR START
PWR/ACCY CIRCUIT BREAKER
30 AMP
PWR
GAGES FUSE
20 AMP
SEE POWER DISTRIBUTION
IGN
FUSE BLOCK
(RED)
(WHT)
SEE POWER DISTRIBUTION
3 ORN/BLK
.8 PNK/BLK
.8 PNK/BLK
A
D
DEFOGGER CONTROL
ON
OFF
INSTRUMENT PANEL LIGHTS
B
C
.8 BLK
.8 PPL
F
E
D
A
B
C
DEFOGGER CONTROL TERMINAL VIEW
TIMER
SOLID STATE
DEFOGGER TIMER
.8 PPL/WHT
3 PPL/WHT
.8 BLK
.8 BLK
G140
C230
3 PPL/WHT
C231
2 BLK
DEFOGGER
2 BLK
G134

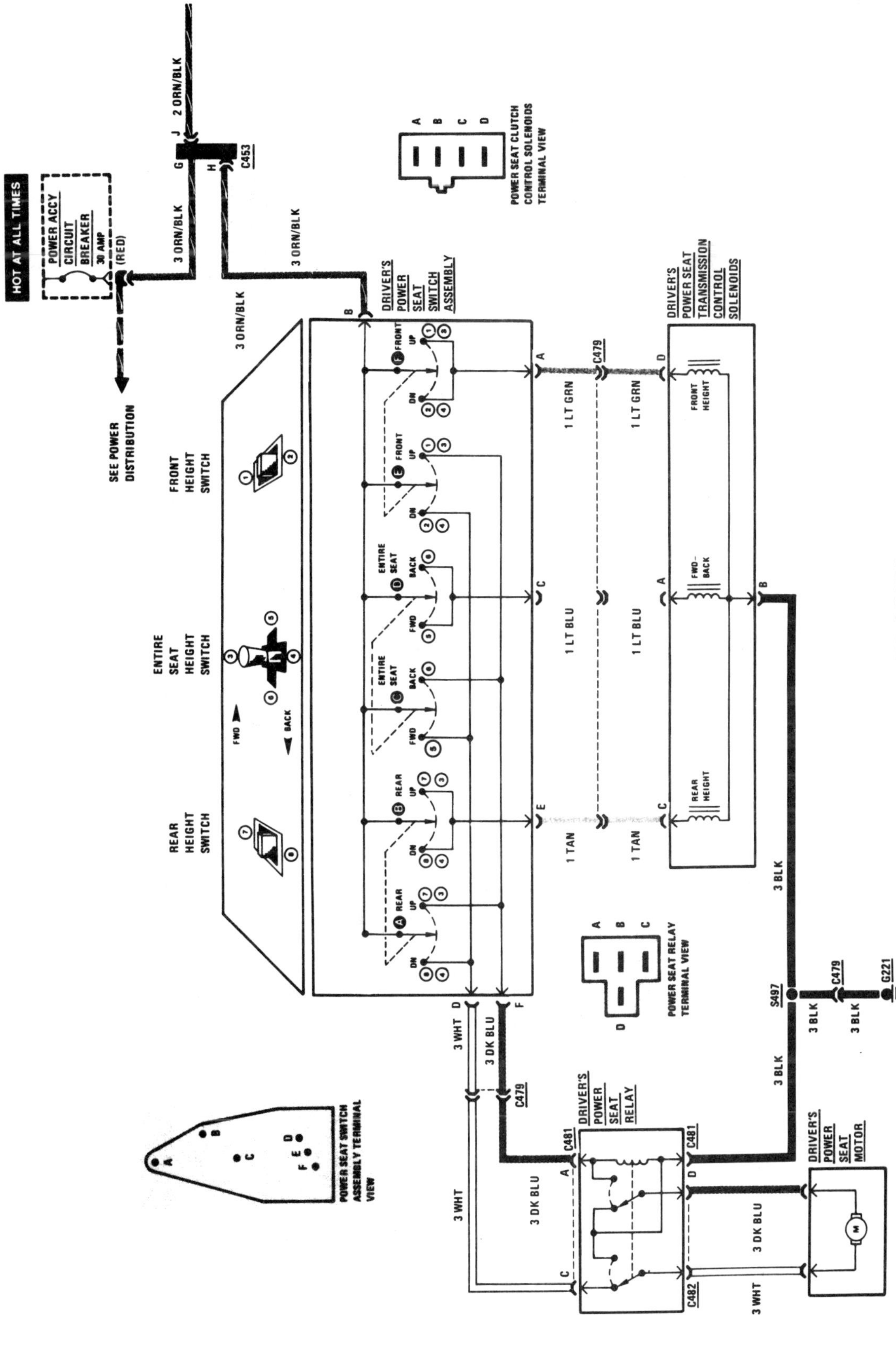

1981 POWER SEATS

15

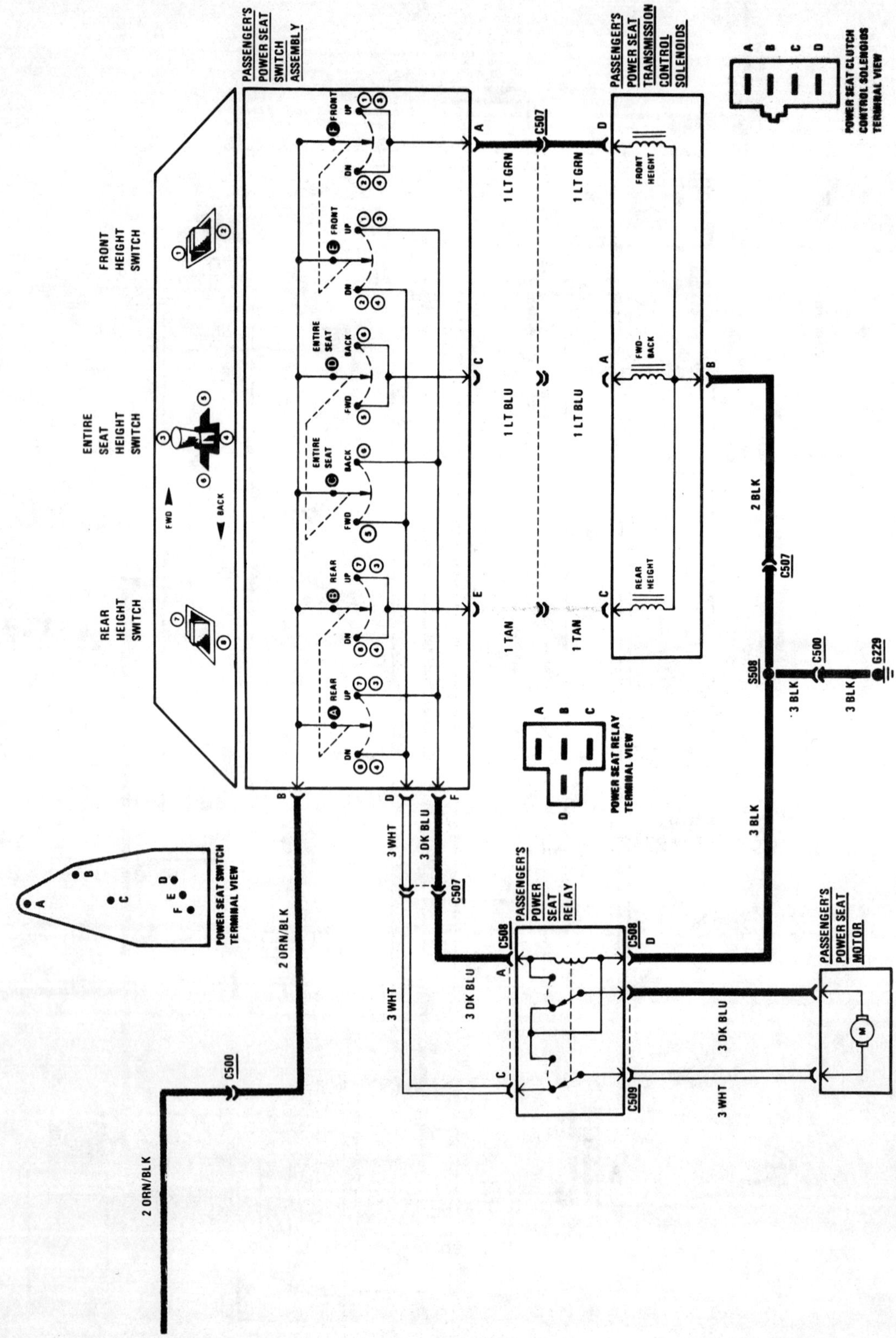

1981 POWER SEATS—Continued

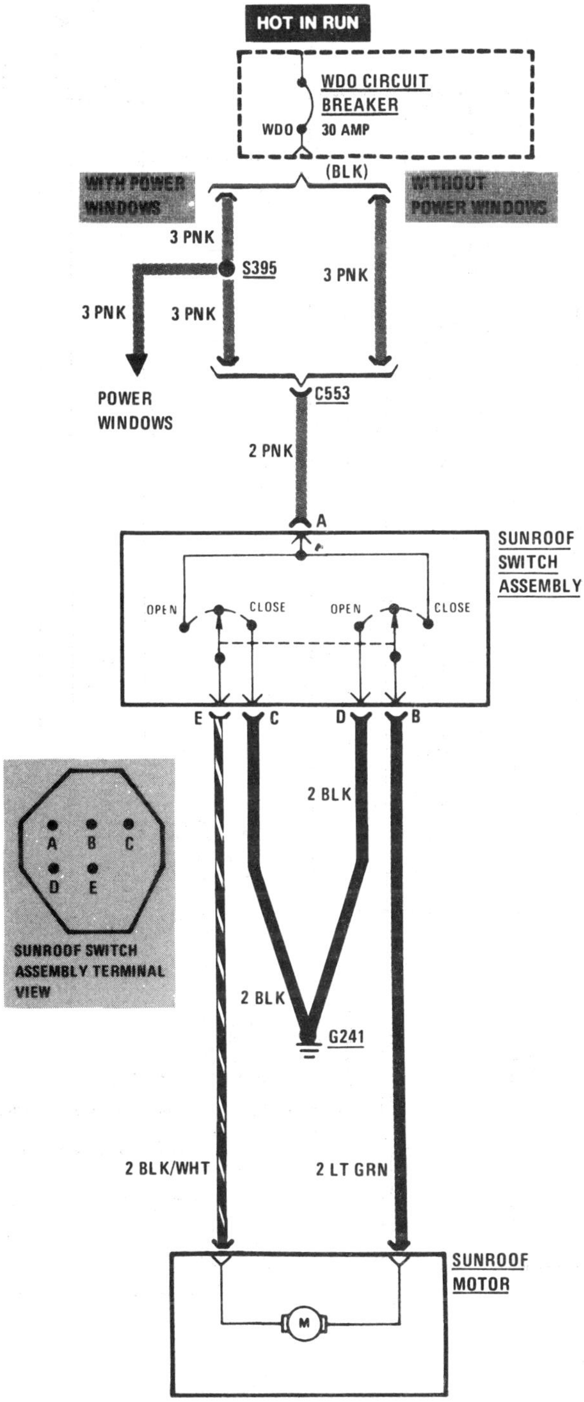

1981 SUNROOF

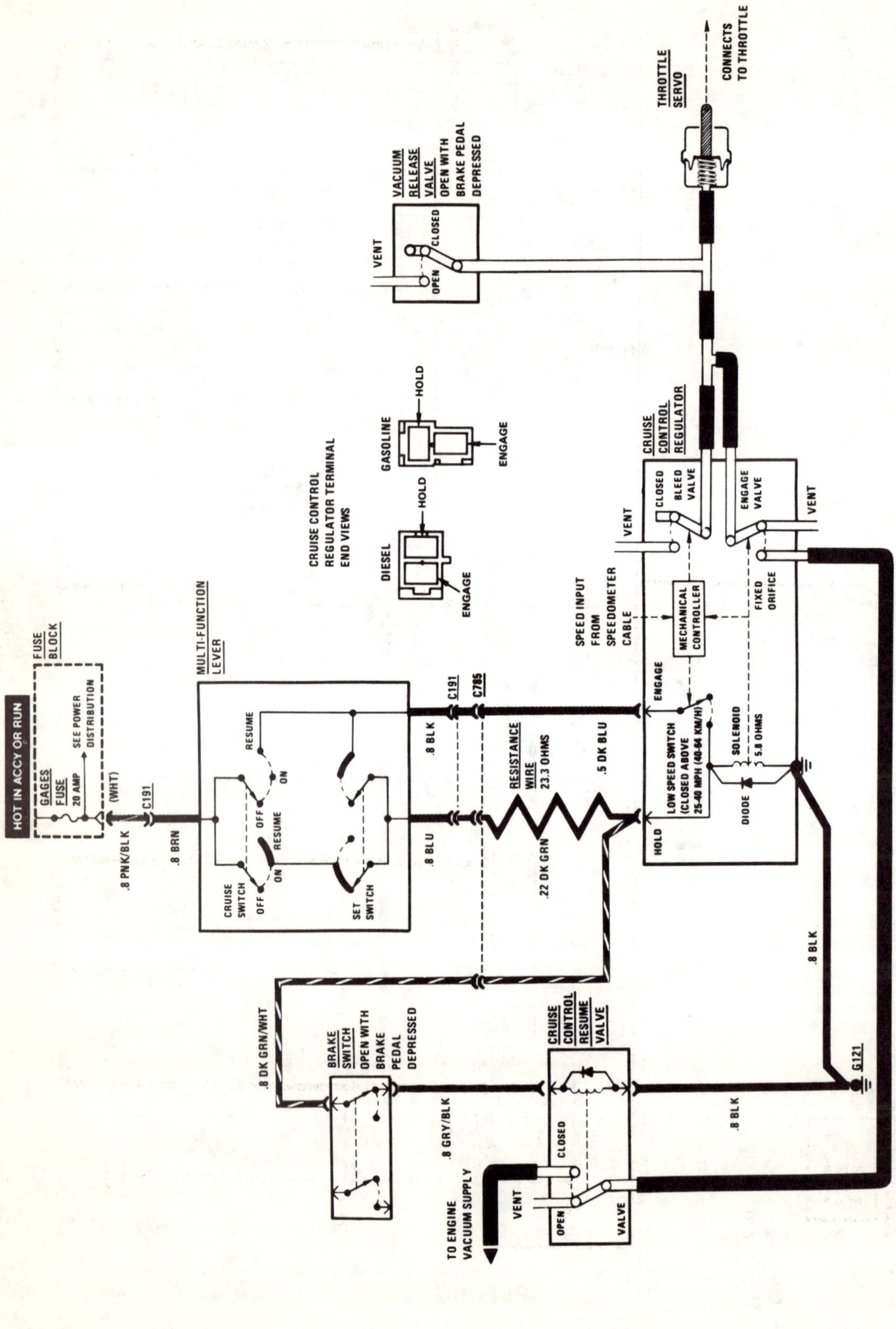

1981 CRUISE CONTROL

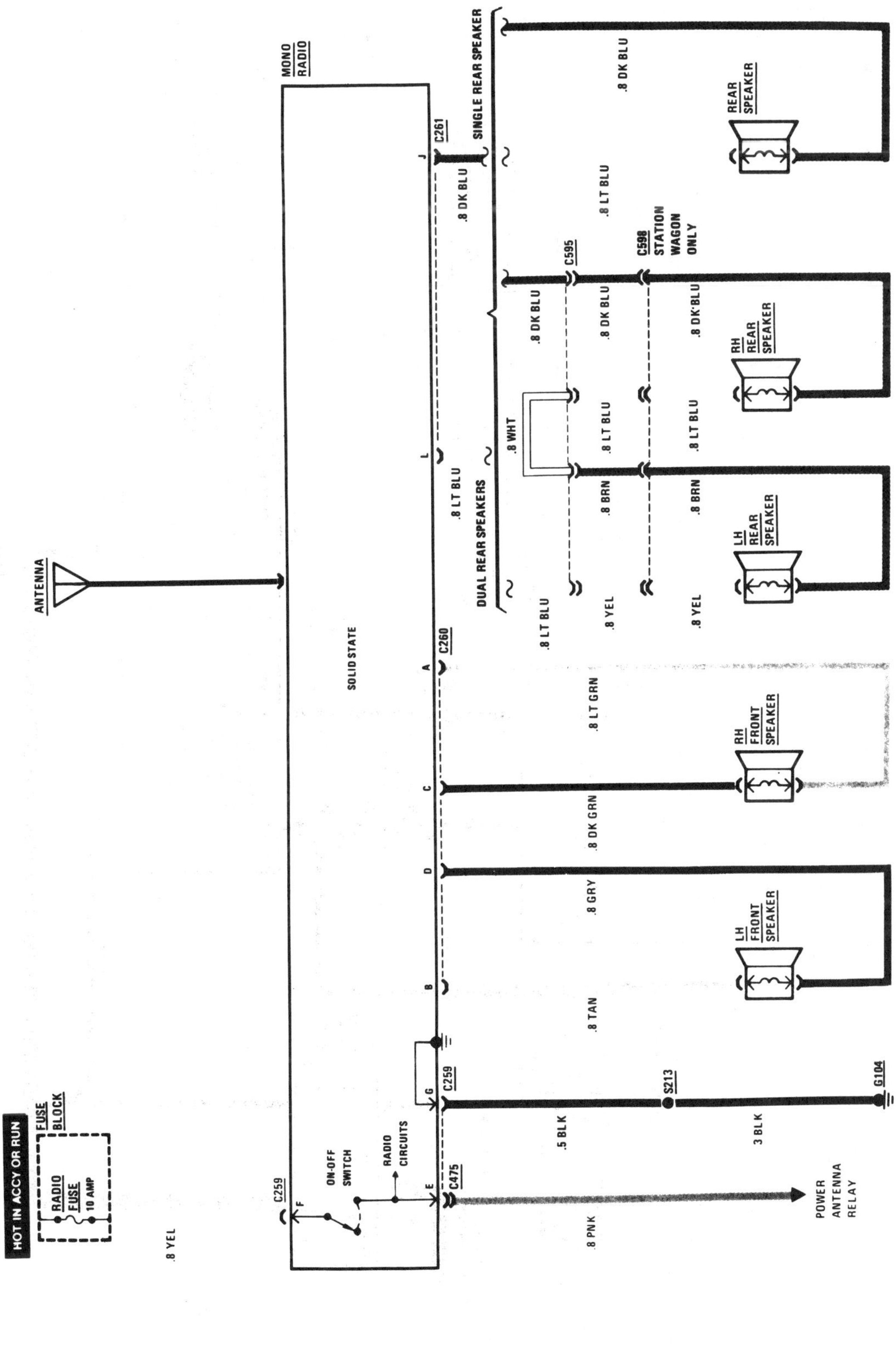

1981 RADIO (MONO)

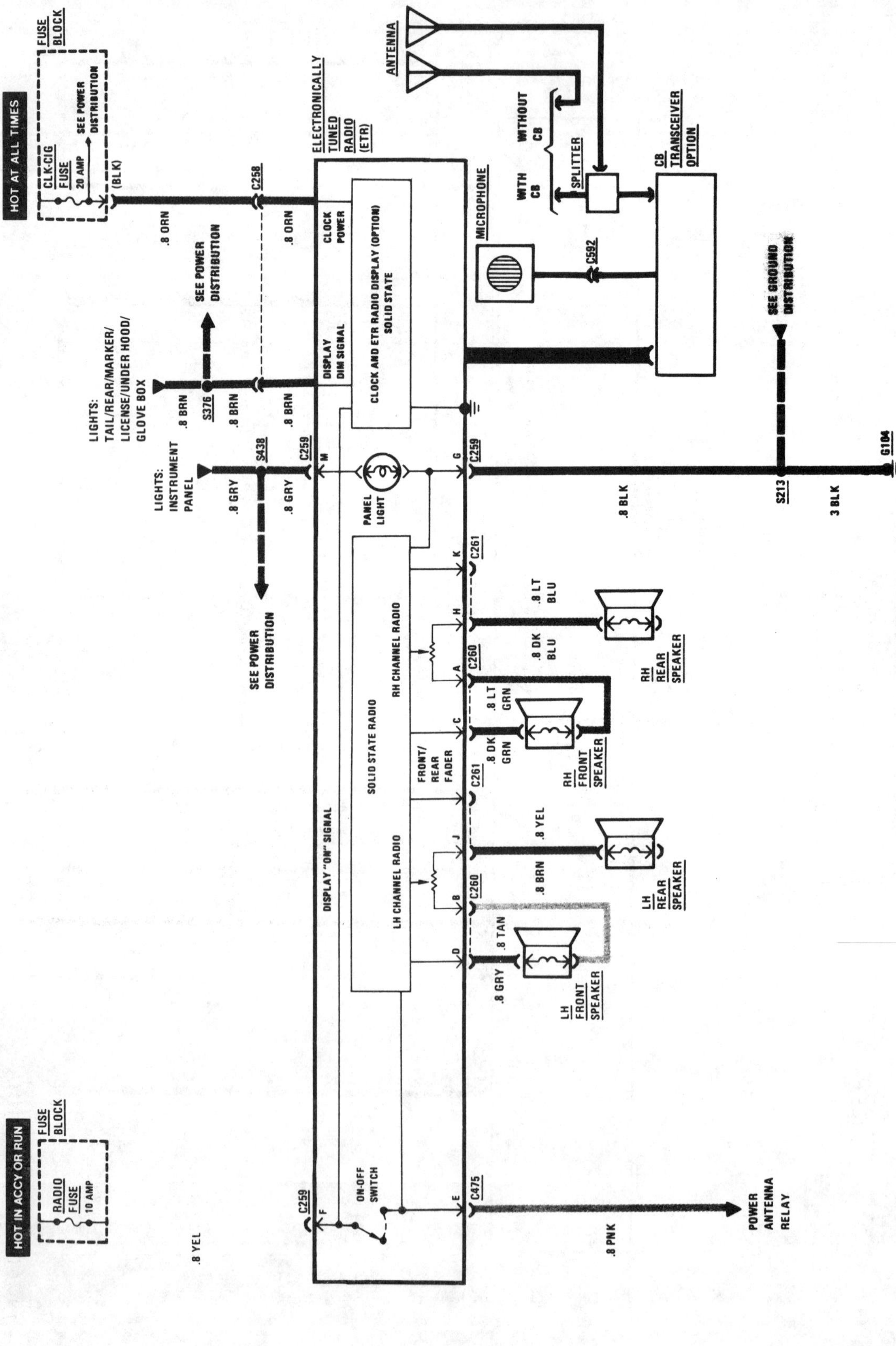

1981 RADIO (STERO)

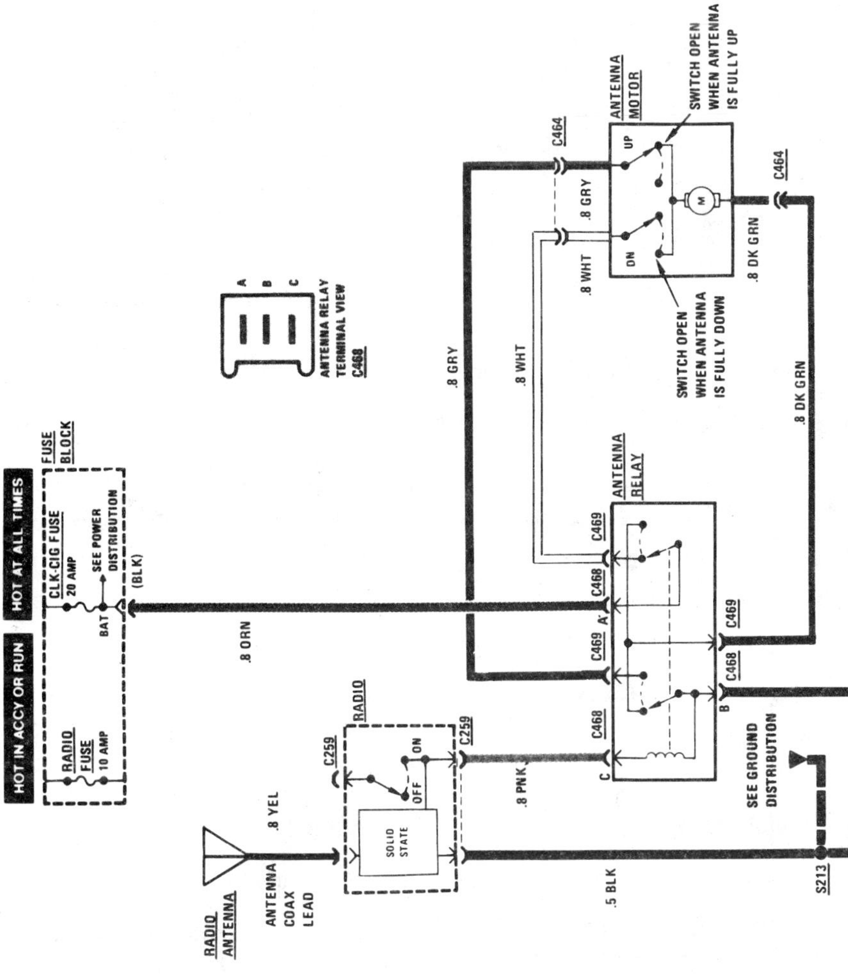

1981 POWER ANTENNA

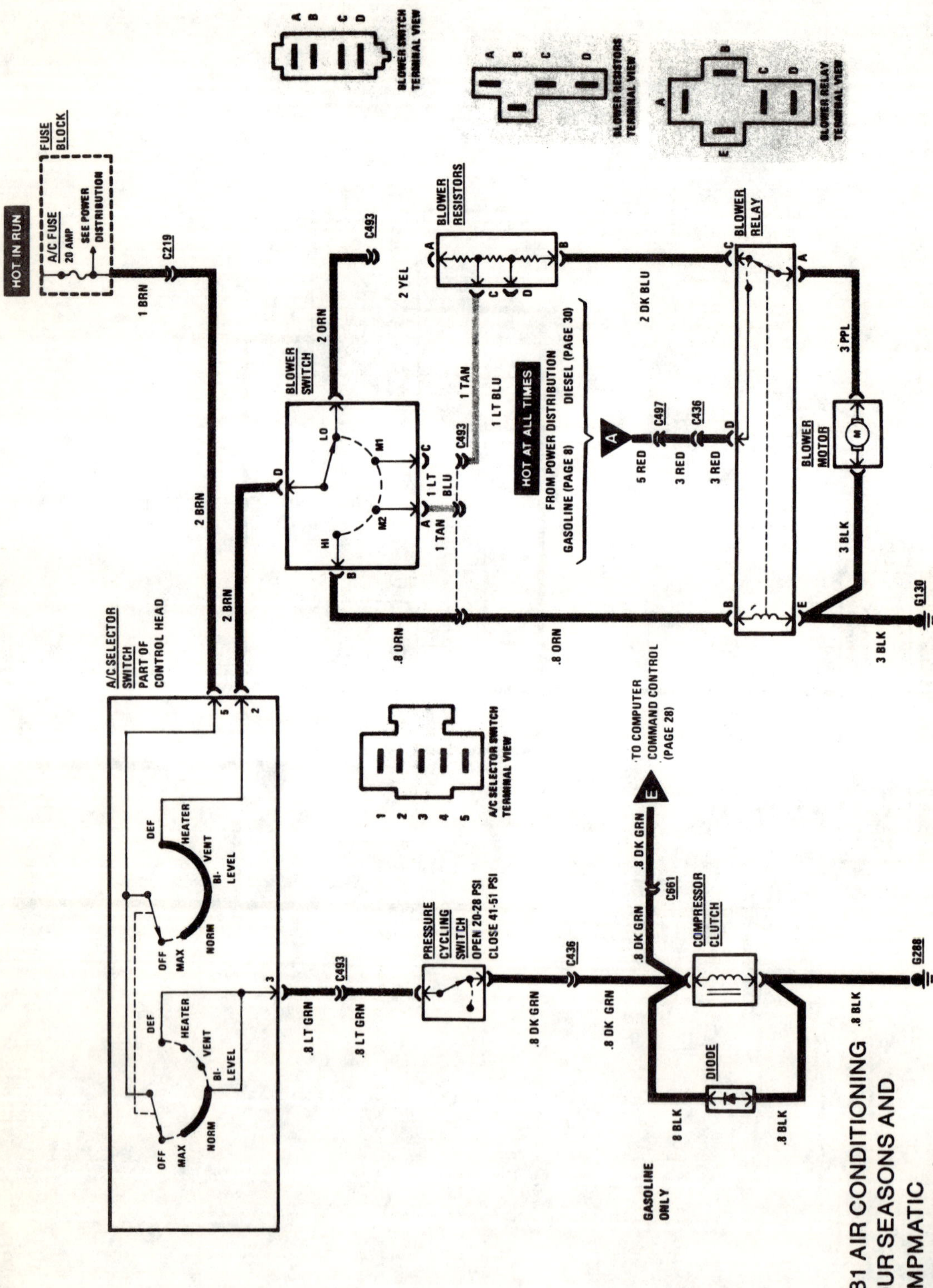

1981 AIR CONDITIONING FOUR SEASONS AND TEMPMATIC (ELECTRICAL)

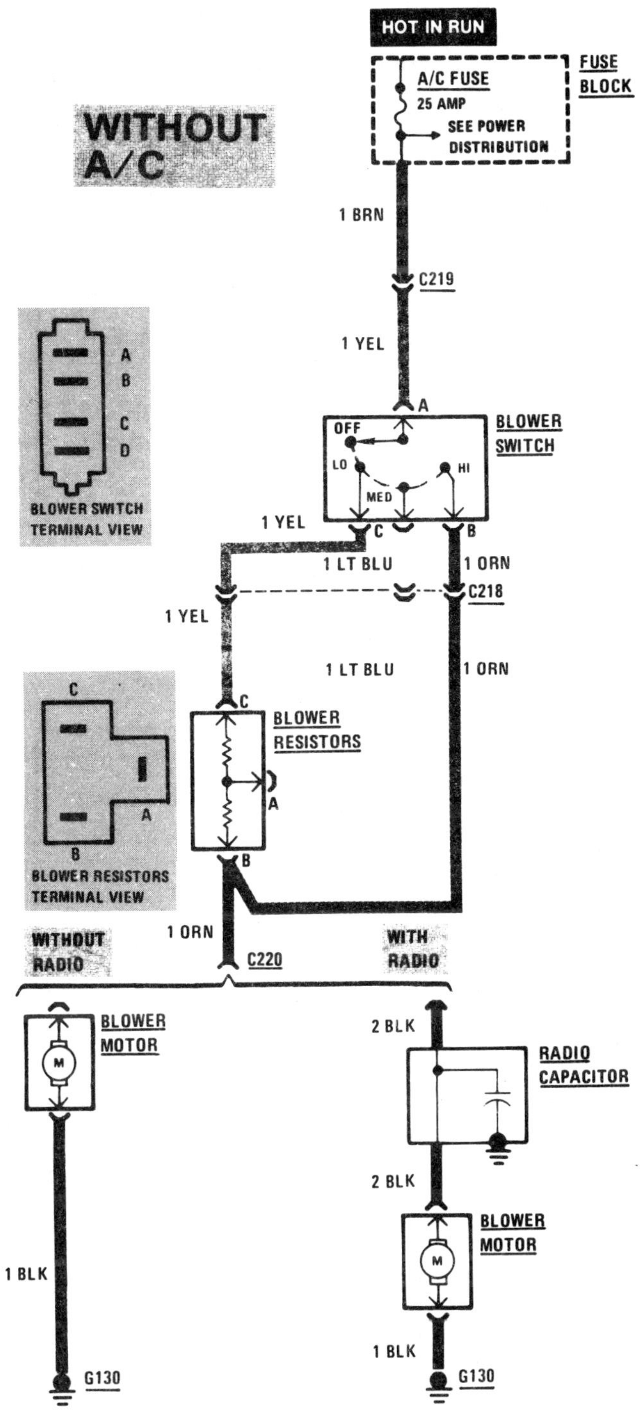

1981 HEATER (WITHOUT A/C)

1982 WIRING DIAGRAM KEY

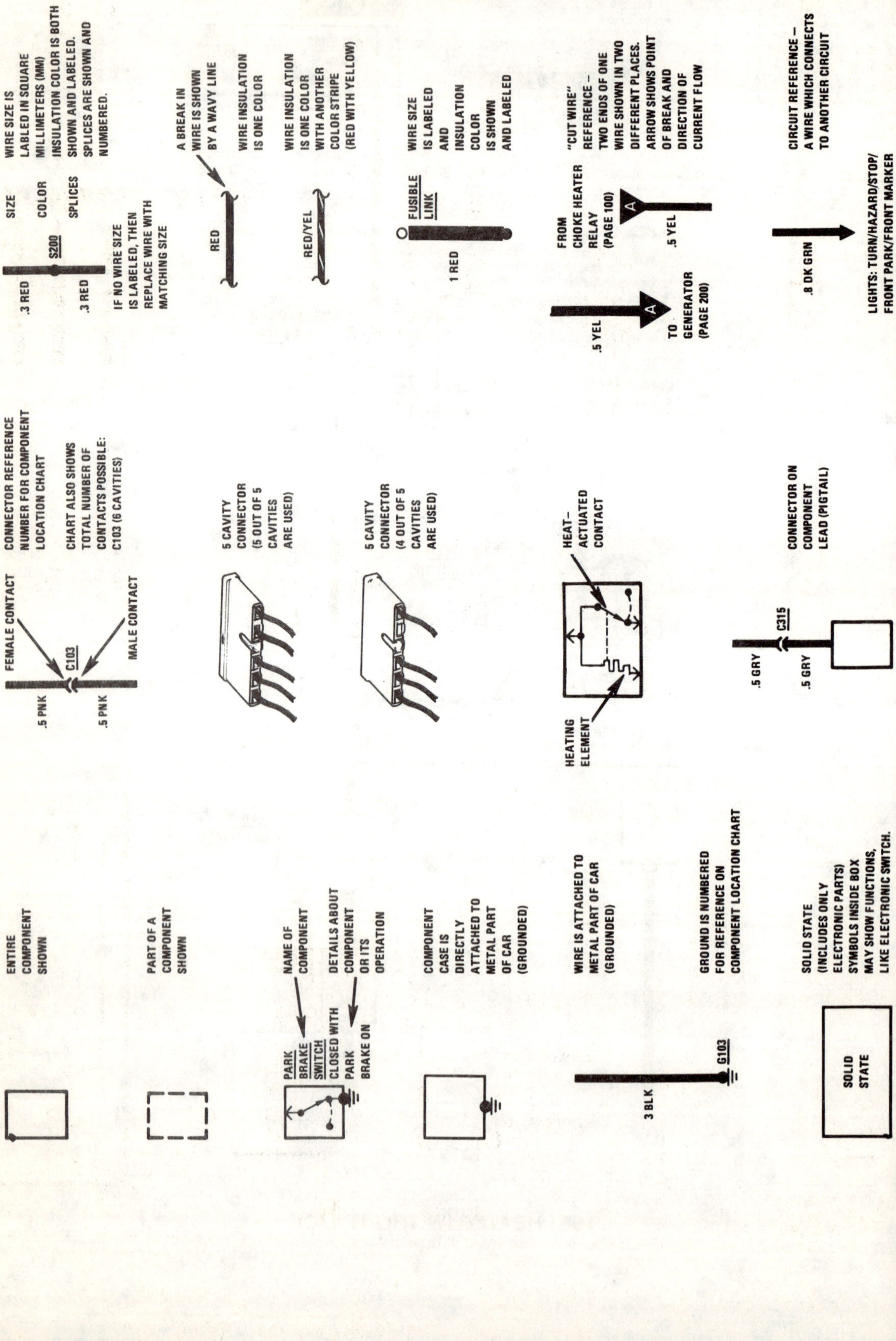

1982 WIRING DIAGRAM KEY

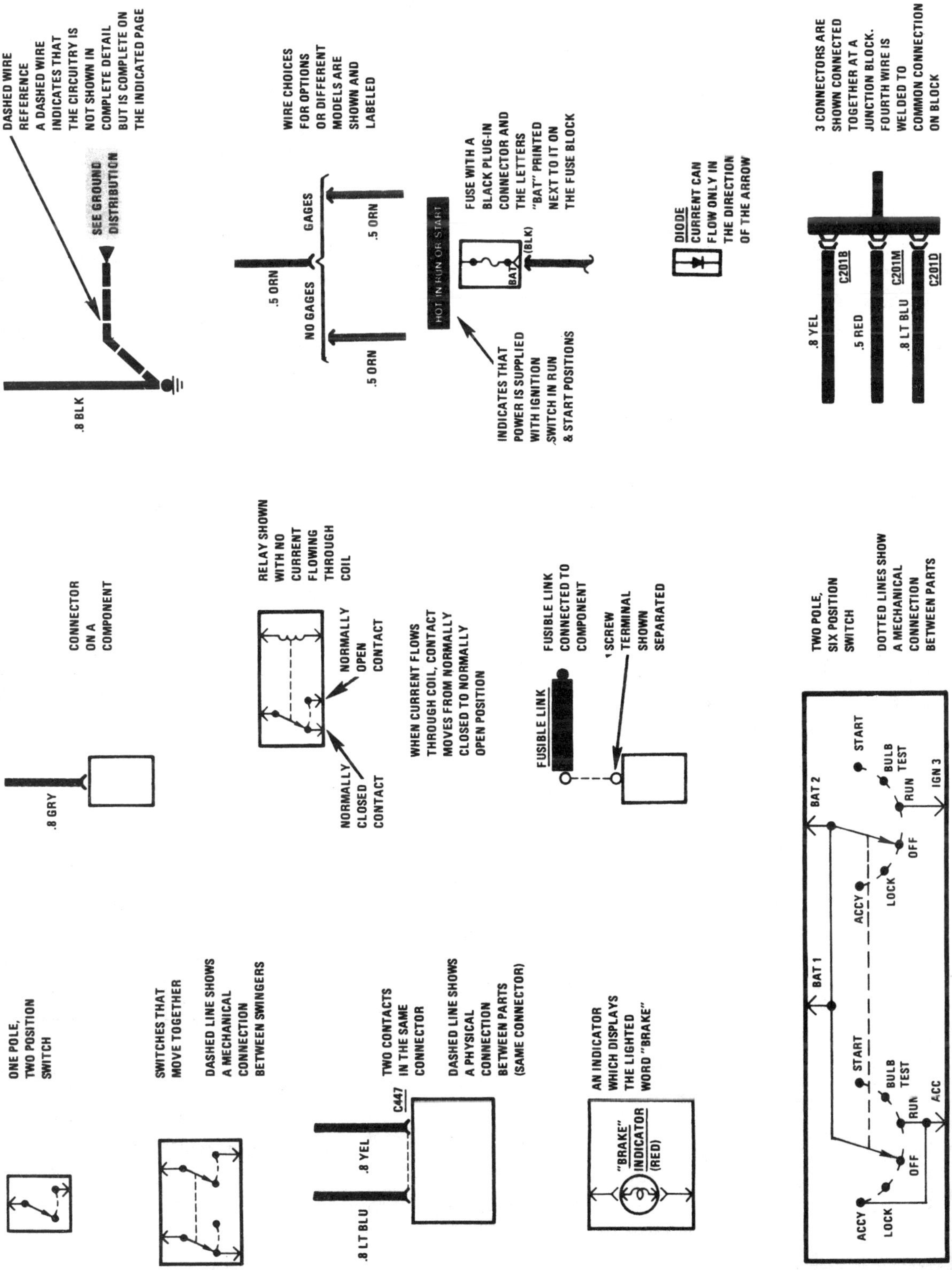

15

1982 POWER DISTRIBUTION (GASOLINE)

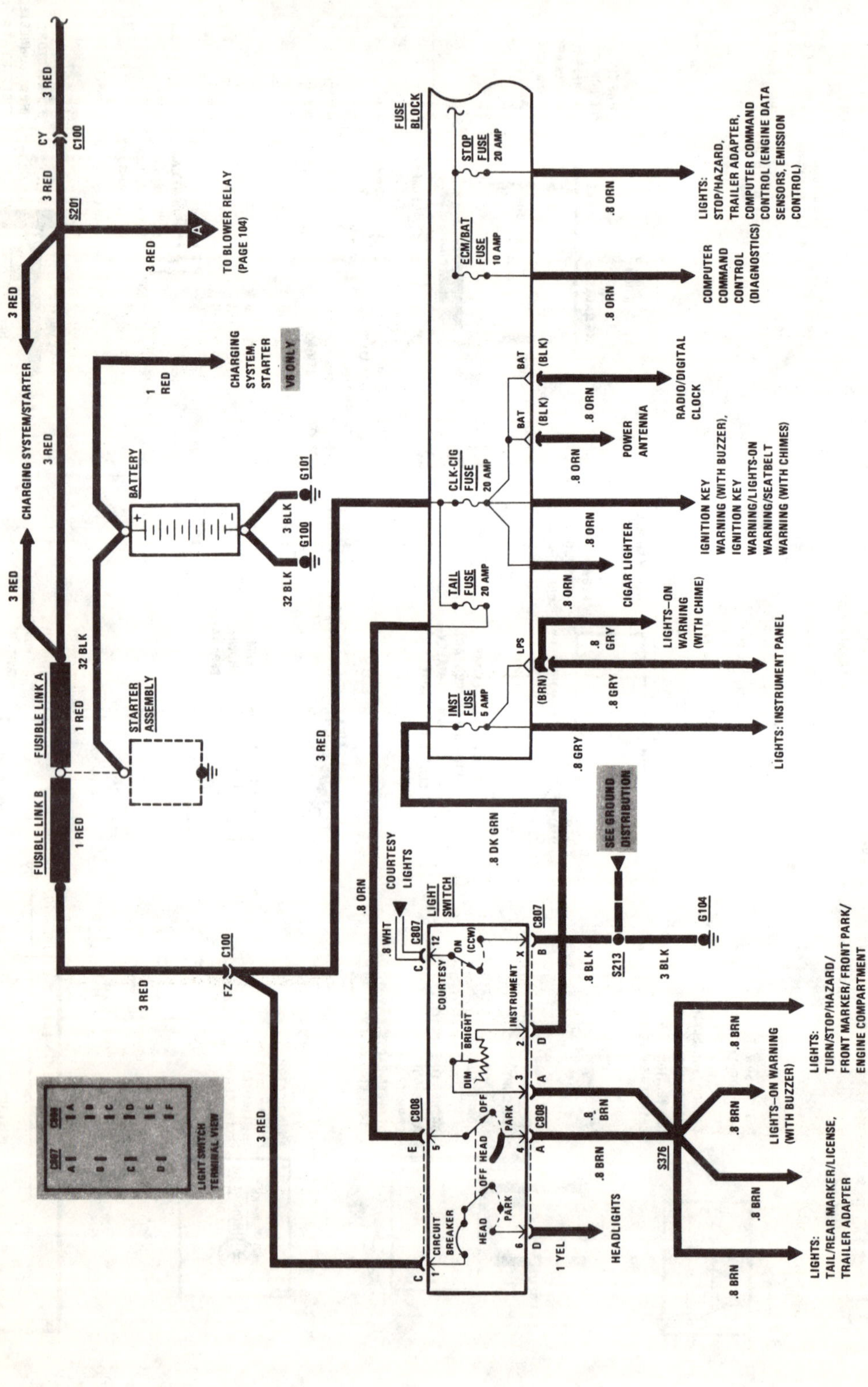

1982 POWER DISTRIBUTION (GASOLINE)

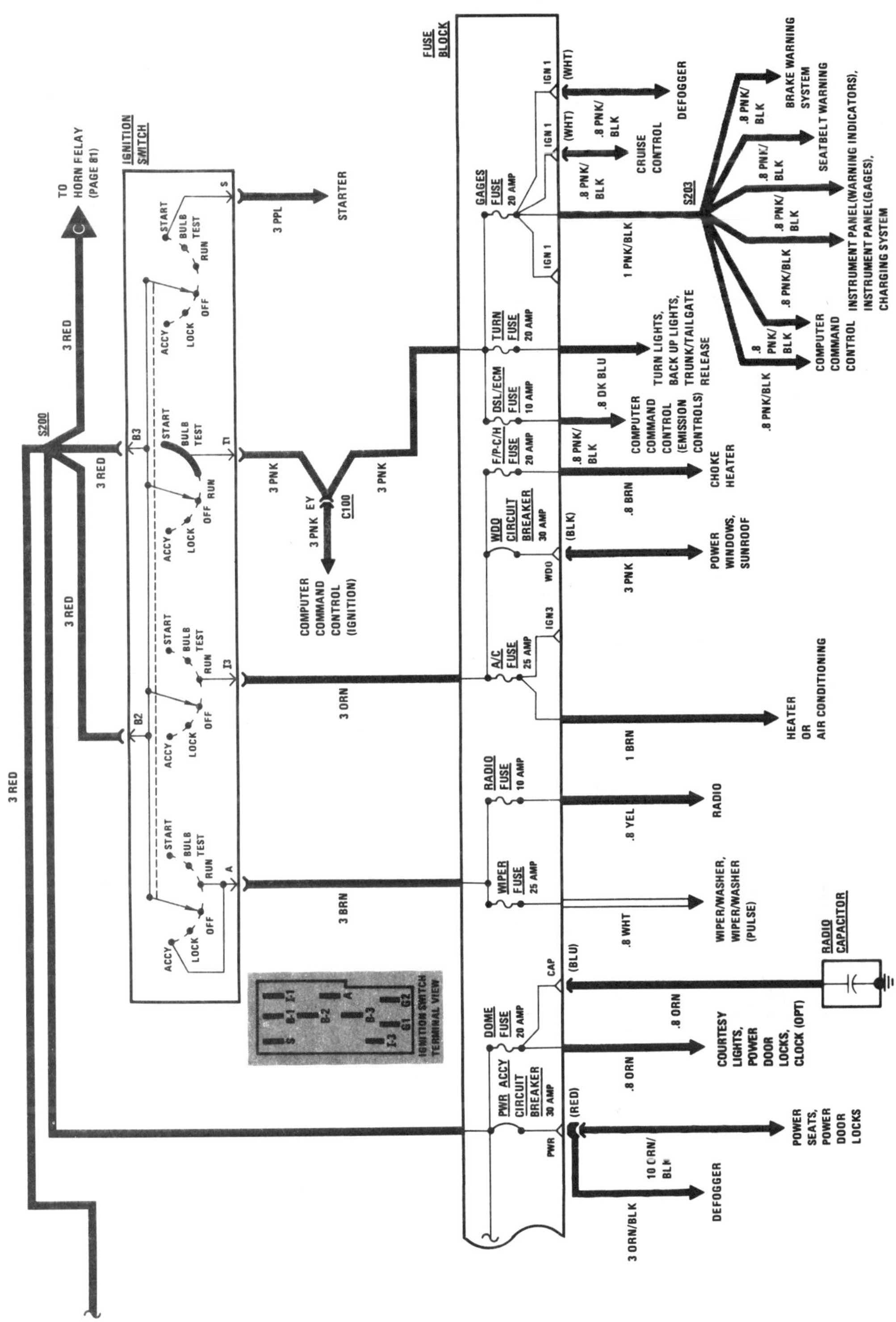

1982 CHARGING SYSTEM/STARTER (GASOLINE)

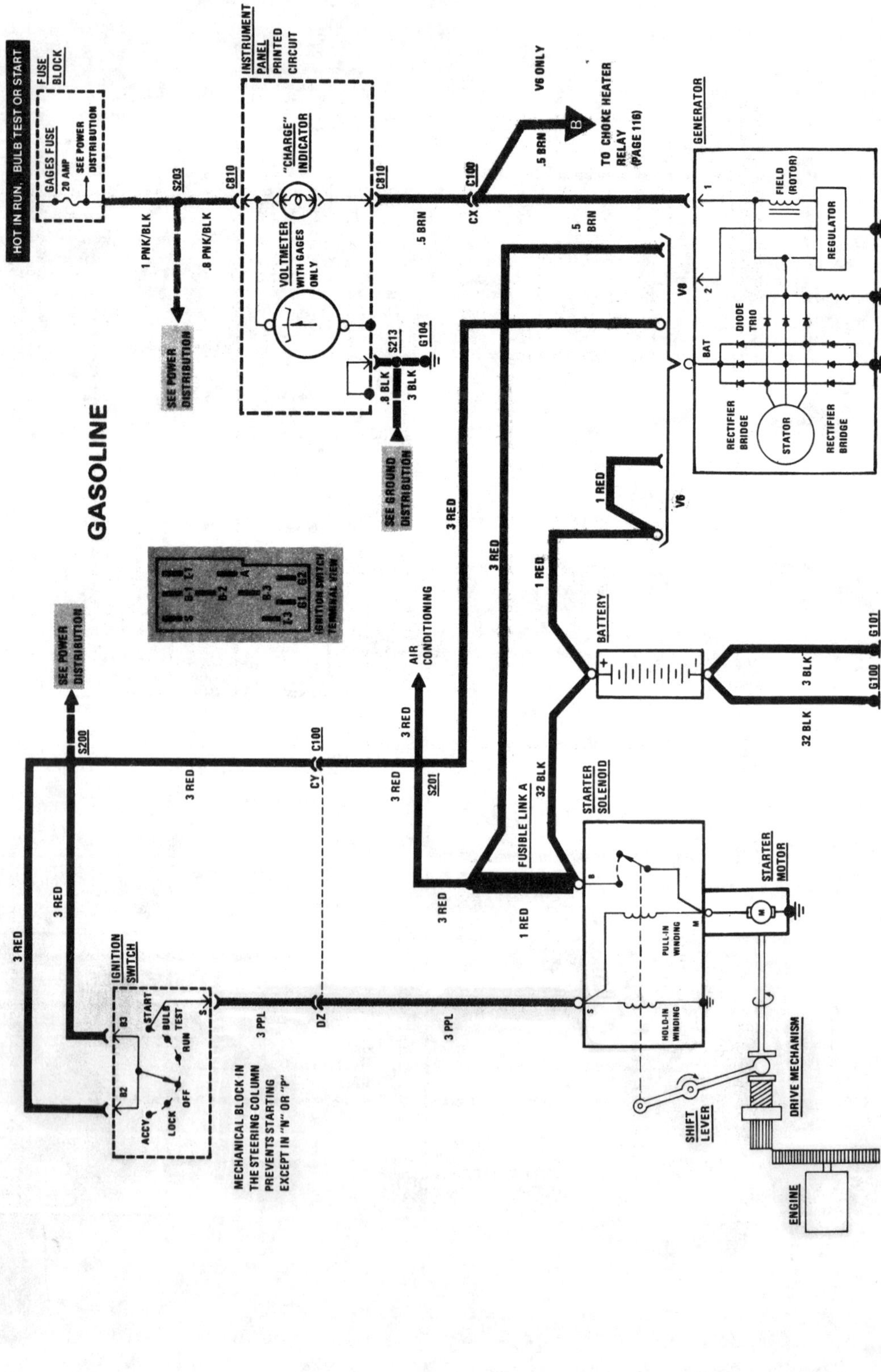

1982 COMPUTER COMMAND CONTROL (DIAGNOSTICS, IGNITION)

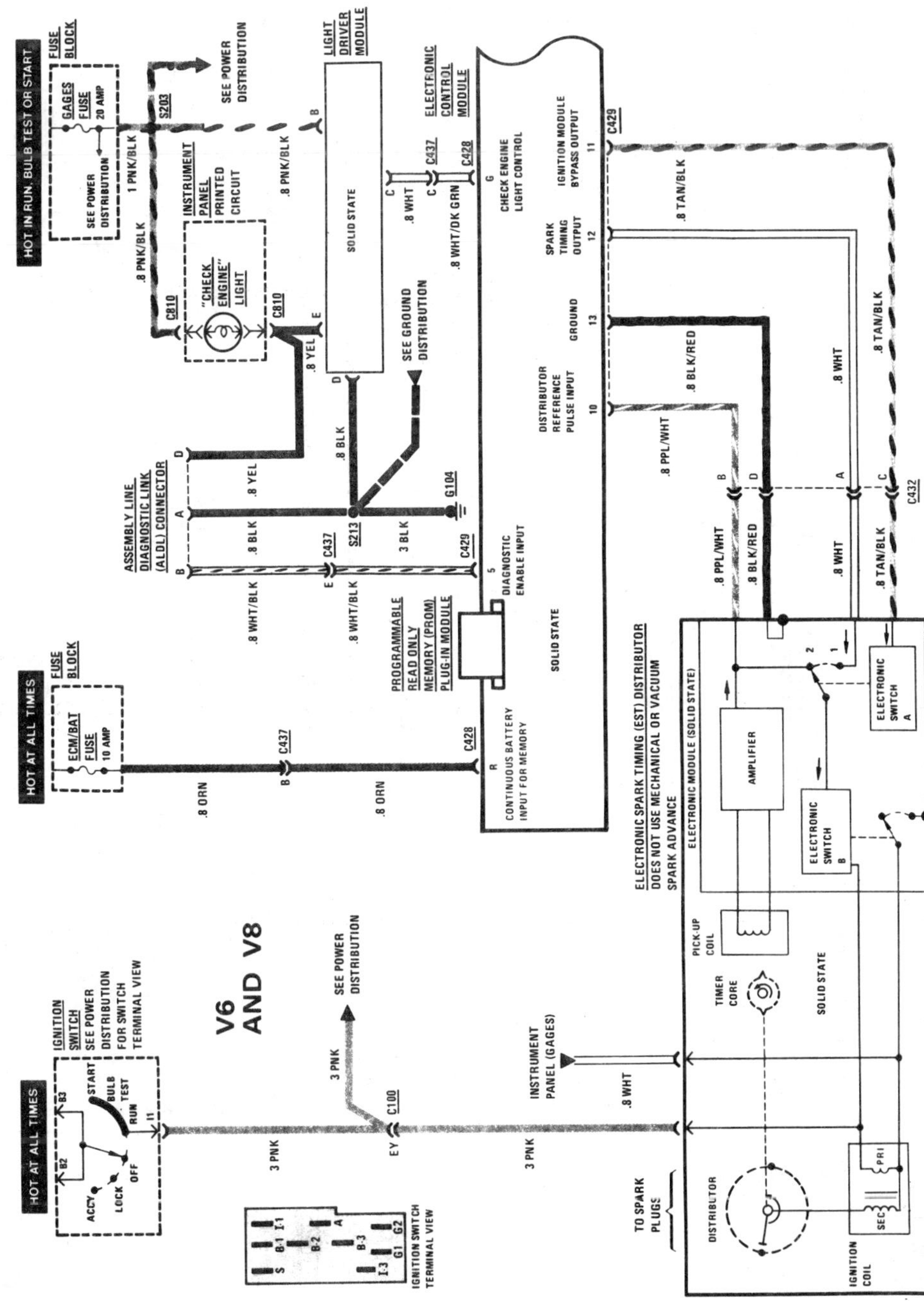

1982 COMPUTER COMMAND CONTROL—V6 (ENGINE DATA SENSORS, EMISSION CONTROLS)

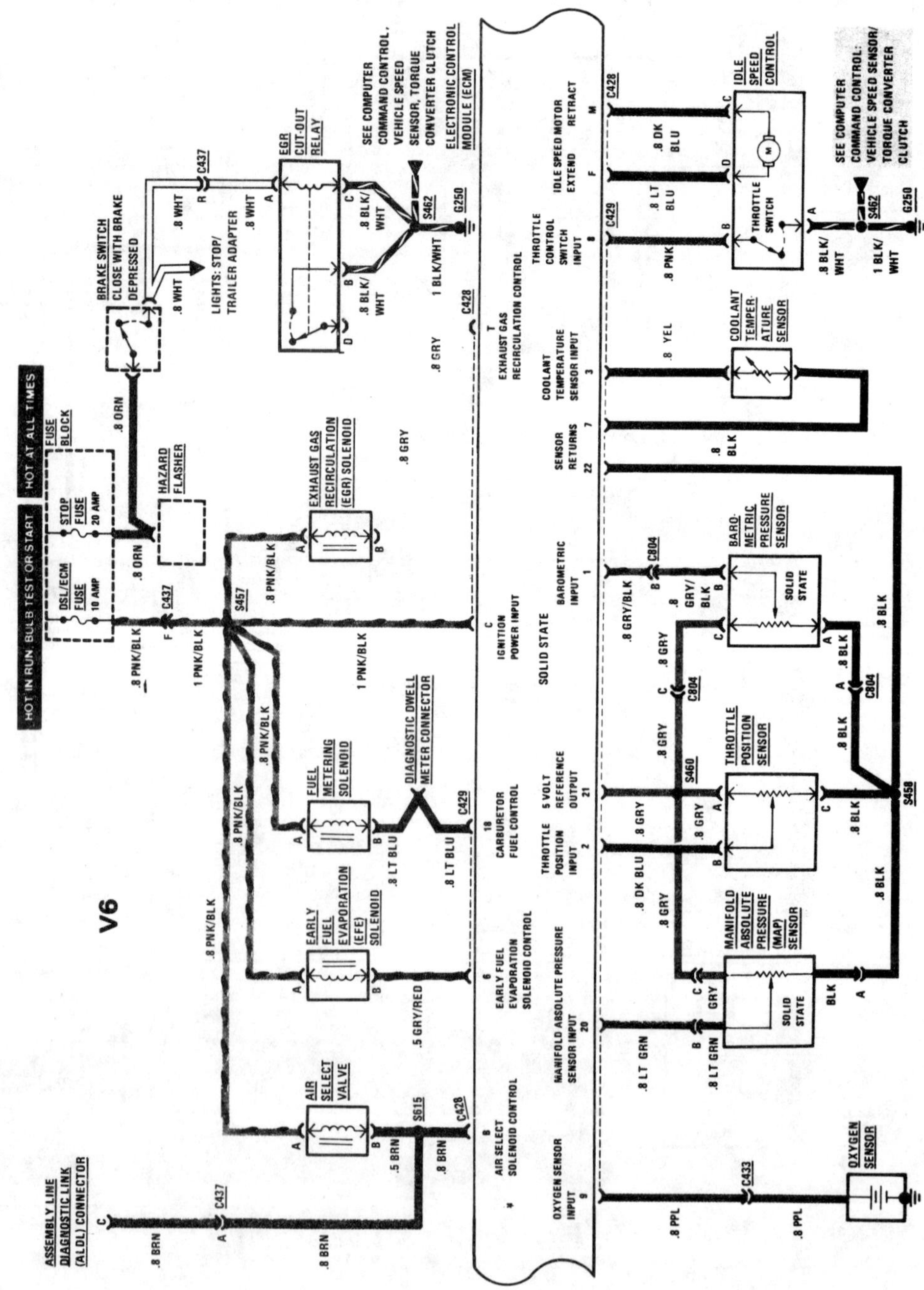

1982 COMPUTER COMMAND CONTROL—V8 (ENGINE DATA SENSORS, EMISSION CONTROLS)

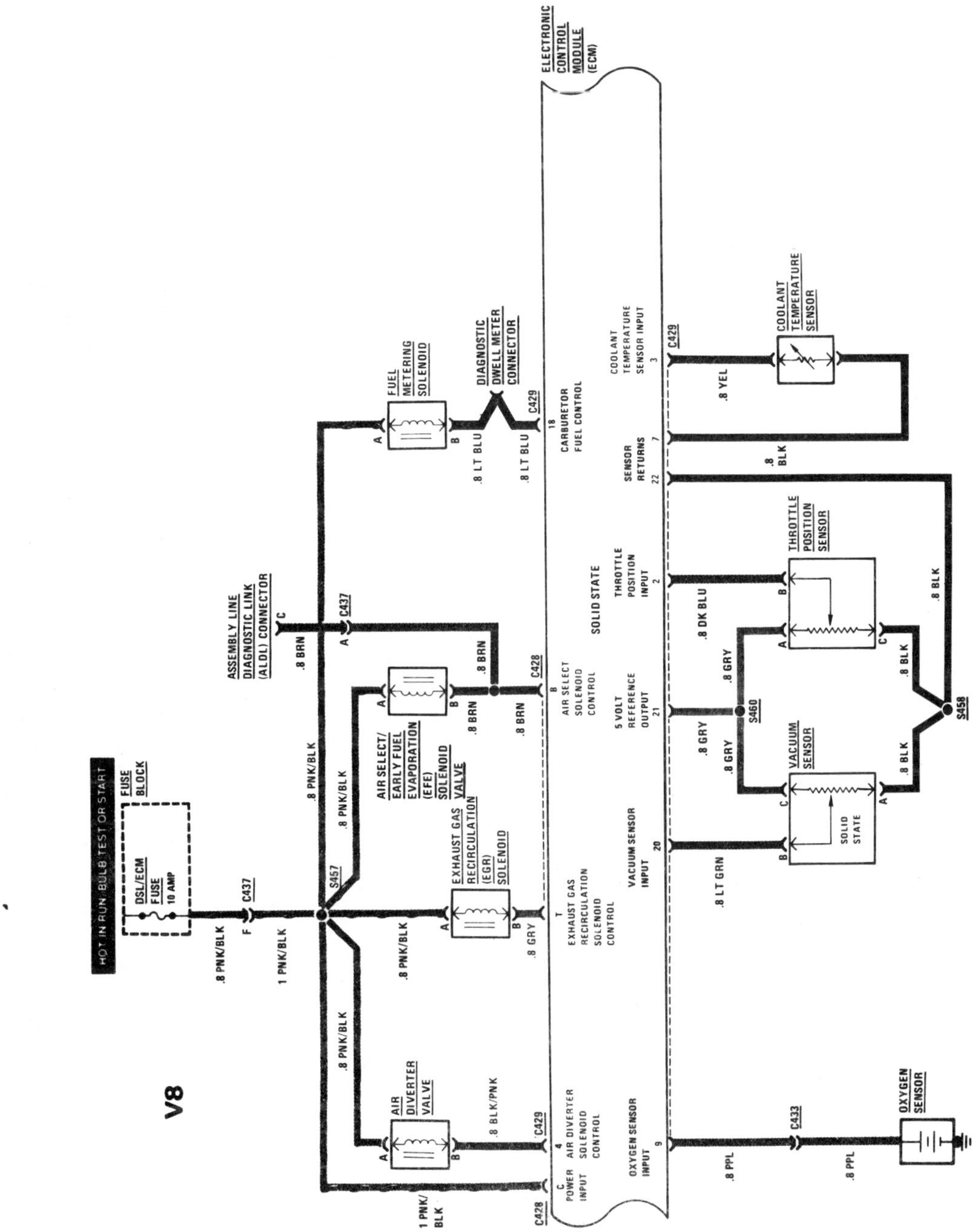

1982 COMPUTER COMMAND CONTROL—V6 (VEHICLE SENSORS, TORQUE CONVERTER CLUTCH)

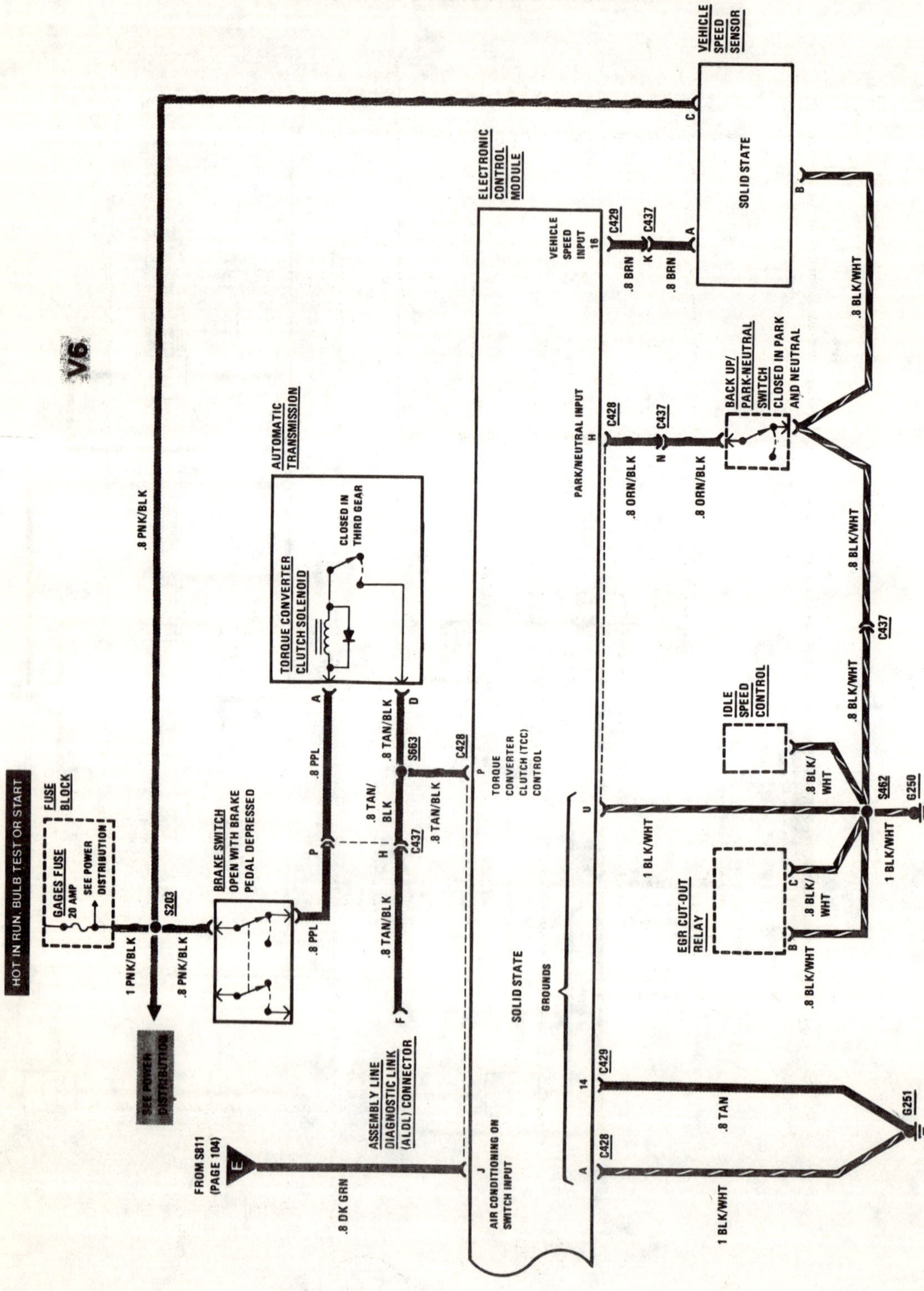

1982 COMPUTER COMMAND CONTROL—V8 (VEHICLE SENSORS, TORQUE CONVERTER CLUTCH)

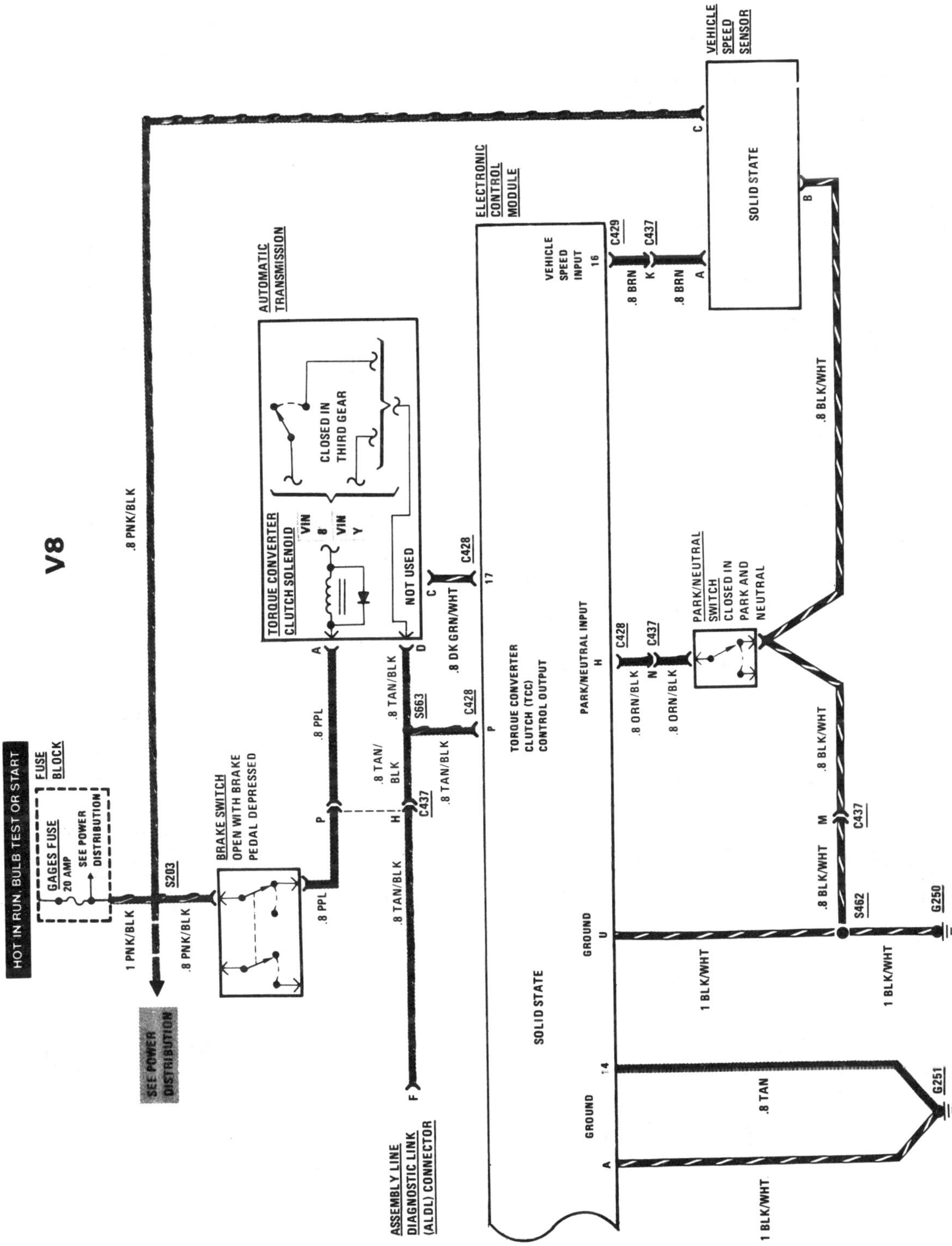

1982 CHARGING SYSTEM/STARTER (DIESEL)

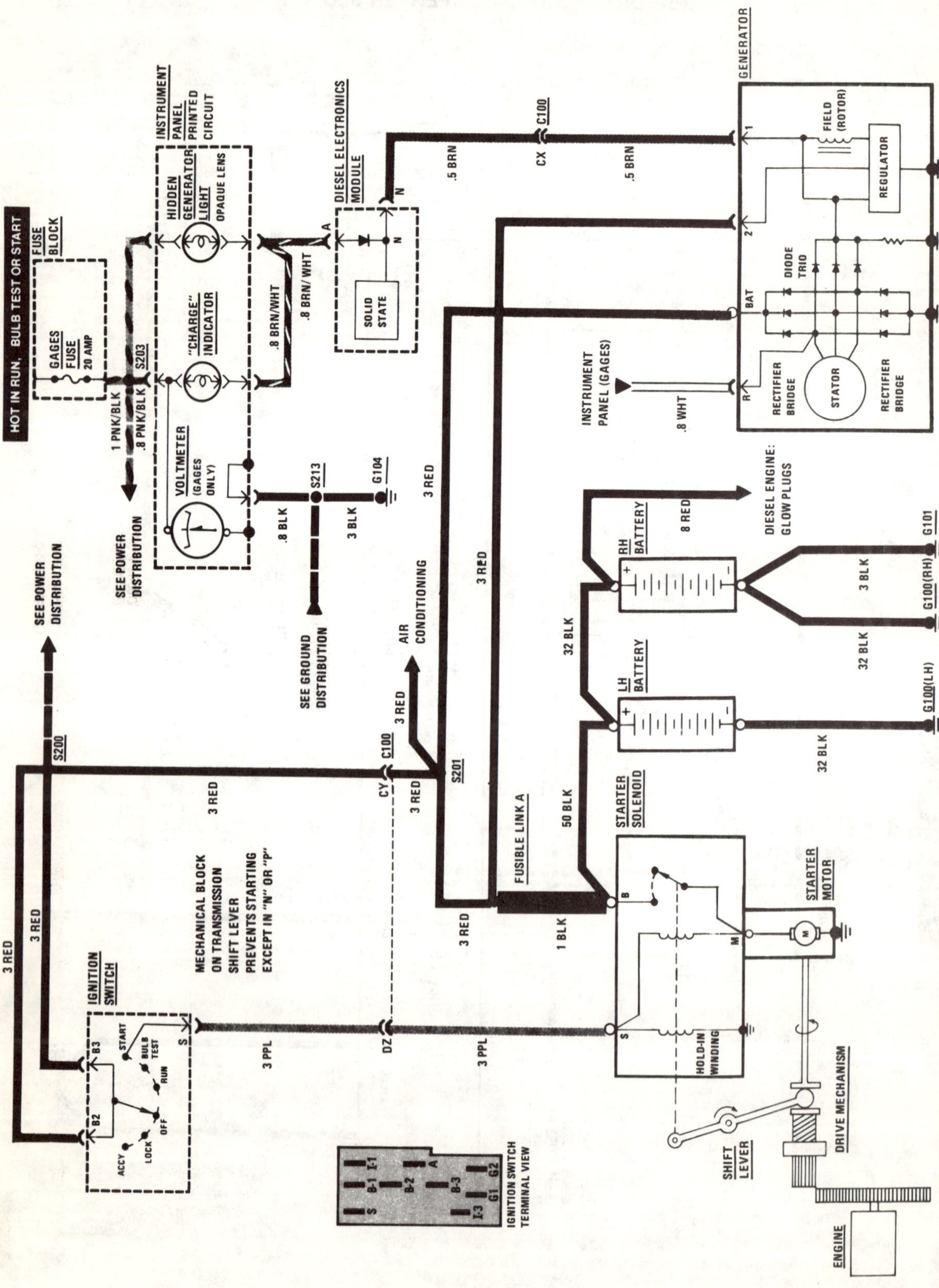

1982 GLOW PLUGS/FUEL CONTROL (V8 DIESEL)

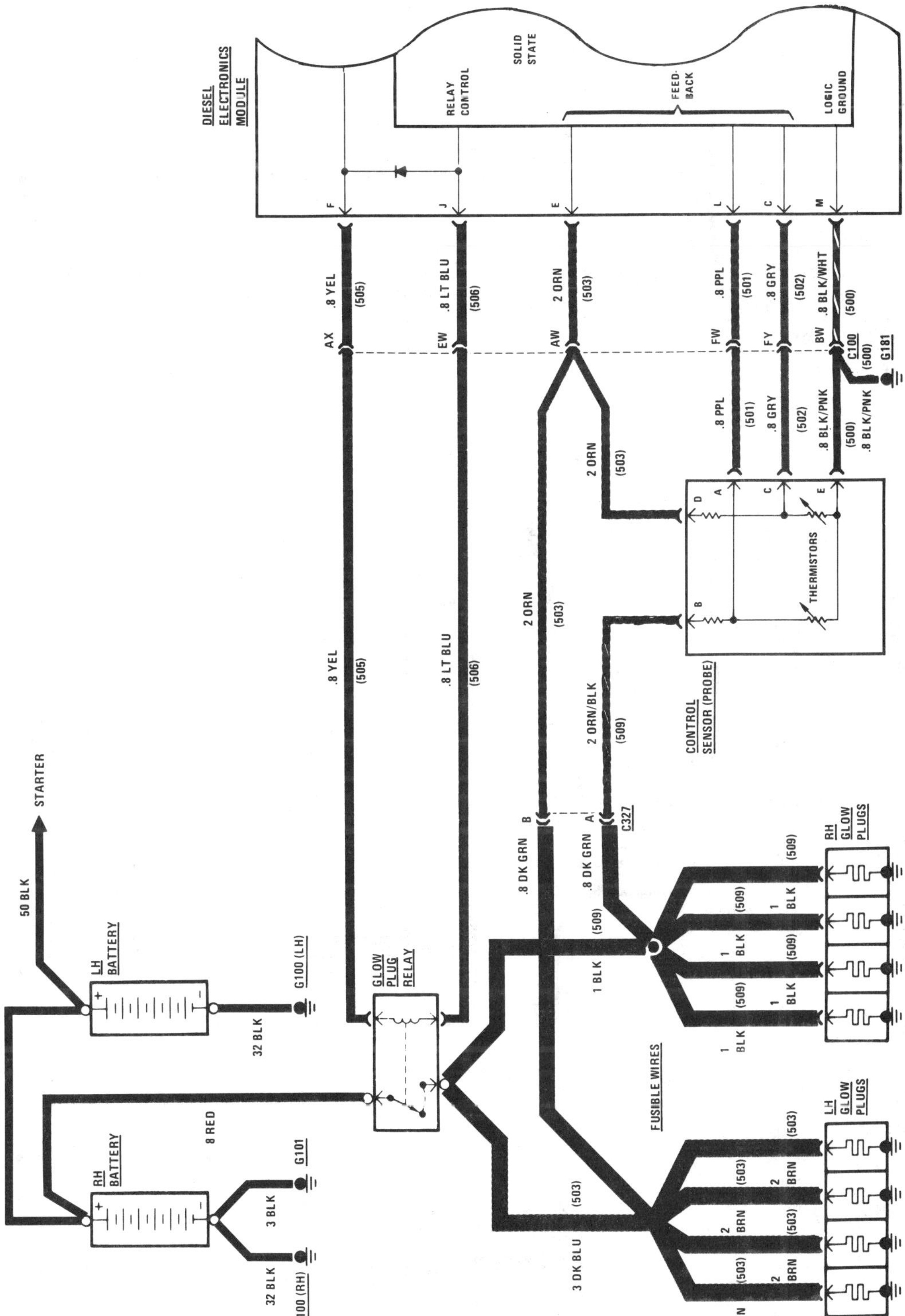

1982 GLOW PLUGS/FUEL CONTROL (V8 DIESEL)

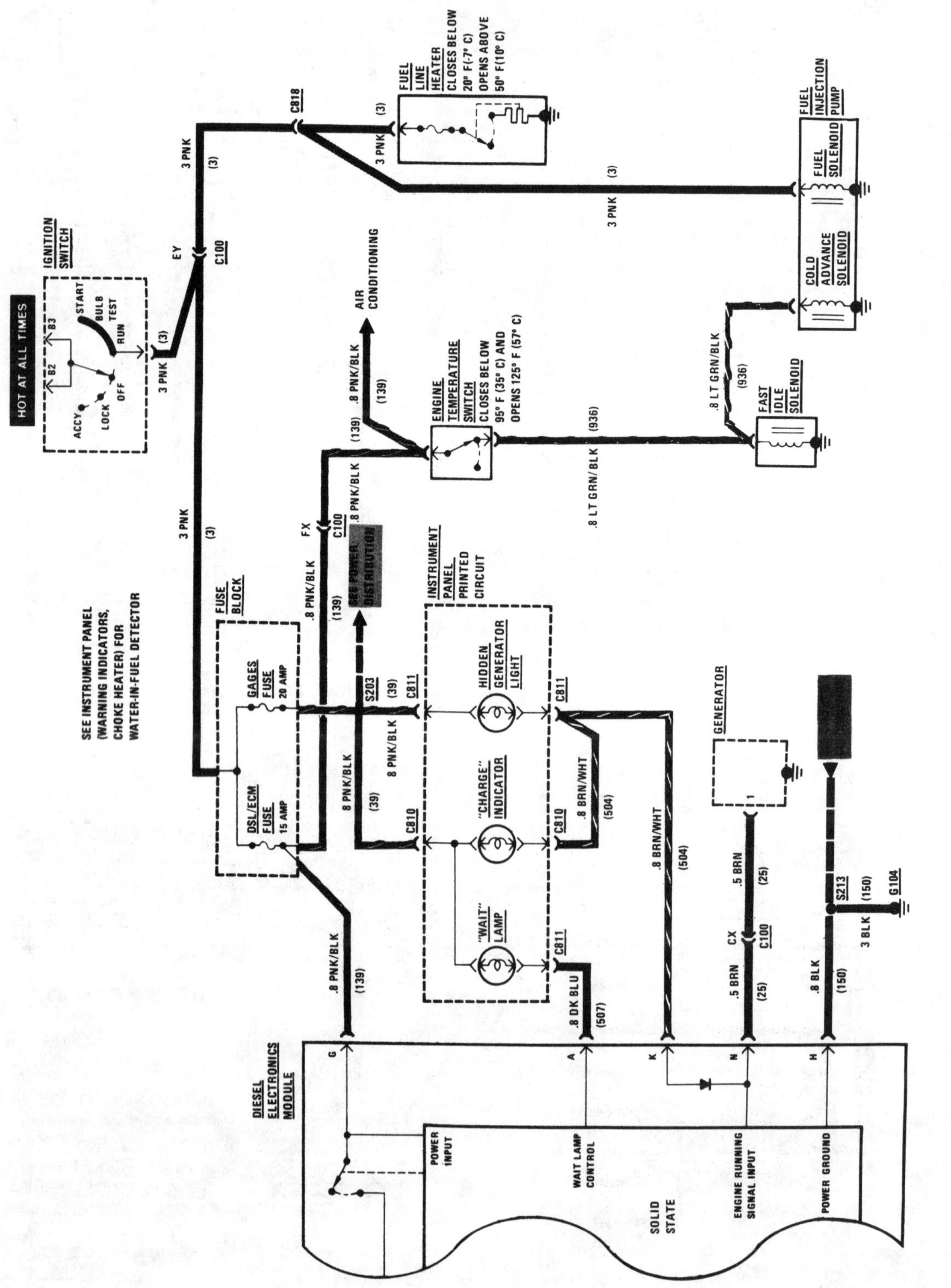

1982 GLOW PLUGS/FUEL CONTROL (V6 DIESEL)

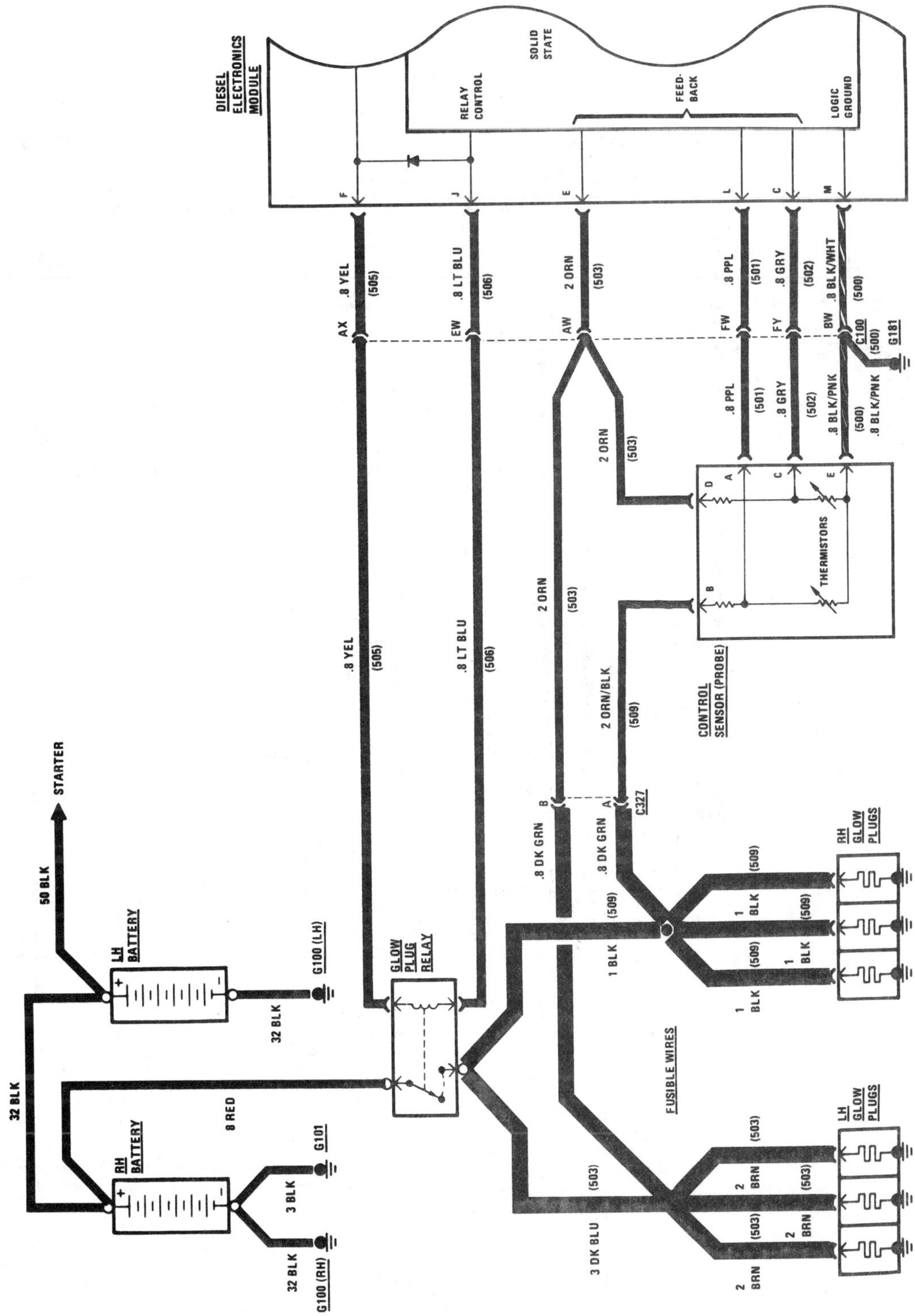

15

1982 GLOW PLUGS/FUEL CONTROL (V6 DIESEL)

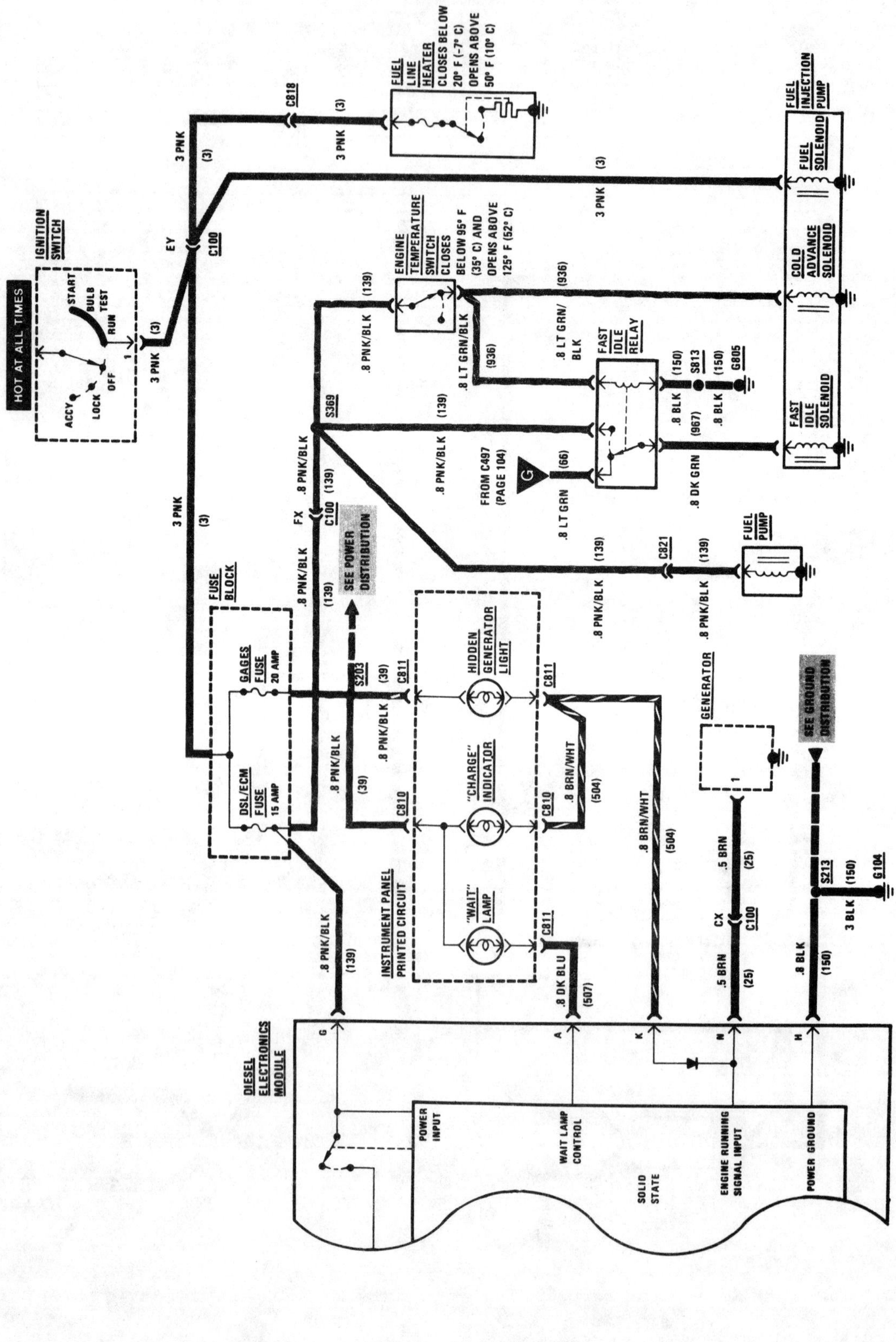

1982 EMISSION CONTROLS/TORQUE CONVERTER CLUTCH (DIESEL)

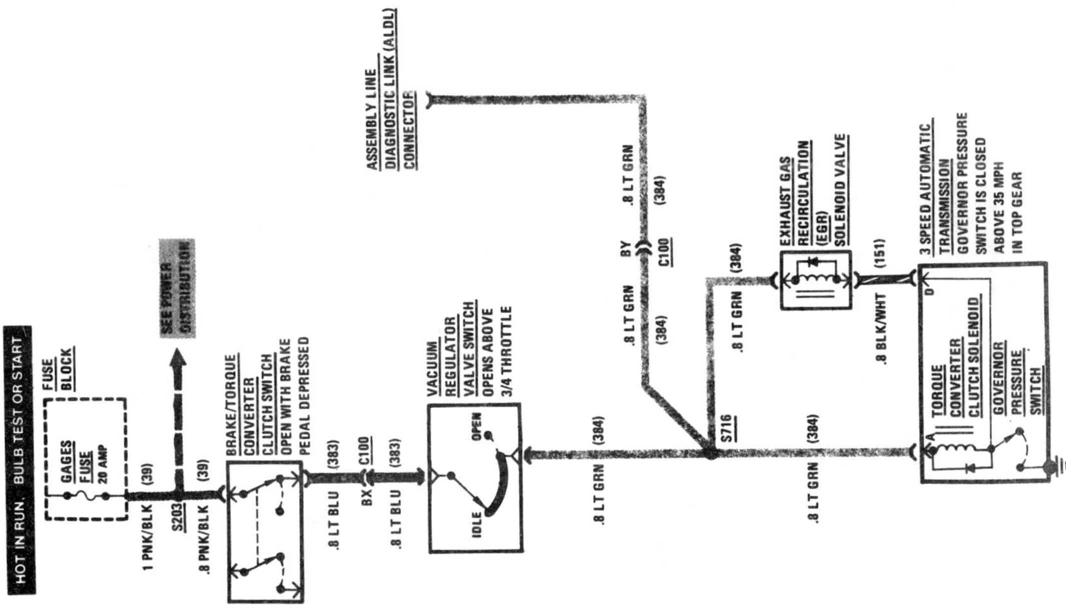

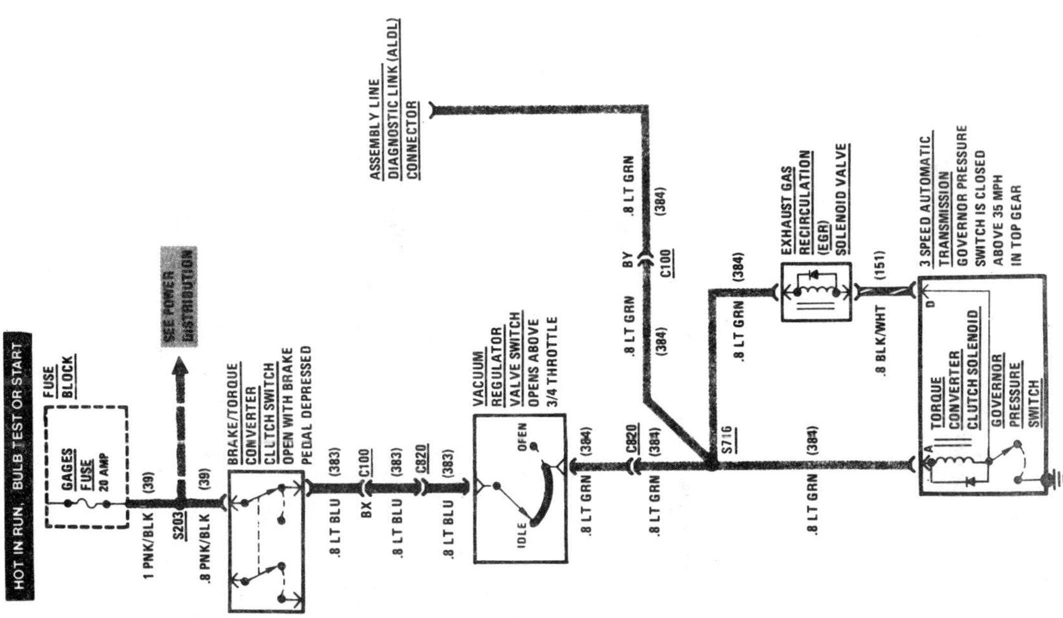

1982 POWER DISTRIBUTION (DIESEL)

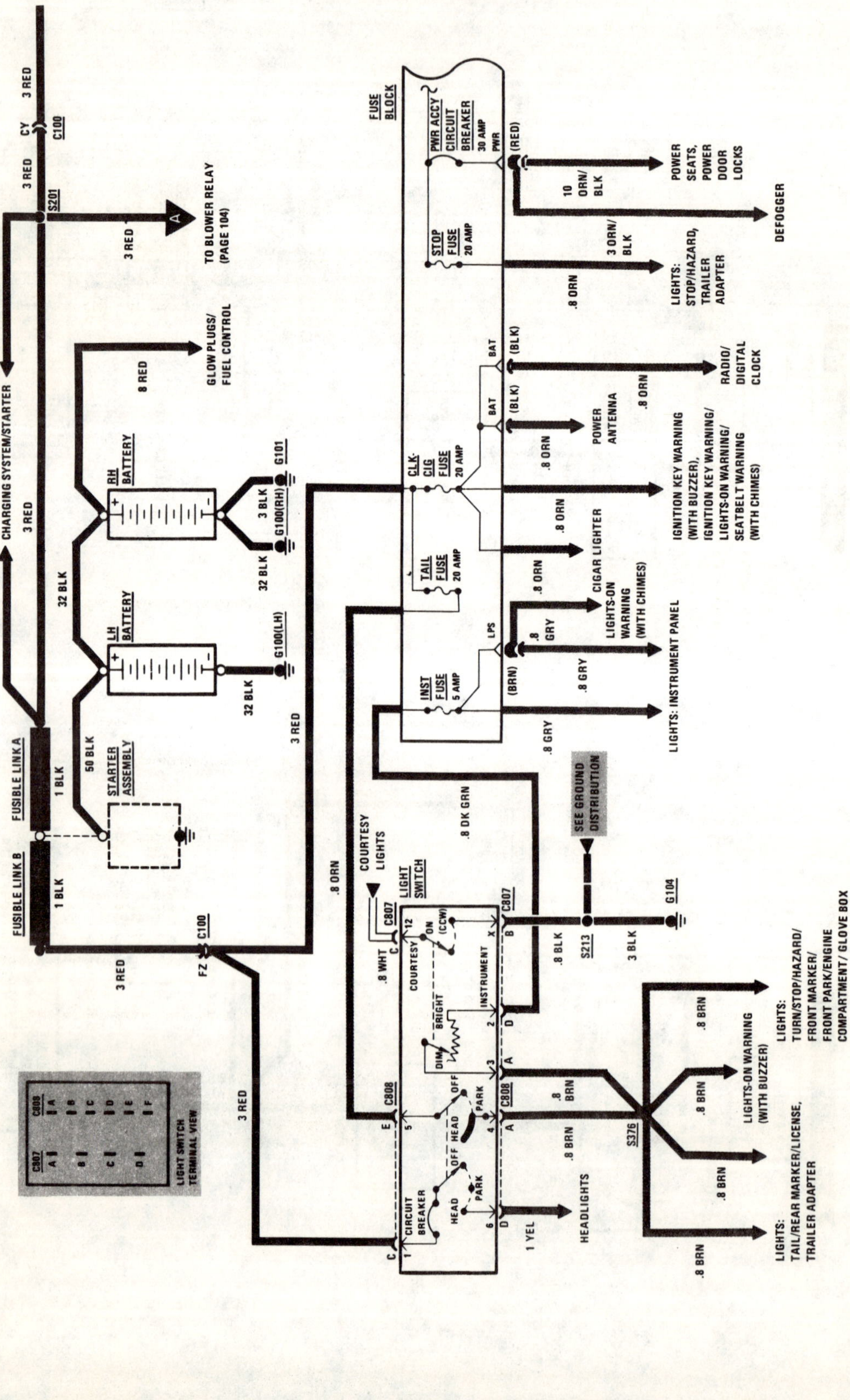

1982 POWER DISTRIBUTION (DIESEL)

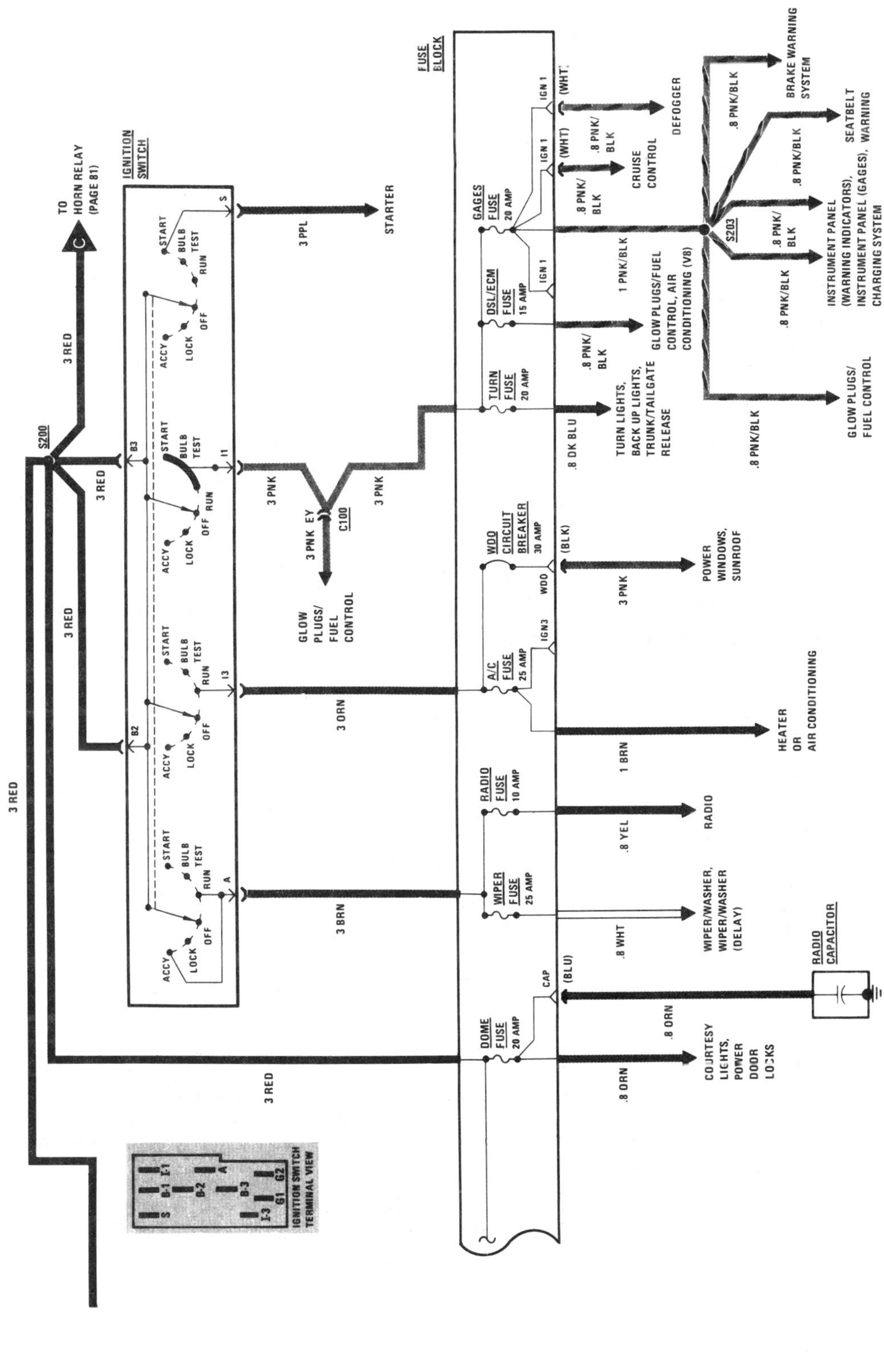

1982 HEADLIGHTS

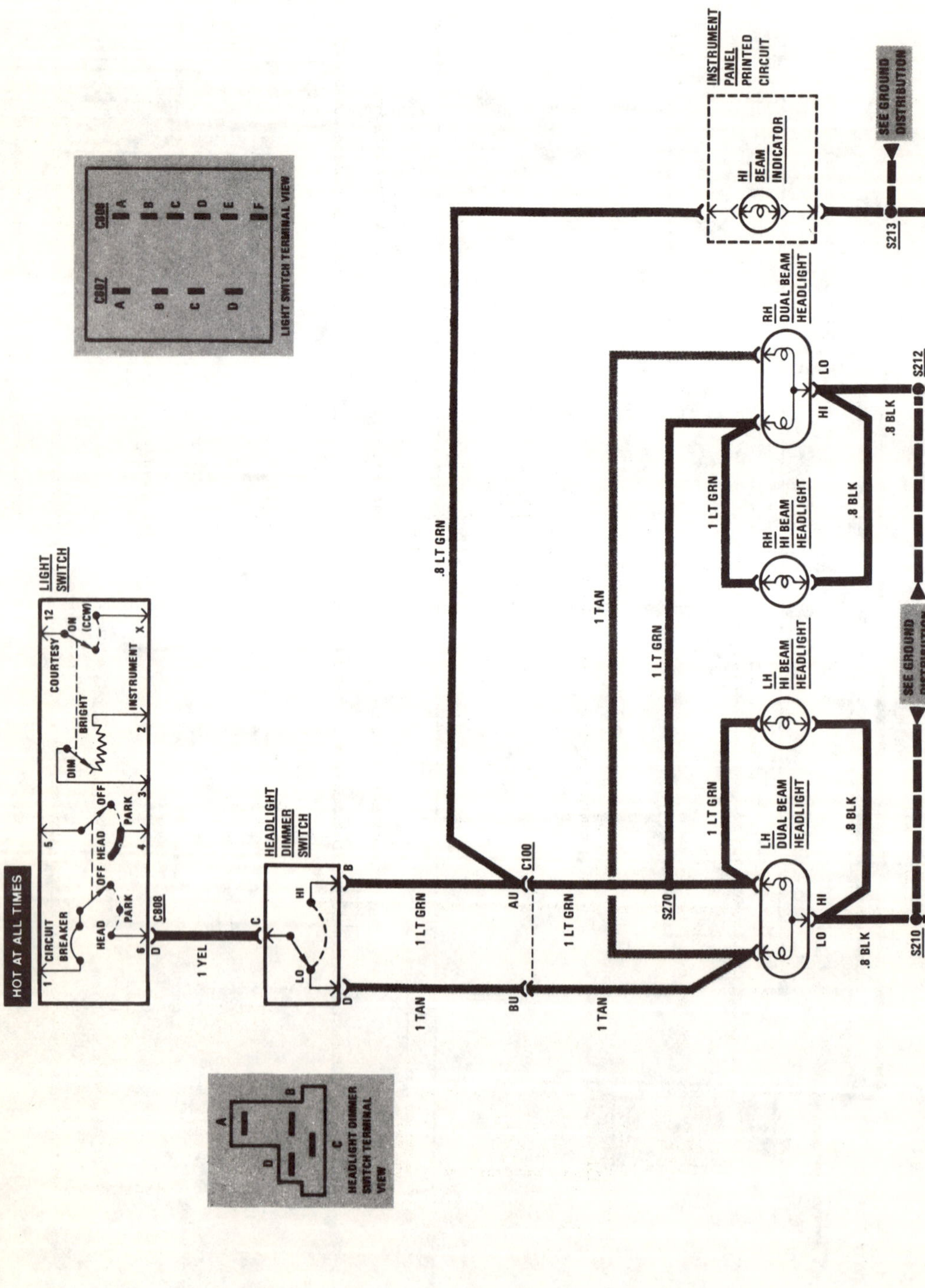

1982 REAR PARK/REAR MARKER AND LICENSE LIGHTS

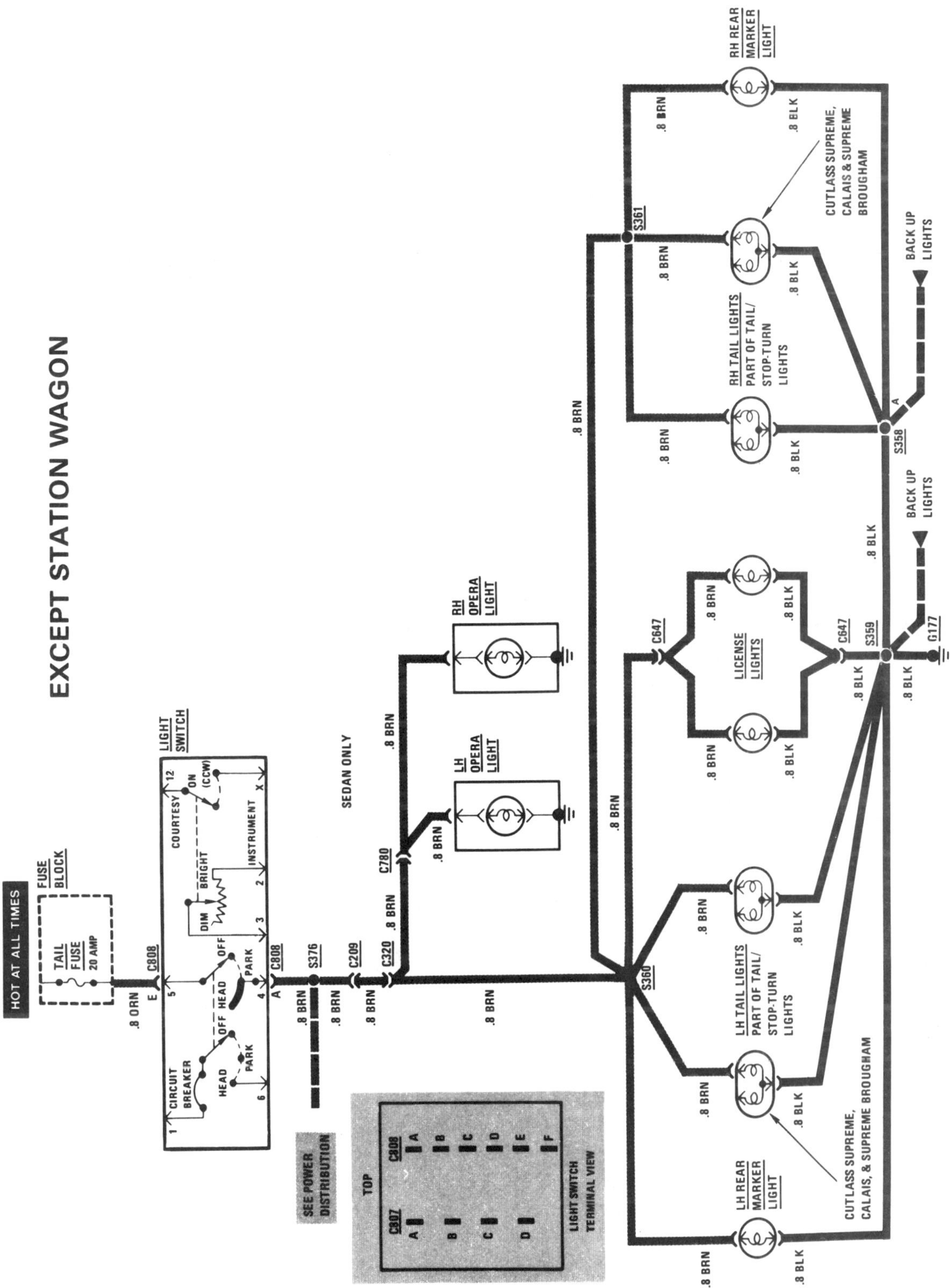

1982 REAR PARK/REAR MARKER AND LICENSE LIGHTS

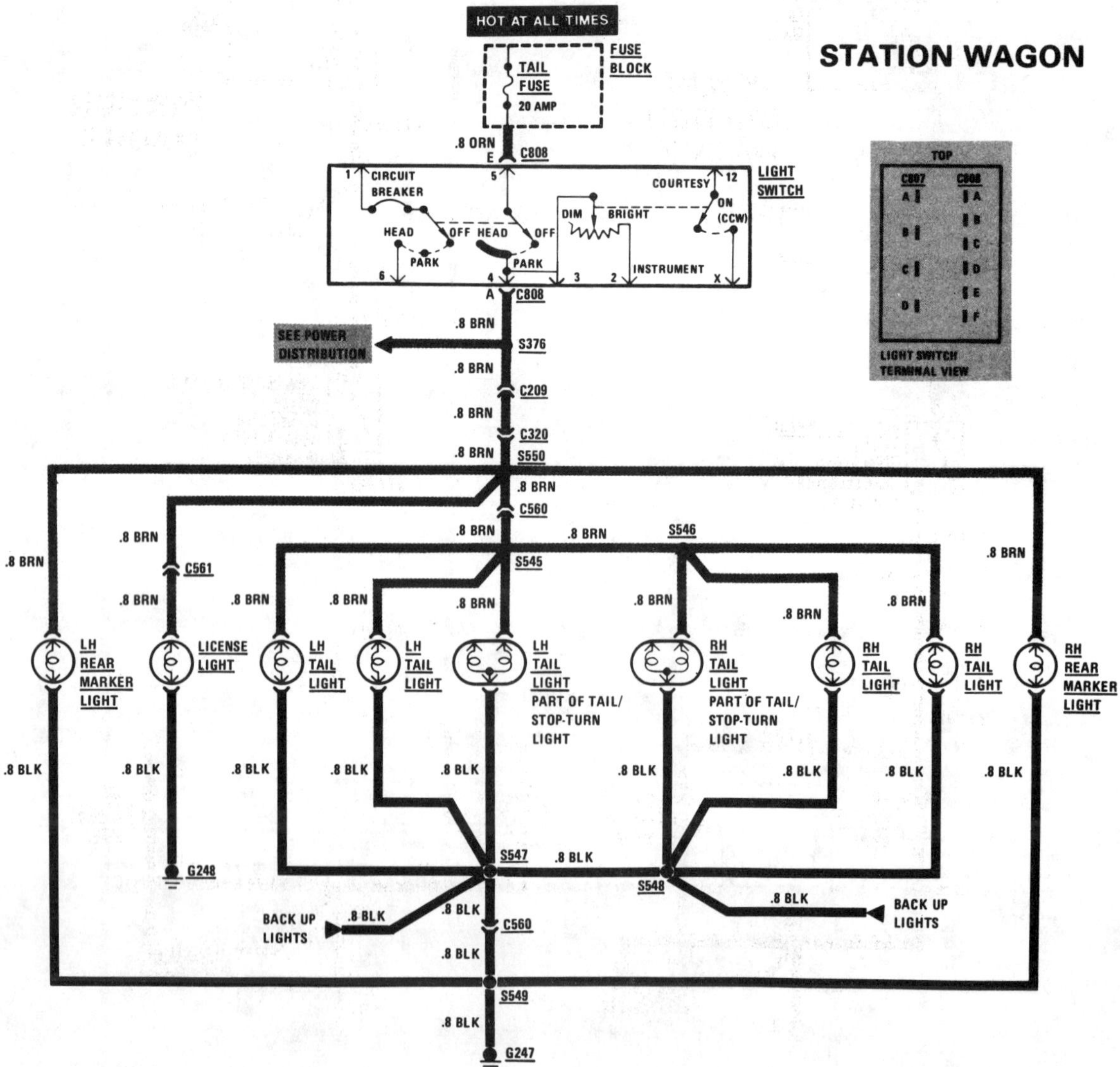

1982 BACK-UP LIGHTS

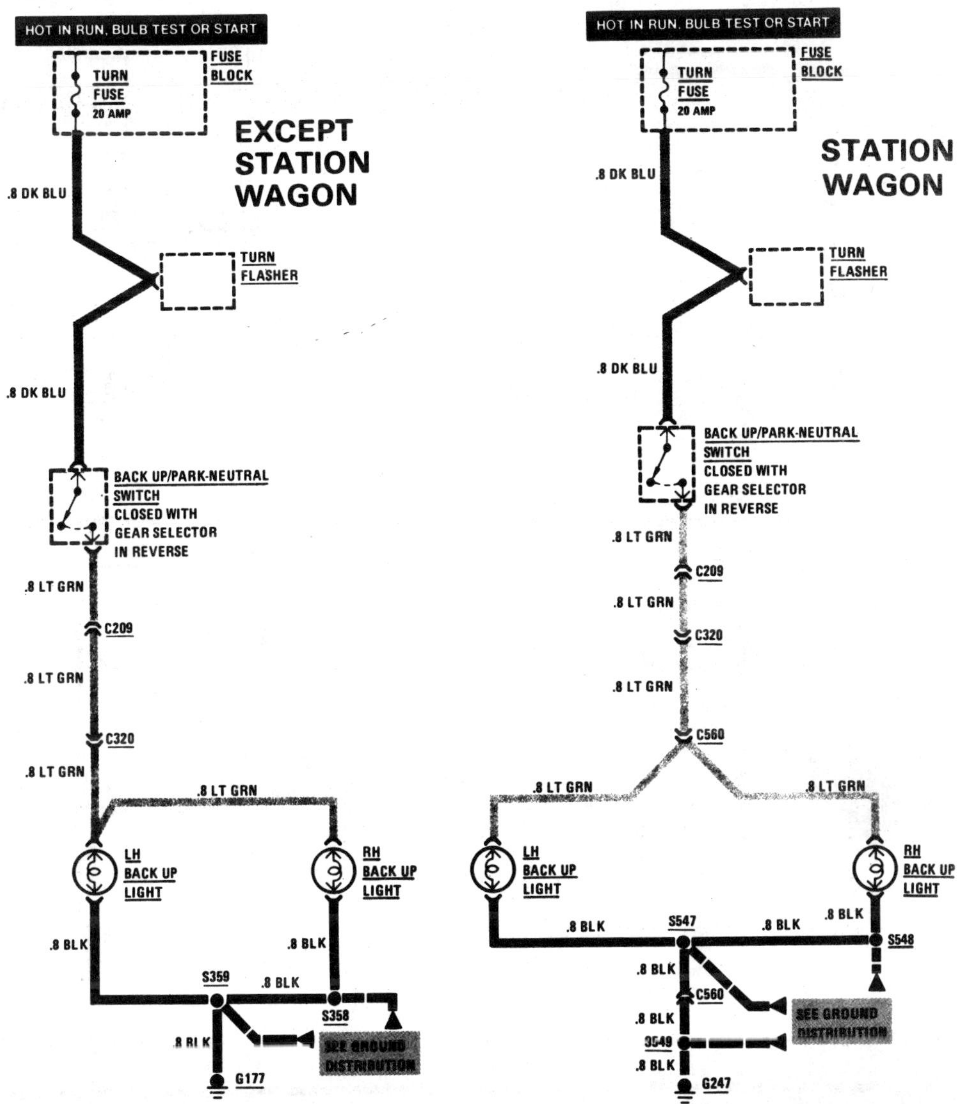

15

1982 COURTESY LIGHTS

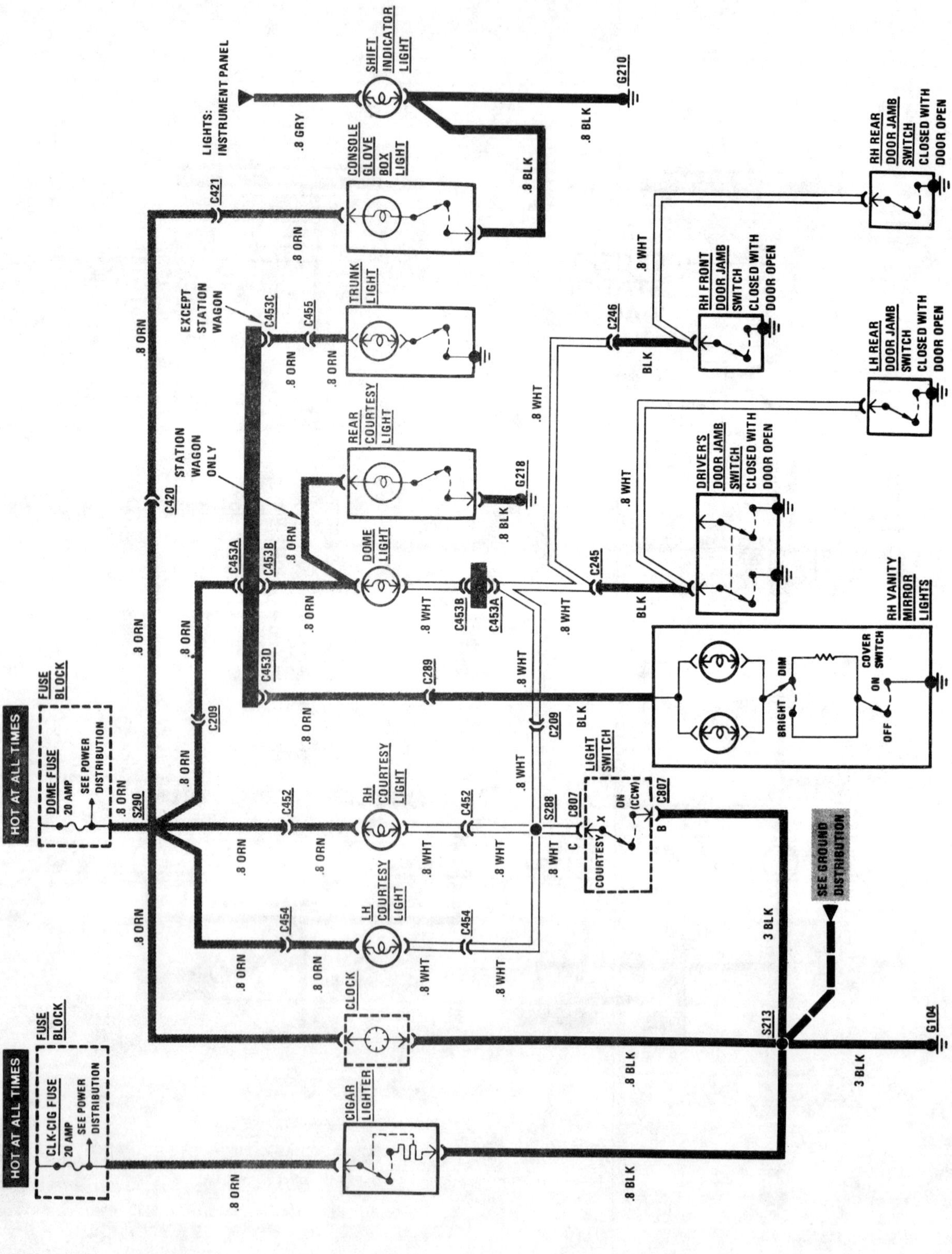

1982 TURN/HAZARD/STOP/FRONT PARK/ FRONT MARKER AND ENGINE COMPARTMENT LIGHTS

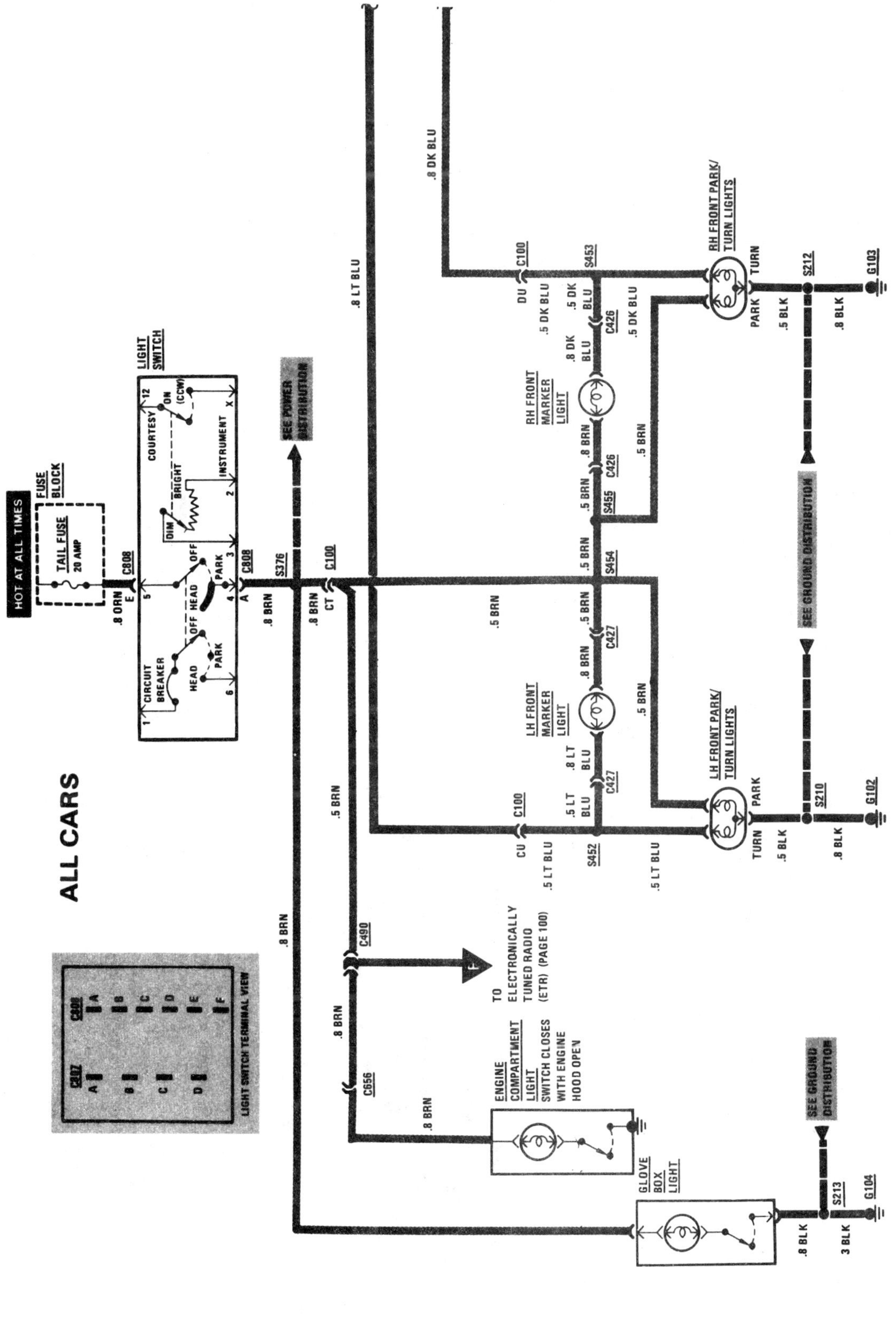

1982 TURN/HAZARD/STOP/FRONT PARK/ FRONT MARKER AND ENGINE COMPARTMENT LIGHTS

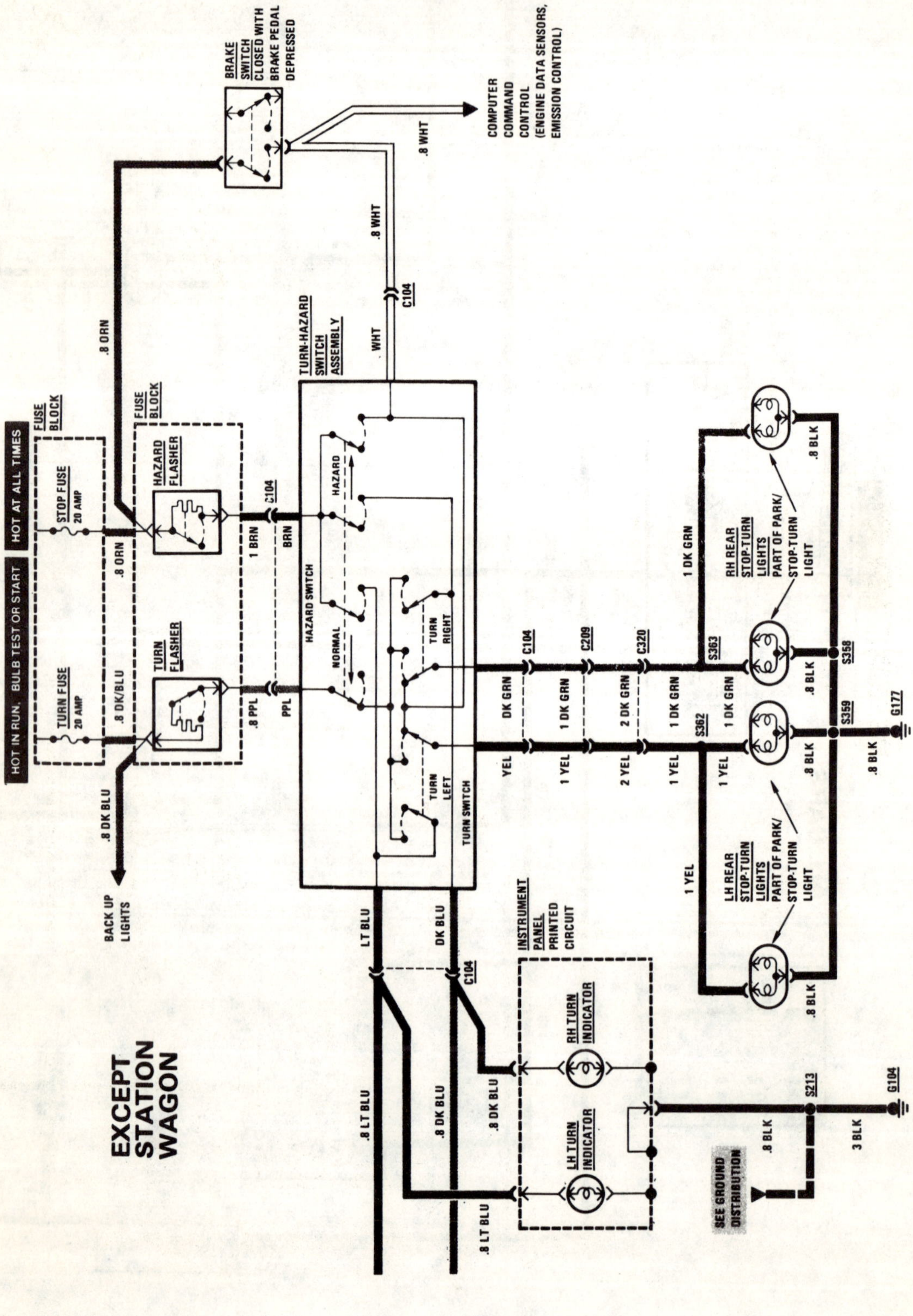

1982 TURN/HAZARD/STOP/FRONT PARK/ FRONT MARKER AND ENGINE COMPARTMENT LIGHTS

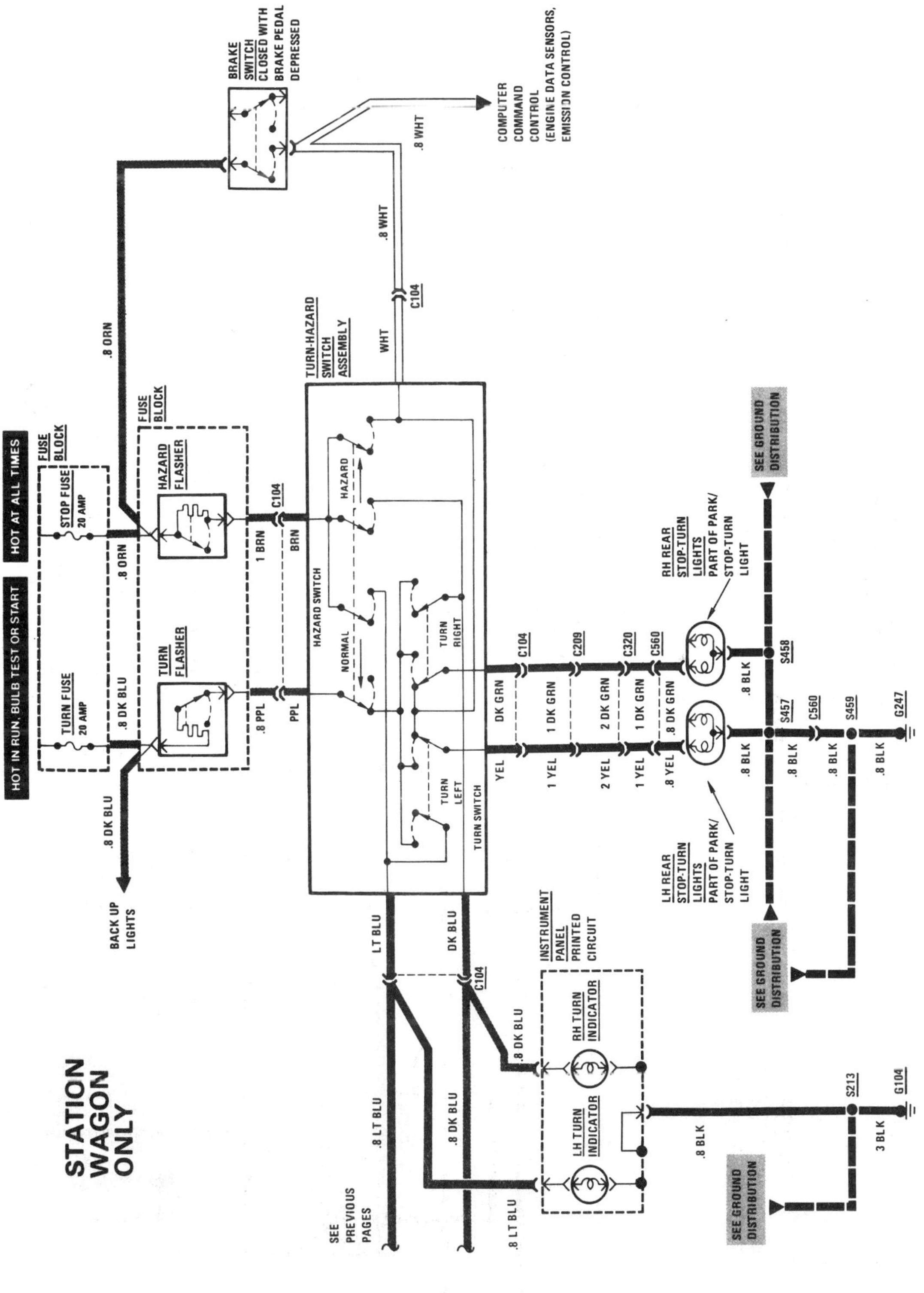

1982 IGNITION KEY WARNING/LIGHTS-ON WARNING (WITH BUZZER)

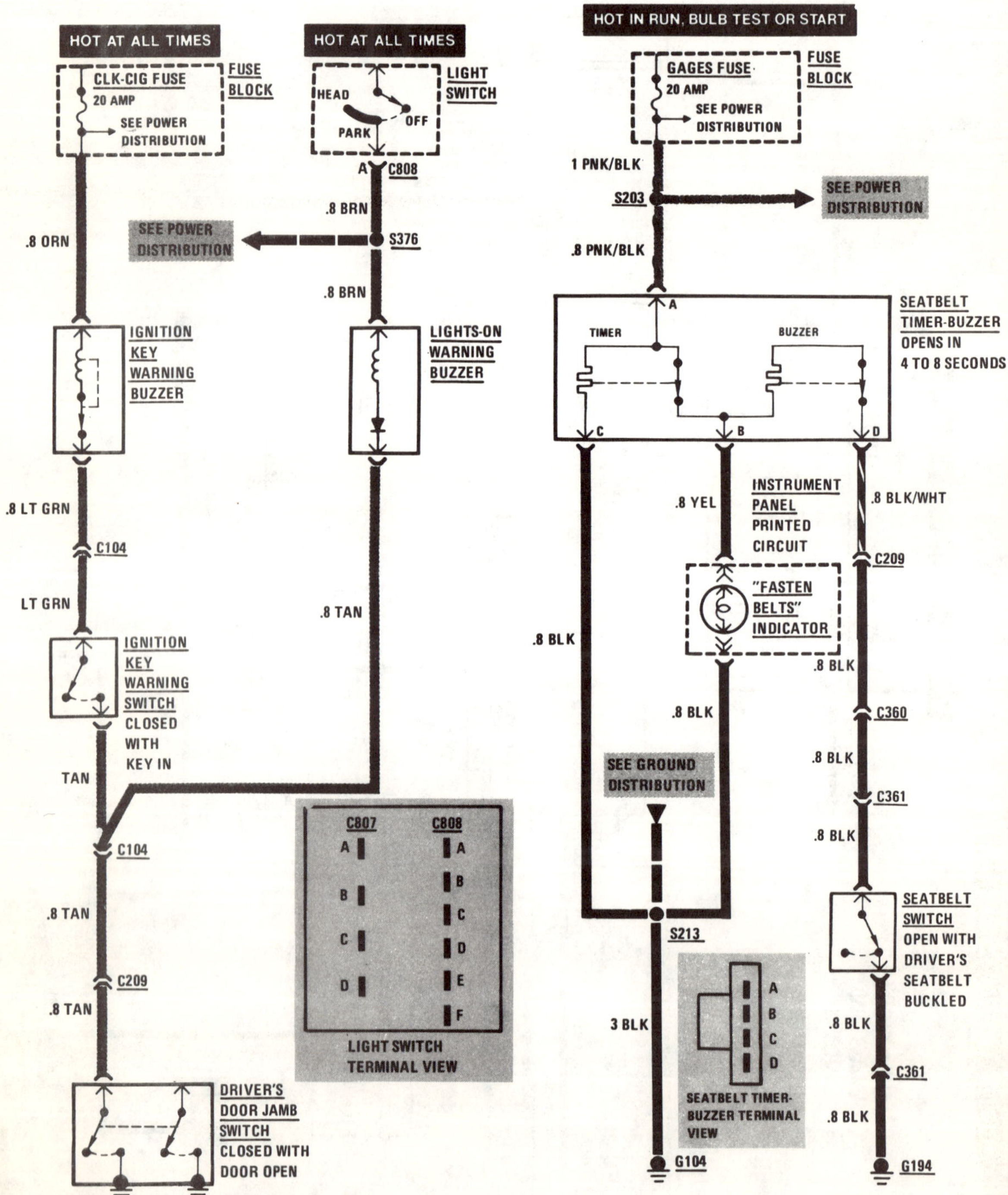

1982 SEATBELT WARNING (WITH BUZZER)

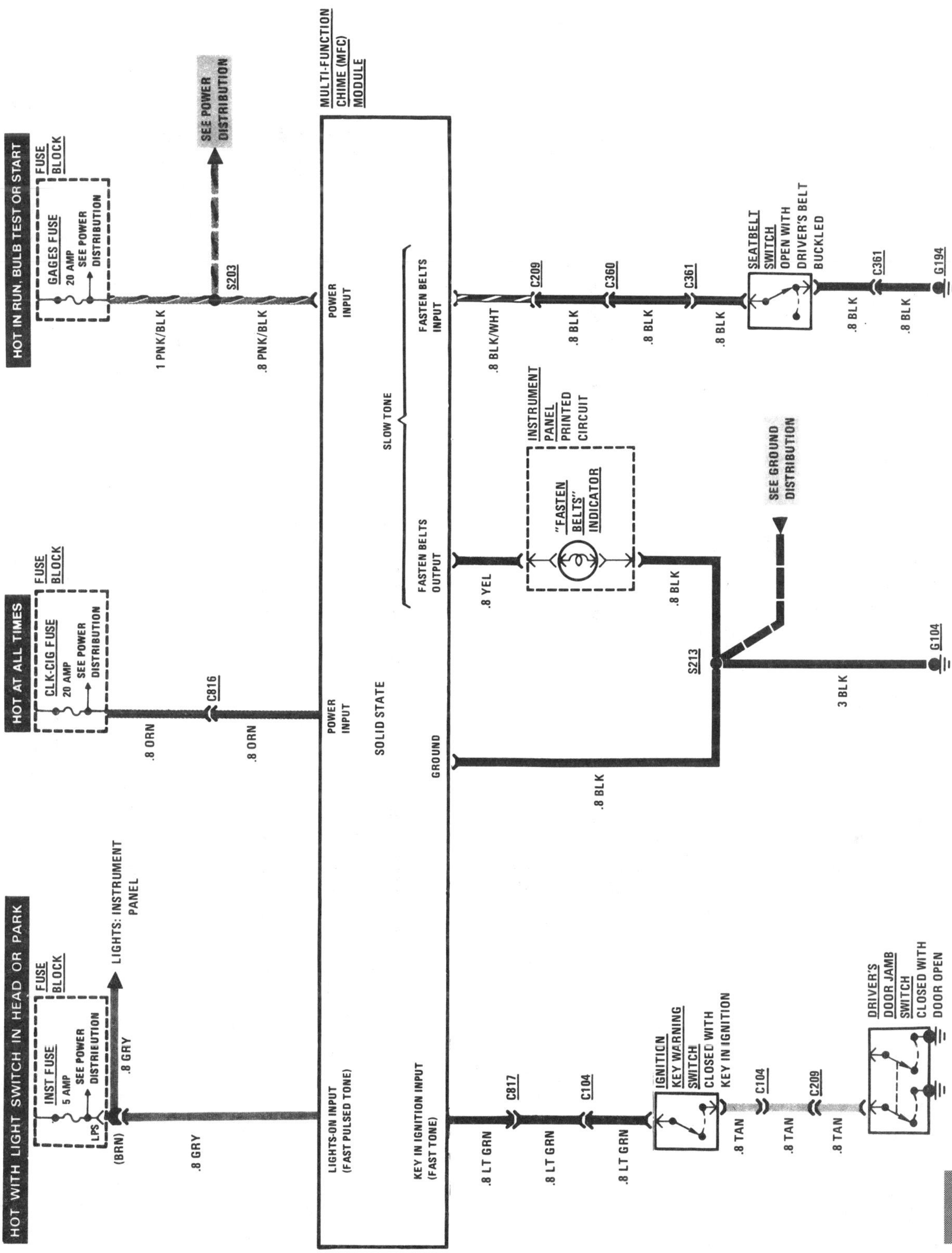

15

1982 IGNITION KEY WARNING/LIGHTS-ON WARNING/ SEATBELT WARNING (WITH CHIMES)

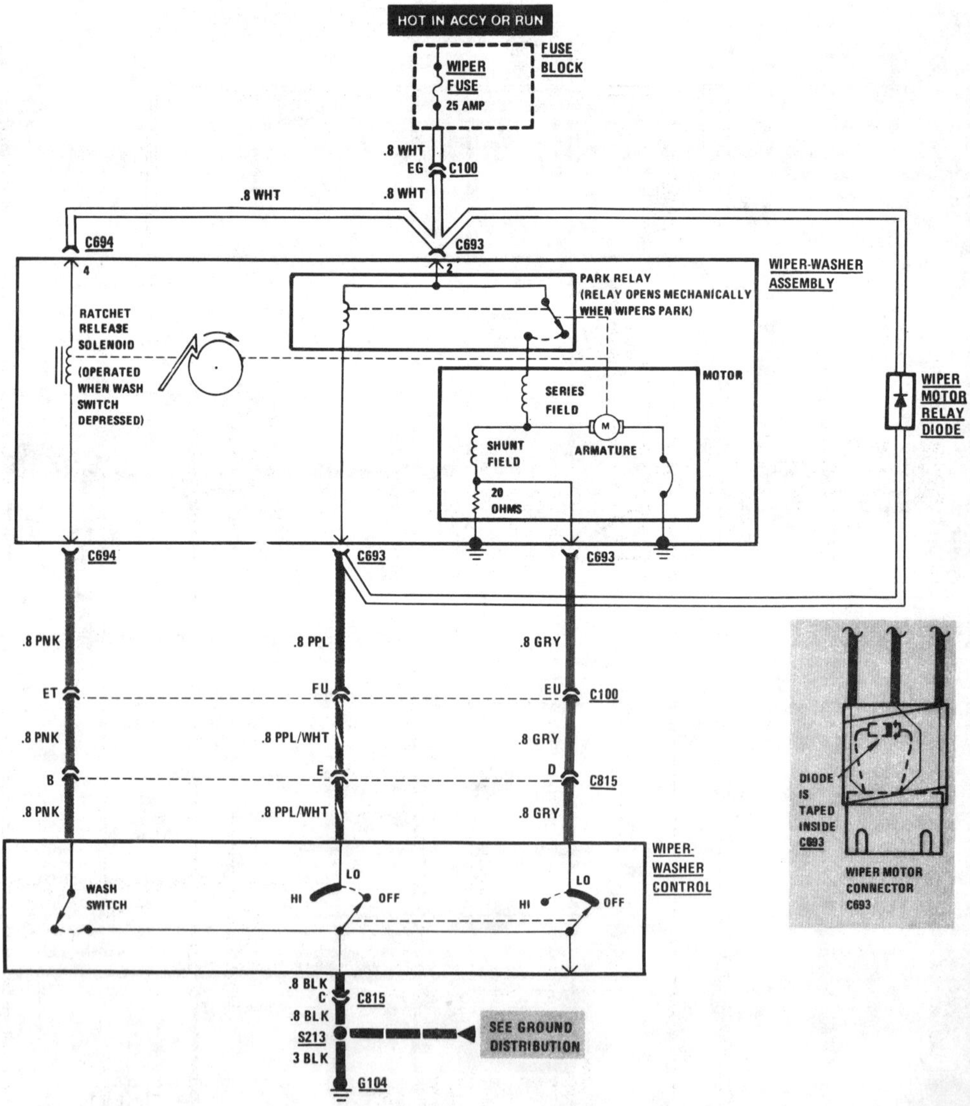

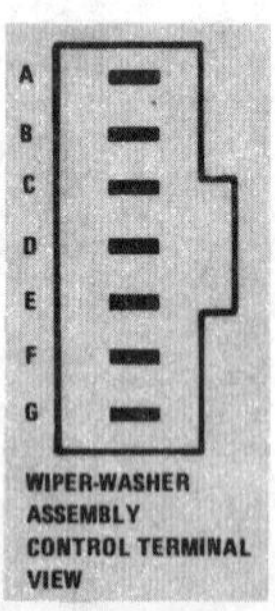

1982 WIPER/WASHER (PULSE)

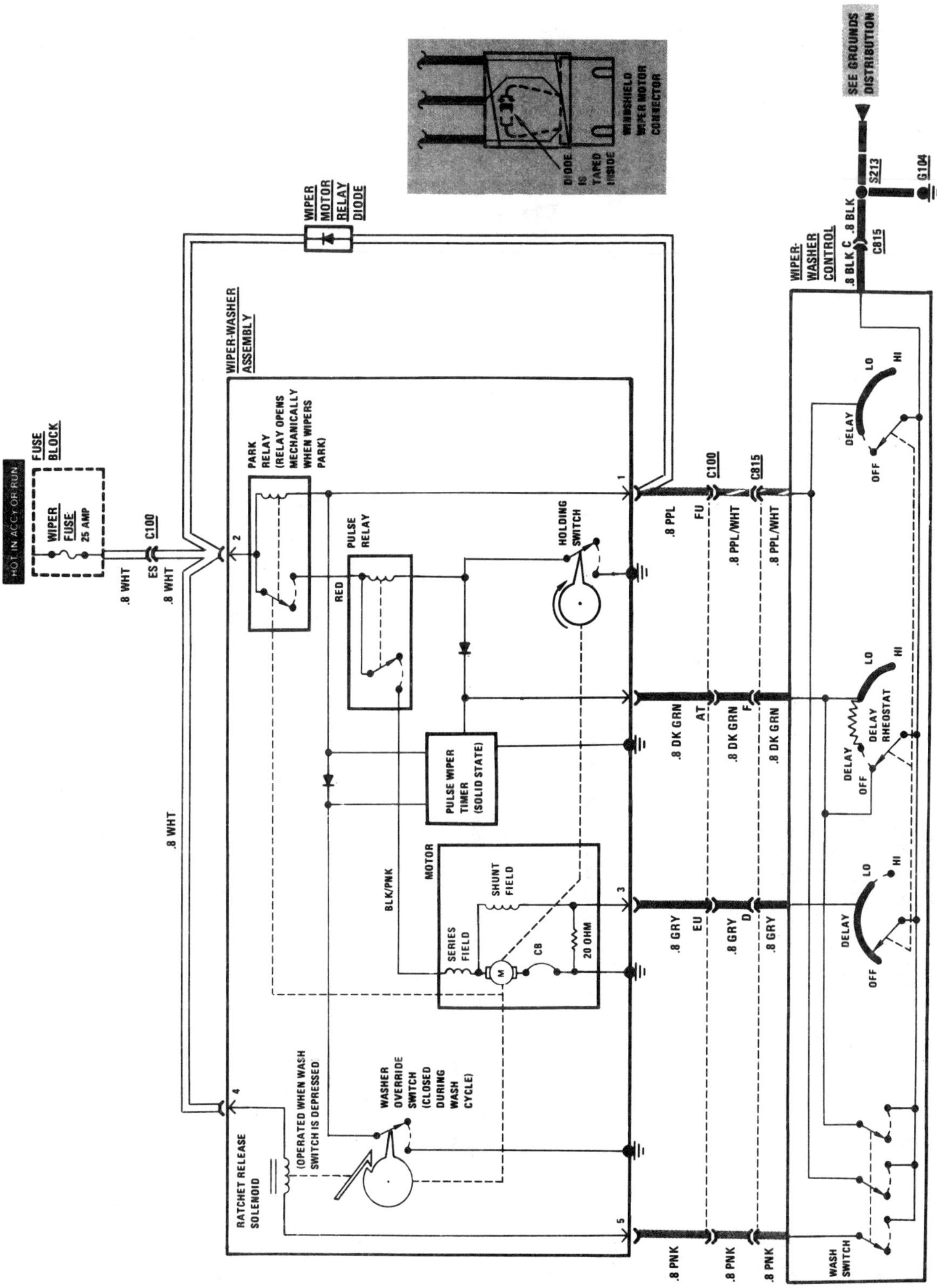

15

1982 HORN

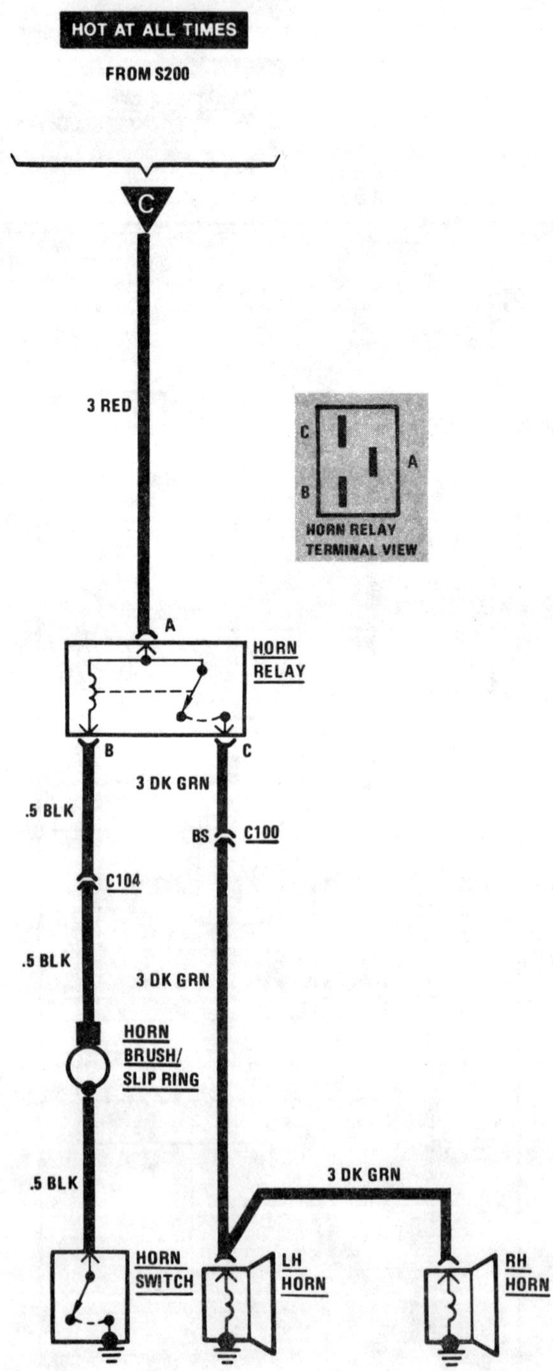
HOT AT ALL TIMES
FROM S200
C
3 RED
C
A
B
HORN RELAY
TERMINAL VIEW
A
HORN
RELAY
B
C
3 DK GRN
.5 BLK
BS
C100
C104
.5 BLK
3 DK GRN
HORN
BRUSH/
SLIP RING
.5 BLK
3 DK GRN
HORN
SWITCH
LH
HORN
RH
HORN

1982 POWER WINDOWS (2-DOOR)

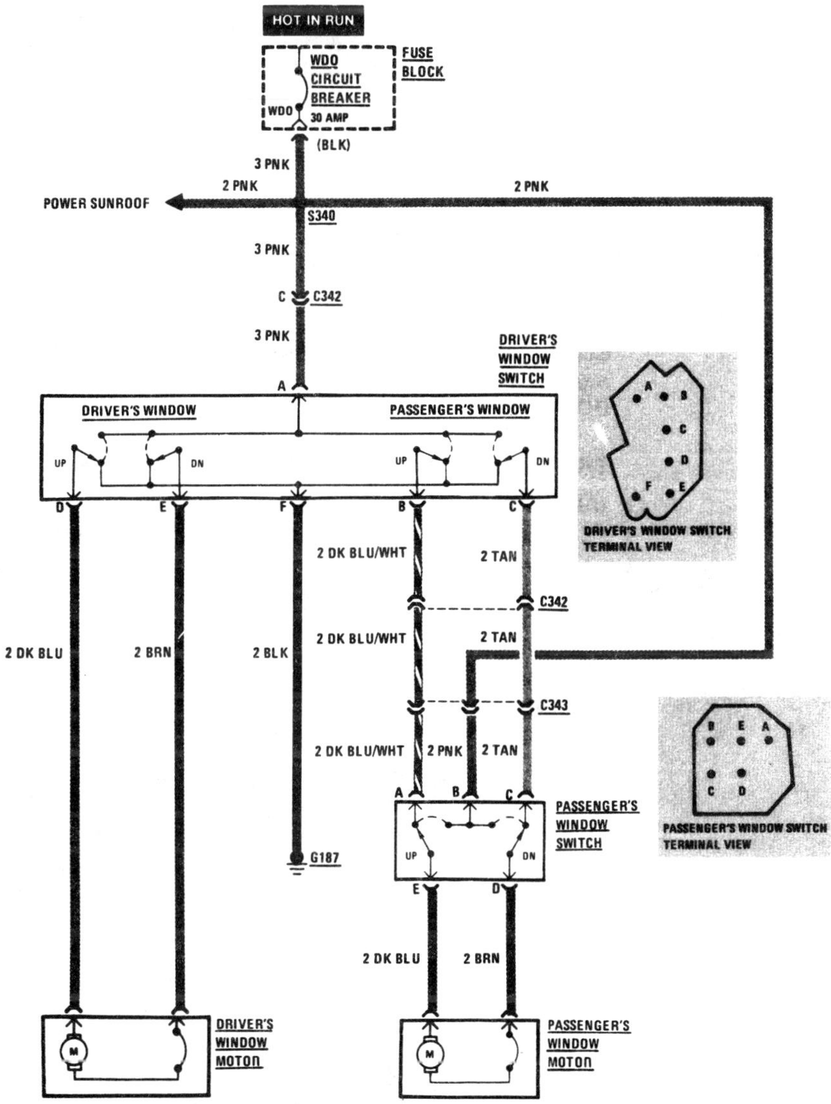

1982 POWER WINDOWS (4-DOOR)

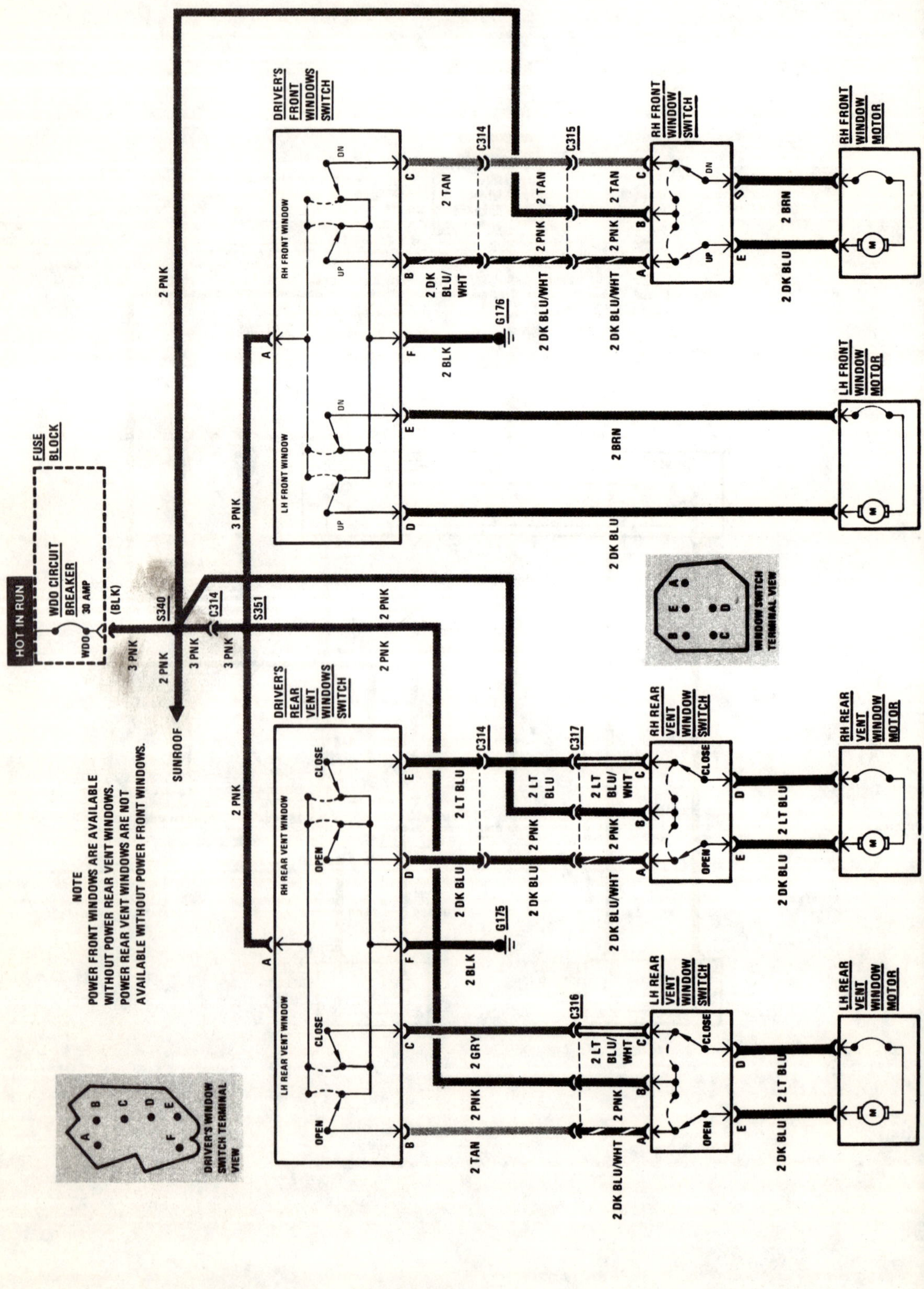

1982 POWER DOOR-LOCKS (2-DOOR)

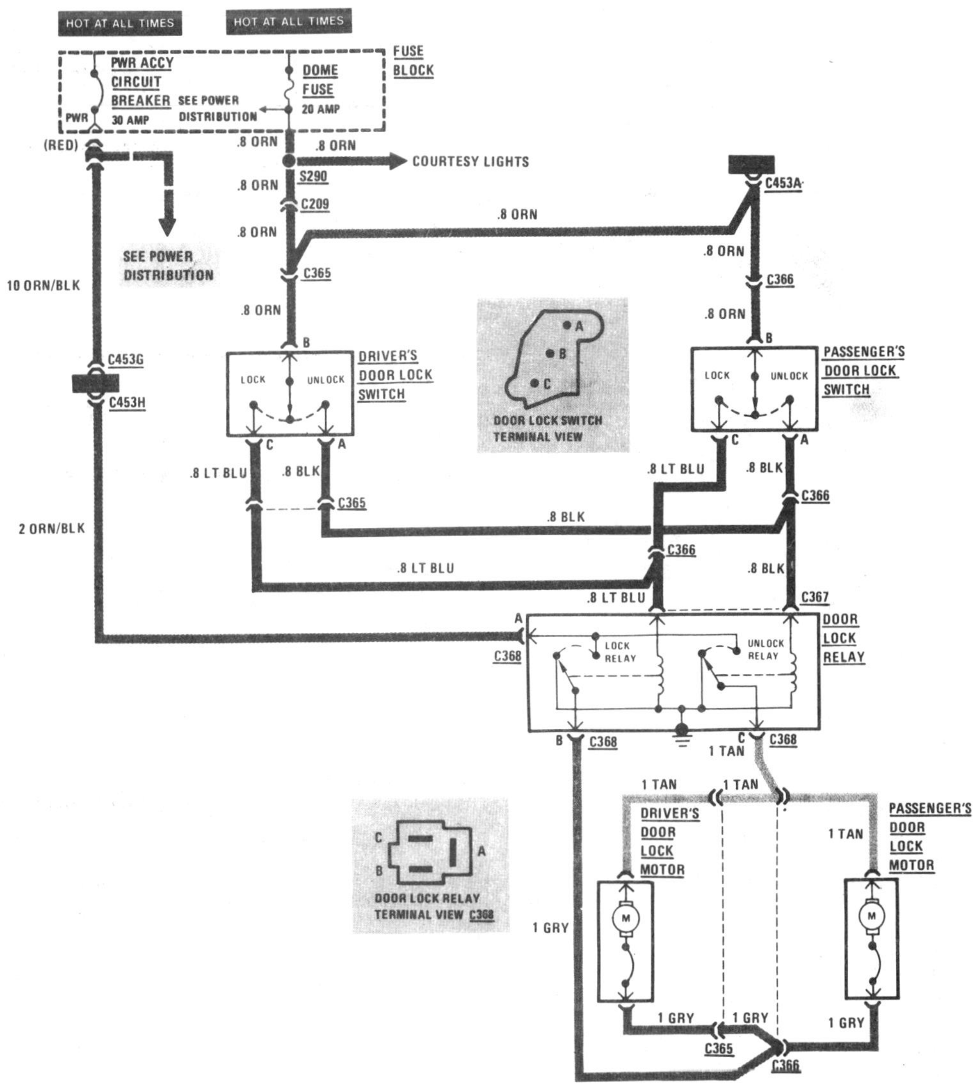

1982 POWER DOOR-LOCKS (4-DOOR)

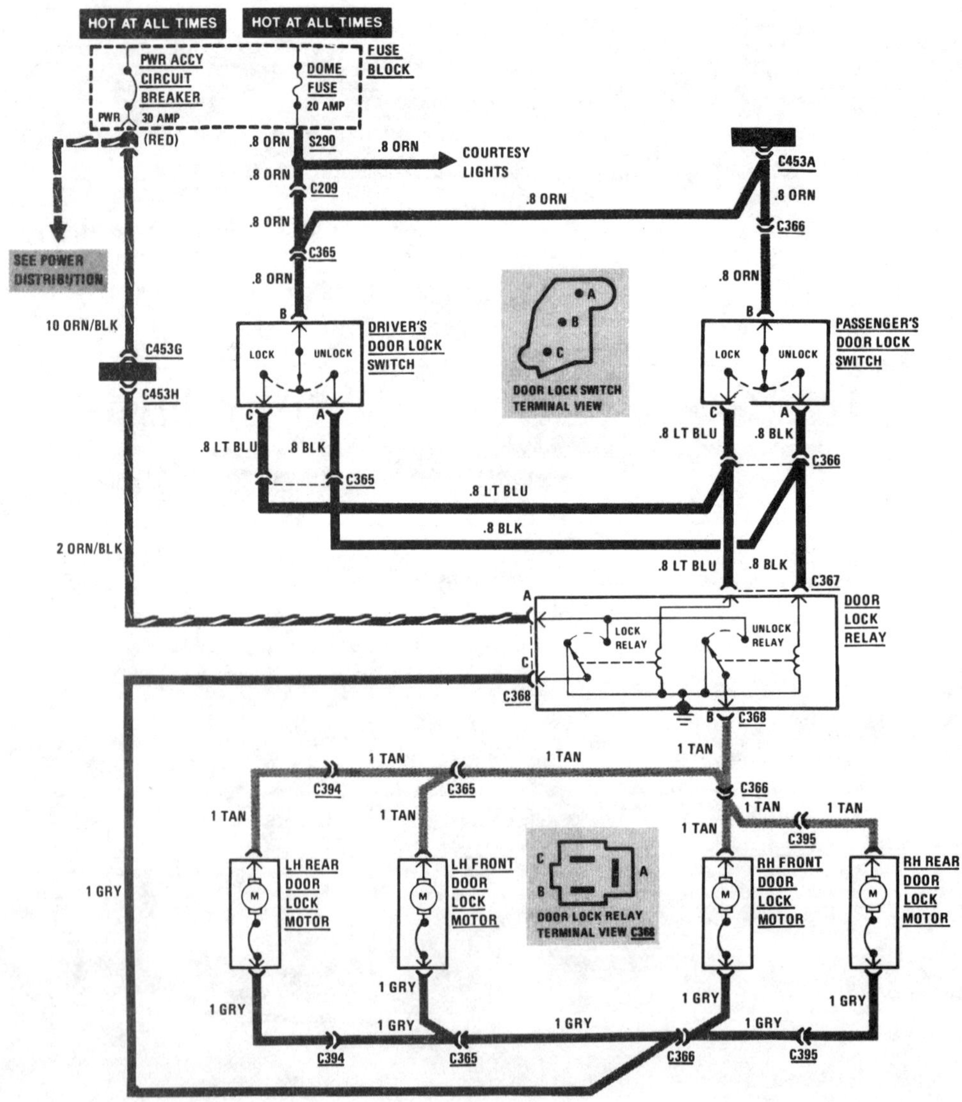

1982 TRUNK/TAILGATE RELEASE

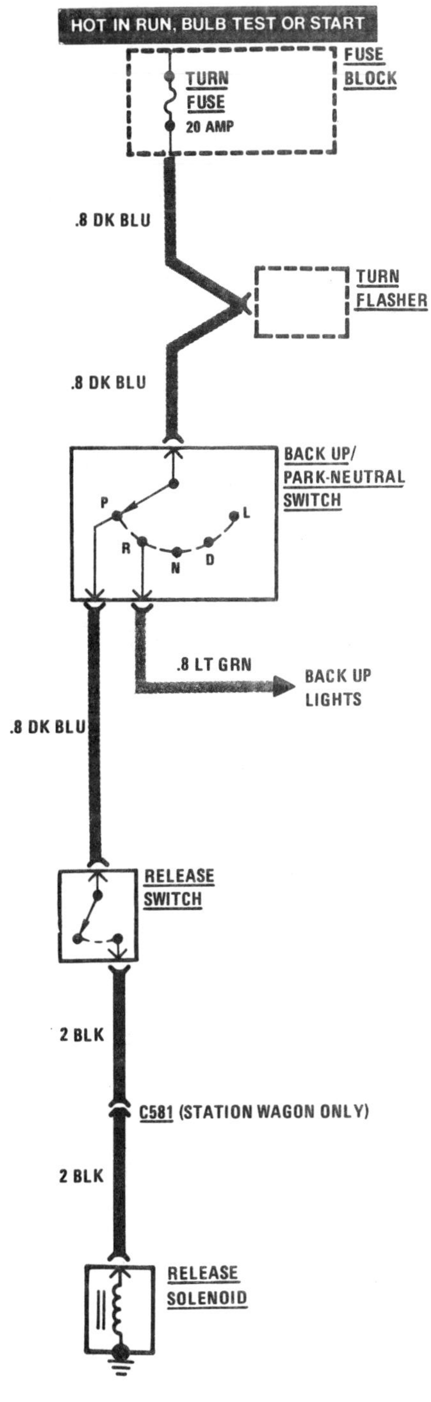

1982 DEFOGGER

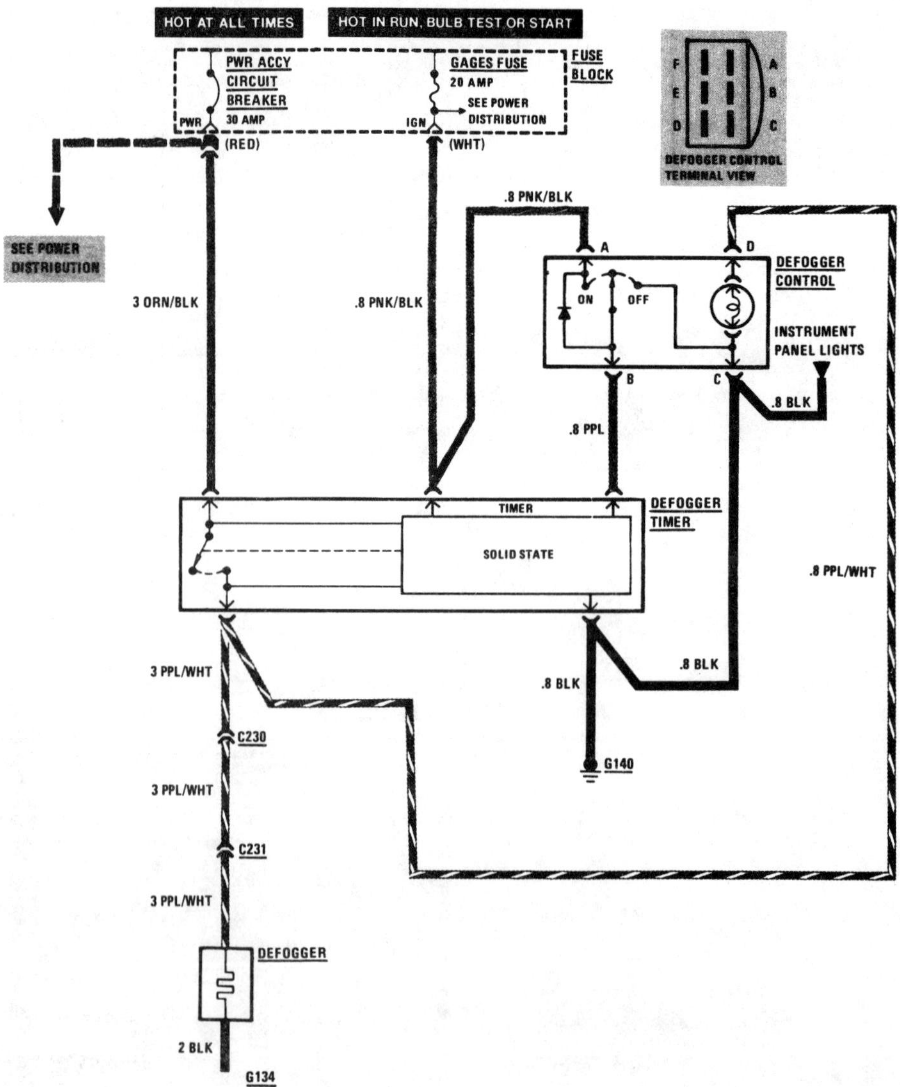

1982 POWER SEATS

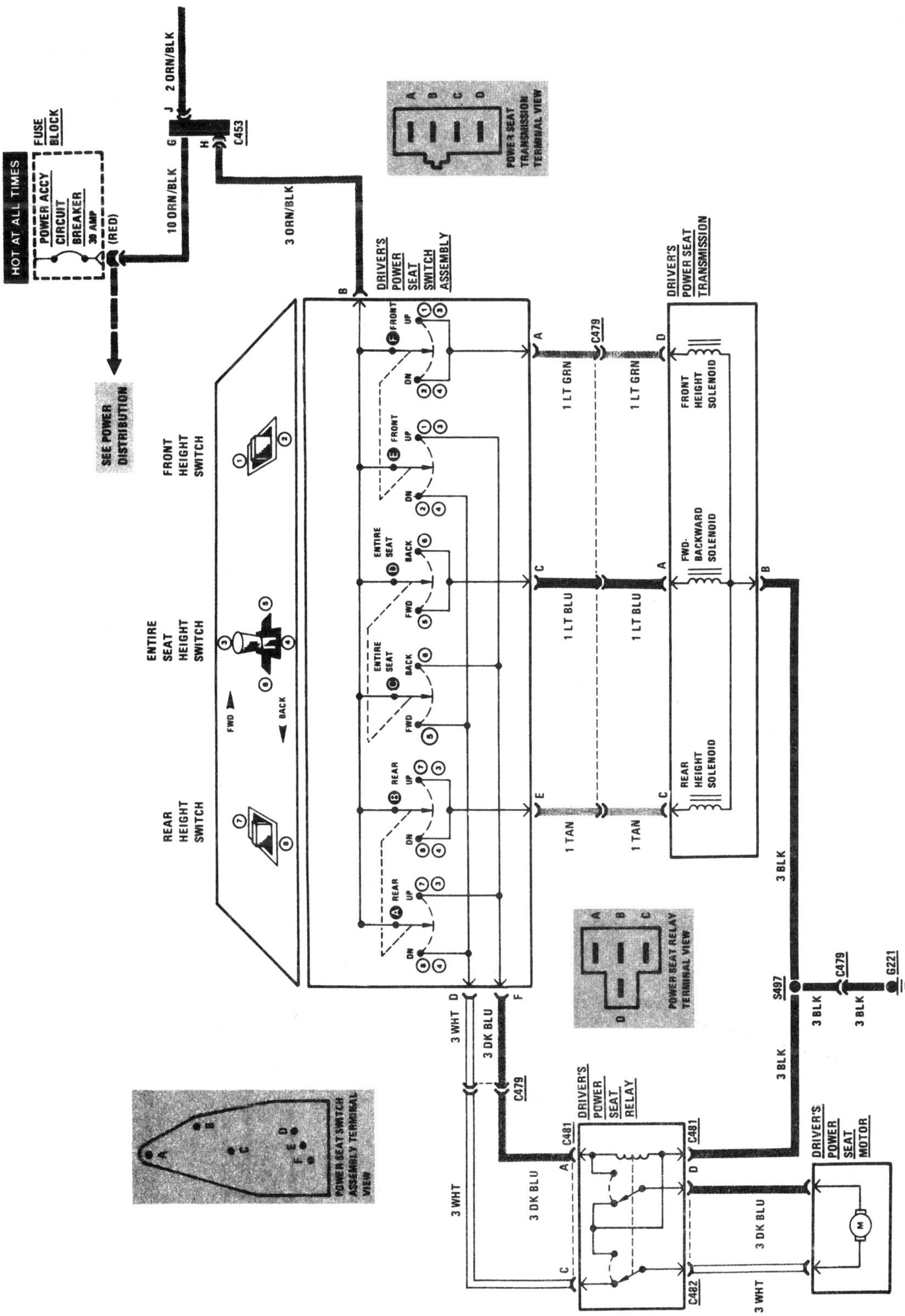

1982 POWER SEATS

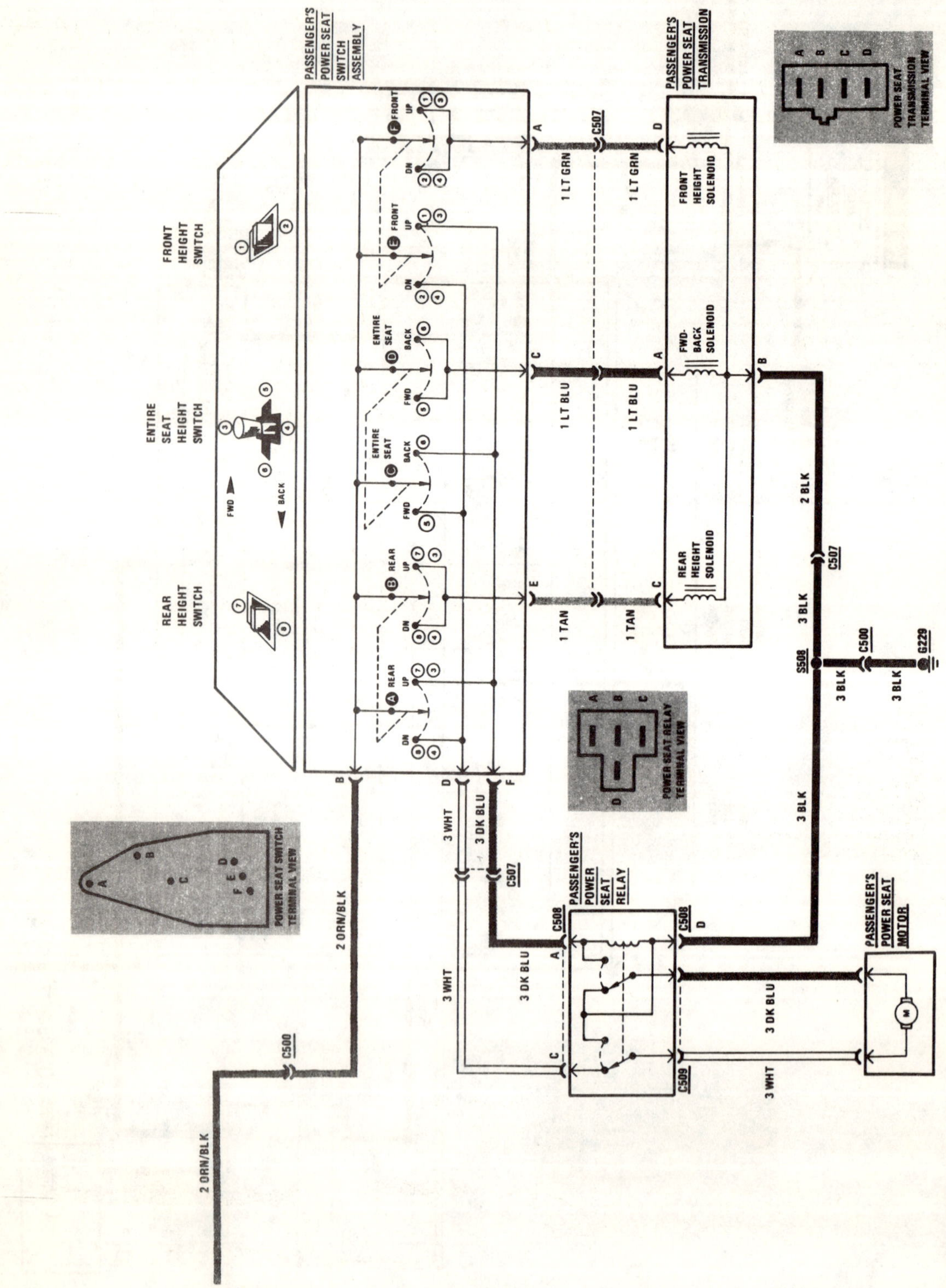

1982 CRUISE CONTROL

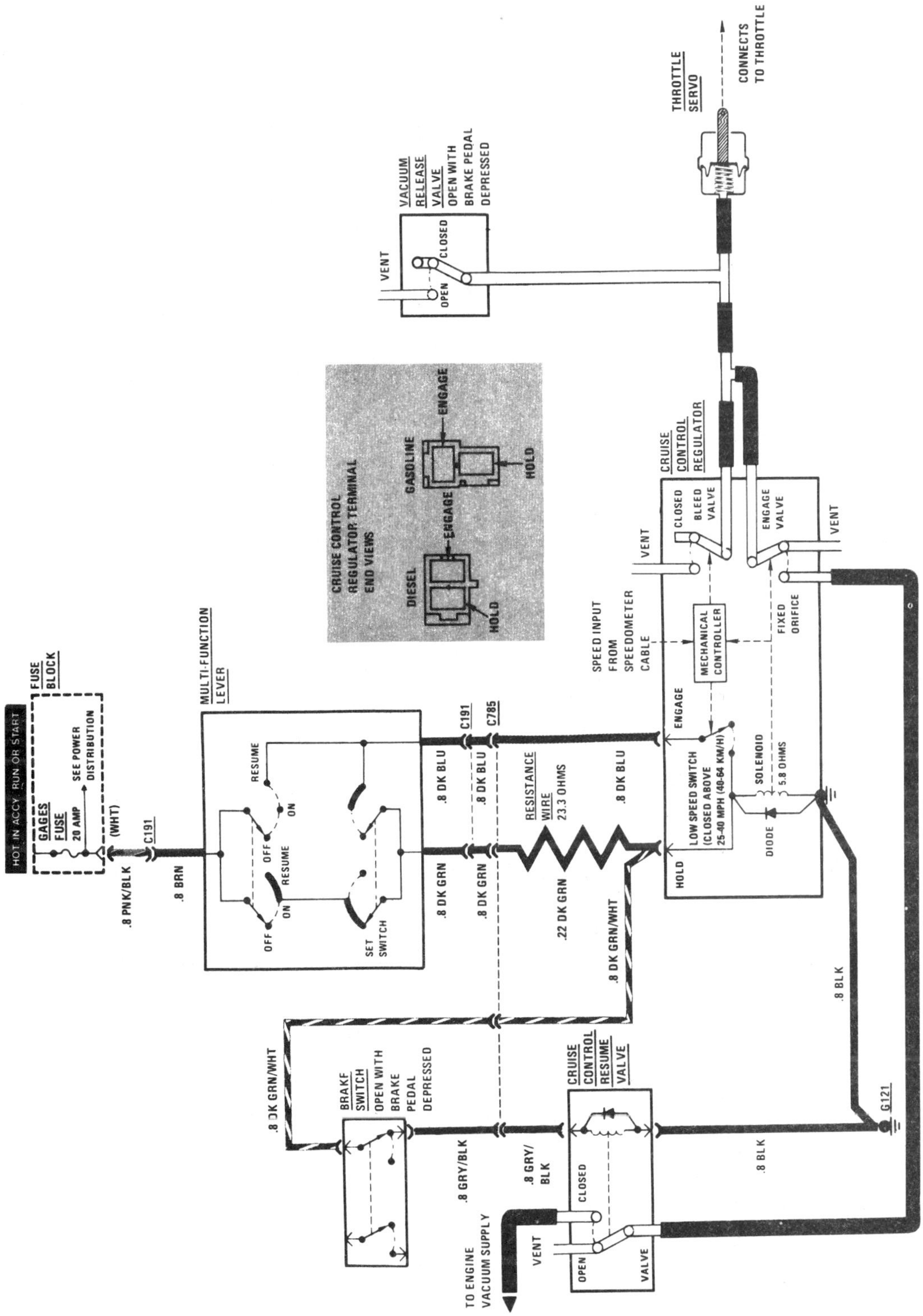

1982 SUN ROOF

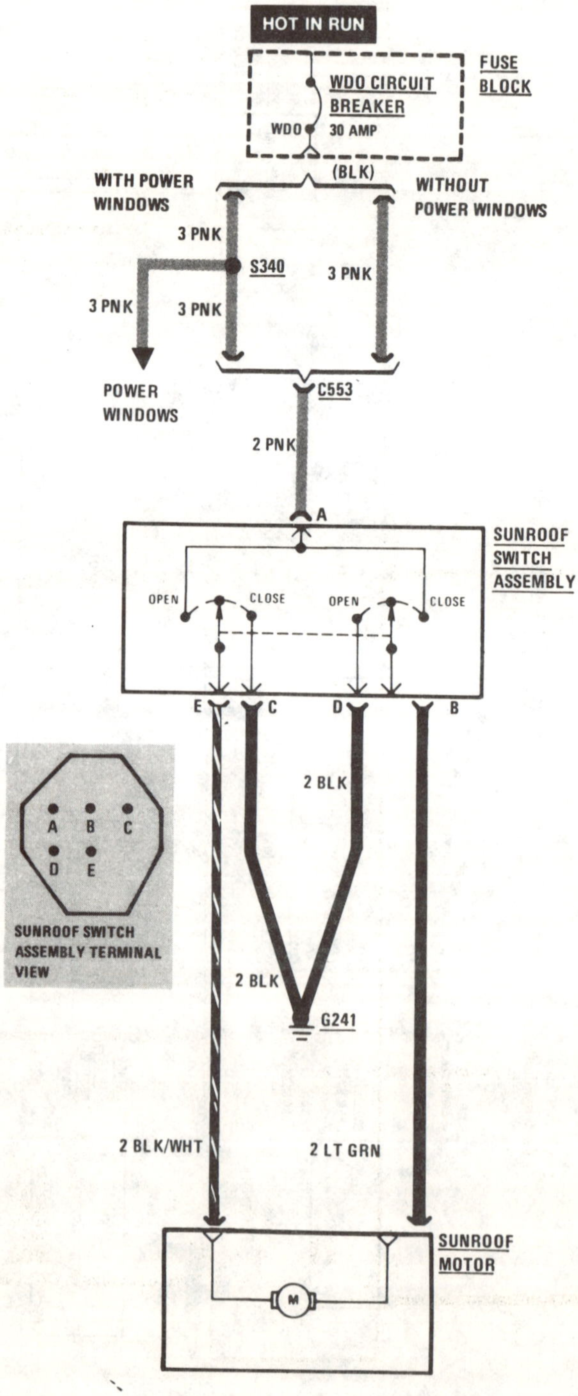

1982 RADIO (MONO)

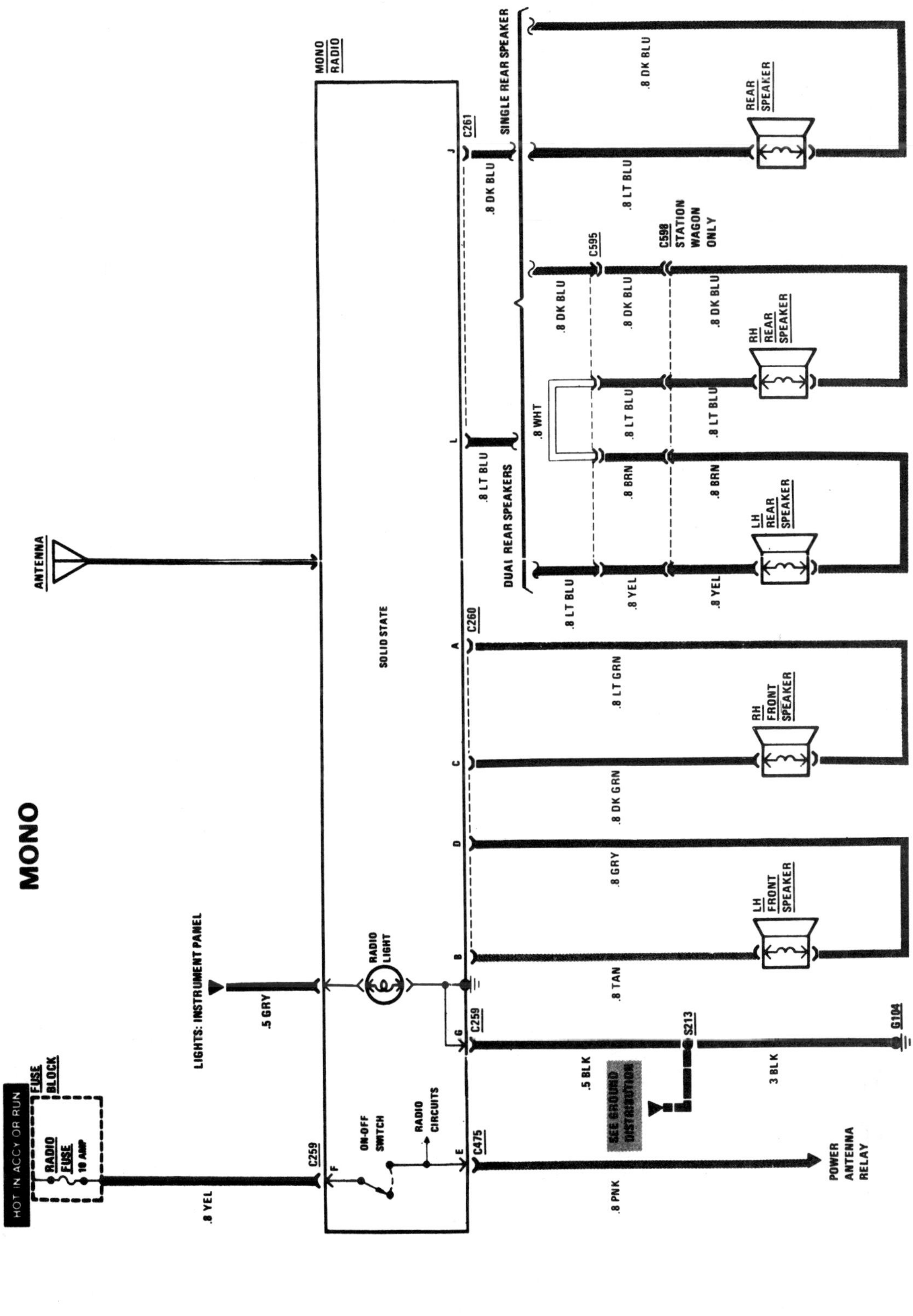

15

1982 RADIO (STEREO)

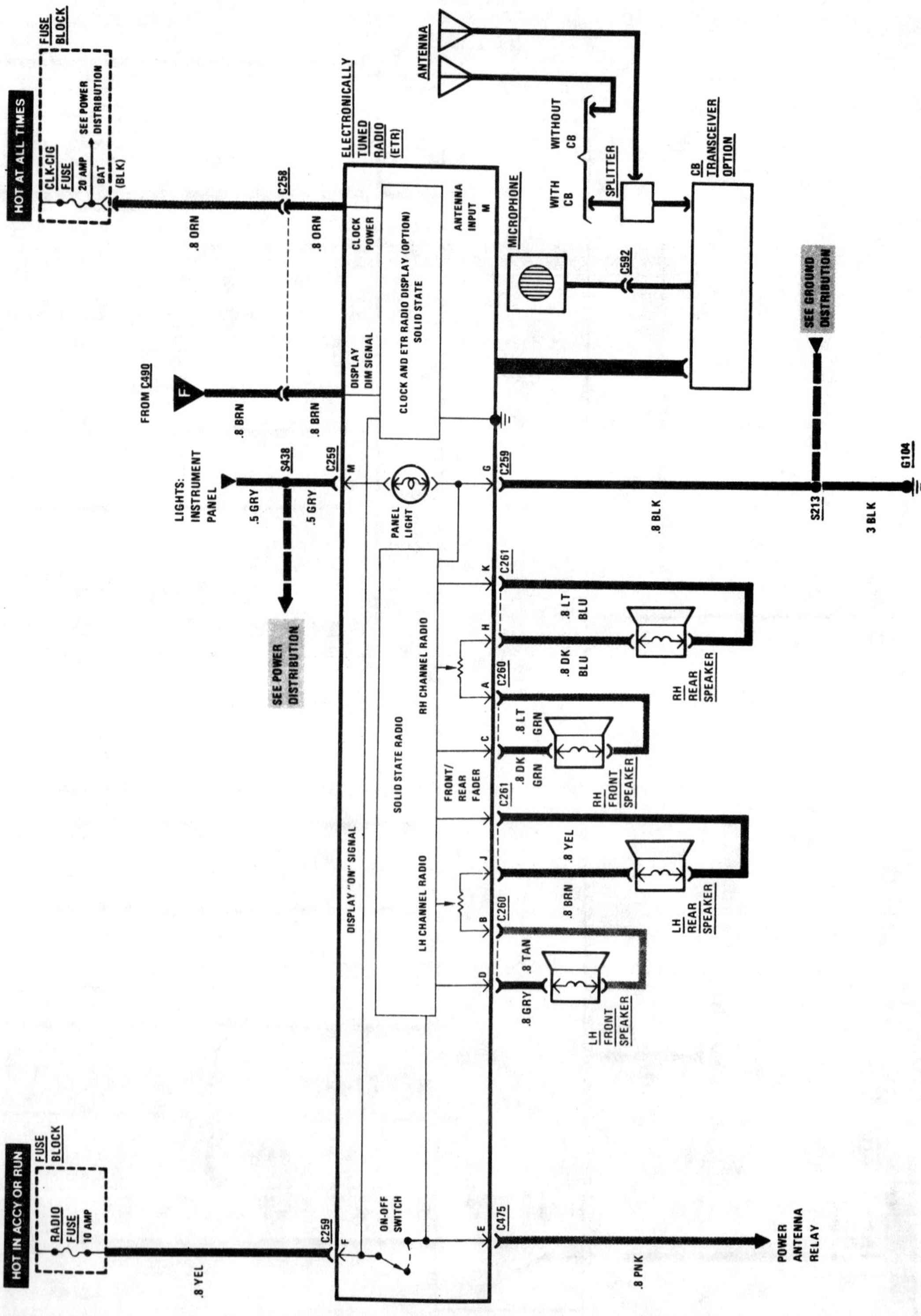

1982 POWER ANTENNA

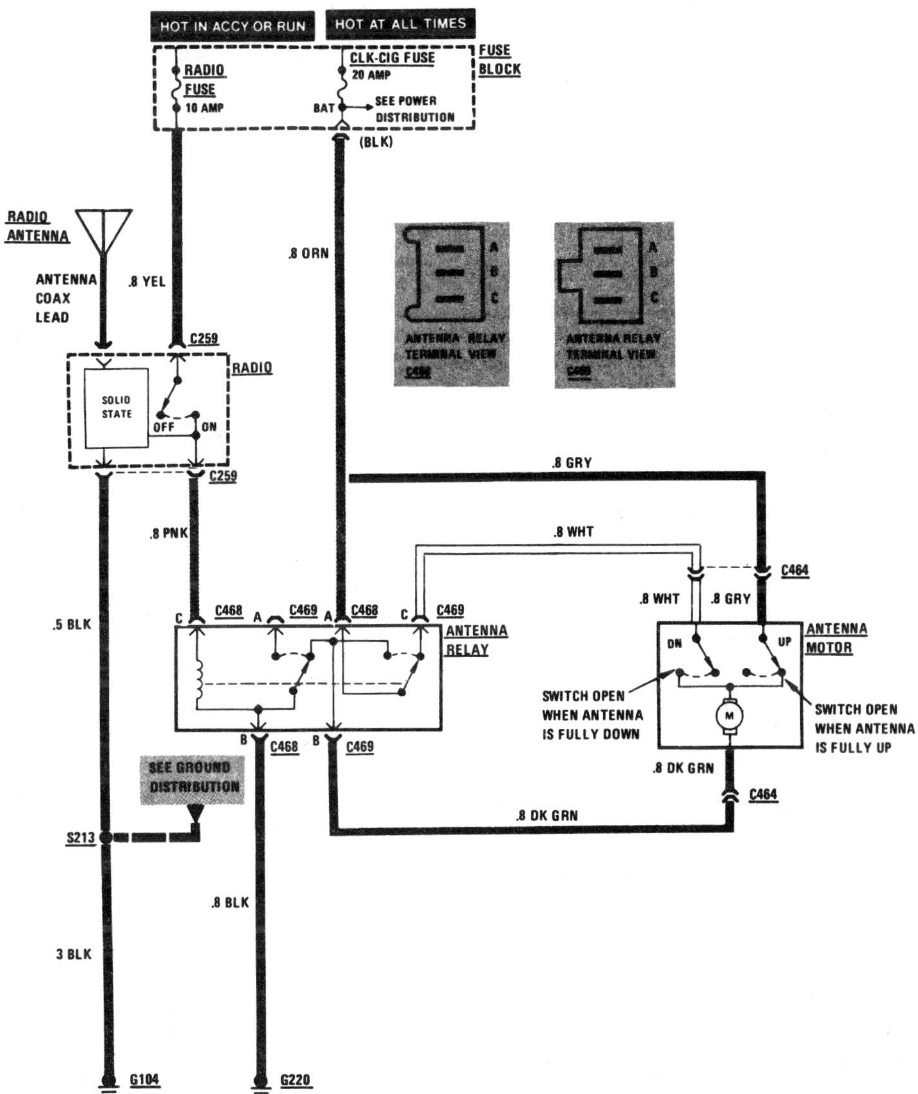

1982 AIR CONDITIONING—FOUR SEASON AND TEMPMATIC (ELECTRICAL)

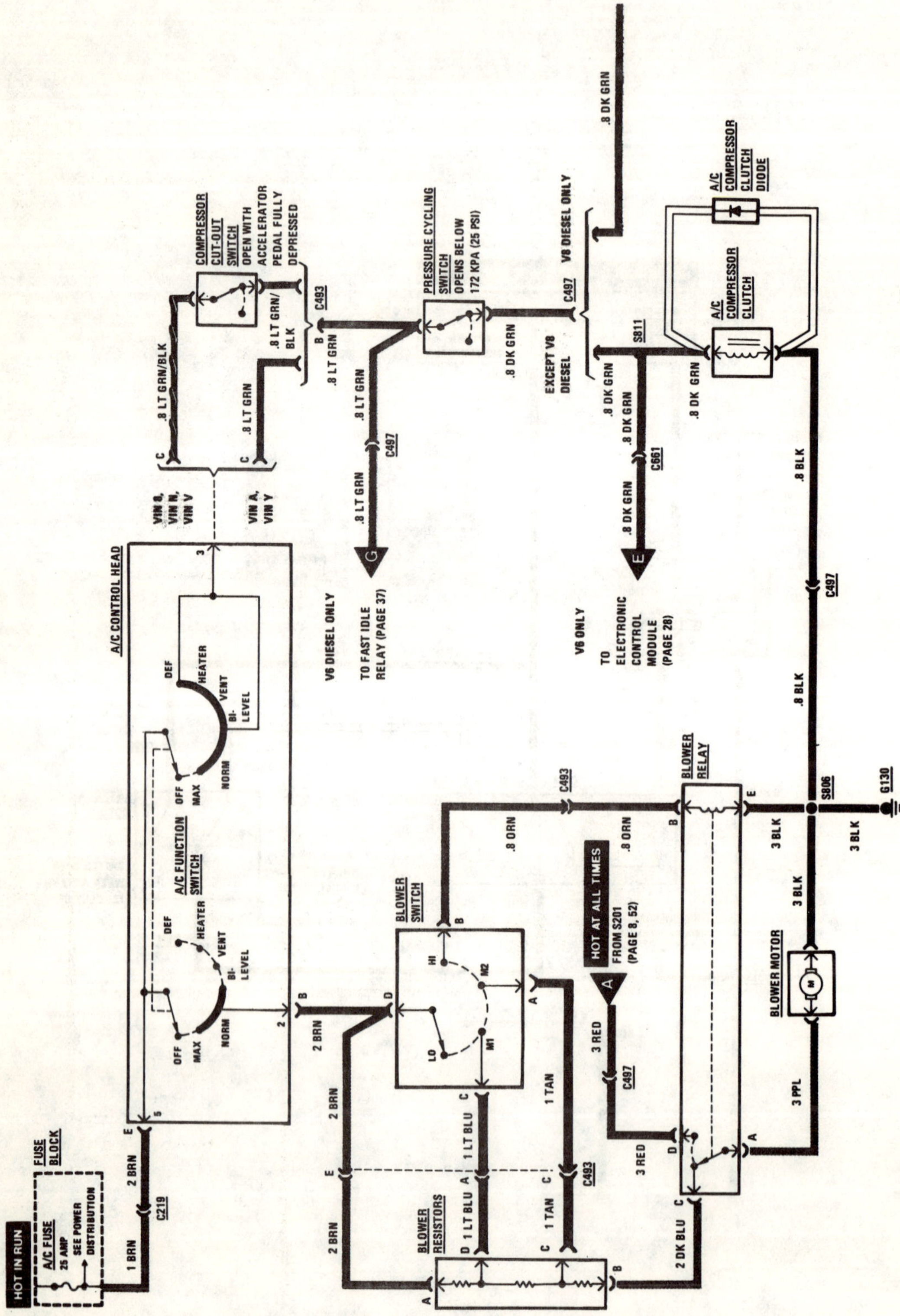

1982 AIR CONDITIONING—FOUR SEASON AND TEMPMATIC (ELECTRICAL)

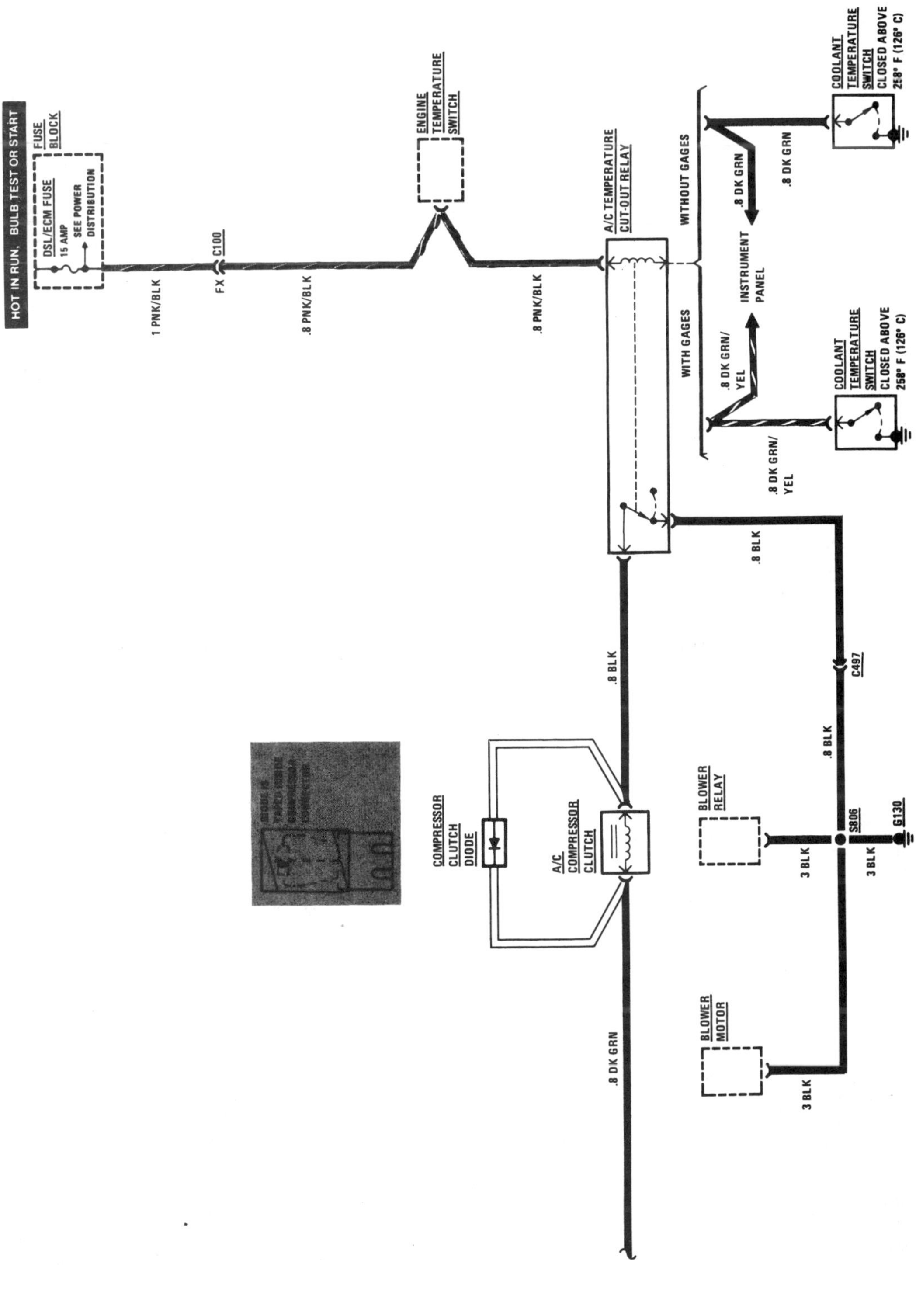

1982 BRAKE WARNING SYSTEM—EXCEPT V6 DIESEL

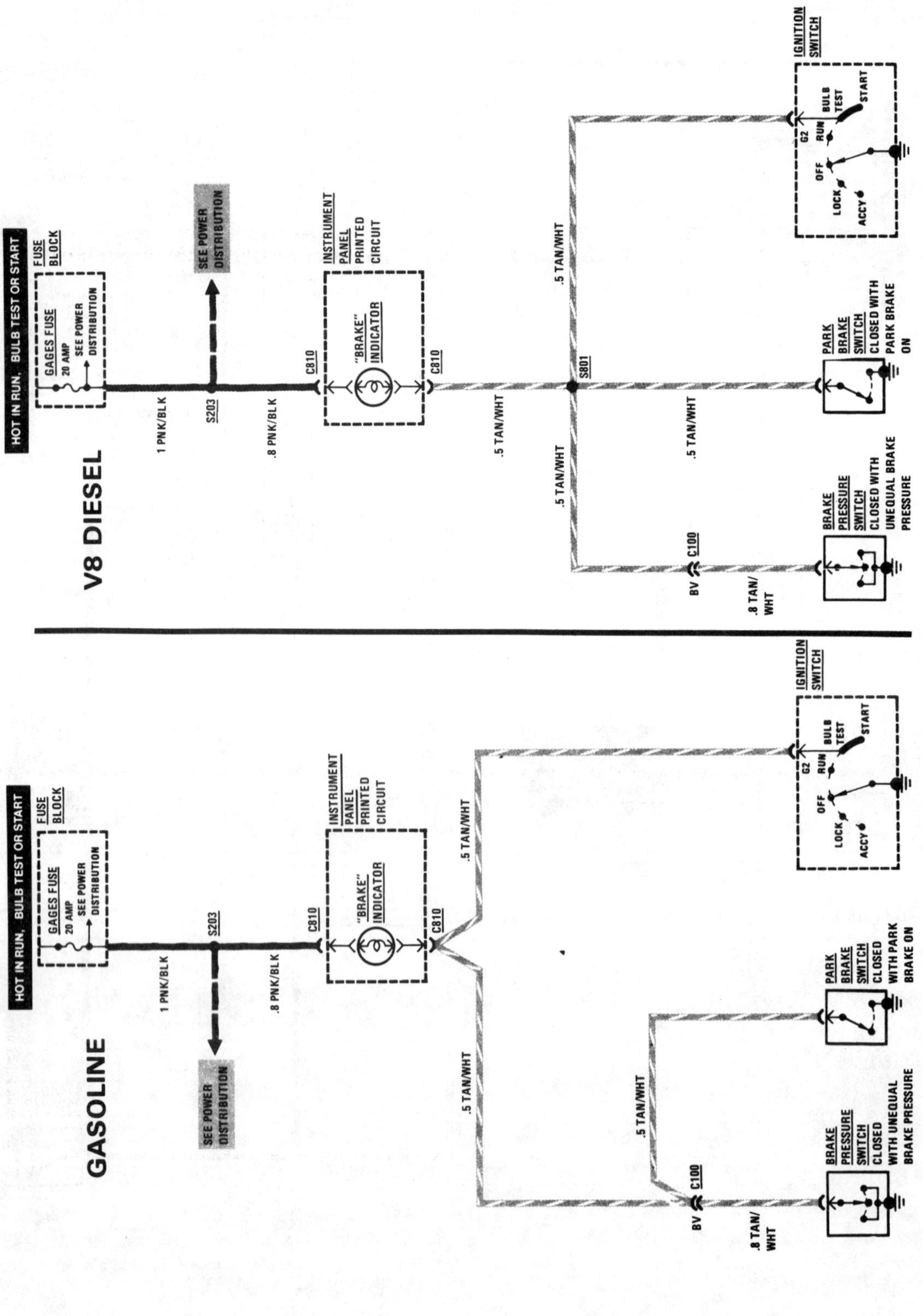

1982 BRAKE WARNING SYSTEM—V6 DIESEL

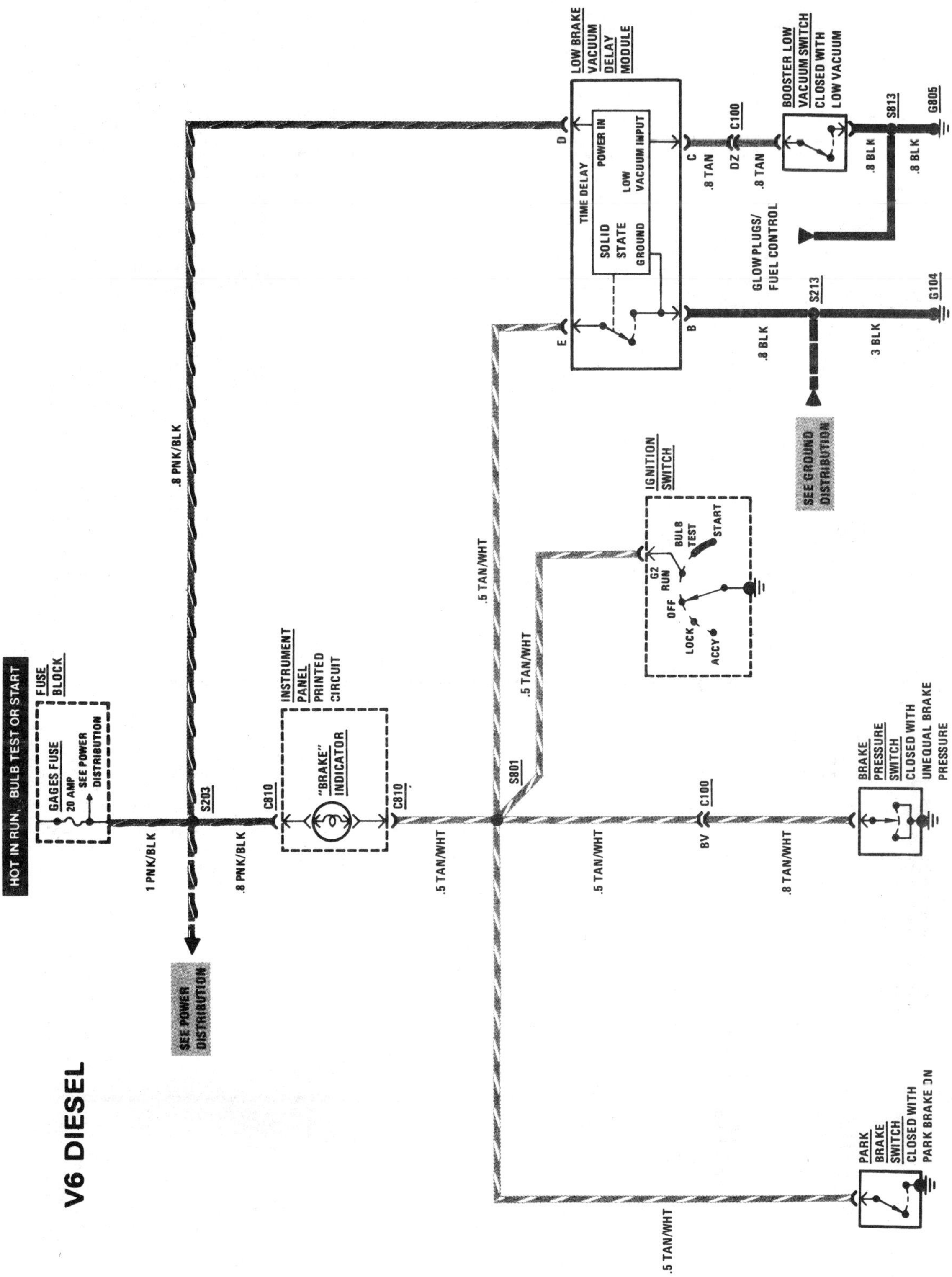

1982 INSTRUMENT PANEL (FUEL GAUGE, GAUGES OPTION, TACHOMETER)

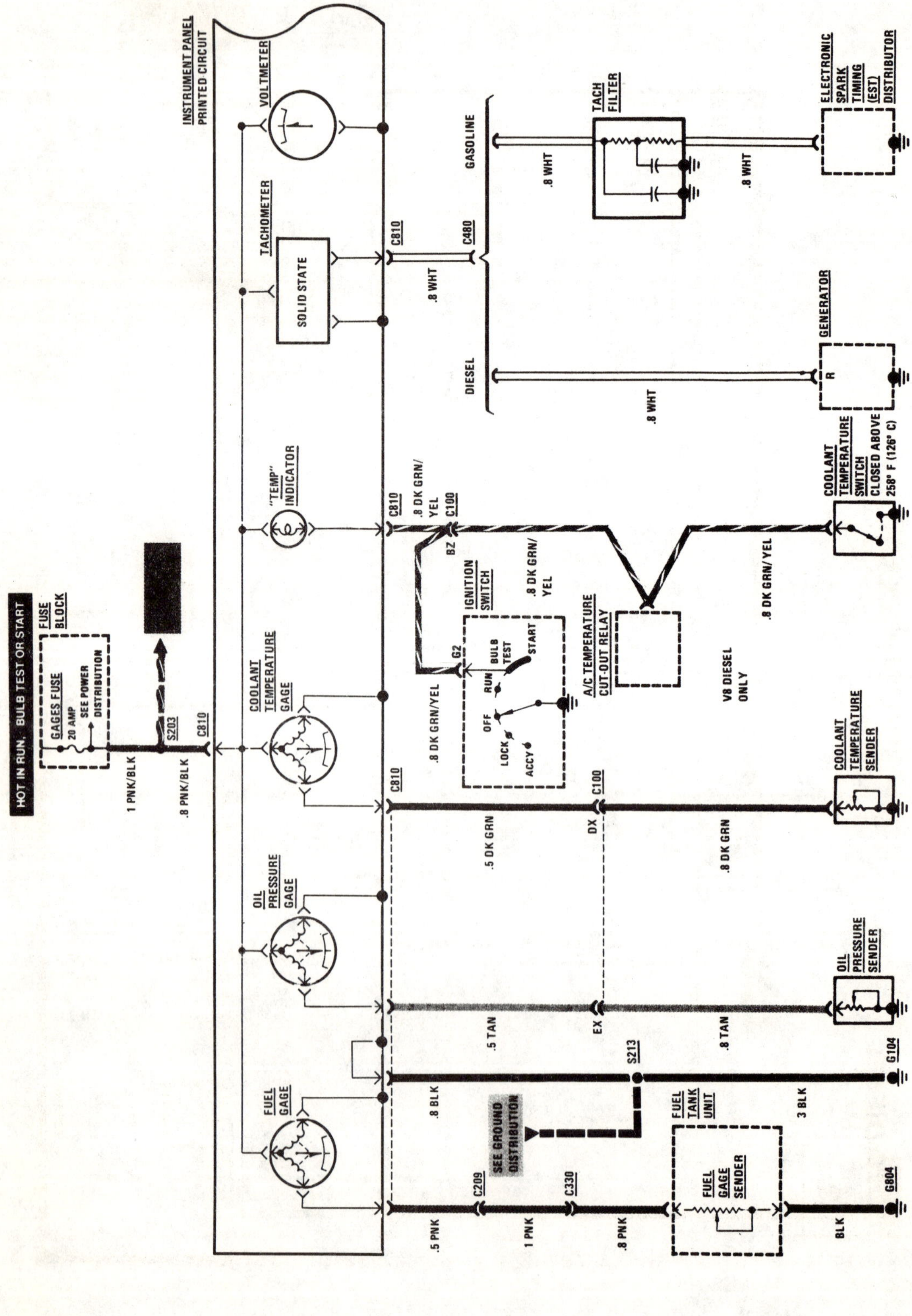

1982 INSTRUMENT PANEL (WARNING INDICATORS CHOKE HEATER)

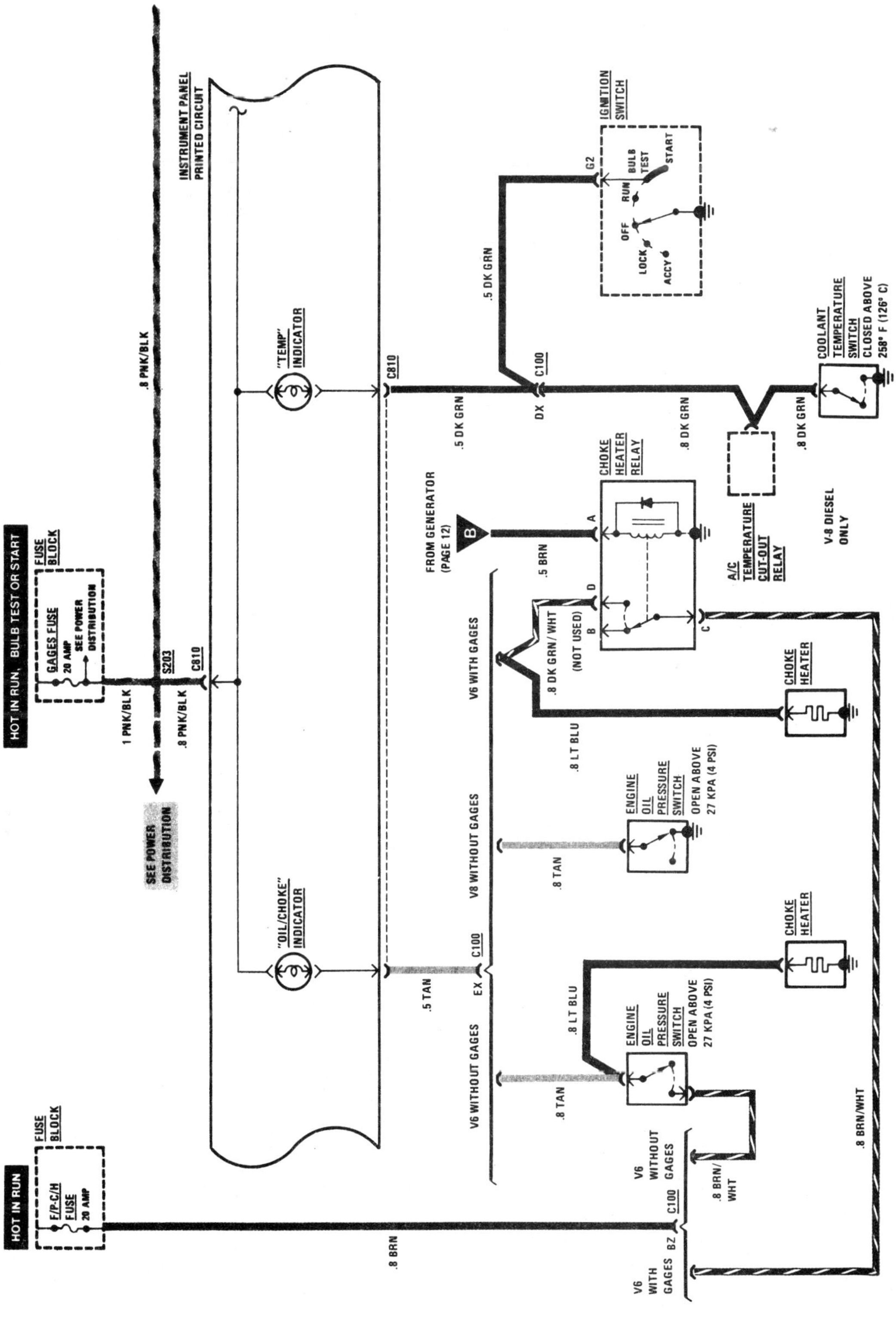

1982 INSTRUMENT PANEL (WARNING INDICATORS CHOKE HEATER)

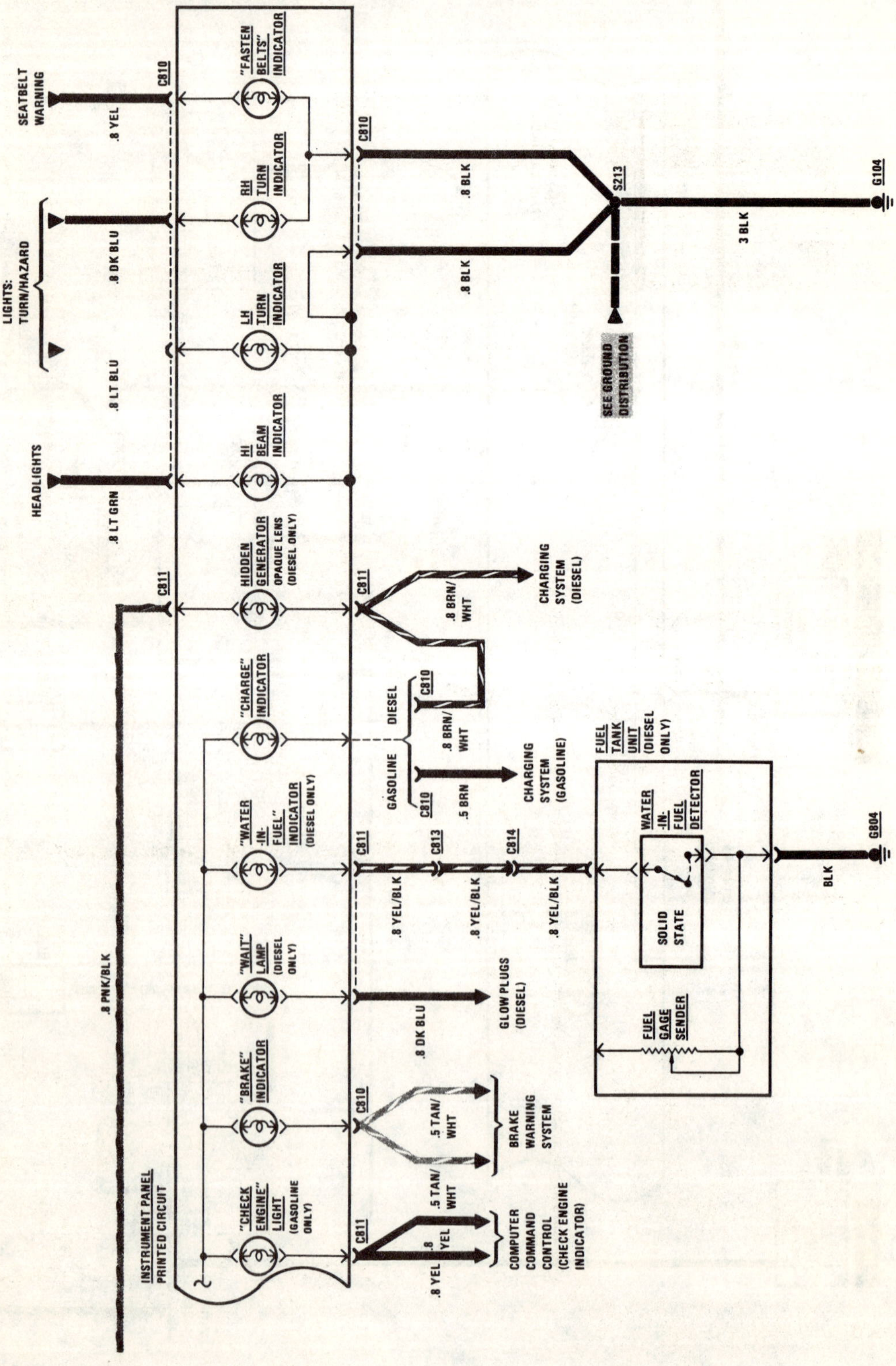

1982 LIGHTS: INSTRUMENT PANEL

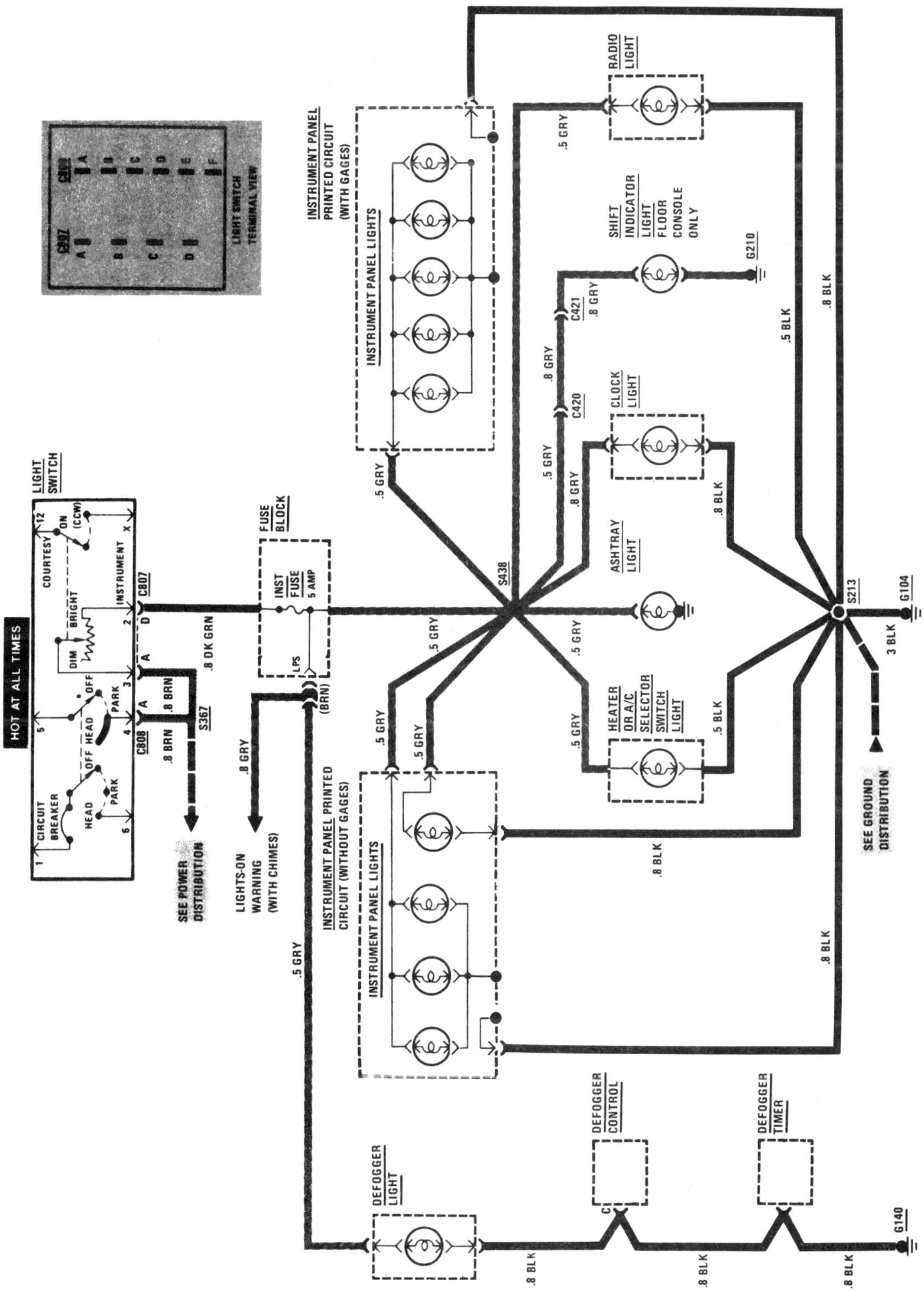

15

1982 HEATER (WITHOUT A/C)

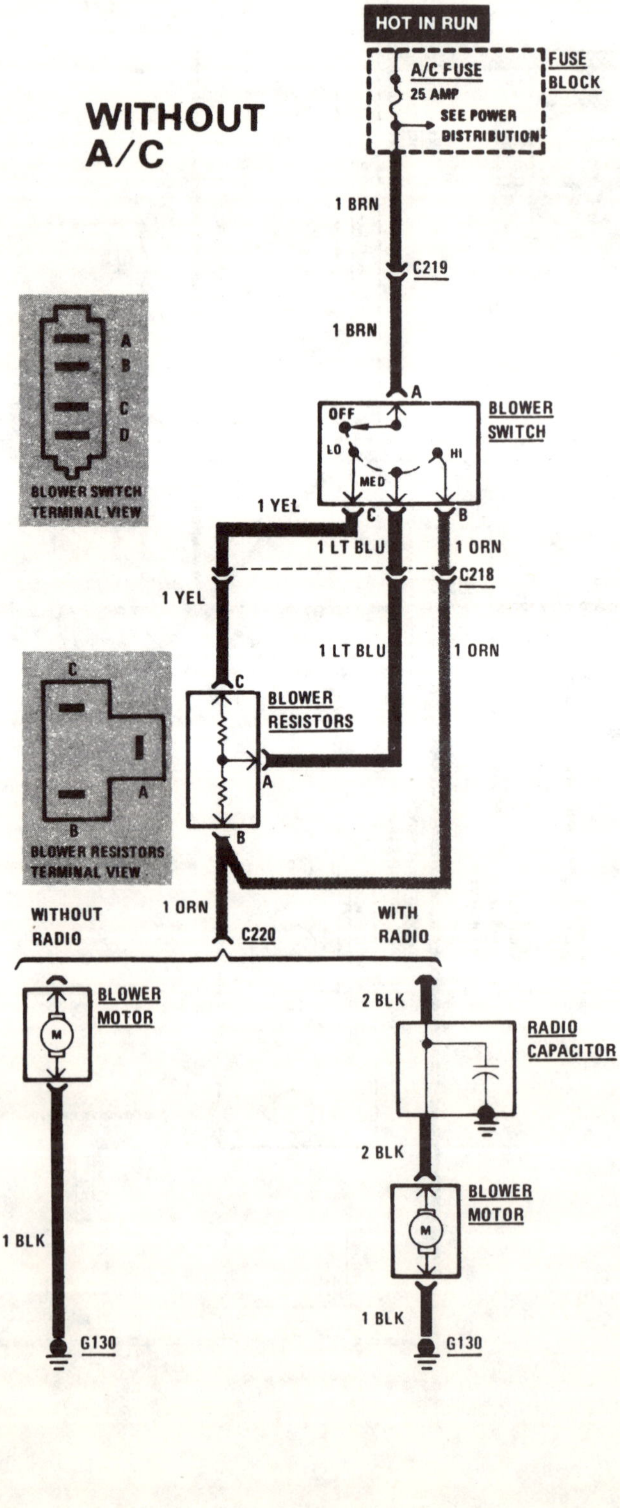

NOTES

NOTES